Process Engineering for CCS Power Plants

Edited by
Detlef Stolten and V. Scherer

Related Titles

Garcia-Martinez, Javier, Serrano-Torregrosa, Elena (eds.)

The Chemical Element
Chemistry's Contribution to Our Global Future

2011
ISBN: 978-3-527-32880-2

Centi, Gabriele, Trifiró, Ferruccio, Perathoner, Siglinda, Cavani, Fabrizio (eds.)

Sustainable Industrial Chemistry

2009
ISBN: 978-3-527-31552-9

Rojey, A.

Energy and Climate
How to achieve a successful energy transition

2009
ISBN: 978-0-470-74427-7

Mitsos, A., Barton, P. I. (eds.)

Microfabricated Power Generation Devices
Design and Technology

2009
ISBN: 978-3-527-32081-3

Morvay, Z., Gvozdenac, D.

Applied Industrial Energy and Environmental Management

2008
ISBN: 978-0-470-69742-9

Georgiadis, M., Kikkinides, E. S., Pistikopoulos, E. (eds.)

Process Systems Engineering
Volume 5: Energy Systems Engineering

2008
ISBN: 978-3-527-31694-6

Schavemaker, P., van der Sluis, L.

Electrical Power System Essentials

2007
ISBN: 978-0-470-51027-8

Wengenmayr, R., Bührke, T. (eds.)

Renewable Energy
Sustainable Energy Concepts for the Future

2008
ISBN: 978-3-527-40804-7

Wilson, E., Gerard, D. (eds.)

Carbon Capture and Sequestration
Integrating Technology, Monitoring, Regulation

2007
ISBN: 978-0-8138-0207-7

Heier, S.

Grid Integration of Wind Energy Conversion Systems

2006
ISBN: 978-0-470-86899-7

IMechE (Institution of Mechanical Engineers)

Hydropower Developments
New Projects, Rehabilitation, and Power Recovery

2005
ISBN: 978-1-86058-479-4

Process Engineering for CCS Power Plants

Edited by
Detlef Stolten and V. Scherer

WILEY-VCH

WILEY-VCH Verlag GmbH & Co. KGaA

The Editors

Prof. Detlef Stolten
Forschungszentrum Jülich GmbH
IEF-3: Fuel Cells
Leo-Brandt-Straße
52425 Jülich
Germany

Prof. Dr.-Ing. Viktor Scherer
Ruhr-Universität Bochum
LS f. Energieanlagen, IB 3/126
Universitätsstr. 150
44780 Bochum
Germany

Library of Congress Card No.: applied for

British Library Cataloguing-in-Publication Data
A catalogue record for this book is available from the British Library.

Bibliographic information published by the Deutsche Nationalbibliothek
The Deutsche Nationalbibliothek lists this publication in the Deutsche Nationalbibliografie; detailed bibliographic data are available on the Internet at http://dnb.d-nb.de.

Composition K+V Fotosatz GmbH, Beerfelden

Printing and Binding Strauss GmbH, Mörlenbach

Cover Design Schulz Grafik-Design, Fußgönheim

Printed in the Federal Republic of Germany
Printed on acid-free paper

ISBN 978-3-527-33002-7

Contents

Materials for CO_2Scrubbing

Processes for CO_2 Scrubbing

Preface

Carbon capture and storage represent one of the major building blocks for CO_2 emissions reduction. It is especially under discussion for fossil-based power plants, which are the main emitters of CO_2 worldwide. Substantial progress has been made in recent years in taking the technology from enhanced oil recovery, where CO_2 has been used for extracting crude oil for about three decades.

On transfer to a mainstream technology for sequestering carbon from fossil fuel-fired power plants, energy-efficient capture, transportation, and storage of the CO_2 become crucial. The search for effective and less energy-intense pathways has led to multiple routes of carbon capture and, hence, different power plant designs. The layout of different capture routes such as scrubbing technologies, oxy-fuel firing, and membrane separation needs the expertise of power plant engineers and process engineers on the one hand and chemists and materials scientists on the other.

The complexity of carbon capture and the wide range of specializations needed gave rise to this book on carbon capture in power plants. It is focused on coal power plants since they represent the major global CO_2 point sources in power production. Moreover, it keeps the requirements and process engineering solutions for the different pathways of carbon capture comparable. Post-combustion, pre-combustion, oxy-fuel, and chemical looping are the categories of carbon capture examined. Each of these topics is subdivided into an article on the process layout and an article on the state of the art of the materials needed for the different categories. Note that it is a special concern of this book to highlight the differences between membrane separation technologies and the other options such as scrubbing, cryogenic oxy-fuel separation, and chemical looping. The different characteristics of the technologies are highlighted with the intention of revealing the advantages and disadvantages. The question of how large the energy efficiency losses associated with this technology are and how we can reduce them was a primary driver for editing this book. A brief overview of the challenges in CO_2 transport and CO_2 storage complements the contents of the book.

This book contains the full texts of the overview presentations of the 2nd ICEPE 2011, Carbon Capture for Coal Power Plants, held in Frankfurt, Germany. DECHEMA is gratefully acknowledged for organizing this conference and supporting this book by making it part of the conference proceedings.

The great efforts that the authors have made in submitting their texts within the very short timeline given is acknowledged and many thanks are due to Michael Weber, who helped us considerably in handling the issues associated with editing this book. Not least, the great efforts of the Wiley team should be mentioned – they made it possible for this book to appear in about 12 months from the first idea to the printed version.

We hope that this book will serve professionals to complement their knowledge and that it will represent a quick but reasonably comprehensive reference for those who need an introduction to this area. It is designed to foster the understanding of the interdisciplinary aspects of carbon capture between plant engineers and process engineers on the one hand and materials scientists and chemists on the other.

Detlef Stolten
Jülich Research Center
and Technical University of Aachen, Germany

Viktor Scherer
Ruhr-Universität Bochum, Germany

February 2011

List of Contributors

Stefan Baumann
Forschungszentrum Jülich GmbH
Institute of Energy and Climate
Research IEK-1
52425 Jülich
Germany

Christoph Bayer
AVT – Chemical Process Engineering
RWTH Aachen-University
Turmstrasse 46
52064 Aachen
Germany

Franz Beggel
RWTH Aachen
Mechanische Verfahrenstechnik
Turmstrasse 46
52056 Aachen
Germany

Erika de Visser
Ecofys
Kanaalweg 16-A
3526 KL Utrecht
The Netherlands

Martin Dittmar
Technische Universität Dortmund
Lehrstuhl Umwelttechnik
Emil-Figge-Strasse 70
44227 Dortmund
Germany

Bernhard Epp
Technische Universität Dortmund
Lehrstuhl Umwelttechnik
Emil-Figge-Strasse 70
44227 Dortmund
Germany

Bernd Epple
Technische Universität Darmstadt
Chair of Energy Systems
and Technology (EST)
Petersenstrasse 30
64287 Darmstadt
Germany

Hans Fahlenkamp
Technische Universität Dortmund
Lehrstuhl Umwelttechnik
Emil-Figge-Strasse 70
44227 Dortmund
Germany

Paul Feron
CSIRO Energy Technology –
Newcastle
Steel River Estate
10 Murray Dwyer Circuit
Mayfield West
NSW 2304
Australia

Peter Michael Follmann
AVT – Chemical Process Engineering
RWTH Aachen-University
Turmstrasse 46
52064 Aachen
Germany

Johannes Franz
Ruhr University of Bochum
Department of Energy Plant Technology
Universitätsstrasse 150
44780 Bochum
Germany

Mark A. Gray
MAG Energy Consulting
Dublin, OH 43017
USA

Marcus Grünewald
Ruhr-University Bochum
Laboratory Fluid Separation
Universitätsstr. 150
44801 Bochum
Germany

Chris A. Hendriks
Ecofys
Kanaalweg 16-A
3526 KL Utrecht
The Netherlands

Mariya Ivanova
Forschungszentrum Jülich GmbH
Institute of Energy and Climate Research IEK-1
52425 Jülich
Germany

Alfons Kather
Hamburg University of Technology (TUHH)
Institute of Energy Systems
Denickestrasse 15 (K)
21073 Hamburg
Germany

Mathias Klostermann
Hamburg University of Technology (TUHH)
Institute of Energy Systems
Denickestrasse 15 (K)
21073 Hamburg
Germany

Joris Koornneef
Ecofys
Kanaalweg 16-A
3526 KL Utrecht
The Netherlands

Ralf Kriegel
Fraunhofer Institute for Ceramic Technologies and Systems
Michael-Faraday-Str. 1
07629 Hermsdorf
Germany

Klaus S. Lackner
Columbia University
The Fu Foundation School of Engineering and Applied Science
Earth and Environmental Engineering
Henry Krumb School of Mines
1038 W. 120th Street
New York, NY 10027
USA

Anders Lyngfelt
Chalmers University of Technology
Dept. Energy and Environment
41296 Gothenburg
Sweden

Tobias Mattisson
Chalmers University of Technology
Dept. Energy and Environment
41296 Gothenburg
Sweden

Thomas Melin
AVT – Chemical Process Engineering
RWTH Aachen-University
Turmstrasse 46
52064 Aachen
Germany

Wilhelm A. Meulenberg
Forschungszentrum Jülich GmbH
Institute of Energy and Climate
Research IEK-1
52425 Jülich
Germany

Michael Modigell
RWTH Aachen
Mechanische Verfahrenstechnik
Turmstrasse 46
52056 Aachen
Germany

Nicolas Nauels
RWTH Aachen
Mechanische Verfahrenstechnik
Turmstrasse 46
52056 Aachen
Germany

Jewgeni Nazarko
Forschungszentrum Jülich
Institute of Energy
and Climate Research
IEK-3: Fuel Cells
Leo-Brandt-Straße
52425 Jülich
Germany

Kitty Nijmeijer
University of Twente
Membrane Science and Technology
P.O. Box 217
7500 AE Enschede
The Netherlands

Jens Potreck
University of Twente
Membrane Science and Technology
P.O. Box 217
7500 AE Enschede
The Netherlands

Graeme Puxty
CSIRO Energy Technology
Newcastle Steel River Estate
10 Murray Dwyer Circuit
Mayfield West
NSW 2304
Australia

Sander R. Reijerkerk
University of Twente
Membrane Science and Technology
P.O. Box 217
7500 AE Enschede
The Netherlands

Ernst Riensche
Forschungszentrum Jülich
Institute of Energy
and Climate Research
IEK-3: Fuel Cells
Leo-Brandt-Straße
52425 Jülich
Germany

Sebastian Schiebahn
Forschungszentrum Jülich
Institute of Energy
and Climate Research
IEK-3: Fuel Cells
Leo-Brandt-Straße
52425 Jülich
Germany

Viktor Scherer
Ruhr University of Bochum
Department of Energy Plant
Technology
Universitätsstrasse 150
44780 Bochum
Germany

Katja Simons
University of Twente
Membrane Science and Technology
P.O. Box 217
7500 AE Enschede
The Netherlands

Hartmut Spliethoff
Technische Universität München
Lehrstuhl für Energiesysteme
Boltzmannstrasse 15
85748 Garching
Germany

Christina Stankewitz
Technische Universität Dortmund
Lehrstuhl Umwelttechnik
Emil-Figge-Strasse 70
44227 Dortmund
Germany

Detlef Stolten
Forschungszentrum Jülich
Institute of Energy
and Climate Research
IEK-3: Fuel Cells
Leo-Brandt-Straße
52425 Jülich
Germany

Jochen Ströhle
Chair of Energy Systems
and Technology (EST)
Technische Universität Darmstadt
Petersenstrasse 30
64287 Darmstadt
Germany

Stefan Telge
Technische Universität Dortmund
Lehrstuhl Umwelttechnik
Emil-Figge-Strasse 70
44227 Dortmund
Germany

Kaj Thomsen
Technical University of Denmark
TU Chemical Engineering
Department of Chemical
and Biochemical Engineering
CERE
Søltofts Plads
2800 Kgs. Lyngby
Denmark

Ruud van den Brink
Energy Research Centre
of The Netherlands ECN
Unit Biomass, Coal
and Environmental Research
Westerduinweg 3
1755 ZG Petten
The Netherlands

Eric van Dijk
Energy Research Centre
of The Netherlands ECN
Unit Biomass, Coal
and Environmental Research
Westerduinweg 3
1755 ZG Petten
The Netherlands

Tim van Gestel
Forschungszentrum Jülich GmbH
Institute of Energy and Climate
Research IEK-1
52425 Jülich
Germany

Peter Viebahn
Wuppertal Institute for Climate,
Environment and Energy
Döppersberg 19
42103 Wuppertal
Germany

Ingolf Voigt
Fraunhofer Institute for Ceramic
Technologies and Systems
Michael-Faraday-Str. 1
07629 Hermsdorf
Germany

Stéphane Walspurger
Energy Research Centre
of The Netherlands ECN
Unit Biomass, Coal
and Environmental Research
Westerduinweg 3
1755 ZG Petten
The Netherlands

Michael Weber
Forschungszentrum Jülich
Institute of Energy
and Climate Research
IEK-3: Fuel Cells
Leo-Brandt-Straße
52425 Jülich
Germany

Matthias Wessling
AVT – Chemical Process Engineering
RWTH Aachen-University
Turmstrasse 46
52064 Aachen
Germany

Matthias Wessling
University of Twente
Membrane Science and Technology
P.O. Box 217
7500 AE Enschede
The Netherlands

Li Zhao
Forschungszentrum Jülich
Institute of Energy
and Climate Research
IEK-3: Fuel Cells
Leo-Brandt-Straße
52425 Jülich
Germany

Part 1
Introduction and Overview

Process Engineering for CCS Power Plants, First Edition. Edited by Detlef Stolten and V. Scherer

1
The Case for Carbon Capture and Storage

Klaus S. Lackner

Abstract

For a world of 10 billion people to achieve a decent standard of living, a prerequisite for political stability and a requirement for curbing population growth, society will need access to far more energy than it has today. Efficiency and conservation can reduce the growth in energy demand, but they cannot stop it. However, concerns over climate change will limit the use of fossil fuels. Fossil fuel consumption is the most important contributor to climate change because of the associated carbon dioxide emissions. While the oxidation of carbon is the unavoidable consequence of extracting energy from fossil carbon, the subsequent emission could be avoided by capturing the carbon dioxide and storing it permanently.

Keywords: carbon capture and storage, excess carbon, carbon reservoirs, storage capacity

1.1
Introduction

Economic growth and human wellbeing require access to plentiful energy. Even if energy efficiency and conservation can reduce the energy requirement for a given economic output, rapid world-wide economic growth is likely to overwhelm these reductions. Indeed, even most business-as-usual scenarios already count on the benefits of efficiency improvements, because they assume that the energy intensity of the world is improving at a rate of 1–1.5% per year [1]. With this assumption, the world's energy consumption, which grew by a factor of 12 in the last century, may only rise by another factor of 3–4 over the course of the twenty-first century.

Universal economic wellbeing is a prerequisite for political stability and apparently also for a gradual decline and ultimate stop of world population growth. Hence the world needs to find a way of providing the energy necessary to assure a decent standard of living for the 10 billion people expected to live

Process Engineering for CCS Power Plants, First Edition. Edited by Detlef Stolten and V. Scherer

on Earth in the second half of the century. This presents an enormous challenge. It is made much more difficult by climate change concerns, which put in question the continued reliance on fossil fuels. Today, fossil fuels provide about 81% of all commercial energy.

The avoidance of climate change demands the stabilization of the CO_2 concentration in the atmosphere [2]. One may debate the precise level at which stabilization will need to occur, but there is little doubt that a continued and unchecked rise in CO_2 concentrations is not sustainable. Stabilizing the atmospheric concentration of CO_2 at any reasonable level can only be accomplished by either abandoning fossil carbon as a source of energy, or by keeping the resulting CO_2 away from the atmosphere. Keeping the fossil carbon, once mobilized, out of the atmosphere requires some form of carbon storage. In effect, for every ton of carbon coming out of the ground, another ton of carbon will have to be returned. This is not a small task, considering that in the process of harnessing the energy of fossil fuels every ton of carbon is converted into 3.7 tons of CO_2.

As the wild swings in oil prices in the last decade have shown, even a small mismatch between supply and demand can lead to drastic price changes. Therefore, a forced abandonment of fossil fuels, which would lead to far more dramatic shortfalls in energy supply, could precipitate a major energy crisis. On the other hand, emission of CO_2 to the atmosphere will have to stop eventually.

Carbon capture and storage (CCS) technologies can assure access to the vast fossil carbon resources known to exist [3]. At a minimum, the world needs sufficient storage capacity to allow for a smooth transition from today's fossil carbon-based energy infrastructure to one that does not depend on fossil fuels. At the other extreme, a large storage capacity could render fossil carbon competitive even in a carbon-constrained world. With sufficient storage capacity, fossil fuel consumption could proceed until other limits, either from additional environmental impacts or from resource limitations, bring it to an end.

1.2
Dilution versus Treatment

The old adage, already criticized in the 1950s, that "the solution to pollution is dilution" [4] tends to work well with small effluents and large reservoirs. As effluent volumes increase, dilution becomes inefficient and waste processing and treatment become necessary. Sewage from cities and sulfur from power plants are two waste streams that have gone through such transitions from dilution to treatment. By contrast, CO_2 is still in the dilution stage. Treatment of this effluent lags behind the others because it is a non-toxic, colorless, odorless gas that occurs naturally in sufficient concentrations to mask emissions. Therefore, it requires a much larger effluent before problems will become visible.

After two centuries of fossil fuel use, dispersion of CO_2 in the atmosphere has found its limits because of climate change. The capacity of the atmosphere is not large enough to dilute the CO_2 to a level that avoids adverse consequences. The CO_2 concentration in the atmosphere is 40% higher than in pre-industrial days, and unchecked it will double or triple before the end of the century. Dilution into the larger system of atmosphere, ocean, and biomass would alleviate the urgency of the problem, but would ultimately run up against the same limits [5].

Options for carbon storage include the dilution of CO_2 into the ocean, or a managed increase in biomass and soil carbon. Both options would buy some time. More permanent storage options include the formation of solid carbonates, which is known as mineral sequestration, or the injection of CO_2 into stable underground geological formations [6].

1.3 Carbon Reservoirs

At present, the concern over fossil fuel use is focused nearly exclusively on climate change, even though excess carbon in the environment also has other negative impacts. For example, excess CO_2 in the air results in higher concentrations of carbonic acid in the surface ocean and this acidification, in turn, is known to stunt coral growth [7, 8].

The air and the ocean are in close contact. Over time, the CO_2 transferred into the ocean migrates to deeper layers, reaching equilibrium throughout the entire water column in a few thousand years [9]. The carbon reservoirs in the ocean and the air are also tightly coupled to the carbon reservoirs comprising biological carbon in the form of living organisms and biological detritus mainly in the soil. The ocean, the world's biomass, and the atmosphere constitute a closely coupled set of carbon reservoirs that are remarkably self-contained. Although there is a large amount of flux between these pools, fluxes in and out of the larger pool are very small.

Emissions of CO_2 into the atmosphere result in a disequilibrium between these reservoirs and the carbon will gradually distribute itself between the ocean, the biomass, and the air. In the course of a century, about half of the CO_2 emitted will leave the atmosphere. On a millennium scale, equilibrium is achieved with 20% or more of the excess CO_2 staying in the atmosphere. The remainder will acidify the ocean. Over tens to hundreds of thousands of years, geological weathering processes will gradually remove excess CO_2 from the mobile carbon pool.

1.4 Excess Carbon

In short, one needs to consider the problem of CO_2 emissions from fossil fuels as a problem of excess in the world's mobile carbon stock. The excess is essentially permanent on a human time scale. On a time scale of less than a century, one can consider the atmosphere alone as a reservoir that will keep about half of the CO_2 that is injected.

The problem with emissions is that they accumulate. The environmental impact does not stop if emissions are halted; the size of the impact only stops growing. To a good approximation, the harm scales with the sum of all emissions over time, not with the rate of emission at a particular time.

Sustainability of fossil fuel consumption demands that the size of the mobile carbon pool is held in a range that avoids harm to the environment. As long as carbon is added to the pool, the CO_2 concentration in the air will increase and the ocean will acidify. Both changes have undesirable environmental consequences [2]. Stabilization of the CO_2 concentration in the atmosphere is possible only if emissions are reduced essentially to zero. The level at which stabilization is achieved will depend on the total amount of carbon that is mobilized and released to the atmosphere.

1.5 The Scale of Carbon Capture and Storage

The scope of the necessary CCS effort is defined by the gap between fossil fuel demand and allowable CO_2 emissions. The maximum acceptable level of CO_2 in the air or the maximum size of the mobile carbon pool is ultimately determined by policy decisions. Once this level has been reached, CO_2 emissions have to fall close to zero. Natural sciences can elucidate the consequences of elevated CO_2, but they cannot provide the value judgment necessary to determine what level is acceptable.

The acceptable upper limit of CO_2 in the air could be well below the current level. It has been argued by Hansen *et al.* [10] that breaching the 350 ppm level will destabilize the Greenland ice sheet and thus will result in unacceptable consequences. It is also possible that the value of 450 ppm, which has been embraced by European politicians, could still be acceptable. The acceptability of even higher levels cannot be entirely ruled out, even though they carry higher risks. Action is necessary, because business-as-usual scenarios could easily approach 1000 ppm before the end of the century [1].

No matter what the limit turns out to be, as the world approaches it, CO_2 emissions will have to be reduced towards zero. Hence continued use of fossil fuels will eventually require storing carbon at the same rate it is extracted from the ground. The question is not whether CO_2 emissions need to stop, but when they need to stop. The inertia in the world's energy infrastructure is such that even stopping at 550 or 650 ppm would suggest immediate action [11].

A stabilization level of 450 ppm would give the world another 60 ppm of uptake capacity in the air. Assuming that 50% of the CO_2 will remain in the air for a long time, this translates to a total carbon budget of 240 Pg (1 $Pg = 10^{12}$ kg or 1 Gt), or slightly more than 30 t for every person on the planet. This should be compared with an annual consumption of 6 t per person in the United States.

The combination of population growth and rapid economic growth in the developing countries suggests that energy demand is likely to grow substantially, even with large improvements in the energy intensity of the world economy [1]. A world of 10 billion people consuming energy at the same rate as the United States, Canada, or Australia would consume energy at roughly 10 times the current rate. Therefore, large new sources of energy will be needed. Whether or not the use of fossil fuels with CCS remains competitive compared with other forms of energy is impossible to predict. The current cost differential between coal-based electricity, nuclear-based electricity, and wind- or solar-based electricity suggests that fossil fuels with CCS could plausibly compete with these technologies.

In summary, a scenario in which an annual storage demand in excess of today's emissions of 30 Pg per year could develop over the next 50 years is not implausible. On the other hand, it is not impossible that, for example, solar energy or nuclear energy will outcompete fossil fuels that are handicapped by cost of capture and the demand for an exceedingly large storage capacity. This, however, would require dramatic improvements in these technologies.

1.6 Storage Capacity Requirements

Storage capacity requirements are large, which makes CCS technology challenging. The amount of CO_2 that may need to be sequestered during the twenty-first century could easily approach 5000 Pg, which is equal to the amount of water in Lake Michigan. The current rate of emission is 30 Pg per year, which equates to 3000 Pg in the course of a century. Economic growth and population growth could easily combine to create a much larger output rate, which would need to be matched by capture if CO_2 is to be stabilized.

The known resource base of fossil fuels is even larger than any reasonable estimate of this century's consumption. Therefore, it is very unlikely that the growth of CO_2 emissions is self-limiting. Instead, it may well be necessary to prepare for the possibility that 5000 Pg of carbon is going to be mobilized over the course of a few centuries. If left unchecked, this would result in the emission of 18 000 Pg of CO_2.

In effect, the required storage capacity is set by the size of the available fossil fuel resource. Conservative planning would assume that given enough time, all available carbon may be used. Only time can tell how much of this carbon is actually accessible or affordable. Advances in technology are notoriously unpredictable [12]. It is not possible to predict the cost of natural gas from hydrates below the ocean floor in 50 years from now. It is equally difficult to predict whether coal seams 2000 m below the ground will remain unmineable. Neither the gas hydrates nor the deep coal seams are counted in today's resource estimate.

1.7 Conclusion

The case for developing CCS and implementing the technology on a large scale rests on a number of premises and observations:

- Energy is critical to human wellbeing. The availability of energy must be increased, even if energy efficiency and energy conservation measures can make the current energy supply go much further than it can go today.
- Energy can be clean. Environmental problems currently associated with energy are not the direct result of the consumption of energy but the unintended consequences of the specific ways in which energy is extracted from the ground, packaged, and delivered to the consumer. Changes in the energy infrastructure can reduce or even eliminate these problems. Today, the most pressing environmental challenge is the elimination of CO_2 emissions, which are associated with all uses of fossil energy.
- CO_2 emissions must stop. CO_2 emissions, or more broadly the problem of excess mobile carbon in the environment, is a stock problem, not a flow problem. Once the concentration of CO_2 in the air reaches the stabilization point, emissions have to be reduced essentially to zero. To a good approximation, the climate change constraint limits the maximum amount of fossil carbon that can safely be mobilized, for example, by releasing it as CO_2 into the atmosphere.

The rate of current CO_2 emissions combined with projections of growth under business-as-usual scenarios could easily triple the preindustrial level of CO_2 in the atmosphere before the century is over, and thus exceed any level that is considered safe and acceptable. Fossil fuel resources are large enough and fossil fuel technology is sufficiently developed that they could play a major role in providing for the world's growing energy demand. Solar energy and nuclear energy may be the only other energy sources large enough plausibly to provide a substantial fraction of world energy demand [11].

The stabilization of CO_2 in the atmosphere will require a revolution in the world's energy infrastructure. All future energy systems will have to operate in a carbon-neutral manner. There are very few options for achieving this goal. CCS is one of them. Because access to energy is vital, it is important to pursue the major options vigorously and not unnecessarily increase the risk of a major energy crisis.

References

1 Lackner, K. S. and Sachs, J. D. A robust strategy for sustainable energy. *Brookings Pap. Econ. Activity*, 2005, **2**(2), 215–284.

2 IPCC. *Climate Change 2007: Synthesis Report. Contribution of Working Groups I, II and III to the Fourth Assessment Report of the Intergovernmental Panel on Climate Change* [Core Writing Team: Pachauri, R.K and Reisinger, A. (eds.)]. IPCC, Geneva, 2007.

3 Rogner, H.-H. An assessment of world hydrocarbon resources. *Annu. Rev. Energy Environ.*, 1997, **22**, 217–262.

4 Lee, D. B. The role of the State Water Pollution Control Authority. *Sewage Ind. Wastes*, 1957, **29**(2), 196–200.

5 Lackner, K. S. Carbonate chemistry for sequestering fossil carbon. *Annu. Rev. Energy Environ.*, 2002, **27**(1), 193–232.

6 Metz, B., *et al. IPCC Special Report on Carbon Dioxide Capture and Storage.* Cambridge University Press, New York, 2005.

7 Langdon, C., Takahashi, T., Sweeny, C., Chipman, D., Goddard, J., Marubini, F., Aceves, H., Barnett, H., and Atkinson, M. J. *Global Biochem. Cycles*, 2000, **14**(2), 639–654.

8 Kleypas, J. A., Feely, R. A., Fabry, V. J., Langdon, C., Sabine, C. L., and Robbins, L. L. *Impacts of Ocean Acidification on Coral Reefs and Other Marine Calcifiers*, Report of a Workshop Sponsored by NOAA, NSF and USGS, June 2006.

9 Archer, D., Kheshgi, H., and Maier-Reimer, E. Multiple timescales for neutralization of fossil fuel CO_2. *Geophys. Res. Lett.*, 1997, **24**, 405–408.

10 Hansen, J., Sato, M., Kharecha, P., Beerling, D., Berner, R., Masson-Delmotte, V., Pagani, M., Raymo, M., Royer, D. L., and Zachos, J. C. *Open Atmos. Sci. J.*, **2**, 217–231.

11 Lackner, K. S. Comparative impacts of fossil fuels and alternative energy sources. In *Carbon Capture Sequestration and Storage. Issues in Environmental Science and Technology*, Vol. 29 (ed. Hester, R. E. and Harrison, R. M.), Royal Society of Chemistry, Cambridge, 2010, pp. 1–40.

12 Smil, V. *Energy at the Cross Roads*, MIT Press, Cambridge, MA, 2005

2
Advanced Power Plant Technology

Hartmut Spliethoff

Abstract

In the field of steam production, recent years have seen the beginning of rapid developments towards enhancing efficiency. The characteristic of this development is the transition to higher live steam conditions with temperatures above 540 °C and pressures above 250 bar. The higher live steam temperatures increase the thermal cycle efficiency because of the higher mean temperature of the heat input. Potential for optimization is also available at the cold end of the steam cycle. In addition to increasing the thermal efficiency, cutting down the auxiliary power demand by the use of low-loss feed pump drives and reducing various other losses also contribute substantially to the enhancement of individual efficiencies. This chapter discusses the options for increasing the efficiency in the design point and also during operation, and describes the steps in the development of coal-fired steam power plants in the past and in the future.

Keywords: coal-fired power plants, steam generation, thermal efficiency, hard coal, brown coal

2.1
Introduction

The generation of electrical power in a steam power plant involves multiple conversion of the primary energy contained within the fuel. In combustion, the fuel is oxidized by the oxygen of the combustion air, thus generating hot flue gas. In this process, the fuel's latent chemical energy is released. The hot flue gas transfers its thermal energy by radiation and convection to the working media (water/steam) via the boiler heat exchanger surfaces.

In the turbine, the thermal energy of the steam is converted into mechanical energy, which in turn is transformed into electrical energy in the generator. These conversions involve certain losses of the fuel energy input, as shown in Fig. 2.1. The major loss of 50% of the fuel heat input can be explained by the cycle efficiency. The waste heat is dissipated to the environment through the

Process Engineering for CCS Power Plants, First Edition. Edited by Detlef Stolten and V. Scherer

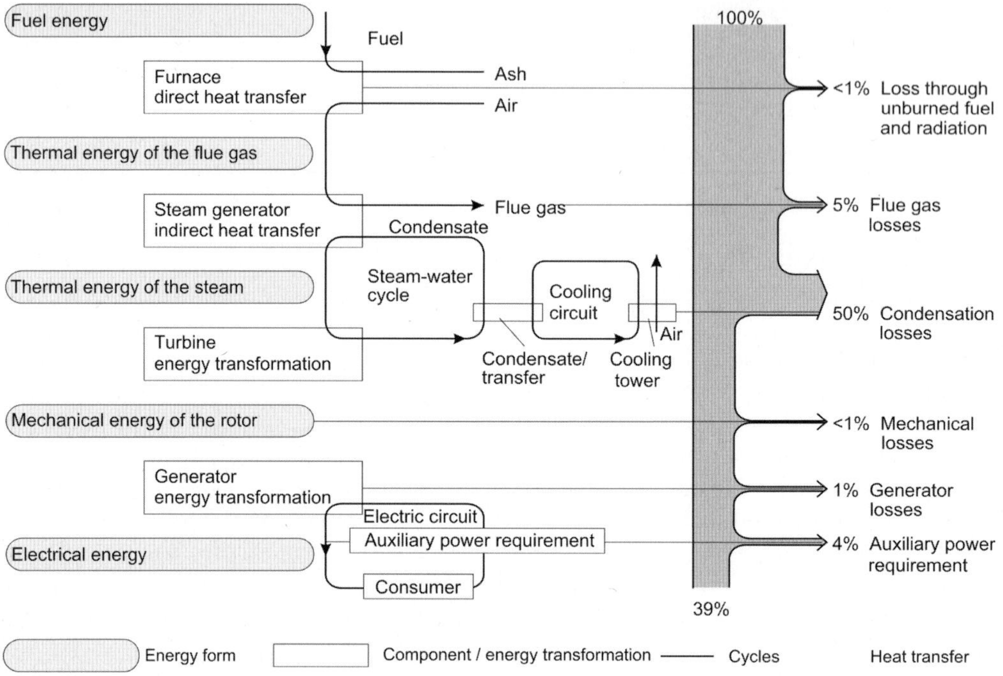

Fig. 2.1 Energy transformation or conversion, circulation of energy-carrying media, and efficiency in a condensation power plant [1].

condenser. Further significant losses occur in the steam generator, mainly as flue gas losses of about 6%. The auxiliary power requirements of about 4% of the fuel energy input add to these losses [1].

2.2
History of the Development of Power Plants – Correlation Between Unit Size, Availability, and Efficiency

The block power station was born out of the need for higher power plant capacities (due to increasing energy demands), changing expectations with respect to lower investment costs and the desire for higher reliability of power supply. In addition to other parameters, particularly the

- unit output,
- efficiency, and
- availability

are the factors that describe the development of the block power station unit.

Given the greater availability of each of the large plant parts, modern hard coal-fired power plants are generally designed as block units, meaning that all the process units are contained together within a single "block". The direct physical interactions of steam generators, turbines, and auxiliary installations involve less investment because of shorter connecting pipelines. In addition, the pressure and the heat losses are lower than in the range-type power stations that were common earlier in the 1900s. From the early 1950s, condensation power plants were built as block units with simple reheating for base and intermediate loads.

The maximum block capacity rose with the maximum capacities of single plant components. Step by step, the power station unit has been supplemented by additional components and plants. Today, the largest unit capacities are 1100 MW in Europe and 1300 MW in the United States (see Fig. 2.2). Conventional live steam conditions proven in operation are 180–250 bar and 540 °C, with the reheater also at 540 °C. Throughout the world, one can see a trend towards higher live steam conditions. Figure 2.3 shows the development of live steam conditions in Germany [1].

With increasing unit capacities and live steam conditions, the efficiency levels also rose (Fig. 2.4) [1]. The power station costs decreased, depending on the capacity, making efficiency-enhancing measures become more cost-effective. The higher efficiencies of large units can also be explained physically: the specific surface heat losses of boilers and losses of rotating machinery due to leakiness diminish with higher capacities.

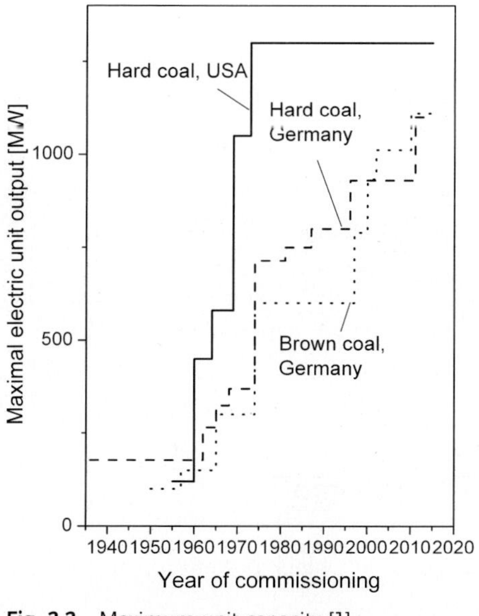

Fig. 2.2 Maximum unit capacity [1].

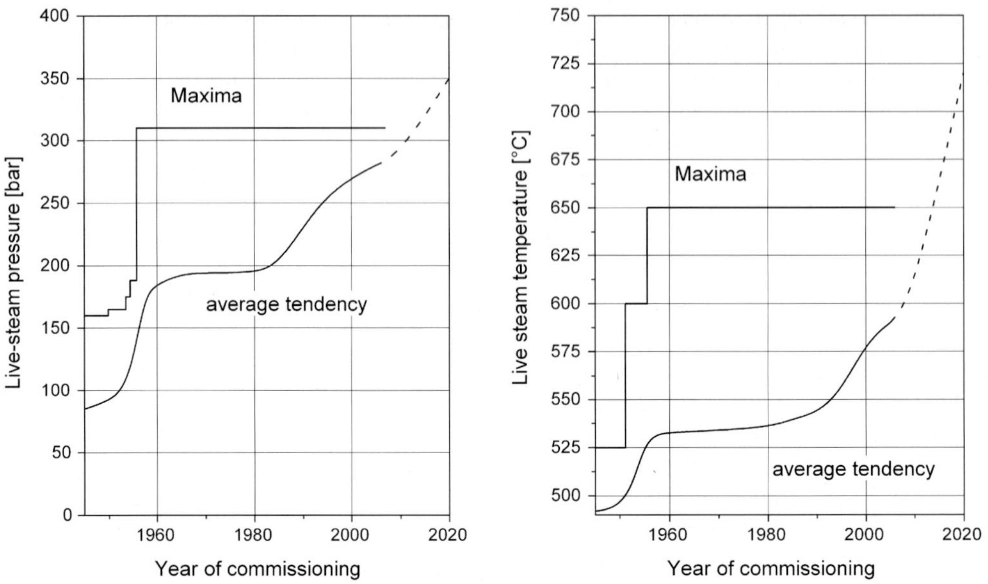

Fig. 2.3 Evolution of live steam conditions in German plants [1].

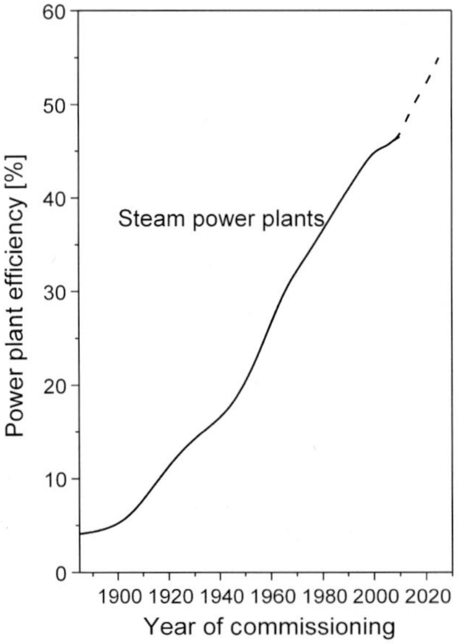

Fig. 2.4 Evolution of the efficiency level of German plants [1].

The availability of technology becomes more important with increasing capacities, the need for more pollution control equipment, and the desire for technical developments towards higher efficiency levels. High availability is desirable for reliable electricity production and a necessary comparative criterion of technical developments. The further development of steam power plants should therefore be based on the comparable availability of proven power plant concepts.

2.3 Possibilities for Efficiency Increases in the Development of a Steam Power Plant

In this section, possibilities for increasing efficiency by different measures will be discussed. The possibilities in this respect are distinguished for stationary operation and are as follows:

- increases in the thermal efficiency of the cycle
- measures to minimize losses
- measures to reduce auxiliary power requirements.

To quantify the contributions of different measures, a newly designed power station will be compared with an existing power station as reference. The reference power plant corresponds to the state of the art from the 1980s in Germany and has a net efficiency of 39%, [steam conditions 200 bar/535 °C (live steam)/ 535 °C (reheat steam)]. The stated efficiency rates usually refer to the rated power. However, the efficiency of the plant in part-load operation and the losses at startup and shutdown should also be taken into account [1].

2.3.1 Increases in Thermal Efficiencies

Improvements in thermal cycles are aimed at attaining a high mean temperature of heat addition and a low mean temperature of heat extraction. High thermal efficiencies are achieved by

- increasing the live steam conditions (temperature and pressure)
- single or double reheating
- regenerative feed water preheating
- reducing reheater spraying
- low exhaust steam temperatures in the condenser.

2.3.1.1 High Temperature of Heat Addition

High mean temperatures of the heat addition contribute to high thermal efficiency. They can be achieved by a high pressure in the high-pressure (HP) steam generator, a high live steam temperature, regenerative feed water preheating, and reheating to high reheater temperatures. Figure 2.5 shows the influence of pressure and temperature on the efficiency of the cycle, given as the rel-

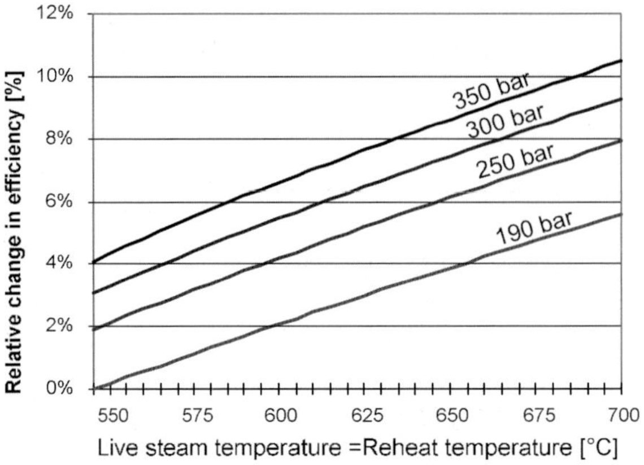

Fig. 2.5 Influence of live steam pressure and temperature on heat rate [1].

ative heat rate gain. For the temperature range up to 600 °C, a rule of thumb is an increase in the net efficiency of 0.02% (absolute) per degree of temperature increase (with the live steam temperature equaling the reheater temperature). In the range from 600 to 700 °C, the increase is about 0.016% per degree of temperature increase. The influence of the temperature increase of the live steam in this process is somewhat higher than that of the reheater temperature. In the pressure range up to 250 bar, an increase in the live steam pressure results in an efficiency improvement of 0.01% bar^{-1}; higher pressures of up to 300 bar result in an improvement of about 0.008% bar^{-1}. With even higher live steam pressures, the gain in net efficiency diminishes again [2–5]. High live steam pressures and temperatures are limited by the construction material available [6, 7].

With increasing live steam conditions in large-scale steam generators, the designs of the membrane wall and final superheater heating surfaces and the HP outlet header approach maximum stress limits of the material [8–11]. Figure 2.6 illustrates how material developments of steels determine and limit the application of higher steam conditions. The development and qualification of new materials for future power plants with steam conditions of up to 350 bar and 750 °C are ongoing in various research projects in Europe, Japan, and the United States [7, 12–20].

Assuming conventional steam conditions, such as those of the reference power plant, introducing double reheating raises the net efficiency rate by up to 2%. Higher live steam pressures increase whereas higher live steam temperatures decrease the gain in efficiency. For a power plant with steam conditions of 280 bar, 585 °C (live steam), 600 °C (reheat steam), the double reheating raises the net efficiency by 0.7% [5]. A drawback is the increased pressure loss. The double reheating can have a disadvantageous effect on the operating regime. Double reheating allows low condenser pressures, because the exhaust steam moisture is reduced, avoiding droplet erosion of the final-stage blades. Double

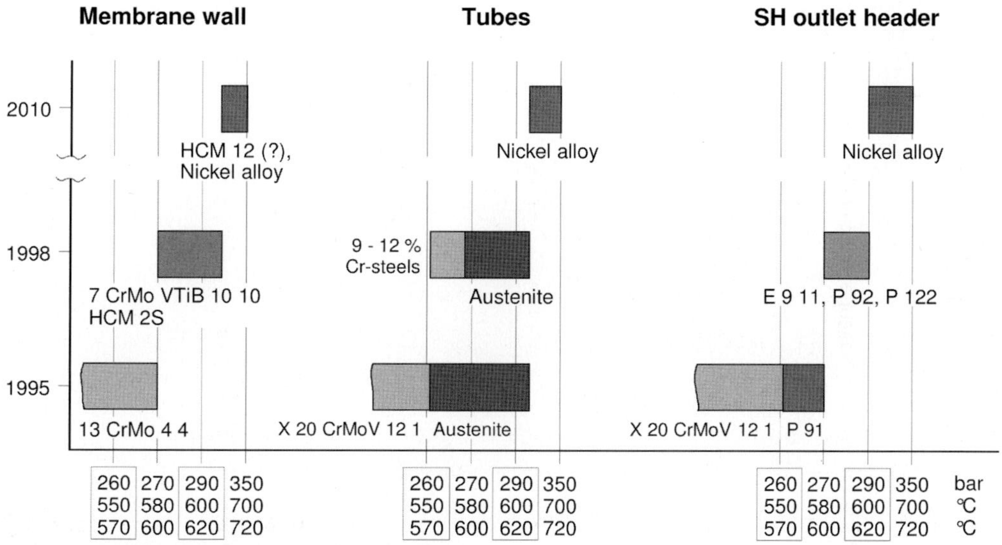

Fig. 2.6 Development of steam conditions and steam generator materials (source: Alstom Power).

reheating, as opposed to single reheating, requires additional investments for the steam generator, the tubes, and the turbine of between 2.5% [21] and 3% [22] of the total investment costs. In Germany, double reheating was installed in 11 plants in total between 1950 and 1983 [23]. Because of the higher costs and the disadvantages for mid-range load operation, this technology was not employed after that time, and it is also not taken into consideration in current projects in Germany.

A new double reheating concept design currently under development is offering promising costs comparable to single reheating. The idea of the so-called Master Cycle is to reduce the exergy loss of the heat transfer of superheated bleed steam from the intermediate pressure (IP) turbine to the feed water. Superheated bleed steam from the IP turbine is used for feed water preheating at the level of its condensation temperature, resulting in high exergy losses. The exergy losses are higher for double than for single reheating. The exergy loss by extracted steam can be reduced by shifting the IP extraction stages to a separate turbine fed by steam from the first cold reheat steam line. The new turbine expands the cold reheat steam to temperatures and pressures of the extraction stages, delivering about 4% of the net power. The steam flow through both reheating stages is reduced, resulting in lower capital costs of the cycle. Calculations give a Master Cycle efficiency improvement of 1.45% (326 bar/592 °C/605 °C/605 °C, 23 mbar) over single reheating [7, 24].

Controlling the reheater temperature by a spray attemperator diminishes the efficiency, because the high-pressure range of the steam generator is bypassed by doing so, and steam is produced at a low pressure and temperature. Figure

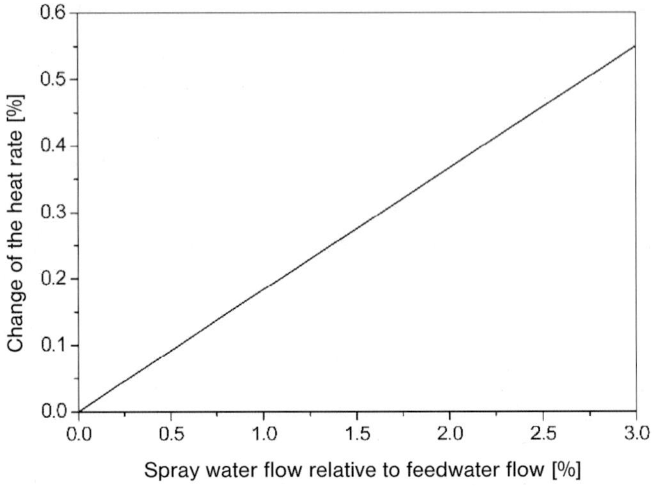

Fig. 2.7 Influence of reheater spraying on the efficiency [25].

2.7 shows the influence of the reheater attemperator mass flow on the efficiency [25]. In the case of the reference power plant, the spraying mass flow at full load is about 0.9% of the feed water mass flow. New power plant designs limit the temperature-controlled spraying mass flow to 0.2% of the feed water mass flow [26].

Regenerative preheating of the feed water raises the feed water temperature via several stages, up to the temperature at which it enters the steam generator. The extracted steam has already produced power in part of the turbine. The final, uppermost HP heater is usually heated by extracted steam from the cold reheat line downstream of the HP turbine. The reheat pressure, derived from optimization calculations for the entire cycle, defines the feed water outlet temperature [27].

The feedwater heating temperature can be further raised by inserting an additional preheater, heated by extraction steam from the HP section of the turbine. Such additional extraction from the HP turbine section uncouples the reheater pressure and feed water outlet temperature. Figure 2.8 shows a heat flow diagram, with stages, where the feed water is preheated to 300 °C.

Nowadays, six to nine feed water heaters with outlet temperatures between 250 and 300 °C are commonly used for large thermal power plants. Higher feed water outlet temperatures are chosen as live steam pressures increase. An increase in the feed water heating temperature from 250 to 290 °C by additional extraction of steam from the HP turbine section results in an efficiency increase of 0.7% [4]; the result of an increase from 270 to 300 °C is an absolute improvement of 0.75% [5]. Figure 2.9 shows that an increase in the feed water temperature results in a relative decrease of the heat rate, which is a dependent on the pressure level [2].

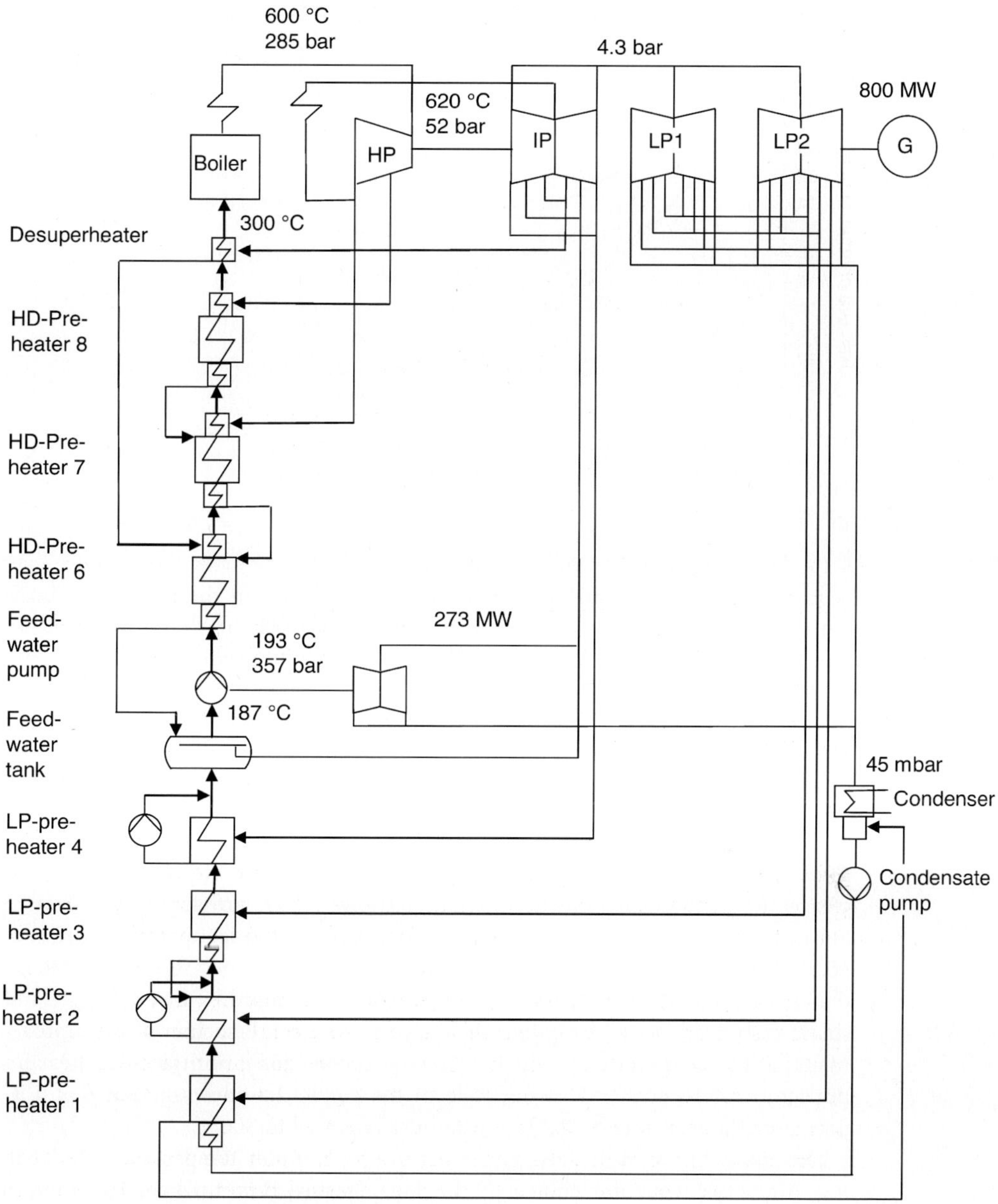

Fig. 2.8 Heat flow diagram of a thermal power plant with advanced steam conditions and nine-stage feed water heating [1].

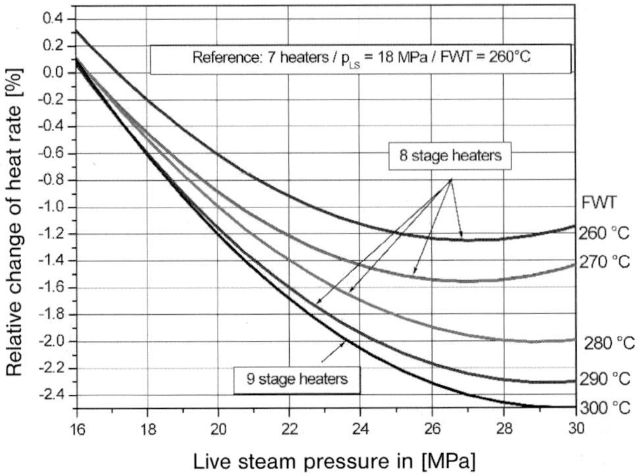

Fig. 2.9 Effect of live steam pressure and feed water temperature on the heat rate [2].

In designing a power plant, after the feed water preheat temperature has been defined, further optimization is possible only within the feed water heating chain. Increasing the number of heaters while maintaining the same outlet temperature results in smaller temperature rises for the individual stages. This helps to achieve a better adaptation of the temperatures of the heat-absorbing to the heat-dissipating heat transfer medium, namely the feed water and extracted steam flows, and hence to minimize the exergy losses.

2.3.1.2 **Lower Heat Dissipation Temperatures – Optimization of the "Cold End"**

The heat dissipation temperature has an impact on the efficiency, which increases when the heat input temperature is lower. This correlation is shown in Fig. 2.10 for the Carnot cycle, with a mean heat dissipation temperature of 30 °C, corresponding to the condensation temperature. Therefore, it is evident that in a pure steam process, in comparison with a combined-cycle (gas and steam turbine) process, a higher efficiency increase can be achieved by improvements at the cold end [28]. For the reference power plant with a mean heat input temperature of 376 °C, a decrease in the condensation temperature of 1 °C decreases the heat rate by 0.29% in the ideal case.

The optimization of the cold end must involve not only the design of the cooling circuit but also the choice of the low-pressure (LP) turbine. In order to make use of the efficiency advantage of low condenser pressures, it is necessary to enlarge the exhaust steam cross-section of the LP turbine. In addition to developing and utilizing greater LP final-stage blade lengths, today final-stage blades are manufactured with lengths up to 1400 mm, for use at 3000 rpm [29]; a larger outlet cross-section can be obtained with two, three, or four LP turbine components mounted on a shaft. Increasing the number of LP turbines, however, entails step changes in the costs for turbines and turbine houses [30].

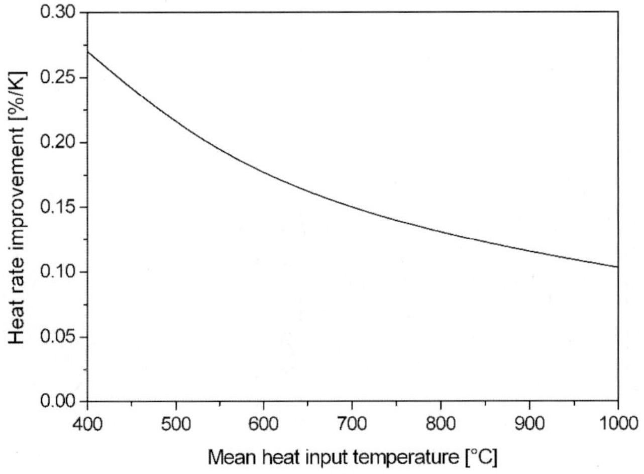

Fig. 2.10 Impact of a heat dissipation temperature reduction of 1 °C [1].

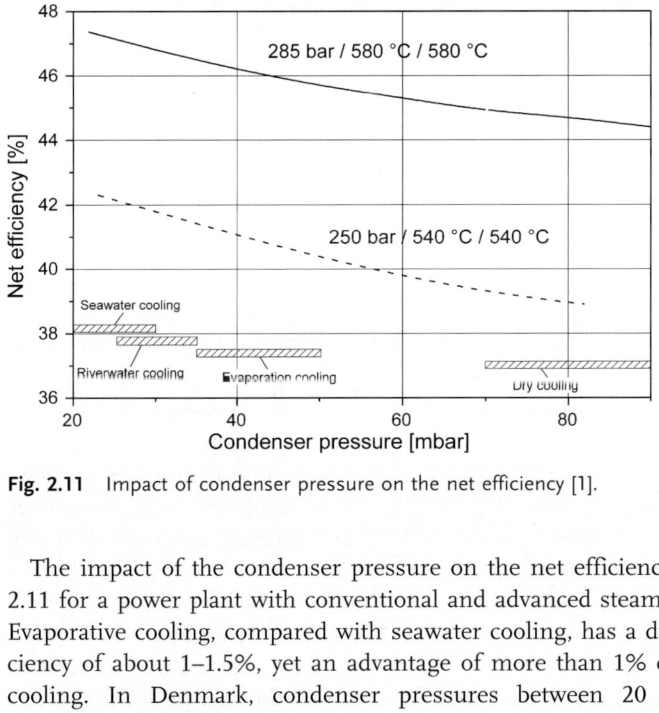

Fig. 2.11 Impact of condenser pressure on the net efficiency [1].

The impact of the condenser pressure on the net efficiency is shown in Fig. 2.11 for a power plant with conventional and advanced steam conditions [3, 21]. Evaporative cooling, compared with seawater cooling, has a disadvantage in efficiency of about 1–1.5%, yet an advantage of more than 1% compared with dry cooling. In Denmark, condenser pressures between 20 and 25 mbar are achieved in advanced steam cycles with seawater cooling at an annual average of about 10 °C. The reference values for river water cooling in Germany range around 30 mbar.

In Germany, natural draft cooling towers for waste heat dissipation and back cooling of the heat carrier, that is, cooling water, have become standard.

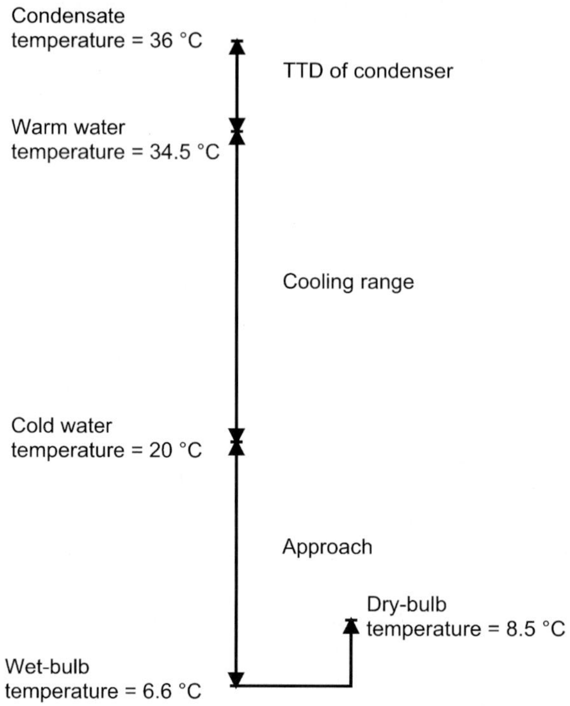

Fig. 2.12 Temperature relations in circuit cooling systems with a wet cooling tower [1].

The wet-bulb temperature of the ambient air is the physical limit for the condensation temperature and hence for the efficiency improvement at the cold end. For plants with wet cooling towers, the difference between the wet-bulb temperature of the ambient air and the economic temperature or the mean condensation temperature results from the following three factors: the cooling range, the terminal temperature difference (TTD) of the condenser, and the approach of the cooling tower (Fig. 2.12).

The economic efficiency of the condenser temperature depends on the degree of utilization of the power plant and the fuel price on the one hand and on the additional investment necessary for an efficiency increase on the other. The condenser pressure for a mid-range plant, with an operation factor of 4000 full-load hours per annum and a unit size of 500 MW_{el}, in the 1980s ranged between 0.04 and 0.05 bar [22]. Recent power station plans provide for values between 45 and 30 mbar [4, 31–35].

2.3.2 Reduction of Losses

2.3.2.1 Internal Turbine Efficiency and Losses

In the expansion process in the turbine, the steam is accelerated and its kinetic energy converted into mechanical work by impulse transfer on to the rotating blades. The measure for the quality of the conversion into mechanical work is the internal turbine efficiency $\eta_{i,T}$, which indicates the difference between real and loss-free isentropic expansion.

About two-thirds of the total losses occur over the blade stages. The losses arise through fluid friction in the channels, friction of the rotating blades in the surrounding steam, steam leakages from rotating and fixed parts, and steam moisture in the final stages [36]. The HP first stages (of the turbine) in particular and the LP final stages (of the turbine) are the areas of the turbine that incur the highest losses.

Turbine improvements have contributed substantially to the increases in efficiency of modern power plants. Three-dimensional calculations (i.e., computer modeling) of flow processes reveal the potential for reducing the flow losses, and modern manufacturing technologies make it possible to build complex blade geometries [37, 38]. The modernization of the turbines of existing power plants is an effective means to increase the efficiency. Figure 2.13 shows the internal turbine efficiencies for existing and planned power plants [4].

2.3.2.2 Steam Generator Losses

In the steam generator or boiler, the chemically bound energy of the fuel is converted into thermal energy of the flue gas and then transferred to the steam–water cycle. The efficiency of the energy conversion is designated the steam generator or boiler efficiency, η_B, and the losses that arise are called the boiler

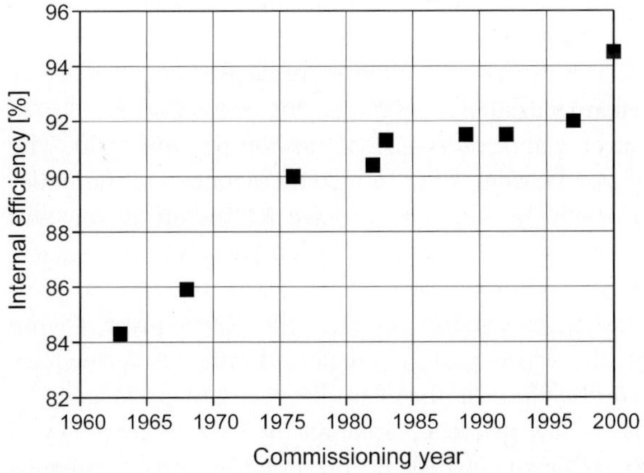

Fig. 2.13 Development of the internal efficiencies of steam turbines [4].

Table 2.1 Boiler losses for the reference power plant and for a new plant [1].

Parameter	Reference power plant	New plant
Air ratio	1.3	1.15
Flue gas temperature (°C)	130 °C	110 °C
Flue gas losses (%)	5.3	3.8
Boiler radiation (%)	0.25	0.3
Loss through unburned matter		
Fly ash (%)	0.2	<0.3
Hopper ash (%)	0.1	<0.2
Sensible heat		
Fly ash (%)	0.02	0.03
Hopper ash (%)	0.04	0.04
Total (%)	5.9	4.6

losses. Referring to the calorific value of the fuel, the steam generator efficiency of modern hard coal-fired furnaces amounts to 94% and that of brown coal-fueled furnaces around 90%.

The losses consist of

- loss through unburned matter (κ_U)
- loss through sensible heat of the slag (κ_S)
- flue gas loss (κ_{FG})
- loss through radiation and convection of the external surfaces of the boiler (κ_{RC}).

Efforts to increase the steam generator efficiency concentrate on reducing the flue gas heat loss. This loss depends both on the flue gas outlet temperature of the steam generator (after the air heater) and on the flue gas mass flow. Table 2.1 compares the boiler losses of the reference power plant and a planned power plant.

2.3.2.3 Advanced Flue Gas Heat Utilization

The designs of advanced power plants make use of low-temperature heat. The usable temperature gradient lies between the flue gas temperature at the boiler exit and the temperature to which the flue gas is cooled for the desulfurization process.

This development utilizes an additional heat exchanger to tap low-temperature heat, which is fed to the thermodynamic cycle of the power plant. Owing to its low temperature level, the waste heat is transferred either to a low-temperature condensate or to combustion air that has not yet been preheated. By preheating a partial condensate flow in the LP area, an increase in efficiency of 0.25% is obtained for hard coal firing – which is countered by considerable expenditure in equipment and therefore not considered economical [35]. For

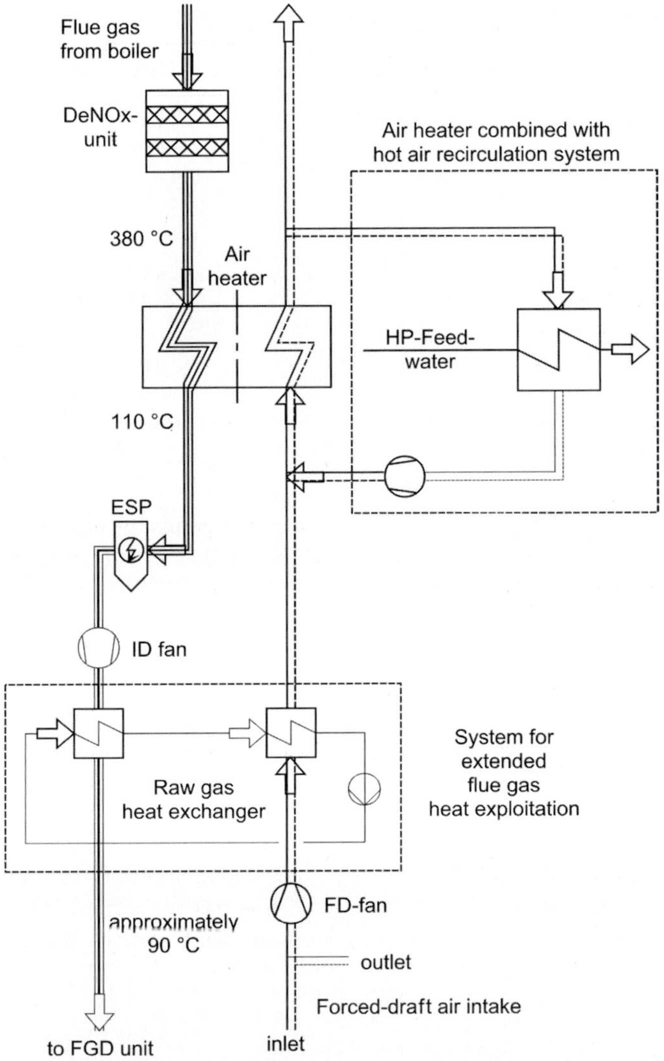

Fig. 2.14 Configuration for extended flue gas heat utilization [4].

brown coal firing, an increase in efficiency of up to 0.7% has been reported [39].

Figure 2.14 shows a schematic diagram of the configuration for transferring low-temperature flue gas heat to the HP feed water for pulverized hard coal firing. The flue gas heat is first transferred to the cold combustion air flow in a heat exchanger and then to the HP feed water via a hot-air recirculation system [4]. A similar configuration is used for the power plant Niederaussem K [32, 40]. Due to the transfer of low-temperature heat of the flue gas to the air, less flue gas heat is required in the regenerative heat exchanger for final air preheating. Therefore, about one-third of the flue gas bypasses the regenerative air preheater, and the heat from this portion of the flue gas is transferred by flue gas/water heat exchangers to the LP and HP feed water preheaters.

Advanced flue gas heat utilization can increase the net efficiency of a hard coal-fired power station by 0.6% [4]. With brown coal firing, given the higher flue gas temperature of about 170 °C and the higher flue gas mass flow, up to 1.6% increases can be achieved [39, 41].

2.3.2.4 **Other Types of Losses**

Pipework Losses

The hot HP steam produced in the steam generator is fed to the HP section, and the hot reheater steam is fed to the IP section of the turbine – both via pipes with lengths of 100–150 m. Even for the highest capacities, it is common to use two live steam pipes for each run. The energetic heat losses of these pipes, which connect the steam generator to the turbine, have an efficiency η_{pipe}, which is around 0.997 [42].

In addition to heat losses, pressure losses also cause a temperature decrease and thus lead to exergy losses of the steam between the steam generator and the turbine inlet. The temperature decrease caused by throttling does not have an effect on the pipe loss, η_{pipe}, but does on the thermal cycle efficiency. The temperature of the live steam decreases by between 5 and 2 °C in total; the pressure loss ranges between 5 and 10 bar. The former values hold for new plants and for power plants in the planning stage.

Generator Losses

In the generator, the mechanical energy is converted into electrical energy. The efficiencies of large generators, in the range 700–900 MVA, with a water-cooled stator and a hydrogen-cooled rotor, are around 98.7%. An efficiency increase to about 99.4% is expected to be possible with a superconducting rotor winding. In addition to the efficiency increase, a smaller space requirement is also expected [43].

Losses of the Main Transformer

Main transformers transform the electrical voltage of the power output supplied to the electrical network, from 21 to 220 or 380 kV or higher, in order to keep the network losses low. They are usually designed as three-phase transformers. The transformer efficiency, which is the ratio of the supplied energy to the energy absorbed, lies, for large main transformers at rated power, in the range 99.6–99.7% [3].

2.3.3
Reduction of the Auxiliary Power Requirements

The auxiliary power requirement of a power plant is the sum of the total electrical and mechanical power demand for driving ancillary and auxiliary systems. The electrical auxiliary power requirement, $P_{aux,el}$, is comprised of the input power of all machines driven by electric motors such as fans [forced draft (FD) and induced draft (ID) fans and primary air fan], pumps (the feed water pump if it has an electrical drive, the condensate, cooling-water, and scrubbing slurry pumps), and mills, and other power electrical devices and equipment such as the transducer and the electrostatic precipitator (ESP) for dust removal. Mechanical power is employed if the feed water pump is driven by a steam turbine instead of an electrical drive.

About 6–10% of the gross electrical power of a power plant is needed for the auxiliary power of the plant, meaning an auxiliary power demand efficiency of 90–94%.

The electrical auxiliary power demand is made of "useful" power plus losses. The useful or net power is roughly between 90 and 99%, whereas the losses are between 1 and 10% of the power demand. Process engineering improvements reduce both the useful power demand and the losses; improvements of the auxiliaries only diminish the losses. In the example of air staging for the reduction of NO_x formation, the useful power demand and losses of auxiliaries are reduced, hence the auxiliary power requirements diminish. The aim of optimization measures for drives and machines is to reduce the losses.

The auxiliary power demand of different main and auxiliary devices are described in the following. For the reference power plant and for a new plant, the auxiliary power values are listed in Table 2.2.

Table 2.2 Auxiliary power requirement breakdown for the reference and a new power plant [1].

Live steam pressure	Reference power plant (%)	New plant (%)
FD fan	0.8	0.5
ID fan	1.2	0.8
Mills and mill fan	0.4	0.4
Coaling, ash removal, dust removal	0.3	0.3
Boiler feed pump	2.7 (260 bar)	2.9 (310 bar)
Condensate pump	0.3	0.2
Cooling water pump	0.7	0.6
FGD unit	2.0	0.8
Other equipment	1.0	0.7
Total	9.4	7.2

2.3.4
Fuel Drying for Brown Coal

The exergetically unfavorable milling and drying processes and the influence of the fuel moisture on the flue gas have resulted in the net efficiency of a brown coal power plant, under comparable steam conditions, being maybe up to 3% lower than the net efficiency of a hard coal power plant, depending on moisture and sulfur contents. The disadvantage of the lower efficiency of brown coal (as opposed to hard coal) can be compensated or even overcompensated by external fuel drying using low-exergy drying media. Higher cycle efficiencies can be achieved if media with low temperature and exergy are utilized for the drying. Since the brown coal drying only needs temperatures of slightly above 100 °C, low-temperature flue gases or extracted steam can be possible drying means.

External drying with separate discharges of water vapor and flue gases can diminish the flue gas losses. Separate discharges allow the use of lower temperatures of both the water vapor and the flue gas due to the low acid dew point.

A further efficiency increase is possible if the evaporation heat of the water vapor can be recovered in the power generating process. In this way, the higher heating value can be tapped for exploitation.

A schematic diagram of the WTA-process (Wirbelschichttrocknung mit Abwärmenutzung; fluidized bed drying with internal waste heat exploitation) is shown in Fig. 2.15 [44]. At a temperature of about 100–120 °C, the bed is fluidized with water vapor from the drying process. To obtain 12% residual moisture of the brown coal, it is necessary to have a fluidized bed temperature of around 110 °C. After a dust removal stage, the water vapor formed during drying is compressed in a vapor compressor to about 4 bar, and then condensed in an in-

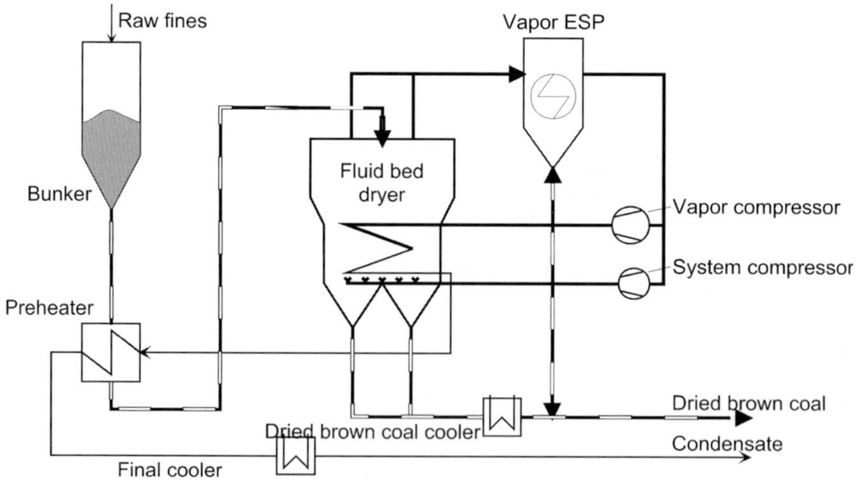

Fig. 2.15 Schematic diagram of WTA-process drying: fluidized bed drying with internal waste heat exploitation [44].

bed heat-transfer surface to dry the fluidized brown coal. In the fluidized-bed heat exchanger, the superheat and the condensation heat of the vapors are transferred to provide the evaporation heat of the brown coal moisture. The WTA-process is reported to have a 5.5% efficiency advantage over brown coal-fired power plants with hot gas drying processes [45].

After several development steps, a full-scale WTA-process prototype with an output of $110\,t\,h^{-1}$ of dried brown coal started operation in 2008. The dried brown coal is fed to the power plant Niederaussem K, delivering up to 30% of the fuel input [45–47].

2.3.5 Losses in Part-Load Operation

2.3.5.1 Impact of the Operating Regime of the Steam Generator and Turbine

The output of the turbine is controlled either by enlarging the turbine inlet cross-section (constant-pressure operation) or by modifying the pressure (sliding-pressure operation). The different control modes diminish the efficiency in part-load operation to varying extents. Figure 2.16 shows the impacts of the control modes on sliding-pressure operation. It has a small advantage compared with constant-pressure operation with governing control. Modified sliding pressure involves a higher heat rate than natural sliding pressure, which is due to the slight throttling of the valves [3].

2.3.5.2 Example of a Reference Power Plant

In part-load operation, the efficiency of the reference power plant diminishes because the efficiencies of the plant components change. In addition to showing the total efficiency, η_{tot}, Fig. 2.17 shows the course of the thermal efficiency, η_{th}, the steam generator efficiency, η_{StG}, the auxiliary power demand efficiency, η_{aux},

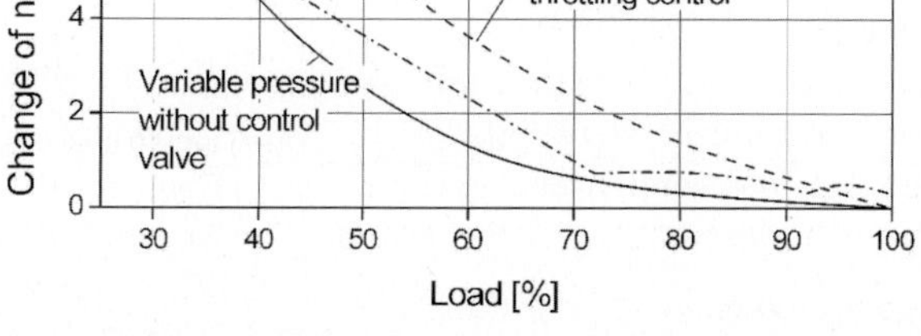

Fig. 2.16 Net heat rate changes with different control modes [3].

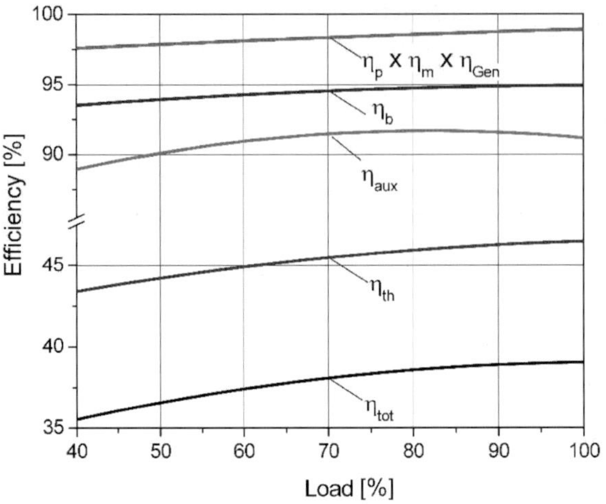

Fig. 2.17 Efficiencies of the reference power plant during part-load operation [1].

and the product of pipework, mechanical, and generator efficiencies, η_{pipe}, η_{mech}, and η_{gen}.

The thermal efficiency depends on the live steam conditions before the turbine, the internal turbine efficiency, and the conditions in the condenser. In the reference power plant, operated at sliding pressure, the mean temperature of the heat input decreases as the pressure decreases, even if the temperatures before the turbine are kept constant. In part-load operation, the pressure in the condenser decreases as a consequence of the diminishing heat flux to be dissipated. This advantage of the lower temperature of heat dissipation, however, cannot offset the disadvantage of the lower temperature of the heat input.

The losses of the steam generator depend only, and slightly, on the load. Although the flue gas temperatures decrease with the shift from convective to radiative heat transfer, the advantage is nullified by the higher air ratio for the reheater temperature control. Since the power requirement of the auxiliary drives, except for the regulated drives, does not decrease with the load, the auxiliary power requirement with respect to the turbine output increases.

For the reference power plant under consideration in Fig. 2.17, the total efficiency decreases from 38.8% at full load to 35.5% at 40% load.

2.3.6
Losses During Startup and Shutdown

Energy supplied to the power plant during start-up heats the steam generator or is lost as waste heat via the stack and condenser. Thermal energy stored in the boiler is lost during shutdown, unless it acts as the standby service until the next (warm) startup.

The startup losses continue at their full extent until the power plant is in parallel connection with the electricity network, and then diminish as the live steam and the reheat steam flows are taken up by the turbine and the steam flow bypasses of the HP, IP, and LP turbines are closed.

The losses occurring during startup and shutdown depend on

- the capacity of the power plant unit
- the construction type of the steam generator
- the period of the preceding outage [3].

The startup losses decrease with increasing unit capacity. The longer the period of an outage, the higher are the losses of the following startup. For a design of a 400 MW unit, for instance, losses of 400 GJ after an 8 h outage have been reported [5].

2.3.7
Efficiency of Power Plants During Operation

In Fig. 2.18, the efficiencies achieved during the operation of German power plants are compared with the respective best design efficiencies for different ages and thermal capacities. The efficiency during operation of power plants is about 2–3% (in absolute terms) lower than the design efficiency. With increasing age of the power plant and decreasing thermal capacity, the net electrical efficiency difference is higher. The lower efficiency during operation is due to the lower part-load efficiency and the losses during startup and shutdown.

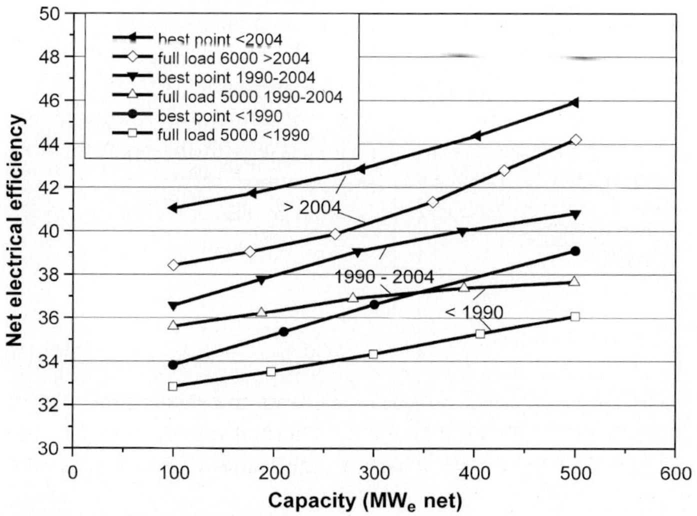

Fig. 2.18 Design and operation efficiencies [1].

2.3.8
Developments – State of the Art and the Future

In the following sections, the stages of development of hard coal-fueled power plants are discussed:

1. power plants with conventional steam conditions (160–200 bar/540 °C/540 °C)
2. power plants with raised steam conditions utilizing known materials (250 bar/540–560 °C/560 °C)
3. power plants with high steam conditions employing austenitic materials for the final superheater and available new steel types for thick-walled parts (270–290 bar/580 °C/600 °C).
4. power plants with the highest steam conditions employing austenitic materials for superheaters and new ferritic steel types for thick-walled parts and the evaporator (300–330 bar/600–630 °C/600–630 °C)
5. power plants with steam conditions up to 400 bar/700–750 °C, with future materials.

Figure 2.19 gives a breakdown of the efficiencies of the individual development stages for hard coal-fired power plants [48]. In the long term, an increase in efficiency to more than 50% can be expected, which will require the development of suitable materials. Table 2.3 describes the materials used and to be used for boilers (numbers 1–5 correspond to the list above).

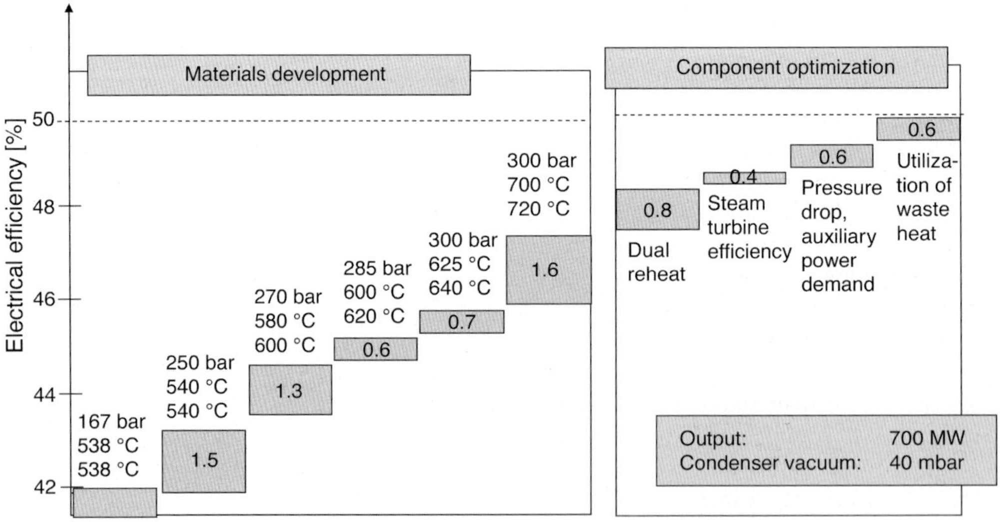

Fig. 2.19 Efficiency development in hard coal-fired power stations [48].

Table 2.3 Materials required for steam generator advancements [1].

No.	Evaporator	Superheater	Thick-walled components	Steam parameters (bar/°C/°C)	η_{el}	Commissioning year
1	13CrMo44	X20	X20	180/540/540	39	1985
2	13CrMo44	X20[a]	X20[a]	250/540/560	43	1993
3	13CrMo44[a]	Austenite	P91[a]	270/580/600	45	2000
4	HCM2S[a] 7CrMMoVTiB 10 10	Austenite[a]	Ferrite (P92, NF616, NF12) or austenite[a]	300/600/620	47	2010
5	HCM12	Nickel alloy	Nickel alloy	350/700/700	50	2020

a) Material stress limit reached.

2.3.8.1 Power Plants with Conventional Steam Conditions

The reference power plant is a typical thermal power plant in Germany with conventional steam conditions operated until the end of the 1980s and was considered as the economic optimum. Typical efficiencies were below 40%, which corresponds to the average efficiency of existing power plants in Germany. Power plants with higher steam conditions have been designed and built first in Germany, Denmark, and Japan, subsequently followed by other countries. Operating data and conclusions drawn from experience are now available.

2.3.8.2 Power Plants with Raised Steam Conditions Utilizing Known Materials (250 bar/540–560 °C/560 °C)

Still employing the known creep-resistant steel types – 13CrMo44 for the evaporator and X20CrMoV121 for the superheater and thick-walled components – the steam conditions are raised through process engineering modifications and by exhausting design reserves. A requirement for this is the homogeneous distribution of steam temperatures over the heating surfaces and of flue gas temperatures over the flue gas cross-sections. Temperature asymmetries of the flue gas are avoided by adequate control of the firing rate and for the steam by adequate configuration of the heating surfaces or by a limited heat-up of the stages. The material limits of thick-walled components and final-stage superheating surfaces are then reached.

Representing this power plant development, Table 2.4 lists several plants: for hard coal, the German Staudinger 5 and Rostock power stations and the Danish Esbjerg 3 power station, and for brown coal, the Schwarze Pumpe power station in Brandenburg, Germany.

The plant currently featuring the highest efficiency amongst the hard coal-fueled thermal power plants of the public power supply in Germany is a 553 MW Rostock power plant unit with 43.2% efficiency that went into operation in 1994. The efficiency is achieved by high live steam conditions (262 bar/545 °C/560 °C), by feed water preheating up to 270 °C, and by seawater cooling.

The 553 MW power plant Staudinger 5, which went into operation in 1992, is similar in design and steam conditions and achieves an efficiency of 42.7%, with an optimized evaporative cooling system. The losses of the boiler and the turbine were diminished and the auxiliary power demand was also reduced by utilizing low-loss drives for the main feed pump and condensate pumps and more effective fans. Still employing conventional materials, the design limitations were set by the final-stage superheater material, X20; a further live steam temperature rise would require the use of austenitic materials [27, 49].

The Danish Vestkraft plant has roughly comparable thermodynamic conditions and reaches a noticeably deeper vacuum due to the once-through cooling with cold seawater. Cooling with seawater helps to achieve a vacuum of 22 mbar at an annual average of 10 °C, whereas inland in Germany, it is only possible to achieve cold water temperatures of 15–18 °C by evaporative cooling, which correspond to a vacuum of 36–42 mbar. By seawater cooling and a rather lower steam-generator flue gas temperature of 104 °C, which, however, requires low-sulfur coal, it is possible to achieve a 2% higher better efficiency.

2.3.8.3 Power Plants with High Steam Conditions Employing Austenitic Materials for the Final-Stage Superheater and Available New Steel Types for Thick-Walled Parts (270–290 bar/580 °C/600 °C)

For power plants designed before 2000, net efficiencies of 45–46% could be achieved [4, 5, 7, 21, 35] . The respective live steam conditions of 270–290 bar and 580 °C and the reheat temperature of 600 °C allowed the use of the approved 13CrMo44 material for the evaporator wall, whereas the superheating surfaces required austenitic materials. Only the development of the ferritic steel P91 for thick-walled components made it possible to employ advanced steam conditions. Helically wound tubing for the evaporators in the furnace section was used.

In the planning of the 700 MW Hessler power plant, the calculated efficiency was in the region of 45% [35].

The Danish Aalborg power plant achieves an efficiency of 47%, applying double reheating, cooling with seawater, and low boiler-outlet temperatures. Today this plant still holds the world record for efficiency of a coal-fired steam power plant.

2.3.8.4 Power Plants with the Highest Steam Conditions Employing Austenitic Materials for Superheaters, and New Ferritic Steel Types for Thick-Walled Parts and the Evaporator (300–330 bar/600–630 °C/600–630 °C)

Power plants employing austenitic materials for superheaters and new martensitic steel types for thick-walled parts and the evaporator are predicted to raise efficiencies to up to 48%. This is made possible by raising the steam conditions up to 300–330 bar and 630 °C/630 °C and advancing the process and the single components [50]. The transition to higher steam conditions requires heavy-duty steel types both for the evaporator wall and for thick-walled components. For the evaporator, new steels such as HCM2 S and 7CrMMoVTiB 10 10 are now available. For thick-walled components, steel P91 becomes unsuitable and the newly developed steel P92 has to be used. The mentioned live steam conditions mean that the stress maximum occurs on the evaporator walls, the final-stage superheater, and the thick-walled components. Turbine shafts and cylinders also need to be built with austenitic materials. Operational data for turbines at live steam conditions of more than 300 bar and temperatures up to 625 °C are available from power plants in the chemical industry.

A number of power plants are currently (2010) under construction in Germany (Datteln, Karlsruhe, Moorburg, and Westfalen). The present maximum achievable steam conditions with currently available steels are considered to be 300 bar and 600 °C/620 °C. The design of these power plants will be similar to that of the Reference Power Plant North Rhine-Westfalia (RPP NRW), which is given in Table 2.4. The RPP NRW achieves an efficiency of 45.9% with live steam conditions of 285 bar and 600 °C and a reheat steam temperature of 620 °C. The design study showed that by increasing the feed water preheating to 320 °C, decreasing the condenser pressure to 35 mbar, and increasing the flue gas heat utilization, it is possible to raise the efficiency to 47.3%. The power plant in Moorburg has a design efficiency of 46.5% [2, 6, 7, 20, 31, 34, 51–57].

2.3.8.5 Power Plants with Steam Conditions up to 350 bar/700–750 °C, with Future Materials

Initiatives have commenced to increase further the live steam temperatures to 700–750 °C and the live steam pressures up to 350 bar [7, 12, 58]. For a power plant with a single reheat cycle cooled by a wet cooling tower, the net efficiencies are in the range 50–51%, and 53–54% if it is based on a double reheat cycle cooled by seawater [59]. The 700 °C technology is expected to be mature after 2010 and the long-term targets are to achieve net efficiencies above 55%, based on maximum steam temperatures in the region of 800 °C [17]. In Germany, E.ON has announced that it is to build a 550 MW power plant with conditions of 365 bar/705 °C/720 °C and an efficiency of about 50% [20, 60, 61].

In order to run such a high-temperature process, new materials and construction procedures have to be developed and approved for the steam generator, turbine, and piping. The testing of the critical components for the 700 °C technology was the reason for launching the European Research Project AD 700 in 1998. Major targets of this research were the development of austenitic steels and nickel-based superalloys for the hottest parts of boilers, steam lines, and turbines, and the development of boiler and turbine designs. The materials now under consideration for the evaporator are HCM12 steel and various nickel-based alloys. For the convective heating surfaces and for thick-walled components, nickel-based alloys are the only choice [7, 12, 15, 58]. When applying the new materials, problems caused by the wall thickness, high-temperature corrosion, and steam oxidation have to be taken into account. In addition to the me-

Table 2.4 Data for various advanced steam power plants [4, 21, 26, 31, 32, 35, 40, 42, 51, 56, 66].

Parameter	Power plant											
	Bexbach I/D	Stau-dinger/D	Esbjerg 3/DK	Hessler/D	Bexbach II/D	Aalborg/ DK	Referenz-kraftwerk NRW/D	Moor-burg/D	Neurath/D	Schwarze Pumpe/D	Lippen-dorf D	Nieder-aussem K/D
Commis-sioned	1983	1992	1992	Planning 1994	Planning 1995	1998	Planning 2004	2010		1997	1999	2002
Coal					*Hard coal*						*Brown coal*	
LHV (MJ kg^{-1})		27					25	25.1	8.5		10.5	9.2
P_{gross} (MW)	742	509	420	700	750	385	600	840	600	740	933	1012
P_{net} (MW)	705	553		732			555.5	779	562		870	965
p_{LS} (bar)	190	250	251	275	250	290	285	276	176	260	268	275
t_{LS} (°C)	530	540	560	580	575	582	600	600	530	545	554	580
p_{RH} (bar)	37		56	50		74	60	51			52	60
t_{RH} (°C)	530	560	560	600	595	580	620	610	530	560	583	600
p_{RH2} (bar) t_{RH2} (°C)	–	–	–	–	–	19/580	–	–	–	–	–	–
$t_{feed\ water}$ (°C)	250	270	275	300	290	290	303	293		270	270	295
n_{stages}	7	7		9	9	9		9		7	8	10
$t_{furnace\ exit}$ (°C)	1110		1260	1250	1200	1300				965	1000	1050
Flue gas exit temperature (°C)	130	125	104	125	110	100	115	115	142	170	170	160
Air ratio	1.3	1.2			1.15	1.17	1.15	1.17			1.15	1.15
Heat shift (°C)	–	–	–	–	110/90	–		–	–	–	170/130	160/100

Table 2.4 (continued)

Parameter Power plant	Bexbach I/D	Stau-dinger/D	Esbjerg 3/DK	Hessler/D	Bexbach II/D	Aalborg/DK	Referenz-kraftwerk NRW/D	Moor-burg/D	Neurath/D	Schwarze Pumpe/D	Lippen-dorf D	Nieder-aussem K/D
Commis-sioned	**1983**	**1992**	**1992**	**Planning 1994**	**Planning 1995**	**1998**	**Planning 2004**	**2010**		**1997**	**1999**	**2002**
Coal					***Hard coal***						***Brown coal***	
Cold water (°C)	16.6	18	10	17.7	15	10		12	21.5	17.5	16.4	14.7
Vacuum (mbar)	55	38/52	23	37	30/43	23	45	27.4	66	42	38	28.5/35.5
Cooling range (°C)	15.9				12			5.9			10.2	10.3
Heat extrac-tion (MW)	–	300	–	–	160	–		225	–	120		
Flue gas extraction	Chimney	Cooling tower	Chimney	Cooling tower	Cooling tower	Chimney	Cooling tower	Chimney	Cooling tower	Cooling tower	Cooling tower	Cooling tower
Efficiency												
Auxiliary	9.4	8.1	7.8		7.1	7.2		7.2			6.8	7.3
η_B (%)	94.0	94.5	95.8	94.4	95.4	95.4	95.0			89.8	90.6	94.4
η_{tot} (%)	38.7	43	45.3	45	46.3	47	45.9	46.5	35.5	40.4	42.3	43.2
Materials												
Evaporator	A [a)]	A [a)]	A [a)]	A [a)]	A [a)]	A [a)]	B [b)]	B [b)]	A [a)]	A [a)]	A [a)]	A [a)]
Final superheater	X20	X20	X20	Aust.	Aust.	Aust.	Aust. TP 347H FG	Aust. Super 304H	X20	X20	Aust.	Aust.

a) 13CRMo44.
b) 7CrMoVTiB1010.

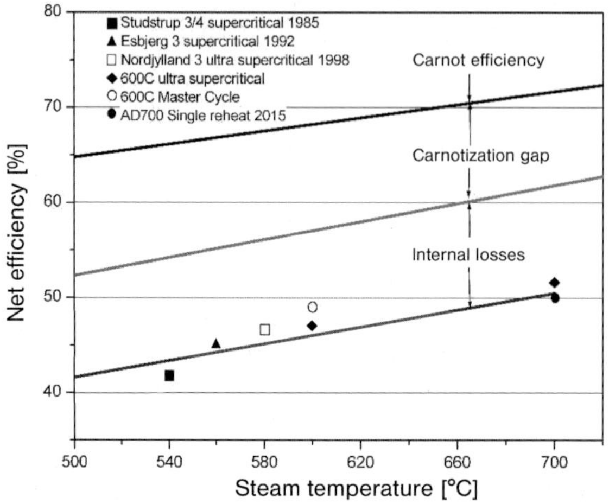

Fig. 2.20 Net efficiency of seawater-cooled supercritical power plants [24].

chanical properties and the workability, the costs have to be borne in mind when choosing the material, considering the large quantities needed for a thermal power plant.

Figure 2.20 shows the net efficiency of a number of seawater-cooled supercritical power plants versus the maximum steam temperature and the ideal Carnot efficiency [24]. It is obvious that higher process temperatures drive the Carnot and the net efficiency upwards. Over the past 20 years, materials developments have resulted in an increase in live steam and reheat temperatures by 60 °C to about 600 °C, corresponding to an improvement in heat rate of 3%. However, further steel-based improvements are not expected for the next 10–15 years as no new candidates replacing P92 are in sight, and at present only nickel-based alloys seem to allow steam conditions of 700 °C.

In Fig. 2.20, the gap between the Carnot and the net efficiency is divided into two gaps of almost equal size, one reflecting the lack of thermodynamic completeness of the supercritical water/steam cycles (the Carnotization gap) and the other internal losses (the internal losses gap). The dividing curve was calculated by setting all equipment efficiencies at 100% and parasitic losses at zero and calculating the net efficiency of the supercritical power plants. Both gaps are about 10% for a steam temperature of 600 °C. Whereas higher steam temperatures raise the Carnot efficiency, optimization of the thermodynamic cycle with a higher live steam pressure, a higher feed water preheating temperature, and double reheating can reduce the Carnotization gap. It is expected that the optimization of the Carnotization gap would provide limited potential for efficiency increases. Efficiency improvements of components and reductions in the auxiliary power demand are already fairly advanced and only minor improvements are likely to be made [24].

2.3.9
Brown Coal

A considerable increase in efficiency could also be achieved for brown coal. After the 600 MW units built in 1972 in the Rhenish mining area, with efficiencies of around 35.5% [62], it was possible to achieve an efficiency of 40% for the Schwarze Pumpe power plant (2 × 800 MW), at live steam conditions comparable to the Staudinger 5 power plant [63]. The Lippendorf power plant (2× 930 MW), in service since 1997, reaches an efficiency of 42.3% with again higher steam conditions [26]. The operating conditions of the Niederaussem power plant, put into operation in 2002 in the Rhenish mining area, are comparable to those of the design of the Hessler plant, and achieves an efficiency of 43.2% [32, 62, 64]. Brown coal-fired power stations with similar steam conditions are under construction [57, 65]. Comparing the power plants in the Rhenish mining area with those in East Germany, the East German brown coal types have a higher fouling tendency, which limits the furnace outlet temperature to between 950 and 1000 °C, whereas the Rhenish brown coal allows furnace outlet temperatures of around 1050 °C, hence favoring the application of higher steam conditions for the former (see Table 2.4).

A predrying stage for the moist brown coal is also a point of discussion. Such a stage would increase the efficiency by 2%, so that brown and hard coal power plants would ultimately feature more or less the same overall efficiency. If the drying method was based on the condensation heat of the water vapor, the efficiency could be increased by about 4–6% to reach 4 eventually 7–49% (280 bar/ 580 °C/600 °C) [46].

2.3.10
Conclusion

In the field of steam production, recent years have seen the beginning of rapid developments towards enhancing the efficiency. The characteristic of this development is the transition to higher live steam conditions with live steam temperatures above 540 °C and live steam pressures above 250 bar. The higher live steam temperatures increase the thermal cycle efficiency. Potential for optimization is also available at the cold end of the steam cycle. Depending on the boundary conditions – once-through water cooling or evaporative cooling – the condensation pressure is reduced when optimizing the cooling circuit. Reducing the auxiliary power demand by the use of low-loss feed pump drives and by reducing the various losses also contribute substantially to the enhancement of the overall efficiency. Measures aimed at reducing the boiler losses focus on diminishing the flue gas temperatures and at reducing the flue gas mass flow. Further developments in turbine engineering also help to advance power plant technology by increasing the efficiency.

For power plants designed before 2000, net efficiencies of 45–46% could be achieved. The respective live steam conditions of 270–290 bar and 580 °C and a reheat temperature of 600 °C allowed the use of the approved 13CrMo44 material for the evaporator wall, whereas the superheating surfaces required austenitic materials. The Danish Aalborg power plant achieves an efficiency of 47%, applying double reheating, cooling with seawater, and low boiler outlet temperatures. Today this plant still holds the world record for the efficiency of a coal-fired steam power plant.

The transition to higher steam conditions requires heavy-duty steel types both for the evaporator wall and for thick-walled components. For the evaporator, new steels such as HCM2 S and 7CrMMoVTiB 10 10 are now available. The current maximum achievable steam conditions with these steels are considered to be 300 bar and 600 °C/620 °C. A number of power plants are currently (2010) under construction in Germany (Datteln, Karlsruhe, Moorburg, and Westfalen). The design of these power plants will be similar to that of the Reference Power Plant North Rhine-Westfalia (RPP NRW), which achieves an efficiency of 45.9% with live steam conditions of 285 bar and 600 °C and a reheat steam temperature of 620 °C.

Initiatives have commenced to increase further the live steam temperatures to 700–750 °C and the live steam pressures up to 350 bar. In Germany, E.ON has announced that it is to build a 550 MW power plant with 365 bar/705 °C/720 °C and an efficiency of about 50%. In order to run such a high-temperature process, new materials and construction procedures have to be developed and approved for the steam generator, turbine and piping.

References

1 Spliethoff, H. *Power Generation from Solid Fuels*, Springer, Berlin, 2010.

2 Klebes, J., High-efficiency coal-fired power plants based on proven technology. *VGB PowerTech*, 2007, **87**(3), 80–84.

3 Adrian, F., Quittek, C., and Wittchow, E. Fossil beheizte Dampfkraftwerke. In *Handbuchreihe Energie*, Band 6 (ed. Bohn, T.), Technischer Verlag Resch, Munich, Verlag TÜV Rheinland, Cologne, 1986.

4 Billotet, T. and Johänntgen, U. Kraftwerk Bexbach II – Fortschrittliche Stromerzeugung aus Steinkohle. *VGB Kraftwerkstechnik*, 1995, **75**(1), 21–25.

5 Kotschenreuther, H., Häuser, U., and Weirich, P.H. Zukünftige wirtschaftliche Kohleverstromung. *VGB Kraftwerkstechnik*, 1993, **73**(1), 22–28.

6 Schütz, M., Hauk, R., and Stamatelopoulos, G.-N. Stand der Planungen für Dampferzeuger der 800-MW-Klasse. *VGB PowerTech*, 2009, **89**(3), 36–41.

7 Blum, R., Bugge, J., and Kjaer, S. USC 700 °C power technology. A European success story. *VGB PowerTech*, 2009, **89**(4), 26–32.

8 Stamatelopoulos, G.-N. and Weissinger, G. Die nächste Generation von Steinkohlekraftwerken. Presented at the VGB-Konferenz "Kraftwerke im Wettbewerb 2005, Ordnungspolitik, Markt und Umweltschutz," Potsdam, 2005.

9 Scheffknecht, G., Stamatelopoulos, G., and Lorey, H., Moderne Kohlekraftwerke. *BWK*, 2002, **54**(6), 46–50.

10 Chen, Q. and Scheffknecht, G. Neue Dampferzeuger- und Rohrleitungswerkstoffe und ihre Anwendungsgrenzen. Presented at the VGB Konferenz "Kraftwerke im Wettbewerb – Technik, Betrieb und Umwelt 2003". Cologne, 2003.

11 Kather, A. and Scheffknecht, G. Neue Werkstoffe und Dampferzeugerbau. *BWK*, 1997, **49**(11/12), 62–67.

12 Blum, R., Kjaer, S., and Bugge, J. Development of a PF fired high efficiency power plant (AD700). In *Proceedings of the Riso International Energy Conference "Energy Solutions for Sustainable Development," Denmark, 2007*, Riso-R-1608 (EN), 2007.

13 Blum, R. and Vanstone, R.W. Materials development for boilers and steam turbines operating at 700 °C. Presented at the 1st International Conference on Super-High Strength Steels, 2–4 November 2005, Rome, Associazione Italiana di Metallurgia.

14 Chen, Q. and Scheffknecht, G. New boiler and piping materials. *VGB PowerTech*, 2003, **83**(11), 91–98.

15 Chen, Q., *et al.* Investigations on oxidation and corrosion characteristics of the advanced boiler materials at steam temperatures up to 720 °C. Presented at the 1st International Conference on Super-High Strength Steels, 2–4 November 2005, Rome, Associazione Italiana di Metallurgia.

16 Husemann, R.-U. Development status of boiler and piping materials for increased steam conditions. *VGB PowerTech*, 2003, **83**(9), 124–128.

17 Kjaer, S., *et al.* The advanced supercritical 700 °C pulverised coal fired power plant. *VGB PowerTech*, 2002, **82**(7), 46–49.

18 Viswanathan, R., *et al.* U.S. program on materials technology for ultra-supercritical coal power plants. *J. Mater. Eng. Performance*, 2005, **14**(3), 281–292.

19 Viswanathan, R., Purgert, R., and Rao, U. Materials technology for advanced coal power plants. Presented at the 1st International Conference on Super-High Strength Steels, 2–4 November 2005, Rome, Associazione Italiana di Metallurgia.

20 Neubronner, M., *et al.* Die Zukunft der Kohleverstromung – Höchste Wirkungsgrade und Dekarbonisierung. *VGB PowerTech*, 2010, **90**(1/2), 31–34.

21 Kjaer, S. Die zukünftigen 400 MW-ELSAM-Blöcke in Aalborg und Skaeraek. *VGB Kraftwerkstechnik*, 1993, **73**(11), 933–940.

22 Steag AG. *Strom aus Steinkohle: Stand der Kraftwerkstechnik*, Springer, Berlin, 1988.

23 VGB. *Kraftwerke 1970 bis 1995 – Wege und Meilensteine. Ein Überblick anläßlich des 75jährigen Bestehens der VGB*. VGB, Essen, 1995.

24 Kjaer, S. and Drinhaus, F. Presentation of a modified double reheat cycle. Presented at the 40th Kraftwerkstechnisches Kolloquium, 14–15 October 2008, Dresden, Vortrag V 5.5.

25 Baehr, R. Konzeption und Aufbau von Dampfkraftwerken. In *Handbuchreihe Energie*, Band 5 (ed. Bohn, T.), Technischer Verlag Resch, Munich, Verlag TÜV Rheinland, Cologne, 1986.

26 Breuer, H., Kehr, M., and Gerstmeyer, R. Darstellung und Entwicklung der neuen Braunkohlekraftwerke der VEAG. Beispiel des Gemeinschaftskraftwerkes Lippendorf. *VGB Kraftwerkstechnik*, 1995, **75**(2), 93–100.

27 Rukes, B., Vollmer, W., and Wittchow, E. Abgestimmte Auslegung von Dampfturbine und Dampferzeuger. *VGB Kraftwerkstechnik*, 1994, **74**(5), 405–411.

28 Johänntgen, U. Kühlwasserseitige Bypass-Schaltung an Kondensationsanlagen. *VGB Kraftwerkstechnik*, 1998, **78**(3), 57–64.

29 Neft, H.G. and Franconville, G. Neue Dampfturbinenkonzepte für höhere Eintrittsparameter und längere Endschaufeln. *VGB Kraftwerkstechnik*, 1993, **73**(5), 409–415.

30 Weber, F., Alzinger, R., and Maschke, U. Optimierungspotential für Kraftwerke zur Stromerzeugung – Maßnahmen am kalten Ende. *VGB PowerTech*, 2005, **85**(9), 85–87.

31 Meier, H.J. Reference Power Plant North Rhine-Westfalia (RPP NRW). *VGB PowerTech*, 2004, **84**(5), 76–89.

32 Lambertz, J. and Gasteiger, G. BoA Niederaußem: Konzeption und Betriebserfahrungen. *VGB PowerTech*, 2003, **83**(5), 82–87.

33 Tremmel, A., *et al.* Modernste Turbinentechnologie mit höchsten Dampftemperaturen für das Kraftwerk Boxberg R. *VGB PowerTech*, 2006, **86**(12), 71–75.

34 Mandel, H. and Schettler, H. Die Neubaublöcke von Vattenfall Europe mit ihren technischen Herausforderungen. Presented at the VGB-Konferenz "Kraftwerke im Wettbewerb," Hamburg, 2007.

35 Eichholtz, A., Houfar, D., and Kübler, D. Der 700-MW-Steinkohleblock der VEBA Kraftwerke Ruhr AG in Gelsenkirchen-Heßler. *VGB Kraftwerkstechnik*, 1994, **74**(1), 25–29.

36 Strauss, K. *Kraftwerkstechnik: zur Nutzung fossiler, nuklearer und regenerativer Energiequellen*, 5th edn., Springer, Berlin, 2006.

37 Nowi, A. and Haller, B.R. Wirkungsgradentwicklung bei Dampfturbinen. *VGB Kraftwerkstechnik*, 1997, **77**(7), 551–555.

38 Oeynhausen, H., Drosdziok, A., and Deckers, M. Dampfturbinen für die neue Kraftwerksgeneration. *VGB Kraftwerkstechnik*, 1996, **76**(12), 974–979.

39 Bauer, G. and Lankes, F. Erhöhung des Blockwirkungsgrades durch verbesserte Abgaswärmenutzung. *VGB Kraftwerkstechnik*, 1997, **77**(5), 376–381.

40 Tippkötter, T. and Scheffknecht, G. Betriebserfahrungen mit dem neuen BOA-Block und Ausblick. *VGB PowerTech*, 2004, **84**(4), 48–55.

41 Bruckmann, W. and Hesel, U. Exergiesteigerung durch Schaltungsänderung der Wärmeauskopplung bei Kohlekraftwerken. Presented at the VDI-Tagung "Entwicklungslinien der Energie- und Kraftwerkstechnik," Siegen, 1996.

42 Köhn, G. Staudinger 5 – Auslegung und Erfahrungen aus Inbetriebnahme und erster Betriebszeit. *VGB Kraftwerkstechnik*, 1993, **73**(11), 927–932.

43 Riedle, K., Rukes. B., and Wittchow. E. Die Erhöhung des Kraftwerkswirkungsgrades in der Vergangenheit und in der Zukunft Presented at VGB Konferenz "Kraftwerkstechnik 2000", 1993, Essen, VGB TB 120.

44 Klutz, H.-J., Klöcker, K.J., and Lambertz, J. Das WTA-Verfahren als Vortrocknungsstufe für moderne Kraftwerkskonzepte auf Basis Braunkohle. *VGB Kraftwerkstechnik*, 1996, **76**(3), 224–229.

45 Schwendig, F., Klutz, H.J., and Ewers, J. Das Trockenbraunkohle befeuerte Kraftwerk. *VGB PowerTech*, 2006, **86**(12), 51–57.

46 Ewers, J., *et al.* The development of predrying and BoA-Plus technology. *VGB PowerTech*, 2003, **83**(11), 60–65.

47 Klutz, H.-J., Moder, C., and Block, D. WTA-Feintrocknung, Baustein für die Braunkohlekraftwerke der Zukunft. *VGB PowerTech*, 2006, **86**(11), 57–61.

48 Rukes, B. Perspectives of fossil power technology. *VGB PowerTech*, 2002, **82**(10), 71–76.

49 E.ON. www.kraftwerk-rostock.de, 2006, accessed 30.11.2006.

50 Kjaer, S. Advanced steam cycles. Presented at the IFRF Conference "Advanced Power Generation," TOTEM 10, Veldhoven, 1994.

51 VGB. *Konzeptstudie Referenzkraftwerk Nordrhein-Westfalen*, VGB, Essen, 2004.

52 Michel, M. Kraftwerksneubau der E.ON Energie AG in Deutschland. *VGB PowerTech*, 2006, **86**(4), 46–51.

53 Schmitz, E. Energiepolitische Perspektiven für Europa und Deutschland sowie die Rolle der Steinkohle im Energiemix. Presented at the VGB-Konferenz "Kraftwerke im Wettbewerb," Hamburg, 25–26 April 2007.

54 Willeke, A. Das Steinkohlekraftwerk Datteln 4 – Start der E.O.N.-Konvoi-Projekte in der 1.100-MW-Klasse. Presented at the VGB-Konferenz "Kraftwerke im Wettbewerb„ Hamburg, 25–26 April 2007.

55 Then, O., Wüllenweber, H.-J., and Kleinhörster, B. Neue Kohlekraftwerke bei RWE Power-Technologien und Herausforderungen. Presented at the VBG-Konferenz "Kraftwerke im Wettbewerb", 25–26 April 2007, Hamburg.

56 Vattenfall. *Antrag auf Genehmigung nach dem Bundes-Immissionsschutzgesetz für die Errichtung und den Betrieb des Kraftwerkes Moorburg 2007*, Vattenfall, Stadt Hamburg, 2007.

57 Elsen, R., Kehr, M., and Nowack, R. Aktuelle Projekte und Strategien für die zukünftige Stromerzeugung im Bereich fossilbefeuerter Kraftwerke der RWE Power. *VGB PowerTech*, 2009, **89**(7), 37–40.

58 Bauer, F., *et al.* Driving coal-fired power plants to over 50% efficiency. *VGB PowerTech*, 2003, **83**(12), 97–100.

59 Hoestgaard-Jensen, P., Bauer, F., and Kjaer, S. Emax: VGB initiative for Clean Coal Technologies. *VGB PowerTech*, 2003, **83**(5), 92–95.

60 Bauer, F., Tschaffon, H., and Hourfar, D. Role of 700 °C technology for the carbon-low power supply. *VGB PowerTech*, 2008, **88**(4), 30–34.

61 Willrodt, A. and Tschaffon, H. 50plus – mit Volldampf zum Kohlekraftwerk der Zukunft. *VGB PowerTech*, 2009,**89**(3), 33–35.

62 Heitmüller, R. J., *et al.* Neue Braunkohleblöcke mit hohem Wirkungsgrad bei RWE Energie AG. *VDI-Ber.*, 1996, **1280**, 77–90.

63 Lauterbach, K., *et al.* Konzepte für 800-MW-Turbinen der VEAG im Kraftwerk Boxberg und im Gemeinschaftskraftwerk der VEAG/ESPAG am Standort Schwarze Pumpe. *VGB Kraftwerkstechnik*, 1993, **73**(6), 500–508.

64 Böcker, D. and Hlubek, W. Umwelt- und Klimaschutzprogramm für die rheinische Braunkohle. *Energiewirtschaftliche Tagesfragen*, 1995, **45**(3), 119–127.

65 Kehr, M., Elsen, R., and Ewers, J. Aktuelle Neubauprodukte und Weiterentwicklungen in der Kohlekraftwerkstechnik. *VGB PowerTech*, 2005, **80**(4), 30–35.

66 Spliethoff, H. and Abröll, G. Das 750 MW-Steinkohlekraftwerk Bexbach. *VGB Kraftwerkstechnik*, 1985, **65**, 346–362.

3
Capture Options for Coal Power Plants

Ernst Riensche, Jewgeni Nazarko, Sebastian Schiebahn, Michael Weber, Li Zhao, and Detlef Stolten

Abstract

This chapter provides an overview of the great variety of options for carbon dioxide capture from coal-fired power plants. It focuses on the question of the energy efficiency of carbon capture. The capture routes are post-combustion, oxyfuel combustion, and pre-combustion – the last with selective CO_2 separation and with selective H_2 separation. The basics of gas separation methods such as CO_2 absorption with liquids, reaction of O_2 and CO_2 with solids, and membrane separation of CO_2, O_2, and H_2 in steam power plants and integrated gasification combined cycles (IGCCs) are outlined. Finally, selected capture processes are briefly described and efficiency potentials are discussed.

Keywords: power plant, coal, carbon dioxide capture, oxyfuel, absorption, chemical looping, membranes

3.1
Introduction

Today's relatively stable and benign climate on Earth provides the basis of modern civilization. Two centuries ago, scientists started to analyze and quantify the natural greenhouse effect and the role of CO_2 in the atmosphere. Intense use of fossil energy represents the backbone of modern civilization and the global economy, creating anthropogenic climate change. Svante Arrhenius was the first scientist to recognize the role of CO_2 in the atmosphere with his review [1] presented as early as 1895. The status of international climate research as of 2007 is documented in the 4th Assessment Report of the International Panel on Climate Change (IPCC) [2].

In 2005, about 26 Gt of CO_2 were emitted. Additional CO_2 emissions are caused by deforestation and amount to 3–7 Gt of CO_2 per annum. In 2007, coal, oil, and natural gas composed 86% of all primary energy [3]. Oil accounted for about 90% of transportation, and coal comprised about 50% of electricity production. In 2007, global CO_2 emissions from fossil fuels were composed of

Process Engineering for CCS Power Plants, First Edition. Edited by Detlef Stolten and V. Scherer

about 12 Gt from coal, about 12 Gt from oil, and about 6 Gt from natural gas. Coal is primarily consumed for electricity production. Hence carbon capture has great potential for future reduction of CO_2 emissions.

An efficiency penalty of 1–2% is attributed today to the removal of pollutants from combustion processes of fossil energy sources, such as nitrogen oxides, particulates, and sulfur dioxide [4, 5]. A much higher additional efficiency penalty of about 10% would arise from application of carbon capture and storage (CCS) technology in coal power plants at the current state of development.

The concentration of CO_2 in the flue gas ranges from 15 mol% for steam power plants (SPPs) to just 4 mol% for coal gas and natural gas combustion in combined cycles – integrated gasification combined cycle (IGCC) and natural gas combined cycle (NGCC). In principle, the total flue gas could be compressed and stored underground. An analysis has shown that the energy required to compress the total volume of flue gas generated by a thermal power plant to 100 bar would amount to ~65% of its power output [6]. This power requirement drops below 8% (about 4% efficiency penalty) when only the contained CO_2 is compressed. Therefore, CO_2 needs to be separated and captured. Subsequently, CO_2 has to be transported to a location where it can be stored for at least hundreds of years.

From a process engineering viewpoint, the total CCS process can be divided into six stages, as depicted in Figure 3.1. Effective and efficient gas separation (2) requires modifications of existing power plants (1). In the case of intolerably high impurity values in the separated CO_2, a subsequent CO_2 cleaning step (3) is necessary. The compression in principle may be carried out in three steps, at the exit of power plant (4), at locations along the pipeline (5) to overcome the pressure losses, and at the storage site (6) [7].

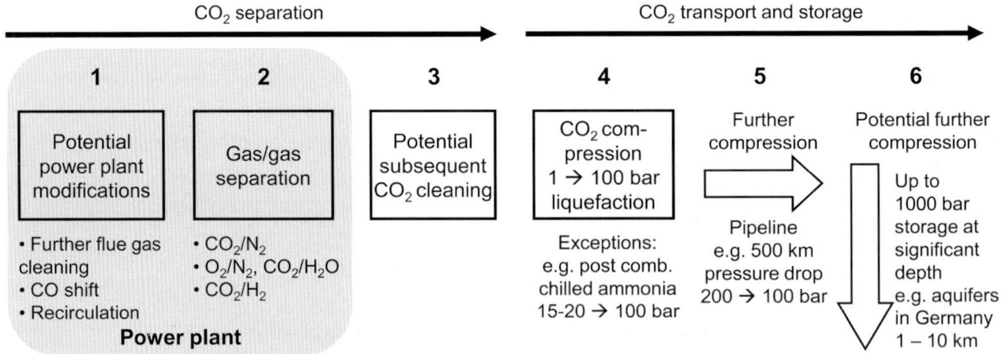

Fig. 3.1 CCS process steps: accumulation of efficiency losses.

3.2 Requirements on CO_2 Capture and Compression

As a starting point of CCS process engineering, the principle requirements on CO_2 capture and compression are to be defined. Efficient transport, injection, and storage of carbon dioxide identify the requirements. The most relevant parameters are

- degree of CO_2 separation
- CO_2 purity
- CO_2 pressure for pipeline transport
- CO_2 pressure for injection
- the additional energy demand for the total CCS process.

With respect to the first parameter, of course, 100% separation is desired. Yet above a degree of CO_2 separation of 90%, the separation devices become overproportionally larger and the process becomes energy-intense [8]. Reasonable levels of relevant parameters are discussed below.

3.2.1 CO_2 Purities and Pressures

The capture process should take the gaseous CO_2 from the combustion process and produce a reasonably clean liquid CO_2 stream, although the final purity required for transport and storage is not yet established. A dense form of the captured CO_2 is needed for effective transport and storage scenarios.

Carbon dioxide is a colorless, non-flammable gas. At temperatures up to the critical point (31.1 °C, 73.9 bar), a pressure increase leads to continuously increasing density, until the saturation line is reached. Only at this line does a two-phase gas–liquid mixture exist. At ambient temperature and pressures above 60 bar, carbon dioxide is a liquid. At temperatures and pressures above the critical point at 31.1 °C and 73.9 bar, CO_2 is in a supercritical state, characterized by liquid-like densities and gas-like mobility. These advantageous physical properties are exploited for transport and storage of CO_2 [9]. Solid and gaseous CO_2 are not suitable for transportation, since it is highly energy consuming [10] to produce solid CO_2 and the gaseous form exhibits too low a density.

Commercialization of CCS requires an infrastructure for transportation of carbon dioxide to the storage sites. Today, pipelines are most commonly used for the transportation of very large quantities of CO_2. In the United States and Canada, about 45×10^6 t of CO_2 are transported annually by a 3100 km long network in 130–200 bar high-pressure pipelines [11, 12]. For pipelines, the supercritical state is preferable, because it is relatively stable compared with the liquid state, minimizing cavitation problems in components such as booster stations and pumps. The CO_2 transport pipelines in the United States are designed to a low pressure drop of up to 10 bar along the total length of the network [12].

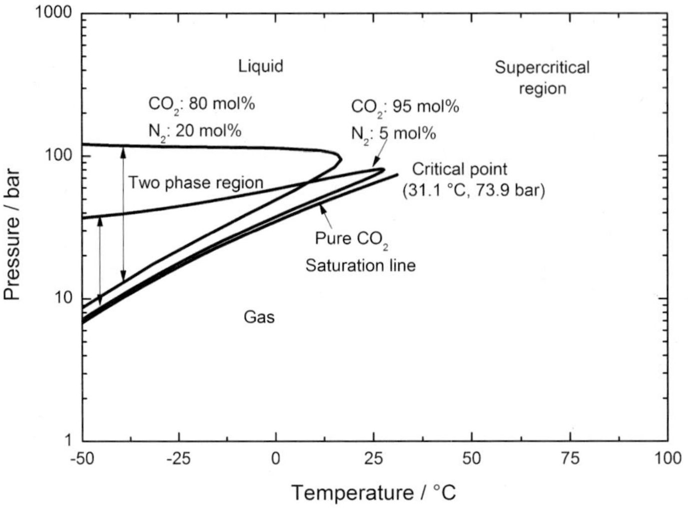

Fig. 3.2 CO_2 phase diagram for pure CO_2 and for CO_2–N_2 mixtures. After Peng-Robinson [20].

In a recent simulation study, the influence of nitrogen impurities on the critical temperature and pressure was studied [13]. Nitrogen as an impurity in CO_2 plays a key role since most flue gases contain nitrogen. This nitrogen may constitute an impurity of the separated CO_2. Figure 3.2 shows a CO_2 phase diagram for pure CO_2 and for CO_2–N_2 mixtures. For pure CO_2, the two-phase state is restricted to the saturation line, ending at the critical point. For 5 mol% N_2 in CO_2 the saturation line is broadened, where inside the envelope two phases exist. For 20 mol% N_2 the two-phase-region ends at about 15 °C. At 0 °C it covers a range between 50 and 110 bar, which is much higher than the critical pressure. Therefore, it can be concluded that for CO_2 containing N_2 as an impurity, the operating pressure of a pipeline should be significantly higher than 100 bar to avoid two-phase flow conditions during transport.

Storage of CO_2 in geological formations is the most favorable option today. In principle, CO_2 is injected via a borehole and stored in an underground rock formation such as saline aquifers, depleted oil and gas reservoirs, and unminable coal beds. The injection involves the same technologies as are applied in the oil and gas exploration and production industry. At depths typically deeper than 800 m, the pressures and temperatures are high enough for CO_2 to remain in a supercritical state, ensuring efficient storage. The storage involves three different mechanisms:

- hydrodynamic trapping as a supercritical fluid under a low-permeability cap rock
- aqueous solubility into underground fluids
- mineralization, that is, reaction with minerals and bacteria and its transformation into minerals in the long term.

Hydrodynamic trapping may dominate in early stages of the injection phase. Here partial displacement of the *in situ* water takes place. Subsequently, CO_2 dissolves slowly into the water. Over thousands of years, this fluid becomes denser than water and sinks deeper into the rock formation [14]. On an even longer timescale, the CO_2 will be converted to solid carbonate minerals.

Deep saline aquifers offer the highest potential in terms of capacity. The lower estimate of the storage capacity indicates 1000 Gt of CO_2, whereas the upper estimate amounts to 10 000 Gt [15]. This would be sufficient to store the CO_2 produced worldwide for over 100 years. Recent studies show a limiting resource for CO_2 storage, however. It was demonstrated that, in many cases, the ability of the geologic formation to dissipate pressure is more limiting than the formation's pore volume [3, 7]. Previously, it had been assumed that a geologically appropriate pore volume was the limiting resource for CO_2 storage. Recently, simulation models of compressible flow in porous media were applied to several formations in the United States. It turned out that in many cases the pore pressures reach levels that would cause large-scale hydraulic fracturing, which may create conduits for buoyant CO_2 to escape through. Hence previous capacity estimates have to be corrected. To date there is no clear understanding of whether larger storage capacities would be available over long periods of time. Figure 3.3 shows the density of pure CO_2 as a function of depth [16]. Below 800 m, high densities between 600 and 700 kg m^{-3} are reached. The second curve shows the density of CO_2 when an impurity of 2.75 mol% N_2 is present. This results in a significant change in the density, especially down to about 2 km depth. For example, at 800 m depth the density reaches merely 300–400 kg m^{-3}, which is just 50% of that of pure CO_2. This behavior seems to be very important with respect

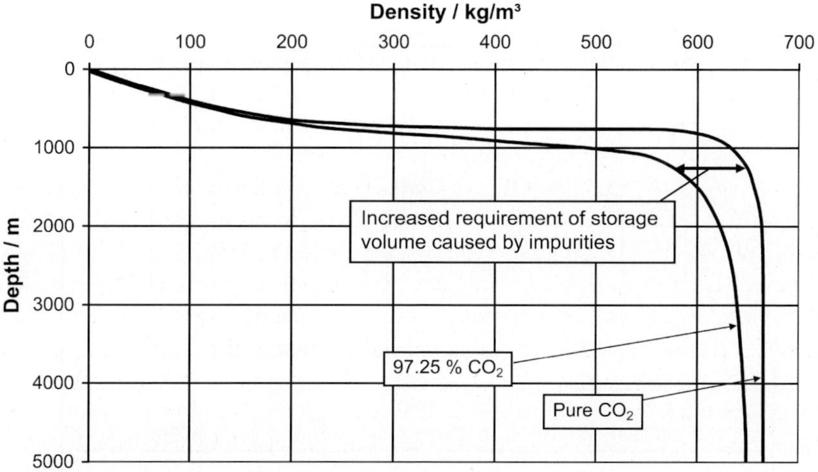

Fig. 3.3 CO_2 density as a function of depth underground, with pressure assumed to be according to that in a water column, namely 0.1 bar m^{-1}. Influence of 2.75% N_2 impurity on the density as a function of depth [16].

to possibly limited storage capacities. As a consequence, high impurities have to be avoided or the CO_2 needs to be stored much deeper.

3.2.2
Basic Considerations on CCS Energy Demand

In this section, the energy demand of CCS is considered. Carbon, coal, and methane are compared with respect to the efficiency penalty that CCS involves. Finally, the question of minimum theoretical work for capture is discussed more in general. A specific answer to this question is fundamental for each capture concept.

3.2.2.1 Relation Between Energy Demand and Efficiency Penalty

Efficiency considerations involve the power plant as such, the CO_2 capture process, and CO_2 compression for transportation and storage. Figure 3.4 displays the most relevant data for comparison of pure carbon, real coal, and methane, on the basis of the lower heating value (LHV). It relates LHV to CO_2 produced. It appears that the LHV per unit mass of CO_2 produced of a typical coal is 17% higher than that of carbon. The reason is that coal also contains hydrogen and other lightweight combustible compounds.

3.2.2.2 Minimum Separation Energy

The minimum specific work required for complete isothermal separation of a mixture of ideal gases with mole fractions y_i and mass fractions x_i into all its components is [19]

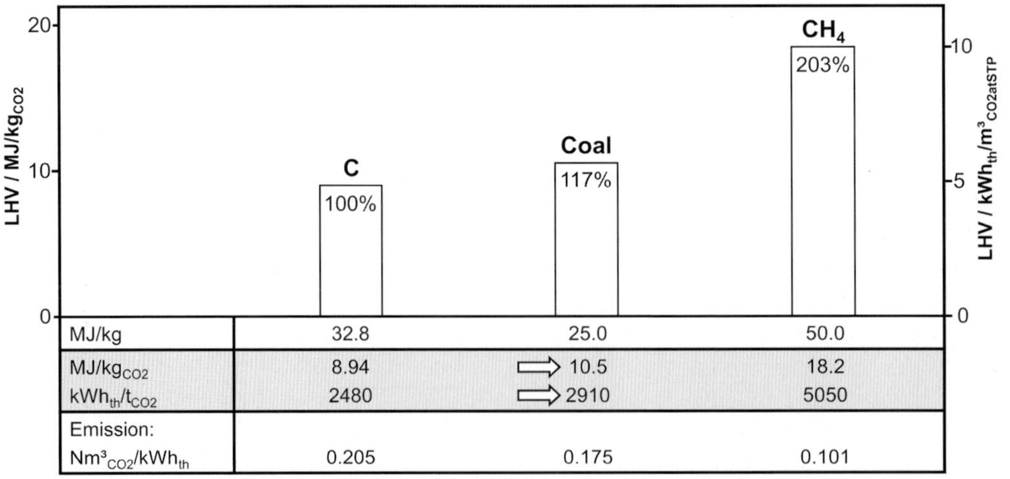

	C	Coal	CH_4
MJ/kg	32.8	25.0	50.0
MJ/kg_{CO2}	8.94	⇨ 10.5	18.2
kWh_{th}/t_{CO2}	2480	⇨ 2910	5050
Emission: Nm^3_{CO2}/kWh_{th}	0.205	0.175	0.101

Fig. 3.4 Lower heating values related to CO_2 produced, for carbon, a typical coal, and methane [17, 18].

$$\tilde{w}_{\text{rev}} = \sum_i y_i RT \ln(p/p_i) [\text{J mol}^{-1}] \quad \text{or} \quad w_{\text{rev}} = \sum_i x_i R_i T \ln(p/p_i) \quad [\text{J kg}^{-1}] \quad (3.1)$$

It represents the energy needed for isothermal recompression of each component back to the initial pressure of the mixture p. The complete separation of just one component out of a mixture is a special case of this approach, if the target gas and the other components are again separated simultaneously out of a stirred volume, such that the partial pressures in the mixture remain constant. The energy is the same as that for complete separation of a two-component mixture:

$$\tilde{w}_{\text{rev}} = y_i RT \ln(p/p_i) + (1 - y_i) RT \ln\left(\frac{p}{p - p_i}\right) \quad [\text{J mol}^{-1}_{\text{mixture}}] \quad (3.2)$$

The same is valid for the mass fraction expression, where R_i and R_{others} replace R.

The partial or full separation of one component out of a passing gas flow, which is closer to real application, needs a different approach because of the decreasing partial pressure of the target gas in the main flow. The separation is to be considered as the integral of infinitesimal separation steps. Each step needs a higher energy for recompression than the previous one. The energy needed for a step (dn_i<0) referred to the initial amount of mixture (n_0, m_0) is

$$d\tilde{w}_{\text{rev}} = \frac{-1}{n_0} RT \ln(p/p_i) dn_i \quad [\text{J mol}^{-1}_{\text{mix},0}] \quad \text{or}$$
$$dw_{\text{rev}} = \frac{-1}{m_0} R_i T \ln(p/p_i) dm_i \quad [\text{J kg}^{-1}_{\text{mix},0}] \quad (3.3)$$

Basic transformation of the molar expression using y_i=p_i/p and $n(1-y_i)$= $n_0(1-y_{i,0})$ – n_{others} yields

$$d\tilde{w}_{\text{rev}} = (1 - y_{i,0}) RT \frac{\ln(y_i)}{1 - y_i} dy_i \quad [\text{J mol}^{-1}_{\text{mixture},0}] \quad (3.4)$$

which needs numerical integration.

In another approach, the energy can be referenced to the amount of the separated component. For any specific stage along the process, this yields

$$\tilde{w}_{\text{rev},i} = RT \ln(p/p_i) \quad [\text{J mol}_i^{-1}] \quad \text{or} \quad w_{\text{rev},i} = R_i T \ln(p/p_i) \quad [\text{J kg}_i^{-1}] \quad (3.5)$$

which is shown as a functional curve in Figure 3.5. Unfortunately, this expression cannot be simply integrated along partial pressure steps, because the mass transferred with each pressure step decreases. A formulation of infinitesimal steps can be done again via changes to the number of moles of the target component in the main flow (dn_i<0), which is now to be divided by the total num-

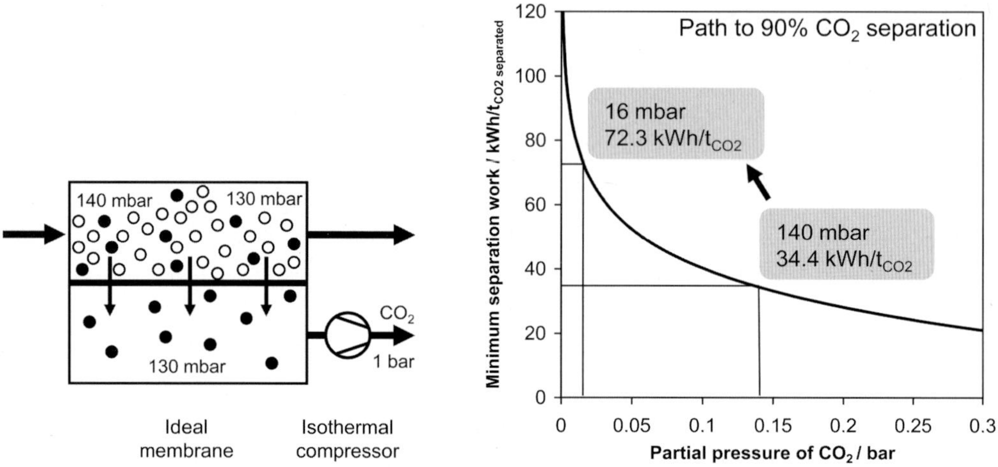

Fig. 3.5 CO_2 separation with membranes and instantaneous minimum separation work at 60 °C.

ber of the moles separated in order to obtain the total energy with respect to the amount separated via integration up to the final state 1:

$$d\tilde{w}_{\text{rev},j} = RT\frac{\ln(p/p_i)}{n_{i,1} - n_{i,0}}dn_i \; [\text{J mole}_i^{-1}] \quad \text{or} \quad dw_{\text{rev},j} = R_iT\frac{\ln(p/p_i)}{m_{i,1} - m_{i,0}}dm_i \; [\text{J kg}_i^{-1}] \tag{3.6}$$

The first equation can be transformed into a similar form as Equation 3.4, with just a different constant. This means that numerical integration is needed again. Simple usage of Equations 3.5 does not yield the correct overall result, as it neglects the decreasing partial pressure. It just gives an instantaneous value for the energy per unit mass of the mixture for a certain stage within the process, and the mass fraction refers to the instantaneous and not constantly to the initial mixture composition.

In the following, an ideal membrane process is considered. It is characterized by a membrane with perfect CO_2 selectivity and by a subsequent isothermal, that is, cooled CO_2 compressor. For a typical CO_2 concentration in the flue gas of 14 mol% at the membrane inlet, the CO_2 partial pressure is 140 mbar. Figure 3.5 shows the first step of a segmented membrane separation process. In the first step (and in each following step), only a small part of the CO_2 is separated to keep the pressure at the secondary side of the membrane (permeate side) as high as possible. The curve for the minimum separation work in Figure 3.5 shows at the beginning of the total membrane process a specific energy demand of 34.4 kWh $t_{CO_2}^{-1}$. When 90% of the initial CO_2 is separated, the specific energy is already 72.3 kWh $t_{CO_2}^{-1}$. The resulting mass-averaged value obtained by numerical integration is 43.8 kWh $t_{CO_2\,\text{separated}}^{-1}$ for gases at a temperature of

60 °C. This is equivalent to 39.4 kWh $t^{-1}_{CO_2\,produced}$, which corresponds to 1.35% net plant efficiency loss, if coal with a thermal energy of 2910 $kWh_{th}\,t^{-1}_{CO_2}$ is used (see Figure 3.4).

3.2.2.3 Minimum Energy Demand for CCS

For isothermal loss-free compression from 1 to 120 bar at 60 °C, a specific energy of 78 kWh $t^{-1}_{CO_2}$ was calculated with a Peng–Robinson gas model [20]. This corresponds to an efficiency penalty of 2.4% – again for 90% CO_2 separation.

Finally, the total energy demand for a degree of CO_2 separation of 90% and compression to 120 bar amounts to 122 kWh $t^{-1}_{CO_2}$, which corresponds to an efficiency penalty of 3.8%. To date, efficiency penalties of real CCS concepts are in the order of 10% [9].

3.3 CO_2 Capture Routes

There are three routes for CO_2 capture from coal-fired power plants:
- post-combustion capture
- oxyfuel combustion
- pre-combustion.

Whereas post-combustion and oxyfuel capture apply to either SPPs or IGCC power plants, pre-combustion just applies to modified IGCC power plants.

Characteristic features of *post-combustion capture* are
- "End-of-pipe" technology, suitable for retrofitting and which is also used to remove other pollutants.
- CO_2 separation from the flue gas, where it is diluted in nitrogen.
- For example, absorption of the CO_2 in aqueous amine solutions and subsequent desorption of the CO_2 by thermal stripping with steam.

In an *oxyfuel* process, oxygen instead of air is used for combustion, rendering flue gas with a high CO_2 concentration and no or little N_2. Characteristic features of the process are
- O_2 separation from air with purities between 95 and 99 mol% (O_2 concentration in air 21 mol%)
- combustion in an O_2–CO_2–H_2O atmosphere and flue gas recycling for adjusting the combustion temperature
- CO_2 separation from the water vapor by cooling for H_2O condensation.

In *pre-combustion* capture, CO_2 is removed prior to combustion. For this purpose, the IGCC power plant concept has to be modified. The syngas mainly consists of carbon monoxide and hydrogen at a pressure of 30–40 bar. CO reacts in a shift reactor with steam to produce CO_2 (concentration about 30 mol%)

and additional hydrogen (about 60 mol%). Then the CO_2 is separated from the syngas. The remaining syngas, mainly composed of hydrogen, is burned in a modified gas turbine to generate power. The characteristic features of pre-combustion *with selective CO_2 separation* are

- modified IGCC power plant design with a CO shift stage
- CO_2 separation from the shifted coal gas; for example, physical absorption of the CO_2 by use of a liquid and subsequent desorption of the CO_2, for example, by expansion
- compression of the atmospheric CO_2 for liquefaction.

In post-combustion and oxyfuel processes, only those separation processes exist today which separate CO_2 and O_2. The alternative task of separating the other main component completely, mostly N_2, has not been accomplished, since there are no practicable methods available for selective N_2 separation. In the case of pre-combustion, the alternative of selective H_2 separation exists, e.g. by membranes. The characteristic features of pre-combustion *with selective H_2 separation* are

- modified IGCC with CO shift
- H_2 separation from the shifted coal gas
- compression of already pressurized CO_2 to the final pressure required for transportation and storage.

3.4 Gas Separation Tasks and Methods

From the three gas separation routes, four different gas separation tasks were identified:

- CO_2 separation from flue gases
- CO_2 separation from shifted coal gas
- O_2 separation from air
- H_2 separation from shifted coal gas.

In principle, many different methods exist for these gas separation tasks, but not all are suitable for large-scale applications, especially considering the efficiency. The promising methods are listed in Table 3.1 and shown in Figure 3.6. Seven principles have been identified, each representing its own class of technology. Each principle (except rectification in cryogenic air separation) is based on materials that either undergo a special type of binding with the separated gas molecules (chemical and physical absorption of CO_2 and also reaction of O_2 and CO_2 with solids) or represent the membrane material in the case of membrane separation. These materials are categorized into 13 material classes.

Table 3.1 Separation routes, tasks, and methods.

Route	Separation tasks	Methods
Post-combustion	CO_2 vs. N_2	Absorption with liquids Carbonate looping Adsorption with solids Membranes
Oxyfuel	O_2 vs. N_2	Cryogenic air separation Chemical looping combustion Membranes
Pre-combustion	CO_2 vs. H_2	Absorption with liquids Carbonate looping Membranes
	H_2 vs. CO_2	Adsorption with solids Membranes

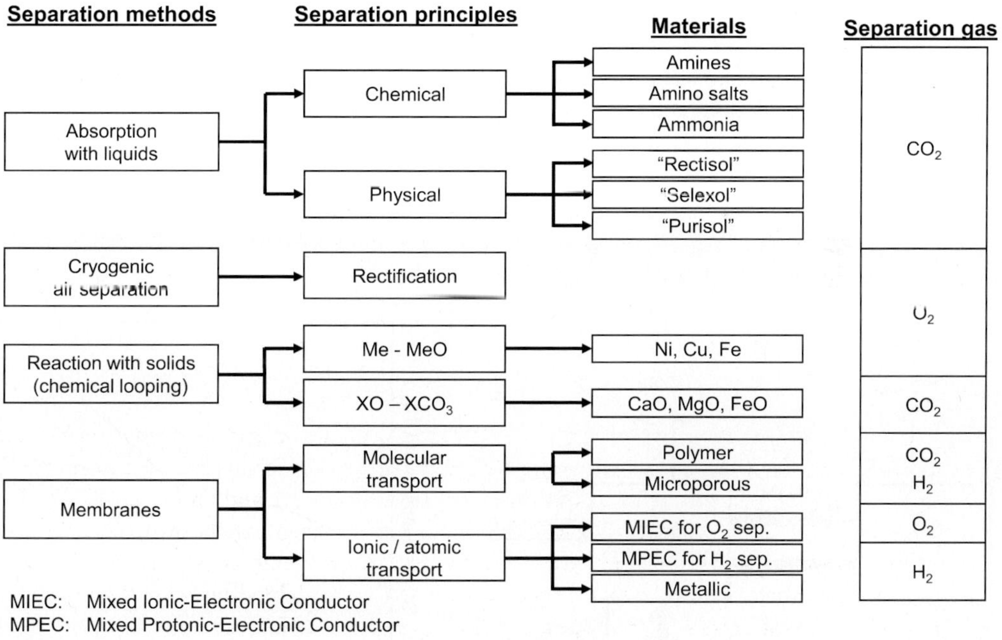

Fig. 3.6 Gas separation methods, principles, and the materials utilized.
x: Ca, Mg, Fe.

3.4.1 Absorption with Liquids

In CO_2 absorption processes, the CO_2 is transferred from the main gas stream, that is, flue gas or coal gas, to a liquid absorbent such as a sorbent or solvent in the absorber (see Figure 3.7). There the CO_2 is selectively absorbed. The absorbent is pumped to a regenerator, in which the CO_2 is released under heat or pressure. The energy input for releasing the CO_2 is strongly dependent on the type of solvent. Then the lean solvent is recycled to the absorber. A make-up flow of fresh absorbent compensates for losses due to deactivation and solvent losses.

CO_2 absorption processes are categorized as either chemical or physical. In chemical absorption, the CO_2 reacts with the solvent. Therefore, chemical solvents have a relatively high loading capacity at low CO_2 partial pressures, qualifying the method for post-combustion. In physical absorption, the amount of CO_2 dissolved in the solvent is proportional to the CO_2 partial pressure in the main gas stream. This is illustrated in Figure 3.7b, where the exact slope and curvature depend on the actual solvent. Consequently, physical absorption is preferred for the treatment of high-pressure gas streams, as in pre-combustion. Hybrid systems combine chemical and physical absorption [12, 15].

3.4.1.1 Chemical Absorption

Chemical absorption is in principle a proven technology, which is commercially available. It is used to purify natural gas, to produce CO_2 from flue gas of gas turbines for commercial applications, and in the chemical industry. None of these applications equals the scale required for coal-based power plant flue gases, however. Ongoing research and development activities are focusing on

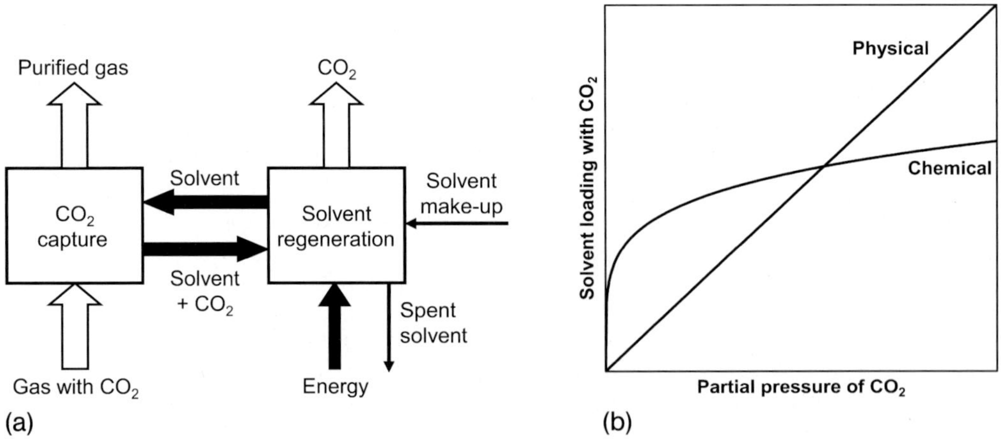

Fig. 3.7 Schematic diagram of CO_2 separation with solvents.

the pilot-plant scale for scaling up the technology. The advantages are as follows:
- 90% degree of CO_2 separation or higher are obtained.
- CO_2 purities above 99% are reached.

Commercially available absorbents are aqueous solutions of alkanolamines such as
- primary amines: monoethanolamine (MEA)
- secondary amines: diethanolamine (DEA)
- tertiary amines: methyldiethanolamine (MDEA)
- hindered amines.

The literature provides values for the heat of reaction between 1.6 and 1.9 MJ $kg_{CO_2}^{-1}$ for MEA. This may be due to several reaction steps, which are based on equilibrium constants and are the subject of simulations. Three main reactions can be identified [21, 22]:
- hydration of CO_2 in the solution: ~25% of the binding energy (~20 kJ $mol_{CO_2}^{-1}$)
- carbamate formation: ~15% of the binding energy (~10 kJ $mol_{CO_2}^{-1}$)
- protonation of the MEA: ~60% of the binding energy (~50 kJ $mol_{CO_2}^{-1}$).

Alternative solvents, showing a lower solvent regeneration energy demand, are being intensively investigated [23, 24]:
- amino acid salts
- chilled ammonia.

In the case of chilled ammonia, the absorber operates at reduced temperatures slightly above 0 °C. A special advantage is that the capture process, which is designed for higher pressure, releases the separated CO_2 with a pressure of 15–20 bar, saving much compression energy for storage [25].

3.4.1.2 **Physical Absorption**

The process of physical absorption utilizes the solubility of acid gases in liquid solvents to separate fractions of sour gases such as CO_2 and H_2S from a gas stream. According to Henry's law, the solvent loading is proportional to the partial pressure of the gas component considered, for example, CO_2:

$$m_{CO_2}/m_{solvent} = kp_{CO_2} \tag{3.7}$$

where k is the Henry's law constant [26]. Hence physical absorption is more suitable for higher partial pressures, making this process applicable for decarbonization and acid gas removal from pressurized syngas and natural gas. Depending on the chosen solvent, the physical absorption process is capable of reducing the amount of acid gases to a value in the ppm range and of producing a CO_2 stream of high purity [27]. Advantages compared with chemical absorption are lower energy consumption for regeneration of the solvent and, since no chemical reactions occur, good stability of the solvent.

The physical solvent is loaded in the absorber and is regenerated in the desorber. This can be done by flashing, stripping, and reboiling or a combination of these measures. During flashing and stripping, the pressure of the gas phase is reduced, so that the equilibrium concentration in the liquid phase decreases proportionally. Reboiling leads to an increase in temperature, which causes a decrease in the acid gas solubility. Therefore, the loading capacity of the solvent is reduced. In all cases, the loaded gases are set free. In comparison with regeneration of chemically bound gases, where chemical bonds have to be broken, these processes require less energy. The main energy consumption is caused by solvent circulation and by refrigeration [9].

The most common absorption processes are Selexol, owned by Union Carbide and licensed by UOP, and Rectisol, licensed by Linde and Lurgi [28]. The Selexol process uses a solvent consisting of a mixture of dimethyl ethers of poly(ethylene glycol) (DMPEG). It is often used for acid gas removal from natural and syngas streams, since it achieves purities of 1 ppm H_2S and COS. The operating temperature lies between 0 and 40 °C [29].

The Rectisol process uses methanol as a solvent. Owing to the high volatility, operating temperatures are between –30 and –60 °C. The Rectisol process is commonly utilized for the purification of hydrogen, ammonia, and methanol syngas with subsequent chemical applications, where very low sulfur contents are required. The Rectisol process is capable of purifying syngas down to 0.1 ppm total sulfur content and CO_2 in the ppm range [28].

Physical–chemical scrubbers use hybrid solvents to combine the characteristics of both chemical and physical solvents and are usually a combination of an amine and an organic physical solvent. The most common hybrid solvent processes are BASF's activated MDEA process, Shell's Sulfinol and Dow's UCARSOL [30].

3.4.2
Cryogenic Air Separation

Today, the single technology for oxygen production on an industrial scale is cryogenic air separation developed by Linde. This technology is applied in the steel and oil industries and in IGCC power plants for coal gasification. Air is compressed and cooled to about –180 °C by means of highly efficient heat exchange with outgoing products. The rectification takes place in a double column with an integrated boiler–condenser.

Up to 95% oxygen purity, only nitrogen and oxygen have to be separated, leaving the predominant impurity argon together with oxygen. For higher purities, the specific energy demand increases, since lower temperatures are required. For purities of 99% and 99.5%, about 10% and 15–20% more energy is required, respectively [9, 31, 32] (Table 3.2).

Table 3.2 Specific energy demand of cryogenic oxygen production [9, 31, 32].

O_2 purity (mol%)	Specific energy demand		Efficiency penalty (%) [a)]
	kWh/(tO_2)	kWh/(tCO_2)[a)]	
95	220	160	5.5
99	240	175	6.0
99.5	260	190	6.5

a) For stoichiometric C combustion.

3.4.3 Reaction with Solids

Figure 3.7 shows the absorption–regeneration principle with liquids, characterized by two reactors and solvent cycling. Reactions with solids can be carried out in similar processes, as shown in Figure 3.8:

- chemical looping combustion (CLC) to separate O_2 from air in an oxyfuel process
- carbonate looping (CL) to separate CO_2 from flue gas in a post-combustion process
- CL to separate CO_2 from shifted coal gas in a pre-combustion process.

3.4.3.1 Chemical Looping Combustion (CLC)

CLC is an indirect combustion method. A solid, usually a metal such as nickel, copper, or iron, transfers oxygen from the air to the fuel. In an air reactor a metal oxide is formed:

$$Me + \frac{1}{2}O_2 \rightarrow MeO \tag{3.8}$$

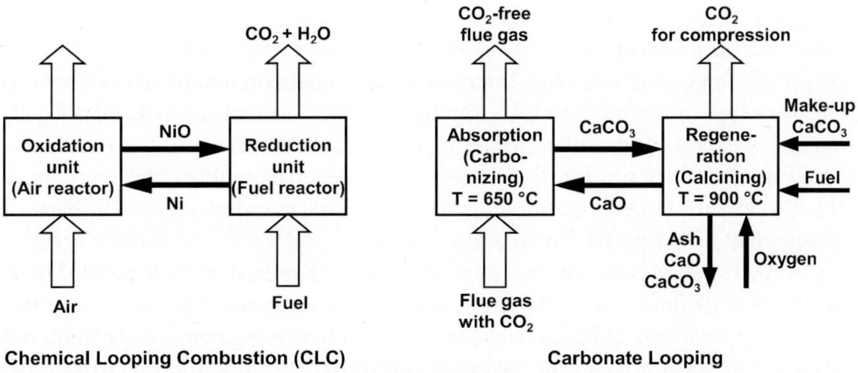

Fig. 3.8 Schemes of chemical looping combustion and carbonate looping [9, 30].

The metal oxide is then transported to the fuel reactor, in which the fuel is oxidized. The reduced metal oxide is then again transported to the air reactor. The temperatures in the reactors are between 800 and 1200 °C [33–35]. The particles have diameters of about 0.1–0.5 mm and are fluidized in the reactors, with air and with fuel, which today is natural gas. There is great interest in using CLC for solid fuels such as coal. By gasification, coal can be converted into a gas, which could react with the metal oxide in the fuel reactor [36–38]. The gas stream from the fuel reactor mainly contains only CO_2 and water vapor, which can be easily separated. The energy released by the two reactions corresponds to the reaction enthalpy of conventional combustion. No transport of pure oxygen is needed in CLC technology. This is a precondition to achieve low energy consumption in oxyfuel processes [39].

3.4.3.2 Carbonate Looping

CO_2 can be adsorbed by dry regenerable sorbents. In a second step, the sorbent is regenerated and recycled. Thereby, a concentrated stream of CO_2 is produced. Several sorbents have been investigated and solid supports impregnated with amines have also been suggested [40, 41]. The sorbents are applicable in post-combustion and pre-combustion. The reaction temperatures are low (absorption at about 50 °C and regeneration at 100–150 °C).

A further possibility is a high-temperature carbonate loop, in which calcined limestone is used [42]. Limestone circulates between a high-temperature calciner, which produces free lime and CO_2, and a high-temperature region of the furnace, where CO_2 is absorbed. The application of high temperatures in the CO_2 separation step has the potential to achieve higher plant efficiencies in comparison with absorption with liquids, because the absorption heat is now released at temperatures suitable for power production [43]. The absorption of CO_2 can also be integrated into a gasification reactor in pre-combustion [4].

3.4.4 Gas Separation with Membranes

Membrane technologies for CO_2 capture in power plants have been widely investigated, since they offer the potential for low energy demand for gas separation processes. In addition to CCS application, they are also investigated for natural gas cleaning. The driving force for CO_2 separation is the CO_2 partial pressure difference between the feed side and the permeate side of the membrane [44]. The world's largest membrane plant, in Pakistan, separates 14×10^6 Nm^3 of CO_2 per day from natural gas in a gas field using 400 000 m^2 of polymeric cellulose acetate membranes [45–47]. As illustrated in Figure 3.9, high natural driving forces exist for the permeation of CO_2 through the membranes, so that under these conditions practically no additional energy is required. During the separation process p_{CO_2} at the feed side and consequently also the driving force decreases continuously, at the maximum until 1 bar is reached, and limits the

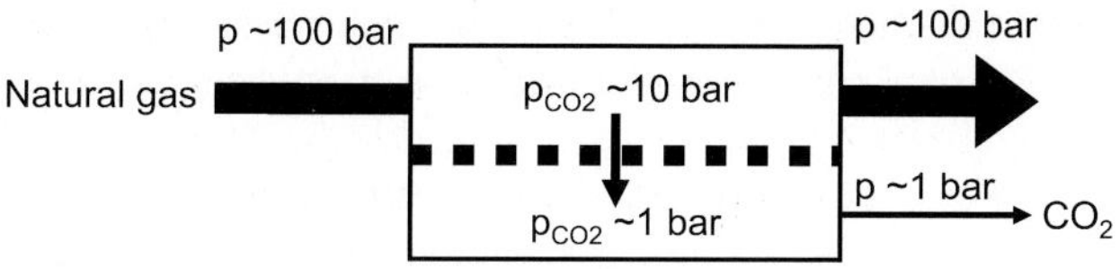

Fig. 3.9 Purification of natural gas by CO_2 separating membranes without additional compressors in the membrane environment.

degree of separation to about 90%. From the data for the plant, a mean driving force of 5.4 bar results, which finally leads to a reasonable CO_2 flux density of 1.5 $Nm^3\,m^{-2}\,h^{-1}$ or 26 t $CO_2\,m^{-2}\,a^{-1}$. Such data for optimal membrane application generate high motivation for concept developments that integrate membranes in power plants, to find similar promising conditions for the different gas separation tasks in CCS.

3.4.4.1 **Polymeric Membranes**

The transport mechanism for a gas component A in polymeric membranes involves solution and diffusion in the material. The driving force for permeation is the partial pressure difference between the feed side and the permeate side. This leads to the transport equation for the flux of component A through the membrane:

$$j_A = (Pb_A/L)(p'_A - p''_A) \tag{3.9}$$

$$Pb_A = S_A D_A \tag{3.10}$$

$$P_A = Pb_A/L \tag{3.11}$$

where Pb_A is the permeability of gas component A, L is the membrane thickness, p'_A and p''_A are the partial pressures at the feed and permeate side, respectively, S_A is the solubility coefficient, D_A is the diffusion coefficient, and P_A is the permeance of gas component A [45].

The selectivity of gas component A against gas component B is defined as

$$\alpha = P_A/P_B \tag{3.12}$$

With regard to polymeric membranes, Powell and Qiao [48] presented a detailed review of different CO_2-selective polymer structures developed for flue gas separation. Two types of polymeric membranes are widely used commercially for gas separations. Glassy membranes are rigid and glass-like, and operate below their glass transition temperatures. On the other hand, rubbery membranes are flexible and soft, and operate above their glass transition temperatures. The original discussion of the "Robeson plot" for both glassy and rubbery membranes has become one of the most often cited papers in the gas membrane separation literature [49]. Current polymeric membranes can be categorized as follows [48, 50, 51].

Polyimides: Polyimides are very attractive materials for gas separation membranes because of their good gas separation and benign physical properties, such as high thermal stability, chemical resistance, mechanical strength, and low dielectric constant. Some polyimides, particularly those incorporating the group 6FDA (Hexafluoroisopropylidene Bisphthalic Dianhydride), possess very high carbon dioxide permeability combined with high selectivity. An extensive review of the gas separation properties of polyimides was published by Langsam in 1996 [52]. However, issues associated with swelling and plasticization of polyimides can limit their application in the gas separation area.

Facilitated transport: These membranes comprise a carrier, typically a metal ion, with a special affinity towards a target gas molecule, and this interaction controls the rate of transport. Facilitated transport membranes can be divided into two general categories: fixed carriers, such as chained carrier, and mobile carriers, such as immobilized liquids. Representative membranes are polyamidoamine (PAMAM) dendrimers [53] and polyvinylamine (PVAm)–poly(vinyl alcohol) (PVA) [54], respectively. Facilitated transport membranes have attracted a lot of attention in gas separations because they offer high selectivity and large fluxes.

Mixed matrix: The microstructure consists of an inorganic material in the form of micro- or nanoparticles as a discrete phase incorporated into a polymeric matrix. Koros and co-workers [55, 56] proposed some criteria for material selection and preparation protocols in order to match the necessary transport characteristics of materials to prepare high-performance mixed matrix membranes for CO_2 capture. However, their performance suffers from defects caused by poor contact at the inorganic/polymer interface.

Poly-ether-oxide (PEO): PEO membranes are considered attractive materials for CO_2 separation because the polymer chain has a strong affinity to the CO_2 molecule from the presence of polar ether oxygen. There have been numerous efforts to design polymers containing PEO for CO_2–N_2 and CO_2–H_2 separations in part because ethylene oxide units have a high concentration of ether oxygen and are relatively easy to fabricate. However, PEO crystallizes easily, which is deleterious for gas permeability. Consequently, various techniques have been investigated to reduce crystallinity in PEO, for example, using low molecular weight liquid PEO or poly(ethylene glycol) (PEG) [57–61].

Polymeric membranes can be used for CO_2 separation in post-combustion and pre-combustion. Some types of polymeric membranes separate H_2 from CO_2 and are potentially applicable to pre-combustion.

3.4.4.2 Inorganic Membranes

Inorganic membranes can be divided into ceramic and metallic membranes. Figure 3.10 shows the main classes of ceramic and metallic membranes. Two types can be identified with respect to their transport mechanism through the membrane:

- Microporous membranes with either crystalline structure such as zeolites or amorphous structure such as sol–gel membranes. Both types work for CO_2–N_2 and H_2–CO_2 separations.
- Dense membranes are either mixed ionic–electronic conductors or metals. Mixed ionic conductors transport oxygen ions for O_2 separation and protons for H_2 separation, as the gas species. Metals transport H atoms for H_2 separation.

The structure of a microporous ceramic membrane is a graded porous multilayer with macroporous, mesoporous, and microporous layers, in which the pores size in the last layer is less than 2 nm. Separation of different gas molecules is achieved by diffusion. The smaller gas molecules permeate through the porous structure of the membrane, whereas larger molecules are sieved out [62]. Therefore, the size of the related gas molecules is important for the development of well-defined nanostructures. The kinetic molecule diameters lie in a very narrow range (H_2 0.289 nm, CO_2 0.330 nm, O_2 0.346 nm, and N_2 0.364 nm) [63].

The transport equation for a gas component A is linear in the driving force:

$$j_A = P_A(p'_A - p''_A) \tag{3.13}$$

where P_A is the permeance of gas component A and p'_A and p''_A are the partial pressures at the feed and permeate side of the membrane, respectively [44]. Permeance is often also called permeability.

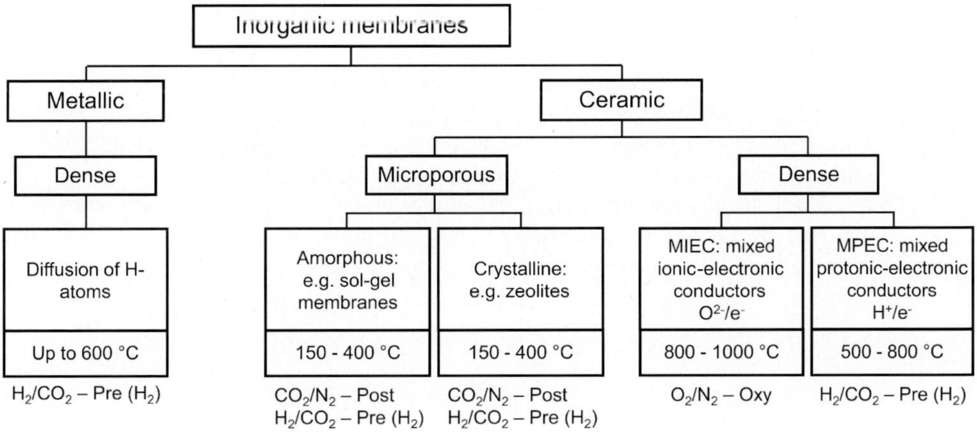

Fig. 3.10 Inorganic membranes: operating temperatures and possible application to the separation tasks within the capture routes.

Amorphous SiO_2 membranes have been investigated and show a potential for high selectivity (e.g., >100 for H_2–CO_2 separation) and for high permeability due to thicknesses of only 50–200 nm reported in the literature.

Mixed ionic electronic conductors (MIECs) transport oxygen ions via vacancies in a crystal lattice. Promising materials are perovskites and fluorites [64]. Barium-containing perovskites show the highest permeability, but are chemically not stable in atmospheres containing CO_2, SO_2, or H_2O. The driving force for permeation is an O_2 partial pressure gradient, which leads to a directed diffusion from the high- to the low-pressure side of the membrane. The O_2 flux density j_{O_2} in the bulk material is described by the Wagner equation:

$$j_{O_2} = (RT/16F^2)(\sigma_{amb}/L)\ln(p'_{O_2}/p''_{O_2}) \tag{3.14}$$

where R is the gas constant, T is the temperature, F is the Faraday constant, σ_{amb} is the ambipolar conductivity, L is the membrane thickness, and p'_{O_2} and p''_{O_2} are the oxygen partial pressure on the oxygen-rich and lean side, respectively [62]. The material-related parameter ambipolar conductivity σ_{amb} is composed of the ionic and electronic conductivities. O_2 permeation increases when the membrane thickness is reduced. However, surface exchange reactions become rate limiting beneath a characteristic thickness L_c [65]:

$$L_c = D/k_s \tag{3.15}$$

where D is the oxygen diffusion coefficient and k_s is the oxygen surface exchange coefficient. Therefore, it seems that membrane development to achieve high permeability by thickness reduction will be successful only if the surface exchange kinetics are accelerated by catalytic activation. Membrane modules for application to oxyfuel are under study, with consideration of several membrane geometries [66, 67].

Ceramic membranes with mixed proton–electron conductivity are promising for pre-combustion capture. The advantages of these membranes are their high thermal, mechanical, and chemical stability. However, the mixed proton–electron conductivity, determining the hydrogen flux, has to be improved. For this purpose, the formation of defects in the material is required, either by variation of the basic component ratios or extrinsically by doping.

Metal membranes are mainly palladium or palladium–alloy membranes supported by porous metallic tubes. Their thickness lies in the range 1–10 μm. Their H_2–CO_2 selectivity is high due to the selective transport of hydrogen atoms through the material. According to Sievert's law, the flux of H_2 through the membrane is

$$j_{H_2} = (S_{H_2}D_{H_2}/L)(\sqrt{p'_{H_2}} - \sqrt{p''_{H_2}} \tag{3.16}$$

where S_{H_2} is the solubility coefficient, D_{H_2} is the diffusion coefficient, L is the membrane thickness, p'_{H_2} and p''_{H_2} and are the hydrogen partial pressure at the feed and permeate side of the membrane, respectively [68, 69]. The permeability is relatively low and can be affected by water, CO, and sulfur compounds in the feed gas (coal gas). They operate at temperatures up to 600 °C [70] and can be applied pre-combustion.

3.5 Plant Concepts for Carbon Capture

3.5.1 Overview

Each of the basic gas separation methods – absorption/reaction, condensation, and membrane separation – can be applied to at least one of the three capture routes: post-combustion, oxyfuel and pre-combustion. Since "pre-combustion" splits into two gas separation tasks, namely selective CO_2 or H_2 separation, a total of four basic concepts exist (Table 3.3). The gray matrix fields in Table 3.3 indicate that the gas separation method is not applicable; in empty fields the method is no longer attractive today. Sixteen matrix elements are filled, indicating that the method is applicable and attractive today. The routes "post-combustion" and "oxyfuel" can be applied to both basic types of coal-based power plants: the steam power plant (SPP) and the combined cycle plant with integrated coal gasification (IGCC). In contrast, pre-combustion capture is applicable only after gasification and additional CO shift (shift of CO to CO_2 by use of steam; see Chapter 3.5.4). There are also eight potentially attractive concepts for such IGCC + shift plants.

As with gas-fired combined cycles, also with IGCCs the gas turbine combustor usually works with excess air. This offers the opportunity for flue gas recycling, which targets either increasing the CO_2 concentration in the flue gas in the case of post-combustion capture, or sweeping the permeate side of H_2 membranes with N_2 in the case of pre-combustion capture. Following these systematics, finally 32 different classes of coal power plants with carbon capture result. These classes may still include different variants, for example, oxyfuel with pressurized air or with permeate vacuum at the membrane.

Figure 3.11 gives a top-level overview of flue gas recycling concepts for large fossil-fuel power plants. It shows on the left-hand side simple schemes for an IGCC and a steam power plant. CO_2 can be separated from the flue gases in post-combustion concepts. The flue gases have CO_2 concentrations of 12–14 mol% in the case of the steam power plant and only about 4 mol% in IGCC power plants. This happens because IGCC plants operate with excess air and consequently with large flue gas streams, extremely diluting the CO_2. The air ratios are in the range 2.2–3 [71].

Burner cooling by flue gas recycling is an option for reducing air and flue gas flow rates, thus increasing the CO_2 concentration [72]. The air ratio is now approaching the value 1, representing the stoichiometric case. A similar advanced

Table 3.3 Capture routes (3), separation tasks (4), and power plant classes (24 + 8).[a)]

Basic concept		Gas separation method								
Route	**Gas separation** task	**Condensation**	**Absorption/reaction**				**Membrane separation**			
		Cryogenic air separation	**Absorption with liquids**		**Reaction with solids**		**Polymer**	**Porous**	**Mixed conducting**	**Metallic**
			Chemical	**Physical**	**Adsorption**	**Reaction**				
Post-combustion	CO_2		SPP IGCC*			SPP IGCC*	SPP IGCC*	SPP IGCC*		
Oxyfuel	O_2	SPP IGCC			SPP IGCC	SPP IGCC			SPP IGCC	
Pre-combustion	CO_2			IGCC + shift		IGCC + shift	IGCC + shift			
	H_2				IGCC + shift		IGCC + shift**	IGCC + shift**	IGCC + shift**	IGCC + shift**

a) Optional modifications: *flue gas recycle for higher CO_2 concentration;
**flue gas recycle for membrane sweep with a large O_2-poor N_2 gas stream.

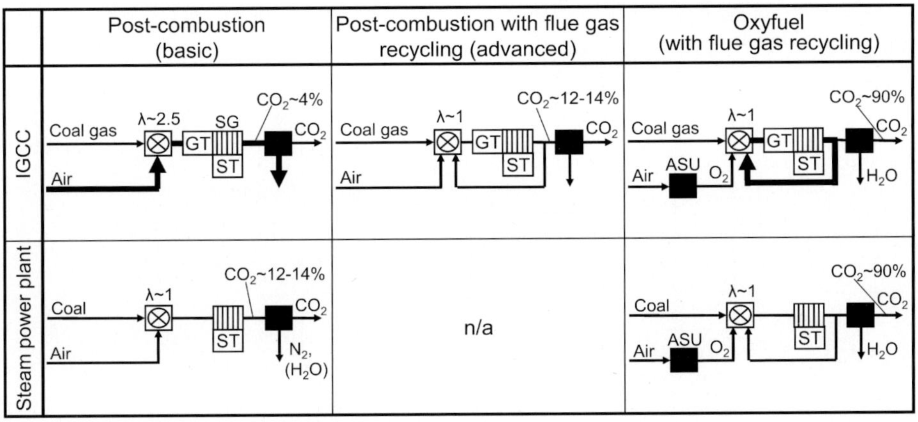

Fig. 3.11 Effect of flue gas recycling on the CO_2 concentration in the flue gases (λ, air ratio, ASU, air separation unit, GT, gas turbine, SG, steam generator, ST, steam turbine).

post-combustion process with flue gas recycling is required for CO_2 capture with chilled ammonia at a natural gas-based NGCC [73]. The modified operating conditions in the combustion system of combined cycles, such as reduced oxygen content, still require basic research on combustion performance and flame stability. This is included in the COORETEC R&D program [72].

The next step to increase the CO_2 concentration further is the elimination of nitrogen in the oxidant by use of an air separation unit. This leads to the so-called oxyfuel process. As shown at the right-hand side of Figure 3.12, this now requires

- flue gas recycling in steam power plants
- increased recycle ratios in IGCCs.

In both oxyfuel plants the CO_2 concentration in flue gas is now of the order of 90 mol%.

3.5.2
Post-Combustion CO_2 Capture

Some important post-combustion capture processes and materials are considered in this section. Enhanced and detailed information can be found in dedicated chapters elsewhere in this book.

3.5.2.1 Post-Combustion Capture by MEA Scrubbing

CO_2 separation by monoethanolamine (MEA) scrubbing is one of the gas separation processes based on chemical absorption [28]. The early development of methods and processes for CO_2 separation by aqueous solutions of MEA was

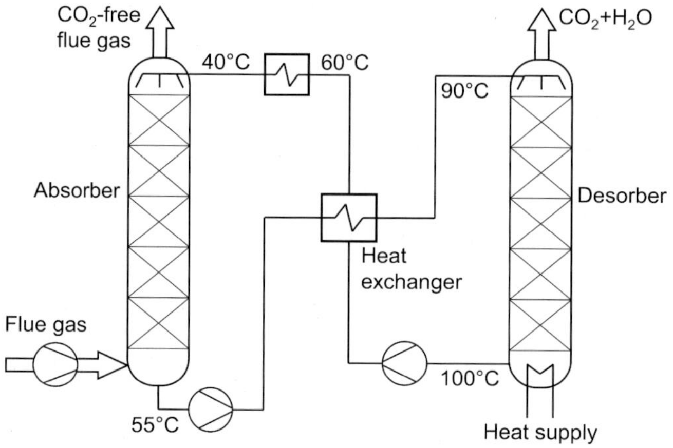

Fig. 3.12 Schematic flow diagram of the amine scrubbing process [72].

done by companies such as Fluor Daniel, ABB Lummus Crest, and Mitsubishi Heavy Industries. The process was applied to flue gases from gas or oil firing. Application in the case of coal firing – with its significantly different flue gas conditions – is currently being tested in pilot plants. The largest realized plant of Fluor Daniel has a capacity of 320 t per day. For use in coal-fired power plants, further scale-up by a factor of about 30 is therefore necessary. The process scheme is shown in Figure 3.12.

In the scrubber column, the flue gas passes through the CO_2-lean MEA solution so that the CO_2 from the flue gas is absorbed by a chemical reaction. The CO_2-rich MEA solution is preheated in a cross-flow heat exchanger, then it enters the desorber. Here the CO_2 is released by warming up the MEA solution to 100–130 °C. The separated gas is then cooled to condense the water vapor. MEA scrubbing is an exothermic process, which is carried out at about 40–50 °C [74].

The total energy demand for the complete CCS process consists of

- regeneration heat demand of the MEA process (steam)
- power demand for flue gas blower and absorbent pump
- power demand of the CO_2 compression for liquefaction.

In theoretical calculations, laboratory tests, and pilot plants, the degree of CO_2 separation ranges from 80 to 96%. CO_2 purity in the dry product is 99 mol% or higher. A large quantity of energy is required for the absorbent regeneration. Typically 4 MJ $kg_{CO_2}^{-1}$ is required without process optimization. The goal of optimization is about 2 MJ $kg_{CO_2}^{-1}$ in the long term [21, 75]. On the other hand, a similar quantity of absorption energy is released. However, due to the low temperature level in the absorber, this energy cannot be used. The development target for amine scrubbing is to discover amines or amine mixtures that require less regeneration heat. In addition, possibilities for waste heat usage are under investigation [76]. In visionary scenarios, solvent development and optimal system integration show promising efficiency penalties of the order of 5% [21]. The other challenge for chemical solvents is the degradation caused by SO_x in flue gas and further environmental impact in dealing with the used solvents [77].

3.5.2.2 Post-Combustion Capture Using Chilled Ammonia

Another important and more complex process makes use of chilled ammonia. In addition to the absorber and desorber, of which the latter works at a pressure of about 20 bar, it needs flue gas cleaning and precooling, flue gas compression followed by cooling to about 0°C, and ammonia recovery by a wash cycle. The process requires five coolers in total [78]. With a capture rate of 100 $kt_{CO_2}\,a^{-1}$ at the Mountaineer power station, the industrial demonstration has reached the same level as that of MEA processes [79]. A level of 100 $kt_{CO_2}\,a^{-1}$ translates into 20 MW power production. A level of 240 MW_e is planned for 2014 [80].

3.5.2.3 Post-Combustion Capture Using CO_2-Selective Polymeric Membranes

As a technology competing with MEA absorption, the CO_2–N_2 gas separation membrane process for post-combustion capture is attracting more and more attention around the world [39, 81, 82]. In comparison with chemical absorption, CO_2 membranes possess the following advantages:

- Modules can be used as add-on equipment in power plants.
- Modules are compact and flexible.

The driving force of gas separation membranes for CO_2 permeation is the partial pressure difference between the feed and the permeate sides. To create this difference, the corresponding measures can be taken:
on the feed side:

- increasing the pressure (using a compressor)
- increasing the CO_2 concentration (recycling an enriched CO_2 stream)

and on the permeate side:

- decreasing the pressure (using a vacuum pump)
- decreasing the CO_2 concentration (using sweep gas).

As the flue gas from coal-fired power plants has a low CO_2 concentration of 13–15 mol% at ambient pressure, a single-stage membrane cannot achieve the separation target, for example, 95 mol% CO_2 purity and 90% degree of CO_2 separation [83]. Current research activities are focused on multi-stage membrane concepts. Two development routes can be observed: three-end membrane (feed, retentate, and permeate streams) and four-end membrane (feed, retentate and sweep gas, sweep gas/permeate). In four-end membrane concepts, steam [84] and air [47] are used as sweep gas, respectively. Figure 3.13 shows a three-end concept [85]. Comparing these concepts, some conclusions can be drawn, as follows: using sweep gas on the permeate side could be an energy-saving concept,

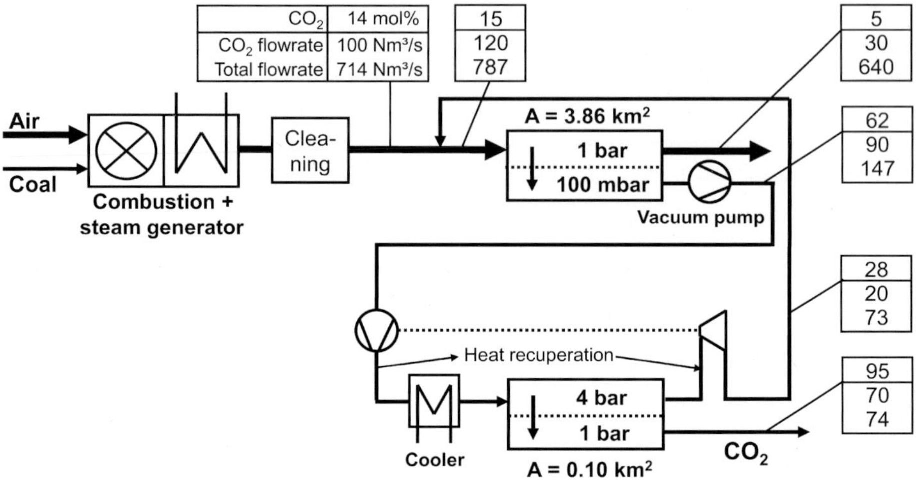

Fig. 3.13 Example calculation of post-combustion capture with CO_2-selective membranes for an 830 MW_e power plant (CO_2–N_2 selectivity 50, CO_2 permeability 3 $Nm^3\ m^{-2}\ h^{-1}\ bar^{-1}$). Three-end cross-flow multi-stage concept: 70% degree of CO_2 separation, CO_2 purity after the first membrane 62 mol% and after the second membrane 95 mol% [85].

but it still consumes energy to separate the captured CO_2 and sweep gas. Furthermore, sweep gas must be recycled within the power plant, for example, the permeate with sweep gas of the first step membrane must be inducted into the boiler [47]. In that case, the above-mentioned advantage will be lost.

A 90% degree of CO_2 separation is a great challenge for any membrane separation concept. This might lead to as much as a 12–14% efficiency loss. An important feature of a three-end membrane capture process is that the specific energy consumption for separation is lower than that of MEA absorption capture. This holds true for 50–70% CO_2 separation. It opens up an application potential for membrane capture as a retrofit for existing power plants.

3.5.3 Oxyfuel Process with Cryogenic Air Separation

In an oxyfuel process, coal is burned with pure oxygen instead of air. Therefore, the conventional power plant process has to be modified. Figure 3.14 shows a schematic process flow diagram. The adiabatic combustion temperature for pure O_2 combustion will be much higher compared with air combustion with the same oxygen excess, because there is no nitrogen to take the heat of reaction. To amend the combustion conditions, about 60% of the flue gas is recycled [86, 87]: burners optimized for air combustion become unstable when operated under a CO_2–O_2 atmosphere. This can be compensated for by increasing the O_2 concentration. Consequently, the O_2 demand is higher and the O_2 impurity in

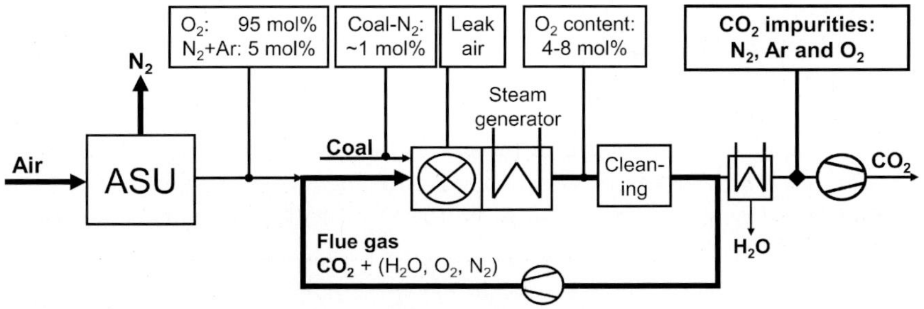

Fig. 3.14 Principle of an oxyfuel process with cryogenic air separation unit (ASU) and flue gas recycling: accumulation of CO_2 impurities from four sources [95].

the final CO_2 stream is higher than anticipated with air combustion. In order to reach lower O_2 concentrations, new burner concepts are required. On the other hand, boilers for oxyfuel processes need smaller heating surfaces and have a smaller size, with benefits in investment costs and heat losses.

As shown in Figure 3.14, a central problem in oxyfuel processes is impurities in the CO_2 stream:

- N_2 and Ar from the cryogenic air separation unit
- non-carbon elements from coal
- O_2 from combustion with excess O_2
- gas components from air leakage.

In order to avoid energy-consuming CO_2 purification [88], it is very important to develop a process that shows low downstream O_2 concentrations and minimum air leakage. New conventional plants have about 4% air leakages related to the flue gas stream. This increases during the operating time up to 10% [87]. However for oxyfuel processes the air leakage into the boiler and the flue gas stream must be reduced to a minimum level to achieve CO_2 concentrations as high as possible.

Efficiency improvement involves integration of waste heat from the compressors of the air separation unit, from the flue gas, and from the CO_2 compression. However, its application is limited owing to the low temperature levels. An alternative is the use of the low-temperature heat in an organic Rankine cycle (ORC) with ammonia, which shows efficiencies of about 20% [87, 89]. For the total oxyfuel CCS process, efficiency penalties between 12 and 8% are reported [9, 90, 91]. Recently, pressurized oxyfuel concepts were proposed [92] in which the CO_2 can be released under pressure. This can save compression energy.

Theoretical analyses of concepts with high-temperature membranes show lower energy requirements for O_2 production and efficiency penalties of only 6% [93, 94]. Since the mixed conducting membranes have selectivities of almost 100%, practically no CO_2 impurities from the oxygen production are expected, opening up a theoretical potential for a CO_2 purity of 90–95 mol%.

3.5.4
Pre-Combustion CO_2 Capture

For pre-combustion capture, coal has first to be gasified. Then CO_2 is captured before the main energy release in the gas turbine combustor takes place. Some important capture processes and materials are mentioned in this section. Enhanced and detailed information can be found in dedicated chapters elsewhere in this book.

3.5.4.1 Pre-Combustion Capture by Physical Absorption

Figure 3.15 shows one possible setup for an IGCC power plant for carbon capture. Coal gas is produced at elevated pressure by addition of steam and sub-stoichiometric oxygen, produced in an air separation unit (ASU). Using oxygen instead of air is essential for pre-combustion capture. The composition of the generated syngas is dominated by large amounts of H_2 and CO, accompanied by small fractions of CO_2, H_2O, Ar, N_2, and other impurities, such as H_2S and fly ash. The main chemical reactions during gasification are

$$C + H_2O \rightarrow CO + H_2 \tag{3.17}$$

$$C + \frac{1}{2}O_2 \rightarrow CO \tag{3.18}$$

After gasification, the syngas is purified in multiple steps and further processed in a water gas shift (WGS) reactor, where CO is converted to CO_2 and H_2 by adding excess steam according to the following exothermic reaction:

$$CO + H_2O \rightarrow CO_2 + H_2 \tag{3.19}$$

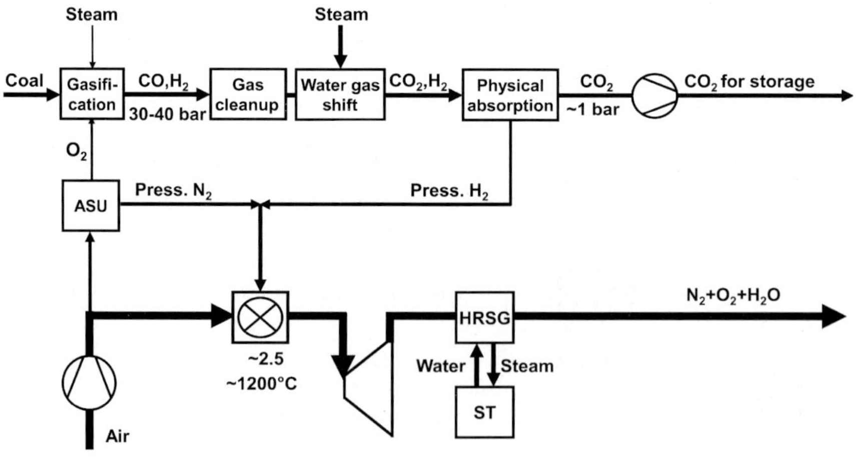

Fig. 3.15 Pre-combustion in an IGCC with CO shift and physical absorption.

The molar concentrations of CO_2 and H_2 in the shifted syngas are approximately 40% and 55%, respectively [96]. This leads to a high partial pressure of CO_2, allowing separation with low energy consumption [30]. Separation is usually done by physical absorption.

The potential for energetic optimization by component development and system integration is high [72]. Efficiency penalties in recent concept studies of about 12% are high, but it is expected that they can be reduced to 7% [97].

For a concept with a CO_2 separating polymer membrane a similar process scheme is useful [98]. To ensure that almost all H_2 flows to the gas turbine, the other part permeating together with the CO_2 through the membrane has to be kept very small, hence high CO_2–H_2 selectivities are required. Sufficiently selective membranes are not yet available. The following section shows that H_2 separating membranes after several development stages may reach degrees of H_2 separation of 98–99%. Hydrogen losses would then contribute only 0.5–1% to the efficiency penalty.

3.5.4.2 Pre-Combustion Capture Using H_2-Selective Membranes

A switch from CO_2 to H_2 separating membranes opens up new ways to design a modified IGCC power plant for advanced pre-combustion with the following three main characteristics:

- reduced excess steam demand for the CO shift reaction
- H_2 flow rate to the gas turbine near to the theoretical maximum
- release of pressurized CO_2 from the H_2 separation step.

Figure 3.16 shows the basic concept, where energy-consuming H_2 recompression to the gas turbine pressure is required [39]. Starting from this basis, the

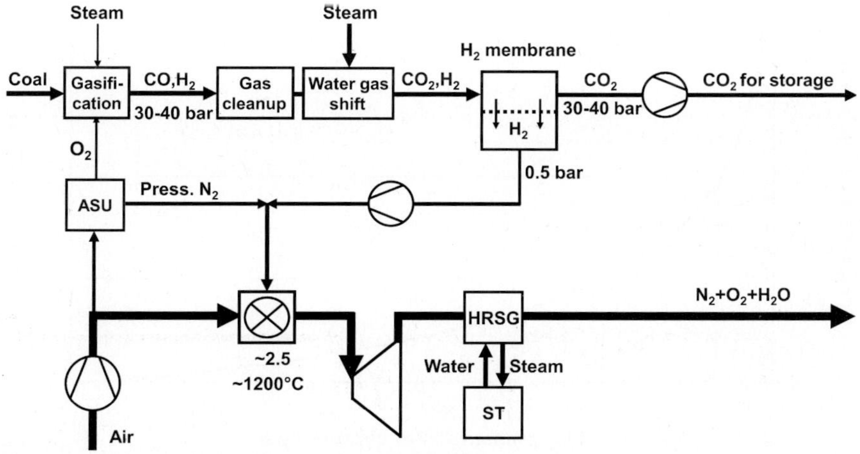

Fig. 3.16 Pre-combustion capture in an IGCC with CO shift, H_2-selective membrane and H_2 recompression [39].

following subsequent optimization and development steps have the potential for a significant increase in the overall CCS plant performance:

- selection of a viable membrane type according to Figure 3.10 (operating temperature, selectivity and permeability, stability)
- selection of an appropriate position of the membrane in the power plant [99]
- integration of shift reaction and H_2 separation into a single membrane reactor [100, 101]
- membrane sweep with pressurized N_2 from the air separation unit, allowing an N_2–H_2 molar ratio of about 1.3 [98, 102, 103]
- membrane sweep with a large stream of pressurized recycled flue gas, allowing an N_2–H_2 molar ratio of about 4 and high driving force [104].

Following this route, the disadvantage arising from the CO shift will be reduced, while the advantages of pre-combustion such as high CO_2 concentration remain.

3.6 Carbon Dioxide Compression

For compression of CO_2 up to a level of 120 bar, which is considered as the transport pipeline inlet pressure, state-of-the-art compressors require an energy of 110 kWh $t_{CO_2}^{-1}$ for pure CO_2 and 115 kWh $t_{CO_2}^{-1}$ for CO_2 diluted with 5 mol% N_2 [4, 39]. The latter corresponds to 4% plant net efficiency loss in the case of complete CO_2 separation. Figure 3.17 shows how the energy demand decreases

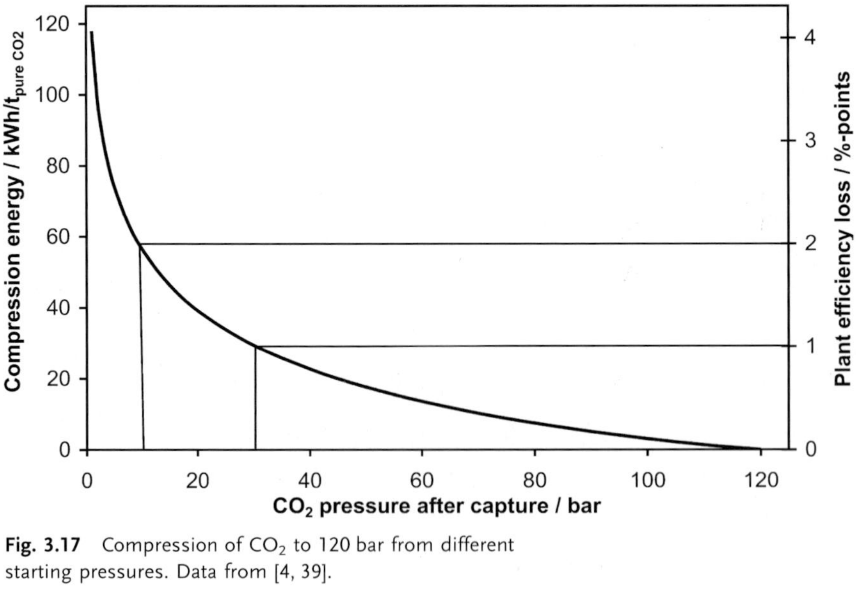

Fig. 3.17 Compression of CO_2 to 120 bar from different starting pressures. Data from [4, 39].

when the capture process delivers CO_2 at pressures higher than 1 bar. A reduction to 2% net efficiency loss can be observed for an initial pressure of 10 bar, and a 1% efficiency loss can be reached with 30 bar.

Various capture process concepts result in CO_2 pressures higher than atmospheric:
- post-combustion with chilled ammonia: 15–20 bar [25]
- oxyfuel pressurized combustion: 10 and 80 bar [92, 105]
- pre-combustion with H_2-selective membranes: 30–40 bar [39]
- pre-combustion in high-pressure gasification systems with H_2-selective water gas shift membrane reactors: about 50 bar [98, 106].

Overall, new compression technologies need to be developed [72]. In that frame, the analysis of the potential of inter-cooling and waste heat use is a key task for cycle calculation and economic process evaluation.

3.7 Conclusion

Recent concepts for CO_2 capture from coal-fired power plants show a broad technological variety in each of the basic routes: post-combustion, oxyfuel, and pre-combustion. Several gas separation methods are currently under investigation at different stages of development. All concepts show potential for further improvement by material development, by process integration of components and apparatus, and by integration of the low-temperature wastes arising from capture and compression systems. On-going R&D is concentrated on the following essential qualities of gas separation:
- minimum plant modifications
- minimum energy demand
- evaluation of tolerable degrees of impurities and separation for all separation tasks
- evaluation of the pressure at which CO_2 is released from the capture process.

Generally, successful development of CCS power plants will require an in-depth dialogue between process engineers and material scientists.

Acknowledgments

This work was supported by the Helmholtz Association, contract HA-104 ("MEM-BRAIN"). The authors thank G. Göttlicher, W.-A. Meulenberg, L. Blum, and R. Menzer for discussions and input.

References

1 Arrhenius, S. On the influence of carbonic acid in the air upon the temperature of the ground. *Philos. Mag. J. Sci.*, 1896, **41**, 237–276.

2 Solomon, S., Qin, D., Manning, M., Chen, Z., Marquis, M., Averyt, K.B., Tignor, M., and Miller, H.L. Climate change 2007: the physical science basis. Contribution of Working Group I to the Fourth Assessment Report of the Intergovernmental Panel on Climate Change, 2007.

3 House, K. *Carbon Dioxide in Terrestrial and Marine Environments – On the Physics and Chemistry of Carbon Dioxide Capture and Storage in Terrestrial and Marine Environments*, Lambert Academic Publishing, Saarbrücken, 2009.

4 Spliethoff, H. *Power Generation from Solid Fuels*, Springer, Berlin, 2010.

5 Reissner, H. and Meschgbiz, A. Flue gas desulfurization as a key requirement of PCC. Presented at the International Symposium on Post Combustion Capture, Düsseldorf/Neuss, 30 July 2009.

6 Herzog, H. Economics. In *Workshop on Carbon Sequestration Science*, May 2001, www.netl.doe.gov/publications/proceedings/01/carbon_seq_wksp/herzog_eco.pdf (last accessed 15 November 2010).

7 Knopf, S., May, F., Müller, C., and Gerling, J.P. Neuberechnung möglicher Kapazitäten zur CO_2-Speicherung in tiefen Aquifer-Strukturen. *Energiewirtsch. Tagesfragen*, 2010, **60**, 76–80.

8 Arlt, W. Thermodynamical optimization of solvents for CO_2 absorption. Presented at the Workshop CO_2-Capture, -Utilization and -Sequestration, Status and Perspectives, PROCESSNET, DECHEMA-Haus, Frankfurt am Main, 21–22 January 2008.

9 VGB. *CO_2 Capture and Storage*, VGB Power Tech, Essen, 2004.

10 Haugen, H.A. and Eide, L.I. CO_2 capture and disposal. *Energy Convers. Manage.*, 1996, **37**, 1061–1066.

11 Race, J.M., Seevam, P.N., and Downie, M.J. Challenges for offshore transport of anthropogenic carbon dioxide. Presented at the 26th International Conference on Offshore Mechanics and Arctic Engineering, San Diego, CA, 10–15 June 2007.

12 Tzimas, E. and Peteves, S. *Controlling Carbon Emissions: the Option of Carbon Sequestration*. Institute for Energy, Petten, The Netherlands, Report EUR 20752 EN, European Commission Joint Research Centre, 2003.

13 Goos, E., Riedel, U., Zhao, L., and Blum, L. Phase diagrams of CO_2 and CO_2–N_2 gas mixtures and their application in compression processes. Presented at the 10th International Conference on Greenhouse Gas Control Technologies, Amsterdam, 19–23 September 2010.

14 Torp, T.A. Underground CO_2 storage – practical experience. Presented at the Workshop on CO_2-Capture, -Utilization and -Sequestration, Status and Perspectives, PROCESSNET, DECHEMA-Haus, Frankfurt am Main, 21–22 January 2008.

15 Metz, B., Davidson, O., de Coninck, H., Loos, M., and Meyer, L. *IPCC Special Report on Carbon Dioxide Capture and Storage*, 2005, www.ipcc.ch (last accessed 15 November 2010).

16 Gerling, J.P. Deutsches Speicherkataster – Planungsgrundlage für dauerhafte CO_2-Speicherung. Presented at the TÜV Nord Akademie CCS-Symposium – CO_2-Abscheidung und -Speicherung, Hamburg, 9–10 June 2009.

17 VGB. *Konzeptstudie: Referenzkraftwerk Nordrhein-Westfalen (RKW NRW)*. VGB PowerTech, Essen, February 2004.

18 Rostrup-Nielsen, J. Reforming and gasification – fossil energy carriers. In *Hydrogen and Fuel Cells – Fundamentals, Technologies and Applications* (ed. Stolten, D.), Wiley-VCH Verlag GmbH, Weinheim, 2010, pp. 291–305.

19 Baehr, H.D. *Thermodynamik*, Springer, Berlin, 1973.

20 Peng, D.-Y. and Robinson, D.P. A new two-constant equation of state. *Ind. Eng. Chem. Fundam.*, 1976, **15**, 59–64.

21 Feron, P.H.M. The potential for improvement of the energy performance of pulverized coal fired power stations with

post-combustion capture of carbon dioxide. *Energy Procedia*, 2009, **1**, 1067–1074.

22 Maeder, M., Wang, X., Conway, W., Fernandes, D., Burns, R., McCann, N., Puxty, G., and Attalla, M. Interactions of CO_2 with aqueous amine solutions – the molecular view. Presented at the 12th Meeting of the International Post-Combustion CO_2 Capture Network, Regina, Canada, 29 September–1 October 2010.

23 Jockenhövel, T. Weniger Verluste als erwartet. *Energie Spektrum*, 2010, **4**, 12–13.

24 Koss, U. Chilled ammonia: status and development. Presented at the Three Countries Workshop on CO_2: CCS and CCR in Germany, Norway, The Netherlands – Challenges and Chances, Essen, 24–25 November 2009.

25 Gray, M. AEP's perspective on carbon capture, status of the Mountaineer demonstration project. Presented at the International Symposium on Post Combustion Capture, Düsseldorf/Neuss, 30 July 2009.

26 Weast, R. C. (ed.). *Handbook of Chemistry and Physics*, 52nd edn., Chemical Rubber Co., Cleveland, OH, 1971.

27 Linde, *Rectisol Wash*, www.linde-engineering.com/en/process_plants/hydrogen_and_synthesis_gas_plants/gas_processing_plants/rectisol_wash/index.html (last accessed 15 November 2010).

28 Kohl, A. and Nielsen, R. *Gas Purification*, 5th edn., Gulf Publishing, Houston, TX, 1997.

29 UOP. *Selexol*, www.uop.com/objects/Selexol.pdf (last accessed 15 November, 2010).

30 US Department of Energy and National Energy Technology Laboratory. *DOE/NETL Advanced Carbon Dioxide Capture R&D Program: Technology Update*, September 2010, www.netl.doe.gov/technologies/coalpower/ewr/pubs/CO2%20Capture%20Tech%20Update%20Final.pdf (last accessed 15 November, 2010).

31 Darde, A., Prabhakar, R., Tranier, J.-P., and Perrin, N. Air separation and flue gas compression and purification units for oxy-coal combustion systems. *Energy Procedia*, 2009, **1**, 527–534.

32 Smith, A. R.and Klosek, J. A review of air separation technologies and their integration with energy conversion processes. *Fuel Process. Technol.*, 2001, **70**, 115–134.

33 Lyngfelt, A., Leckner, B., and Mattisson, T. A fluidized-bed combustion process with inherent CO_2 separation: application of chemical loop combustion. *Chem. Eng. Sci.*, 2001, **56**, 3101–3113.

34 Ryden, M., Lyngfelt, A., and Mattisson, T. Chemical-looping combustion and chemical looping reforming in a circulating fluidized-bed reactor using Ni-based oxygen carriers. *Energy Fuels*, 2008, **22**, 2585–2597.

35 Anthony, E. J. Solid looping cycles: a new technology for coal conversion. *Ind. Eng. Chem. Res.*, 2008, **47**, 1747–1754.

36 Berguerand, N. and Lyngfelt, A. Design and operation of a 10 kW(th) chemical-looping combustor for solid fuels – testing with South African coal. *Fuel*, 2008, **87**, 2713–2726.

37 Leion, H., Mattisson, T., and Lyngfelt, A. Solid fuels in chemical-looping combustion. *Int. J. Greenhouse Gas Control*, 2008, **2**, 180–193.

38 Cao, Y., Casenas, B., and Pan, W. P. Investigation of chemical looping combustion by solid fuels, 2. Redox reaction kinetics and product characterization with coal, biomass, and solid waste as solid fuels and CuO as an oxygen carrier. *Energy Fuels*, 2006, **20**, 1845–1854.

39 Göttlicher, G. The energetics of carbon dioxide capture in power plants, Dissertation, 1999. Fortschr.-Ber. VDI Reihe 6 Nr. 421, Düsseldorf, VDI Verlag, 1999. National Energy Technology Laboratory (NETL), US Department of Energy, English translation at www.netl.doe.gov/publications/carbon_ seq/refshelf.html, 2004 (last accessed 15 November 2010).

40 Contarini, S., Barbini, M., Piero, G. D., Gambarotta, G., Mazzamurro, G., Riocci, M., and Zappelli, P. Solid sorbents for the reversible capture of carbon dioxide. Presented at the 6th International Conference on Greenhouse Gas Control Technologies, Kyoto, 1–4 October 2002.

41 Kato, M., Yoshikawa, S., and Nakagawa, K. Carbon dioxide absorption by lithium

orthosilicate in a wide range of temperature and carbon dioxide concentration. *J. Mater. Sci. Lett.*, 2001, **21**, 485–487.

42 Abanades, C., Oaky, J., and Hämäläinen, D. Novel combustion cycles incorporating capture of CO_2 with CaO. Presented at the 6th International Conference on Greenhouse Gas Control Technologies. Kyoto, 1–4 October 2002.

43 Griffin, T., Bill, A., Marion, J. L., Ya Nsakala, N. CO_2 control technologies: Alstom power approach. Presented at the 6th International Conference on Greenhouse Gas Control Technologies, Kyoto, 1–4 October 2002.

44 Melin, T. and Rautenbach, R. *Membranverfahren*, 2nd edn., Springer, Berlin, 2003.

45 Peinemann, K.-V. Overview of polymeric gas separation membranes. Presented at the 9th Jülicher Werkstoffsymposium, Jülich, 15–16 November 2007.

46 Nunes, S. P. and Peinemann, K. V. *Membrane Technology in the Chemical Industry*, 2nd edn., Wiley-VCH Verlag GmbH, Weinheim, 2006.

47 Merkel, T. C., Lin, H., Wei, X., and Baker, R. Power plant post-combustion carbon dioxide capture: an opportunity for membranes. *J. Membr. Sci.*, 2010, **359**, 126–139.

48 Powell, C. E. and Qiao, G. G. Polymeric CO_2/N_2 gas separation membranes for the capture of carbon dioxide from power plant flue gases. *J. Membr. Sci.*, 2006, **279**, 1–49.

49 Robeson, L. M. Correlation of separation factor versus permeability for polymeric membranes. *J. Membr. Sci.*, 1991, **62**, 165–185.

50 Brunetti, A., Scura, F., Barbieri, G., and Drioli, E. Membrane technologies for CO_2 separation. *J. Membr. Sci.*, 2010, **359**, 115–125.

51 Shekhawat, D., Luebke, D. R., and Pennline, H. W. *A Review of Carbon Dioxide Selective Membranes*, 1 December 2003, www.osti.gov/bridge/product.biblio.jsp?osti_id=819990 (last accessed 15 November 2010).

52 Langsam, M. Polyimides for gas separation. *Plast. Eng.*, 1996, **36**, 697–741.

53 Kouketsu, T., Duan, S., Kai, T., Kazama, S., and Yamada, K. PAMAM dendrimer composite membrane for CO_2 separation: Formation of a chitosan gutter layer. *J. Membr. Sci.*, 2007, **287**, 51–59.

54 Deng, L., Kim, T.-J., and Hägg, M.-B. Facilitated transport of CO_2 in novel PVAm/PVA blend membrane. *J. Membr. Sci.*, 2008, **340**, 154–163.

55 Zimmerman, C. M., Singh, A., and Koros, W. J. Tailoring mixed matrix composite membranes for gas separations. *J. Membr. Sci.*, 1997, **137**, 145–154.

56 Mahajan, R. and Koros, W. J. Factors controlling successful formation of mixed matrix gas separation materials. *Ind. Eng. Chem. Res.*, 2000, **39**, 2692–2696.

57 Car, A., Stropnik, C., Yave, W., and Peinemann, K. V. Tailor-made polymeric membranes based on segmented block copolymers for CO_2 separation. *Adv. Funct. Mater.*, 2008, **18**, 2815–2823.

58 Car, A., Stropnik, C., Yave, W., and Peinemann, K. V. PEG modified poly(amide-b-ethylene oxide) membranes for CO_2 separation. *J. Membr. Sci.*, 2008, **307**, 88–95.

59 Car, A., Stropnik, C., Yave, W., and Peinemann, K. V. Pebax/polyethylene glycol blend thin film composite membranes for CO_2 separation. *Sep. Purif. Technol.*, 2008, **62**, 110–117.

60 Yave, W., Car, A., Funari, S. S., Nunes, S., and Peinemann, K. V. CO_2-philic polymer membrane with extremely high separation performance. *Macromolecules*, 2010, **43**, 326–333.

61 Yave, W., Car, A., and Peinemann, K. V. Nanostructured membrane material designed for carbon dioxide separation. *J. Membr. Sci.*, 2010, **350**, 124–129.

62 Meulenberg, W. A., Baumann, S., Ivanova, M., van Gestel, T., Bram, M., and Stöver, D. Ceramic membranes for gas separation in advanced fossil power plants. Presented at the 9th Liège Conference on Materials for Advanced Power Engineering, Liège, 27–29 September 2010.

63 Pauling, L. *Nature of Chemical Bond*, Cornell University Press, Ithaca, NY, 1960.

64 Sunarso, J., Baumann, S., Serra, J. M., Meulenberg, W. A., Liu, S., Lin, Y. S., and

Diniz da Costa, J. C. Mixed ionic–electronic conducting (MIEC) ceramic-based membranes for oxygen separation. *J. Membr. Sci.*, 2008, **320**, 13–41.

65 Bouwmeester, H. J. M. and Burggraaf, A. J. Dense ceramic membranes for oxygen separation. *Membr. Sci. Technol.*, 1996, **4**, 435–528.

66 Vente, J. F., Haije, W. G., Ijpelaan, R., and Rusting, F. T. On the full-scale module design of an air separation unit using mixed ionic electronic conducting membranes. *J. Membr. Sci.*, 2005, **278**, 66–71.

67 Carolan, M. F., Stein, V. E., and Armstrong, P. A. Development of planar ceramic membranes for tonnage oxygen production in advanced energy appliations. Presented at the 10th Jülicher Werkstoffsymposium, Jülich, 15–16 June 2009.

68 Bhargav, A. and Jackson, G. S. Thermo-kinetic modeling and parameter estimation for hydrogen permeation through $Pd_{0.77}Ag_{0.23}$ membranes. *Int. J. Hydrogen Energy*, 2009, **34**, 5164–5173.

69 Bhargav, A., Jackson, G. S., Ciora Jr., R. J. C., and Liu, P. T. K. Model development and validation of hydrogen transport through supported palladium membranes. *J. Membr. Sci.*, 2010, **356**, 123–132.

70 Kreutz, T. G. and Law, C. K. Production of hydrogen and electricity from coal with CO_2 capture. Presented at the 6th International Conference on Greenhouse Gas Control Technologies. Kyoto, 1–4 October 2002.

71 Bolland, O. CO_2 capture technologies – an overview. Presented at the Second Trondheim Conference on CO_2 Capture, Transport and Storage, Trondheim, 25–26 October 2004.

72 Cooretec. *Cooretec Lighthouse Concept – the Path to Fossil-Fired Power Plants for the Future.* Federal Ministry of Economics and Technology (BMWi), 2008, www.cooretec.de/publikationen (last accessed 15 November 2010).

73 Winkler, D., Brautsch, A., and Wirsum, M. Gasturbinenprozess mit chemischer CO_2-Abtrennung aus dem Abgas. Presented at the 39th Kraftwerkstechnisches Kolloquium. Dresden, 11–12 October 2007.

74 Chapel, D. G., Mariz, C. L., and Ernest, J. Recovery of CO_2 from flue gases: commercial trends. Presented at the Canadian Society of Chemical Engineers Annual Meeting, Saskatoon, Saskatchewan, 4–6 October 1999.

75 Mangalapally, H. P., Notz, R., Hoch, S., Asprion, N., Sieder, G., Garcia, H., and Hasse, H. Pilot plant experimental studies of post combustion CO_2 capture by reactive absorption with MEA and new solvents. *Energy Procedia*, 2009, **1**, 963–970.

76 Tigges, K.-D. Challenges for the plant manufacturing. Presented at the Three Countries Workshop on CO_2: CCS and CCR in Germany, Norway, The Netherlands – Challenges and Chances, Essen, 24–25 November 2009.

77 Eide-Haugmo, I., Brakstad, O. G., Hoff, K. A., Sørheim, K. R., da Silva, E. F., and Svendsen, H. F. Environmental impact of amines. *Energy Procedia*, 2009, **1**, 1297–1304.

78 Görner, K. 700 °C-Technologie: Voraussetzung für CCS. Presented at Anwenderforum Power Plant Technology, Hannover, 21–23 April 2009.

79 Alstom. *Carbon Capture and Sequestration CCS Project*, www.alstom.com/power/news-and-events/press-releases/AEP-and-Alstom-com mission-first-of-Its-kind-Carbon-Capture-and-Sequestration-CCS-project-20091030/(last accessed 15 November 2010).

80 Sullivan, C. Global warming: climate change and the law. *New York Times Online*, 5 March 2010.

81 Herzog, H. J. What future for carbon capture and sequestration? *Environ. Sci. Technol.*, 2001, **35**, 148–153.

82 Koros, W. J. Evolving beyond the thermal age of separation processes: membranes can lead the way. *AIChE J.*, 2004, **50**, 2326–2334.

83 Zhao, L., Riensche, E., Menzer, R., Blum, L., and Stolten, D. A parametric study of CO_2/N_2 gas separation membrane processes for post-combustion capture. *J. Membr. Sci.*, 2008, **325**, 284–294.

84 Hussain, A. and Hägg, M.-B. A feasibility study of CO_2 capture from flue gas by a facilitated transport membrane. *J. Membr. Sci.*, 2010, **359**, 140–148.
85 Zhao, L., Riensche, E., Blum, L., and Stolten, D. Multi-stage gas separation membrane processes with post-combustion capture: energetic and economic analyses. *J. Membr. Sci.*, 2010, **359**, 160–172.
86 Hellfritsch, S., Gonschorek, S., Wilhelm, R., Löser, J., Klemm, M., Weigl, S., and Gampe, U. Entwicklungsstand des Oxyfuel-Prozesses für Braunkohlekraftwerke. Presented at the 38th Kraftwerkstechnisches Kolloquium, Dresden, 24–25 October 2006.
87 Kather, A., Hermsdorf, C., and Klostermann, M. Der kohlebefeuerte Oxyfuel-Prozess. *VGB PowerTech*, 2007, **4**, 84–91.
88 Ritter, R., Holling, B., Altmann, H., and Biele, M. Konzepte und Ausblick für eine CO_2-Anlage eines Oxyfuel-Kraftwerkes am Beispiel Schwarze Pumpe. Presented at the 39th Kraftwerkstechnisches Kolloquium, Dresden, 11–12 October 2007.
89 Gericke, B. Konzeptüberlegungen bezüglich der Auswirkung des Oxyfuel-Prozesses auf konventionelle Kraftwerksanlagen. *VGB PowerTech*, 2006, **10**, 64–72.
90 Dunker, R. Brennertechnik mit neuem Freiheitsgrad. *BWK*, 2007, **59**, 51–54.
91 Anderson, K. and Johnsson, F. Process evaluation of an 865 MWe lignite fired O_2/CO_2 power plant. *Energy Convers. Manage.*, 2006, **47**, 3487–3498.
92 Renz, U. Ein Oxyfuel-Prozess mit Druckfeuerung. Presented at Ebsilon-Anwendertagung, Bensheim, 8–9 October 2009.
93 Beggel, F., Engels, S., Modigell, M., and Nauels, N. Oxyfuel combustion by means of high temperature membranes for air separation. Presented at the 4th International Conference on Clean Coal Technologies (CCT2009), Dresden, 18–21 May 2009.
94 Nazarko, J., Riensche, E., Blum, L., and Stolten, D. Optimierung der Oxyfuel-Kraftwerkskonzepte mit der Sauerstoffbereitstellung durch Hochtemperaturmem-branen. Presented at the 42nd Kraftwerkstechnisches Kolloquium, Dresden, 12–13 October 2010.
95 Kass, H. Inbetriebnahme der 0,5 MWth – Testanlage zum Oxyfuel-Prozess. Presented at the Innovations forum CO_2-freies Kraftwerk, Cottbus, 26–27 April 2007.
96 Blomen, E., Hendriks, C., Neele, F. Capture technologies: improvements and promising developments. *Energy Procedia*, 2009, **1**, 1505–1512.
97 Prins, M., van der Ploeg, R., van den Berg, R., Patil, C., van Dorst, E., Geuzeboek, F., and Paul, D. Technology advances in IGCC with CO_2 capture. Presented at the 4th International Conference on Clean Coal Technologies. Dresden, 18–20 May 2009.
98 Franz, J. and Scherer, V. An evaluation of CO_2 and H_2 selective polymeric membranes for CO_2 separation in IGCC processes. *J. Membr. Sci.*, 2010, **359**, 173–183.
99 Marano, J. J. and Ciferino, J. P. Integration of gas separation membranes with IGCC Identifying the right membrane for the right job. *Energy Procedia*, 2009, **1**, 361–368.
100 Bracht, M., Alderliesten, P. T., Kloster, R., Pruschek, R., Haupt, G., Xue, E., Ross, J. R. H., Koukou, M. K., and Papayannakos, N. Water gas shift membrane reactor for CO_2 control in IGCC systems: techno-economic feasibility study *Energy Convers. Manage.*, 1997, **38**, 159–164.
101 Carbo, M. C., Jansen, D., Haije, W. G., and Verkooijen, A. H. M. Advanced membrane reactors for fuel decarbonisation in IGCC: H_2 or CO_2 separation? Presented at the 5th Annual Conference on Carbon Capture and Sequestration, Alexandria, VA, 8–11 May 2006.
102 Chiesa, P., Kreutz, T. G., and Lozza, G. G. CO_2 sequestration from IGCC power plants by means of metallic membranes. *J. Eng. Gas Turbines Power*, 2007, **129**, 123–134.
103 Battersby, S., Tasaki, T., Smart, S., Ladewig, B., Liu, S., Duke, M. C., Rudolph, V., and Diniz da Costa, C. J. Performance of cobalt silica membranes

in gas mixture separation. *J. Membr. Sci.*, 2009, **329**, 91–98.

104 Blum, L., Riensche, E., Nazarko, J., Menzer, R., and Stolten, D. Overview on 1st and 2nd generation coal-fired membrane power plants (with and without turbo machinery in the membrane environment). Presented at the 4th International Conference on Clean Coal Technologies (CCT2009), Dresden, 18–21 May 2009.

105 Hong, J., Chaudhry, G., Brisson, J.G., Field, R., Gazzino, M., and Ghoniem, A. F. Analysis of oxy-fuel combustion power cycle utilizing a pressurized coal combustor. *Energy*, 2009, **34**, 1332–1340.

106 Schiebahn, S., Riensche, E., Nazarko, J., Blum, L., and Stolten, D. IGCC-Kraftwerk mit einem Wassergas-Shift-Membranreaktor (WGS-MR) sowie Verfahren zum Betreiben eines solchen IGCC-Kraftwerks mit Spülgas. Deutsche Patentanmeldung 102010036056.2, 2010.

4
Life Cycle Assessment for Power Plants with CCS

Peter Viebahn

Abstract

Several recent life cycle assessment (LCA) studies for power plants with carbon capture and storage (CCS) are summarized and compared. The first joint assessment of all the three capture routes was undertaken in 2007 for Germany. Its findings were qualitatively confirmed by later ones, which partially considered more details. If the entire process chain is considered, the greenhouse gas emissions will only be reduced in total by around 68–87% compared with a power plant without CCS. In exceptional cases this might amount to up to 95%. This was calculated for CCS power plants operational in 2020, including the upstream chains of substances and energies used for capture. However, other environmental impacts should be considered in addition to greenhouse gas emissions. Most studies have concluded that for the post-combustion process virtually all of the other environmental impacts increase by 26–250%. The individual processes cannot yet be modeled in detail for pre-combustion and oxyfuel processes. Nevertheless, rough estimates for IGCC show 20–66% increases for other environmental impacts. In contrast, for oxyfuel processes 22–80% decreases in other environmental impacts are observed. The analysis also revealed a need to harmonize LCA for CCS technology since a wide range of assumptions for capture, transport, and storage, the timing of the CCS process, the type of reference power plant, the choice of parameters, and the varying environmental impact categories were found in the LCA studies analyzed.

Keywords: carbon capture and storage, life cycle assessment, power plants

4.1
Introduction

In the RECCS report [1], containing an ecological, economic, and structural comparison of renewable energy technologies (RE) with carbon capture and storage (CCS), life cycle assessments (LCAs) were carried out for the first time for the three conventional carbon dioxide capture routes for Germany. An LCA

Process Engineering for CCS Power Plants, First Edition. Edited by Detlef Stolten and V. Scherer

illustrates a “cradle-to-grave” approach, in which all energy and material flows that occur during manufacture, use, and disposal of products, are modeled. The resulting emissions are calculated and their environmental impacts are assessed. The individual processes involved in the capture of CO_2 were considered in detail only for post-combustion plants, and for pre-combustion and oxyfuel only the additional energy consumption was included. Since then, several new comprehensive LCAs covering all prevalent capture routes applied to lignite-, hard coal-, and natural gas-fired power plants have been presented by a number of institutions. Most of these studies were compiled in 2008. The selected studies, however, are restricted to those in which LCAs of the *entire CCS chain* were performed. Another criterion was whether the respective ISO Standards for LCAs were followed. An analysis is made of the precision with which the individual steps in the process, such as capture, compression, transport, and storage, are modeled and also of what assumptions are made in the process.

This chapter gives a short overview of the methodology of LCA. Four individual studies published in 2008 and 2009 are analyzed, and the results are compared with the earlier findings of the RECCS report. The overall results of the analysis are then presented and some constraints of using LCA as an assessment tool are illustrated. The development of fossil technologies, including CCS, is compared with LCAs for renewable electricity production. In conclusion, needs of action in terms of harmonization and data availability are considered.

4.2
Life Cycle Assessment as an Assessment Method

For the ecological evaluation of selected systems configurations, often the method of LCA according to ISO Standards 14.040 and 14.044 is used: “LCA addresses the environmental aspects and potential environmental impacts, e.g., use of resources and environmental consequences of releases, throughout a product's life cycle from raw material acquisition through production, use, end-of-life treatment, recycling, and final disposal” (ISO 14.040).

The *first step* of an LCA, the goal and scope definition, addresses the systems boundary, the level of detail, the subject of the LCA study, and the intended use (Fig. 4.1).

The *second step* compares the material and energy flows that enter a system, are converted there and leave it in a different form. This is also termed input/output balance or life cycle inventory analysis. A product LCA thus examines all the material and energy flows caused by a single product, beginning with the extraction and processing of the raw materials and following the process through manufacturing and use to the product's eventual disposal. This is what is meant by “cradle-to-grave” approach.

The *third step* is to calculate the environmental impact of the assessed flows. During this life cycle impact assessment, it is necessary to “weigh up, aggregate, or generalize flows of different materials in different environmental media with different environmental impacts” [2]. Such impact categories are, for exam-

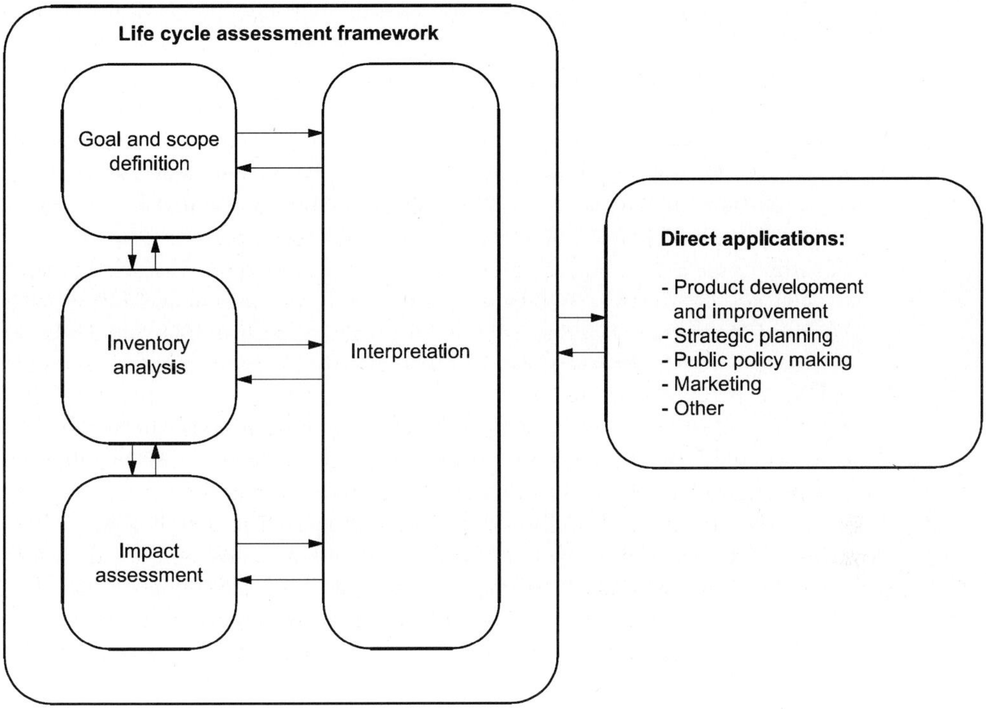

Fig. 4.1 Stages of an LCA (source: ISO 14040).

ple, consumption of limited resources, greenhouse effect, acidification of soils and bodies of water, eutrophication of soils and bodies of water, photochemical smog, particles, and dust [3]. By scaling the results to a functional unit, for example, 1 kWh electricity output at the power plant's stack, different production processes can be compared and the best technology with regard to an impact category can be selected.

In the *final step*, called life cycle interpretation, the results are discussed and serve as a basis for recommendations and political consultation.

4.3
Review of Life Cycle Assessments Along the Whole CCS Chain

4.3.1
Overview

The comparison includes LCAs along the whole CCS chain. Studies that included either only greenhouse gas emissions or only the power plant without the upstream and downstream chains were not included in this exercise. The minimum criteria were as follows:

- LCA according to, or in line with, the relevant LCA ISO Standards 14040 and 14044
- consideration of relevant environmental impact categories
- modeling the whole chain: recovery and transport of primary energy, power plant processes with and without CO_2 capture, transport of CO_2, storage of CO_2
- inclusion of second- and third-order processes, that is, the upstream chains for individual materials, such as the solvent monoethanolamine (MEA), and plant infrastructure, for instance, the construction of the power plant.

Table 4.1 provides a comprehensive overview of the most important assumptions of the studies described in the following sections. Note, however, that only the key parameters were analyzed in the direct comparison.

The following studies have been compared:

- Utrecht University (The Netherlands): *Hard coal-fired steam power plant* in The Netherlands, including post-combustion using MEA capture, current situation, 50 km pipeline, natural gas storage facility.
- Forschungszentrum (FZ) Jülich (Germany): *Hard coal-fired steam power plants*, including post-combustion reference case, situation in 2020
  - steam power plant from 2010, retrofitted with MEA capture in 2020
  - steam power plant in 2020 based on MEA capture (new construction).
- Institute for Energy and Environmental Research, Heidelberg (Germany): *Lignite-fired power plants* in eastern Germany (Lusatia), situation in 2020, 325 km pipeline, natural gas storage facility
  - steam power plant with post-combustion technology based on MEA capture
  - steam power plant with oxyfuel technology
  - IGCC with pre-combustion technology based on Selexo.
- Paul Scherrer Institute (PSI), Villigen (Switzerland)/Institute of Energy Economics and the Rational Use of Energy (IER), University of Stuttgart (Germany): *Hard coal-, lignite- and natural gas-fired power plants*, 200 and 400 km pipeline, 2500 m natural gas storage facility and 800 m aquifer
  - hard coal with post-combustion, pre-combustion, and oxyfuel technology
  - lignite with post-combustion, pre-combustion, and oxyfuel technology
  - natural gas with post-combustion technology.

The following sections give an overview of the individual studies.

4.3.2
Hard Coal (Utrecht University)

A detailed and very well documented LCA analysis for *a hard coal-fired steam power plant*, including post-combustion and subsequent transportation and storage of the CO_2, was published in 2008 by workers at Utrecht University [4]. Three cases were explored:

1. reference case: average hard coal-fired steam power plant operating in The Netherlands in 2000 (subcritical, 460 MW_{el})

Table 4.1 Comprehensive overview of assumptions of life cycle assessments along the whole CCS chain.

	Units	Wuppertal Institute, 2007			University of Utrecht, 2008	FZ Jülich, 2009	IFEU, 2009	PSI and IER, 2008		
		Coal	Lignite	NG	Coal	Coal	Lignite	Coal	Lignite	NG
CCS in general										
Type of capture		Post, Oxy, Pre	Post	Post	Post	Post	Post, Oxy, Pre	Post, Oxy, Pre	Post, Oxy, Pre	Post
Year of installation		2020	2020	2020	2008	2020	2020	2025/2050	2025/2050	2025/2050
New/retrofit		Newly built	Newly built	Newly built	Newly built	Retrofit Newly built	Newly built	Newly built	Newly built	Newly built
Reference year		2020	2020	2020	(a) 2000, (b) 2008	(a) 2010, (b) 2020	2020	2025/2050	2025/2050	2025/2050
Capture model										
Infrastructure considered		–	–	–	X	–	X	(X)	(X)	(X)
Capture process modeled		(X), –, –	(X)	–	X	X	X, (X), (X)	(X)	(X)	(X)
Dismantling considered		–	–	–	–	n.a.	n.a.	X	X	X
Upstream energy demand		Hard coal mix D	Lignite mix D	NG mix D	Hard coal mix NL	Hard coal mix D	Lausitz lignite	Hard coal mix EU	Lignite mix D	NG mix EU
Upstream materials		(X), –, –	(X)	(X)	X[b)]	X	X, (X), (X)	X	X	X
Compression										
Compression to	MPa	11	11	11	11	n.a.	18.7	n.a.	n.a.	n.a.
Compression energy		Only electricity consumption			Gas turbine	n.a.		Only electricity consumption		
CO_2 leakage rate	$t\ MW^{-1}a^{-1}$	–	–	–	23	0	0	n.a.	n.a.	n.a.

Table 4.1 (continued)

	Units	Wuppertal Institute, 2007			University of Utrecht, 2008	FZ Jülich, 2009	IFEU, 2009	PSI and IER, 2008		
		Coal	Lignite	NG	Coal	Coal	Lignite	Coal	Lignite	NG
Transport										
CO_2 capacity of pipeline	$Mt\ a^{-1}$	5	5	5	30	n.a.		7.9	7.9	7.9
Pipeline diameter	cm	95	95	95	95	n.a.		n.a.	n.a.	n.a.
Distance	km	300	300	300	50	300/400	325	200/400	200/400	200/400
Construction pipeline		NG pipeline	NG pipeline	NG pipeline	NG pipeline	NG pipeline	X	NG pipeline	NG pipeline	NG pipeline
CO_2 leakage rate		–	–	–	$2.32\ t\ km^{-1}\ a^{-1}$	0	0	$0.26\ g\ t^{-1}\ km^{-1}$	$0.26\ g\ t^{-1}\ km^{-1}$	$0.26\ g\ t^{-1}\ km^{-1}$
Energy demand		X	X	X	0	0	X	X	X	X
Storage										
Energy consumption		50% surcharge on transport				50% surcharge on transport				
Storage formation					NG field			(a) Aquifer at a depth of 800 m (b) Depleted NG field at a depth of 2500 m		
Infrastructure considered					As NG storage		–[a)]	–	–	–
CO_2-Injection capacity	$Mt\ a^{-1}$				7.3 (6 wells, 3 km length)			7.9 (2 wells)	7.9 (2 wells)	7.9 (2 wells)
Pressure at end of pipe	MPa				10.7		15	n.a.	n.a.	n.a.
Compression to	MPa				15		–	(a) 10.8, (b) 20	(a) 10.8, (b) 20	(a) 10.8, (b) 20
Energy demand					Electricity mix NL			Electricity mix	Electricity mix	Electricity mix
Leakage rate	%	0	0	0	X		0	0	0	0

Table 4.1 (continued)

	Units	Wuppertal Institute, 2007			University of Utrecht, 2008	FZ Jülich, 2009	IFEU, 2009	PSI and IER, 2008		
		Coal	Lignite	NG	Coal	Coal	Lignite	Coal	Lignite	NG
Life cycle assessment										
Method		UBA	UBA	UBA	CML	CML, 2001	n.a.	–	–	–
Normalization		–	–	–	NL	D	–	–	–	–
Impact categories[c)]		6	6	6	10	6	5	–	–	–
Contribution analysis		X	X	X	X	–	–	X	X	X
Sensitivity analyses		X	–	–	X	X	X	–	–	–

a) Assumption: use of existing EOR infrastructure as available in the Weyburn EOR project in USA/Canada.
b) Characterization factors for MEA added in CML.
c) For details, see Table 4.2.
Abbreviations: X = applied, available; (X) = partly applied; – = not considered; n.a. = not available; NG = natural gas; Oxy = oxyfuel; Post = post-combustion with MEA; Pre = pre-combustion; D = Germany; NL = The Netherlands; EU = European Union.
Sources: Wuppertal Institute [1]; University of Utrecht [4]; FZ Jülich [6]; IFEU Institute [7]; PSI and IER [10].

2. state-of-the-art hard coal-fired steam power plant (ultra-supercritical, 660 MW_{el})
3. as in Case 2, but fitted with MEA-based CO_2 separation (post-combustion, capture rate 90%, net 450 MW_{el}, capture of 3.6 Mt a^{-1} of CO_2).

Rather than modeling a future situation in 2020 as done in the RECCS study, an analysis was made of what a state-of-the-art coal-fired power plant, with and without CCS, could achieve compared with actual reality. The model focused on the capture of CO_2. To this end, existing LCA modules for hard coal-fired power plants were first updated using as their model the state-of-the-art system in flue gas cleaning in The Netherlands to create a starting point for Case 3. As an upstream chain, the average hard coal mix for The Netherlands from 2004 was used.

With regard to *CO_2 capture*, the equipment required for the infrastructure was modeled according to manufacturers' data. This model excluded energy expenditure for construction and dismantling, energy consumption, and material costs for maintenance and later disposal. In addition, the demand for heat and electricity was determined. The capture process was modeled in detail using data from published work such as consumption of MEA and degradation to salts, reaction with other emissions in addition to CO_2, or consumption of other materials.

The *compression process* was modeled using LCA data for a gas turbine. The energy expenditure for compression with an initial pipeline pressure of 11 MPa was calculated in detail. Based on the IPCC method [5], a leakage rate in the compression of 23 t MW^{-1} a^{-1} of CO_2 was established. With an assumed capture quantity of 3.6 Mt a^{-1} (or 0.011 Mt MW^{-1} a^{-1}) of CO_2, this constitutes a 0.4% loss of CO_2.

A distance of 50 km was assumed for the *transport*. This distance was considered representative of a pipeline from a coastal location to one of several possible onshore storage sites in northern Holland. The infrastructure was modeled in line with the natural gas pipeline. A leakage of 2.32 t km^{-1} a^{-1} of CO_2 was assumed. With a total transport of 30 Mt a^{-1}, for which the pipeline was designed, this produces a leakage rate of 0.0000077% per kilometer transported.

When assessing *CO_2 storage* in a depleted natural gas storage site, the infrastructure was modeled based on experience gained from storing natural gas. In doing so, both energy expenditure for construction and later dismantling, energy expenditure, and material costs for maintenance and subsequent disposal were excluded. In addition, the installation and operating expenses for six 3 km boreholes were included. It was taken into account that the pressure at the end of the pipeline is insufficient for injection and that a pressure boost from 10.7 to 15 MPa is required, using electricity from the grid. Instead of modeling leakage, a leakage rate of zero or an "insignificant amount" was assumed.

The *life cycle impact assessment* was carried out using the CML process, one of the most commonly used methods of analysis. Ten different environmental impact categories were considered and standardized for The Netherlands. In addition, sensitivity analyses were performed for six critical parameters.

The *key conclusions* drawn from the study are as follows. State-of-the-art hard coal-fired power plants (Case 2) improve all impact categories, compared with the average power plant fleet in The Netherlands (Case 1) by 23–83%. It is not possible to make such unequivocal statements for CCS-based power plants because there are a number of trade-offs: if applied today, CCS could reduce the *greenhouse gas emissions* of a hard coal-fired power plant to 243 g (CO_2-equiv.) kWh_{el}^{-1}. Compared with Cases 1 and 2, this constitutes a 78 and 71% reduction, respectively. This is much less than expected for a CO_2 capture rate of 90% because of the additional emissions from the hard coal upstream chain and other processes. These figures correspond roughly to the results in the RECCS study. There, a reduction of 67% for power plants in 2020 was determined, for which a 49% initial degree of utilization and a CO_2 capture rate of 88% were assumed.

In this context, the authors introduced the term *avoidance efficiency*. The avoidance efficiency, 68% in this example, implies that only 68 t of CO_2 would be avoided if 100 t of CO_2 were stored. They correctly pointed out that this should be taken into account in the calculation of CO_2 allowances.

The impact of infrastructure expenses such as capture, transport, and storage on the total greenhouse gas emissions is very low at 0.3%.

If we compare Case 3 (CCS) and Case 2 (state of the art) with the other impact categories, the authors concluded that only "marine aquatic ecotoxicity" is reduced (by 27%). This was caused by the reduced emissions of hydrogen fluoride. All other impact categories deteriorate by 27–181%, although the direct emissions of dust, SO_2, hydrogen chloride, and hydrogen fluoride also decline, due to CO_2 scrubbing.

Despite the detailed modeling of the various stages in the process, many uncertainties remain. For this reason, the authors called their study "advanced screening LCA." Thus the MEA upstream chain taken from the ecoinvent database is considered to be highly uncertain. It was also not possible to model explicitly the entire capture process, for example, to determine exactly the quantity of MEA used. Sensitivity analyses show that the variation for some of the parameters has a considerable impact on the computed results. This concerns, for example, the CO_2 capture rate and the capture efficiency of other chemicals. For this reason, the authors recommended the introduction of a comprehensive measurement program for the first CCS power plants to obtain information about the actual emissions caused or reduced due to capture, and about their interaction with other scrubbing technologies. In this sense, the aim of their study was also to ascertain which areas have an urgent need for action.

4.3.3
Hard Coal (Forschungszentrum Jülich)

Hard coal-fired steam power plants, including post-combustion, were also the focus of an analysis published by workers at Forschungszentrum Jülich in 2009 [6]. Five cases were explored:

1. reference case: a hard coal-fired steam power plant (500 MW_{el}) installed in the 1990s, but still in operation in 2005
2. a state-of-the-art hard coal-fired steam power plant, installed in 2010 (552 MW_{el})
3. a hard coal-fired steam power plant projected to standards for 2020 (697 MW_{el})
4. as in Case 2, but retrofitted in 2020 with MEA scrubbing, applying a 90% capture rate (426.5 MW_{el})
5. as in Case 3, but fitted with MEA scrubbing, applying a 90% capture rate (592 MW_{el}).

As in the previous study, a comparison was made between a new, state-of-the-art coal-fired power plant, both with and without CCS, and a more realistic, older plant. To this end, the power plant was divided into its various modules. The individual units, such as the generator, flue gas scrubbing, and dust scrubbing, were projected and supplemented by CO_2 scrubbing, including CO_2 compression for Cases 4 and 5. The average hard coal mix for Germany was used as the upstream chain. Neither transport nor storage was considered in either of the basic variants for CCS. The presentation of results focused on the separation of emissions and consumption in the actual power plant process and all upstream and downstream processes.

The additional need for heat and electricity was determined with regard to *CO_2 capture*[1]. In Case 5 (CCS new construction), the SO_x concentration in the flue gas was reduced from 150 to 29 mg m^{-3} for the capture process. The consumption of NH_3 and limestone, and also MEA and caustic soda, which is used for the capture process, was varied. The infrastructure was not modeled. The energy demand for compression was included, but not explained in further detail. Leakages were not taken into account.

The *life cycle impact assessment* was also carried out using the CML method. Six different environmental impact categories were considered and standardized for Germany. The impact of other hard coal upstream chains, the addition of transport over 300 km and storage, a longer transportation distance of 400 km, and a higher CO_2 absorption rate of the MEA were investigated in sensitivity analyses.

The *key conclusions* drawn from the study were as follows. State-of-the-art (Case 2) or future state-of-the-art (Case 3) hard coal-fired power plants, not applying CO_2 capture, improve all impact categories compared with power plant generation from the 1990s.

Retrofitting with CO_2 capture (Case 4) and the construction of a new CCS power plant (Case 5) generated a $27^{(*)}$ and $20^{(*)}$% higher *cumulated energy consumption* than for reference power plants 2 and 3[2]. *The greenhouse gas emissions* were reduced from around $800^{(*)}$ and $770^{(*)}$ to $210^{(*)}$ and $190^{(*)}$ g (CO_2-equiv.) kWh_{el}^{-1},

1) No sources were given for this, however.

2) All of the values marked with $^{(*)}$ can only be estimated since they are not available as numerical values, but have to be measured from diagrams.

that is, by 74$^{(*)}$ and 75%$^{(*)}$. The relatively low additional consumption and the relatively high greenhouse gas reduction in Case 5 are due to the low efficiency loss of just 7%. The assumption of losses of just 10.5% in the case of retrofitting was probably also calculated too generously because these values have already been exceeded by other authors for newly constructed power plants.

If Case 4 (CO_2 retrofitting) is compared with Case 2 (state of the art) with regard to the other impact categories, the authors concluded that all categories would worsen by 6$^{(*)}$–210$^{(*)}$%. If, on the other hand, Case 5 (CCS-based new construction in 2020) is compared with Case 3 (new construction in 2020), the summer smog category improves by 13$^{(*)}$%, whereas the other categories deteriorate by 5$^{(*)}$–90$^{(*)}$%.

Sensitivity analyses show that the consequences of transport and storage [3] have only a minor effect on all impact categories. Varying the solvent's absorbency does not cause any noticeable change. Only the use of Australian or South African coal instead of the German hard coal mix leads to a 40$^{(*)}$–120$^{(*)}$% reduction in acidification, summer smog, and eutrophication.

4.3.4 Lignite (Institute for Energy and Environmental Research, Heidelberg)

Shortly afterwards, workers at the Institute for Energy and Environmental Research published a detailed LCA for various CCS-based *lignite power plants* [7]. The analysis was based on the work of Henkel [8] and Idrissova [9]. The analysis refers to 2020 and lignite mined in Lusatia. The following were analyzed:

1. a steam power plant with and without post-combustion technology based on MEA capture at a rate of 90% CO_2 capture
2. an oxyfuel-based steam power plant at a rate of 92% CO_2 capture
3. an IGCC with and without pre-combustion technology based on Selexo at a rate of 90% CO_2 capture

all of which were designed with a capacity of between 500 and 800 MW_{el}. As in the RECCS study, the current state-of-the-art example was first projected to 2020 for the steam power plant and IGCC, regarding future efficiency and emissions. Based on this projection, the capture of CO_2 was modeled. In addition to these reference conditions, "slower development" and "faster development" routes were considered as a sensitivity analysis, both with and without CCS. Lusatian lignite was used as the upstream chain.

The CO_2 capture was modeled in different ways:

- Steam power plant with post-combustion capture: MEA manufacture, consumption, and degradation, CO_2 capture rate and energy consumption, capture of other emissions, and infrastructure were modeled. Compared with all

3) Storage is modeled in accordance with the approach in the RECCS study, in which, for cost estimates, 50% of the energy consumption and the emissions balanced for transport were applied for storage.

other studies, the very high efficiency loss of 18.2%, which leads to an increased energy consumption of 66%, is conspicuous.
- Steam power plant with oxyfuel combustion: oxygen requirement and energy consumption for air separation, CO_2 capture rate, SO_x separation, 92 and 100% retention of all other emissions in sensitivity analyses, and infrastructure were modeled.
- IGCC with pre-combustion: capture rate and energy consumption, SO_x emissions, and infrastructure were modeled.

For *compression* and *transportation*, the following data were imported from the Weyburn EOR project (United States/Canada), as this was deemed to be representative for Germany also: 325 km pipeline transportation and a compression to 18.7 MPa at the power plant using electricity. This ensures that a pressure of 15 MPa, sufficient for the injection, is achieved at the end of the pipeline. No CO_2 losses were assumed during compression and transportation. The infrastructure was not modeled.

No expenditure was assumed for *CO_2 storage* in a depleted natural gas storage site. Instead, it was assumed that the existing infrastructure from the previous recovery of natural gas could continue to be used. Since the pipeline pressure is adequate, no energy consumption is required for storage either. No leakages were modeled.

Six different environmental impact categories were considered when carrying out the *life cycle impact assessment*. Emissions in water and the soil were not included. All results were given for the reference case and two sensitivity cases, "slower development" and "faster development." .

The *key conclusions* drawn from the study were as follows. Depending on the technology, between 24% additional consumption of *primary energy* in the case of the IGCC and 66% in the case of steam power plant is required for the CCS-based power plants.

The *greenhouse gas emissions* can be reduced from approximately 940 to 190$^{(*)}$ (post-combustion) and 120$^{(*)}$ g (CO_2-equiv.) kWh_{el}^{-1} (oxyfuel), or around 80$^{(*)}$ and 87$^{(*)}$%.[4] For the IGCC power plant the reduction was from 880 to 140$^{(*)}$ g (CO_2-equiv.) kWh_{el}^{-1}, or 84$^{(*)}$%. The difference from the assumed CO_2 reduction rates of 90–92% is revealed by the expense incurred by the additional energy consumption. In contrast to hard coal-fired power plants, most emissions occur directly at the power plant (79% with post-combustion and oxyfuel and 98% with pre-combustion) and not in the upstream fuel chain.

The values for the steam power plant using post-combustion correspond to those of the RECCS findings; the other two routes were only considered for hard coal and not for lignite in RECCS.

Potential leakages were not modeled. Nevertheless, it is assumed that CO_2 is released ("it is clear that leakage will not be zero"), albeit at a much later stage. How-

4) All of the values marked with $^{(*)}$ can only be estimated since they are not available as numerical values, but have to be measured from diagrams.

ever, it was pointed out that some research suggests that a delayed release of CO_2 emissions could have a positive impact on climate change, expecting a slow escape of small quantities instead of higher rates at the present time. It was also stressed that the lower the permeability of the storage site, the more CO_2 is dissolved, which is then unable to reach any leakage points. This, however, contradicts the catalog of available criteria for the selection of suitable storage sites. One of these criteria is that a sufficiently high level of permeability is required to be able to store huge quantities of CO_2 from power plants within a short space of time.

At 2.3–2.6%, the proportion of infrastructure expenses within the overall remaining greenhouse gas emissions of the whole CCS chain is low. However, it is still one order of magnitude higher than the values offered by Koornneef *et al.* [4], even though transport and storage infrastructure have not been reconciled. If compression is assigned to the power plant, transport and storage expenses, which mainly refer to the energy consumption, constitute 0.1–2.8% of the total remaining greenhouse gas emissions.

As with the study by Koornneef *et al.* [4], there is a wide variety of trade-offs in the case of *other impact categories*. Depending on the technology, very different results are produced, which were analyzed in detail by the authors and can only be given here for the reference case (Table 4.2):

- Steam power plant with post-combustion: with the steam power plant, the environmental impacts in the reference case increase more or less substantially, with the exception of acidification. The main causes are the additional energy consumption, the CO_2 capture process, and the manufacture of the solvent.
- Steam power plant with oxyfuel combustion: in contrast, oxygen combustion generates the lowest emissions, considering the reference case and 92% retention of all emissions at the power plant's stack.
- IGCC with pre-combustion: the environmental impacts of an IGCC *without* CCS are 33–66% lower in the reference case than for a steam power plant *without* CCS. With CO_2 capture, the environmental impacts increase in the reference case, as with the steam power plant, but nevertheless remain far below the values for the steam power plant *without* CCS in all categories.

Table 4.2 Trade-off for individual environmental impact categories in the LCA study of lignite-fired power plants [7].

	Units	Summer smog	Eutrophication	Acidification	Health effects
SSP [a] post-combustion	%	+250	+98(*) [c]	–5(*)	+26(*)
SPP [b] oxyfuel combustion	%	–56(*)	+80(*)	–76(*)	–22(*)
IGCC with pre-combustion	%	+40(*)	+20(*)	+66(*)	+27(*)

a) Steam power plant with post-combustion.
b) Steam power plant with oxyfuel combustion.
c) Values marked with (*) can only be estimated since they are not available as numerical values, but have to be measured from diagrams.

In the case of the steam power plant, the results are comparable to those computed in the RECCS study; the other two routes were only modeled with regard to energy consumption in RECCS.

The authors pointed to the considerable uncertainties that must be taken into account when interpreting the results. This is in particular because IGCC and oxyfuel combustion are technologies that are still in their infancy or have not yet been constructed. In addition, the results for the oxyfuel power plant are heavily reliant on whether non-CO_2 emissions can be fully retained, as with CO_2, or not at all. In the former case, realistically an oxyfuel power plant would be a "near zero"-emission power plant. In the latter case, it would be assessed as being similar to, or worse than, the post-combustion process.

4.3.5
Hard Coal, Lignite, Natural Gas (Paul Scherrer Institute, Villigen, and Institute of Energy Economics and the Rational Use of Energy, University of Stuttgart)

The most extensive study, also published in 2008, was carried out within the EU NEEDS[5)] project, part of a global framework to derive new "external cost" factors for future power generation systems in Europe. The potential of various CCS technologies was modeled within the context of long-term scenarios considering 2005, 2025, and 2050. Based on the scenario development, costs were computed and their life cycle inventories determined [10]. In contrast to the other studies, no life cycle impact assessment was carried out. Three main scenarios representing potential developments from the perspective of CCS technology were considered: "pessimistic," "optimistic–realistic," and "very optimistic." The "optimistic–realistic" option, analyzed below, was taken to be the most likely, as with the other technologies explored in the NEEDS study. In addition, driving and constraining forces, together with the general role of fossil electricity generation in future energy systems, were analyzed.

The dynamics applied in NEEDS are its main distinguishing feature compared with all other published studies: by linking life cycle inventories to development scenarios, relevant "learning effects" on the technical side, and hence the material balance, could be taken into account. Double dynamics are created because the background processes are also adjusted decade by decade by including, for instance, higher recycling rates for steel or an altered energy mix based on European energy scenarios in manufacture. The path for the climate protection scenario at 440 ppm was chosen for the following situation.

In this study, the three processes of post-combustion, pre-combustion, and oxyfuel were considered for hard coal- and lignite-fired power plants; the post-combustion process was applied for the natural gas combined cycle (NGCC). The year of application for pre-combustion IGCC was assumed to be 2015. The complete life cycle, including upstream chains of the energy sources and also the manufacture, operation, and dismantling of the power plants and the stor-

5) NEEDS = New Energy Externalities Developments for Sustainability; http://www.needs-project.org/.

age part, was calculated. Life cycle inventory modules from the ecoinvent database were used throughout. These modules describe the actual situation in Europe or the transportation there. Future periods were modeled in accordance with ESU and IFEU [11]. All results were divided into the manufacture, operation, fuel, and dismantling of the plant.

The following assumptions were made for the reference power plants, not including CCS:

1. hard coal-fired steam power plant reference: Rostock power plant (ultra-supercritical, 350, 600, 800 MW_{el})
2. hard coal-fired IGCC reference: Puertollano power plant (Spain), projected (450 MW_{el})
3. lignite-fired steam power plant reference: Niederaussem K power plant, Bergheim (lignite-fired power plant with optimized plant technology, 950 MW_{el})
4. lignite IGCC reference: Vrespva power plant (Czech Republic) (400 MW_{el})
5. natural gas reference: Mainz–Wiesbaden power plant (400 MW_{el}).

The power plants were projected according to three different technology scenarios, "pessimistic," "optimistic–realistic," and "very optimistic," to 2025 and 2050. Several CCS variants were also modeled.[6] The degree of utilization, the CO_2 capture rate, and the use of chemicals for CO_2 scrubbing were considered to be the key parameters for *CO_2 capture*. The infrastructure of the capture plants was not modeled because it generally has only a minor impact on the results of the LCA.

- The *degree of utilization* was projected according to literature references. Depending on the source of primary energy, the technology, and the scenario, efficiency losses between 6 and 10% were assumed for 2025 and between 4 and 10% for 2050. If the values for 2025 are compared with those of the RECCS study, between 25 and 40% lower losses than in the power plants modeled for 2020 in RECCS were assumed.
- For the *CO_2 capture rates*, 90% was assumed with post- and pre-combustion, 99.5% with coal-based oxyfuel, and 100% in case of natural gas oxyfuel (RECCS assumed 88% for post- and pre-combustion and 99.5% for coal-based oxyfuel).
- The consumption of MEA, caustic soda (NaOH), and activated carbon was modeled for the post-combustion capture process, but not projected to the future, owing to a lack of data. Assumptions about the behavior of NO_x and SO_x emissions were made. For the oxyfuel combustion, only the energy consumption of the air separation was computed.

Two different distances of 200 and 400 km were assumed for the *transport*. Compression at 200 km was calculated for the 400 km pipeline. The resulting

6) Although hard coal- and lignite-fired steam power plants were modeled differently for the acutal situation, no differentiation is made any longer between lignite and hard coal for future steam power plants. This means that lignite has to develop more rapidly than hard coal, in particular with regard to the degree of utilization.

power requirement was taken from two reports [12,13]. The infrastructure model was based on natural gas pipelines. A leakage rate of 0.26 g per 1000 km was assumed, which corresponds to a value of 0.0052% per 200 km, or 0.000026% per kilometer transported.

To model *CO_2 storage,* the power requirement to establish the necessary injection pressure, as calculated [12, 13], was included for two different storage formations. Both an aquifer at a depth of 800 m and a depleted natural gas storage site at a depth of 2500 m were considered. Leakages were not modeled. Instead, a leakage rate of zero was adopted, and it was assumed that other storage sites would not be authorized and that tests and monitoring would be able to detect any leakages.

Unlike "complete" LCAs, the life cycle impact assessment was not carried out in NEEDS. The greenhouse gas potential was merely summarized from the impact factors of the IPCC. All life cycle inventory results can be retrieved from the NEEDS website and can therefore be further processed as required.

The *key conclusions* drawn from the study were as follows. Depending on the fuel, the technology used, and the scenarios considered, the application of CCS by 2050 can reduce greenhouse gases emissions from fossil fuel-fired power plants by 70–95%. Between 26 and 192 g (CO_2-equiv.) kWh_{el}^{-1} were computed as resultant values for all scenarios. In the "optimistic–realistic" scenario, the lowest value in 2050 was achieved for the oxyfuel steam power plant fired by lignite, 28 g (CO_2-equiv.) kWh_{el}^{-1}. By contrast, NGCC achieved 77, lignite IGCC 118 and the hard coal-fired post-combustion based steam power plant 168 g (CO_2-equiv.) kWh_{el}^{-1}. Transport and storage comprise only a fraction of the total emissions.

The low greenhouse gas emissions must be seen in the context of an increased primary energy expenditure, which amounts to 10–20% in 2050, and a corresponding increase in all emissions caused by the recovery and transportation of energy sources. As also shown in RECCS, this particularly concerns the hard coal and natural gas upstream chain. A higher burden then results in the total amount with numerous emissions than without CCS. It is not possible to draw more detailed conclusions from this because the study did not calculate any environmental impact parameters.

Compared with the RECCS study, a greater reduction in emissions and a lower energy expenditure are generally obtained. This is due, in particular, to the ~25–40% lower efficiency losses assumed in 2025 and further improvements in efficiency by 2050. The energy expenditure of the lignite-fired power plants modeled was too low: although the capture expenditure for lignite is much higher than that for hard coal-fired power plants, due to the higher CO_2 emissions, the same efficiency losses were used for both.

4.4 Results

4.4.1 Development of emissions

The findings of the RECCS study were mainly confirmed for post-combustion processes; even if in the latest studies, in particular by Koornneef *et al.* [4], the capture and, in part, transport and storage were modeled in more detail. Substantial new findings were generated for pre-combustion and oxyfuel, despite the fact that the capture processes have not yet been considered in detail. All of the results of the emissions calculations are presented in Table 4.3.

The only value given in all of the studies is that of *greenhouse gas emissions*, consisting mainly of CO_2, CH_4, and N_2O emissions. They are particularly relevant in the case of hard coal- and natural gas-fired power plants because consid-

Table 4.3 Comprehensive overview of increase in environmental impact categories along the whole CCS chain evaluation of four different studies.

Impact category [a]	Units	Wuppertal Institute, 2007					University of Utrecht, 2008	FZ Jülich, 2009		IFEU Institute, 2009		
		Coal			Lignite	NG	Coal	Coal		Lignite		
		Steam Post	Steam Oxy	IGCC Pre	Steam Post	CC Post	Steam Post	Steam Post	Steam Post	Steam Oxy	Steam Pre	IGCC
ADP	%	28 [b]			44 [b]		33	27 [b]	20 [b]	66 [b]	66 [b]	24 [b]
GWP	%	–67	–78	–68	–78	–67	–71	–74	–75	–80	–87	–84
ODP	%						55					
HTP	%						181	210	157	26	–22	27
FWAETP	%						46					
MAETP	%						–29					
TEP	%						57					
POP	%	94			524		27	6	–13	250	–56	40
AP	%	–10			–3		46	13	5	–5	–76	66
EP	%	36			40		80	80	66	98	–80	20
PM10	%	2			24							

a) Environmental impact categories: ADP = abiotic depletion potential; GWP = global warming potential; ODP = ozone depletion potential; HTP = human toxicity potential; FWAETP = fresh water aquatic ecotoxicity potential; MAETP = marine aquatic ecotoxicity potential; TEP = terrestrial ecotoxicity potential; POP = photochemical oxidation potential; AP = acidification potential; EP = eutrophication potential.

b) Only cumulated energy demand (CED) calculated.

Abbreviations: NG = natural gas, Oxy = oxyfuel, Post = post-combustion with MEA; Pre = pre-combustion.

Sources: Wuppertal Institute 2007 [1]; University of Utrecht [4]; FZ Jülich [6]; IFEU Institute [7]. In the EU NEEDS report [10], no impact assessment was done; emissions are available in the NEEDS database for calculating environmental impact categories.

Table 4.4 Reduction of greenhouse gas emissions from CCS power plants operational in 2020/2025.

Primary energy	Units	Steam power plant	Oxyfuel	IGCC
Hard coal	%	67–72	78–85	67–68
Lignite	%	78–81	87–95	83–84
Natural gas	%	67–75		

erable methane (CH_4) emissions are created here in the upstream chains. This effect is intensified by the increased energy consumption of CCS power plants. If the entire process chain is considered, the greenhouse gas emissions from CCS power plants operational in 2020/2025 will be reduced in total by only around 67–87% (Table 4.4). In exceptional cases it might amount to up to 95%.

Figure 4.2 presents an overview of the development of greenhouse gas emissions in the various studies. It should be noted that the sensitivity analyses contained in all studies are not shown. The capture rates and initial emissions in 2020 were generally assumed to be identical in all studies. Considerable deviations are visible, however, in the reduction of greenhouse gas emissions. The reason for this is that varying assumptions were made about the additional energy consumption caused by capture, and hence the reduction in the degrees of utilization. In the NEEDS study [10], in particular, significantly lower losses than in other studies were assumed.

However, *other environmental impacts* should be considered in addition to greenhouse gas emissions. The higher energy consumption required in all of the processes and the materials used in the capture processes can be perceived in direct proportion to the various impact categories of the LCA. Depending on the assumptions made in the studies, the various interactions in the capture processes lead to many trade-offs in the individual environmental impact categories. In some studies, all other emission impacts increase in accordance with the additional energy consumption. Other studies, however, modeled trade-offs that arise from the simultaneous reduction of other emissions in the course of the CO_2 capture process.

Most of the studies concluded that for the post-combustion process, increases are observed with virtually all of the other environmental impacts in the range 26–250%, compared with a power plant without CCS. The individual processes cannot yet be modeled in detail for pre-combustion and oxyfuel. Nevertheless, rough estimates for *IGCC* show 20–66% increases for all other environmental impacts. In contrast, for *oxyfuel* processes, 22–80% decreases in all other environmental impacts are observed.

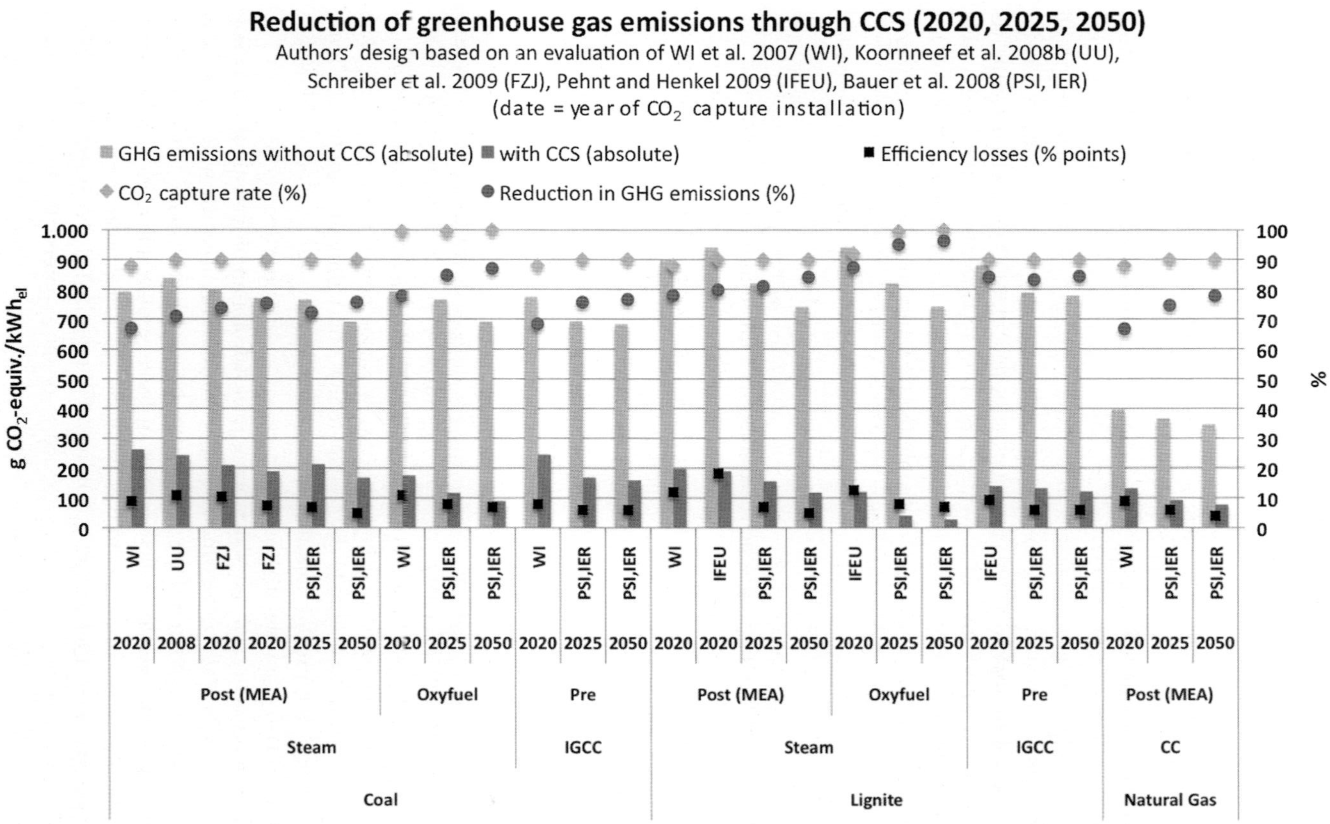

Fig. 4.2 Development of greenhouse gas emissions with CO_2 capture. Evaluation of five different studies each for 2020/2025 and for 2050.

4.4.2
Share of Infrastructure, Transport, and Storage

The proportion related to the manufacture of the infrastructure, that is, the plant required to capture, transport, and store the gas, is analyzed as being as low as 0.3–2.6% in all of the studies. Transportation of the CO_2 is modeled more or less uniformly, even if assumptions regarding the transport distance vary. Leakages of CO_2 in the compression and transportation processes were only partially modeled. Leakages at the CO_2 storage site were neglected in all studies. It was assumed in some studies that the storage site would otherwise not have been approved. Other studies assumed that CO_2 would indeed be released, albeit with a long delay, which would be significantly better for the environment than the current high rates of emission. The injection was either not modeled at all, or it was modeled only for the purpose of power requirements or for the required infrastructure.

4.5
Constraints of LCA Regarding an Assessment of CCS

Several environmental and social impacts cannot be evaluated using LCAs. Some direct and indirect environmental impacts of CCS and also possible impacts of CO_2 storage on subterranean ecosystems have to be considered during an environmental assessment.

4.5.1
Direct Environmental Impact Outside LCA

The consumption of water and cooling water was not included in the CML method. The use of land, caused by the additional infrastructure installation and in particular increased coal mining, was not taken into account [4]. The risks that could occur due to the transportation and temporary storage of chemicals such as MEA or the captured carbon dioxide were also neglected. These are subject to risk assessment [4].

The consequences of salt water being released offshore following the injection of CO_2, dramatically increasing the local concentration of salt water, have also not yet been considered. It would have to be examined whether this would directly affect the area around the point of leakage or whether its rapid dilution with the sea water would mean that its impact was negligible. If CO_2 is injected to enhance oil recovery, large quantities of contaminated salt water would also be released into the oceans [14]. Applied to onshore potentials, this argument is even more far-reaching. If it were necessary to extract salt water on the same scale as CO_2 is injected to keep the increase in pressure under control in aquifers, it would be extremely difficult to utilize the onshore storage capacity.

The spatial and long-term impacts of balanced emissions have not yet been considered in any of the LCAs. For example, SO_2 and NO_x emissions caused by the transport of coal by sea, which have a considerable effect on the acidification and eutrophication category [4], should be assessed differently from the same emissions released straight from the power plant.

Furthermore, the extent to which amines, used in the CO_2 scrubbing process, have an impact on health and the environment has not yet been adequately clarified. The gaps in our knowledge on this subject were highlighted in a study carried out by the Norwegian environmental organization Bellona [15]. Bellona proposed seven points that need to be clarified before CCS technology is used commercially in order to minimize or to avoid altogether the risk of damage being caused to the environment by amines. The most important ones are

- Closing the gaps in knowledge: determining the degradation routes, degradation quantities, retention times of the decay products of amines in the atmosphere, toxic contamination thresholds for humans, and the degree of ecotoxicology of amines with respect to terrestrial and aquatic ecosystems.
- Developing amines with a low environmental impact.
- Ensuring that the waste products from the amines are dealt with properly, because enormous quantities of hazardous waste would be created if this method were to be used globally.

4.5.2
Indirect Environmental Impact of Coal Mining and Social Aspects

Coal mining is generally linked to drastic, extensive changes to the landscape. In most cases, the consequences of such a restructuring of the landscape are a lowering of the water table, contamination of the water by mine drainage, and the creation of enormous slag heaps that have a negative impact on groundwater supply for agriculture and the surrounding ecosystems. In addition, enormous quantities of water are consumed to wash the coal and to cool coal-fired power plants. In some areas, this has led to significant water shortages in rivers and streams.

Due to the resettlement or displacement of the population, cultivable land is lost and homes and entire village communities are destroyed, resulting in social and cultural problems. Often residents are unwilling to leave voluntarily, and forced evictions occur, often with the threat or use of violence, or even, in some extreme cases, murder.

In the mining process itself, sulfur, methane, and dust emissions are released. Coal miners are often directly exposed to such emissions, because there are no or only inadequate safety precautions in many coal-producing countries. These emissions lead to acid rain and the formation of smog. Workers are at risk from gas and dust explosions, and also from flooding. In the regions affected by coal mining, increasing occurrences of respiratory diseases and skin rashes are diagnosed among workers and local residents. Once mining has been abandoned, subsidence and collapses often occur, creating devastating damage to the existing infrastructure and the surrounding houses [16]. When coal is combusted, residuals are created that contain significant concentrations of heavy metals, radioactive substances, and other substances that are harmful to Nature, soil and bodies of water.

4.5.3
Potential Impact of CO_2 Storage on Subterranean Ecosystems

It is not yet apparent whether the transportation of large quantities of CO_2 will have a biogeochemical impact on the microbial biota in deep rock formations. Drilling to depths of 3.5 km has revealed bacteria, viruses, and fungi. These microbes can be found in quantities of up to 10^8 per gram of sediment or ground water in tiny cracks and pores in the rock [17].

This field of research is very much in its infancy. Many of these types of bacteria found in deep rock formations are completely unknown. The research into their role within this ecosystem has hardly begun. It is known, however, that the metabolism of some better known types of bacteria changes chemical compounds in deep rock. Some gain their energy by converting manganese, sulfur, nitrogen, phosphorus, iron, and carbon compounds [17].

During a pilot storage test in a saline aquifer (Frio Brine) near Houston, TX, USA, research results on the mobilization of metals and organic compounds before, during, and after the injection of CO_2 were compiled between 2004 and 2008 [18]. It was shown that the injection of CO_2 led to considerable changes in the chemical and isotopic composition, including a dramatic shift of the pH value from 6.3 to 3.0. In addition, iron, manganese, lead, and aromatic hydrocarbons reacted to the increased concentration of CO_2, and their concentrations in the formation water rose sharply.

4.6
Comparison of Electricity from CCS and from Renewable Energies

The development of fossil technologies, including CCS, will now be compared with renewable energies. Only greenhouse gas emissions are taken into account because no standardized data exists for the other environmental impact categories for CCS.

We use the results from the aforementioned NEEDS study, in which not only future fossil energy technologies, but also a number of renewable energies, were investigated in terms of their future development. LCAs for three stages of development were also generated, namely current technology, the situation in 2025, and the situation in 2050. As mentioned previously, the background processes were simultaneously adjusted to energy mixes arising from the various pan-European energy scenarios. From these scenarios, we use the 440 ppm climate protection scenario here. The minimum and maximum values from the three technology scenarios "pessimistic," "optimistic–realistic," and "very opti-

Table 4.5 Greenhouse gas emissions from solar thermal power plants, photovoltaics, and offshore wind at the current situation, 2025 and 2050[a)].

	Solar thermal			Photovoltaics			Offshore wind		
	Current	**2025**	**2050**	**Current**	**2025**	**2050**	**Current**	**2025**	**2050**
Value	30.56			68.94			14.60		
Minimum		19.83	16.99		16.65	5.55		5.71	7.67
Maximum		20.90	17.55		36.64	7.39		11.81	6.66

a) Data are given in g CO_2-equiv./kWh_{el} (mix from different technologies).
Sources of basic values for CO_2, CH_4, N_2O: Solar thermal systems, Viebahn *et al.*, 2010 [21]; photovoltaic systems, Frankl *et al.*, 2008 [20]; offshore wind, DONG Energy, 2008 [19].
Source of conversion rates from CH_4 and N_2O to CO_2 equivalents: IPCC, 2007 [22].

mistic" are selected. Offshore wind power [19], photovoltaics [20], and solar thermal power plants [21] are used for the comparison with CCS power plants. Their greenhouse gas emissions (CO_2, CH_4, and N_2O) are shown in Table 4.5. The values describe a mix of various technologies, which can be referred to in the sources cited.

In Fig. 4.3, electricity generation from renewable energies is compared with CCS power plants in 2020/2025 and 2050. The minimum and maximum values and the mean of all options considered are given. The range given for fossil fuel-fired power plants results from combining steam power plants and IGCC from Table 4.3. For renewable energies, the range stems from the minimum and maximum values presented in Table 4.5.

Even compared with CCS power plants, renewable energies create only a fraction of greenhouse gas emissions. Most of them originate from the construction of the plant. The chart shows that in 2025, offshore wind will create only 5–8%, solar thermal energy 11–18%, and photovoltaics 14–24% of the emissions of CCS power plants. By 2050, photovoltaics in particular will improve, releasing only 7–12% of the emissions of CCS power plants; offshore wind will create 9–15% and solar thermal energy 13–23%.

4.7
Conclusion on Needs for Action

In conducting our analysis, we were surprised about the wide range of assumptions for capture, transport and storage, timing of the CCS process, the type of reference power plant, the choice of parameters, and the varying environmental

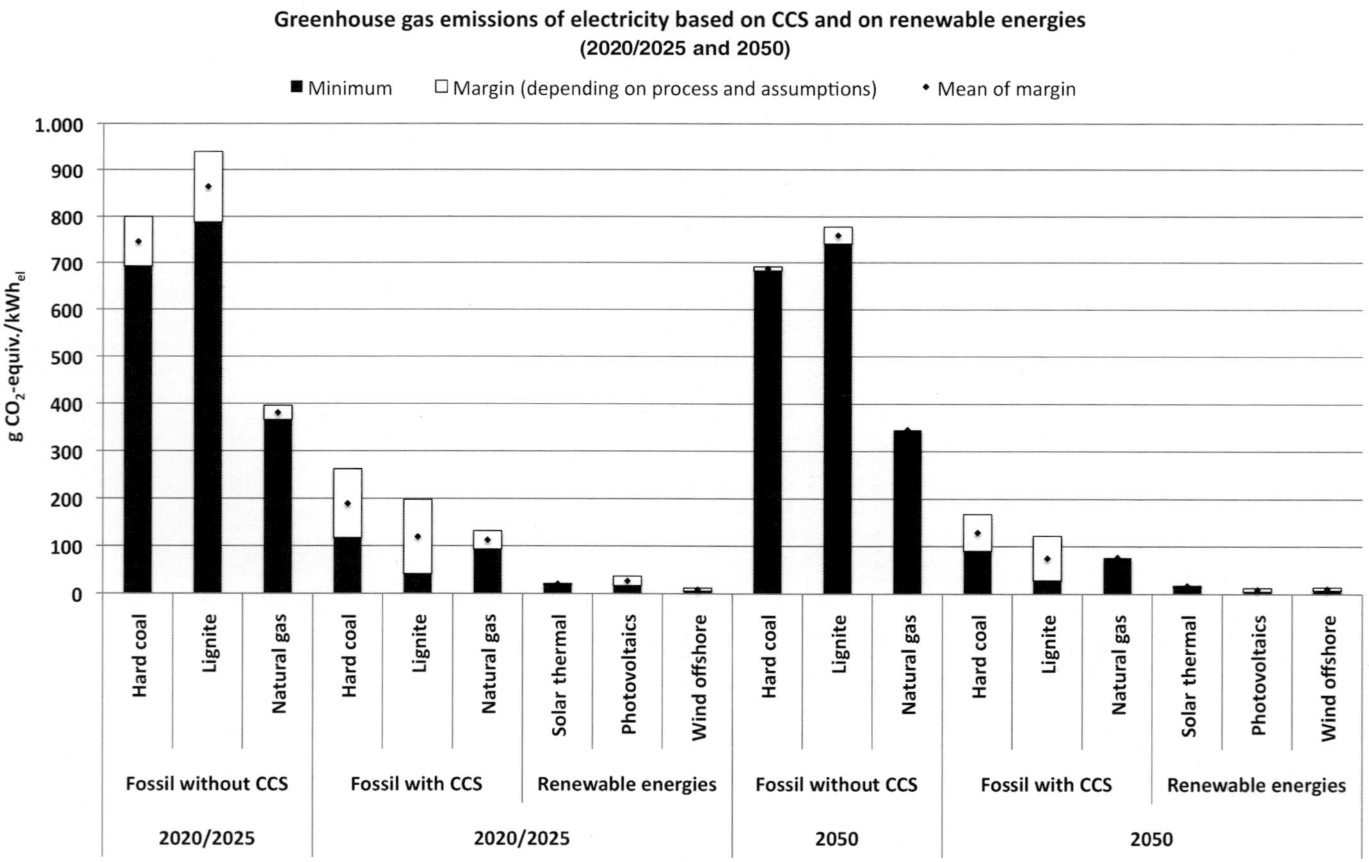

Fig. 4.3 Greenhouse gas emissions from fossil fuel-fired power plants with CCS in comparison with electricity from renewable energies (solar thermal power plants, photovoltaics and offshore wind, each for 2020/2025 and for 2050).

impact categories. As in many other LCAs, this reveals a need for action to harmonize LCA for CCS technology. Following the American example of assessing all State plans according to standard "Carbon Capture and Sequestration Systems Analysis Guidelines" [23], LCA guidelines should be drawn up and used in all investigations. In Germany, the "Network on Life Cycle Inventory Data" would lend itself to this aim of uniformity. The objective of this network is to provide harmonized LCAs, including in the energy sector, in collaboration with relevant stakeholders [24]. On the European level the European Platform of Life Cycle Thinking and Assessment [25] and on the international level the UNEP/SETAC Life Cycle Initiative [26] have similar goals.

Some authors recommend installing a comprehensive monitoring program for the first CCS power plants. In this way, we could gain information about the actual emissions caused or reduced by the capture of CO_2. Such information would considerably improve our understanding of the individual chemical processes, and how to model them.

Most studies have neglected to take into account that the commercial use of CCS, including transport and storage, is only expected to become available in 2025, or more likely in 2030. Only the NEEDS study included a balance for the periods 2025 and 2050, albeit based on the commercial introduction of CCS in 2020.

Even though in scientific studies the captured quantities are differentiated from the quantities of CO_2 that are actually avoided, this is not common practice. For this reason, the *avoidance efficiency coefficient* [4] should be stated in all CO_2 avoidance calculations. This coefficient plays a particularly important role when CO_2 allowances are later offset within the meaning of the amended European Union Emissions Trading Scheme (ETS) Directive.

Acknowledgment

The work this chapter is based on was financially supported by the German Federal Ministry for the Environment, Nature Conservation and Nuclear Safety.

References

1 WI (Wuppertal Institute for Climate, Environment and Energy), DLR (German Aerospace Centre), PIK (Potsdam Institute for Climate Impact Research), ZSW (Centre for Solar Energy and Hydrogen Research). *RECCS – Ecological, Economic and Structural Comparison of Renewable Energy Technologies (RE) with Carbon Capture and Storage (CCS) – an Integrated Approach.* Study on behalf of the German Federal Ministry for the Environment, Nature Conservation and Nuclear Safety, 2007, http://www.wupperinst.org/projekte/proj/index.html?projekt_id=25&bid=155, last accessed 20. Jan. 2011.

2 Schmidt, M. and Häuslein, A. *Ökobilanzierung mit Computerunterstützung.* Springer, Berlin, 1996.

3 Guinée, J.B., Gorree, M., Heijungs, R., Huppes, G., Kleijn, R., Sleeswijk, A.W., Haes, H.A.U.D., Bruijn, J.A.D., Duin,

R.V., and Huijbregts, M.A.J. *Handbook on Life Cycle Assessment: Operational Guide to the ISO Standards.* Kluwer, Dordrecht, 2002.

4 Koornneef, J., van Keulen, T., Faaij, A., and Turkenburg, W. Life cycle assessment of a pulverized coal power plant with post-combustion capture, transport and storage of CO_2. *Int. J. Greenhouse Gas Control*, 2008, *2*(4), 448–467.

5 Intergovernmental Panel on Climate Change (IPCC). Fugitive emissions. In: *2006 IPCC Guidelines for National Greenhouse Gas Inventories, Vol. 2, Energy* (ed. Eggleston H.S., Buendia L., Miwa K., Ngara T., and Tanabe K.), prepared by the National Greenhouse Gas Inventories Programme. Institute for Global Environmental Strategies (IGES), Hayama, on behalf of the IPCC, 2006, Chapter 4.

6 Schreiber, A., Zapp, P., and Kuckshinrichs, W. Environmental assessment of German electricity generation from coal-fired power plants with amine-based carbon capture. *Int. J. Life Cycle Assess.*, 2009, **14**(6), 547–559.

7 Pehnt, M. and Henkel, J. Life cycle assessment of carbon dioxide capture and storage from lignite power plants. *Int. J. Greenhouse Gas Control*, 2009, **3**(1) 49–66.

8 Henkel, J. Life cycle assessment of carbon dioxide capture and storage. Diploma thesis. Institute for Energy and Environmental Research and Berlin University of Technology, Heidelberg, 2006.

9 Idrissova, F. Life cycle assessment of selected carbon dioxide capture, transportation and storage options from coal-fired power plants. Master Thesis, Faculty of Business Administration and Economics, Technische Universität Freiberg, 2004.

10 Bauer, C., Heck, T., Dones, R., Mayer-Spohn, O., and Blesl, M. *Final Report on Technical Data, Costs, and Life Cycle Inventories of Advanced Fossil Power Generation Systems.* Deliverable 7.2, NEEDS (New Energy Externalities Developments for Sustainability), http://www.needs-project.org/, 2008 (last accessed 6. 11. 2009).

11 ESU and IFEU *LCA of Background Processes. Deliverable 15.1.* NEEDS (New Energy Externalities Developments for Sustainability), http://www.needs-project.org/, 2008 (last accessed 23 November 2009).

12 Wildbolz C. Life cycle assessment of selected technologies for CO_2 transport and sequestration. Diploma thesis, No. 2007 MS05. Department Bau, Umwelt und Geomatik, Institute of Environmental Engineering (IfU), ETHZ, Zurich, 2007.

13 Doka, G. Critical review of life cycle assessment of selected technologies for CO_2 transport and sequestration of the Diploma thesis No. 2007 MS05 by C. Wildbolz, ETHZ, Zurich, 2007.

14 Lindeberg, E., Vuillaume, J., and Ghaderi, A. Determination of the CO_2 storage capacity of the Utsira formation. Energy Procedia, 2009, 1(1), 2777–2784.

15 Bellona. *Amines Used in CO_2 Capture – Health and Environment Impacts.* Bellona Report, September 2009, http://www.bellona.org/files/fil_Bellona_report_September_2009_Amines_used_in_CO2_capture.pdf (last accessed 13 January 2010).

16 Greenpeace. False Hope. Why Carbon Capture and Storage Won't Save the Climate, http://www.greenpeace.org/raw/content/international/press/reports/false-hope.pdf , 2008 (last accessed 13 January 2010).

17 Scinexx. *Fremde Welt im Untergrund – Wimmelndes Leben und jede Menge Fragen*, http://www.scinexx.de/dossier-detail-7-5.html, 2004 (last accessed 13 January 2010).

18 Kharaka, Y., Thordsen, J., and Kakouros, E. Potential environmental impacts of geologic storage of carbon dioxide: Mobilization of metals and organic compounds. Presented at Int. Conf. on Deep Saline Aquifers for Geological Storage of CO_2 and Energy, IFP, Paris, 27–29 May 2009.

19 DONG Energy. *Final Report on Offshore Wind Technology.* EU NEEDS, www.needs-project.org, 2008 (last accessed 8 February 2010).

20 Frankl, P., Menichetti, E., and Raugei, M. *Final Report on Technical Data, Costs*

and Life Cycle Inventories of PV applications. Deliverable 11.2-RS Ia. EU NEEDS, www.needs-project.org, 2008 (last accessed 8 February 2010).

21 Viebahn, P., Lechon, Y., and Trieb, F. The potential role of concentrated solar power (CSP) in Africa and Europe – a dynamic assessment of technology development, cost development and life cycle inventories until 2050. *Energy Policy*, 2010, doi:10.1016/j.enpol.2010.09.026.

22 Intergovernmental Panel on Climate Change (IPCC). *Climate Change 2007: Synthesis Report. Contribution of Working Groups I, II and III to the Fourth Assessment Report of the Intergovernmental Panel on Climate Change* [Core Writing Team, Pachauri, R.K. and Reisinger, A. (eds)], IPCC, Geneva, 2007.

23 National Energy Technology Laboratory (NETL). *Carbon Capture and Sequestration Systems Analysis Guidelines*, http://www.netl.doe.gov/technologies/carbon_seq/Resources/Analysis/, 2005 (last accessed 11 November 2009).

24 Bundesministerium für Bildung und Forschung (BMBF). *Netzwerk Lebenszyklusdaten. Daten für die Innovationen von Morgen*, http://www.netzwerk-lebenszyklusdaten.de/, 2008 (last accessed 20 December 2010).

25 European Commission, Joint Research Center. *Life Cycle Thinking and Assessment*, http://lct.jrc.ec.europa.eu/assessment, 2010 (last accessed 20 December 2010).

26 United Nations Environment Programme (UNEP). *The Life Cycle Initiative. International life Cycle Partnerships for a Sustainable World.* Launched by The United Nations Environment Programme (UNEP) and the Society for Environmental Toxicology and Chemistry (SETAC), http://lcinitiative.unep.fr/, 2010 (last accessed 20 December 2010).

Part 2
CO_2 Scrubbing

Process Engineering for CCS Power Plants, First Edition. Edited by Detlef Stolten and V. Scherer

5
Physics and Chemistry of Absorption for CO_2 Capture to Coal Power Plants

Paul Feron and Graeme Puxty

Abstract

CO_2 capture from power stations is essentially a gas separation challenge where the goal is to produce from combustion/energy conversion processes a pure CO_2 stream that can be compressed, transported, and stored. The leading processes to perform this separation are the chemical and/or physical absorption of CO_2 using a liquid absorbent, and subsequent regeneration of the absorbent to release CO_2. In this chapter, the engineering aspects of an absorption process are described, followed by detailed discussion of the physical and chemical processes that occur during absorption using commercially available state-of-the-art absorbents. Absorbents are characterized in terms of absorption rate, absorption capacity, and absorption enthalpy. Included in the discussion is the relationship between physical and chemical properties of the absorbent and process energy requirements and equipment size.

Keywords: absorption, carbon dioxide capture, power stations, physical absorption, chemical absorption, liquid absorbents

5.1
Gas Separation for CO_2 Capture

Carbon dioxide (CO_2) capture processes can be divided into three main categories or general process routes [1]:

- *Post-combustion processes.* CO_2 is captured from a flue gas at low pressure (101.3 kPa) and low CO_2 content (3–20%), in general. The separation task is to remove CO_2 from a mixture of mainly nitrogen (N_2) and oxygen (O_2), but the impact of flue gas impurities (sulfur oxides, nitrogen oxides, particulates) also needs to be taken into account.
- *Pre-combustion processes.* CO_2 is captured from a gas mixture with predominantly H_2 gas at high pressure (1.5–4.0 MPa) and medium CO_2 content (15–40%). Apart from the CO_2–H_2 separation, the feed gases also contain CO, H_2S, and sometimes other sulfur components.

Process Engineering for CCS Power Plants, First Edition. Edited by Detlef Stolten and V. Scherer

- *Denitrogenation processes.* A concentrated stream of CO_2 can be produced by the separation of O_2 from air and then carrying out the combustion/conversion process in an O_2–CO_2 atmosphere. This is in essence separation of O_2 from N_2. This results in a flue gas concentrated in CO_2, eliminating the need for CO_2 separation. An additional advantage is that the resulting flue gas is free of nitrous oxides and any other impurities resulting from the components of air.

Capture of CO_2 from energy conversion processes generally means that at some point in the process CO_2 will need to be separated, which is the case for post- and pre-combustion capture. Alternatively, the energy conversion process can produce a concentrated CO_2 stream, provided that N_2 is not present during CO_2 formation. This is the principle of the denitrogenation process route, which is more often described as oxyfuel. In any case, it is clear that overall the CO_2 capture process will in general consist of a combination of energy or fuel conversion steps and separation steps.

The separation of CO_2 from gas streams can be achieved by several physical–chemical separation methods, which can be seen as the technological platforms. These methods are as follows:

- membranes, using selective barrier materials with different gas permeability to separate gases in a continuous process
- absorption, using liquids with a strong affinity to one or more of the components to separate gases in a cyclic operation
- adsorbents, using solids with a strong affinity to one or more of the components to separate gases in a cyclic operation
- cryogenic, using differences in points of condensation or solidification to separate gases.

A technology matrix can be built up, linking the three capture process routes with the different technological platforms. The matrix is filled with the specific technologies that form part of the overall CO_2 capture process. The resulting CO_2 capture toolbox is shown in Table 5.1. It is a useful instrument for both mapping the relevant separation technologies and guiding future research initiatives in this area.

In Table 5.1, the leading separation processes are indicated in bold. For capture of CO_2 in both pre- and post-combustion, absorption-based separation processes are currently preferred. This is despite the large difference in CO_2 partial pressures and gas consumption between post- and pre-combustion capture. They have been extensively utilized for the production of CO_2 from flue gases [3, 4] under low-pressure and oxidizing conditions, and for CO_2 removal from natural gases and synthesis gases [5] under high-pressure and reducing environments. Historically, the development of absorption-based gas separation processes for CO_2 removal has been spurred by the industrial need, for example, to meet natural gas specification requirements for transport or gas purity specifications as needed for further processing of the product (H_2, CO, CH_4, and other

Table 5.1 CO_2 capture toolbox, adapted from Table 3.1 in [1] and also [2]. The leading separation technologies are indicated in bold.

Capture method	Post-combustion capture	Pre-combustion capture	Denitrogenation
Targeted separations	CO_2 from N_2/O_2	CO_2 from H_2–CO–CH_4	O_2 from N_2
Technology platform			
Membranes	Polymeric membranes Ceramic membranes Facilitated transport membranes Carbon molecular sieve membranes Membrane contactors	Ceramic membranes Polymeric membranes Palladium membranes Membrane contactors	High temperature O_2-conducting membranes Facilitated transport membranes
Adsorption	Lime carbonation/ calcinations Carbon-based sorbents Amine-functionalized sorbents	Dolomite, hydrotalcites and other carbonates Zirconates Carbon-based sorbents	Carbon-based sorbents High-temperature adsorbents, e.g., perovskites
Absorption	**Alkanolamine solutions** Amino acid solutions and other amines Carbonate solutions and slurries Emulsions	**Alkanolamine solutions** **Non-aqueous physical solvents** Amino acid solutions and other amines Carbonate solutions and slurries Emulsions Water	Absorbents with O_2 carriers (artificial blood)
Cryogenic	Anti-sublimation	CO_2 liquefaction	**Distillation for air separation**

alkane gases). The need for removal of other acid gases, in particular H_2S, with very stringent requirements has further influenced the development of absorption processes. In the application of these processes for CO_2 capture from power stations, the following aspects need to be considered, some of which are very different from the current practice of absorption processes used for acid gas removal:

- *Bulk removal of CO_2*
  The required removal of CO_2 is most likely determined by an economical optimization, with 80–90% removal being typical. Complete removal of CO_2 is not required, but removing less than 50% from the CO_2 present in the feed gas will have a much reduced impact on the CO_2 emission reduction goals in power stations.

- *High CO_2 purity*
  As the CO_2 product is likely to be compressed and/or liquefied before transport or storage in underground reservoirs, purity in excess of 95% is desired. Further CO_2 concentration and purification are to be avoided.
- *No stringent retentate gas specifications*
  The retentate gas, lean in CO_2, will either be vented (post-combustion capture) and required to meet atmospheric emission limits, or be used as a fuel (pre-combustion capture), required to meet gaseous fuel composition requirements for combustion processes.
- *Low energy requirement*
  The energy consumption of the separation process will have a direct impact on the main product of a power station, electricity, and is therefore most critical.
- *Scale of process*
  Although absorption processes have been used at scales similar to the requirements of a pre-combustion capture operation, there is a considerable challenge in scaling up absorption processes for post-combustion.
- *Integration*
  The capture process will have to be integrated with the power station to bring about optimal performance. It is therefore not a stand-alone operation with its own utility provisions, such as heat, water, and electricity.

Absorption processes make use of a liquid which removes CO_2 from a gas stream either by a chemical reaction with alkaline components, or physical affinity imparted through specific molecular interactions. Table 5.2 describes the main differences between chemical and physical absorbents. Some absorption processes exhibit characteristics of both physical and chemical absorption and the dividing line between the two modes of interaction is somewhat blurred.

One of the distinguishing features between chemical and physical absorbents is their achievable absorption capacity at a given partial pressure of CO_2, which is determined by the equilibrium partial pressure over the solution. In Fig. 5.1, the equilibrium partial of some representative absorbents is given. The operating ranges for post- and pre-combustion capture are also represented in Fig. 5.1. It shows that achievable loadings at CO_2 partial pressures around 1 MPa, typical of pre-combustion, are similar for both chemical and physical absorbents. Because part of the CO_2 is bound chemically, the energy requirement for solvent regeneration will be higher for a chemical absorbent. For partial pressures of 10 kPa, typical of post-combustion, only chemical absorbents will be suitable.

Perhaps the single most important aspect of CO_2 capture in power stations is its impact on the generation efficiency of the power station. To permit its assessment, it is helpful to estimate the minimum energy requirement for separation of CO_2 from flue gas or fuel gas and subsequent compression of CO_2 to the transport and/or storage pressure (10–20 MPa). The thermodynamic minimum energy requirement (W_{min}/J) for the complete separation of a binary gas mixture into its pure components can be determined from the following equation [6], which represents the mixing energy for two ideal gases:

Table 5.2 Summary of the main distinguishing characteristics of chemical and physical absorbents.

Type of absorption process	Chemical	Physical
Partial pressure CO_2 range	Typically <0.5 MPa	Typically >0.5 MPa
Separation task specification	Deep removal and bulk removal	Bulk removal
Absorption enthalpy	30–200 kJ mol^{-1} CO_2	10–30 kJ mol^{-1} CO_2
Operating pressure	0.1–10 MPa	1–10 MPa
Operating temperature	Absorption 303–343 K Regeneration 373–423 K	Absorption 243–313 K Regeneration 293–373 K
Absorbent flow rate	10–40 m^3 t^{-1} CO_2	7–70 m^3 t^{-1} CO_2
Absorbent regeneration	Typically thermal regeneration with flash regeneration possible at high feed pressure	Typically flash regeneration augmented with thermal regeneration
Active components	Amines in aqueous solution Carbonates/phosphates in aqueous solution Hydroxide solutions	Ethers Ketones Alcohols
Capture application	Post-combustion – coal/gas Pre-combustion – gas	Pre-combustion – coal
Absorption rate	Determined by gas solubility and reaction kinetics	Determined by gas solubility and diffusion rates
Energy consumption	2–5 MJ kg^{-1} CO_2 (thermal) 0.1–0.5 MJ kg^{-1} CO_2 (electrical)	0.06–0.07 MJ kg^{-1} CO_2 (thermal) 0.1–0.25 MJ kg^{-1} CO_2 (electrical)

$$W_{min} = nRT\,[X_{CO_2}\ln(X_{CO_2}) + (1 - X_{CO_2})\ln(1 - X_{CO_2})] \tag{5.1}$$

where n is the total moles of gas, R the gas constant (J mol^{-1} K^{-1}), T the temperature (K), and X_{CO_2} the mole fraction of CO_2 in the mixture.

The specific energy requirement can be obtained by dividing Equation 5.1 by the amount of desired product, nX_{CO_2}. It can be understood as the compression energy needed to raise the individual partial pressure to the feed gas pressure. Amel'kin et al. [7] provided more general equations where the separation is incomplete, that is, where the binary mixture is separated into two streams which are still mixtures of the original components. The CO_2 product will need to be compressed from atmospheric pressure to pressures suitable for transport and storage, typically 15 MPa. The minimum energy requirement for compression can be derived from the difference in the Gibbs free energy (G/J) between the end state and the start state (taken as 0.1 MPa and 313 K). The Gibbs free energy is a function of both the enthalpy (H/J) and entropy (S/J) of the system:

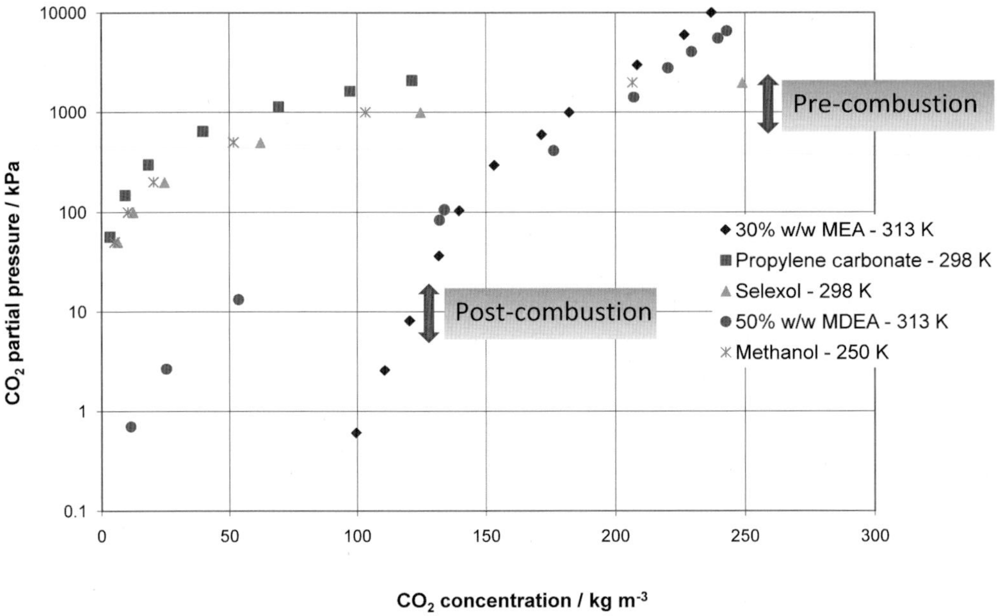

Fig. 5.1 Equilibrium partial pressure for a given absorbent CO_2 loading.

$$G = H - TX \tag{5.2}$$

The Gibbs free energy for CO_2 under a large range of conditions can be calculated from the thermodynamic data given by Span and Wagner [8].

The overall minimum energy requirement for complete CO_2 separation and compression can now be determined by adding the minimum energy requirement for compression to the minimum energy for separation. The minimum energy requirement at a given temperature is determined only by the partial pressure of CO_2 in the feed gas, the feed gas total pressure, and the required end-pressure of the CO_2 product. Figure 5.2 shows the total minimum energy requirement for complete separation and compression to 15 MPa as a function of CO_2 mole fraction in the feed gas at different total pressures at a temperature of 313 K.

Figure 5.2 shows that the minimum energy requirement for complete separation and compression of CO_2 present in an atmospheric flue gas, that is, at 0.1 MPa and 0.04–0.15 mole fraction, is typically between 400 and 500 kJ kg^{-1} CO_2. This value, representative for the post-combustion capture route, is higher than the value for a pre-combustion capture route: at 25 MPa and a 0.3–0.4 mole fraction the minimum energy requirement for complete separation and compression is typically around between 150 and 200 kJ kg^{-1} CO_2. This illustrates the advantage in separation energy efficiency when starting with a higher pressure and higher CO_2 content feed gas. It should be mentioned that for the pre-combustion route the fuel conversion of the hydrocarbon fuels to gaseous

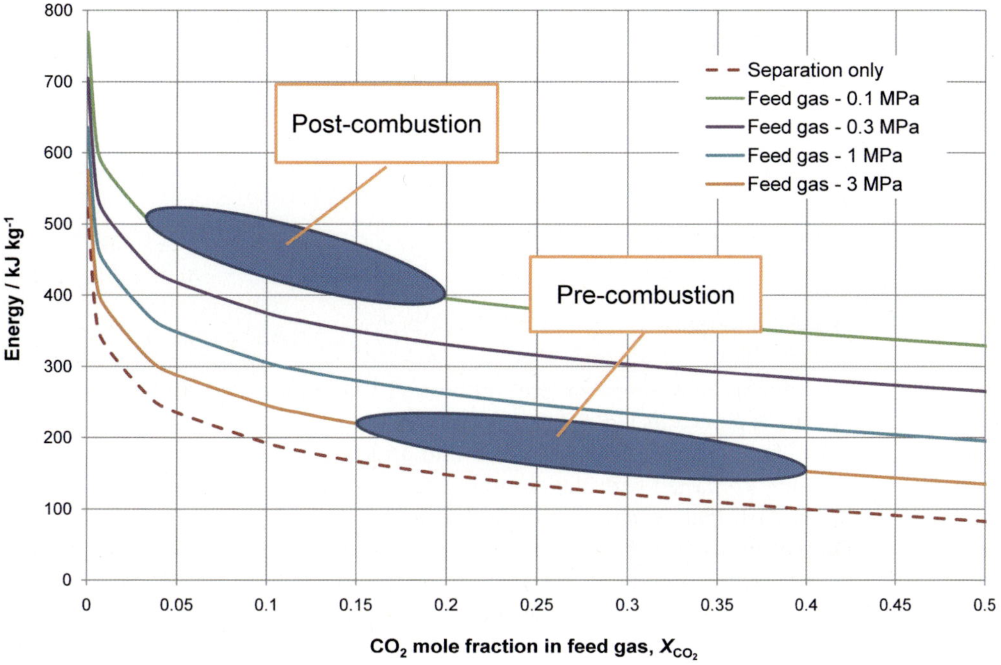

Fig. 5.2 Overall minimum specific energy requirement for complete CO_2 separation and compression as a function of CO_2 mole fraction in the feed gas with compression to 15 MPa.

fuel leads to an additional loss in efficiency. This should be taken into account for a full comparison between pre- and post-combustion, which is beyond the scope of this chapter.

The energy consumption of an absorption process normally consists of two components:

- thermal energy for regeneration of the liquid absorbents extracted from the power plant
- electrical energy for rotating equipment, such as fans and pumps.

Bolland and Undrum [9] provided a generalized methodology by which the three capture routes can be compared. The comparison of different capture process routes for power plants has to be expressed by the reduction in power output as determined by the capture of CO_2, as electricity is the main product [10]. This involves integration of the CO_2 capture process into the power plant. The level of integration will also influence the reduction in power plant efficiency, but in general a tighter integration will result in a loss of flexibility in power plant operation. Several studies have focused on an overall techno-economic assessment of different power plants with CO_2 capture [11–15], effectively providing overall data for the electricity consumption and costs per tonne of CO_2 captured. A recent report by the National Energy Technology Laboratory in the

United States places the cost of pre-combustion CO_2 capture (90%) in an IGCC power plant at US$ 59 t^{-1} CO_2 and US$ 57–64 t^{-1} post-combustion CO_2 capture (90%) in a pulverized coal power plant [12]. Note that these values include CO_2 compression, transport, and storage costs.

5.2 Process Engineering and Performance

5.2.1 Physical Absorption

The process flow sheet for pre-combustion CO_2 capture for a physical solvent process is shown in Fig. 5.3.

The high-pressure feed gas is fed into an absorber, where the gas is countercurrently contacted with the absorbent. CO_2 is absorbed and the liquid is fed to a first flash tank, where co-absorbed hydrogen and less soluble components are flashed off and fed back in the feed stream after recompression. In a second flash tank, the pressure is reduced further and the bulk of the CO_2 is recovered. Further pressure reduction, down to vacuum pressures, will lead to a greater recovery of CO_2. Additionally, an air or N_2 stripper can be included to give very

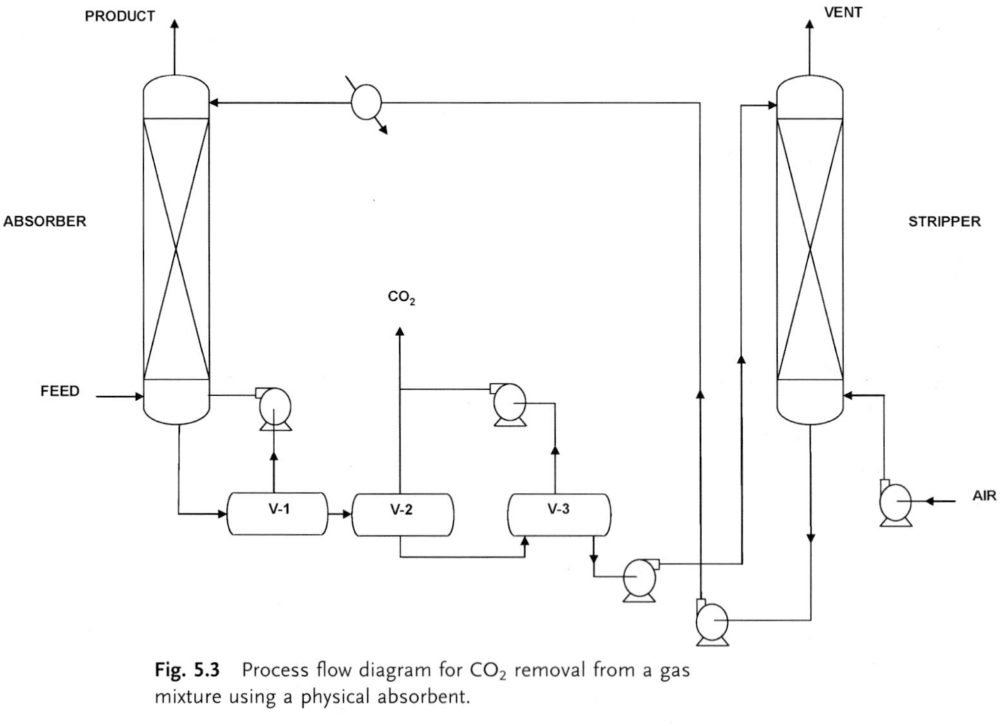

Fig. 5.3 Process flow diagram for CO_2 removal from a gas mixture using a physical absorbent.

deep removal of CO_2. After regeneration, the solvent will be brought up to the absorber pressure and fed back to the top of the column. The number of regeneration stages is dependent on the CO_2 removal required, but also on the process economics. The repressurization of the solvent is the main energy requirement of this process, with some solvents also requiring cooling to achieve high CO_2 solubility. In pre-combustion capture, a sizeable fraction of the CO_2 can be available at pressures higher than atmospheric by the staged regeneration process. This reduces the requirement for compression of CO_2. The choice of a particular physical absorbent is primarily determined by the solvent loading and the selectivity. Other critical issues in the selection of physical absorbents are the vapor pressure (to reduce losses), stability, and viscosity. The overall energy requirement of a physical absorption process applied to pre-combustion capture will typically be in the range 0.2–0.4 MJ kg^{-1} CO_2 [10], excluding compression of the product CO_2.

5.2.2 Chemical Absorption

The typical flow sheet of a chemical absorption process is shown in Fig. 5.4.

The flue gas from the power station is first cooled, after which it is brought into contact with the liquid absorbent in the absorber. A blower is required to pump the gas through the absorber. At temperatures typically between 313 and 333 K, CO_2 is then bound by the chemical absorbent. Next, the flue gas passes through a water wash section to balance water in the system and a demister section. This will remove any liquid droplets or vapor carried over before it leaves the absorber. It is possible to the reduce CO_2 concentration in the feed gas to very low values as a result of the chemical reaction in the liquid absorbent. However, lower exit concentrations tend to increase the height of the absorption vessel. The "rich" absorbent, which contains the chemically bound CO_2, is then pumped to the top of a stripper, via a heat exchanger. The regeneration of the chemical absorbent is carried out in the stripper at elevated temperatures (373–413 K) and pressures typically between 100 and 200 kPa. Steam is supplied externally to the reboiler to maintain the regeneration conditions, thus providing the required desorption heat for removing the chemically bound CO_2 and for steam production from the liquid absorbent which acts as a stripping gas. This steam is recovered as water in the condenser and fed back to the stripper, whereas the CO_2 product gas leaves the condenser. The CO_2 product is relatively pure (>99%), with water vapor being the main other component. Due to the selective nature of the chemical absorption process, the concentration of inert gases is low. The CO_2 product might contain trace components, such as volatile solvent decomposition products or components carried over from the flue gas. A further CO_2 purification step makes it possible to bring the CO_2 quality up to food-grade standard. The "lean" liquid absorbent, containing far less CO_2, is then pumped back to the absorber via the lean-rich heat exchanger and a cooler that cools it to the absorber temperature level.

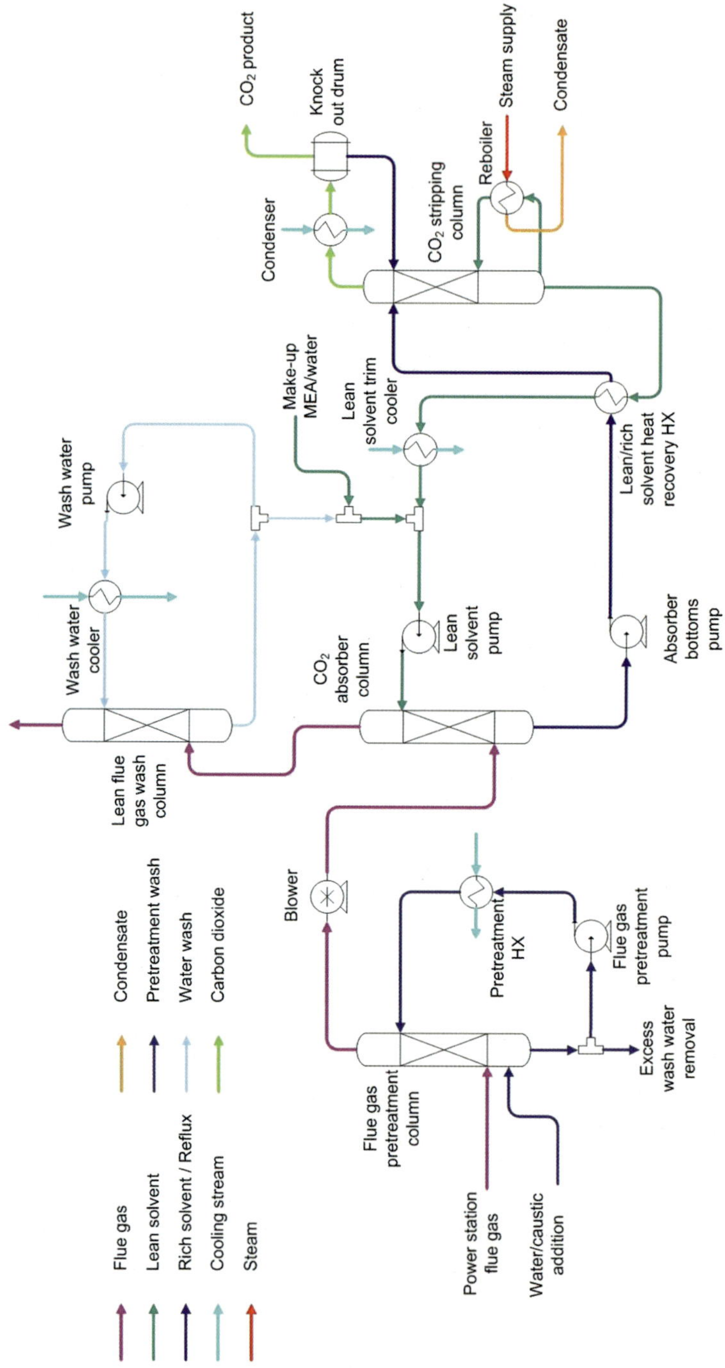

Fig. 5.4 Process flow diagram for CO_2 recovery from flue gas with a chemical absorbent [monoethanolamine (MEA)].

5.2.3
Energy Consumption

The power requirement in the chemical absorption process is in general determined by

- The power needed for the blower to guide the flue gases through the absorber. This is determined by the volumetric flow rate, the pressure drop over the absorption column, and the percentage removal of CO_2.
- The power needed for the liquid absorbent pumps. This is determined by the achievable liquid absorbent loading in the absorption process and the level of liquid absorbent regeneration. This is the difference in the absorbent CO_2 content between the lean absorbent at the absorber liquid inlet and the rich absorbent at the liquid outlet (Δa).

The energy use for liquid absorbent regeneration is described in Section 5.4.2.3 and Equations 5.38–5.41. Steam is needed to provide the thermal energy for the regeneration process and the extraction of this steam will lead to a reduction in the power plant output. The power factor (0.2–0.3) converts the steam usage into lost electrical output. The energy required for a chemical absorption process applied to post-combustion will be in the range 0.8–1.3 MJ kg^{-1} CO_2 [6, 10], excluding compression.

5.2.4
Process Flow Sheet Modifications

The energy requirement of absorption processes can be reduced by consideration of modifications to the process flow sheet given in Fig. 5.4. Although process flow sheet modifications have been applied in the treatment of high-pressure gas streams similar to pre-combustion capture applications, there is potential for such process modifications also to be applied in post-combustion capture. Typical alterations to the standard flow sheet involve:

- *Use of intercooling.* In the standard flow sheet, the absorption process operates adiabatically, which results in a temperature increase in the liquid absorbent due to the exothermic reaction of CO_2 with the liquid absorbent. This will limit the achievable CO_2 loading of the absorbent. Through cooling at a point in the absorber, the achievable loading can be increased, resulting in a higher partial pressure of CO_2 in the desorption process [16].
- *Split flow.* Two or more liquid flow circuits, with liquid absorbent regeneration occurring at different positions in the stripper, can be used if high removal is required [17]. This results in only part of the liquid absorbent having to undergo complete regeneration, as the semi-lean solution will be adequate for bulk removal of CO_2.
- *Vapor compression.* As the product CO_2 needs to be available at high pressure, compression of the overall vapor stream from the stripper with heat recovery for the regeneration of liquid absorbents will be of interest. The system oper-

ates basically as an integrated heat pump. Process flow configurations along these lines have been developed along with strippers operating at different pressures [18, 19]. Furthermore, the vapor coming from the lean solvent lends itself to compression and heat recovery. As this modification requires the use of electricity, its full impact can only be determined by considering the capture process integrated with the power plant.

Several of these process modifications have been assessed for post-combustion capture applications and shown to have potential to reduce the heat duty for a standard 4.9 mol dm^{-3} (30% w/w) MEA process by nearly 20% [20].

5.2.5 Gas–Liquid Contactors

The gas–liquid contactors used for both absorption and desorption form a large part of the capital costs in an absorption process. The use of a countercurrent flow pattern ensures that uniformity of driving forces for gas transfer is maintained over the column. It also allows a maximum loading of the liquid absorbent to be achieved, which is beneficial for the regeneration of the absorbent. For selective removal of gases from off-gas, flue gas, natural gas, synthesis gases, and other process gases, four different types of contactors can be used: packed columns, plate columns, spray columns, and bubble columns. An overview of the different characteristics of these columns is given in Table 5.3. Guidance for the selection of the most appropriate contactor for a given duty can be found in several textbooks [21, 22]. These and other references [23, 24] provide generalized methods for the design of chemical and physical absorption processes. For applications in post-combustion capture of CO_2 using chemical absorption, packed columns are the most suitable, as they provide the highest mass transfer and lowest pressure drop. Spray columns are also considered because of their simplicity and low cost [25].

In pre-combustion capture applications using either physical or chemical absorption, both packed columns and plate columns can be used. For gas treatment conditions, the absorption rates are sufficiently large to result in small diffusion layers. This rules out bubble columns, which are normally used for gas absorption involving slow reactions.

5.3 Physical Absorption

5.3.1 Process Principle

The physical absorption of CO_2 into a liquid absorbent involves the solvation and diffusion of the gas in a liquid, without any subsequent chemical reactions occurring. Absorption is driven by the difference in the liquid phase concentra-

Table 5.3 Overview of gas–liquid contactors [26–29].

Parameter	Column type			
	Packed column	**Plate column**	**Spray column**	**Bubble column**
Features	Packing provides contact area Low liquid hold-up Dumped and structured packing materials	Intermediate liquid hold-up Long residence time Complete mixing on tray	Liquid is dispersed phase Suitable for gases with high solubility One spray bank is equilibrium stage	Large liquid hold-up Long residence time
Advantages	Low gas-side pressure drop Able to handle foaming liquids	Wide operational gas/liquid flow range Can accommodate heat exchangers on trays	Can handle gases with solids present Low gas-side pressure drop	Simple construction Low capital cost Low maintenance
Disadvantages	Not suitable for liquids with solids present Operation constrained by gas/liquid flow ratios Less suitable for inclusion of heat exchangers for thermal control	Vulnerable to foaming liquids High gas-side pressure drop High capital costs	Energy required for spray formation Liquid coalescence Liquid entrainment in gas outlet stream	High gas-side pressure drop Bubble coalescence
Mass transfer limitations	Gas side limited preferred but suitable for both gas and liquid limited mass transfer conditions	Liquid side limited preferred but suitable for both gas and liquid limited mass transfer conditions	Gas side limited preferred	Liquid side limited preferred
Gas/liquid ratio ($m^3\ m^{-3}$)	100–1000	100–1000	500–2000	10–100
$k_l a$ (s^{-1})	0.005–0.02	0.01–0.05	0.007–0.015	0.005–0.01
$k_g a$ (mol s^{-1} kPa^{-1} m^{-3})	0.5–2.0	0.5–1.0	0.1–0.8	0.4–1.0
Specific surface area ($m^2\ m^{-3}$)	100–300	100–400	20–80	10–30
Gas hold-up	>0.95	>0.8	>0.8	<0.2
Liquid hold-up	<0.05	<0.2	<0.2	>0.8
Ratio of bulk liquid to diffusion layer	~10	~10	~10	1000–10 000
Gas-side pressure drop (kPa m^{-1})	0.2–0.8	>10 (determined by filled plate height and pressure drop through orifices)	0.05–0.1 per spray bank	~10 (determined by filled column height and pressure drop through orifices)

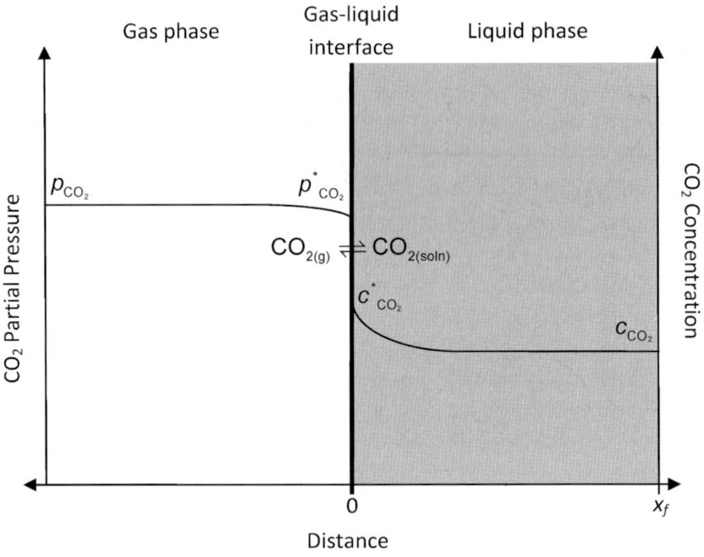

Fig. 5.5 Graphical representation of CO_2 physical absorption in a liquid. Gaseous CO_2 diffuses towards and crosses the gas–liquid interface. It then diffuses from the interface into the film.

tion of CO_2 and its solubility at a given temperature and gas-phase partial pressure. Graphically this can be represented as in Fig. 5.5, which shows CO_2 gas [$CO_{2(g)}$] in contact with a liquid absorbent as a film of thickness x_f (m). CO_2 dissolved in the liquid is given by $CO_{2(soln)}$. In the gas phase, CO_2 diffuses from the bulk where it has partial pressure p_{CO_2} (kPa) to the gas–liquid interface. Near the interface its partial pressure falls to $p^*_{CO_2}$ as it is absorbed into the liquid. In the liquid phase, the concentration of CO_2 at the gas–liquid interface $c^*_{CO_2}$ (mol dm^{-3}) is defined by the solubility of CO_2 in the liquid, which is a function of temperature and the CO_2 partial pressure at the interface. The dissolved CO_2 moves via diffusion from the gas–liquid interface into the liquid bulk, where it has concentration c_{CO_2}. If the gas and liquid remain in contact for long enough, eventually they will reach equilibrium when $p_{CO_2} = p^*_{CO_2}$ and $c_{CO_2} = c^*_{CO_2}$.

The driving force for CO_2 absorption is the difference between the CO_2 concentration in the film, c_{CO_2}, and its solubility at the gas–liquid interface, $c^*_{CO_2}$. CO_2 mass transfer across the interface expressed as a flux, N_{CO_2} (mol s^{-1} m^{-2}), can be defined according to this difference and an overall mass transfer coefficient, K_l (m s^{-1}) [30]:

$$N_{CO_2} = K_l(c^*_{CO_2} - c_{CO_2}) \tag{5.3}$$

The value of K_l depends on the resistance to diffusion on both the gas side and liquid side of the gas–liquid interface. A mass transfer coefficient can be defined for CO_2 in the gas phase (k_g/mol s^{-1} kPa^{-1} m^{-2}) and the liquid phase k_l^*/m s^{-1}). The * in k_l^* is used to indicate mass transfer without chemical reaction in the liquid phase. In physical absorbents where no chemical reactions occur, the magnitude of k_l^* is defined by the diffusion coefficient of CO_2 (D_{CO_2}/m^2 s^1). A discussion of diffusion is given in Section 5.5.

$$\frac{1}{K_l} = \frac{1}{k_{H,CO_2} k_g} + \frac{1}{k_l^*} \tag{5.4}$$

The solubility of CO_2 in a liquid is generally defined using a Henry coefficient, k_{H,CO_2}. This coefficient is the ratio of gas-phase CO_2 partial pressure and the respective liquid phase CO_2 concentration, and is generally expressed as a function of temperature. In physical solvents, it is commonly expressed using MPa and mole fraction (X_{CO_2}):

$$k_{H,CO_2} = \frac{p_{CO_2}}{X_{CO_2}} \tag{5.5}$$

An important characteristic of physical absorption illustrated by the form of Equation 5.5 is that CO_2 solubility changes linearly with CO_2 partial pressure. This is why physical absorption is suited to applications with a large CO_2 partial pressure in the gas phase such as pre-combustion capture, but performs poorly in post-combustion capture applications where the CO_2 partial pressure is typically less than 20 kPa. As will be seen in the next section, this is not the case for chemical absorption.

Something worth noting, and often done in practice, is that the relationship in Equation 5.3 can also be written in terms of the gas-phase partial pressures and a mass transfer coefficient in gas-phase units, K_g (mol s^{-1} kPa^{-1} m^{-2}):

$$N_{CO_2} = K_g(p_{CO_2} - p_{CO_2}^*) \tag{5.6}$$

$$\frac{1}{K_g} = \frac{1}{k_g} + \frac{k_{H,CO_2}}{k_l^*} \tag{5.7}$$

Equations 5.3 and 5.5 or 5.6 and 5.5 can be related by using Equation 5.5 to give the value of $c_{CO_2}^*$, the solubility of CO_2 at the gas–liquid interface. If Equation 5.5 is rearranged and substituted into Equation 5.3, Equation 5.8 results:

$$N_{CO_2} = K_l\left(\frac{p_{CO_2}^*}{k_{H,CO_2}} - c_{CO_2}\right) \tag{5.8}$$

The equations given so far represent steady-state mass transfer. They can be derived from Fick's first law [31], which is given below for CO_2:

$$N_{CO_2} = D_{CO_2} \frac{\partial c_{CO_2}}{\partial x} \tag{5.9}$$

Fick's laws can be applied to both the gas and liquid phases; however, the discussion from here on will focus mainly on the liquid phase. The reason is that in real-world applications where a liquid flowing as a film over a packing is countercurrently contacted with a gas, the gas phase can be considered well mixed and of infinite thickness, in which case $p^*_{CO_2} = p_{CO_2}$. Application of Fick's law requires knowledge of the diffusion coefficient of CO_2, which can be measured [32] or estimated [33].

Fick's second law, which describes non-steady-state mass transfer, can be derived from the first law by applying a mass balance and using the method of "combination of variables" [31]:

$$\frac{\partial c_{CO_2}}{\partial t} = D_{CO_2} \frac{\partial^2 c_{CO_2}}{\partial x^2} \tag{5.10}$$

Integration of this partial differential equation yields the concentration profile of CO_2 in the liquid film as a function of both time (t/s) and distance (x/m) from the gas–liquid interface. An example of this is shown in Fig. 5.6, which is a plot of the diffusion of CO_2 into a 1×10^{-4} m film of methanol at 263 K. At the gas–liquid interface, the film is exposed to a constant CO_2 partial pressure of 100 kPa for 5 s. Using the Henry coefficient for CO_2 in methanol at 263 K

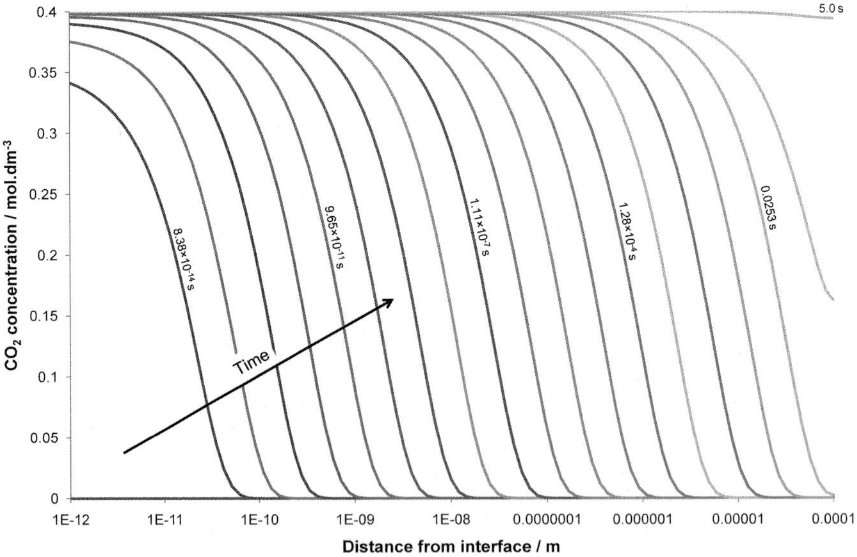

Fig. 5.6 Diffusion of CO_2 into a 1×10^{-4} m liquid film of methanol exposed to a constant CO_2 partial pressure of 100 kPa for 5 s.

(k_{H,CO_2} = 8.2 MPa [34]) results in an interfacial CO_2 concentration, or boundary condition at $x=0$, of 0.40 mol dm^{-3}. The boundary condition at $x=x_f$ is zero CO_2 flux. The diffusion coefficient of CO_2 in methanol at 263 K is 3.6×10^{-9} m^2 s^{-1} [32]. The progression of CO_2 through the film as a function of time can be clearly seen. After 5 s, CO_2 has diffused from the gas–liquid interface through the entire film.

If gas phase diffusion was limiting, and it was necessary to incorporate it in a description of diffusion in the liquid film, it can be easily done. Rather than assuming that the concentration of CO_2 at the gas–liquid interface is defined by the Henry coefficient alone, the driving force and gas-side mass transfer coefficient k_g can be used to define the interfacial flux [35]:

$$D_{CO_2}\frac{\partial c_{CO_2}}{\partial x} = k_{H,CO_2}k_g(c^*_{CO_2} - c_{CO_2}) \tag{5.11}$$

at x=0.

5.3.2 Range of Physical Absorbents

Liquids used as physical absorbents are based on organic compounds. Any aqueous system, even those such as the Benfield process which has only potassium carbonate ions in solution, is chemically reactive and will be discussed in the next section. This is because CO_2 is an acid gas and absorption in water involves chemical conversion to carbonic acid/bicarbonate/carbonate species and acid–base equilibria between these species.

For physical absorbents, the two key parameters to be considered are the solubility of CO_2 and selectivity in the presence of other gases such as nitrogen and hydrocarbons. CO_2 solubility is governed by solvent–solvent interactions and solvent–CO_2 interactions [36]. The most desirable situation is weak solvent–solvent interactions, allowing CO_2 to incorporate into the solvent, and strong solvent–CO_2 interactions to keep it there. Water is a poor solvent for CO_2 (chemical reactivity aside) because its high polarity and hydrogen bonding between water molecules mean that solvent–solvent interactions are strong. CO_2 is most soluble in weakly polar solvents due their weak solvent–solvent interactions. In terms of solvent–CO_2 interactions, polar interactions will be weak. However, strong and specific interactions with CO_2 occur with solvents that exhibit some Lewis base character. The molecules in these solvents contain O or N atoms with a free lone pair of electrons able to interact with the C in CO_2. An attractive feature of physical absorbents is that CO_2 is only weakly bound in solution and has not undergone physical transformation. This means that CO_2 can be removed and the solvent regenerated via mild heating and/or pressure reduction.

The commercially available processes based on physical absorbents used for CO_2 capture are listed in Table 5.4. The table gives the Henry coefficient for

Table 5.4 Commercially available CO_2 capture processes based on physical absorbents [5].

Process name	Absorbent	Company	k_{H,CO_2} at 298 K (MPa)	$-\Delta H_{abs}$ (kJ mol^{-1})
Fluor Solvent	Propylene carbonate	Fluor Daniel	8.5 [37]	12.2
SELEXOL	Dimethyl ether of poly(ethylene glycol)	Union Carbide	3.6 [37]	14.3
Sepasolv MPE	Methyl isopropyl ether of poly(ethylene glycol)	BASF	Similar to SELEXOL [5]	Similar to SELEXOL
Purisol	N-Methyl-2-pyrrolidone	Lurgi	6.6 [38]	14.8
Rectisol	Methanol	Lurgi and Linde	14.5 (4.3[a)]) [39]	15.1
Ifpexol	Methanol	IFP		
Methyl cyanoacetate	Methyl cyanoacetate	Unocal	8.9 [37]	12.3

a) Value at 250 K for comparison as methanol-based processes generally operate in the range 243–273 K.

each absorbent at 298 K and the enthalpy of absorption, ΔH_{abs}. ΔH_{abs} was determined by plotting values of $\ln k_{H,CO_2}$ as a function of inverse temperature (T^{-1}) and using linear regression to determine the slope according to the van't Hoff equation (Equation 5.12). As only physical absorption is occurring, this then provides a good estimate of the absorption enthalpy.

$$\frac{d \ln k_{H,CO_2}}{d\left(\frac{1}{T}\right)} = -\frac{\Delta H_{abs}}{R} \tag{5.12}$$

These compounds all share the key characteristics of weak solvent–solvent interactions and strong solvent–CO_2 interactions. The choice of physical absorbent from those available is mainly driven by the selectivity of the solvent for CO_2 relative to the other components in the gas stream. For example, propylene carbonate and methyl cyanoacetate show good selectivity for CO_2 in the presence of hydrocarbons whereas methanol is suited to the simultaneous removal of CO_2 along with ecologically damaging impurities such as hydrogen cyanide and organic sulfur compounds [5]. Of the absorbents listed, methanol has a lower molecular weight than the others and hence a much higher vapor pressure at a given temperature. For this reason, it is commonly used at low temperature (243–273 K) whereas the others are used at ambient temperature or above.

5.4
Chemical Absorption

5.4.1
Process Principle

As explained above, physical absorption of CO_2 in a liquid involves the solvation of CO_2 by the liquid. When chemical absorption occurs, in addition to solvation CO_2 reacts chemically with components of the liquid and is converted into other compounds. If we again consider gaseous CO_2 being contacted with a liquid film, chemical reaction modifies what occurs in the liquid phase. Once CO_2 has crossed the gas–liquid interface and entered the film, it undergoes both diffusion and chemical reaction. This is illustrated in Fig. 5.7, which shows solvated CO_2 [$CO_{2(soln)}$] reacting with chemical species A to form product P. The result of this reaction is to lower the concentration of solvated CO_2 in the film (c_{CO_2}) and deplete it more rapidly than would occur by diffusion alone.

Chemical reaction of CO_2 has two main effects. One is to increase the amount of CO_2 that can be absorbed by converting it to other chemical forms. The other is to enhance mass transfer. Mass transfer is enhanced for two reasons. One is that by lowering the concentration of $CO_{2(soln)}$ in the liquid, the difference ($c^*_{CO_2} - c_{CO_2}$) given in Equation 5.3 is increased. This difference is the driving force for absorption. The other reason is that mass transfer in the

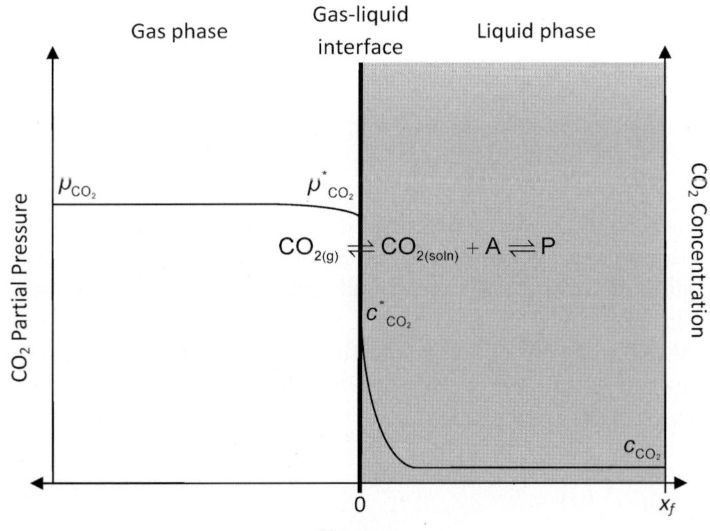

Fig. 5.7 Graphical representation of chemical absorption of CO_2 in a liquid. Gaseous CO_2 diffuses towards and crosses the gas–liquid interface. It then diffuses from the interface into the film while simultaneously undergoing chemical reaction with A.

liquid phase is enhanced, because CO_2 is being depleted from the gas–liquid interface more rapidly than would occur by diffusion alone. Recall that in a liquid mass transfer without chemical reaction is defined by the mass transfer coefficient k_l^* (Equation 5.4). When diffusion and reaction occur, the mass transfer coefficient is written k_l (m s^{-1}) with k_l and k_l^* related by a dimensionless enhancement factor E [30]:

$$k_l = Ek_l^* \tag{5.13}$$

$$\frac{1}{K_l} = \frac{1}{k_{H,CO_2}k_g} + \frac{1}{k_l} \tag{5.14}$$

Under conditions where CO_2 reacts rapidly in a pseudo-first-order regime, the CO_2 flux can be written in terms of the reaction and diffusion of CO_2 only. Pseudo-first-order conditions are obtained when the following conditions are satisfied: the reactant A is present in large excess; CO_2 reacts rapidly with A via an overall second-order reaction with rate constant k_A like that given in Fig. 5.7; and the reverse reaction is negligible. The equation for k_l under such conditions is

$$k_l = \sqrt{D_{CO_2}k_A c_A} \tag{5.15}$$

If we consider conditions with no gas-side limitation to mass transfer (k_g is large so K_l=k_l and $p^*_{CO_2} = p_{CO_2}$) and no CO_2 is present in solution, Equation 5.8 simplifies to Equation 5.16. Such conditions are often used by in wetted-wall reactors to determine the value of the rate constant k_A.

$$N_{CO_2} = \frac{p_{CO_2}}{k_{H,CO_2}}\sqrt{D_{CO_2}k_A c_A} \tag{5.16}$$

For reactions that are of greater complexity, a range of other relationships can be derived for specific cases [30, 31, 40, 41]. However, as was the case for physical absorption, all of these equations are simplifications that represent steady-state mass transfer.

A more general description of diffusion and chemical reaction is found by extending Fick's second law as given in Equation 5.10 to include the chemical reaction rate of CO_2 (r_{CO_2}/dm^3 mol^{-1} s^{-1}):

$$\frac{\partial c_{CO_2}}{\partial t} = D_{CO_2}\frac{\partial^2 c_{CO_2}}{\partial x^2} + r_{CO_2} \tag{5.17}$$

An equivalent equation is required for each chemical species in the liquid to describe its diffusion and reaction in the film. The general form of Equation 5.17 for the ith chemical species is

$$\frac{\partial c_i}{\partial t} = D_i \frac{\partial^2 c_i}{\partial x^2} + r_i \tag{5.18}$$

The reaction rate equations consist of ordinary differential equations (and possibly coupled algebraic equations for equilibria) that can be derived according to standard theories of chemical kinetics [42–44] and integrated as part of the partial differential equation (5.18).

5.4.2
Chemical Reaction Pathways

All processes for the chemical absorption of CO_2 that are in use or at the later stages of development are based on aqueous solutions of carbonate, amines, or mixtures thereof. The chemistry of CO_2 absorption will therefore be discussed in two sections: first the carbonate chemistry of CO_2 absorption in water; followed by the additional reactions introduced when an amine or other base is present in the absorbent.

5.4.2.1 CO_2–H_2O Chemistry and Carbonate Absorbents

The chemical reactions that occur between CO_2 and H_2O have been extensively studied and are well documented. They can be fully described by the following set of reactions [45].

$$CO_{2(soln)} \xrightleftharpoons{k_{H,CO_2}} CO_{2(g)} \tag{5.19}$$

$$CO_{2(soln)} + H_2O \xrightleftharpoons[k_{-1}]{k_1,\, K_1} H_2CO_3 \tag{5.20}$$

$$CO_{2(soln)} + OH^- \xrightleftharpoons[k_{-2}]{k_2,\, K_2} HCO_3^- \tag{5.21}$$

$$H^+ + CO_3^{2-} \xrightleftharpoons{K_3} HCO_3^- \tag{5.22}$$

$$H^+ + HCO_3^- \xrightleftharpoons{K_4} H_2CO_3 \tag{5.23}$$

$$H^+ + OH^- \xrightleftharpoons{K_5} H_2O \tag{5.24}$$

Rate and equilibrium constants used for these chemical reactions are given in Table 5.5. The first equation is the physical solvation of $CO_{2(g)}$ by water to give dissolved CO_2 [$CO_{2(soln)}$]. This is described by a Henry coefficient with pressure in kPa and concentration in molarity (mol dm^{-3}) for chemical absorbents. The next two reactions are $CO_{2(soln)}$ with water or hydroxide (OH^-) to form carbonic acid (H_2CO_3) and bicarbonate (HCO_3^-), respectively. Where the value of a rate or equilibrium constant was unknown, it was calculated using the relationship be-

Table 5.5 Rate constant and equilibrium constant values for CO_2–H_2O reactions [a].

Reaction	Parameter	Ref.
$CO_{2(soln)} \xrightleftharpoons{k_{H,CO_2}} CO_{2(g)}$	$k_{H,CO_2} = 2.82 \times 10^6 e^{-17000/RT}$ (kPa dm^3 mol^{-1})	46
$CO_{2(soln)} + H_2O \underset{k_{-1}}{\overset{k_1, K_1}{\rightleftharpoons}} H_2CO_3$	$k_1 = \dfrac{e^{(22.66-7799/T)}}{c_{H_2O}}$ (mol dm^{-3} s^{-1}), $k_{-1} = e^{(30.15-8018/T)}$ (s^{-1}), $K_1 = \dfrac{k_1}{k_{-1}}$ (mol dm^{-3})	47
$CO_{2(soln)} + OH^- \underset{k_{-2}}{\overset{k_2, K_2}{\rightleftharpoons}} HCO_3^-$	$k_2 = 3.279 \times 10^{13} e^{-6613/T}$ (mol dm^{-3} s^{-1}), $k_{-2} = \dfrac{k_2}{K_2}$ (s^{-1}), $K_2 = \dfrac{K_1 K_5}{K_4}$ (mol dm^{-3})	48
$H^+ + CO_3^{-2} \overset{K_3}{\rightleftharpoons} HCO_3^-$	$\log K_3 = \dfrac{-2902.39}{T} - 0.02379T + 6.4980$	49
$H^+ + HCO_3^- \overset{K_4}{\rightleftharpoons} H_2CO_3$	Calculated as $\dfrac{K_1}{K_1'}$, where $K_1' = \dfrac{c_H c_{HCO_3}}{c_{CO_2} c_{H_2O}}$ and $\log K_1' = \dfrac{-5251.43}{T} - 36.781 \log T + 102.2685 - \log c_{H_2O}$	50
$H^+ + OH^- \overset{K_5}{\rightleftharpoons} H_2O$	$\log K_5 = \dfrac{-5839.48}{T} - 22.4773 \log T + 61.2060 - \log c_{H_2O}$	50

a) The values are at infinite dilution and zero ionic strength. c_{H_2O} is the concentration of water in dilute solution (55.6 mol dm^{-3}). All equations originally given as natural logarithms have been converted to base 10 logarithms.

tween an equilibrium constant and the ratio of the forward and reverse reaction rate constants [44] as follows:

$$K_i = \frac{k_i}{k_{-i}} \tag{5.25}$$

The remaining three reactions are the first and second protonations of carbonate (CO_3^{2-}) followed by water dissociation.

These reactions represent all those that occur in water and aqueous carbonate systems such as the hot potassium carbonate or Benfield process [5]. In the hot potassium carbonate process, CO_2 is absorbed in a solution of K_2CO_3 at ~80 °C. Without CO_2 present, when K_2CO_3 is dissolved in water an aqueous solution of OH^-, CO_3^{2-} and K^- ions results. Upon absorption of CO_2, protons and HCO_3^- are produced and the carbonate–bicarbonate protonation equilibrium shifts in favor of HCO_3^-. Absorption ceases when sufficient CO_2 is absorbed to deplete the CO_3^{2-} concentration. Heating and/or pressure reduction result in CO_2 release and a shift in the equilibrium back to CO_3^{2-}. The K^+ ions present in solution play no role in the reactions that occur; however, their presence has an influence on the ionic strength of the solution and the activity of the chemical species [51].

The mass transfer of CO_2 into potassium carbonate solutions is slow in comparison with aqueous amine solutions, because absorption occurs via reaction of CO_2 with OH^- and H_2O. At high pH (approximately >10), absorption occurs predominantly via reaction with OH^-. Although the rate constant of this reaction is comparable to that of amines which react rapidly with CO_2, the concentration of OH^- is low. In a pH 12 solution, which can be achieved in a CO_2-free carbonate solution, the OH^- concentration is only 0.01 mol dm^{-3}. Below pH 10 absorption is dominated by the slower reaction with H_2O.

To enhance mass transfer in aqueous carbonate solutions, an additional rate promoter can be added. Two types of rate promoter are commonly used. The first is a weak acid such as arsenious acid, phosphoric acid, or boric acid [52]. These weak acids promote the rate of CO_2 absorption by acting as a pH buffer. Some weak acids such as arsenious acid have been shown to enhance the absorption rate to a greater extent than would be observed by pH buffering alone. It has been postulated that these weak acids also act as Lewis bases, with a free lone pair of electrons that allows them to catalyze the water hydration reaction of Equation 5.20 [52, 53]. The second class of rate promoters is primary and secondary amines that are known to react rapidly with CO_2 [41, 54, 55]. These amines, such as MEA, diethanolamine (DEA), or piperazine (PZ), behave according to the CO_2–amine chemistry described in the following section. The synergy that is found upon addition to a carbonate solution is that the CO_3^{2-} ion is a stronger base than most amines and acts to maintain the amine in its conjugate base form and thus active form as CO_2 is absorbed.

5.4.2.2 CO_2–Amine Chemistry

Amines are organic compounds containing a basic nitrogen atom with a lone pair of electrons [56]. Amines are derivatives of ammonia (NH_3) in which one or more hydrogen atoms have been replaced by an alkyl or aryl substituent. In aqueous amine solutions, a number of additional reactions between CO_2, primary and secondary amine groups, and amine protonation need to be considered. Analogous reactions occur for solutions of amino acids (amines which also contain carboxylic acid functionality) and ammonia. Normally, one of two reaction mechanisms is used to describe the direct reaction between CO_2 and a primary or secondary amine (R_1R_2NH) to form a carbamate ($R_1R_2NCOO^-$), known as the zwitterion and termolecular mechanisms [46]. They are both derived from a detailed mechanism in which CO_2 and an amine react to form an intermediate carbamate complex from which a proton is then removed via a base. This mechanism is illustrated in Fig. 5.8.

The detection of the intermediate has never been reported in the literature. If it is assumed that the concentration of the intermediate complex is low and at steady state, and this steady-state assumption is applied, then the rate of reaction of CO_2 can be described by the following equation:

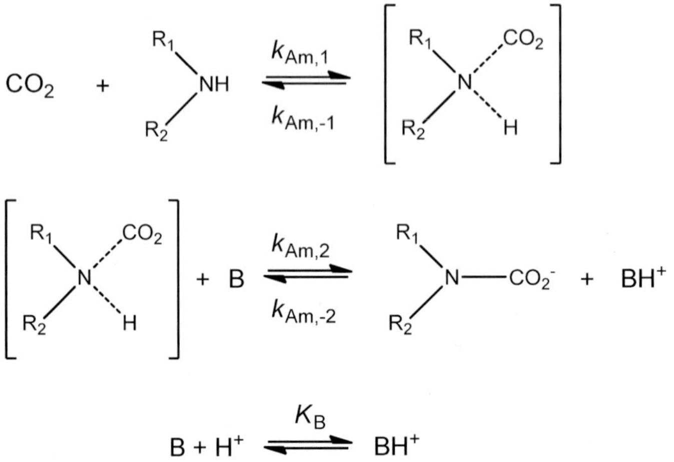

Fig. 5.8 The detailed mechanism for the direct reaction of a primary or secondary amine (R_1R_2NH) with CO_2 in the presence of a base (B).

$$\frac{dc_{CO_2}}{dt} = -\frac{k_{Am,1}k_{Am,2}c_B}{k_{Am,-1} +_{Am,2}c_B}c_{CO_2}c_{R_1R_2NH} + \frac{k_{Am,-1}k_{Am,-2}}{k_{Am,-1}+k_{Am,2}c_B}c_{R_1R_2NCOO}c_{BH} \tag{5.26}$$

This equation has a number of limiting cases depending on the relative magnitude of the reaction rate constants. Given that the base B is a reasonably good base, as is the case for an amine, the single most likely limiting case is $k_{Am,2} \gg k_{Am,-1}$. $k_{Am,2}$ is the rate constant for a proton transfer reaction which is effectively instantaneous in aqueous media (of the order 10^{11}–10^{12} $dm^3\ mol^{-1}\ s^{-1}$ [57, 58]). $k_{Am,-1}$ will be relatively small in magnitude if the intermediate is to form, which is necessary if the reaction is to proceed to carbamate formation. When this limiting case is applied, Equation 5.26 reduces to Equation 5.27:

$$\frac{dc_{CO_2}}{dt} = -k_{Am,1}c_{CO_2}c_{R_1R_2NH} + \frac{k_{Am,-1}k_{Am,-2}}{k_{Am,2}c_B}c_{R_1R_2NCOO}c_{BH} \tag{5.27}$$

If the equilibrium constant K_B is substituted into equation (5.27) then equation (5.28) results.

$$\begin{aligned}\frac{dc_{CO_2}}{dt} &= -k_{Am,1}c_{CO_2}c_{R_1R_2NH} + \frac{k_{Am,-1}k_{Am,-2}K_B}{k_{Am,2}}c_{R_1R_2NCOO}c_H \\ &= -k_{Am,1}c_{CO_2}c_{R_1R_2NH} + k'_{Am,-1}c_{R_1R_2NCOO}c_H\end{aligned} \tag{5.28}$$

where $K_B = \frac{c_{BH}}{c_Bc_H}$ and $k'_{Am,-1} = \frac{k_{Am,-1}k_{Am,-2}K_B}{k_{Am,2}}$.

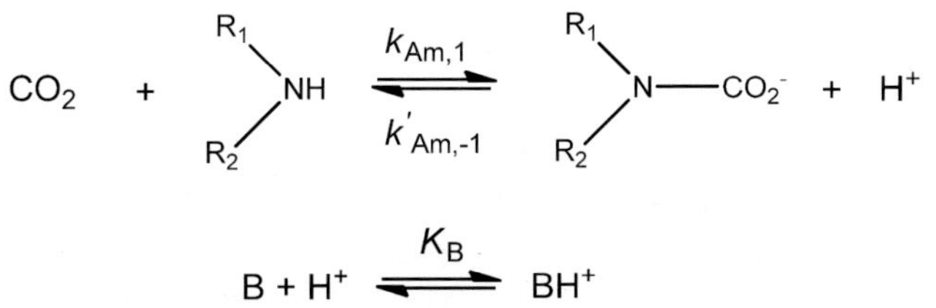

Fig. 5.9 The overall mechanism for the direct reaction of a primary or secondary amine (R_1R_2NH) with CO_2 in the presence of a good base (B).

According to Equation 5.28, in the presence of a good base this reaction should be described by the overall mechanism given in Fig. 5.9. The value of the forward reaction rate constant is unaffected by the properties of the base B; however, both the value of $k'_{\mathrm{Am},-1}$ and the concentration of H^+ will be influenced by the magnitude of K_B. In a concentrated solution of a single type of amine this does not create an issue as B will be the amine itself. Hydroxide ions may also contribute, but their low concentration means they can generally be neglected. In binary amine mixtures, B may be either amine, complicating the analysis. In reality, this effect can be neglected because at pH values and concentrations relevant for CO_2 capture the product $K_B c_H$ is essentially invariant [59].

5.4.2.3 **Characteristics of Commonly Used Amines and Ammonia**

A range of primary, secondary, and tertiary amines (generally alkanolamines), ammonia, and mixtures thereof have found application in aqueous solution as CO_2 absorbents. The performance of amines as CO_2 absorbents varies considerably as a function of amine chemical structure, including carbon chain length, the position and chain length of any alcohol and alkyl substituents, and whether the structure is open chain or cyclic [60–62]. In terms of the rate of absorption, the single most important property is the second-order reaction rate constant $k_{\mathrm{Am},1}$ of the direct reaction between CO_2 and the amine to form a carbamate. In conjunction with diffusion, this rate constant is generally the determining factor for the rate of mass transfer of CO_2.

In terms of absorption capacity, both carbamate and bicarbonate formation need to be considered. Based on the reactions of Equations 5.19–5.24 and Fig. 5.9, these can be defined in terms of two overall equilibria and their corresponding equilibrium constants K_{AmCOO} and $K_{\mathrm{HCO_3}}$:

$$\mathrm{CO_2 + Am + B} \xrightleftharpoons{K_{\mathrm{AmCOO}}} \mathrm{AmCOO^- + BH^+} \tag{5.29}$$

$$K_{\mathrm{AmCOO}} = \frac{c_{\mathrm{AmCOO}}c_{\mathrm{BH}}}{c_{\mathrm{CO_2}}c_{\mathrm{Am}}c_{\mathrm{B}}} \tag{5.30}$$

$$\mathrm{CO_2 + H_2O + B} \xrightleftharpoons{K_{\mathrm{HCO_3}}} \mathrm{HCO_3^- + BH^+} \tag{5.31}$$

$$K_{HCO_3} = \frac{c_{HCO_3} c_{BH}}{c_{CO_2} c_{H_2O} c_B} \tag{5.32}$$

In the case of solutions of a single monoamine, the base B is the amine Am. In the case of diamines that form a dicarbamate, or amine mixtures, B is the sum of all possible bases. The other important property in terms of capacity is the basicity of B as defined by the equilibrium constant K_B:

$$B + H^+ \overset{K_B}{\rightleftharpoons} BH^+ \tag{5.33}$$

$$K_B = \frac{c_{BH}}{c_B c_H} \tag{5.34}$$

For a solution of a single monoamine the equilibrium constants K_{AmCOO} ($dm^3\ mol^{-1}$) and K_{HCO_3} ($dm^3\ mol^{-1}$) are both linked to K_B via the following relationships. For diamines and mixtures the relationship is more complex:

$$K_{AmCOO} = \frac{c_{AmCOO} c_H}{c_{CO_2} c_{Am}} K_B \tag{5.35}$$

$$K_{HCO_3} = \frac{c_{HCO_3} c_H}{c_{CO_2} c_{H_2O}} K_B \tag{5.36}$$

The final important property is the enthalpy of CO_2 absorption, ΔH_{abs} ($kJ\ mol^{-1}$). This enthalpy is a combination of the proportion of CO_2 absorbed that becomes carbamate versus bicarbonate, and the enthalpies associated with the equilibrium constants K_{AmCOO} and K_{HCO_3}, ΔH_{AmCOO} ($kJ\ mol^{-1}$) and ΔH_{HCO_3} ($kJ\ mol^{-1}$), respectively. ΔH_{CO_2} is the enthalpy of CO_2 solvation (17.0 $kJ\ mol^{-1}$) derived from the Henry coefficient in Table 5.5.

$$\Delta H_{abs} = \left(\frac{c_{AmCOO}}{c_{CO_2,tot}}\right) \Delta H_{AmCOO} + \left(\frac{c_{HCO_3}}{c_{CO_2,tot}}\right) \Delta H_{HCO_3} + \Delta H_{CO_2} \tag{5.37}$$

where $c_{CO_2,tot} = c_{CO_2} + c_{CO_3} + c_{HCO_3} + c_{AmCOO}$.

Along with the absorption capacity, this enthalpy is a major factor in determining the energy efficiency of the capture process. The three main energy requirements during CO_2 desorption (q_{des}/kJ) are the heating of the absorbent to desorption temperature (q_{heat}/kJ g^{-1}), which is governed by the liquid circulation rate and heat capacity, water evaporation (q_{vap}/kJ g^{-1}), and the energy required to reverse the chemical reactions and release CO_2 (q_{abs}/kJ g^{-1}) [63]. Note that overall the physical solvation of CO_2 and its chemical reaction in the absorber is an exothermic process. The reverse process in the desorber is endothermic and requires energy input.

$$q_{des} = q_{heat} + q_{vap} + q_{abs} \tag{5.38}$$

The value of each of these terms depends on the process operating parameters but can be expressed simply as follows [63]:

$$q_{\text{heat}} = \frac{C_{\text{p}}(T_{\text{des}} - T_{\text{feed}})}{\Delta a} \frac{M_{\text{sol}}}{M_{CO_2}} \frac{1}{X_{\text{Am}}} \tag{5.39}$$

$$q_{\text{vap}} = \Delta H_{\text{vap}} \frac{p_{H_2O}}{p_{CO_2}} \frac{1}{M_{CO_2}} \tag{5.40}$$

$$q_{\text{abs}} = \frac{\Delta H_{\text{abs}}}{M_{CO_2}} \tag{5.41}$$

where C_p is the heat capacity of the solution (kJ g^{-1} K^{-1}), T_{des} is the desorber temperature (K), T_{feed} is the temperature of the liquid entering the desorber (K), Δa is the difference in the CO_2 loading between the absorber liquid inlet (lean) and outlet (rich), M_{sol} is the molecular weight of the solution (g mol^{-1}), M_{CO_2} is the molecular weight of CO_2 (g mol^{-1}), ΔH_{vap} is the enthalpy of water vaporization (kJ mol^{-1}), X_{Am} is the mole fraction of amine in the solution, p_{H_2O} is the vapor pressure of water at the desorber temperature (kPa), and p_{CO_2} is the vapor pressure of CO_2 at the desorber temperature.

Generally, for a given CO_2 removal efficiency, q_{des} is minimized by maximizing Δa and p_{CO_2}, which reduces q_{heat} and q_{vap}, respectively. With regard to the chemistry, these two terms increase with increase in ΔH_{abs}. Hence there is a correlation and associated trade-off between all three terms and there are also circumstances, such as when desorption is carried at low temperature and pressure, when reducing ΔH_{abs} is favorable to minimize q_{des} [63].

The parameters described above are listed in Table 5.6 for five commonly used amines, ammonia, and potassium carbonate. Determination of ΔH_{abs} according to Equation 5.37 requires determination of the enthalpies ΔH_{AmCOO} and ΔH_{HCO_3} using the van't Hoff equation:

$$\frac{\mathrm{d} \ln K_{\text{AmCOO/HCO}_3}}{\mathrm{d}\left(\frac{1}{T}\right)} = -\frac{\Delta H_{\text{AmCOO/HCO}_3}}{R} \tag{5.42}$$

The system of simultaneous algebraic equations that represent the mass balance and equilibrium constants of Equations 5.29–5.34 were then solved at 313 K using the Newton–Raphson method implemented in Matlab [42]. This yielded the speciation and from this the proportion of CO_2 as carbamate and bicarbonate.

If we first consider the sterically free primary amine MEA, it has the largest value of K_{AmCOO} ($\log K_{AmCOO} = 4.06$), which is consistent with the fact that it favors the formation of carbamate. This is in contrast to the other primary amine, 2-amino-2-methyl-1-propanol (AMP), which has steric hindrance around the amino functional group ($\log K_{AmCOO} = 1.52$). AMP essentially forms no stable carbamate. This is also reflected in the rate constant $k_{Am,1}$, which is 11 times smaller for AMP.

Table 5.6 Parameters defining the CO_2 capture performance of a range of commonly used primary (p), secondary (s), and tertiary (t) amines, ammonia, and potassium carbonate. All data are given at 313 K and a CO_2 partial pressure of 15 kPa. Some measured enthalpies of absorption are also listed for comparison [64–66].

Compound	Concentration (mol dm^{-3})	$k_{Am,1}$ (dm^3 mol^{-1} s^{-1})	Log K_{AmCOO}	Log K_{HCO_3}	Log K_B	$-\Delta H_{abs}$ (kJ mol^{-1})	Ref.
2-Amino-2-methyl-1-propanol (AMP, p)	3.0 (30% w/w)	1286	1.52	1.27	9.31	73.8 (73)	67, 68
Ammonia (NH_3)	1.5 (2.5% w/w)	1337	2.78	0.716	8.80	78.9 (78)	69, 70
Diethanolamine (DEA, s)	3.8 (40% w/w)	333	2.85	0.452	8.52	87.8	69, 71
Methyldiethanolamine (MDEA, t)	2.9 (35% w/w)	n/a	n/a	0.151	8.22	58.8 (58)	72
Monoethanolamine (MEA, p)	4.9 (30% w/w)	14270	4.06	0.991	9.07	83.6 (84)	46, 73
Piperazine (PZ, s)	1.0 (9% w/w)	114880 0.0019$^{d)}$	2.64 1.62$^{a, b)}$	1.35$^{b)}$	9.46 9.20$^{c)}$	59.3 (60)	74–76
Potassium carbonate (K_2CO_3)	1.5 (20% w/w)		n/a	2.14	10.2	27.2 (32)	47

a) The second equilibrium constant is defined in terms of the formation of the dicarbamate species $PZ(COO)_2^{2-}$ from the carbamate $PZCOO^-$.
$$K = \frac{c_{PZ(COO)_2} c_{BH}}{c_{CO_2} c_{PZCOO} c_B}.$$
b) In the case of PZ the base $B = PZ + PZCOO^-$.
c) The second equilibrium constant is defined in terms of the protonation of the PZ carbamate species. $K = \dfrac{c_{HPZCOO}}{c_{PZCOO} c_H}$.
d) Reaction rate constant for the reaction between CO_2 and H_2O (Equation 5.20), which dominates the reaction kinetics for potassium carbonate.

The CO_2 absorption rate is maximized by maximizing $k_{Am,1}$, and capacity is maximized by maximizing K_{HCO_3}. With this in mind, it is apparent why PZ has been found to perform well. It has the largest value for both $k_{Am,1}$ and K_{HCO_3} of the amines listed. The main issue with PZ is its solubility, which is limited to approximately 1.7 mol dm^{-3} (15% w/w). A correlation exists between the value of K_{HCO_3} and the amine basicity as expressed by K_B. The larger the value of K_B, the larger is the value of K_{HCO_3}. $k_{Am,1}$ loosely shows a similar trend with K_B, but with more variability. AMP and PZ have values of $k_{Am,1}$ much smaller and larger, respectively, than K_B alone would predict. This is because the reaction kinetics, and K_{AmCOO}, are strongly influenced by the accessibility of CO_2 to the amino group and its reactive lone pair of electrons. The electronic effects of substituents on the amino group have also been proposed to play a role in these differences in reactivity [77].

No single amine provides the ideal combination of fast absorption rate, large absorption capacity at low temperature, large CO_2 partial pressure at elevated temperature, and solubility. To address this issue, modern amine-based absorbents are often formulations of two or more amines. The ultimate goal is to combine the amines in such a way as to produce a formulation with all the desired properties. In terms of amine mixtures, the parameters in Table 5.6 provide a good guide as to which amines to mix. For example, PZ is a good choice in terms of absorption rate. Even though its solubility is limited, the fact that $k_{\mathrm{Am},1}$ is eight times larger than for MEA leads to higher absorption rates even though it is present at lower concentration. The optimal choice in terms of a more soluble amine to achieve a large absorption capacity (i.e., large value for $K_{\mathrm{HCO_3}}$), which also has a large ΔH_{abs} to achieve a high CO_2 partial pressure during desorption, is AMP. The mixture of PZ and AMP has proven in practice to be a very good solvent formulation leading to better absorption rates and reductions in energy requirement relative to MEA [78–80], the most commonly used amine for CO_2 capture.

Figure 5.10 is a plot of the concentration profiles of PZ and its reaction products in a thin liquid film determined by integrating Equation 5.18 for each chemical species according to the method and using the model given by Puxty and Rowland [59]. The conditions were a solution of 1 mol dm^{-3} PZ exposed to

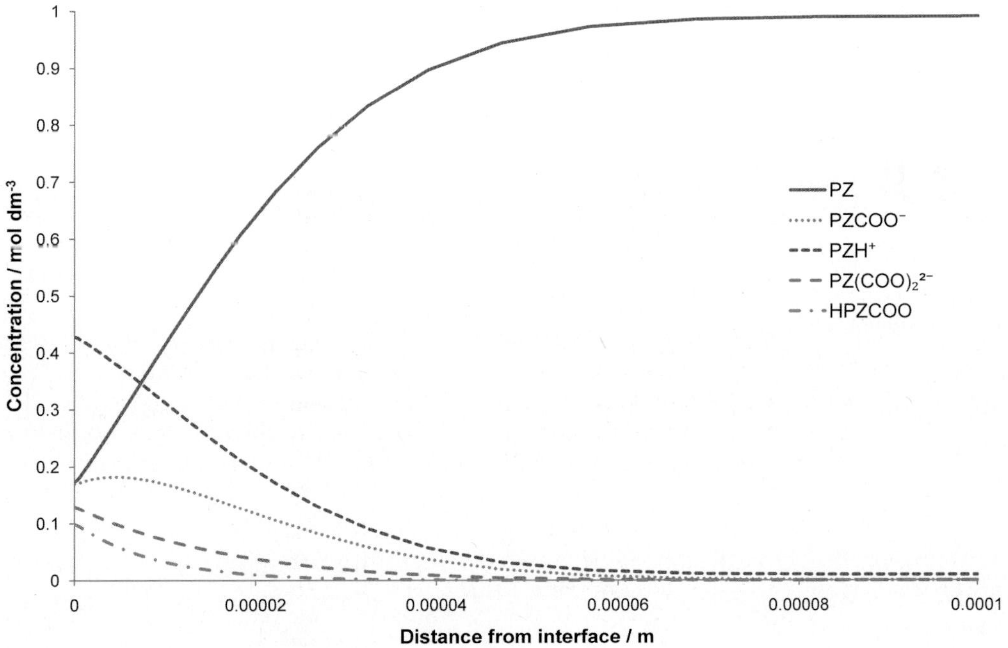

Fig. 5.10 Plot of the concentration profiles as a function of distance from the gas–liquid interface in a 1×10^{-4} m film of 1 mol dm^{-3} PZ exposed to 15 kPa CO_2 for 0.25 s at 313 K.

15 kPa CO_2 for 0.25 s at 313 K. Clearly, the concentration of PZ has fallen dramatically at the gas–liquid interface. With such a kinetically rapid amine under such conditions, the diffusion of PZ and $PZCOO^-$ becomes a major factor controlling mass transfer. The large value of K_{HCO_3} in Table 5.6 suggests that PZ should also form significant amounts of HCO_3^-, but it is absent from the plot, the reason being that within the 0.25 s timeframe the system is far from equilibrium and the formation of the PZ carbamates is kinetically favored.

Ammonia shares the same chemistry as amines in terms of CO_2 absorption, but has the benefit of not undergoing oxidative or thermal degradation [81–84]. The main challenge of using ammonia is that being a small molecule, it has a high vapor pressure [85]. At ambient temperature in aqueous solution it is a dissolved gas rather than a liquid. This presents additional process challenges, requiring absorption at reduced temperature and the use of slurries and/or gas-phase ammonia scrubbing to avoid excessive ammonia loss.

5.4.2.4 **Characteristics of Amino Acids**

Amino acids are a major subject of research investigation as an alternative to amines [86–90]. The chemical behavior of amino acids is essentially the same as that described for amines, with the main difference that amino acids when used as absorbents are present in solution as their overall negatively charged conjugate base. Neutral amino acids generally have a low solubility and exist in aqueous solution as zwitterions (overall neutral but carrying a positive and a negative charge on different atoms) with the amino group protonated and inactive as a base. The nominal and zwitterionic forms of glycine are shown in Fig. 5.11. Amino acids are ionized either by mixing the amino acid with a strong base such as sodium hydroxide or by using a salt of the amino acid. This increases their solubility and renders them non-volatile, eliminating issues around process emissions. Other benefits of amino acids are their stability towards oxidative and thermal degradation and low toxicity. A number of organizations, such as Siemens, are currently completing pilot-scale trials of amino acid-based capture processes [91].

5.4.2.5 **Hybrid Absorbents**

A number of commercial processes are based on hybrid absorbents consisting of a mixture of physical and chemical absorbents [5]. In terms of process perfor-

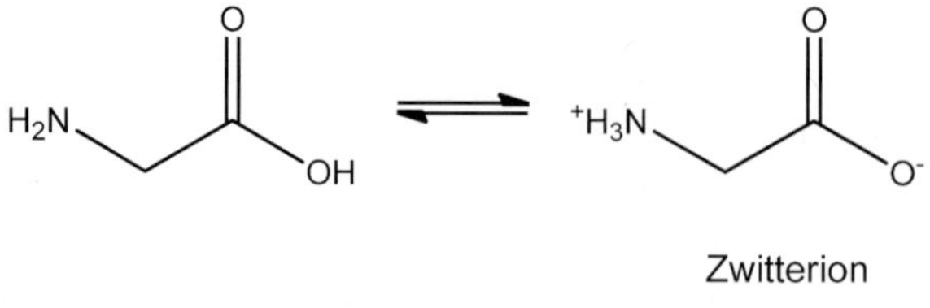

Fig. 5.11 Glycine in its nominal and zwitterionic forms.

mance they fall between a physical and a chemical absorbent. The physical absorbent allows bulk CO_2 removal from high-pressure gas streams while the presence of the chemical absorbent allows removal down to low CO_2 partial pressures. Hybrid absorbents are suited to high-pressure gas streams such as gasification with pre-combustion capture, where it is necessary to achieve low levels of CO_2 emissions. They are not suited to post-combustion capture with low-pressure flue gas due to little physical absorption occurring, making the physical absorbent component of no benefit. Commercial hybrid absorbents include the Sulfinol process (Shell Oil), which uses tetrahydrothiophene dioxide (sulfolane) and diisopropanolamine (DIPA) or *N*-methyldiethanolamine (MDEA) the Amisol process (Lurgi), which employs methanol as the physical absorbent with either MEA, DEA, diisopropylamine (DIPAM) or diethylamine (DETA), and the Selefining process (Snamprogetti), which uses an undisclosed physical absorbent mixed with a tertiary amine.

5.4.2.6 **Absorbent Stability and Corrosion**

As CO_2 capture moves from laboratory and pilot-scale research to demonstration, understanding the long-term stability of absorbents is becoming increasingly important. Stability in this context refers to the following properties: thermal stability, particularly under desorption conditions (up to ~433 K); oxidative stability, as power station flue gas contains around 5% residual oxygen; and volatility of the absorbent and degradation products and any associated environmental and health impacts of process emissions. Stability of and corrosion caused by physical absorbents are generally not an issue and are not discussed here.

As MEA has been in use for some decades, its stability has been studied in most detail, although even for this amine understanding of what are chemically complex degradation processes is limited. Thermal degradation of MEA has been shown to be a function of concentration, CO_2 content, temperature, and time. The major degradation products under desorption conditions have been identified as 2-oxazolidone, *N,N′*-di(2-hydroxyethyl)urea, 1-(2-hydroxyethyl)-2-imidazolidone, and *N*-(2-hydroxyethyl)ethylenediamine [92]. These compounds form via a combination of cyclization and dimerization reactions. Amines that constitute more recently developed absorbents, such as PZ and AMP, show much greater thermal stability than MEA [93]. Thermal degradation products are generally larger non-volatile molecules and do not pose a significant emission risk.

Interest in oxidative degradation is twofold: from the perspective of absorbent stability and ongoing cost, and also more importantly from the perspective of oxidation products that may form a part of the process emissions. A study of the components in the reclaimer of an operating MEA-based facility capturing CO_2 from coal-fired flue gas found carboxylic acids and ammonia [94] to be the dominant oxidation products. A laboratory study found formate, hydroxyethylformamide, and hydroxyethylimidazole as the dominant products [95]. However, there was significant variability depending upon the metal ions present in solution and the reaction conditions. Developing a thorough understanding of oxidation pathways and extent is an ongoing research challenge.

The process emissions and any potential environmental and health effects associated with them have recently become important research topics as the first commercial-scale CO_2 capture demonstration projects move closer to reality. Part of this interest has been sparked by the detection of very low concentrations of nitrosamines in lean MEA samples from an operating CO_2 capture facility [94]. Nitrosamines may form from the reaction between amines and nitrogen oxides and are carcinogenic. A significant number of assessments of possible environmental and health impacts are yet to be completed. However, one of the first indicates that the most significant source of environmental and health impacts will be from emission of the absorbent itself [96]. Impacts of the emission of degradation products were found to be negligible.

Pure aqueous amine solution are generally non-corrosive [5]. The effect of CO_2 absorption is to increase the ion concentration and thus the conductivity and reduce the pH, resulting in a more corrosive solution. This is due to corrosion being a pH-dependent redox process. MEA concentrations are typically limited to 3–5 mol dm^{-3} in flue gas applications due to corrosion concerns even when corrosion inhibitors are used [94]. The ability of amines to chelate metal ions in solution also contributes to corrosion. It is also thought that some degradation products may increase corrosion by further acidifying the absorbent or acting as metal chelating agents.

It should also be mentioned that the presence of other acid gases, such as sulfur and nitrogen oxides, in flue gas is a potential issue. Although in the United States and Europe practically all power stations have flue gas desulfurization and denitrification units installed, some less populous regions such as Australia do not. Chemical absorbents such as aqueous amines that react with CO_2 via acid–base reactions will also react with other acid gases. NO_2 is not a major issue as it is present in flue gas at ppm concentrations. The solubility of NO_2 in aqueous solution is significantly less than that of either CO_2 or SO_2 [97], with only 100–300 ppm of nitrate/nitrite being detected in an MEA solution operating with a coal-fired flue gas [94]. SO_2, the main sulfur oxide, is more soluble than CO_2 and NO_2 in aqueous solution and is a stronger acid than CO_2 [50]. Furthermore the reaction between SO_2 and water is more rapid than that between CO_2 and an amine [98]. Hence the accumulation of sulfates/sulfites will occur [94] even when SO_2 is present in flue gas at only a few ppm.

5.5 Physical Properties

The single most important physical property common to physical and chemical absorption is the diffusion coefficient of CO_2 ($D_{CO_2}/m^2\,s^{-1}$). In the case of physical absorption, the value of D_{CO_2} defines the magnitude of the mass transfer coefficient. During chemical absorption, the rate of chemical reaction and the diffusion of chemically reactive species to the gas–liquid interface also play a role.

For physical absorbents the value of D_{CO_2} can be directly determined from mass transfer measurements using an apparatus such as a wetted-wall reactor. For chemical absorbents it cannot be measured directly because CO_2 mass transfer is influenced by both diffusion and reaction. To determine D_{CO_2} the diffusion coefficient of an unreactive gas, N_2O, is measured and related to D_{CO_2} by the N_2O analogy. The N_2O analogy states that the ratio of diffusion coefficients between N_2O and CO_2 in water ($D_{N_2O,\,H_2O}$ and $D_{CO_2,\,H_2O}$ [32, 99]) is equal to that in an aqueous amine solution ($D_{N_2O,\,Am}$ and $D_{CO_2,\,Am}$). $D_{N_2O,\,Am}$ can be measured directly or estimated using a correlation such as that given by Cullinane and Rochelle [100] in which $D_{N_2O,\,Am}$ is related to the viscosity of an amine-based absorbent (μ_{Am}/cP). The following four equations show the calculation of the diffusion coefficients:

$$D_{CO_2,\,H_2O} = 1.81 \times 10^{-6} \exp\left(\frac{-2032.7}{T}\right) \tag{5.43}$$

$$D_{N_2O,\,H_2O} = 5.07 \times 10^{-6} \exp\left(\frac{-2371}{T}\right) \tag{5.44}$$

$$D_{N_2O,\,Am} = 4.4974 \times 10^{-11} \exp\left(\frac{T}{\mu_{Am}^{0.545}}\right) \tag{5.45}$$

$$D_{CO_2,\,Am} = D_{N_2O,\,Am} \frac{D_{CO_2,\,H_2O}}{D_{N_2O,\,H_2O}} \tag{5.46}$$

If Equation 5.18 is used to describe absorption by chemical reaction, the diffusion coefficients of all chemical species are required. Their values can be measured but are often estimated using the Wilkie–Chang correlation [33]. This correlation relates the diffusion coefficient of a chemical species at infinite dilution ($D_{i,\infty}$) to the molecular volume ($V_m/cm^3\,mol^{-1}$) and viscosity of water (μ_{H_2O}). The diffusion coefficient in the absorbent is then given by the ratio of μ_{H_2O} to the viscosity of the absorbent:

$$D_{Am,\infty} = \frac{7.4 \times 10^{-12} T \sqrt{2.6 \times 18.02}}{V_m \mu_{H_2O}} \tag{5.47}$$

$$D_i = D_{i,\infty} \left(\frac{\mu_{H_2O}}{\mu_{Am}}\right)^{0.6} \tag{5.48}$$

5.6 Outlook

As implied by the breadth and depth of information presented above, CO_2 capture using a physical absorbent or an aqueous amine/aqueous carbonate/ammonia chemical absorbent is reasonably well understood. The chemistry has been characterized to the extent necessary, and the engineering processes are well understood and have been extensively used in practice. The next step is for the technology in this form to be demonstrated on a commercial scale on-site at a coal- or gas-fired power station as part of a capture and storage chain. It is only through this scale-up and practical operation that the technology can be proven and significant additional learning be achieved rapidly to drive further improvement and large-scale deployment.

This is not to imply that deployment would be the final chapter, as the additional generation costs for CO_2-capture continue to provide a formidable challenge for further deployment. Potentially game-changing liquid absorbent-based capture technologies are currently under development and it is hoped that some of these will form the next generation of more efficient, safer, and lower cost capture processes. These include:

- *Room temperature ionic liquids.* These are salt solutions that are liquids at ambient temperature. Many ionic liquids that are potential CO_2 absorbents are based on imidazolium salts [101, 102]. The major advantage of ionic liquids is that they have no vapor pressure and therefore no process emissions. They are also thermally stable and the properties of the ionic liquid can be tuned to provide a physical, chemical, or hybrid-type interaction with CO_2 [103].
- *Phase change absorbents.* These are absorbents that undergo a phase separation upon absorption of CO_2 [104, 105]. One phase contains chemically absorbed CO_2 and can be separated to give a concentrated CO_2 phase for regeneration. If we refer back to Equations 5.38–5.41, this will potentially lead to an increase in $\Delta\alpha$, the difference in the lean-rich CO_2 loading, and the CO_2 partial pressure in the desorber, resulting in a reduction in energy requirement.
- *Enzyme-promoted processes.* Enzymes such as carbonic anhydrase that facilitate CO_2 exchange between a gas and a liquid in living systems can also be used for CO_2 capture [106–109]. As is the role of enzymes, they have the potential to increase greatly the rate of CO_2 hydration in aqueous solution and thus mass transfer, resulting in significant size reductions of absorber and desorber columns and their associated costs. They are also being used in capture processes leading to solid carbonate products.
- *Hyperbranched polymers.* These polymers are highly branched and dispersed macromolecules with tree-like structures. They have the potential to be used as physical absorbents with greater absorption capacity than those currently in use [110]. Through modulation of their properties, it is hoped to produce hyperbranched polymers that have a large enough capacity at low CO_2 partial pressures to make then suitable for post-combustion capture, while retaining the low energy regeneration of a physical absorbent.

CO_2 capture and storage from power stations is currently at a crucial stage. Current state-of-the-art processes are ready for demonstration and then deployment. To maximize the impact of this technology on the mitigation of greenhouse gas emissions and subsequent climate effects, the sooner deployment can be achieved the better. However, as this is a pollution control technology, there is still a strong imperative for further research to improve the process efficiency, reduce the costs, and eliminate degradation and harmful process emissions. Such improvements would then form the basis of next-generation capture processes that will allow capture to occur with a smaller parasitic energy demand, lower capital and operating costs, and a smaller environmental footprint.

Acknowledgments

The authors acknowledge the support given through CSIRO's Advanced Coal Technology portfolio and the financial contribution from ANLEC R&D in the preparation of this chapter.

Symbols and Nomenclature

a	Effective surface area of gas–liquid contact ($m^2\ m^{-3}$)
$\Delta\alpha$	Difference in molar CO_2 loading (α=moles of CO_2/moles of amine) between the absorber liquid inlet and outlet
$CO_{2(g)}$	Gas-phase CO_2
$CO_{2(soln)}$	CO_2 dissolved in solution
c_{CO_2}	Dissolved CO_2 concentration in the liquid bulk ($mol\ dm^{-3}$)
c_i	Liquid-phase concentration of chemical species i ($mol\ dm^{-3}$)
$c^*_{CO_2}$	Dissolved CO_2 concentration at the gas–liquid interface ($mol\ dm^{-3}$)
D_{CO_2}	Diffusion coefficient of CO_2 ($m^2\ s^{-1}$)
$D_{CO_2,Am}$	Diffusion coefficient of CO_2 in an aqueous amine ($m^2\ s^{-1}$)
D_{CO_2,H_2O}	Diffusion coefficient of CO_2 in water ($m^2\ s^{-1}$)
$D_{N_2O,Am}$	Diffusion coefficient of N_2O in an aqueous amine ($m^2\ s^{-1}$)
D_{N_2O,H_2O}	Diffusion coefficient of N_2O in water ($m^2\ s^{-1}$)
D_i	Diffusion coefficient of chemical species i ($m^2\ s^{-1}$)
$D_{i,\infty}$	Diffusion coefficient of chemical species i at infinite dilution ($m^2\ s^{-1}$)
G	Gibbs free energy (J)
H	Total enthalpy (J)
ΔH_{CO_2}	Enthalpy of CO_2 solvation ($kJ\ mol^{-1}$)
ΔH_{abs}	Enthalpy of CO_2 absorption ($kJ\ mol^{-1}$)
ΔH_{AmCOO}	Enthalpy of CO_2 carbamate formation ($kJ\ mol^{-1}$)
ΔH_{HCO_3}	Enthalpy of CO_2 bicarbonate formation ($kJ\ mol^{-1}$)
ΔH_{vap}	Enthalpy of water vaporization ($kJ\ mol^{-1}$)
M_{CO_2}	Molecular weight of CO_2 (44.01 $g\ mol^{-1}$)
M_{sol}	Average molecular weight of a solution ($g\ mol^{-1}$)

N_{CO_2}	CO_2 flux across the gas–liquid interface (mol s^{-1} m^{-2})
$k_{Am,1}$	Reaction rate constant to form carbamate intermediate (dm^3 mol^{-1} s^{-1})
$k_{Am,-1}$	Reaction rate constant for dissociation of carbamate intermediate (s^{-1})
$k_{Am,2}$	Reaction rate constant of proton abstraction from carbamate intermediate (dm^3 mol^{-1} s^{-1})
$k_{Am,-2}$	Reaction rate constant of proton abstraction from base by carbamate intermediate (dm^3 mol^{-1} s^{-1})
K_{AmCOO}	Overall equilibrium constant of carbamate formation (dm^3 mol^{-1})
K_B	Base protonation constant (dm^3 mol^{-1})
K_{HCO_3}	Overall equilibrium constant of bicarbonate formation (dm^3 mol^{-1})
k_i	Reaction rate constant of reaction i
K_i	Equilibrium constant of reaction i
k_l	Liquid-phase mass transfer coefficient (m s^{-1})
k_g	Gas-phase mass transfer coefficient (mol s^{-1} kPa^{-1} m^{-2})
K_g	Overall mass transfer coefficient in gas units (mol s^{-1} kPa^{-1} m^{-2})
K_l	Overall mass transfer coefficient in liquid units (m s^{-1})
k_{H,CO_2}	Henry coefficient of CO_2 solubility (MPa or kPa dm^3 mol^{-1})
n	Number of moles
p_{CO_2}	$CO_{2(g)}$ bulk partial pressure (kPa)
$p^*_{CO_2}$	$CO_{2(g)}$ partial pressure at the gas–liquid interface (kPa)
p_{H_2O}	Vapor pressure of water (kPa)
q_{abs}	Energy consumed to overcome the enthalpy of absorption during desorption (kJ g^{-1})
q_{des}	Total energy consumed for CO_2 desorption (kJ g^{-1})
q_{heat}	Energy consumed to heat absorbent to desorption temperature (kJ g^{-1})
q_{vap}	Energy consumed by water evaporation during desorption (kJ g^{-1})
R	Ideal gas constant, 8.314 J mol^{-1} K^{-1}
S	Total entropy (J K^{-1})
T	Temperature (K)
T_{des}	Temperature of desorption (K)
T_{feed}	Temperature of liquid feed into desorber (K)
μ_{Am}	Viscosity of an aqueous amine (cP)
μ_{H_2O}	Viscosity of water (cP)
V_m	Molecular volume (cm^3 mol^{-1})
W_{min}	Thermodynamic minimum energy requirement to separate CO_2 from a binary mixture
x	Distance from the gas–liquid interface (m)
x_f	Liquid film thickness (m)
X_{Am}	Mole fraction of amine
X_{CO_2}	Mole fraction of CO_2

References

1 IPCC. *IPCC Special Report on Carbon Dioxide Capture and Storage*, Cambridge University Press, Cambridge, 2005.

2 Feron, P. H. M. and Hendriks, C. A. CO_2 capture process principles and costs. *Oil Gas Sci. Technol. Rev. Instit. Fr. Pet.*, 2005, **60**(3), 451–459.

3 Rushing, S. A. CO_2 recovery provides environmental and economical benefits to power and combustion projects. *UTA Int.*, 1999, **3**, 8–13.

4 Thomsen, H. B. and Geleff, S. *Recovery of CO_2 from Exhaust Gases – Another Way to Improve the Economical and Environmental Aspects of CHP Plants*, Union Engineering, Fredericia, Denmark, 2002, www.union.dk/media/ICCI_2002.pdf, last accessed Dec. 6th, 2010.

5 Kohl, A. and Nielson, R. *Gas Purification*, 5th edn., Gulf Professional Publishing, Houston, TX, 1997.

6 Feron, P. H. M. Exploring the potential for improvement of the energy performance of coal fired power plants with post-combustion capture of carbon dioxide. *Int. J. Greenhouse Gas Control*, 2010, **4**(2), 152–160.

7 Amel'kin, S. A., Burtzler, J. M., Hoffmann, K. H., and Tsirlin, A. M. Evaluating the efficiency frontier of separation processes. *Theor. Found. Chem. Eng.*, 2001, **35**(3), 217–223

8 Span, R. and Wagner, W. A new equation of state for carbon dioxide covering the fluid region from the triple-point temperature to 1100 K at pressures up to 800 MPa. *J. Phys. Chem. Ref. Data*, 1996, **25**(6), 1509–1596.

9 Bolland, O. and Undrum, H. A novel methodology for comparing CO_2 capture options for natural gas-fired combined cycle plants. *Adv. Environ. Res.*, 2003, **7**(4), 901–911.

10 Goetlicher, G. *The Energetics of Carbon Dioxide Capture in Power Plants*, National Energy Technology Laboratory, Pittsburgh, PA, 2004.

11 Davison, J. Performance and costs of power plants with capture and storage of CO_2. *Energy*, 2007, **32**(7), 1163–1176.

12 Grol, E. *Assessment of Power Plants that Meet Proposed Greenhouse Gas Emission Performance Standards*, DOE/NETL-401/110509, National Energy Technology Laboratory, Pittsburgh, PA, 2009.

13 Kanniche, M., Gros-Bonnivard, R., Jaud, P., Valle-Marcos, J., Amann, J. M., and Bouallou, C. Pre-combustion, post-combustion and oxy-combustion in thermal power plant for CO_2 capture. *Appl. Thermal Eng.*, 2010, **30**(1), 53–62.

14 Metz, B., Davidson, O., de Coninck, H., Loos, M., and Meyer, L. *IPCC Special Report on Carbon Dioxide Capture and Storage*, Cambridge University Press, Cambridge, 2005.

15 Rao, A. B. and Rubin, E. S. A technical, economic, and environmental assessment of amine-based CO_2 capture technology for power plant greenhouse gas control. *Environ. Sci. Technol.*, 2002, **36**(20), 4467–4475.

16 Aroonwilas, A. and Veawab, A. Heat recovery gas absorption process. World Patent WO 2007/07004 A1, 2007.

17 Shethna, H. K. and Towler, G. P. Design of mixed-solvent processes for chemisorption with ultrahigh recovery. *Ind. Eng. Chem. Res.*, 1997, **36**(12), 5307–5320.

18 Jassim, M. S. and Rochelle, G. T. Innovative absorber/stripper configurations for CO_2 capture by aqueous mono ethanolamine. *Ind. Eng. Chem. Res.*, 2006, **45**(8), 2465–2472.

19 Oyenekan, B. A. and Rochelle, G. T. Alternative stripper configurations for CO_2 capture by aqueous amines. *AIChE J.*, 2007, **53**(12), 3144–3154.

20 Cousins, A., Wardhaugh, L. T. and Feron, P. H. M. Preliminary analysis of process flow sheet modifications for energy efficient CO_2 capture from flue gases using chemical absorption, In *Distillation Absorption 2010, Eindhoven, The Netherlands* (ed. de Haan, A. B., Kooijman, H., and Górak, A.), 2010.

21 Kister, Z. H. *Distillation Design*, McGraw-Hill, New York, 1992.

22 Perry, R.H. and Green, D. *Perry's Chemical Engineer's Handbook*, 6th edn., McGraw-Hill, New York, 1984.

23 Coulson, J.M., Richardson, J.F., and Sinott, R.K. *Chemical Engineering. Vol. 6. An Introduction to Chemical Engineering Design*. Pergamon Press. Oxford, 1983.

24 Strigle, R.F. *Packed Tower Design and Applications*, 2nd edn., Gulf Publishing, Houston, TX, 1994.

25 Kuntz, J. and Aroonwilas, A. Performance of spray column for CO_2 capture application. *Ind. Eng. Chem. Res.*, 2008, **47**(1), 145–153.

26 Beek, W.J., Muttzall, K.M.K., and van Heuven, J.W. *Transport Phenomena*, 2nd edn., John Wiley & Sons Ltd., Chichester, 1999.

27 Bratzler, K. and Doerges, A. Gasreinigung and Gastrennung durch Absorption. In *Ullmann Encykopaedie der Technische Chemie*, 4 edn., Verlag Chemie, Weinheim, 1972, pp 575–599, by Barthholomé, E. *et al.*

28 Laurent, A. and Charpentier, J.C. Aires Interfaciales et Coefficients de Transfert de Matière dans les Divers Type d'Absorbeurs et de Reacteurs Gaz–Liquide. *Chem. Eng. J.*, 1974, **8**, 85–101.

29 Lee, S.Y. and Tsui, Y.P. Succeed at gas/liquid contacting. *Chem. Eng. Prog.*, 1999, **95**(7), 23–49.

30 Danckwerts, P.V. *Gas–Liquid Reactions*, McGraw-Hill, New York, 1970.

31 Cussler, E.L. *Diffusion – Mass Transfer in Fluid Systems*, 3rd edn., Cambridge University Press, Cambridge, 2009.

32 Frank, M.J.W., Kuipers, J.A.M., and van Swaaij, W.P.M. Diffusion coefficients and viscosities of CO_2+H_2O, CO_2+-CH_3OH, NH_3+H_2O, and NH_3+CH_3OH liquid mixtures. *J. Chem. Eng. Data*, 1996, **41**, 297–302.

33 Wilkie, C.R. and Chang, P. Correlation of diffusion coefficients in dilute solutions. *AIChE J.*, 1955, **1**, 264–270.

34 Xia, J.Z., Jodecke, M., Kamps, A.P.S., and Maurer, G. Solubility of CO_2 in (CH_3OH + H_2O). *J. Chem. Eng. Data*, 2004, **49**(6), 1756–1759.

35 van Swaaij, W.P.M. and Versteeg, G.F. Mass-transfer accompanied with complex reversible chemical-reactions in gas-liquid systems – an overview. *Chem. Eng. Sci.*, 1992, **47**(13–14), 3181–3195.

36 Gwinner, B., Roizard, D., Lapicque, F., Favre, E., Cadours, R., Boucot, P., and Carrette, P.L. CO_2 capture in flue gas: semiempirical approach to select a potential physical solvent. *Ind. Eng. Chem. Res.*, 2006, **45**(14), 5044–5049.

37 Xu, Y.M., Schutte, R.P., and Hepler, L.G. Solubilities of carbon dioxide, hydrogen sulfide and sulfur dioxide in physical solvents. *Can. J. Chem. Eng.*, 1992, **70**(3), 569–573.

38 Melzer, W.M., Schrodter, F., and Knapp, H. Solubilities of methane, propane and carbon dioxide in solvent mixtures consisting of water, *N,N*-dimethylformamide, and *N*-methyl-2-pyrrolidone. *Fluid Phase Equil.*, 1989, **49**, 167–186.

39 Miyano, Y., and Fujihara, I. Henry's constants of carbon dioxide in methanol at 250–500 K. *Fluid Phase Equil.* 2004, **221**(1–2), 57–62.

40 Astarita, G. *Mass Transfer with Chemical Reaction*, Elsevier, Amsterdam, 1967.

41 Astarita, G., Savage, D.W., and Bisio, A. *Gas Treating with Chemical Solvents*, John Wiley & Sons, Inc., New York, 1983.

42 Maeder, M., and Neuhold, Y.-M. *Practical Data Analysis in Chemistry*, Elsevier, Amsterdam, 2007.

43 Maeder, M., Neuhold, Y.M., Puxty, G., and King, P. Analysis of reactions in aqueous solution at non-constant pH: no more buffers? *Phys. Chem. Chem. Phys.*, 2003, **5**(13), 2836–2841.

44 Wilkins, R.G. *Kinetics and Mechanism of Reactions of Transition Metal Complexes*, 2nd edn., VCH Verlag GmbH, Weinheim, 2002.

45 McCann, N., Phan, D., Attalla, M., Puxty, G., Fernandes, D., Conway, W., Wang, X.G., Burns, R., van Altena, I., Lawrance, G., and Maeder, M. Molecular interactions between amine and carbonate species in aqueous solution – kinetics and thermodynamics. *Energy Procedia*, 2009, **1**(1), 995–1002.

46 Versteeg, G.F., van Dijck, L.A.J., and van Swaaij, W.P.M. On the kinetics between CO_2 and alkanolamines both in aqueous and non-aqueous solutions. An

overview. *Chem. Eng. Commun.*, 1996, **144**, 113–158.
47 Soli, A.L. and Byrne, R.H. CO_2 system hydration and dehydration kinetics and the equilibrium CO_2/H_2CO_3 ratio in aqueous NaCl solution. *Mar. Chem.*, 2002, **78**, 65–73.
48 Kucka, L., Müller, I., Kenig, E.Y., and Górak, A. On the modelling and simulation of sour gas absorption by aqueous amine solutions. *Chem. Eng. Sci.*, 2003, **58**, 3571–3578.
49 Harned, H.S., Samuel, R., and Scholes, J. The ionization constant of HCO_3^- from 0 to 50 °C. *J. Am. Chem. Soc.*, 1941, **63**, 1706–1709.
50 Edwards, T.J., Maurer, G., Newman, J., and Prausnitz, J.M. Vapour–liquid equilibria in multicomponent aqueous solutions of volatile weak electrolytes. *AIChE J.*, 1978, **24**, 966–976.
51 Knuutila, H., Juliussen, O., and Svendsen, H.F. Kinetics of the reaction of carbon dioxide with aqueous sodium and potassium carbonate solutions. *Chem. Eng. Sci.*, 2010, **65**, 6077–6088.
52 Meldon, J.H., Smith, K.A., and Colton, C.K. Effect of weak acids upon transport of carbon dioxide in alkaline solutions. *Chem. Eng. Sci.*, 1977, **32**(8), 939–950.
53 Astarita, G., Savage, D.W., and Longo, J.M. Promotion of CO_2 mass-transfer in carbonate solutions. *Chem. Eng. Sci.*, 1981, **36**(3), 581–588.
54 Cullinane, J.T. and Rochelle, G.T. Carbon dioxide absorption with aqueous potassium carbonate promoted by piperazine. *Chem. Eng. Sci.*, 2004, **59**(17), 3619–3630.
55 Tseng, P.C., Ho, W.S., and Savage, D.W. Carbon dioxide absorption into promoted carbonate solutions. *AIChE J.*, 1988, **34**(6), 922–931.
56 Fox, M.A. and Whitesell, J.K. *Organic Chemistry*, Jones and Bartlett Publishers, Boston, 1994.
57 Atkins, P.W. *Physical Chemistry*, 5th edn., Oxford University Press, Oxford, 1995.
58 Mohammed, O.F., Adamczyk, K., Pines, D., Pines, E., and Nibbering, E.T.J. Aqueous proton transfer pathways in bimolecular acid–base neutralization. In *Ultrafast Phenomena XVI* (ed. Corkum, P., Silvestri, S.D., Nelson, K.A., Riedle, E., and Schoenlein, R.), Springer, New York, 2009, 622–624.
59 Puxty, G. and Rowland, R. Modelling CO_2 mass transfer in amine mixtures – PZ-AMP and PZ-MDEA. *Environ. Sci. Technol.*, 2011, doi: 10.1021/es1022784.
60 Puxty, G., Rowland, R., Allport, A., Yang, Q., Bown, M., Burns, R., Maeder, M., and Attalla, M. Carbon dioxide postcombustion capture: a novel screening study of the carbon dioxide absorption performance of 76 amines. *Environ. Sci. Technol.*, 2009, **43**(16), 6427–6433.
61 Singh, P. and Versteeg, G.F. Structure and activity relationships for CO_2 regeneration from aqueous amine-based absorbents. *Process Safety Environ. Protect.*, 2008, **86**(B5), 347–359.
62 Tomizaki, K., Shimizu, S., Onoda, M., and Fujioka, Y. An acid dissociation constant (pK(a))-based screening of chemical absorbents that preferably capture and release pressurized carbon dioxide for greenhouse gas control. *Chem. Lett.*, 2008, **37**(5), 516–517.
63 Oexmann, J. and Kather, A. Minimising the regeneration heat duty of post-combustion CO_2 capture by wet chemical absorption: the misguided focus on low heat of absorption solvents. *Int. J. Greenhouse Gas Control*, 2010, **4**, 36–43.
64 Kim, I. and Svendsen, H.F. Comparative study of the heats of absorption of post-combustion CO_2 absorbents. *Int. J. Greenhouse Gas Control*, 2010, doi:10.1016/j.ijggc.2010.05.003.
65 Qin, F., Wang, S.J., Kim, I., Svendsen, H.F., and Chen, C.H. Study of the heat of absorption of CO_2 in aqueous ammonia: comparison between experimental data and model predictions. *Ind. Eng. Chem. Res.*, 2010, **49**(8), 3776–3784.
66 Tosh, J.S., Field, J.H., Bensen, H.E., and Haynes, W.P. *Equlibrium Study of the System Potassium Carbonate, Potassium Bicarbonate, Carbon Dioxide, and Water*, Report 5484, United States Bureau of Mines, Washington, DC, 1959.
67 Saha, A.K., Bandyopadhyay, S.S., and Biswas, A.K. Kinetics of absorption of CO_2 into aqueous-solutions of

2-amino-2-methyl-1-propanol. *Chem. Eng. Sci.*, 1995, **50**(22), 3587–3598.

68 Silkenbaumer, D., Rumpf, B., and Lichtenthaler, R.N. Solubility of carbon dioxide in aqueous solutions of 2-amino-2-methyl-1-propanol and *N*-methyldiethanolamine and their mixtures in the temperature range from 313 to 353 K and pressures up to 2.7 MPa. *Ind. Eng. Chem. Res.*, 1998, **37**(8), 3133–3141.

69 NIST. *NIST Critically Selected Stability Constants of Metal Complexes*, National Institute of Standards and Technology, Gaithersburg, MD, 2001.

70 Puxty, G., Rowland, R., and Attalla, M. Comparison of the rate of CO_2 absorption into aqueous ammonia and monoethanolamine. *Chem. Eng. Sci.*, 2010, **65**(2), 915–922.

71 Barth, D., Tondre, C., and Delpuech, J.J. Stopped-flow investigations of the reaction kinetics of carbon dioxide with some primary and secondary alkanolamines in aqueous solutions. *Int. J. Chem. Kinet.*, 1986, **18**(4), 445–457.

72 Bishnoi, S. and Rochelle, G.T. Thermodynamics of piperazine/methyldiethanolamine/water/carbon dioxide. *Ind. Eng. Chem. Res.*, 2002, **41**(3), 604–612.

73 Bates, R.G. and Pinching, G.D. Acidic Dissociation constant and related thermodynamic quantities for monoethanolammonium ion in water from 0-degrees-C to 50-degrees-C. *J. Res. Natl. Bur. Stand.*, 1951, **46**(5), 349–352.

74 Bishnoi, S. and Rochelle, G.T. Absorption of carbon dioxide into aqueous piperazine: reaction kinetics, mass transfer and solubility. *Chem. Eng. Sci.*, 2000, **55**, 5531–5543.

75 Ermatchkov, V., Kamps, A. P.-S., and Maurer, G. Chemical equilibrium constants for the formation of carbamates in (carbon dioxide + piperazine + water) from ^{1}H-NMR spectroscopy. *J. Chem. Thermodyn.* 2003, **35**, 1277–1289.

76 Samanta, A. and Bandyopadhyay, S.S. Kinetics and modeling of carbon dioxide absorption into aqueous solutions of piperazine. *Chem. Eng. Sci.*, 2007, **62**, 7312–7319.

77 Chakraborty, A.K., Bischoff, K.B., Astarita, G., and Damewood, J.R. Molecular-orbital approach to substituent effects in amine–CO_2 interactions. *J. Am. Chem. Soc.*, 1988, **110**(21), 6947–6954.

78 Aronu, U.E., Svendsen, H.F., Hoff, K.A., and Juliussen, O. Solvent selection for carbon dioxide absorption. *Energy Procedia*, 2009, **1**(1), 1051–1057.

79 Inoue, Y., Taura, M., Yoshiyama, T., Oishi, T., Iijima, M., Mimura, T., Yagi, Y., Yoshiyama, R., Tanoura, M., Iljima, M., and Minura, T. Absorption liquid used in carbon dioxide or hydrogen sulfide removal apparatus comprises linear or cyclic amine compound with primary amino group, and linear or cyclic amine compounds with secondary amino group. Patents JP 2008307519-A, WO 2008156085-A1, CA 2692154-A1, EP 2189207-A1, and US 2010180764-A1, 2009.

80 Samanta, A. and Bandyopadhyay, S.S. Absorption of carbon dioxide into aqueous solutions of piperazine activated 2-amino-2-methyl-1-propanol. *Chem. Eng. Sci.*, 2009, **64**(6), 1185–1194.

81 Diao, Y.F., Zheng, X.Y., He, B.S., Chen, C.H., and Xu, X.C. Experimental study on capturing CO_2 greenhouse gas by ammonia scrubbing. *Energy Conversion Manage.*, 2004, **45**(13–14), 2283–2296.

82 Kim, Y.J., You, J.K., Hong, W.H., Yi, K.B., Ko, C.H., and Kim, J.N. Characteristics of CO_2 absorption into aqueous ammonia. *Sep. Sci. Technol.*, 2008, **43**(4), 766–777.

83 Yeh, A.C. and Bai, H.L. Comparison of ammonia and monoethanolamine solvents to reduce CO_2 greenhouse gas emissions. *Sci. Total Environ.*, 1999, **228**(2–3), 121–133.

84 Yeh, J.T., Resnik, K.P., Rygle, K., and Pennline, H.W. Semi-batch absorption and regeneration studies for CO_2 capture by aqueous ammonia. *Fuel Process. Technol.*, 2005, **86**(14–15), 1533–1546.

85 Edwards, T.J., Newman, J., and Prausnitz, J.M. Thermodynamics of vapor–liquid equilibria for ammonia–

water system. *Ind. Eng. Chem. Fundam,* 1978, **17**(4), 264–269.

86 Hamborg, E. S., Niederer, J. P. M., and Versteeg, G. F. Dissociation constants and thermodynamic properties of amino acids used in CO_2 absorption from 293 to 353 K. *J. Chem. Eng. Data,* 2007, **52**(6), 2491–2502.

87 Hook, R. J. An investigation of some sterically hindered amines as potential carbon dioxide scrubbing compounds. *Ind. Eng. Chem. Res.,* 1997, **36**(5), 1779–1790.

88 Munoz, D. M., Portugal, A. F., Lozano, A. E., de la Campa, J. G., and de Abajo, J. New liquid absorbents for the removal of CO_2 from gas mixtures. *Energy Environ. Sci.,* 2009, **2**(8), 883–891.

89 Portugal, A. F., Sousa, J. M., Magalhaes, F. D., and Mendes, A. Solubility of carbon dioxide in aqueous solutions of amino acid salts. *Chem. Eng. Sci.,* 2009, **64**(9), 1993–2002.

90 van Holst, J., Versteeg, G. F., Brilman, D. W. F., and Hogendoorn, J. A. Kinetic study of CO_2 with various amino acid salts in aqueous solution. *Chem. Eng. Sci.,* 2009, **64**(1), 59–68.

91 Jockenhoevel, T., Schneider, R., and Rode, H. *Validation of a Second-Generation Post-Combustion Capture Technology – Results from POSTCAP Pilot Plant Operation,* Power-Gen Europe, Amsterdam, 2010.

92 Davis, J. and Rochelle, G. Thermal degradation of monoethanolamine at stripper conditions. *Energy Procedia,* 2009, **1**(1), 327–333.

93 Freeman, S. A., Davis, J., and Rochelle, G. T. Degradation of aqueous piperazine in carbon dioxide capture. *Int. J. Greenhouse Gas Control,* 2010, **4**(5), 756–761.

94 Strazisar, B. R., Anderson, R. R., and White, C. M. Degradation pathways for monoethanolamine in a CO_2 capture facility. *Energy Fuels,* 2003, **17**(4), 1034–1039.

95 Sexton, A. J. and Rochelle, G. T. Catalysts and inhibitors for oxidative degradation of monoethanolamine. *Int. J. Greenhouse Gas Control,* 2009, **3**(6), 704–711.

96 Veltman, K., Singh, B., and Hertwich, E. G. Human and environmental impact assessment of postcombustion CO_2 capture focusing on emissions from amine-based scrubbing solvents to air. *Environ. Sci. Technol.,* 2010, **44**(4), 1496–1502.

97 Lee, Y. N. and Schwartz, S. E. Reaction kinetics of nitrogen dioxide with liquid water at low partial pressure. *J. Phys. Chem.,* 1981, **85**(7), 840–848.

98 Eigen, M., Kustin, K., and Maas, G. Die Geschwindigkeit der Hydratation von SO_2 in wässriger Lösung. *Z. Phys. Chem.,* 1961, **30**, 130–136.

99 Versteeg, G. F. and van Swaaij, W. P. M. Solubility and diffusivity of acid gases (CO_2, N_2O) in aqueous alkanolamine solutions. *J. Chem. Eng. Data,* 1988, **33**, 29–34.

100 Cullinane, J. T. and Rochelle, G. T. Kinetics of carbon dioxide absorption into aqueous potassium carbonate and piperazine. *Ind. Eng. Chem. Res.,* 2006, **45**, 2531–2545.

101 Bara, J. E., Camper, D. E., Gin, D. L., and Noble, R. D. Room-temperature ionic liquids and composite materials: platform technologies for CO_2 capture. *Acc. Chem. Res.,* 2010, **43**(1), 152–159.

102 Huang, J. and Rüther, T. Why are ionic liquids attractive for CO_2 absorption? An overview. *Aust. J. Chem.,* 2009, **62**, 298–308.

103 Teague, C. M., Dai, S., and Jiang, D. E. Computational investigation of reactive to nonreactive capture of carbon dioxide by oxygen-containing Lewis bases. *J. Phys. Chem. A,* 2010, **114**(43), 11761–11767.

104 Jung, E. Carbon dioxide removal from gases – using absorption liquid containing ethanolamine and a cyclotetramethylene sulphone and regenerating two liquid phases separately.Patents DE 1544122-B and NL 148648-B, 1974.

105 Rojey, A., Cadours, R., Carrette, P. L., Boucot, P., and Carrette, P. Deacidifying effluent gas containing hydrogen sulphide and/or carbon dioxide, e. g. natural gas or combustion fumes, by absorption in a solution which forms two phases on heating and then regen-

erating the high-acid phase. Patents FR 2898284-A1, WO 2007104856-A1, EP 1996313-A1, AU 2007226446-A1, US 2009199709-A1, JP 2009529420-W, and FR 2898284-A1, 14 September 2007, B01D-053/14 200829, 2008.

106 Dilmore, R., Griffith, C., Liu, Z., Soong, Y., Hedges, S.W., Koepsel, R., and Ataai, M. Carbonic anhydrase-facilitated CO_2 absorption with polyacrylamide buffering bead capture. *Int. J. Greenhouse Gas Control*, 2009, **3**(4), 401–410.

107 Favre, N., Christ, M.L., and Pierre, A.C. Biocatalytic capture of CO_2 with carbonic anhydrase and its transformation to solid carbonate. *J. Mol. Catal. B*, 2009, **60**(3–4), 163–170.

108 Trachtenberg, M.C., Cowan, R.M., Smith, D.A., Horazak, D.A., Jensen, M.D., Laumb, J.D., Vucelic, A.P., Chen, H.B., Wang, L.P., and Wu, X.Q. Membrane-based, enzyme-facilitated, efficient carbon dioxide capture. *Energy Procedia*, 2009, **1**(1), 353–360.

109 Ondrey, G. Enzyme-based method could make CO_2 capture economically viable. *Chem. Eng.* 2010, 1 September, 11.

110 Rolker, J., Seiler, M., Mokrushina, L., and Arlt, W. Potential of branched polymers in the field of gas absorption: experimental gas solubilities and modeling. *Ind. Eng. Chem. Res.*, 2007, **46**(20), 6572–6583.

Materials for CO_2Scrubbing

6
Chemical Absorption Materials for CO_2 Capture

Kaj Thomsen

Abstract

Chemical absorption materials that potentially can be used for post-combustion carbon dioxide capture are discussed. They fall into five groups: alkanolamines, alkali metal carbonates, ammonia, amino acid salts, and ionic liquids. The chemistry of the materials is discussed and advantages and drawbacks are mentioned. References to review papers, papers with experimental data, and papers describing the thermodynamic modeling of the systems are given.

Keywords: chemical absorption materials, carbon dioxide capture, alkanolamines, alkali metal carbonates, chilled ammonia, amino acid salts, ionic liquids

6.1
Introduction

Flue gases from large point sources such as fossil-fueled power plants usually contain less than 15 vol.% of carbon dioxide and small amounts of sulfur dioxide and nitrous oxide. The remaining part of the flue gas consists of nitrogen, water vapor, and some oxygen. Modern power plants have facilities for removing sulfur dioxide and nitrous oxide. It is not considered an option to transport and store all the remaining flue gas. Current methods under investigation for possible carbon dioxide capture and storage therefore involve the selective absorption of carbon dioxide in an absorption column.

Possible chemical absorption materials for selective carbon dioxide capture include alkanolamines, alkali metal carbonates, ammonia, and amino acid salts. These chemical absorption materials are all dissolved in water when used for carbon dioxide capture. Ionic liquids are in a separate category because they do not require any solvent such as water. In the following general discussion, these aqueous mixtures or liquids will be denoted solvents.

Process Engineering for CCS Power Plants, First Edition. Edited by Detlef Stolten and V. Scherer

When the solvent leaves the absorption column, it is to a certain extent saturated with carbon dioxide. The term "loading" is used to describe the degree of saturation of such solutions. The loading is defined as the number of moles of carbon dioxide per mole of chemical absorption material. When the solvent leaves the absorption column, it has a "rich" loading. The solvent with a rich loading is led to a desorption column, where it is heated. The heating causes the evaporation of a certain fraction of the absorbed carbon dioxide. After desorption, the solvent now has a "lean" loading and is returned to the absorption column.

Carbon dioxide is an acidic gas. The chemical absorption materials of relevance are basic. When carbon dioxide is dissolved in basic aqueous solutions or in ionic liquids, exothermic chemical reactions take place. These chemical reactions are acid–base reactions.

During carbon dioxide removal in the desorption column, the chemical reactions are reversed by the addition of heat. The amount of heat required for the reversal of these reactions is of the same magnitude as the amount of heat released in the exothermic absorption process. After the production of pure carbon dioxide, it is pressurized and liquefied before transportation to the storage site. The compression required for pressurizing and liquefying the carbon dioxide consumes a significant amount of energy. Chemical absorption materials that can release the absorbed carbon dioxide at a high pressure are therefore preferred.

A relevant criterion for the selection of chemical absorption materials is the amount of heat required per kilogram of carbon dioxide for the absorption–desorption cycle. It is also important to consider the temperatures at which the absorption and desorption take place. The absorption might require cooling to maintain a suitable low temperature during the exothermic absorption process. A low desorption temperature is preferred because excess heat is often available at relatively low temperature. The amount of solvent required to absorb a certain amount of carbon dioxide varies widely between different chemical absorption materials. The effect of this property is included in the amount of heat required per kilogram of carbon dioxide for the absorption–desorption cycle. It is difficult to make a comparison of this heat requirement for different chemical absorption materials. The optimal operation parameters such as absorber/desorber temperatures, lean and rich loadings, and concentration of the chemical absorption material are different for each type of chemical absorption material, making comparisons difficult. An accurate, rigorous process simulation or actual pilot plant data are required in order to perform the comparisons.

Another relevant criterion for the selection of chemical absorption materials is the rate of absorption of carbon dioxide in the solvent. The rate of absorption determines the contact area between the solvent and gas required to transfer a certain amount of gas between the phases. A low rate of absorption results in large equipment size. During absorption, carbon dioxide is transported from the bulk gas to the gas–liquid interface. After passing this interface, carbon dioxide is transported to the bulk liquid. The transport in the liquid phase may be enhanced by chemical reactions between carbon dioxide, water, and the chemical absorption material.

It is also important to consider the stability of the chemical absorption material. In the absorber, the absorption material is in contact with flue gas containing a few percent of oxygen. Oxidative degradation of the material might therefore take place. These materials also have to be able to withstand repeated heating in the desorption column. Thermal degradation might take place.

6.2 Alkanolamines

The first paper on the solubility of carbon dioxide in alkanolamines was published in 1931 by Bottoms [1]. At that time, carbon dioxide was usually removed from gas mixtures using potassium or sodium carbonate solutions. Alkanolamines were considered a good alternative to carbonates because of the limited solubility of sodium and potassium hydrogencarbonate that formed during absorption. A solution of alkanolamines had a larger capacity for carbon dioxide absorption than a carbonate solution.

6.2.1 Experimental Data

Alkanolamines have been known and used for carbon dioxide absorption for many years. Considerable amounts of experimental data have therefore been published. These experimental data include vapor liquid equilibrium data such as the solubility of carbon dioxide in solutions of alkanolamines at various temperatures and pressures and vapor pressures of aqueous alkanolamine solutions and of pure alkanolamine [1–51], solid–liquid equilibrium data such as the freezing point of alkanolamine solutions [52, 53], heat of absorption data such as the heat evolved per mole of carbon dioxide dissolved in the solution and experimentally measured heat capacity data [54–63], and speciation data such as the distribution of species in an aqueous solution of alkanolamines with various amounts of carbon dioxide [63–66].

All these experimental data are important for the thermodynamic modeling of the solvent system. Thermodynamic models are required for designing and optimizing the carbon dioxide absorption and desorption process. With this amount of experimental data, thermodynamic models can be made very accurate and cover all relevant ranges of temperatures and pressures [66–69].

The alkanolamines commonly considered today as chemical absorption materials for carbon dioxide absorption are mainly the primary amine monoethanolamine (MEA), the secondary amine diethanolamine (DEA), and the tertiary amine methyldiethanolamine (MDEA), or mixtures of these. There has also been some interest in the primary amine diglycolamine (DGA) [15, 49, 50, 58] and the secondary amine diisopropanolamine (DIPA) [51–54]. The tertiary amine diethylethanolamine (DEEA) and the secondary amine ethylethanolamine (EEA) have also been considered [55]. Piperazine and 2-amino-2-methyl-1-propanol (AMP) have been considered kinetic promoters together with alkanolamines [14, 32, 36–38, 40, 66, 70].

A 30 wt% MEA solution is often used as a benchmark solvent for comparing different chemical absorption materials. Such a solution corresponds roughly to 8 mol of MEA per kilogram of water.

6.2.2 Chemistry of Aqueous Carbon Dioxide–Alkanolamine Solutions

When carbon dioxide is dissolved in basic solutions, several reactions take place until equilibrium is reached. The first reaction is the dissociation of water:

$$H_2O \rightleftharpoons H^+ + OH^- \tag{6.1}$$

The next reaction is the reaction of gaseous carbon dioxide to form aqueous carbon dioxide. This reaction is equivalent to a reaction that could be written by adding water on both sides of the reaction equation and giving the end product carbonic acid, H_2CO_3 (aq.). As the two reactions are equivalent, the short form is preferred:

$$CO_2(g) \rightleftharpoons CO_2(aq.) \tag{6.2}$$

Carbon dioxide reacts with hydroxide ions to form hydrogencarbonate ions:

$$CO_2(aq.) + OH^- \rightleftharpoons HCO_3^- \tag{6.3}$$

Depending on the amount of base, part of the hydrogencarbonate will react with hydroxide ions to form carbonate ions:

$$HCO_3^- + OH^- \rightleftharpoons H_2O + CO_3^{2-} \tag{6.4}$$

Primary and secondary alkanolamines are able to react with aqueous carbon dioxide to form carbamate in certain pH ranges. The carbamate ion, $R_1R_2NCOO^-$, can be formed in the reaction:

$$R_1R_2NH + HCO_3^- \rightleftharpoons H_2O + R_1R_2NCOO^- \tag{6.5}$$

R_1R_2NH is the generic expression for a primary or a secondary alkanolamine. If R_1=–CH_2CH_2OH and R_2=H, the alkanolamine is MEA.

Speciation studies such as those performed by Jakobsen *et al.* [68], Böttinger *et al.* [69], and Hilliard [70] show that the amount of hydrogencarbonate, HCO_3^-, is lower than expected at low loadings of carbon dioxide whereas the amount of carbamate present is relatively high. This indicates that hydrogencarbonate that would usually be formed in the basic solution reacts with the alkanolamine to form carbamate ions. Speciation studies showed that at a certain loading, the amount of carbamate formed reaches a maximum. By further increasing the loading, the amount of carbamate decreases and the amount of hydrogencarbonate increases correspondingly. For a 30 wt% MEA solution, the maximum amount of carbamate is formed at a loading of ~ 0.5 at both 293 and at 333 K according to [69]. The same is the case for a 20 wt% DEA solution at 313 K [69]. Bailey and Feron [74] reported that a typical reaction rate of carbon dioxide with an MEA solution is more than 1000 times higher than the corresponding reaction rate of carbon dioxide with MDEA. MEA is a primary amine and therefore forms carbamate. MDEA as a tertiary amine cannot form carbamate and therefore has a low reaction rate. The typical carbon capture process with MEA will operate with a maximum rich loading of 0.5 in order to benefit from the high reaction rate obtained during carbamate formation.

If the lean loading achieved in the desorber is 0.2 mol of carbon dioxide per mole of MEA, a 30 wt% MEA solvent (8 molal) is able to transfer (0.5–0.2) $\times 8$ mol = 2.4 mol of CO_2 per kilogram of water in the solution.

In addition to forming carbamate ions, alkanolamines react with water to form protonated alkanolamines:

$$R_1R_2NH + H_2O \rightleftharpoons R_1R_2NH_2^+ + OH^- \qquad (6.6)$$

This process takes place for primary, secondary, and tertiary amines. The hydroxide ions produced contribute to creating a driving force for the dissolution of carbon dioxide and the formation of hydrogencarbonate.

Alkanolamine evaporation can be described by the equilibrium

$$R_1R_2NH(aq.) \rightleftharpoons R_1R_2NH(g) \qquad (6.7)$$

The vapor pressures of the alkanolamines used as chemical absorption materials are usually low. Some evaporation of the alkanolamine takes place, especially in the desorber where the temperature is usually above 100 °C.

6.2.3
Alkanolamine Degradation

It is well known that alkanolamines degrade in contact with carbon dioxide, oxygen, and sulfur dioxide, and the degradation chemistry was discussed in the scientific literature in the 1950s, as reviewed by Blake in 1967 [75]. Blake mentioned the following degradation products from the degradation of MEA, DEA, and DIPA: oxazolidinones, imidazolidinones, piperazine, and the so-called "heat-stable" salts [75].

The heat-stable salts are not ordinary salts. According to Blake [75], the term "heat-stable salt" is used to describe the protonated alkanolamine that is produced according to Equation 6.6 in the presence of carboxylic acids or another compound (acid) more willing to donate a proton than water. No hydroxide ion is produced by the formation of the heat-stable salt because a stronger acid replaces water in the reaction in Equation 6.6. Instead, an acetate ion or another anion that is a much weaker base than the hydroxide ion is formed. The amount of alkanolamine involved in the formation of a heat-stable salt is therefore not active as a chemical absorption material. The term "heat stable" is used because the protonated alkanolamine cannot be regenerated by heating, as the vapor pressure of the acid corresponding to the anion is too low. The heat-stable salt can only be regenerated to an active alkanolamine by the addition of a stronger base, such as sodium carbonate. This converts the heat-stable salt to an alkanolamine and the sodium salt of the acid that caused the formation of the heat-stable salt. By the addition of an extra piece of equipment in the carbon dioxide desorption process, a reclaimer, the alkanolamine and water are distilled under vacuum and at high temperature from the salt solution, which is discarded together with other low-boiling degradation products. Without the reclaimer, the concentration of salt and other degradation products would continue to increase and cause problems.

In their overview of CO_2 capture technologies, MacDowell *et al.* [76] referred to relevant newer literature dealing with the degradation chemistry of alkanolamines. The chemistry will not be repeated here owing to its complexity. Alkanolamine degradation in contact with industrial flue gases containing oxygen, carbon dioxide, and sulfur dioxide is complex and not yet fully understood. In addition to the degradation products considered by Blake [75], McDowell *et al.* [76] mentioned amides, amines, amino acids, aldehydes, amino acid derivatives, carboxylic acids, and nitrosamines (potent carcinogens) as some of the currently considered alkanolamine degradation products.

The thermal and oxidative degradation of alkanolamines is a challenge for processes with alkanolamines. Plants need to be designed so that emissions of hazardous degradation products are reduced to a minimum.

6.3 Sodium and Potassium Carbonates

Sodium and potassium carbonates were in use as chemical absorption materials before alkanolamines. When CO_2 dissolves in aqueous carbonate solutions such as Na_2CO_3 or K_2CO_3 solutions, only the chemical reactions in Equations 6.1–6.4 take place. The formation of carbamate in primary and secondary alkanolamine solutions was found to be an advantage for the reaction rate of carbon dioxide in these solutions. At the same time, the formation of carbamate creates a stronger chemical bond between the chemical absorption material and carbon dioxide. The formation of this chemical bond is exothermic, which means that heat is released during the absorption. A correspondingly larger amount of heat has to be added in the desorber to release carbon dioxide from alkanolamine solutions again.

According to Knuutila *et al.* [77], the renewed interest in carbonates as chemical absorption materials for carbon dioxide stems from the potentially low desorption energy demand in carbonate processes. The paper by Knuutila *et al.* [77] contains references to several recent papers on sodium and potassium carbonates as chemical absorption materials for carbon dioxide capture.

Owing to the absence of carbamate formation, the reaction rate of carbon dioxide in carbonate solutions is expected to be correspondingly lower. This problem can be solved by the addition of small amounts of a primary or secondary amine. It was demonstrated by Yi *et al.* [78] that the reaction rate of carbon dioxide with a potassium carbonate solution could be increased significantly by the addition of small amounts of the secondary amine DEA. Yi *et al.* [78] used a rotating packed bed which supposedly has a high gas–liquid mass transfer efficiency. At the same time, rotating packed beds are able to handle slurry. The limited solubility of potassium and sodium hydrogencarbonate can be seen as another drawback of carbonate processes. If the handling of slurry is technically possible, the precipitation of hydrogencarbonate is not a drawback, but could be an advantage of the process.

The formation of solids increases the possible loading range considerably. The solubility of potassium carbonate is around 40 wt% or 8.6 mol of K_2CO_3 per kilogram of water at 40 °C. Only a small amount of carbon dioxide can be absorbed by a saturated potassium carbonate solution before potassium hydrogencarbonate starts to precipitate. If slurry can be handled, this solvent is able to absorb 8.6 mol of carbon dioxide before it is fully loaded. Solutions of the alkanolamine MEA are not used at loadings above 0.5 due to a low reaction rate above this loading. If the reaction rate of potassium carbonate solutions can be increased and maintained at a high level in a wider loading range, potassium carbonate certainly has an advantage.

Pérez-Salado Kamps *et al.* [79] gave a short review of absorption processes using potassium carbonate as chemical absorption material. Processes usually operate at an absorption temperature of 60 °C and higher. Desorption takes place at temperatures up to 130 °C. The potassium carbonate solutions used are relatively dilute, 1.8–4.8 mol kg^{-1} water. The absorption pressure is relatively high, up to 3 bar partial pressure of carbon dioxide. The desorption pressure is relatively low, 3–4 bar total pressure. A low absorption pressure and a high desorption pressure would have been preferred. In addition to DEA, piperazine and ammonia have also been used in the process to promote the absorption of carbon dioxide through carbamate formation.

Absorption processes using potassium carbonate are often called "hot potassium carbonate" processes owing to the relatively high temperatures applied. Another term is "Benfield" processes, derived from the names of two key figures in the development of this process, Benson and Field [80–82]. The Benfield process was developed in the 1950's for removing carbon dioxide from synthesis gas for the Fischer–Tropsch synthesis. The partial pressure of carbon dioxide in the synthesis gas was typically up to 3 bar. The partial pressure of carbon dioxide in the flue gas from coal-fired power plants is less than 0.15 bar. The Benfield process therefore needs to undergo some adaptation before it can be applied to carbon dioxide removal in post-combustion carbon capture plants.

6.3.1
Experimental Data for Sodium and Potassium Carbonate Systems

Knuutila *et al.* [77] measured and modeled vapor–liquid equilibria of the carbon dioxide-loaded sodium carbonate–water system in the temperature range 40–80 °C for sodium carbonate concentrations of 8–12 wt%. They also referred to and discussed the quality of experimental data found in the literature for this system.

Pérez-Salado Kamps *et al.* [79] provided new measurements for the equilibrium between carbon dioxide and relatively dilute potassium carbonate solutions. They also modeled the system and referred to and discussed previous published data. Experimental data for the potassium carbonate system can also be found in a paper by Tosh *et al.* [82].

Although carbonate systems have been used for carbon dioxide absorption for many years, the experimental data for these systems are still limited. Also, according to the discussions in the papers cited above [77, 79], there is some scatter in the available data. Data for concentrated potassium carbonate solutions seem to be lacking.

6.3.2
Absence of Degradation Problems for Carbonate Solutions

Potassium and sodium carbonates are both inorganic substances and cannot undergo thermal or oxidative degradation. Sodium and potassium hydrogencarbonates release carbon dioxide when heated. The use of sodium hydrogencarbonate as baking powder is based on this thermal degradation process. By releasing carbon dioxide, the hydrogencarbonates are converted to carbonates, and can again be used for absorbing carbon dioxide.

6.4
Ammonia

The use of ammonia for capturing carbon dioxide is based on a patent by Gal from 2006 [83]. Like amines, ammonia is able to form carbamate. The chemical reactions in Equations 6.1–6.7 that were mentioned for amines all apply to ammonia also. In the reactions in Equations 6.5–6.7, R_1 and R_2 should both be replaced with H to reflect the chemistry of carbon dioxide in ammonia solutions. In addition to the reactions in Equations 6.1–6.7, the formation of several solid phases in the ammonia–carbon dioxide–water system has to be considered. The formation of ammonium carbamate will only take place in solutions with higher ammonia concentrations than considered for carbon dioxide capture:

$$NH_4^+ + NH_2COO^- \rightleftharpoons NH_2COONH_4(s) \tag{6.8}$$

The formation of ammonium carbonate monohydrate takes place at low temperature in solutions with sufficient ammonia concentration and with a lean loading:

$$2NH_4^+ + CO_3^{2-} + H_2O \rightleftharpoons (NH_4)_2CO_3 \cdot H_2O(s) \quad (6.9)$$

The formation of the double salt sesquicarbonate takes place in solutions with sufficient ammonia concentration in the temperature range 10–85 °C in a narrow, intermediate loading range:

$$4NH_4^+ + CO_3^{2-} + 2HCO_3^- \rightleftharpoons (NH_4)_2CO_3 \cdot 2NH_4HCO_3(s) \quad (6.10)$$

The formation of ammonium hydrogencarbonate takes place at a high carbon dioxide loading in solutions with sufficient ammonia concentration:

$$NH_4^+ + HCO_3^- \rightleftharpoons NH_4HCO_3(s) \quad (6.11)$$

Different processes using ammonia as chemical absorption material are being developed. In Gal's patent [83], absorbing carbon dioxide at low temperature, 0–20 °C, is suggested. The process is therefore called the chilled ammonia process. By using this low temperature in the absorber, the partial pressure of ammonia is kept at a low level. This reduces the ammonia slip from the absorber. In order to perform the carbon dioxide absorption at this low temperature, not only does the flue gas have to be chilled, but also the heat evolved from the exothermic reactions of absorbing carbon dioxide needs to be removed. In the patent, it is suggested that the process is operated at ammonia concentrations sufficiently high that an ammonium hydrogencarbonate slurry is formed. This slurry is pumped to the desorber and releases carbon dioxide when it is heated.

An alternative process scheme using ammonia as chemical absorption material is being developed. This process is operated at lower concentrations of ammonia and the absorption takes place at temperatures above 20 °C. Such process schemes are being developed both at CSIRO (Australian Commonwealth Scientific and Industrial Research Organization) [84] and at KIER (Korea Institute of Energy Research) [85].

The ammonia-based carbon dioxide capture processes all promise a significant reduction in the heat required for regenerating the solvent in the desorber.

According to an Electric Power Research Institute (EPRI) report [86], a capture efficiency of carbon dioxide higher than 90% can be achieved in a chilled ammonia plant. The chilling of the flue gas reduces its volume significantly by chilling and by the condensation of water. The temperature bulge experienced in absorbers using alkanolamines is decreased because the reactions taking place in absorbers using ammonia are less exothermic and the heat of reaction is continually removed. Absorbers for chilled ammonia processes can therefore be reduced in size compared with the absorbers used with alkanolamines. Much less heat is used for regenerating ammonia and releasing carbon dioxide than in alkanolamine processes. According to Darde *et al.* [87], the heat requirement for desorbing carbon dioxide is less than 2000 kJ kg^{-1} CO_2 in the ammonia process, whereas it is 3700 kJ kg^{-1} CO_2 in the MEA process. The saving stems from the fact that less heat is required to reverse the reactions of carbon dioxide absorption. In addition, less water is evaporated in the desorber due to the increased pressure. Carbon dioxide is regenerated at high pressure, which also reduces the construction and operating costs. In addition, it is expected that the cooling/condensation of the flue gas will remove practically all sulfur oxides, nitrous oxides, and fly ash from the flue gas before entering the absorber. If some remaining sulfur oxides enter the absorber, they will automatically be converted to ammonium sulfate rather than causing any harm. In fact, Yu *et al.* [84] reported that 95% of the sulfur oxides was removed from the flue gas in test runs at the Munmorah pilot plant. Another advantage is that ammonia is stable under the operating conditions of the process. The solubility of ammonia in water is very high. An up to 28 wt% ammonia solution was recommended in the chilled ammonia process [83]. This concentration corresponds to 22.8 mol of ammonia per kilogram of water. In comparison, the typical 30 wt% MEA solution contains only 8 mol of MEA per kilogram of water, and potassium carbonate solutions used in the carbonate process contain up to 4.8 mol of potassium carbonate per mole of water. These three chemical absorption materials all react 1:1 with carbon dioxide. The ammonia solution therefore has the capacity to absorb much more carbon dioxide than any of the other solutions, especially if the process is operated with the formation of slurry consisting of ammonium hydrogencarbonate in the absorber.

6.4.1
Formation of Solids in Aqueous Ammonia Solutions

Owing to the formation of carbamates, the reaction rate of carbon dioxide with aqueous ammonia is expected to be similar to that of MEA. Most chemical processes proceed more slowly at low temperature. It can therefore be expected that the reaction kinetics and the absorption rate of carbon dioxide in aqueous ammonia will be slow at low temperatures, and this has been verified experimentally [88]. Preliminary tests at the CSIRO pilot plant using 2–5 wt% ammonia solutions at 15–30 °C showed mass transfer coefficients to be only half of those measured in the laboratory [84].

A factor other than temperature that can impede the rate of absorption of carbon dioxide in aqueous ammonia is the formation of solid phases. If a 27 wt% ammonia solution is used for the absorption of carbon dioxide at 10 °C at a total pressure of 1 bar, the formation of solids will start at a relatively low loading. The lean solution recommended [83] has a loading of about 0.5 mol of carbon dioxide per mole of ammonia. Figure 6.1 shows the amount of solid phases formed as a function of loading. It can be seen that the solution is oversaturated with ammonium carbonate at a loading of 0.5. The amount of ammonium carbonate precipitated is actually at its maximum at a loading of 0.5. The graph in

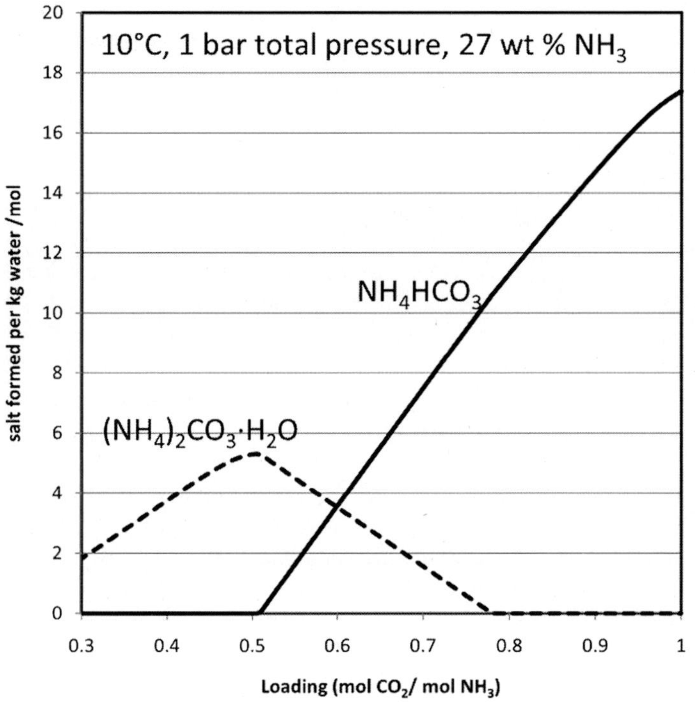

Fig. 6.1 Amount of salt formed in a 27 wt% ammonia solution at 10 °C and 1 bar as a function of loading. The graph was calculated with the Extended UNIQUAC model of Thomsen and Rasmussen [89].

Fig. 6.1 was calculated using the Extended UNIQUAC thermodynamic model for electrolytes [89]. If the process is operated between loadings of 0.5 and 0.9, the amount and type of salt formed can be followed in Fig. 6.1. At a loading of 0.5. a certain amount of ammonium carbonate is already present. It will probably precipitate during the chilling of this lean stream. As the loading increases, the amount of ammonium hydrogencarbonate increases and the amount of ammonium carbonate decreases. This means that ammonium carbonate dissolves and at the same time ammonium hydrogencarbonate forms. The overall process taking place is

$$CO_2(g) + (NH_4)_2CO_3 \cdot H_2O(s) \rightleftharpoons 2NH_4HCO_3(s) \tag{6.12}$$

The rate of this reaction has probably never been studied, but it is expected that it is relatively slow as it involves a solid-phase transition.

In Fig. 6.2, the calculated partial pressures of carbon dioxide and of ammonia during the absorption are depicted. This figure was also obtained using the Extended UNIQUAC thermodynamic model for electrolytes [89]. The partial pressures of carbon dioxide and of ammonia are both constant in the loading range

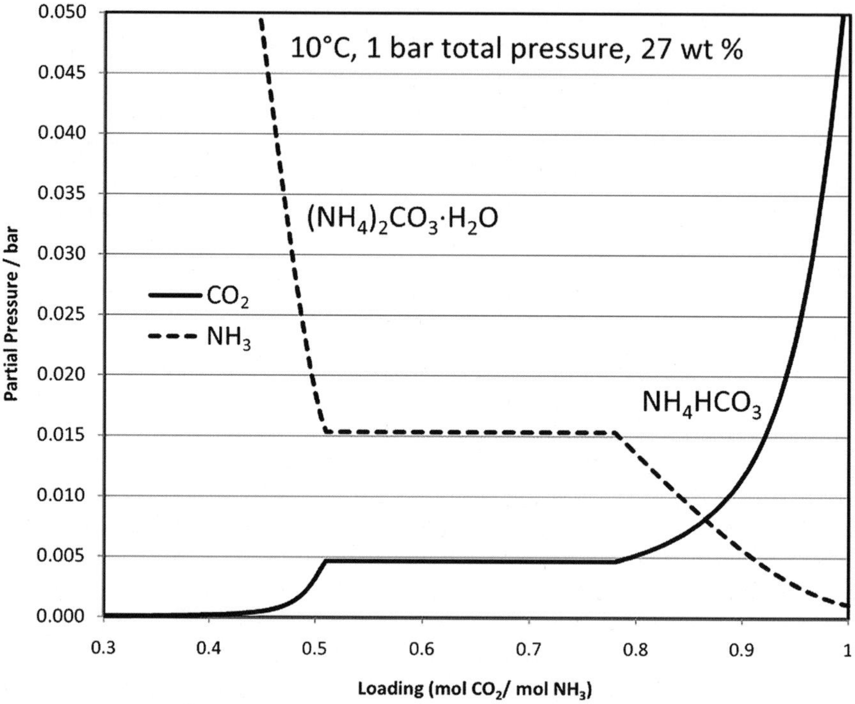

Fig. 6.2 Partial pressure of carbon dioxide and ammonia in equilibrium with a 27 wt% ammonia solution at 10 °C and 1 bar as a function of loading. The graph was calculated with the Extended UNIQUAC model of Thomsen and Rasmussen [89].

0.51–0.78. The reason is that the temperature and the pressure are both constant as they would ideally be during carbon dioxide absorption. The system is a three-component system. According to Gibbs' phase rule, the number of degrees of freedom is five minus the number of phases. In the loading range considered, there are two solid phases, a liquid phase, and a gas phase. The system therefore only has one degree of freedom. When carbon dioxide is added to the system, the liquid phase or the gas phase cannot change composition. That would result in a change of pressure and the pressure is held constant at 1 bar. The only thing that can change is the composition of the solid phase, which is the change in solid composition that was seen in Fig. 6.1. By comparing Fig. 6.1 and 6.2 it can be seen that in loading ranges where only one salt (one solid phase) is present, both partial pressures vary.

Figures 6.1 and 6.2 were both calculated using the Extended UNIQUAC thermodynamic model for electrolytes of Thomsen and Rasmussen [89]. New parameters have been determined for this model by Darde *et al.* [90]. This new parameter set improved the ability of the model to reproduce the phase behavior of the system at temperatures above 110 °C. Other thermodynamic models for this system include the model of Mathias *et al.* [91], who included a quantitative evaluation of the aqueous ammonia process for carbon capture. An evaluation of the process has also been published by Darde *et al.* [87].

Many experimental data are available for the thermodynamic modeling of the ammonia process for carbon capture. Most of the data available in the open literature can be found in three papers [89–91]. Data on the heat of mixing, heat capacity, and other thermal properties of the system are scarce. This type of data is generally in short supply for all the chemical absorption systems considered here. Thermodynamic models could probably be improved further if such data were available. The accurate calculation of the energy consumption of these processes is often the focus of process simulation.

6.5
Amino Acid Salts

Amino acid salts are interesting alternatives to alkanolamines, carbonates, and ammonia. Amino acid salts have properties that combine the best from these three groups. Amino acid salts have the same functional groups as alkanolamines and are able to form carbamates, they do not evaporate, and some of them are highly soluble in water. Some amino acid salts have interesting acid–base properties. It was found, for example, that the potassium salt of methionine [2-amino-4-(methylthio)butanoic acid] is very basic at low temperature and therefore has good properties for the absorption of carbon dioxide. When the temperature is increased, the acid–base properties change so that carbon dioxide is only sparingly soluble in the solution, according to Sengeløv [92]. Unfortunately, the solubility of methionine is low (56 g kg^{-1} water) and it could precipitate as the amino acid in the system even though the isoelectric point at a pH of 5.74 is not reached during a carbon capture process. Experimental vapor–liquid equilibrium data for this system were measured by Kumelan *et al.* [93]

Much more experimental work needs to be done in order to evaluate the use of amino acid salts as chemical absorption materials for carbon capture. Kumar *et al.* measured the properties of potassium taurate [94], the potassium salt of taurine (2-aminoethanesulfonic acid). A process for using potassium taurate for carbon dioxide was patented by Versteeg *et al.* [95] in 2003. Aronu *et al.* [96] measured vapor–liquid equilibrium data for potassium sarcosinate, the potassium salt of sarcosine [2-(methylamino)acetic acid]. Portugal *et al.* [97] measured the solubility of carbon dioxide in potassium glycinate, the potassium salt of glycine (aminoacetic acid). They found that the solubility of carbon dioxide did not change much with temperature for this amino acid salt. This is a surprising finding and is contrary to what was found for the potassium salt of methionine by Sengeløv [92].

Amino acid salts have the same functional groups as alkanolamines. They might also have inherited the tendency to undergo oxidative and thermal degradation from the alkanolamines. This is one of the topics that still need to be examined.

6.6
Ionic Liquids

Ionic liquids are compounds consisting of ions like salts, but are liquid at relatively low temperatures compared with salts. The size difference between the cations and anions are so large that they cannot form lattices like other salts. Ionic liquids can be tailored for specific purposes. The amine groups that are active in alkanolamines can, for example, be built into ionic liquids and might give them some properties that are useful as chemical absorption materials for carbon capture. Many ionic liquids are hydrophobic, which could be a good property for carbon capture materials. Perhaps ionic liquids that are specific to carbon dioxide can be constructed. In 2004, Anthony *et al.* [98] published a feasibility study using ionic liquids for carbon capture. The study focused on the ionic liquid [bmim][PF_6] with some ability to dissolve carbon dioxide at high pressure. This ionic liquid was not particularly designed for the purpose of absorbing carbon dioxide. The ionic liquid was not functioning as a chemical absorption material for carbon dioxide but was apparently able physically to contain some carbon dioxide in cavities between the ions. It appeared from the study that 5914 kg of ionic liquid was required to absorb 1 kg of carbon dioxide.

In a recent review, Vega *et al.* [99] reported on recent advances and perspectives in the modeling of ionic liquids and the solubility of gases in them. The solubility of carbon dioxide in ionic liquids is one of the main topics of their paper.

6.7
Conclusion

The choice of chemical absorption materials for carbon dioxide capture is not easy. Processes based on alkanolamines and alkali metal carbonates represent the old, well-known technology. However, both of these processes were developed for the absorption of carbon dioxide at a much higher partial pressure than is found in the flue gas from fossil-fueled power plants, and they both need to be adapted to the process conditions feasible for a post-combustion carbon capture plant. Problems concerning the oxidative and thermal degradation of alkanolamines have to be addressed. Additional equipment might need to be added to the processes to avoid the emission of carcinogens and other degradation products. To the extent that alkanolamines are used as promoters or activators in the alkali metal carbonate processes, this problem concerns both of these two groups of processes.

Processes based on ammonia do not suffer from the problem of oxidative and thermal degradation of the chemical absorption material. They are very promising in terms of energy consumption and loading capacity. The rate of absorption of carbon dioxide into these solutions might be a problem for this process. In the chilled ammonia process, the volume of the flue gas is reduced considerably due to the low temperature and to the condensation of moisture during the chilling. This factor contributes to a more compact absorber size as less gas volume has to pass. The low rate of absorption results in the requirement for a larger contact area, which results in a larger absorber. A significant advantage of the ammonia process might be its ability to absorb sulfur dioxide simultaneously with carbon dioxide. In addition, it can operate with high pressure in the desorber, reducing the size of the desorber column and reducing the energy consumption for compressing carbon dioxide. Problems include the high volatility of ammonia, the need for the technology required for handling slurry, and the chilling of flue gas and process streams.

Amino acids and ionic liquids are interesting alternatives to processes based on alkanolamines, alkali metal carbonates, and ammonia. However, much more research needs to be done in order to develop competitive processes using these absorption materials.

References

1 Bottoms, R. R. Organic bases for gas purification. *Ind. Eng. Chem.*, 1931, **23**(5), 501–504.

2 Mason, J. W. and Dodge, B. F. Equilibrium absorption of carbon dioxide by solutions of the ethanolamines. *Trans. Am. Inst. Chem. Eng.*, 1936, **32**, 27–48.

3 Jones, J. H., Froning, H. R., and Claytor, E. E. Jr. Solubility of acidic gases in aqueous monoethanolamine. *J. Chem. Eng. Data*, 1959, **4**, 85–92.

4 Lee, J. I., Otto, F. D., and Mather, A. E. Solubility of carbon dioxide in aqueous diethanolamine solutions at high pressures. *J. Chem. Eng. Data*, 1972, **17**, 465–468.

5 Lee, J. I., Otto, F. D., and Mather, A. E. The solubility of mixtures of carbon dioxide and hydrogen sulphide in aqueous diethanolamine solutions: supplementary unpublished material. *Can. J. Chem. Eng.*, 1974, **52**, 125–7.

6 Lee, J. I., Otto, F. D., and Mather, A. E. Solubility of Mixtures of Carbon Dioxide and Hydrogen Sulfide in 5M Monoethanolamine Solution. *J. Chem. Eng. Data*, 1975, **20**, 161–163.

7 Lee, J. I., Otto, F. D., and Mather, A. E. Measurement and prediction of the solubility of mixtures of carbon dioxide and hydrogen sulphide in a 2. 5 n monoethanolamine solution. *Can. J. Chem. Eng.*, 1976, **54**, 214–219.

8 Lee, J. I., Otto, F. D., and Mather, A. E. Equilibrium between carbon dioxide and aqueous monoethanolamine solutions. *J. Appl. Chem. Biotechnol.*, 1976, **26**, 541–549.

9 Lawson, J. D. and Garst, A. W. Hydrocarbon gas solubility in sweetening solutions: methane and ethane in aqueous monoethanolamine and diethanolamine. *J. Chem. Eng. Data*, 1976, **21**, 20–30.

10 Isaacs, E. E., Otto, F. D., and Mather, A. E. Solubility of mixtures of hydrogen sulfide and carbon dioxide in a monoethanolamine solution at low partial pressures. *J. Chem. Eng. Data*, 1980, **25**, 118–120.

11 Jou, F. Y., Mather, A. E., and Otto, F. D. Solubility of H_2S and CO_2 in aqueous methyldiethanolamine solutions. *Ind. Eng. Chem. Process Des. Dev.*, 1982, **21**, 539–544.

12 Nath, A. and Bender, E. Isothermal vapor–liquid equilibria of binary and ternary mixtures containing alcohol, alkanolamine, and water with a new static device. *J. Chem. Eng. Data*, 1983, **28**(4), 370–375.

13 Lal, D., Otto, F. D, and Mather, A. E. The solubility of H_2S and CO_2 in a diethano-

lamine solution at low partial pressures. *Can. J. Chem. Eng.*, 1985, **63**, 681–685.

14 Chakma, A. and Meisen, A. Solubility of CO_2 in aqueous methyldiethanolamine and *N,N*-bis(hydroxyethyl)piperazine solutions. *Ind. Eng. Chem. Res.*, 1987, **26**(12), 2461–2466.

15 Maddox, R. N., Bhairi, A. H., Diers, J. R., and Thomas, P. A. *Equilibrium Solubility of Carbon Dioxide or Hydrogen Sulfide in Aqueous Solutions of Monoethanolamine, Diglycolamine, Diethanolamine and Methydiethanolamine, GPA Research Report RR-104*, Gas Processors Association, Tulse, OK, 1987.

16 MacGregor R. J. and Mather A. E. Equilibrium solubility of H_2S and CO_2 and their mixtures in a mixed solvent. *Can. J. Chem. Eng.*, 1991, **69**, 1357–1366.

17 Austgen, D. M., Rochelle, G. T., and Chen, C. C. Model of vapor–liquid equilibria for aqueous acid-gas–alkanolamine systems. 2. Representation of H_2S and CO_2 solubility in aqueous MDEA and CO_2 solubility in aqueous mixtures of MDEA with MEA or DEA. *Ind. Eng. Chem. Res.*, 1991, **30**, 543–555.

18 Xu, S., Qing, S., Zhen, Z., Zhang, C., and Carroll, J. J. Vapor pressure measurements of aqueous *N*-methyldiethanolamine solutions. *Fluid Phase Equil.*, 1991, **67**, 197–201.

19 Kling, G. and Maurer, G. Solubility of hydrogen in aqueous ethanolamine solutions at temperatures between 323 and 423 K. *J. Chem. Eng. Data*, 1991, **36**, 390–394.

20 Shen, K.-P. and Li, M.-H. Solubility of carbon dioxide in aqueous mixtures of monoethanolamine with methyldiethanolamine. *J. Chem. Eng. Data*, 1992, **37**, 96–100.

21 Jou, F.-Y., Carroll, J. J., Mather, A. E., and Otto, F. D The solubility of carbon dioxide and hydrogen sulfide in a 35 wt% aqueous solution of methyldiethanolamine. *Can. J. Chem. Eng.*, 1993, **71**, 264–268.

22 Dawodu, O. F. and Meisen, A. Solubility of carbon dioxide in aqueous mixtures of alkanolamines. *J. Chem. Eng. Data*, 1994, **39**, 548–552.

23 Jou, F.-Y., Mather, A. E., and Otto, F. D. The solubility of CO_2 in a 30 mass percent monoethanolamine solution. *Can. J. Chem. Eng.*, 1995, **73**, 140–147.

24 Cai, Z.-Y., Xie, R.-J., and Wu, Z.-L. Binary isobaric vapor–liquid equilibria of ethanolamines plus water. *J. Chem. Eng. Data*, 1996, **41**, 1101–1103.

25 Kuranov, G., Rumpf, B., Smirnova, N. A., and Maurer, G. Solubility of single gases carbon dioxide and hydrogen sulfide in aqueous solutions of *N*-methyldiethanolamine in the temperature range 313–413 K at pressures up to 5 MPa. *Ind. Eng. Chem. Res.*, 1996, **35**, 1959–1966.

26 Rogers, W. J., Bullin, J. A., Davison, R. R. Frazier, R. E, and Marsh, K. N. FTIR method for VLE measurements of acid-gas–alkanolamine systems. *AIChE J.*, 1997, **43**(12), 3223–3231.

27 Mathonat, C., Majer, V., Mather, A. E., and Grolier, J.-P. E Enthalpies of absorption and solubility of CO_2 in aqueous solutions of methyldiethanolamine. *Fluid Phase Equil.*, 1997, **140**, 171–182.

28 Rho, Seung-Wook, Yoo, Ki-Pung, Lee, Jong Sup, Nam, Sung Chan, Son, Jae Ek, Min, Byoung-Moo. Solubility of CO_2 in Aqueous Methyldiethanolamine Solutions. *J. Chem. Eng. Data*, 1997, **42**(6), 1161–1164.

29 Haji-Sulaiman, M. Z., Aroua, M. K., and Benamor, A. Analysis of equilibrium data of CO_2 in aqueous solutions of diethanolamine (DEA), methyldiethanolamine (MDEA) and their mixtures using the modified Kent Eisenberg model. *Trans. IChemE*, 1998, **76**, 961–996.

30 Tochigi, K., Akimoto, K., Ochi K., Liu, F., and Kawase, Y. Isothermal vapor–liquid equilibria for water + 2-aminoethanol + dimethyl sulfoxide and its constituent three binary systems. *J. Chem. Eng. Data*, 1999, **44**, 588–590.

31 Pérez-Salado Kamps, A., Balaban, A., Joedecke, M., Kuranov, G., Smirnova, N. A., and Maurer, G. Solubility of single gases carbon dioxide and hydrogen sulfide in aqueous solutions of *N*-methyldiethanolamine at temperatures from 313 to 393 K and pressures up to 7.6 MPa: new experimental data and

model extension. *Ind. Eng. Chem. Res.*, 2001, **40**(2), 696–706.

32 Dang, H. and Rochelle, G.T. CO_2 absorption rate and solubility in monoethanolamine/piperazine/water. Presented at the First National Conference on Carbon Sequestration, Washington, DC, 14–17 May 2001.

33 Pérez-Salado Kamps, A., Rumpf, B., Maurer, G., Anoufrikov, Y., Kuranov, G., and Smirnova, N.A. Solubility of CO_2 in H_2O + *N*-methyldiethanolamine + (H_2SO_4 or Na_2SO_4). *AIChE J.*, 2002, **48**, 168–177.

34 Sidi-Boumedine, R., Horstmann, S., Fischer, K., Provost, E., Furst, W., and Gmehling, J. Experimental determination of carbon dioxide solubility data in aqueous alkanolamine solutions. *Fluid Phase Equil.*, 2004, **218**, 85–94.

35 Voutsas, E., Vrachnos, A., and Magoulas, K. Measurement and thermodynamic modeling of the phase equilibrium of aqueous *N*-methyldiethanolamine solutions. *Fluid Phase Equil.*, 2004, **224**(2), 193–197.

36 Ali, B.Si., and Aroua, M.K. Effect of piperazine on CO_2 loading in aqueous solutions of MDEA at low pressure. *Int. J. Thermophys.*, 2004, **25**(6), 1863–1870.

37 Rebolledo-Libreros, M.E. and Trejo, A. Gas solubility of CO_2 in aqueous solutions of N-methyldiethanolamine and diethanolamine with 2-amino-2-methyl-1-propanol. *Fluid Phase Equil.*, 2004, **218**, 261–267.

38 Daneshvar, N., Zaafarani Moattar, M.T., Abedinzadegan Abdi, M., and Aber, S. Carbon dioxide equilibrium absorption in the multi-component systems of CO_2 + TIPA + MEA + H_2O, CO_2 + TIPA + Pz + H_2O and CO_2 + TIPA + H_2O at low CO_2 partial pressures: experimental solubility data, corrosion study and modeling with artificial neural network. *Sep. Purif. Technol.*, 2004, **37**(2), 135–147.

39 Kundu, M. and Bandyopadhyay, S.S. Modelling vapour–liquid equilibrium of CO_2 in aqueous *N*-methyldiethanolamine through the simulated annealing algorithm. *Can. J. Chem. Eng.*, 2005, **83**, 344–353.

40 Hosseini Jenab, M., Abedinzadegan Abdi, M., Najibi, S.H., Vahidi, M., and Matin, N.S. Solubility of carbon dioxide in aqueous mixtures of *N*-methyldiethanolamine + piperazine + sulfolane. *J. Chem. Eng. Data*, 2005, **50**, 583–586.

41 Ma'mun, S., Nilsen, R., and Svendsen, H.F. Solubility of carbon dioxide in 30 mass% monoethanolamine and 50 mass% methyldiethanolamine solutions. *J. Chem. Eng. Data*, 2005, **50**(2), 630–634.

42 Ermatchkov, V., Pérez-Salado Kamps, A., and Maurer, G. Solubility of carbon dioxide in aqueous solutions of *N*-methyldiethanolamine in the low gas loading range. *Ind. Eng. Chem. Res.*, 2006, **45**, 6081–6091.

43 Kundu, M., and Bandyopadhyay, S.S. Solubility of CO_2 in water+ diethanolamine+ *N*-methyldiethanolamine. *Fluid Phase Equil.*, 2006, **248**, 158–167.

44 Barreau, A., Blanchon le Bouhelec, E., Habchi Tounsi, K.N., Mougin, P., and Lecomte, F. Absorption of H_2S and CO_2 in alkanolamine aqueous solution: experimental data and modelling with the electrolyte-NRTL model. *Oil Gas Sci. Technol. – Rev. IFP*, 2006, **61**(3), 345–361.

45 von Niederhausern, D.M., Wilson, G.M., and Giles, N.F. Critical point and vapor pressure measurements for 17 compounds by a low residence time flow method. *J. Chem. Eng. Data*, 2006, **51**, 1990–1995.

46 Kierzkowska-Pawlak, H. Enthalpies of absorption and solubility of CO_2 in aqueous solutions of methyldiethanolamine. *Sep. Sci. Technol.*, 2007, **42**, 2723–2738.

47 Huttenhuis, P.J.G., Agrawal, N.J, Hogendoorn, J.A., and Versteeg, G.F. Gas solubility of H_2S and CO_2 in aqueous solutions of *N*-methyldiethanolamine. *J. Pet. Sci. Eng.*, 2007, **55**, 122–134.

48 Kim, I., Svendsen, H.F., and Børresen, E. Ebulliometric determination of vapor–liquid equilibria for pure water, monoethanolamine, *N*-methyldiethanolamine, 3-(methylamino)propylamine, and their binary and ternary solutions. *J. Chem. Eng. Data*, 2008, **53**(11), 2521–2531.

49 Dingman, J.C., Jackson, J.L., Moore, T.F., and Branson, J.A. Equilibrium data

for the H_2S–CO_2–diglycolamine agent–water system. In *Proceedings of the 64th Annual Convention of the Gas Processors Association*, Gas Processors Association, Tulsa, OK, 1983, pp. 256–268.

50 Martin, J. L., Otto, F. D., and Mather, A. E. Solubility of hydrogen sulfide and carbon dioxide in a diglycolamine solution. *J. Chem. Eng. Data*, 1978, **23**(2), 163–164.

51 Isaacs, E. E., Otto, F. D., Mather, and A. E. Solubility of hydrogen sulfide and carbon dioxide in an aqueous diisopropanolamine solution. *J. Chem. Eng. Data*, 1977, **22**, 71–73.

52 Dell'Era, C., Uusi-Kyyny, P., Rautama, E.-L., Pakkanen, M., and Alopaeus, V. Thermodynamics of aqueous solutions of methyldiethanolamine and diisopropanolamine. *Fluid Phase Equil.*, 2010, **299**, 51–59.

53 Kim, J. H., Dobrogowska, C., and Hepler, L. G. Thermodynamics of ionization of aqueous alkanolamines. *Can. J. Chem.*, 1987, **65**, 1726–1728.

54 Littel, R. J., Bos, M., and Knoop, G. J. Dissociation constants of some alkanolamines at 293, 303, 318,and 333 K. *J. Chem. Eng. Data.*, 1990, **35**, 276–277.

55 Vaidya, P. D. and Kenig, E. Y. A study on CO_2 absorption kinetics by aqueous solutions of *N,N*-diethylethanolamine and *N*-ethylethanolamine. *Chem. Eng. Technol.*, 2009, **32**, 556–563.

56 Chang, H. T., Posey, M., and Rochelle, G. T. Thermodynamics of alkanolamine–water solutions from freezing point measurements. *Ind. Eng. Chem. Res.*, 1993, **32**, 2324–2335.

57 Fosbøl, P. L., Pedersen, M. G., and Thomsen, K. Freezing point depressions of aqueous MEA, MDEA, and MEA–MDEA measured with a new apparatus. *J. Chem. Eng. Data*, 2011, 56(4), p. 995–1000.

58 Touhara, H., Okazaki, S., Okino, F., Tanaka, H., Ikari, K., and Nakanishi, K. Thermodynamic properties of aqueous mixtures of hydrophilic compounds. 2. Aminoethanol and its methyl derivatives. *J. Chem. Thermodyn.*, 1982, **14**, 145–156.

59 Hayden, T. A., Smith, T. G. A., and Mather, A. E. Heat-capacity of aqueous methyldiethanolamine Solutions. *J. Chem. Eng. Data*, 1983, **28**, 196–197.

60 Dohnal, V., Roux, A. H., and Hynek, V. Limiting partial molar excess enthalpies by flow calorimetry: some organic solvents in water. *J. Solution Chem.*, 1994, **23**, 889–900.

61 Posey, M. L. Thermodynamic model for acid gas loaded aqueous alkanolamine solutions. Ph.D. Thesis, University of Texas at Austin, 1996.

62 Weiland, R. H., Dingman, J. C., and Cronin, D. B. Heat capacity of aqueous monoethanolamine, diethanolamine, *N*-methyldiethanolamine, and *N*-methyldiethanolamine-based blends with carbon dioxide. *J. Chem. Eng. Data*, 1997, **42**, 1004–1006.

63 Chiu, L.-F. and Li, M.-H. Heat capacity of alkanolamine aqueous solutions. *J. Chem. Eng. Data*, 1999, **44**, 1396–1401.

64 Maham, Y., Mather, A. E., and Mathonat, C. Excess properties of (alkyldiethanolamine + H_2O) mixtures at temperatures from 298.15 to 338.15 K. *J. Chem. Thermodyn.*, 2000, **32**(2), 229–236.

65 Chen, Y.-J., Shih, T.-W., and Li, M.-H. Heat capacity of aqueous mixtures of monoethanolamine with *N*-methyldiethanolamine. *J. Chem. Eng. Data*, 2001, **46**, 51–55.

66 Zhang, K., Hawrylak, B., Palepu, R., and Tremaine, P. R. Thermodynamics of aqueous amines: excess molar heat capacities, volumes, and expansibilities of {water + methyldiethanolamine (MDEA)} and {water + 2-amino-2-methyl-1-propanol (AMP)}. *J. Chem. Thermodyn.*, 2002, **34**, 679–710.

67 Hawrylak, B., Palepu, R., and Tremaine, P. R. Thermodynamics of aqueous methyldiethanolamine (MDEA) and methyldiethanolammonium chloride ($MDEAH^+Cl^-$) over a wide range of temperature and pressure: apparent molar volumes, heat capacities, and isothermal compressibilities. *J. Chem. Thermodyn.*, 2006, **38**, 988–1007.

68 Jakobsen, J. P., Krane, J., and Svendsen, H. F., Liquid-phase composition determination in CO_2–H_2O–alkanolamine systems: an NMR study. *Ind. Eng. Chem. Res.*, 2005, **44**, 9894–9903.

69 Böttinger, W., Maiwald., M, and Hasse, H. Online NMR spectroscopic study of species distribution in MEA–H_2O–CO_2 and DEA–H_2O–CO_2. *Fluid Phase Equil.*, 2008, **263**(2), 131–143.

70 Hilliard, M. D. A predictive thermodynamic model for an aqueous blend of potassium carbonate, piperazine, and monoethanolamine for carbon dioxide capture from flue gas. Dissertation, University of Texas at Austin, 2008.

71 Faramarzi, L., Kontogeorgis, G. K., Thomsen, K, and Stenby, E. H. Extended UNIQUAC model for thermodynamic modeling of CO_2 absorption in aqueous alkanolamine solutions. *Fluid Phase Equil.*, 2009, **282**, 121–132.

72 Hessen, E. T., Haug-Warberg, T., and Svendsen, H. F. The refined e-NRTL model applied to CO_2–H_2O–alkanolamine systems. *Chem. Eng. Sci.*, 2010, **65**(11), 3638–3648.

73 MacDowell, N., Llovell, F., Adjiman, C. S., Jackson, G., and Galindo, A. Modeling the fluid phase behavior of carbon dioxide in aqueous solutions of monoethanolamine using transferable parameters with the SAFT-VR approach. *Ind. Eng. Chem. Res.*, 2010, **49**, 1883–1899.

74 Bailey, D. W. and Feron, P. H. M. CO_2 capture – post-combustion decarbonisation processes. *Oil Gas Sci. Technol. – Rev. IFP*, 2005, **60**, 461.

75 Blake, R. J. How acid-gas treating processes compare: aqueous alkanolamine still offers inherent advantages that should not be overlooked. *Oil Gas J.*, 1967, **65**, 105–8.

76 MacDowell, N., Florin, N., Buchard, A., Hallett, J., Galindo, A., Jackson, G., Adjiman, C. S., Williams, C. K., Shah, N., and Fennell, P. An overview of CO_2 capture technologies. *Energy Environ. Sci.*, 2010, **3**, 1645–1669.

77 Knuutila, H., Hessen, E. T., Kim, I., Haug-Warberg, T., and Svendsen, H. F. Vapor–liquid equilibrium in the sodium carbonate–sodium bicarbonate–water–CO_2-system. *Chem. Eng. Sci.*, 2010, **65**, 2218–2226.

78 Yi, F., Zou, H.-K., Chu, G.-W., Shao, L., and Chen, J.-F. Modeling and experimental studies on absorption of CO_2 by Benfield solution in rotating packed bed. *Chem. Eng. J.*, 2009, **145**, 377–384.

79 Pérez-Salado Kamps, A., Meyer, E., Rumpf, B., and Maurer, G. Solubility of CO_2 in aqueous solutions of KCl and in aqueous solutions of K_2CO_3. *J. Chem. Eng. Data*, 2007, **52**(3), 817–832.

80 Benson, H. E., Field, J. H., and Jimeson, R. M. CO_2 absorption employing hot potassium carbonate solutions. *Chem. Eng. Prog.*, 1954, **50**, 357–364.

81 Benson, H. E., Field, J. H., and Haynes, W. P. Improved process for CO_2 absorption uses hot carbonate solutions. *Chem. Eng. Prog.*, 1956, **52**, 433–438.

82 Tosh, J. S., Field, J. H., Benson, H. E., and Haynes, W. P. Equilibrium study of the system potassium carbonate, potassium bicarbonate, carbon dioxide, and water. *US Bur. Mines Rep. Invest.*, 1959, No.5485.

83 Gal, E. Ultra cleaning combustion gas including the removal of CO_2. World Patent WO 022885, 2006.

84 Yu, H., Morgan, S., Allport, A., Do, T., Cottrell, A., McGregor, J., and Feron, P. Update on aqueous ammonia based post combustion capture pilot plant at Munmorah. In *Proceedings of Distillation Absorption 2010* (ed. de Haan, A. B., Kooijman, H., and Górak, A.), 2010.

85 Park, S. Y., Yi, K. B., Ko, C. H., Park, J.-H., Kim, J.-N., and Hong, W. H. Selection of optimal operating conditions for a continuous CO_2-capture process using an aqueous ammonia solution. *Energy Fuels*, 2010, **24**, 3704–3709.

86 Rhudy, R. and Gal, E. *Chilled-Ammonia Post Combustion CO_2 Capture System – Laboratory and Economic Evaluation Results. Report 1012797*, EPRI, Palo Alto, CA, 2006.

87 Darde, V., Thomsen, K., van Well, W. J. M., and Stenby, E. H. Chilled ammonia process for CO_2 capture. *Int. J. Greenhouse Gas Control*, 2010, **4**, 131–136.

88 Puxty, G., Rowland. R., and Attala, M. Comparison of the rate of CO_2 absorption into aqueous ammonia and monoethanolamine. *Chem. Eng. Sci.*, 2010, **65**, 915–922.

89 Thomsen, K. and Rasmussen, P. Modeling of vapor–liquid–solid equilibrium in gas–aqueous electrolyte systems. *Chem. Eng. Sci.*, 1999, **54**, 1787–1802.

90 Darde, V., van Well, W. J. M., Stenby, E. H., and Thomsen, K. Modeling of carbon dioxide absorption by aqueous ammonia solutions using the extended UNIQUAC model. *Ind. Eng. Chem. Res.*, 2010, **49**(24), 12663–12674.

91 Mathias, P. M., Reddy, S., and O'Connell, J. P. Quantitative evaluation of the aqueous-ammonia process for CO_2 capture using fundamental data and thermodynamic analysis. *Energy Procedia*, 2009, **1**, 1227–1234.

92 Sengeløv, L. W. Thermodynamic modeling of CO_2 absorption systems. Bachelor thesis, DTU Chemical Engineering, Technical University of Denmark, 2010.

93 Kumelan, J., Pérez-Salado Kamps, A., and Maurer, G. Solubility of CO_2 in aqueous solutions of methionine and in aqueous solutions of (K_2CO_3 + methionine). *Ind. Eng. Chem. Res.*, 2010, **49**(8), 3910–3918.

94 Kumar, P. S., Hogendoorn, J. A., Timmer, S. J., Feron, P. H. M., and Versteeg, G. F. Equilibrium solubility of CO_2 in aqueous potassium taurate solutions: Part 2. Experimental VLE data and model. *Ind. Eng. Chem. Res.*, 2003, **42**, 2841–2852.

95 Versteeg, G. F., Kumar, P. S., Hogendoorn, J. A., and Feron, P. H. M. Method for absorption of acid gases. World Patent WO 03/095071 A1, 2003.

96 Aronu, U. E., Hessen, E. T., Haug-Warberg,T., Hoff, K. A., and Svendsen, H. F. Equilibrium absorption of carbon dioxide by amino acid salt and amine amino acid salt solutions. *Energy Procedia*, 2010, (4), p. 109–116.

97 Portugal, A. F., Sousa, J. M., Magalh~aes, F. D., and Mendes, A. Solubility of carbon dioxide in aqueous solutions of amino acid salts. *Chem. Eng. Sci.*, 2009, **64**(9), 1993–2002.

98 Anthony, J. L., Aki, S. N. V. K., Maginn, E. J., and Brennecke, J. F. Feasibility of using ionic liquids for carbon dioxide capture. *Int. J. Environ. Technol. Manage.*, 2004, **4**, 105–115.

99 Vega, L. F., Vilaseca, O., Llovell, F., and Andreu, J. S. Modeling ionic liquids and the solubility of gases in them: recent advances and perspectives. *Fluid Phase Equil.*, 2010, **294**, 15–30.

7
Physical Absorption Materials for CO_2 Capture

Sebastian Schiebahn, Li Zhao and Marcus Grünewald

Abstract

Pre-combustion capture with physical absorption in integrated gasification combined cycles (IGCCs) offers a feasible opportunity for separating carbon dioxide with a relatively low efficiency penalty. This chapter provides an introduction to the IGCC process and the integration of physical scrubbing to separate CO_2. General considerations for process and material selection for physical absorption are explained. Commercially available physical and mixed chemical–physical solvents are described and characterized with respect to the relevant application. Additionally, current research activities regarding ionic liquids as alternative physical solvents are presented.

Keywords: pre-combustion, physical absorption, integrated gasification combined cycles, ionic liquids, carbon capture

7.1
Introduction

The search for energy-efficient separation of carbon dioxide from point sources such as coal power plants has led to the investigation of different separation methods and technologies, which could be divided into post-, oxy- and pre-combustion capture routes. Depending on the characterizing process conditions, different CO_2 separation techniques are applicable and economically reasonable.

Within this context the main characteristic of integrated gasification combined cycle (IGCC) power plants is the elevated partial pressure offered for capturing CO_2 prior to the final combustion of CO_2. Due to this, physical adsorption and physical absorption are attractive process alternatives in terms of energy efficiency. Hence pre-combustion capture offers the potential for separating CO_2 with the theoretically lowest reachable energetic effort, because only physically ad-/absorbed CO_2 has to be removed from the ad-/absorbent.

In this chapter, we focus on physical absorption media and processes for carbon capture. We discuss the opportunities for integration of physical absorption

Process Engineering for CCS Power Plants, First Edition. Edited by Detlef Stolten and V. Scherer

units in the complex IGCC process and give hints for the choice of the optimal solvent.

7.2 Pre-Combustion Capture in IGCC

In IGCC processes, fuel is gasified at elevated pressure. This provides an opportunity to develop an energy-efficient pre-combustion route in comparison with other capture approaches. For a better understanding of the integration of physical absorption processes, it is essential to look at the IGCC process as a whole, including the complexity and the variety of possible configurations.

7.2.1 IGCC without CO_2 Separation

Figure 7.1 shows a simplified IGCC process without carbon capture. The first step of the IGCC process consists of the gasification of a solid fuel such as coal or bio-waste to produce a combustible gas under elevated pressure:

$$C + \frac{1}{2}O_2 \rightarrow CO \tag{7.1}$$

$$C + H_2O \rightarrow CO + H_2 \tag{7.2}$$

Therefore, several different types of gasifiers are viable choices, such as moving bed, fluid bed, and entrained flow gasifiers [1]. Their major distinguishing features are their operating temperature, their residence time for the fuel used, their oxygen demand, and the composition of the gas produced. The preferred choice for IGCCs for highly efficient power production is entrained flow gasifiers, since they work at very high temperatures (1250–1600 °C), which leads to a low residence time, high carbon conversion rate (>99%), low reactor volume, and syngas consisting mainly of CO and H_2. For gasification pressurized coal, steam and oxygen are required. The coal can be pressurized and fed to the process either by dry-coal feeding via lock-hoppers or by wet-coal feeding as a

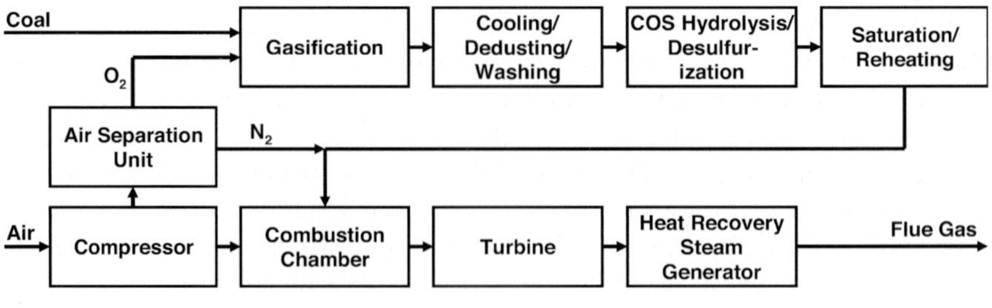

Fig. 7.1 IGCC without carbon capture.

water–coal slurry. The latter is easier to handle and can reach higher pressures, but at the cost of a lower efficiency in the gasifier, since the additional water has to be vaporized and heated additionally, thus increasing the oxygen demand. The oxygen can be provided in a diluted form by air or as a highly concentrated stream (x_{O_2}=95–99 mol%), which is produced by an air separation unit (ASU). Most often, an ASU is used, because using pure oxygen reduces the amount of inert gases, which have to be heated during gasification and carried through the whole process, thereby increasing the volume flow. Hence the utilization of an ASU leads to higher efficiency of the gasifier and a decrease in the size of the subsequent pipes and processing facilities.

After gasification, the syngas, containing molten slag, has to be cooled in order to clean it from fly ash and impurities such as sulfur-containing components (H_2S, COS), halogen compounds (HCl, HF), or ammonia (NH_3). The temperature reduction can be executed by partial or full water quench, by circulated cold syngas, or by radiative coolers. Subsequently, dust can be separated by cyclones or ceramic filters. Venturi scrubbers can provide better purification; removing dust particles in addition to cleaning the water-soluble components such as HCl and NH_3.

The syngas has to be cleaned further to remove sulfur-containing components. Therefore, chemical scrubbing, for example with *N*-methyldiethanolamine (MDEA), is applied in order to separate H_2S for further processing in a Claus plant. Since chemical scrubbing selectively removes only H_2S, the COS present in the syngas is catalytically converted with H_2O to H_2S and CO_2 prior to scrubbing. The desulfurization is usually performed at low temperatures, so that the syngas is heated again before combustion to increase the efficiency and saturated with water to lower thermal NO_x production during combustion.

The purified and saturated syngas, consisting mainly of H_2 and CO, is burned in the gas turbine and expanded to ambient pressure to produce mechanical and thereby electric power. The flue gas is further cooled in a downstream heat recovery steam generator (HRSG) to use the remaining heat for steam and power generation.

7.2.2
IGCC with CO_2 Separation by Physical Absorption

Before separating the carbon content of the syngas prior to combustion, the chemically bound energy of the CO in the syngas has to be transferred into a different carbon-free burnable gas, to maintain the high exergetic value of the syngas. The separation of carbon in the form of CO would cause a very large reduction in the heating value, thus reducing the overall efficiency considerably. The conversion of CO to CO_2 can be achieved by the water gas shift (WGS) reaction, which involves the catalytic reaction of carbon monoxide with steam to yield carbon dioxide and hydrogen (Equation 7.3). The latter now carries a large amount of the chemical energy, so that it is not lost or completely transferred into exergetically less valuable heat.

$$CO + H_2O \rightleftharpoons CO_2 + H_2 \tag{7.3}$$

The WGS reaction is an exothermic equilibrium reaction, which means that the lower the reaction temperature, the more the chemical equilibrium lies on the side of the products. For a high CO conversion, the WGS reaction has to be performed in multiple steps with intermediate cooling including additional steam to shift the equilibrium. Common operating temperatures are between 300 and 500 °C for high-temperature shift and between 200 and 270 °C for low-temperature shift [1]. The additional steam required causes a major fraction of the overall efficiency loss regarding pre-combustion capture [2, 3]. Hence, concerning the amount of additional steam, a compromise has to be made between a high CO conversion rate and a low efficiency loss. A high conversion rate allows for a high CO_2 capture rate.

The positioning of the WGS reactor depends on the sulfur tolerance of the chosen catalyst. Since physical absorption is capable of removing all acid gas components such as H_2S and CO_2 simultaneously, a sour WGS permits the utilization of this feature, since thereby desulfurization and CO_2 removal can be carried out at the same place. To reduce the amount of additional steam, a location with a high water content is favorable, for instance, after the Venturi scrubber. A possible process design is shown in Fig. 7.2.

The subsequent reheating, saturation, and combustion in the combined cycle are the same as those of the IGCC without carbon capture.

The composition of the syngas produced prior to the WGS reactor depends on many factors, such as the feedstock used and the type of coal feeding (wet/dry). The composition after the WGS reactor depends on the CO conversion rate and the outlet temperature. Table 7.1 shows a typical composition of syngas after gasification of Pittsburgh #8 coal in a Shell entrained flow gasifier with dry coal feeding [4].

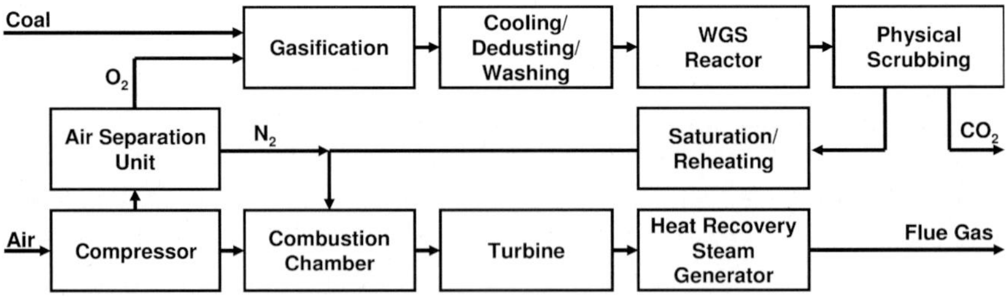

Fig. 7.2 IGCC with carbon capture by physical scrubbing.

Table 7.1 Gas composition prior to WGS reactor at $p = 26.8$ bar and $T = 133\,^{\circ}C$ [4].

Component	Mole fraction
CO	53.3%
H_2	30.4%
H_2O	11.8%
CO_2	0.4%
N_2	2.6%
Ar	0.7%
H_2S	7350 ppm
COS	550 ppm

7.3 Physical Absorption Materials and Processes

7.3.1 Process Description and Design

The process of physical absorption is based on the solubility of acid gases in liquid solvents to separate fractions of sour gases such as CO_2 and H_2S from a gas stream. According to Henry's law, the solvent loading is proportional to the partial pressure of the gas component considered, for example, CO_2:

$$x_{CO_2} \times H_{CO_2,L} = y_{CO_2} \times p \tag{7.4}$$

where $H_{CO_2,L}$ is the Henry's law constant of carbon dioxide in the solvent L [5]. In the literature, many different expressions for solubility, which describes the amount of a specific gas that can be dissolved in the solvent, exist, such as Henry constants (MPa) [6], absorption coefficients λ (mol $kg^{-1}_{solvent}$) [7], and mole fractions x (mol $mol^{-1}_{solvent}$) [8] at a given temperature and pressure. Since the circulation rate of the solvent mainly dominates the economics, regarding both the investment costs and the operating costs in the form of pumps [9], this chapter expresses the solubility, needed to calculate the loading c, as the ratio of volume gas per unit volume of solvent ($m^3\ m^{-3}_{solvent}$ [10, 11]. However, in combination with the molecular weight and density of the solvent and the molecular weight of the examined gas component, it is easy to convert them into each other.

Figure 7.3 shows characteristic equilibrium lines for chemical and physical absorption at a given temperature. As illustrated, the influence of the positive effects dominates as the partial pressure of the considered gas component increases, since the cyclic loading capacity Δc_{ph} rises continuously, whereas c_{ch} is limited by chemical boundaries.

Hence physical absorption is more suitable for high partial pressures, making this process applicable to decarbonization and acid gas removal of pressurized

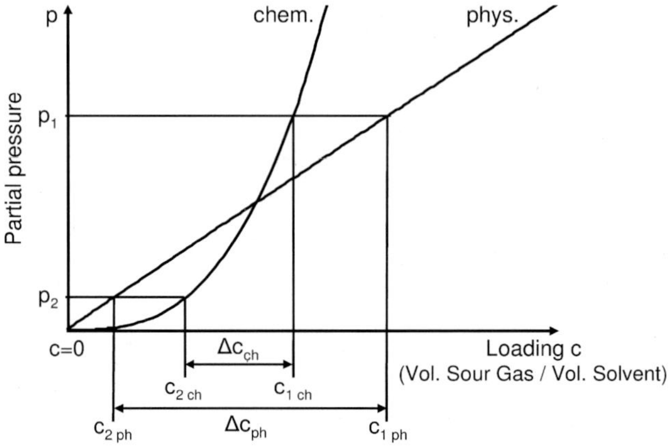

Fig. 7.3 Equilibrium lines for chemical and physical absorption [12].

syngas and natural gas. Depending on the solvent chosen, the physical absorption process is capable of reducing the amount of acid gases to a value in the ppm range and of producing a CO_2 stream of high purity [13].

The benefits of physical scrubbing used for high-pressure acid gas removal compared with chemical scrubbing are as follows:

- Lower circulation rate due to the high cyclic loading capacity.
- Lower electric energy and heat consumption for regeneration, since most of it can be done by flashing. In addition, no chemical bonds have to be broken, which reduces the heat of reaction or solution to an insignificant factor [14].
- Good stability of the solvent, because no reactions occur.
- Capable of separating acid gases selectively.

The process design of physical scrubbers is very flexible and can be adapted to the given task to meet the required targets of degree of separation and purity of the product streams.

The physical solvent is loaded in the absorber by falling from the head of the column in counter-current flow to the rising gas stream. This is basically the same for all physical scrubbers, although an additional inlet of solvent can be arranged at a medium height. This can be reasonable if the column is divided into a part for primary cleaning and a part for final purification, whereby regeneration costs are saved, since the part of the solvent for the primary cleaning does not have to be regenerated at very high purity.

Regeneration of the solvent can be done by three different methods, which are illustrated in Fig. 7.4. The simplest means of regeneration is flash desorption (Fig. 7.4 a), which means that the pressurized, loaded solvent from the bottom of the absorber column is fed to a desorber with a lower operating pressure. According to Fig. 7.3, the solvent releases dissolved gas until equilibrium is reached again. This can be done in multiple steps to gain the first released gas under pressure, reducing the later levels even to vacuum. The amount of re-

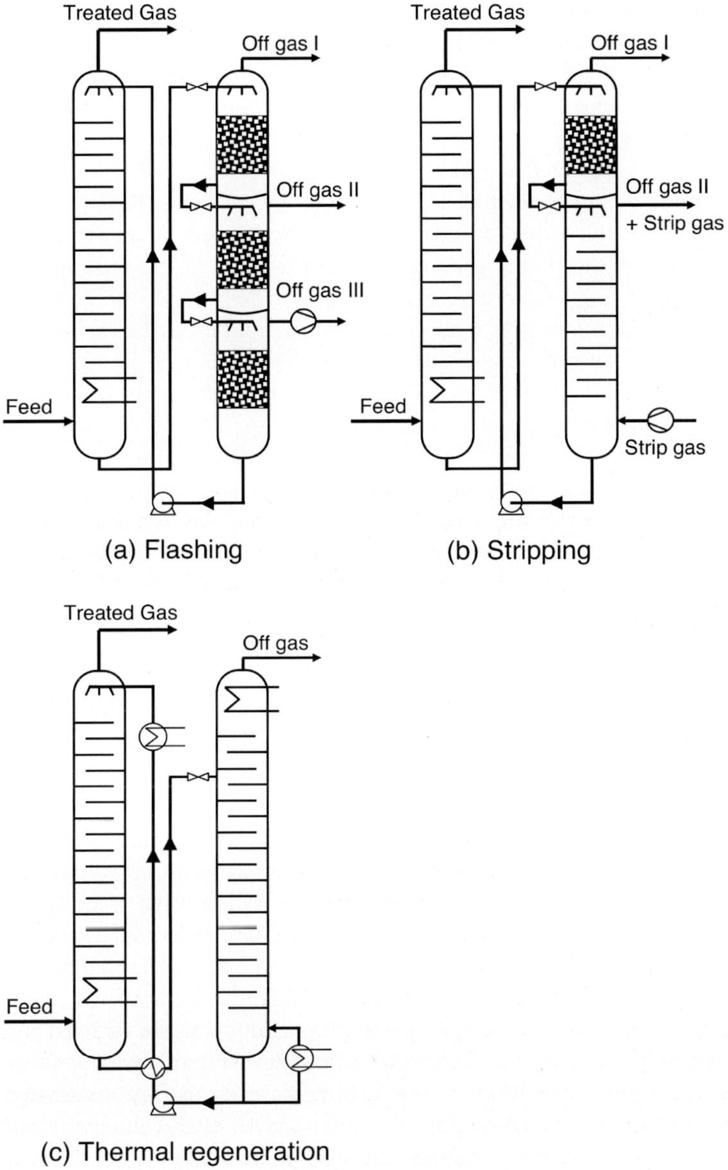

Fig. 7.4 Different basic methods of solvent regeneration [12].

leased gases and their purity depend on the particular operating pressure and the solvent chosen. Sometimes, if valuable gases such as H_2, CO, or methane are also dissolved to a significant extent, it may be expedient to redirect the product of the first flash to the entry of the absorber column to minimize loss of valuable gas components.

Regeneration by stripping (Fig. 7.4b) is basically similar. The first step consists of pressure reduction to release a large amount of the gases. Subsequently, the solvent is fed to a desorber column, where it is stripped by an inert gas. The inert gas reduces the partial pressure of the dissolved gases, thus having the same effect as if the total pressure was reduced. Because of the very low partial pressure at the inlet of the inert gas, the regeneration can be carried out close to purity of the solvent. The disadvantage is the dilution of the separated gases by the inert gas. In the case of CO_2 separation, these novel impurities could be a problem when storing the CO_2.

The energetically most expensive but also the most effective regeneration is done thermally (Fig. 7.4c). Here, the solution is heated to its boiling point, which has two effects. First, the solubility decreases rapidly with increase in temperature (see Fig. 7.5, later). Second, the boiling of the solution results in the production of stripping gas. The necessary additional expenditure for this is reflected in higher investment costs for heat exchangers and higher operating costs due to the additional heat necessary for boiling and supplying the desorption energy of the dissolved gases. In an IGCC process, the additional heat can be supplied by low-pressure steam. Analyses are necessary to assess whether it is favorable to utilize the steam for electricity production or for solvent regeneration.

Basically, every degree of purity of the syngas and also of the separated streams can be reached by a combination of the described methods of regeneration. Nevertheless, high-purity grades have to be achieved by complex physical scrubbing processes with an intricate design of regeneration, thus resulting in high expenditure for compounds and operation. In addition, a complex process is always accompanied by more difficult handling and reduced availability of the facility.

When designing a physical absorption process, in addition to the targeted purities of the product streams, the aim is to reduce investment and operating costs. In addition to the complexity of the facility, these factors are dominated by the solvent circulation rate, which influences the size of nearly all equipment and also the energy demand for pumping and regeneration [9].

A reduction in the amount of a certain solvent circulated can primarily be accomplished by reducing the operating temperature, since the absorption capacity increases with decrease in temperature. Using heat exchangers to maintain the refrigeration of process streams leaving the physical scrubber is common practice. If further temperature reduction is reasonable, this can be done by mechanical or absorption refrigeration [10]. The minimum process temperature is limited primarily by the increasing viscosity of the solvent, which causes a decrease in the solvent's heat and mass transport capabilities.

Another commonly used approach to save power for circulation is the use of hydraulic turbines to regain power, when the rich solvent is flashed. Thereby, up to almost 50% of the solvent pumping energy can be saved [10].

However, when implementing a physical scrubber process into an IGCC, the plant has to be considered as a whole. This is the only way to evaluate what efforts, such as for external refrigeration or complex heat exchanger systems, are energetically and economically reasonable.

7.3.2
Process Selection

Several physical solvents and the corresponding processes have been developed, licensed, and introduced commercially for applications of bulk CO_2 and H_2S removal (see Table 7.2). When selecting and designing the appropriate physical absorption process, several parameters have to be taken into account [10, 14]:

- Framework conditions, such as syngas flow rate and composition, especially concerning CO_2, H_2S, and COS, and also process pressure and temperature.
- Extent of acid gas removal required from the syngas. On the one hand, the degree to which CO_2 bulk removal is required and energetically and economically reasonable; on the other hand, the total sulfur removal required, for example, for environmental specification.
- Required product purity of the separated CO_2 for sequestration or further processing and also the purity of the sulfur removal stream in terms of suitability for further treatment (e.g., for a Claus plant feed). The integration of the physical absorption process into the whole plant has to be considered to evaluate whether additional expenditure in the physical scrubbing process is favorable in comparison with additional treatment.
- Capital costs in the form of column sizes and also auxiliary equipment (heat exchanger, pumps, compressors, etc.).
- Operating costs and energy consumption, due to heat requirements for solvent regeneration and electric power demand for pumps, compressors, and refrigeration, if used. In the case of CO_2 sequestration, it is important to take the CO_2 release pressure into account, since a higher outlet pressure reduces the need for further compression, thus saving energy elsewhere.
- Acceptable loss of valuable constituents, typically in the form of light and heavy hydrocarbons.
- Cost of initial solvent charge and solvent losses, which have to be replaced.
- Process complexity, influencing the availability, and flexibility, regarding adaptation to different feedstocks, new purity requirements or disturbances.
- Experience with regard to corrosion, foaming, or other operating problems.
- Process royalty costs.

7.3.3
Commercially Available Solvents

Water was the first physical solvent used to absorb carbon dioxide, particular in ammonia synthesis [15], but the solubilities of CO_2 and H_2S in water are too low to establish water washing as a commercially practicable physical absorption process [10]. Organic solvents have replaced water, because of their much higher solvent loading capacity.

Although a large number of organic solvents seem to be applicable for physical scrubbing, their number is limited by a series of requirements that have to be fulfilled to make the solvent economically usable. The solubility for acid gases, especially CO_2 and H_2S, has to be several times that of water (see Table 7.3), combined with a high selectivity compared with valuable components such as hydrogen and carbon monoxide. A very low vapor pressure at operating temperatures is advantageous to reduce solvent losses. Additionally, a low viscosity is necessary to ensure high heat and mass transport capabilities. The solvent must be non-corrosive towards metals and chemically inert towards all components in the syngas stream. Finally, it has to be commercially available at an acceptable price for the particular application.

Table 7.2 lists commercially available organic physical and mixed solvents that which fulfill these criteria the best, with the corresponding process name and licensor.

Table 7.2 Commercially established physical and mixed solvents [10, 11, 16, 17].

Component	Solvent	Process licensor	Ref. a)
Physical solvents			
Rectisol	Methanol	Lurgi and Linde	8–10, 12, 14, 15, 18–23
Selexol	Dimethyl ethers of poly(ethylene glycol) (DMPEG)	Union Carbide/UOP	8, 10, 11, 14, 15, 19, 20, 23–25
Purisol	*N*-Methyl-2-pyrrolidone (NMP)	Lurgi	6, 8–12, 15, 20, 23, 26
Fluor Solvent	Propylene carbonate (PC)	Fluor Daniel	6, 10, 11, 15, 20, 23, 25–27
Sepasolv MPE	Methyl isopropyl ether of poly(ethylene glycol) (MPE)	Badische (BASF)	10, 11, 23, 28
Ifpexol	Methanol	Institut Français du Pétrole (IFP)	8, 10, 19, 21, 23
Mixed physical–chemical solvents			
Sulfinol	Sulfolane + DIPA or MDEA	Shell Global Solutions	6, 10, 14, 15, 26, 29
Amisol	Methanol + secondary alkylamine	Lurgi	10, 15, 30

a) For process description and solvent parameters.

Table 7.3 Physical properties of solvents [11, 20].

Property	Solvent			
	Methanol	DMPEG	NMP	PC
Viscosity at 25 °C (cP)	0.6	5.8	1.65	3
Specific gravity at 25 °C (kg m^{-3})	785	1030	1027	1195
Molecular weight (g mol^{-1})	32	280	99	102
Vapor pressure at 25 °C (Pa)	16.7×10^3	0.097	53	11.3
Freezing point (°C)	–92	–28	–24	–48
Boiling point at 1 atm (°C)	65	275	202	240
Thermal conductivity (W m^{-1} K^{-1})	0.21	0.19	0.16	0.21
Maximum operating temperature (°C)	–	175	–	65
Specific heat at 25 °C (kJ kg^{-1} K^{-1})	2.37	2.05	1.68	1.4
CO_2 solubility[a] at 25 °C ($m^3_{CO_2}$ $m^{-3}_{solvent}$)	3.18	3.63	3.57	3.40

a) For comparison, the solubility of CO_2 in water at 35 °C and 1 atm CO_2 partial pressure is 0.6 $m^3_{CO_2}$ m^{-3}_{water} [9].

Since the process layout and the proper composition of the solvents are the property of the respective licensors, the exact performances and solvent characteristics are valuable know-how, and thus unavailable for the public [11]. Nevertheless, a series of open publications are available, sufficient for a qualitative characterization. Table 7.3 summarizes the properties of methanol, dimethyl ethers of poly(ethylene glycol) (DMPEG), *N*-methyl-2-pyrrolidone (NMP), and propylene carbonate (PC).

The corresponding solubilities, expressed as volume of dissolved gas per unit volume of solvent, and selectivities related to carbon dioxide, are presented in Table 7.4.

A general examination of all physical solvents reveals some common characteristics. As shown in Table 7.3, the amount of carbon dioxide dissolved at a giv en temperature does not vary very much between the different solvents. Even so, the actual quantity dissolved under operating conditions differs considerably, since the solubility is strongly dependent on the temperature. On the basis of methanol, Fig. 7.5 shows the dependence of the solubility on the operating temperature. For example, the capability to dissolve carbon dioxide rises to about 60 $m^3(STP)_{CO_2}$ $m^{-3}_{methanol}$ at –60 °C. With the assumption that Henry's law is applicable over the whole pressure range, at a CO_2 partial pressure of 10 bar this is equal to a molar loading of $x = 0.99$ mol_{CO_2} $mol^{-1}_{methanol}$.

Regarding the selectivity, all physical solvents show the same tendency. The heavier the hydrocarbons, the better they are absorbed. However, due to the high gasification temperature, the concentration of heavy hydrocarbons in the syngas is very low. Hence the high solubility of heavy hydrocarbons does not result in an expedient loss of valuable gases. Another characteristic is that among the important fractions to be separated such as CO_2, H_2S and COS, H_2S is always the most soluble gas, followed by COS.

Table 7.4 Absolute and relative solubilities of physical solvents [11, 20].

Gas	Solvent							
	Methanol (Rectisol)		DMPEG (Selexol)		NMP (Purisol)		PC (Fluor Solvent)	
	Solubility at 25 °C[a] ($m^3\ m^{-3}$)	Selectivity relative to CO_2 at −25 °C	Solubility at 25 °C ($m^3\ m^{-3}$)	Selectivity relative to CO_2 at 25 °C	Solubility at 25 °C ($m^3\ m^{-3}$)	Selectivity relative to CO_2 at 25 °C	Solubility at 25 °C ($m^3\ m^{-3}$)	Selectivity relative to CO_2 at 25 °C
Hydrogen	0.017	0.0054	0.047	0.013	0.023	0.0064	0.027	0.0078
Nitrogen	0.038	0.012	0.073	0.020	–	–	0.029	0.0084
Oxygen	0.064	0.020	–	–	0.12	0.038	0.089	0.026
Carbon monoxide	0.064	0.020	0.102	0.028	0.075	0.021	0.071	0.021
Methane	0.16	0.051	0.24	0.066	0.26	0.072	0.13	0.038
Ethane	1.33	0.42	1.52	0.42	1.36	0.38	0.58	0.17
Ethylene	1.46	0.46	1.70	0.47	1.96	0.55	1.19	0.35
Carbon dioxide	3.18	1.0	3.63	1.0	3.57	1.0	3.40	1.0
Propane	7.47	2.35	3.66	1.01	3.82	1.07	1.74	0.51
n-Butane	–	–	8.59	2.37	12.4	3.48	5.95	1.75
Carbonyl sulfide	12.5	3.92	8.34	2.30	9.70	2.72	6.40	1.88
Hydrogen sulfide	22.4	7.06	32.0	8.82	36.4	10.2	11.2	3.29
Ammonia	73.7	23.2	17.4	4.80	–	–	–	–
n-Pentane	–	–	19.8	5.46	–	–	17.0	5.0
Methanethiol	–	–	81.2	22.4	121	34.0	92.5	27.2
Sulfur dioxide	–	–	334	92.1	–	–	233	68.6
Water	–	–	2647	730	14266	4000	1021	300

a) Solubilities in MeOH at 25 °C calculated with selectivities of MeOH at −25 °C.

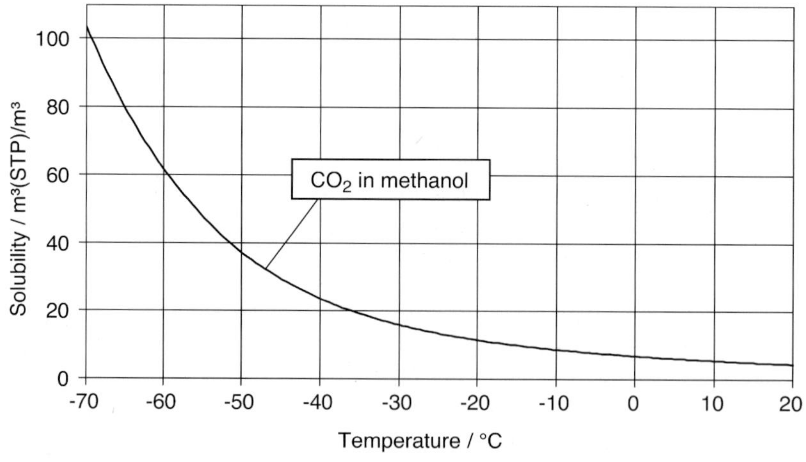

Fig. 7.5 Solubility of carbon dioxide in methanol as a function of temperature at 1 atm CO_2 partial pressure [18].

An obvious feature of all solvents is the high attraction of water, which will cause an accumulation in the solvent. The water content of the physical absorbent is an important factor, since large amounts influence the solubility characteristics and generate extra pumping costs [11]. Another effect of dissolving water is the high heat of absorption, which is set free, resulting in an appreciable increase in temperature.

Specific features of the individual physical solvents are discussed in the following sections.

7.3.3.1 Rectisol

The Rectisol process was invented by Lurgi around 1950 and further developed in cooperation with Linde [10]. Rectisol wash units are operated worldwide in coal- and oil-based facilities for the purification of hydrogen, ammonia, and methanol syngas and also the production of pure carbon monoxide and oxo alcohols [13]. The Rectisol process is very flexible and can be adapted to meet the product specifications and process objectives. Hence many different flow schemes are possible, depending on the optimized configuration.

The Rectisol process uses methanol as solvent. Due to the high vapor pressure at ambient temperature, the operating temperature has to be reduced drastically, typically to around –40 to –62 °C. At these temperatures, the viscosity is still low enough that it remains sufficient for heat and mass transfer capabilities. As shown in Fig. 7.5 and 7.6, the solubility increases with decrease in temperature. Additionally, the solubilities of hydrogen, nitrogen, and carbon monoxide are nearly independent of the temperature [15]. Hence, additionally to the overall solubility, the selectivity under the required operating conditions is improved compared with ambient temperature.

The high selectivities of methanol for H_2S and COS over CO_2 means that the Rectisol process can achieve deep sulfur removal (<0.1 ppmv H_2S plus COS)

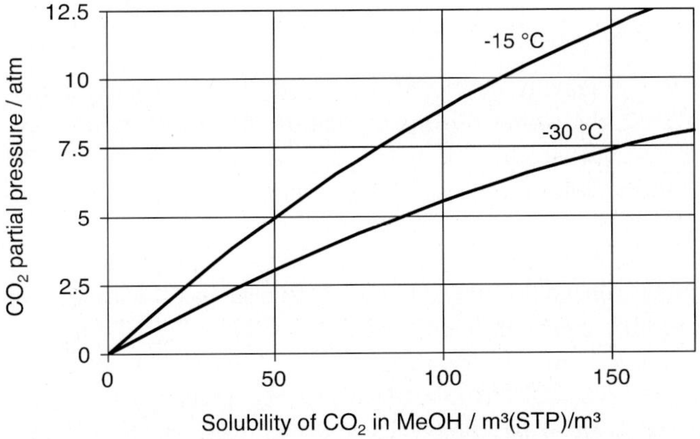

Fig. 7.6 Effect of partial pressure on the solubility of carbon dioxide in methanol [12].

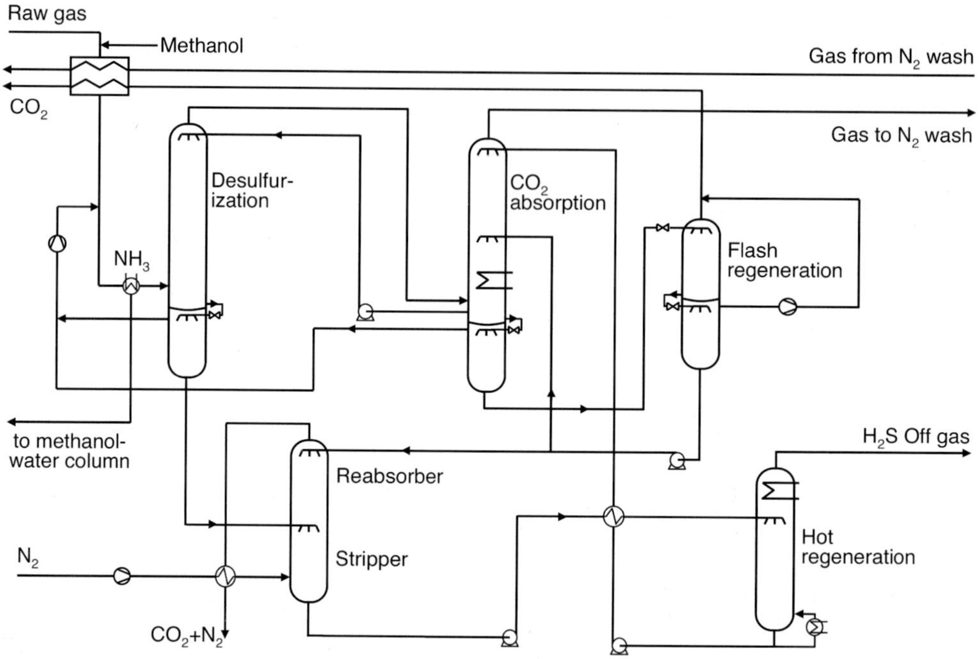

Fig. 7.7 Selective Rectisol process for the separate removal and recovery of H_2S and CO_2 [15].

without prior COS hydrolysis. On top of this, chilled methanol also absorbs other impurities such as HCN, NH_3, and iron and nickel carbonyls [14].

Although a major part of the cooling is done by internal heat exchange, the process requires refrigeration to operate, which is very expensive and energy intensive. This and the complex process scheme (Fig. 7.7) are the major disadvantages for the Rectisol process. A detailed description of different Rectisol process designs and applications can be found elsewhere [10, 15].

7.3.3.2 **Selexol**

The Selexol process was originally developed by Allied Chemical Corporation over 30 years ago. In 1982, the Norton Company acquired the rights to this process. In 1990, Union Carbide acquired it from Norton, and since 1993, UOP has owned the rights of the Selexol process and is responsible for process licensing.

The Selexol process has established itself for a variety of utilizations. Most commercial applications consist of natural and synthesis gas treatment, in which the feed gas has high concentrations of H_2S and CO_2 or CO_2 alone.

The solvent used is a mixture of dimethyl ethers of poly(ethylene glycol) (DMPEG), which has the chemical formula $CH_3(CH_2CH_2O)_nCH_3$ with n in the range 3–9 [14]. It has proven to be chemically and thermally stable and also non-toxic and biodegradable. The low vapor pressure reduces solvent losses and

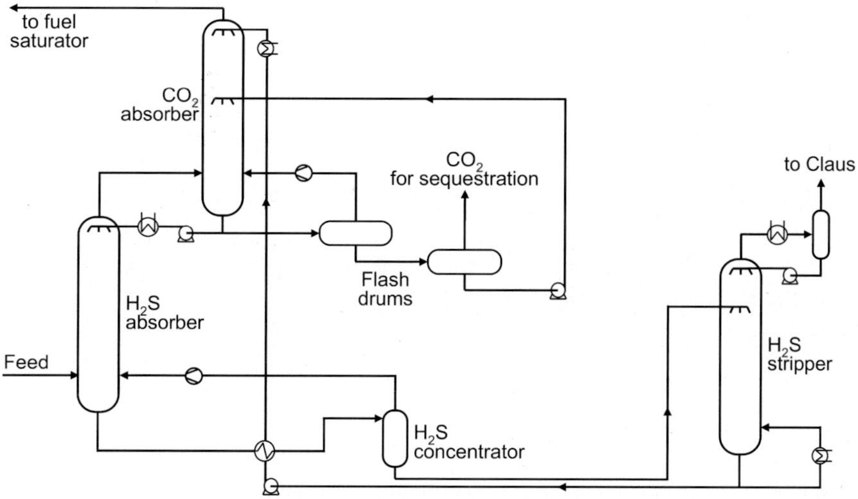

Fig. 7.8 Two-stage Selexol process for selective hydrogen sulfide [10] and CO_2 removal [31].

makes it applicable for use at elevated temperatures (up to 175 °C), although this feature has to be compensated by an increased solvent circulation rate.

DMPEG has a very good selectivity for hydrogen sulfide over carbon dioxide, making it favorable for the selective separation of these acid gases. On the other hand, the solubility of carbonyl sulfide is only twice that of carbon dioxide, which results in a dilution of the acid gas stream for the Claus plant, if COS is also meant to be separated. In this case, COS hydrolysis to convert the carbonyl sulfide to hydrogen sulfide may be reasonable.

The Selexol process can be configured in many ways. Owing to the high selectivities, a two-stage process, as shown in Fig. 7.8, is expedient for the acquisition of two high-purity product streams. The major part of H_2S is absorbed in the first column. The solvent is regenerated thermally in the H_2S stripper and recirculated to the top of the CO_2 absorber, where it serves for final gas cleaning of the syngas due to its high purity. The CO_2 separation occurs in the second column, where the major fraction of the lean solvent is inserted at an intermediate stage in the column, it is then regenerated by flashing at different pressure levels. To increase the purity of the product stream and minimize loss of valuable gases, the gas of the first flash is re-fed to the insert of the absorber.

7.3.3.3 **Purisol**

The Purisol process was developed and commercialized by Lurgi. It is utilized for acid gas removal from natural gas, fuel gas, and syngas and is advantageous if high CO_2 contents have to be separated or bulk acid gas removal is required [15].

The Purisol process uses a mixture of *N*-methyl-2-pyrrolidone (NMP) and water as solvent. It can be operated either at ambient temperature or, with refrigeration, at temperatures down to –15 °C. NMP has the highest selectivity of the solvents considering H_2S over CO_2 and also a very good selectivity of CO_2 over H_2 and CO. Owing to this quality, the Purisol process is able to produce highly purified product streams, whereby most of the regeneration can be achieved by simple flashing, thus reducing energy consumption. The solubility of COS compared with CO_2 is only a factor of 2.7, but COS is partly hydrolyzed to H_2S by the NMP solvent, so that 60–80% COS removal is achieved indirectly.

The design of the Purisol process is similar to that of the Selexol process. However, the vapor pressure of NMP is high in comparison with PC and DMPEG and water washing of the all product streams is recommended to reduce solvent losses.

7.3.3.4 **Fluor Solvent**

The Fluor Solvent process was commercialized in the late 1950s by the cooperative efforts of El Paso Natural Gas Company and Fluor and was first applied to remove carbon dioxide from natural gas at mildly sub-ambient temperatures [10].

The solvent used in the Fluor Solvent process is propylene carbonate (PC), which has a good selectivity of carbon dioxide over all light hydrocarbons. Thereby, loss of valuable gases is reduced and the regeneration can be accomplished by simple flashing. However, the solubilities of H_2S and COS are relatively low compared with CO_2, making this process unattractive for selective H_2S and CO_2 removal. This makes the application of the Fluor Solvent process expedient only if H_2S is present in very low concentration.

PC is not completely soluble in water and reacts slowly but irreversibly with water and CO_2 at around 90 °C. This chemical instability makes it impossible to regenerate PC thermally. The operating temperature is limited to a minimum of –18 °C and a maximum of 65 °C [20].

7.3.3.5 **Other Solvents**

In addition to the solvents considered above, a wide variety of other solvents exist, although their general characteristics are often similar to those described above. For example, the Sepasolv MPE process, devolved by BASF, uses a mixture of poly(ethylene glycol) methyl isopropyl ethers (MPE) and resembles in process design and solvent properties the Selexol process [11]. Ifpexol, introduced by the Institut Français du Pétrole, utilizes methanol and is applied for natural gas treatment regarding dehydration, natural gas liquids recovery, and acid gas removal [10].

Ongoing research is also investigating novel solvents such as perfluorinated compounds (PFCs) for pre-combustion capture of CO_2 at elevated pressures and temperatures up to 227 °C [32].

7.3.4 Mixed Solvents

A number of processes use a mixture of amines and physical solvents to take advantage of the best characteristics of the individual compounds. They are termed physical–chemical absorption or mixed solvent processes. In comparison with pure physical solvents, mixed solvents show higher loading capacities, especially at low partial pressures, due to the additional chemical bonding with the amine [10]. Thus, the physical solvent absorbs the major fraction of the acid gases at high partial pressures, whereas the chemical solvent guarantees the final absorption to a high degree of purification. The process flow schemes resemble that of a conventional amine treating unit, although less steam is required for regeneration and additional flashing is integrated to set free part of the dissolved gas. Among the best known are the Shell Sulfinol and the Lurgi Amisol processes, which are described below.

7.3.4.1 Sulfinol

The Sulfinol process was developed by Shell in 1964 and was first applied for hydrogen sulfide and carbon dioxide removal from gasification gases in steam reforming, and also from natural gases. Since then, it has become established and widely used for acid gas treatment of natural, refinery, and synthesis gases. The solvent used is a solution containing an alkanolamine, sulfolane (tetrahydrothiophene dioxide), and water. The physical solvent provides a solution with a high bulk removal capacity. Especially all sulfur compounds, such as H_2S and COS, are very soluble in sulfolane, providing the Sulfinol process with excellent selectivity for H_2S and COS over CO_2.

There are different variants of Sulfinol solvents, characterized by their chemical constituents. Sulfinol-D utilizes diisoproanolamine (DIPA) as the alkanolamine and is favorable when essentially complete removal of both H_2S and CO_2 and also deep removal of COS are required [10].

The chemical component of Sulfinol-M is methyldiethanolamine (MDEA). Sulfinol-M is utilized when a higher selectivity of H_2S over CO_2 is required, and is capable of almost complete removal of COS in addition to reducing the total amount of all sulfur components to less than 40 ppmv [14].

Recent developments at Shell resulted in the Sulfinol-X solvent, containing sulfolane, water, MDEA, and piperazine. The piperazine functions as an accelerator, enhancing CO_2 reaction kinetics and COS hydrolysis. This is reported to result in higher CO_2 and COS removal rates for the same absorber size [29].

7.3.4.2 Amisol

The Amisol process was introduced by Lurgi. It is suitable for treating raw gases, produced by gasifying coal, peat, or heavy oil, with selective desulfurization or complete separation of CO_2, H_2S, and COS along with trace components.

Table 7.5 Properties of amines used in the Amisol process [30].

Property	DIPAM	DETA	DEA
Chemical formula	$[(CH_3)_2CH]_2NH$	$(C_2H_5)_2NH$	$(HOC_2H_4)_2NH$
Molecular weight	101.2	73.14	105.14
Boiling point (°C)	84	56	269
Vapor pressure at 20 °C (Pa)	1×10^4	2.53×10^4	10
Density at 20 °C (kg m^{-3})	0.716	0.704	1.092

The Amisol process uses a combination of physical and chemical solvents for acid gas removal (Table 7.5). Like the Rectisol process, it utilizes methanol as the physically absorbing component. The first plants in the early 1970s utilized alkanolamines, typically MEA and DEA, as chemical solvents to separate CO_2, H_2S, and COS. When selective absorption of sulfur components versus carbon dioxide is required, aliphatic alkylamines such as diisopropylamine (DIPAM) and diethylamine (DETA) are preferred rather than alkanolamines. In comparison, they have greater stability, higher acid gas loading capacity, easier regeneration at a lower reboiler temperature, and higher volatility [15].

Whereas the absorption takes place at ambient temperature, the regeneration is performed only at 80 °C, since the boiling point of methanol is very low [10]. At these process temperature levels, the volatility of the solvent is very high, so water wash stages at the top of each column are required to reduce solvent losses.

7.3.5
Ionic Liquids

Ionic liquids (ILs) are salts in the liquid state. They are composed of cations and anions, and melt at or below 100 °C [33]. All ionic liquids possess low volatility because of their salt-like character. Moreover, the physico-chemical properties, such as solubility, viscosity, electrical conductivity, and stability, can vary widely depending on the types of the ions. Any salt that melts without decomposing or vaporizing usually yields an IL. Conversely, when an IL is cooled, it often forms an ionic solid, which may be either crystalline or glassy. The ionic bond is usually stronger than the van der Waals forces between the molecules of ordinary liquids [34].

Research on ILs began in 1914 with the observation of the melting point of ethylammonium nitrate $[C_2H_5][NH_3][NO_3]$ at 13–14 °C by Walden [35]. In an authoritative review [33], the history of the development of ILs was well illustrated. A public, verified, web-based database of physical, thermodynamic and related data for ILs is now available [36].

Table 7.6 Comparison of organic solvents with ionic liquids[a)] [33].

Property	Organic solvents	Ionic liquids
No. of solvents	$>10^3$	$>10^6$
Catalytic ability	Rare	Common and tuneable
Vapor pressure	Obeys the Clausius–Clapeyron equation	Negligible vapor pressure under normal conditions
Flammability	Usually flammable	Usually nonflammable
Solvation	Weakly solvating	Strongly solvating
Polarity	Conventional polarity concepts apply	Polarity concept questionable
Tuneability	Limited range of solvents available	Virtually unlimited range means "designer solvents"
Cost	Normally cheap	Between 2 and 100 times the cost of organic solvents
Viscosity (cP)	0.2–100	22–40 000
Density (g cm^{-3})	0.6–1.7	0.8–3.3

a) The data summarized in this table are not comprehensive, nor do they represent outliers; they are meant to give a brief visual comparison of typical values. Detailed values can be found elsewhere [38].

7.3.5.1 General Properties

Unlike the ~600 conventional solvents that are extensively used in industrial processes, ILs represent a different paradigm. As there are at least one million simple ILs, and so up to one trillion ternary systems, they can be designed to optimize a specific reaction, for example, to deliver maximum yield and purity of the isolated product. This is why ILs have been named "designer solvents" [37]. Table 7.6 compares the properties of ILs and organic solvents.

ILs can be used in a great variety of applications, for example, as solvents, catalysts, and electrolytes in the chemical, technical, biochemical, and pharmaceutical industries. Obviously the important issues for industry are security of supply and cost. With regard to supply, this is no longer an issue as there are now many suppliers of ILs, from the gram to the multi-ton scale [33, 39].

7.3.5.2 Using ILs for CO_2 Absorption

ILs represent a platform for the development of new processes and materials aimed at the capture of CO_2 from flue gas and syngas. The physical properties of ILs may be modified and adjusted by employing different cation–anion combinations [40–44]. For these reasons, a large number of studies on the solubility of CO_2 in ILs have been performed [44–51].

Room-temperature ionic liquids (RTILs) are defined as those salts whose melting points are below ambient temperature [44]. Owing to their extremely low vapor pressure, such as $\sim 10^{-10}$ Pa at 25 °C for 1-butyl-3-methylimidazolium hexafluorophosphate [36], CO_2 can be desorbed without loss of solvent. RTILs

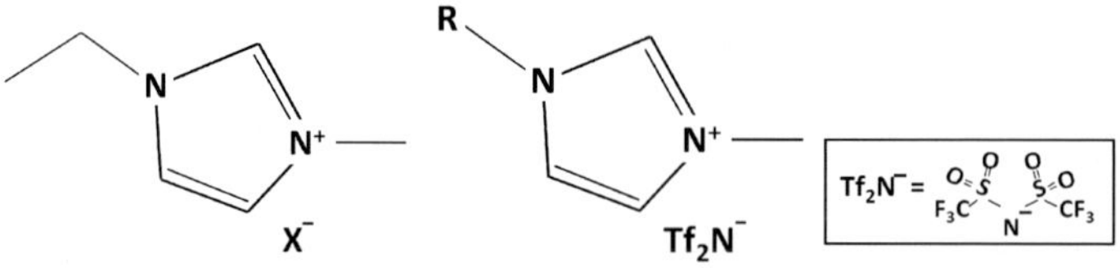

Fig. 7.9 General structures of 1-ethyl-3-methylimidazolium ($C_6H_{11}N_2$) [emim][X] and [Rmim][Tf_2N] [36, 44].

can be used in a pressure-swing configuration, where CO_2 is preferentially absorbed, while H_2, CH_4, or N_2 and other gases are swept away. Such a scheme can operate similarly to common physical absorption processes, such as the Selexol process shown in Fig. 7.8 [44]. The ideal single gas solubility of CO_2 in RTILs at temperatures near ambient and low pressures is comparable to some of the lower values measured for common organic solvents [52]. The ideal solubility selectivities, that is, the ratio of pure-component solubilities of CO_2 relative to N_2 and CH_4 in RTILs [53–55], are also similar to those of common organic solvents. RTILs offer unprecedented control over solvent composition and functionality. This attractive feature, coupled with their unique physical properties, has made CO_2 separations one of the most prominent areas for the development of RTILs.

Imidazolium-based RTILs (see Fig. 7.9) have been the systems of choice for CO_2 absorption, as they are commercially available or readily synthesized and highly tuneable. Although other classes of RTILs, such as phosphonium [47, 54], ammonium [49, 50], and pyridinium [40,56], have certainly attracted attention in CO_2 separations, imidazolium salts dominate the literature [43, 44, 46, 48, 57–59]. A series of 1-ethyl-3-methylimidazolium molten salts and closely related systems were the first set of RTILs to be widely examined [40, 45, 56, 58, 59]. In some publications, the imidazolium-based RTILs are represented using the cation–anion structures as [$C_n$mim][X]. Predominant anion species (X) include triflate (OTf), dicyanamide (dca), tetrafluoroborate (BF_4), bis(trifluoromethane)sulfonimide (Tf_2N), and other molecular anions. The variations in the length of alkyl substituent (C_n) of cations can be distinguished as 1-ethyl-3-methylimidazolium [C_2mim] or [emim], 1-butyl-3-methylimidazolium [C_4mim] or [bmim], 1-hexyl-3-methylimidazolium [C_6mim] or [hmim], and so on [44].

A rough summary of the Henry's constant H and solubility S of different gas components in RTILs is presented in Table 7.7. Generally, imidazolium-based RTILs still exhibit lower CO_2 solubility than common physical solvents [44]. The significant property of RTILs is the extremely low vapor pressure near their triple points, due to the combination of strongly ionic character and of low melting temperatures [60]. The other two potential application fields are supported ionic liquid membranes (SILMs) and in combination with amines for CO_2 capture [44].

Although considerable progress has been made in understanding CO_2-based separations by RTILs, a number of obstacles exist before serious consideration

Table 7.7 Henry's constant and solubility for CO_2, N_2, CH_4, H_2, and H_2S in related RTILs.

Gas	RTIL	*T* (°C)	*H* (atm)	S ($Nm^3\ m^{-3}\ atm^{-1}$)	Ref.
CO_2	[emim][BF_4]	25	80	1.83	44, 45
	[emim][Tf_2N]	25	37	2.40	44, 45
	[emim][dca]	40	96	1.44	44, 61
	[hmim][Tf_2N]	25	34	2.08	44, 45
	[bmim][BF_4]	25	56	2.16	44, 61
	TEGO IL K5®	27	27	–	50
N_2	[emim][BF_4]	25	7100	0.020	44, 45
	[emim][Tf_2N]	25	1400	0.062	44, 45
	[emim][dca]	40	4900	0.028	44, 61
	[hmim][Tf_2N]	25	1000	0.069	44, 45
CH_4	[emim][BF_4]	25	2900	0.050	44, 45
	[emim][Tf_2N]	25	580	0.149	44, 45
	[emim][dca]	40	2000	0.068	44, 61
	[hmim][Tf_2N]	25	350	0.196	44, 45
H_2	[emim][BF_4]	25	8100	0.018	44, 45
	[emim][Tf_2N]	25	2600	0.033	44, 45
	[hmim][Tf_2N]	25	1800	0.038	44, 45
H_2S	TEGO IL K5	27	23	–	50

can be given to the implementation of RTILs in commercial CO_2 scrubbing applications:

- absorption behavior of the other gas components, arising from the related CO_2 capture process, such as steam and SO_x
- process design and optimization related to RTILs and absorbed gas components
- the expense of producing significant volumes of RTIL solvents.

7.4 Conclusions and Outlook

A detailed examination of the IGCC process reveals its feasibility for pre-combustion capture of carbon dioxide and demonstrates the advantages compared with post- and oxy-combustion in terms of the specific energy demand resulting from high operating pressures and CO_2 concentrations. Physical absorption processes, such as the Rectisol or Selexol process, have been applied to high-pressure CO_2 removal of syngas for years, making them feasible for carbon capture and storage.

The high flexibility of the process design permits the adaptation of the physical scrubbing process to the required specifications, such as degree of CO_2 separation and product purity, under the restrictions of minimized energy demand and complexity. In addition, long-term experience in designing and operating physical scrubbers reduces the risks in the design and construction of novel power plant concepts in IGCC with CO_2 removal.

A variety of commercially available solvents exist, each having its own characteristic parameters such as optimal operating temperature, acid gas solubility, and selectivity. The proper choice is dependent on the process conditions, such as total pressure and carbon dioxide and sulfur concentrations, and also the separation task, with regard to degree of CO_2 removal, parallel desulfurization, and required product purity.

All of the presented physical absorption processes operate at temperatures below that of the WGS reactor. Since the different temperature levels require the extensive and expensive use of heat exchangers, resulting in high investment costs and exergetic heat losses, new developments are focused on the development of physical solvents, capable of working at higher temperatures. ILs are a promising alternative, offering the opportunity to separate carbon dioxide at elevated temperatures with almost no solvent losses.

Further developments could make use of higher operating temperatures to integrate the CO_2 removal into the WGS process, for example, by reducing the carbon dioxide concentration between the different shift steps, thus reducing the demand for additional steam and increasing the overall efficiency. However, in addition to the high temperature, the water balance in the scrubbing process is a crucial factor in achieving this target, since a highly hydrophilic solvent would dehydrate the syngas and thus reduce the amount of water available for the WGS reaction.

References

1 Higman, C. and van der Burgt, M. *Gasification*, 2nd edn, Gulf Professional Publishing, Elsevier, Burlington, MA, 2008.
2 Franz, J. and Scherer, V. An evaluation of CO_2 and H_2 selective polymeric membranes for CO_2 separation in IGCC processes. *J. Membr. Sci.*, 2010, **359**, 173–183.
3 Prins, M., van der Ploeg, R., van den Berg, R., Patil, C., van Dorst, E., Geuzeboek, F., and Paul, D. Technology advances in IGCC with CO_2 capture. Presented at the 4th International Conference on Clean Coal Technologies, Dresden, 18–20 May 2009.
4 Kloster, R. Thermodynamische Analyse und Optimierung von Gas-/Dampfturbinen-Kombi-Kraftwerken mit integrierter Kohlevergasung. Dissertation. *Fortschr.-Ber. VDI*, 1998, No. 409, Reihe 6, Energietechnik.
5 Kriebel, M. *Ullmann's Encyclopedia of Industrial Chemistry, Absorption, 2. Design of Systems and Equipment*, Wiley-VCH Verlag GmbH, Weinheim, 2008.
6 Rivas, O.R. and Prausnitz, J.M. Sweetening of sour natural gases by mixed-solvent absorption: solubilities of ethane, carbon dioxide, and hydrogen sulfide in mixtures of physical and chemical solvents. *AIChE J.*, 1979, **25**, 975–984.
7 Kriebel, M. *Ullmann's Encyclopedia of Industrial Chemistry, Gas Producion*, VCH Verlag GmbH, Weinheim, 1989.
8 Sweeney, C.W., Ritter, T.J., and McGinley, E.B. A strategy for screening physical solvents. *Chem. Eng.*, 1988, **95**, 119–125.
9 Hochgesand, G. Rectisol and Purisol. *Ind. Eng. Chem.*, 1970, **62**, 37–43.

10 Kohl, A. and Nielsen, R. *Gas Purification*, 5th edn, Gulf Publishing, Houston, TX, 1997.

11 Bucklin, R.W. and Schendel, R.L. Comparison of Fluor Solvent and Selexol processes. *Energy Prog.*, 1984, **4**, 137–142.

12 Hochgesand, G. Anwendung von Absorptionsverfahren für die CO_2-Entfernung aus Natur- und Synthesegasen. *Chem.-Ing.-Tech.*, 1968, **40**, 432–440.

13 Linde Engineering. *Rectisol Wash*, http://www.linde-engineering.com/en/process_plants/hydrogen_and_synthesis_gas_plants/gas_processing_plants/rectisol_wash/index.html, 2011, last accessed Feb. 24th, 2011.

14 Korens, N., Simbeck, D.R., and Wilhelm, D.J. *Process Screening Analysis of Alternative Gas Treating and Sulfur Removal for Gasification. Prepared for US Department of Energy by SFA Pacific, Inc. Revised Final Report*, http://www.netl.doe.gov/technologies/coalpower/gasification/pubs/pdf/SFA%20Pacific _Process%20Screening%20Analysis_Dec%2020, 02.pdf, 2002, last accessed 3 January 2011.

15 Kriebel, M. *Ullmann's Encyclopedia of Industrial Chemistry, Gas Producion*, Wiley-VCH Verlag GmbH, Weinheim, 2007.

16 Öhrman, O. *Syngas Cleaning Using Absorption and Adsorption*, https://ciweb.chydenius.fi/project_files/HighBio-Projektiseminaari%2018, 0510/Olov%20%C3%96hrman%2018, 05-2010.pdf, 2010 last accessed 3 January 2011.

17 Munder, B., Grob, S., and Fritz, P.M. Selection of wash systems for sour gas removal. Presented at the 4th International Freiberg Conference on IGCC & XtL Technologies, 5 May 2010.

18 Herbert, W. Das Rectisol-Verfahren zur Reinigung von Druckgasen. *Erdöl Kohle*, 1956, **9**, 77–81.

19 Nassar, V.L., Bullin, J.A., and Lyddon, L.G. *Solubility of Hydrocarbons in Physical Solvents*, http://www.bre.com/portals/0/technicalarticles/Solubility%20of%20-Hydrocarbons%20in%20Phys ical%20-Solvents.pdf, 2000, last accessed 3 January 2011.

20 Burr, B. and Lyddon, L. Which physical solvent is best for acid gas removal? *Hydrocarbon Process.*, 2009, January, 43–50.

21 Esteban, A., Hernandez, V., and Lunsford, K. *Exploit the Benefits of Methanol*, http://www.bre.com/portals/0/technicalarticles/Exploit%20the%20Benefits%20of%20Methanol.p df, 2000, last accessed 3 January 2011.

21 Lurgi. *The Rectisol Process*, http://www.lurgi.com/website/fileadmin/user_upload/1_PDF/1_Broshures_Flyer/englisch/0308e_Rectisol.pdf, 2002, last accessed 3 January 2011.

23 Zawacki, T.S., Duncan, D.A., and Macriss, R.A. Process optimized for high pressure gas cleanup. *Hydrocarbon Process.*, 1981, April, 143–149.

24 UOP. *Selexol Process*, http://www.uop.com/objects/97%20Selexol.pdf, 2002, last accessed 3 January 2011.

25 Xu, Y., Schutte, R.P., and Hepler, L.G. Solubilities of carbon dioxide, hydrogen sulfide and sulfur dioxide in physical solvents. *Can. J. Chem. Eng.*, 1992, **70**, 569–573.

26 Murrieta-Guevara, F., Romero-Martinez, A., and Trejo, A. Solubilities of carbon dioxide and hydrogen sulfide in propylene carbonate, *N*-methylpyrrolidone and sulfolane. *Fluid Phase Equilib.*, 1988, **44**, 105–115.

27 Mak, J., Wierenga, D., Nielsen, D., and Graham, C. New physical solvent treating configurations for offshore high pressure CO_2. Presented at the Offshore Technology Conference 5–8 May 2003.

28 Wölfer, W. Helpful hints for physical solvent absorption. *Hydrocarbon Process.*, 1982, November, 193–197.

29 Nikolic, D.L. and Claessen, M. *Sulfinol-X for Acid Gas Removal: a Solution for LNG Production from Contaminated Gas*, 2009, www.kgu.or.kr, last accessed 3 January 2011.

30 Kriebel, M. Improved Amisol process for gas purification. *Energy Prog.*, 1984, **4**, 143–146.

31 Chiesa, P. and Consonni, S. UOP. Selexol^{TM} Process. http://www.uop.com/objects/Selexol.pdf, 2007, last accessed 3. Jan. 2011.

32 Heintz, Y. J., Sehabiague, L., Morsi, B. I., Jones, K. L., and Pennline, H. W. Novel physical solvents for selective CO_2 capture from fuel gas streams at elevated pressures and temperatures. *Energy Fuels*, 2008, **22**, 3824–3837.

33 Plechkova, N. V. and Seddon, K. R. Applications of ionic liquids in the chemical industry. *Chem. Soc. Rev.*, 2007, **37**, 123–150.

34 Wikipedia. *Ionic Liquid*, http://en.wikipedia.org/wiki/Ionic_liquid, 2011, last accessed 24. Feb. 2011

35 Walden, P. Über die Molekulargrösse und elektrische Leitfähigkeit einiger geschmolzenen Salze. *Bull. Acad. Imper. Sci., St. Petersburg*, 1914, **8**, 405–422.

36 IL Thermo. *Ionic Liquids Database*, http://ilthermo.boulder.nist.gov/ILThermo/mainmenu.uix, 2006, last accessed 3 January 2011.

37 Freemantle, M. Designer solvents – ionic liquids may boost clean technology development. *Chem. Eng. News*, 1998, **76**, 32–37.

38 Merk. http://www.merck-chemicals.com/applications/c_d7Ob.s1Odr UAAAEdDmcme9RK, 2011, last accessed 24. Feb. 2011.

39 Lide, D. R. (ed.). *CRC Handbook of Chemistry and Physics*, 73rd edn, CRC Press, Boca Raton, FL, 1992.

40 Anderson, J. L., Dixon, J. K., and Brennecke, J. F. Solubility of CO_2, CH_4, C_2H_6, C_2H_4, O_2, and N_2 in 1-hexyl-3-methylpyridinium bis(trifluoromethylsulfonyl)imide: comparison to other ionic liquids. *Acc. Chem. Res.*, 2007, **40**, 1208–1216.

41 Muldoon, M. J., Aki, S. N. V. K., Anderson, J. L., Dixon, J. K., and Brennecke, J. F. Improving carbon dioxide solubility in ionic liquids. *J. Phys. Chem. B*, 2007, **111**, 9001–9009.

42 Baltus, R. E., Culbertson, B. H., Dai, S., Luo, H., and DePaoli, D. W. Low-pressure solubility of carbon dioxide in room-temperature ionic liquids measured with a quartz crystal microbalance. *J. Phys. Chem. B*, 2004, **108**, 721–727.

43 Mutelet, F., Revelli, A.-L., Jaubert, J.-N., Sprunger, L. M., Acree, W. E. A., and Baker, G. A. Partition coefficients of organic compounds in new imidazolium and tetralkylammonium based ionic liquids using inverse gas chromatography. *J. Chem. Eng.* Data, 2010, **55**, 234–242.

44 Bara, J. E., Carlisle, T. K., Gabriel, C. J., Camper, D., Finotello, A., Gin, D. L., and Noble, R. D. Guide to CO_2 separations in imidazolium-based room-temperature ionic liquids. *Ind. Eng. Chem. Res.*, 2009, **48**, 2739–2751.

45 Finotello, A., Bara, J. E., Camper, D., and Noble, R. D. Room-temperature ionic liquids: temperature dependence of gas solubility selectivity. *Ind. Eng. Chem. Res.*, 2008, **47**, 3453–3459.

46 Jalili, A. H., Mehdizadeh, A., Shokouhi, M., Ahmadi, A. N., Hosseini-Jenab, M., and Fateminassab, F. Solubility and diffusion of CO_2 and H_2S in the ionic liquid 1-ethyl-3-methylimidazolium ethylsulfate. *J. Chem. Thermodyn.*, 2010, **42**, 1298–1303.

47 Carvalho, P. J., Alvarez, V. H., Marrucho, I. M., Aznar, M., and Coutinho, J. A. P. High carbon dioxide solubilities in trihexyltetradecylphosphonium-based ionic liquids. *J. Supercrit. Fluids*, 2010, **52**, 258–265.

48 Revelli, A.-L., Mutelet, F., and Jaubert, J.-N. High carbon dioxide solubilities in imidazolium-based ionic liquids and in poly(ethylene glycol) dimethyl ether. *J. Phys. Chem. B*, 2010, **114**, 12908–12913.

49 Pennline, H. W., Luebke, D. R., Jones, K. L., Myers, C. R., Morsi, B. I., Heintz, Y. J., and and Ilconich, J. B. Progress in carbon dioxide capture and separation research for gasification-based power generation point sources. *Fuel Process. Technol.*, 2008, **89**, 897–907.

50 Heintz, Y. J., Sehabiague, L., Morsi, B. I., Jones, K. L., Luebke, D. R., and Pennline, H. W. Hydrogen sulfide and carbon dioxide removal from dry fuel gas streams using an ionic liquid as a physical solvent. *Energy Fuels*, 2009, **23**, 4822–4830.

51 Bara, J. E., Camper, D. E., Gin, D. L., and Noble, R. D. Room-temperature ionic liquids and composite materials: platform technologies for CO_2 capture. *Acc. Chem. Res.*, 2009, **43**, 152–159.

52 Lin, H. and Freeman, B. D. Materials selection guidelines for membranes that remove CO_2 from gas mixtures. *J. Mol. Struct.*, 2005, **739**, 57–74.

53 Morgan, D., Ferguson, L., and Scovazzo, P. Diffusivities of gases in room-temperature ionic liquids: data and correlations obtained using a lag-time technique. *Ind. Eng. Chem. Res.*, 2005, **44**, 4815–4823.

54 Ferguson, L. and Scovazzo, P. Solubility, diffusivity, and permeability of gases in phosphonium-based room temperature ionic liquids: data and correlations. *Ind. Eng. Chem. Res.*, 2007, **46**, 1369–1374.

55 Bara, J. E., Gabriel, C. J., Carlisle, T. K., Camper, D., Finotello, A., Gin, D. L., and Noble, R. D. Gas separations in fluoroalkyl-functionalized room-temperature ionic liquids using supported liquid membranes. *Chem. Eng. J.*, 2009, **147**, 43–50.

56 Hou, Y. and Baltus, R. E. Experimental measurement of the solubility and diffusivity of CO_2 in room-temperature ionic liquids using a transient thin-liquid-film-method. *Ind. Eng. Chem. Res.*, 2007, **46**, 8166–8175.

57 Bara, J. E., Carlisle, T. K., Gabriel, C. J., Camper, D., Finotello, A., Gin, D. L., and Noble, R. D. A guide to CO_2 separations in imidazolium-based room-temperature ionic liquids. *Ind. Eng. Chem. Res.*, 2009, **48**, 2739–2751.

58 Kilaru, P. K., Condemarin, R. A., and Scovazzo, P. Correlations of low-pressure carbon dioxide and hydrocarbon solubilities in imidazolium-, phosphonium-, and ammonium-based room-temperature ionic liquids. Part 1. Using surface tension. *Ind. Eng. Chem. Res.*, 2008, **47**, 900–909.

59 Kilaru, P. K. and Scovazzo, P. Correlations of low-pressure carbon dioxide and hydrocarbon solubilities in imidazolium-, phosphonium-, and ammonium-based room-temperature ionic liquids. Part 2. Using activation energy of viscosity. *Ind. Eng. Chem. Res.*, 2008, **47**, 910–919.

60 Bier, M. and Dietrich, S. Vapour pressure of ionic liquids. *Mol. Phys.*, 2010, **108**, 211–214.

61 Camper, D., Bara, J., Koval, C., and Noble, R. D. Bulk-fluid solubility and membrane feasibility of Rmim-based room-temperature ionic liquids. *Ind. Eng. Chem. Res.*, 2006, **45**, 6279–6283.

Processes for CO_2 Scrubbing

8
CO_2 Removal in Coal Power Plants via Post-Combustion with Absorbents

Hans Fahlenkamp and Martin Dittmar

Abstract

One of the most preferred techniques for CO_2-capture is the so called post combustion capture or tail-end capture, as the coal is, in contrast to other processes, fired in a state of the art furnace chamber. The resulting flue gas stream is treated after the FGD-Unit downstream the furnace chamber. Hence to the location of separation of CO_2 in the flue gas stream results the name "post combustion".

This chapter covers the post combustion process and its absorption/desorption technique. Furthermore, different solvents for the CO_2-capture in post combustion capture plants are introduced with their advantages and disadvantages to this specific process. A separate section is referring to arising problems in the absorption/desorption process as a tail-end-capture and due to this, various standards for flue gas components other than CO_2, like O_2, dust and SO_2 are reported.

This chapter also contains a section about pilot plants and demo plants for the CO_2-postcombustion-capture technique that are either planned or already in operation. The reader is introduced to first test results from various pilot plants. Also, the necessity of pilot plants on the way to full scaled post combustion CO_2-capture from power plants is shown. In addition, the problem of foam formation which occurs with amine solutions in the absorption/desorption unit is examined closer so that the reader will have a first overview of the relevant impact factors which lead to this undesired phenomenon.

Keywords: post-combustion, absorbents, carbon dioxide capture, demonstration plants, pilot plants

Process Engineering for CCS Power Plants, First Edition. Edited by Detlef Stolten and V. Scherer

8.1
Tail-End CO_2 Capture

8.1.1
Amine-Based Absorption

The chemical absorption of CO_2 with primary amines – primarily by monoethanolamine (MEA) – is one of the classic processes in the chemical treatment of natural gases (Fig. 8.1). In particular, "sour gas" or "acidic gas" (natural gas which contains H_2S and CO_2 and with a relevant percentage of hydrocarbons heavier than methane) is separated through this absorption process. MEA allows a high final purity in relation to the natural gas being purified. In comparison with physical absorption, chemical absorption does not require high pressure but the wash solution must be thermally regenerated. This regeneration occurs through "boiling out," that is, heating the MEA solution and thereby reversing the chemical reaction and releasing CO_2, which is removed from the solution through its own stripping steam. The disadvantage of this process is its high energy consumption (Ullmann, 1938), which is composed of the reaction enthalpy, the energy for heating the solution, and the evaporation enthalpy of the evaporated water and the absorption solution (Göttlicher, 1999).

The common feature of CO_2 absorption from both natural gas and synthetic gas is absence of oxygen. Therefore, sulfur is present as H_2S and nitrogen as NH_3 (ammonia) or HCN (hydrogen cyanide, prussic acid). The circumstances are therefore important because free oxygen and sulfur are available as SO_2 in a flue gas, where they generate irreversible competing reactions with the wash solution.

Amines are derivatives of ammonia (NH_3). In primary amines, one hydrogen atom is replaced by a functional group (R^1NH_2), in secondary amines, two func-

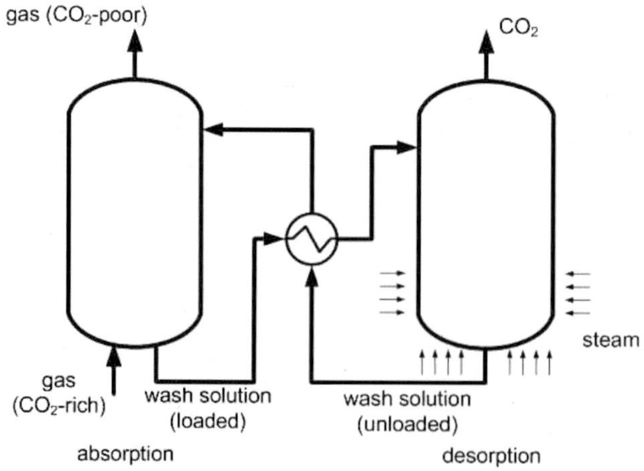

Fig. 8.1 Depiction of an absorption–desorption process.

tional groups are bound to the nitrogen atom (R^1R^2NH), and in tertiary amines, all three hydrogen molecules are replaced by functional groups ($R^1R^2R^3N$).

Primary (and also secondary) amines react with CO_2 to form a carbamate; the reaction is formulated through the following two steps (for MEA, $R^1{=}CH_2CH_2OH$):

$$R^1NH_2 + CO_2 \rightleftharpoons R^1HNCOOH \tag{8.1}$$

$$R^1HNCOOH + R^1NH_2 \rightleftharpoons R^1HNCOO^- + R^1NH_3^+ \tag{8.2}$$

that is, the overall reaction uses 2 mol of amine per mole of CO_2:

$$2R^1NH_2 + CO_2 \rightleftharpoons R^1HNCOO^- + R^1NH_3^+ \tag{8.3}$$

The reaction of CO_2 is a second-order reaction, that is, the reaction rate r_{CO_2} depends on the concentrations of MEA and CO_2 (Table 8.1). The reaction rate constant k contains the temperature dependence of the reaction:

$$r_{CO_2} = -k[CO_2][MEA] \tag{8.4}$$

The reaction order of other primary and secondary amines lies between 2 and 3 (Schubert, 2004).

8.1.1.1 Primary Amines

So far, the primary amine monoethanolamine (Fig. 8.2) has been the main amine considered as a wash solution for tail-end CO_2 absorption because of long experience in its use for the chemical treatment of natural gases.

The reaction enthalpy is about –85 kJ mol^{-1} CO_2. The reaction with MEA leads to a comparatively stable carbamate in the wash solution. The theoretical

Table 8.1 Kinetic data for the CO_2–amine reaction (Schubert, 2004).

Amine	k (m^3 mol^{-1} s^{-1})	T (°C)	c_{amine} (wt%)
MEA	7.74	30	12.2
	10.4	45	19.5
	25.7	60	19.5
MDEA	0.0072	30	26.2
	0.0170	45	26.2
	0.0410	60	26.2

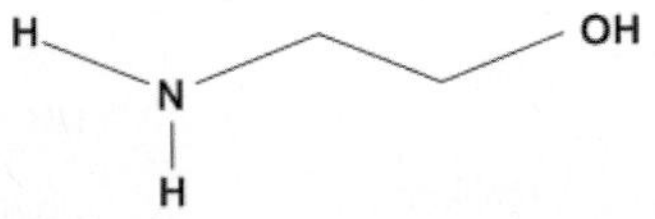

Fig. 8.2 Structural formula of monoethanolamine.

loading of the aforementioned reaction is 0.5 mol CO_2 per mole MEA. An additional hydrolysis of the carbamate may occur and leads to a higher load, although this happens very slowly. The theoretical load will generally not be used up. Normally, the starting load is already at 0.2 mol CO_2 per mole MEA after regeneration and reaches 0.4–0.5 mol CO_2 per mole MEA in the absorber. Due to the high reaction enthalpy and the marginal load, the energy necessary for regeneration is very high in MEA. The entire wash solution must be heated to 120–140 °C at 1 bar positive pressure and subsequently cooled to the absorption temperature. Another reason for energy requirements of MEA regeneration is that the aqueous MEA solution may contain only a maximum of 30 wt% MEA and would cause intense corrosion at higher concentration.

8.1.1.2 **Secondary Amines**

Diethanolamine (DEA) (Fig. 8.3) is the most often used secondary amine. The reaction with CO_2 is similar to that of primary amines. Since it is resistant towards COS and CS_2, it is especially used in the absorption of CO_2 and H_2S from refinery gases. There are limited possibilities to apply DEA apart from refinery gas treatment, since absorption is slower with DEA than with MEA and DEA can degrade under certain conditions with high CO_2 concentration (McKetta, 1978; Kohl and Nielsen, 1997).

8.1.1.3 **Tertiary Amines**

The tertiary amine methyldiethanolamine (MDEA) (Fig. 8.4) is especially used for selective H_2S absorption from natural gases; triethanolamine is used for CO_2 capture in the Haber–Bosch process. Tertiary amines are especially suitable for atmospheric CO_2 absorption since they react differently to primary and sec-

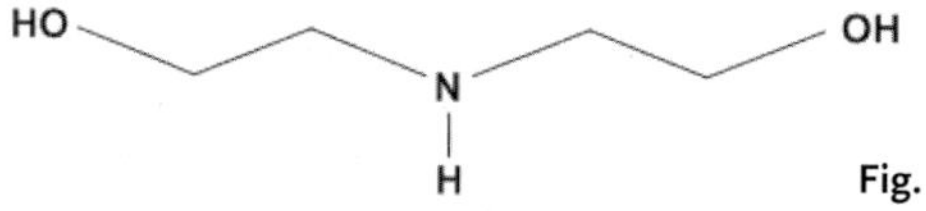

Fig. 8.3 Structural formula of diethanolamine.

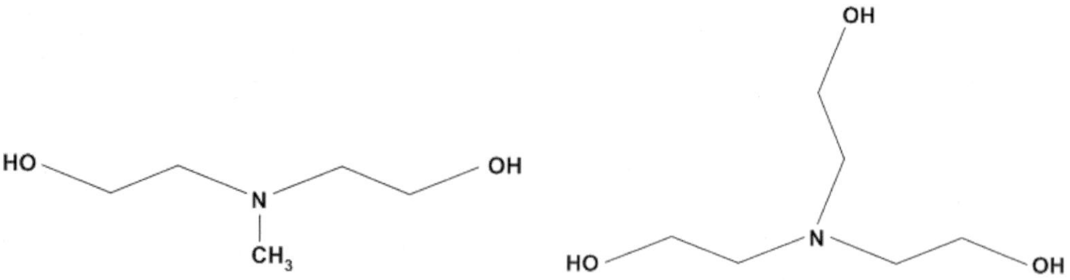

Fig. 8.4 Structural formulae of methyldiethanolamine and triethanolamine.

ondary amines: there is no direct reaction with CO_2; the amines catalyze the reaction of CO_2 with water to give the hydrogencarbonate ion (Schubert, 2004):

$$R^1R^2R^3N + CO_2 \rightleftharpoons R^1R^2R^3NH^+ + HCO_3^- \quad (8.5)$$

Due to stoichiometry, the theoretical load of tertiary amines is 1 mol CO_2 per mole amine; since the reaction kinetics are considerably slower, such a high load is not reached (Table 8.1). The reaction enthalpy of tertiary amines with CO_2 is lower than that of MEA; therefore, the energy consumption for the regeneration is lower. The absorption by MDEA, similarly to MEA absorption, is a second-order reaction:

$$r_{CO_2} = -k[CO_2][MDEA] \quad (8.6)$$

Tertiary amines have much less corrosiveness compared with MEA; furthermore, MDEA is less poisonous and less harmful to the environment than MEA.

8.1.1.4 Mixed Amines

Amine blends, such as MEA + MDEA and MDEA + piperazine, utilize the advantages of primary and secondary amines (rapid reaction with CO_2) and also those of the tertiary amines (high molar load, mol CO_2 per mole amine). The optimal mix ratio needs to be established on an individual basis so that the advantages of the particular amines outweigh the disadvantages, such as corrosion and a slow reaction. With a favorable ratio of MEA to MDEA, the energy required for regeneration is reduced from over 3000 kJ kg^{-1} CO_2 in an MEA solution to 1200–2400 kJ kg^{-1} CO_2 (Aroonwilas and Veawab 2006).

8.1.1.5 Sterically Hindered Amines

2-Amino-2-methyl-1-propanol (AMP) (Fig. 8.5) is a sterically hindered amine, with typical bonding of the amino group to a tertiary carbon atom (in primary amines) or the bonding of the amino group to a secondary or tertiary carbon atom (in secondary amines). Similarly to tertiary amines, sterically hindered amines can reach a maximum load of 1 mol mol^{-1}.

Mitsubishi Heavy Industries sells a sterically hindered amine blend named KS-1; this amine can be used in a CO_2 absorption unit with only slight corrosion and degradation problems and little energy requirement for regeneration.

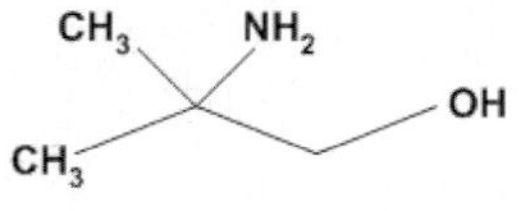

Fig. 8.5 Structural formula of 2-amino-2-methyl-1-propanol.

8.1.1.6 **Amino Acid Salts**

A further promising wash solution is represented by amino acid salt-based solutions (Fig. 8.6). Amino acids have the same functional group as alkanolamines, they react similarly with CO_2, and they do not degrade in the presence of oxygen as fast as primary amines. An advantage is that the amino acids can introduce salt formation. The carboxylate group can be neutralized with potassium hydroxide in order to preserve the potassium salt in the amino acid. The salt formation leads to nonvolatility of the amino acid salts; the implementation of an additional scrubber in the clean gas duct can be avoided.

Furthermore, amino acid salts can be used in membrane contactors with simple polyolefin membranes (e.g., polypropene), whereas MEA can be used only by means of costly membranes because of wetting difficulties (Van Holst *et al.*, 2006).

An alternative process for CO_2 capture with amino acid salts can be controlled when the product can be precipitated, thereby maintaining a nearly constant equilibrium partial pressure of CO_2 above the slurry. This produces an increased liquid load and decreased energy consumption for solvent regeneration (Majchrowicz *et al.*, 2006).

8.1.1.7 **Diamines**

A frequently used diamine is piperazine (Fig. 8.7). Piperazine is a cyclic compound and exists as a solid at room temperature. It is generally used in combination with or as an activator in other wash solutions, for example, in a potash solution or in the aMDEA process of BASF (Kohl and Nielsen, 1997).

8.1.2
Inorganic Chemical Approaches

8.1.2.1 **Potash**

A process that has only recently attracted attention is the use of the well-known potash solution for CO_2 scrubbing. Potash liquor absorbs CO_2 during the formation of potassium hydrogencarbonate and then desorbs CO_2 through thermal regeneration (Park and Sung, 2006; Benson *et al.*, 1954). Absorption and desorption can also be carried out at the same temperature if the gaseous phase is put under pressure, although this is not the case in conventional coal and natural

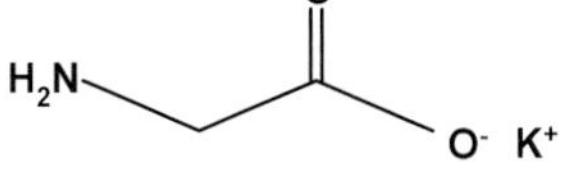

Fig. 8.6 Structural formula of the potassium salt of glycine.

Fig. 8.7 Structural formula of piperazine.

gas fuel power stations. As a downstream operating increase in flue gas flow pressure would cause disproportionately high additional energy consumption, this method is no longer pursued. Desorption occurs through a decrease in pressure. Potassium carbonate (potash) binds CO_2 stoichiometrically and reversibly during the formation of hydrogencarbonate (potash):

$$CO_2 + K_2CO_3 + H_2O \rightarrow 2\,KHCO_3 \qquad \Delta H = -1215\ \text{kJ m}^{-3}\ \text{N}\ CO_2 \tag{8.7}$$

Since the load capacity depends on a stoichiometric chemisorption process, it is proportional to the solubility of the less soluble salt. In the case of the potassium carbonate process, the less soluble salt is potassium hydrogencarbonate. Due to the equilibrium characteristics of potassium carbonate, absorption at higher temperatures is still practicable. This means that absorption and desorption can be executed at proximate temperatures and expensive heat exchanges can be avoided. In hot potash scrubbing, absorption occurs under pressure at ~105 °C, and desorption occurs through the decrease in pressure only. However, it is also possible to perform absorption at ~40–60 °C and desorption at ~120 °C. The mechanical layout of the potash process is similar to that of the amine-based processes. The solubilities given in Table 8.2 illustrate the binary salt–H_2O system at 25 °C as percent by weight anhydrous salt.

The salt precipitated first in the potassium carbonate solution is potassium sulfate, which can be formed via SO_2 in the flue gas and which must then be separated from the process, followed by potassium hydrogencarbonate. The potash process is never operated in the slurry of hydrogencarbonate. This is ensured with an adequately high liquor rate, that is, there is a low load of CO_2 in the wash solution and therefore lower hydrogencarbonate concentrations. The CO_2 absorption is due to stoichiometrically running chemisorption. The load of the liquor is limited by the solubility limit of the slightly soluble hydrogencarbonate, the loaded salt. The attainable load is ~0.28 mol CO_2 per mole potassium carbonate. The entire process is controlled not only by the CO_2 capture but also by the separation of impurities and other unwanted reaction products. Potash liquor binds to CO_2 and H_2S as carbonate (hydrogencarbonate) or sulfide (including SO_2 and SO_3) under the influence of oxygen in the form of less soluble sul-

Table 8.2 Solubilities at 25 °C.

Salt	Solubility (wt%)
K_2CO_3	52.83
$KHCO_3$	26.47
K_2SO_3	51.55
$KHSO_3$	33.6
K_2SO_4	10.7
KNO_2	75.7
KNO_3	27.7

fite and sulfate, and then must be locked out of the process. Degradation of the solution through oxygen or SO_2 as occurs in amine solutions is not achieved in potash liquor, although the potassium balance has to be held steady by adding, for example, caustic potash to the solution. The absorptive solution of alkali metal carbonate does not corrode any material chemically. There is no known damaging effect of oxygen on alkali metal carbonates. In spite of these indisputable advantages, there has not yet been a significant effort to analyze the process of CO_2 separation from flue gases intensively. The reasons lie in the limited capacity of the potash liquor (0.28 mol CO_2 per mole potassium carbonate) and the slow absorption rate in comparison with amine solutions. In order to increase the yield of the CO_2 wash, some processes utilize activators (Rochelle and Hilliard, 2005; Derks *et al.*, 2006; Ghosh and Kentish, 2009).

8.1.2.2 **Ammonia**

A further wash solution used in downstream functioning CO_2 capture is chilled ammonia as an aqueous solution. As with hot potash liquor, the carbon dioxide exists in the form of carbonates and hydrogencarbonates in the equilibrium reaction of absorption and desorption:

$$(HN_4)_2CO_3 + CO_2 + H_2O \underset{90-130\,^\circ C}{\overset{0-10\,^\circ C}{\rightleftharpoons}} 2NH_4HCO_3 \tag{8.8}$$

The energy providers We Energies, Alstom, and the Electric Power Research Institute (EPRI) are conducting a research project with a pilot installation at a power plant in Pleasant Prairie, WI, USA. The pilot plant operates with a partial flow of <1% of the generating unit (corresponding to a flue gas stream of 5 MW_{el}). The CO_2 separation efficiency of flue gas is specified as ~90%.

As depicted in Fig. 8.8, the flue gas branch current leaving the desulfurization unit is cooled to 0–5 °C. In the absorber, aqueous ammonium hydrogencarbonate liquor absorbs CO_2 from the flue gas and binds it as hydrogencarbonate. The resulting CO_2-depleted flue gas is released into the atmosphere. The loaded wash solution is sent through a high-pressure pump at 20 bar into the regenerator, where the addition of heat at 120 °C causes the re-formation of carbonate and the release of CO_2; the procedural advantage of this is that a considerable proportion of the compression work for the desorbed CO_2 is shifted to the liquid phase. Furthermore, the elevated pressure prevents water and ammonia from entering the gas phase and therefore maintains the purity of the CO_2.

A further 30 MW_{th} pilot plant on an AEP power plant site in Mountaineer, WV, USA should validate the findings of the pilot installation in Pleasant Prairie and will store the detached CO_2 geologically. By 2011, there will be a commercial operation at the AEP Northeastern Plant in Oklahoma. With an electrical power of 200 MW_{el}, 1.5 million tons of CO_2 will be removed per year and then be either stored geologically or used for oil transfer according to the enhanced oil recovery process. Since 2008, Alstom and E.ON have been carrying

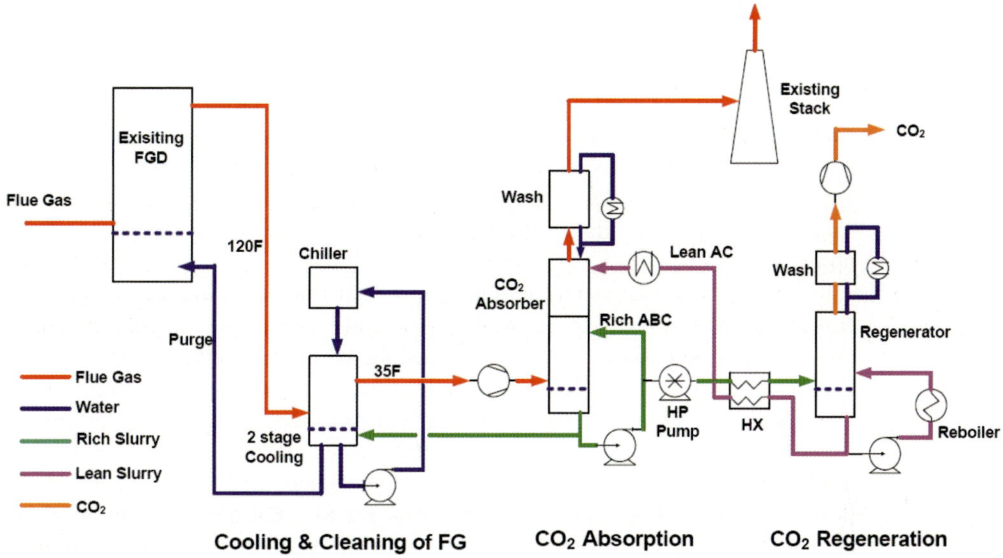

Fig. 8.8 Schematic diagram of the chilled-ammonia process (EPRI, 2006).

out a further research project at the Karlshamn power plant site, Sweden (Rhudy, 2007).

A disadvantage of ammonia cleaning is the volatility and toxicity of ammonia. Care must be taken to ensure that the ammonia level in clean gas does not lead to unacceptable emissions.

Advantages of ammonia absorption in comparison with MEA cleaning are its lower sensitivity to impurities (dust, SO_2) and its lower demand for energy for regeneration; a disadvantage is the necessary cooling. According to a study in which the absorption system was integrated into the power generating process, CO_2 capture with chilled ammonia had an efficiency advantage of over 8% with 1140 MW_{th} power compared with MEA absorption (MEA absorption, 28.9%; chilled ammonia, 37%; without CO_2 capture, 40.5%) (Black, 2006; Braine 2007; Rhudy, 2007).

In interpreting these quantities, it should be noted that steam power plants with supercritical parameters are not as accepted in the United States and Canada as they are in Europe, which can lead to the unfavorable omission of the efficiency and therefore the CO_2 output per kWh.

8.1.3 Advantages and Disadvantages of Various Processes

Although the CO_2 wash is established among chemical processes, it cannot be simply applied to flue gases of power plants. In comparison with the chemical treatment of natural gases, oxidizing conditions exist in the flue gas; this means that there is oxygen in the flue gas. Some of the wash solutions commonly used in the chemical treatment of natural gases become degraded by the reaction with flue gas components such as oxygen and traces of sulfur dioxide and nitric oxide.

Another significant difference between application in power plants and in the chemical treatment of natural gases is that a load distributor regulates the power output of the power plant and therefore the amount of coal burned. The amount of flue gas to be treated consequently fluctuates. Solutions need to be sought in order to configure the CO_2 absorber flexibly to the standards of the power plant.

In this context, it is also important to note that flue gas is available at the cold end of the power plant at atmospheric pressure, whereas considerably higher pressure may be present in chemical processes or the treatment of natural gas.

As can be seen in Fig. 8.9, in the case of post-combustion decarbonization a CO_2 wash is installed downstream of the existing flue gas cleaning equipment. The CO_2 washing unit exists as an absorber, which runs the flue gas in a countercurrent flow to the wash solution. The CO_2 is absorbed as the reaction heat is released and the loaded wash solution is led into the desorber. In the desorber, thermal energy separates the CO_2 from the wash solution, which is then fed back into the absorber. Since the wash solution is kept cool in the absorber for kinetic reasons and has to be heated in the desorption unit, a heat exchanger is installed between the columns (Fig. 8.1).

The energy consumption used in "boiling out" is so considerable that the net efficiency of a power plant is noticeably reduced. The energy consumption of the solution alone can be influenced by the choice of the absorption liquid. An energetic integration of the solution into the power plant process is necessary for the minimization of the energy consumption of the entire process.

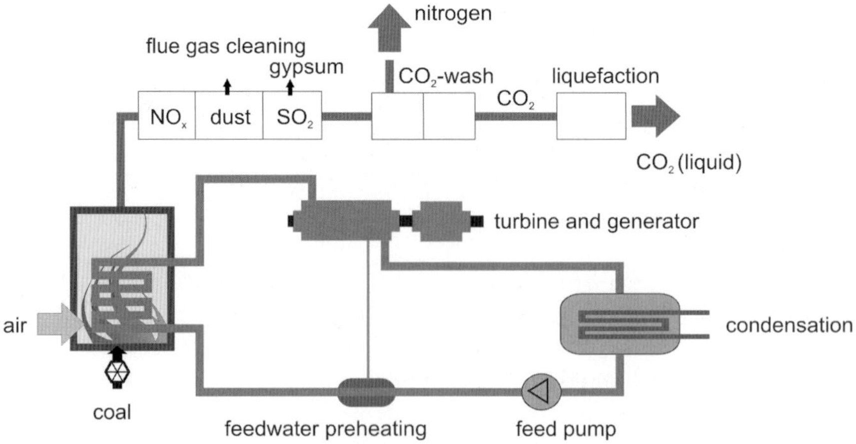

Fig. 8.9 Design of a coal power plant with post-combustion decarbonization (Epp, 2011).

8.1.3.1 **Degradation**

This section considers the degradation problems that can occur with the example of MEA.

MEA can degrade because of heat, oxidation, polymerization, and acid formation due to NO_x and SO_x. Degradation is undesirable for the following reasons:

- Solvents destroyed through degradation must be replaced.
- Degradation products have different material characteristics to the starting material:
  - Degradation products can be harmful to health and the environment.
  - Degradation products can lead to corrosion.

Heat degradation occurs at temperatures higher than 200 °C, but this type of degradation can be disregarded, as such high temperatures do not occur during regeneration. In degradation through polymerization, high molecular weight carbamate compounds inevitably form during regeneration.

With regard to power plant practices, the focus is turned to degradation through oxidation, since flue gases – in contrast to natural gases – obtain an oxidizing atmosphere containing up to 5% oxygen. Oxidative degradation occurring through two mechanisms, electron abstraction and hydrogen abstraction, was investigated by Goff and Rochelle (2004). Ammonia is produced by both mechanisms, serving as an indicator of degradation during the use of a MEA–CO_2 solution (Strazisar *et al.*, 2003).

As described above, a further problem in MEA implementation is its high corrosiveness, present at MEA concentrations of over 20% and especially in loaded solutions. MEA is used as a corrosion inhibitor in aqueous solutions that do not contain CO_2, hence it seems that carbamates and their degradation products cause severe corrosion effects (Kittel *et al.*, 2009). Corrosivity can be reduced through the addition of corrosion inhibitors such as cuprous salts and vanadium salts. However, it should be noted that copper catalyzes oxidative degradation. There is a great need for research to find the optimal corrosion inhibitor for an oxidizing atmosphere that does not also lead to degradation (Goff and Rochelle, 2004; Schubert, 2004).

The acidifiers NO_x and SO_x form nitrate/nitrite salts and sulfate/sulfite salts with amines. On the one hand, they can precipitate as solid matter; on the other, they deactivate the solution through salt formation through NO_x or SO_x. In the solution, sulfur dioxide forms almost entirely into a salt and can be removed in the so-called reclaimer through the addition of NaOH or $NaHCO_3$ (Epp *et al.*, 2010).

These problems of corrosion, degradation, and energy consumption should not hide the fact that MEA is a good absorption liquid for CO_2. Most data and negative operational experiences with MEA have been compiled exactly for the reason that it is a very good washing liquid.

8.1.4
Standards for the Power Plant Process

Every power plant is designed for a specific purpose; there are therefore power plants for base load, medium load, and peak load operations. The use of CO_2 scrubbers in base load operations helps in the prevention of large-scale CO_2 emissions since they universally run on invariable power. For the insertion of CO_2 scrubbers in medium and peak load operations, the design should be elaborated so that they can follow the dynamic behavior of the power plant. Research on this topic was reported by Suenson *et al.* (1985) and Lawal *et al.* (2008). Current analysis and software-based simulations of the dynamic behavior of CO_2 scrubbing devices were discussed by Kubacz (2011).

As a basic principle, there is the possibility of operating both absorber and desorber columns in parallel. This allows for the shutting down of one or more absorber and desorber under a partial load. Base load power stations also have the possibility of parallel design, as the diameter of an absorption column for a power plant with 600 MW_{el} gross output is about 15–20 m. These column dimensions are disadvantageous from a hydrodynamic point of view. Because of this, today there are only columns with a diameter of up to 14 m (Geuzebroek, 2007).

Since there are still no CO_2 scrubbers that can clean the entire flue gas stream of a power plant, the so-called capture-ready design presents the possibility for new power plants to keep the changeover time as short as possible through a design which is prepared for a CO_2 scrubber retrofit including pre-installed connection fittings and extra space for future equipment.

Since the scrubbing of the wash solution requires a great amount of energy and therefore steam flow, it is not possible to extract the steam from a tapping point in the turbine. The duct to the low-pressure part of the power station turbine is the only reasonable extraction point (Alie, 2004). In order to lower the efficiency of the power plant with CO_2 scrubbing as little as possible, Stöver *et al.* (2010) suggested carrying out modifications on the steam turbine. Of especial interest is that the low-pressure turbine must be in a position to work with a strongly fluctuating steam flow.

A rough estimate of the site area required for a CO_2 scrubbing plant was specified by Stankewitz (2011).

The diameter of the columns is calculated with the help of the load factor of inserted packings. The rate of the gas can then be calculated based on the flow rate used. The intended load factor for the built-in components of the columns is generally given by the manufacturer. The equation for calculating the rate is

$$F_{gas} = u_{gas}\sqrt{\rho_{gas}} \tag{8.9}$$

where F_{gas} is the load factor, u_{gas} the gas velocity, and ρ_{gas} the gas density. According to the continuity equation, the surface area and diameter of the columns can be calculated:

$$A_{col} = \frac{\dot{G}}{u_{gas}} \tag{8.10}$$

$$d_{col} = 2\sqrt{\frac{A_{col}}{\pi}} \tag{8.11}$$

where $\dot{G}$ is the flow rate, A_{col} the surface area, and d_{col} the column diameter. If Mellapak 250Y is inserted into the equation, the required diameter of the absorber would be 18 m and that of the desorber would be 11 m. For the aforementioned reasons, parallel operated absorber and desorber columns are employed in real implementation, increasing the required site area. Next to the actual columns there must also be space for enormous heat exchangers, which in the case of an MEA solution for a 600 MW_{el} power plant must be able to chill and heat the wash solution circulating at $\sim 3000\ kg\ s^{-1}$. (Stankewitz, 2011). Furthermore, pipelines and the CO_2 compressor location need to be taken into account. A CO_2 wash will raise the amount of heat exchange water required considerably, as it was heated in the desorber sump but must be chilled in the desorber head, in the wash solution flow, in the flue gas stream, and in CO_2 compression.

For the future implementation of a CO_2 scrubber in a power plant, this section will consider the potential for improvement of flue gas scrubbing. Amine washes, specifically pertaining to MEA, will then be discussed, as this combination is the best developed wash solution.

Amines form traces of SO_x and NO_x and also degradation products with oxygen that are non-regenerative. These degradation products cause corrosion of the equipment. Fly ash, which may pass an upstream-installed electrostatic precipitator (ESP), and a flue gas desulfurization (FGD) device can accumulate in a closed loop absorber–desorber system and cause clogging risks. These challenges and possible approaches will be discussed in the following section.

8.1.4.1 Limitation of the Input Values of Amine-Based CO_2 Scrubbers

The flue gas scrubbing of a typical coal-fueled power plant consists of a denitrification unit (DeNOx or SCR stage), an ESP, and a, FGD unit. The legal limitations of the equipment do not suffice for problem-free implementation of amine based CO_2 scrubbers. SO_x values of 10 ppm and NO_2 values of <65 ppm are considered to be acceptable (Knudsen *et al.*, 2006). More than 95% of NO_x consists of NO and less than 5% of NO_2. NO does not react with the amine solution and develops amine degradation products. As NO_2 is up to $\sim 30\%$ absorbed in the FGD, it can be concluded that the NO_x concentrations in flue gas already suffice for conventional flue gas scrubbing standards (Feraud *et al.*, 2006).

8.1.4.2 Improvements in Flue Gas Desulfurization

SO_x is mostly composed of SO_2 with a small amount of SO_3 ($\sim$2%). Both compounds create strong acids with water. It is therefore assumed that both substances lead to the neutralization of amines, so that heat-resistant salts form as irreversible degradation of amines occurs. Consequently, SO_x separation must be improved. Possible approaches are the following:

- Performance amplification of the liquid/gas ratio in the FGD scrubber by internals (i.e., nozzle adjustment, additional trays, additional buffles).
- Introduction of a two-stage absorber (double loop system of FISIA Babcock or Babcock-Noell Group).
- A "turbulent bed absorber" (Flowpac, from Alstom).
- An aerosol eliminator, such as a wet ESP; such a device in particular can lead to a reduction in SO_3 emissions since SO_3 combined with water forms sulfuric acid aerosols, of which only up to 30–50% can be precipitated by a conventional spray tower.

A further possibility is the installation of an additional scrubber. In doing so, a so-called prescrubber, operated with NaOH solution, would be taken into account before CO_2 scrubbing. The NaOH concentration in this solution should not exceed the pH level of 8 of the scrubbing solution. Otherwise significant amounts of CO_2 would be absorbed in the prescrubber than in the absorber. Prescrubbers are mainly installed in pilot plants to feed them with a "capture-ready" flue gas from a ordinary brachn current of flue gas (see also Chapter 2).

8.1.4.3 Prevention of Oxygen-Related Operation Problems

While NO_x and SO_x occur only in very low concentrations in flue gas downstream of a conventional flue gas treatment, the "cold end" of the power plant has more than a 3.5% proportion of oxygen; when a power plant uses a partial load, a higher oxygen concentration is calculated. Since amines also enter into irreversible reactions with oxygen, action must be taken related to downstream CO_2 scrubbers operated with amine liquors.

Oxygen ends up in the flue gas through various channels. Combustion is achieved by an excess of air of $\sim$20% in order to burn the coal completely and eliminate the emission of carbon monoxide. Before the hot combustion air is freed from dust, it passes an air preheater (APH) in order to heat the fresh air to increase the efficiency. As the APH has operational leaks, oxygen can enter the flue gas here. Oxygen also ends up in the flue gas in the ESP when dust discharges over a rotary feeder The functionality of FGD is based on the absorption of sulfur oxide in water, oxidation, gas–water separation, the dissolution of limestone, and gypsum crystallization. The entry of oxygen at this juncture occurs due to oxidation and can be avoided through structurally separate oxidation. Oxygen can also potentially be eliminated at the chilled end with the help of flameless combustion with natural gas; in doing so, the CO emissions must be considered. The energy freed in the catalytic post-combustion can be used for the production of steam (Nsakala *et al.*, 2001).

A further possibility would be the introduction of amines with corrosion inhibitors which form a passivating protective coating on the metal (Chapel, 1999).

PRAXAIR owns two patents for oxygen-tolerant processes. In the first, a loaded amine solution is heated to 60–90 °C and is then degassed in a flash container through a decrease in pressure. The absorption in this process occurs above atmospheric pressure, and the degassing proceeds at a low pressure of 0.35–0.7 bar. In the second process, oxygen is stripped with an oxygen scavenger gas (Chakravarty *et al.*, 2001).

8.1.4.4 **Improvement of Dust Removal**

The wet precipitator behind the FGD unit is recognized as an opportunity for improvement of flue gas desulfurization. This can reduce the dust emissions, especially of respirable dust. An improvement in dust removal by means of larger separation surfaces in existing dry precipitators and the introduction of more modern regulators is not impossible, although it takes a separate dust filter function as a part of the FGD solution (Engelking, 2007).

8.1.5 Prevention of Environmental Pollution

There are various safety aspects and regulations to be respected according to the post-combustion technique. It must be taken into consideration that in amine-based CO_2 scrubbers, for example, the nitrogen stream leaving the absorber contains more amines than the German Technical Instructions on Air Quality ("Technische Anleitung zur Reinhaltung der Luft/TA-Luft") allow. In other countries, similar regulations must be taken into account. In the CAFE Directive ("Clean Air For Europe," Directive 2008/50/EC of the European Parliament and of the Council of 21 May 2008) (Amann *et al.*, 2005), maximum permissible values for ammonia emissions, one of the main degradation products, are discussed. Ammonia emissions must also be kept in mind in the chilled ammonia process.

It is therefore recommended to install an additional scrubbing stage, operated with water, at the head of the absorber column in order to reduce amine and ammonia emissions. This also has a certain economic advantage, as the absorption liquid is kept in the process.

A further possibility for reducing emissions is absorption with non-volatile absorption liquids. There are no emissions in CO_2 capture with potash solution, although there is also a lower CO_2 capture expected compared with amines. Other possible wash solutions are amino acid salts, whose reaction mechanism is similar to that of the primary amines and, like potash, they do not possess vapor pressure because of their salt character.

As both amines and potash are slightly hazardous towards water (in Germany specified as "class 1 water hazard"), measures must be taken to prevent any amine leakages from ending up in ground water.

8.2
Demonstration Plants and Pilot Plants

CO_2 absorption as a counter measure to greenhouse gas emissions has not been realized on a large scale. It should be mentioned that some long-standing plants exist that separate CO_2 from branch currents of power plant flue gas, predominantly from gas and oil combustion systems. The CO_2 capture is used in these cases for the production of CO_2 for the food processing industry, carbamide production, or methanol synthesis and not as a counter measure to greenhouse gas emissions. The degradation of the wash solution through the components of the power plant's flue gas, O_2, SO_x and NO_x, is known in this systems, although it is accepted because of the economically driven utilization. At the same time, SO_x and NO_x are found at much lower concentrations in the flue gas from gas and oil combustion systems than in the exhaust of coal-fired power plants. Hence a direct transfer of the established process of CO_2 absorption to the scale of power plants cannot be accomplished. The degradation of the scrubber liquor increases considerably and can no longer be disregarded. To complicate matters, equipment for the recovery of CO_2 as a resource is left behind because of quantity, as they are needed on a commercial scale for CO_2 absorption for the reduction of greenhouse gas emissions. In a 600 MW_{el} coal-fired power plant, the flue gas flow rate is $\sim 2\times10^6\ m^3\ h^{-1}$ (STP), from which $\sim 500\ t\ h^{-1}$ of CO_2 must be removed. It is beneficial that substantial information and data on the reactions of many alkanolamines with CO_2 are known due to the long-term operational experience.

The problems that accompany the degradation of the wash solution and the plant capacity can be studied only at experimental plants. These plants should clarify the degree to which a cost-effective CO_2 capture is possible under the conditions of a power plant. The test plants are therefore differentiable by their integration in the power plant process and also by their size. Pilot plants are generally operated with a flue gas branch current and are laid out for diverse test series.

Table 8.3 gives a summary of worldwide pilot plants using a partial flue gas flow of a coal-fired power station. Because of the great number of planned and newly constructed pilot plants, it makes no claim to be complete, but it reflects the plants that the authors found to be particularly worth mentioning. The table refers to the scrubbing liquor (i.e., amine-based wash solutions or ammonia solutions). A further breakdown is not provided.

Demonstration plants in comparison with pilot plants are much larger and can be seen as precursors to the commercial application of the techniques used there. At this juncture, the process of CO_2 capture is integrated in the flue gas pathway (which can be bypassed to some extent) so that the entire flue gas flow from the boiler of the plant can be treated. Most demonstration plants include the CO_2 conditioning and pressing into underground geologic formations. For this reason, in comparison with pilot plants, it is not only the CO_2 capture that is investigated, but rather the entire process to eliminate the global warming effect of CO_2 on the climate. In demonstration plants there is generally long-term

Table 8.3 Overview of CO_2 capture pilot plants worldwide with capacity and solvents used (AAS, amino acid salts; THF, tetrahydrofuran; DME, ethylene glycol dimethyl ether; PC, potassium carbonate); decision upon appearance by author.

Start-up	Country	Location	Captured Co_2	Solvents	Participants
2005	Canada	Boundary Dam Power Plant	$4\ t\ d^{-1}$	MEA–MDEA	ICT University of Regina, Sask Power, Babcock & Wilson, E.ON, nPower (RWE)
2008	Germany	Hamburg Reitbrook	125 kg/d	THF, DME	E.ON Hanse
2008	Australia	New South Wales	$3–14\ t\ d^{-1}$	MEA	CO2CRC (27 partners): e.g. BP, Shell, CSIRO, Australian Government, etc.
2009	Germany	Power Station Staudinger, Unit 5	n.a.	AAS	E.ON, Siemens
2009	China	Beijing	$8\ t\ d^{-1}$	MEA based	CSIRO, China Huaneng Group, Thermal Power Research Institute (TPRI)
2009	Germany	Heyden	$90\ t\ d^{-1}$	MEA	Cansolv, E.ON
2009	Germany	Power Plant Niederaussem	$7\ t\ d^{-1}$	MEA	RWE, Linde, BASF
2009	USA	South Charleston, West Virginia	$5\ t\ d^{-1}$	n.a.	Alstom, Dow Chemical
2010	Italy	Brindisi	$14\ t\ d^{-1}$	MEA	Enel, Eni
2010	USA	Yates, Georgia	$500\ t\ d^{-1}$	KS-1	Southern Company, MHIA
2010	China	Shanghai	$250\ t\ d^{-1}$	n.a.	Huaneng Power international
2010	Germany	n.a.	$100\ t\ d^{-1}$	KS-1	MHI, E.ON
2010	Germany	Wilhelmshaven	$8\ t\ d^{-1}$	MEA	Fluor Daniels, E.ON
2010	Norway	Trondheim	n.a.	n.a.	Sintef, NTNU, Aker Clean Carbon (30.03.2009 E.ON, ScottishPower, Statkraft)
2010/2011	Germany	Duisburg	$6\ t\ d^{-1}$	MEA	LUAT, IUTA, Ef-Ruhr, E.ON, Hitachi
2011	Wales	Aberthaw Power Station/South Wales	$50\ t\ d^{-1}$	MEA, KS-1, PC	RWE nPower, Cansolv
2011	USA	Barry Power Plant, Alabama	$500\ t\ d^{-1}$	KS-1	MHI, Southern Company
2015	Poland	Belchatow Power Plant	270 t/a	n.a.	Alstom, PGE Elektrownia Belchatow S.A., Fortum
2008	USA	Shadyside, Ohio RE Burger Plant	$20\ t\ d^{-1}$	Ammonia	Siemens, Powerspan, FirstEnergy Corp.
2008	Australia	Delta Electricity's Munmorah powerstation	$1\ t\ d^{-1}$	Ammonia	Asian Pacific Partnership funding
2009	USA	New Haven, West Virginia	$270–820\ t\ d^{-1}$	Ammonia	AEP Mountaineer, Alstrom, EPRI
2012	Canada	Alberta, Keephills Power Plant	$2700\ t\ d^{-1}$	Ammonia	Alstom, TransAlta

Table 8.4 Overview of planned CO_2 capture demonstration plants in Europe with estimated capacity and storage options for CO_2; decision upon appearance by author.

Start-up	Country	Location	Captured CO_2	Storage option	Participants
2010	Germany	Wilhelmshaven	0.57 Mt a^{-1}	Saline aquifer	E.ON CE
2011	Netherlands	Pistoolhaven, Rotterdam	2 Mt a^{-1}	n.a.	ENECO, International Power
2011	Netherlands	Barendrecht	0.38 Mt a^{-1}	Depleted gas field/coal seams	Shell
2012	Norway	Bergen	2 Mt a^{-1}	Post saline aquifer	StatoilHydro, Gasnova
2012	Italy	Porto Tolle	1.5 Mt a^{-1}	Saline aquifer/ depleted oil field	ENEL
2012	Norway	Husnes	2.5 Mt a^{-1}	Post saline aquifer	Tinfos, Sor-Norge, Eramet, Sargas
2013	Denmark	Aalborg		stopped because of citizen protest	
2013	Norway	Hammerfest	0.56 Mt a^{-1}	Post saline aquifer	Hammerfest Energi, Sargas, Siemens
2014	Norway	Mongstad	1.2 Mt a^{-1}	Post – pre saline aquifer	BKK
2014	UK	Kingsnorth, South East England	2 Mt a^{-1}	Post depleted gas field	E.ON UK
2014	UK	Scotland	n.a.	n.a.	Scottish Power
2015	Czech Republic	Hodonin, SE	0.3 Mt a^{-1}	Post depleted oil and gas field	CEZ
2015	Czech Republic	Ledvice, N	0.9 Mt a^{-1}	Post saline aquifer	CEZ
2015	Netherlands	Eemshaven	0.19 Mt a^{-1}	Post depleted oil and gas field	RWE Power, BASF, Linde
2015	Netherlands	Maasvlakte, Rotterdam	1 Mt a^{-1}	Post depleted gas field	E.ON Benelux
2015	Poland	Belchatow	1.7 Mt a^{-1}	Post saline aquifer	PGE, ICPC, CMI, PGI
2016	Denmark	Kalundborg	3.4 Mt a^{-1}	Post saline aquifer	DONG Energy
2016	Italy	Saline Joniche (RC)	3.94 Mt a^{-1}	Post saline aquifer	SEI (Rätia Energie & Partners)
2016	Poland	Warsaw	2.87 Mt a^{-1}	n.a.	Vattenfall
2016	Spain		n.a.	Post saline aquifer	Union Fenosa
2016	UK	Tilbury, Thames Estuary	9 Mt a^{-1}	Post saline aquifer	RWE nPower

operational experience (>1 year) in focus. An overview of planned demonstration plants in Europe can be found in Table 8.4. Interested readers will be referred to further literature on the topic.

In the following section, two selected pilot plants will be introduced and the first test results of the completed campaigns are presented. Based on these test results, the phenomenon of foam formation within CO_2 capture is also addressed. The possible factors influencing foam formation and their repercussions are outlined.

8.2.1
Pilot Plants and CO_2 Capture Results

The overview in Tables 8.3 and 8.4 shows that different scrubber liquors are being investigated in CO_2 capture. It is not possible to distinguish a trend, as there are often unspecified or abbreviated wash solutions (e.g., KS-1), which give no clue to the chemical composition of the wash solution used. The most that can be known about these solutions is that they deal with sterically hindered amines or other general properties. Because of its availability, MEA is the most often used wash solution, especially for the first trials with a pilot plant. Absorption with potash liquor, which is limited by slow kinetics, has been investigated in only a few test plants (Chen and Seibert, 2006).

Absorption via chilled ammonia is still being developed; so far no significant test studies have been published. It is not yet clear how the process will be developed. In order to develop it, the first test results, especially from long-term experiments, must exist. Alstom has asserted that the chilled ammonia process is a cost-effective CO_2 capture option (Black, 2006).

In this section, two pilot plants are examined. First, the plant from the CASTOR Project at the Esbjerg power station in west Denmark has been operating since 2005 under the direction of DONG Energy Generation. The pilot plant receives its flue gas from a flue gas branch current from a coal-fired 400 MW power station, which has been equipped with a DeNOx stage, an electrostatic precipitator, and an FGD unit.

Second, there are two plants in Regina, Canada, operated by the International Test Centre for CO_2 Capture (ITC). One plant is located on the campus of the University of Regina (UR plant) and serves to create an assortment of wash solutions, which are then utilized in the pilot plant on the Boundary Dam power station (BD plant). The UR plant is supplied with flue gas from a gas-fired boiler and a gas-fired turbine, whereas the BD plant utilizes a partial flue gas flow of the coal-fired furnace.

The selection of the following pilot plants is based on the data available in the literature mentioned. It should be noted that it is generally difficult to obtain exact results for the composition (properties and concentrations) of the scrubber liquors used. Further, not all necessary results are present in the literature, making it hard to compare the results of different plants and to interpret the operational experience. The results of the two pilot plants are summarized along with the experience that the plants have obtained. It will be revealed that pilot plants are an obligatory intermediate stage in the process of developing the large-scale application of CO_2 absorption.

8.2.1.1 Esbjerg Pilot Plant

The pilot plant in Esbjerg, which was designed for the EU-funded CASTOR project in the Elsam power station at Esbjerg, is currently the largest existing pilot plant for operational experience. The plant capacity is specified as $1\,t\,h^{-1}$ CO_2 with one absorber column.

Layout of the Plant

The absorber column consists of four full-bodied sections; each section is 4.25 m high and has its own separate feed. The level of packing can therefore be used flexibly. A washing stage is installed on the head of the absorber in order to precipitate the entrainment of the wash solution. This wash stage is 3 m tall and equipped with the structured packing Mellapack 252Y. All of the other stages are equipped with the unstructured packing IMTP50. The inner diameter of the absorber is 1.1 m. The total height of the absorption unit is 33.4 m.

The two-tiered desorber column has a total height of 27.4 m and is equipped with the same packing as the absorber column (IMPT50). These two stages each have a height of 5 m. On top of the two stages, there is a 3 m tall wash section installed with Mellapack 250Y. The stripper also has an inner diameter of 1.1 m. The wiring of the absorber, stripper, and the other existing peripheral apparatus can be seen in Fig. 8.10. It should be noted that pilot plants which utilize amine-based wash solutions and have similar constructions and wirings to that shown in Fig. 8.10. Generally, only a couple of the peripheral units, for example a flue gas pretreatment unit or a reclaimer are added or left out.

In the process in Esbjerg, there is an additional reclaimer available beside the filtration unit. This separates the heat-stable byproducts of the cyclical process through heating and the feeding of alkaline media and regenerates the partly degraded wash solution components. After the absorber, the loaded wash solution ("rich MEA") runs through the filters containing active carbon in order to separate entrained solid components such as gypsum and flue ash, which may be carried over from the upstream desulfurization unit.

In Table 8.5, the design specifications for the test plant are shown.

Test Results

The absorber was kept at atmospheric pressure and was supplied with flue gas held at a constant temperature. In order to prevent excessive degradation of the wash solution, the plant was turned off at an SO_x content above 65 ppm. An additional prewash for the reduction of SO_x content was not installed.

During 2006 and 2007, multiple test runs were made. The first trial was performed with a 30 wt% MEA solution and ran for over 1000 h. In the first 500 h, the unit had to be shut down 18 times, producing a downtime of 20 h. The

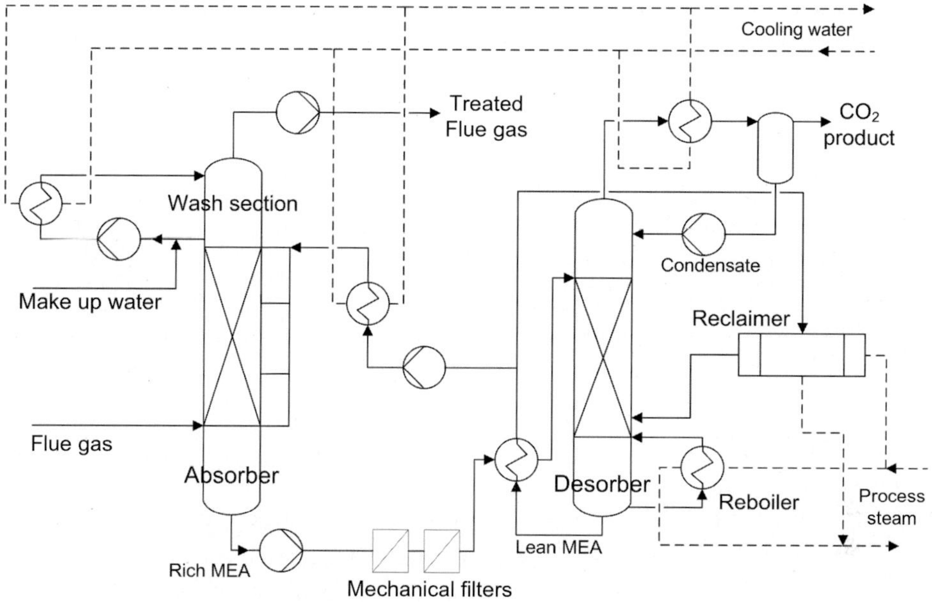

Fig. 8.10 Flow chart for the CASTOR pilot plant at Esbjerg power station (Knudsen *et al.*, 2006).

Table 8.5 Design specifications for the CASTOR pilot plant (Knudsen *et al.*, 2006).

Parameter	**Design value**
Flue gas capacity	5000 Nm^3 h^{-1} = 0.5% of total flue gas flow
CO_2 production (at 12 vol.% CO_2)	1000 kg h^{-1}
Absorption degree	90%
Maximum solvent flow	40 m^3 h^{-1}
Maximum reboiler steam flow	2500 kg h^{-1} (3.5 bar)
Maximum stripper pressure	2 bar (g)
Flue gas conditions	47 C (sat.), <10 ppm SO_2, <65 ppm NO_x, <10 mg Nm^{-3} dust

problems that led to the shutdowns were an excessive SO_x concentration in the flue gas, clogging of the filters, and liquid levels in the stripper that were too high. To overcome these downtimes, the desorber pressure and desorber temperature were raised after the first 500 h. The desorber pressure was raised from 1.25 to 2 bar, which consequentially raised the sump temperature from 109 to 120 °C. Through the subsequently higher wash solution regeneration rate, the wash solution flow could be reduced to 20 m^3 h^{-1}. Optimised operating parameters for the second 500 h are given in Table 8.6.

In addition to the given operating parameters, studies of the change in volume flow in relation to the flue gas flow were undertaken. Load fluctuations

Table 8.6 Optimal operating conditions in the CASTOR 500 h test (Knudsen *et al.*, 2007 a, b).

Setting	Average value, 2nd 500 h test	Performance	Average value, 2nd 500 h test
Flue gas	5020 $Nm^3\,h^{-1}$, wet	CO_2 production rate	1000 $kg\,h^{-1}$
Solvent flow	15.5 $m^3\,h^{-1}$	CO_2 removal efficiency	88%
Reboiler temperature	120 °C	Steam consumption	3.7 $GJ\,t^{-1}\,CO_2$
Stripper pressure	1.89 bar (g)	Cooling water	3.4 $GJ\,t^{-1}\,CO_2$
MEA concentration	30 wt%	Water accumulation	19 $kg\,h^{-1}$
CO_2 loading	Lean, 0.23; rich, 0.36 $mol\,mol^{-1}$	Solvent consumption	1.4 $kg\,t^{-1}\,CO_2$

and the corresponding changes in the gas volume flow and gas composition are to be expected in daily power plant operation. Special interest was taken in the behavior of the reboiler in reaction to these fluctuations. The results of this can be seen in Fig. 8.11.

Figure 8.11 shows a linear relationship between the wash solution required and the input flue gas. In addition, the tests showed that the pilot plant is able to process the load gas stream fluctuations and at the same time safeguard the high CO_2 capture of around 90%.

In Fig. 8.12, the specific steam consumption for varying solvent flow rates of the wash solution is shown. The optimum for the CASTOR pilot plant is revealed to lie at 15 $m^3\,h^{-1}$. At this rate, there is a balance between the desorption energy expended and the energy required to heat the wash solution; this is the optimal operating point for the pilot plant. In longer-lasting operations, the degradation behavior of the wash solution was also studied.

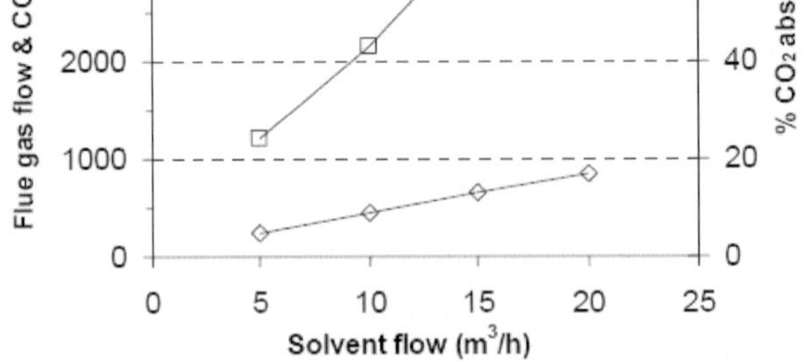

Fig. 8.11 Change in flue gas volume flow in the CASTOR pilot plant (Knudsen *et al.*, 2006).

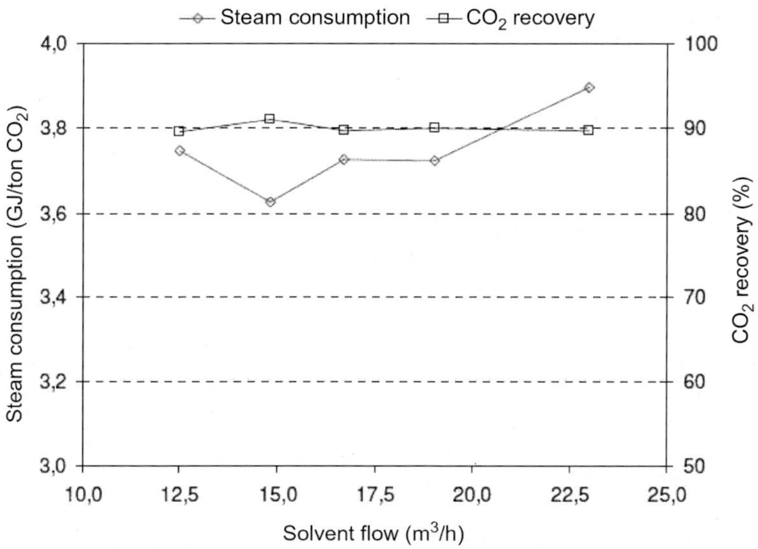

Fig. 8.12 Optimal solvent flow rate in the CASTOR pilot plant (Knudsen *et al.*, 2007a,b).

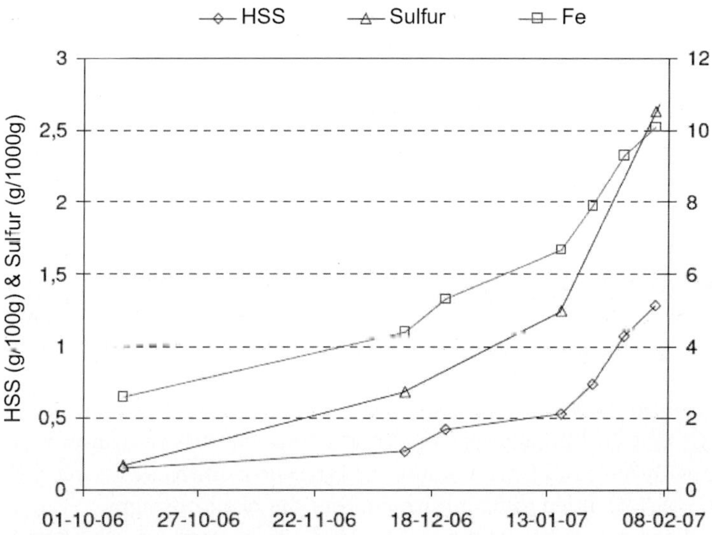

Fig. 8.13 Accumulation of various degradation products over more than 1000 h (Knudsen *et al.*, 2006).

Figure 8.13 shows the accumulation of heat-stable salts (HSS), sulfur, and Fe ions in the wash solution.

After 14 January 2007, according to the adjustment of the optimal parameters as shown in Table 8.6, a sharp increase in the degeneration products and the Fe concentration was registered during the second uninterrupted 500 operational hours. Nearly all of the SO_2 emitted after the FGD unit was retained in

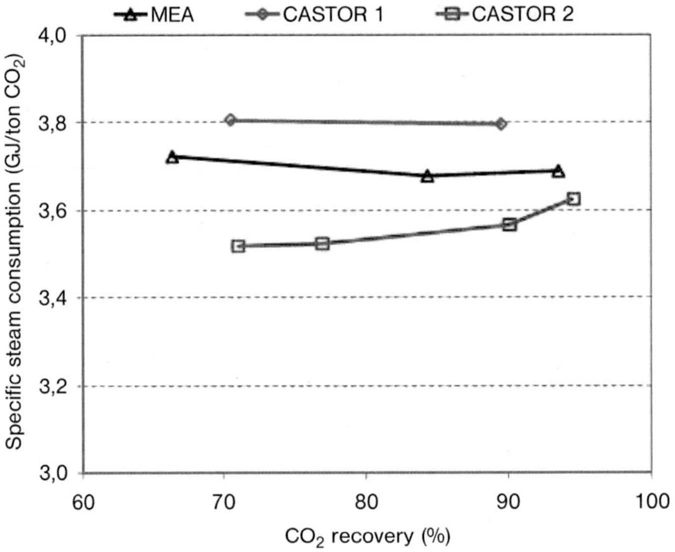

Fig. 8.14 Comparison of reboiler steam consumption of different solvents (Knudsen *et al.*, 2009).

the CO_2 absorber column. This decidedly leads to degradation of the wash solution.

Because of this, in 500 operating hours, 720 kg of MEA were required in order to maintain the wash solution concentration at 30 wt%. At the same time, 503 t of CO_2 were captured with the optimal setting, leading to a wash solution consumption of 1.4 kg t^{-1} CO_2.

Further tests with wash solutions named CASTOR-1 and CASTOR-2 were carried out. These wash solutions were tested under laboratory conditions. Both wash solutions were amine blends of unstated composition. Figure 8.14 shows the performance of the new wash solutions in comparison with an MEA solution.

Figure 8.14 can be disregarded insofar as the steam consumption can be reduced to 3.55 GJ t^{-1} CO_2 through the use of CASTOR-2 for 90% CO_2 absorption in comparison with the use of MEA solution. CASTOR-1 exhibits a poorer performance than MEA, as it led to foam formation in the test campaigns.

The trend to use wash solutions other than MEA increased considerably with the number of pilot plants constructed. The advantages of the fast reaction kinetics of MEA stand in contrast to disadvantages of the corrosion tendency and desorption energy, which make up a significant proportion of the plant and operational costs in the case of large-scale implementation, which should be lowered through the use of a different scrubbing liquor. The list of researched wash solutions studied is extensively varied and has led to the common use of identifiers and pseudonyms or synonyms such as Castor-1 and-2, KS-1 and -2, and Ecoamine.

8.2.1.2 **Boundary Dam Pilot Plant and University of Regina Test Plant**

Foam formation which occurs in CO_2 absorption with amine-based wash solutions can significantly disturb the process (cf., Idem *et al.*, 2006). These effects that appear include a reduced product quality of the separated CO_2 stream and also a decrease in capacity, as this often led to a preemptive flooding of the absorber column. This also diminishes the overall performance of the absorber. Furthermore, the foam may cause an increased discharge of wash solution and a large amount of wash solution that must be added supplementally.

In spite of the serious problems caused by foam formation, it is comparatively seldom reported in operations. Reclaimers, filtration units with active carbon, or antifoam solution are generally used for the suppression of foam formation. Antifoam additives only prevent the formation of foam and do not alleviate the cause, so that during operation more and more antifoam additive will probably be necessary. However, the occurrence of foam formation in the absorber results in ever-increasing investment and operational costs, which can be enormous for a large-scale commercial application. Because of this, the group of Thitakamol and Aroonwilas was the first to investigate systematically the various factors in foaming tendency and the foam volumes formed from amine-based wash solutions (Thitakamol and Aroonwilas, 2008, 2009). The dominating characteristics identified for the foaming tendencies of wash solutions were surface tension, solution density, solution viscosity, and surface viscosity.

The following sections summarize the results for foaming tendencies in MEA-based wash solutions. For more detailed information, the original reports can be consulted (Thitakamol and Aroonwilas, 2008, 2009). It should further be noted that the following test results were achieved under laboratory conditions. The behavior under normal circumstances, with power plant flue gas and the impurities that it contains, can only be approximated and therefore do not need to correspond to those listed here. However, the underlying tendencies, causes, and contexts of foaming tendencies are briefly summarized and can give a first hint about this problem.

Impact of the Gas Volume Flow

Figure 8.15 shows the influence of the gas volume flow on the accumulated amount of foam. As can be seen, the foam formation decreases nonlinearly in relation to the increasing gas volume flow and leads to a stationary final value. The turbulence brought about by the higher gas flow leads to lower stability of the foam, so that it is quickly destroyed.

Impact of amine Concentration

In Fig. 8.16, the foaming tendency at different wash solution amine concentrations can be seen. The trend is not uniform; in the areas with an increasing foaming tendency, the surface tension decreases due to the increase in amine concentration. In the regions of decreasing foaming tendency, the increase in surface viscosity correlates with the increase in amine concentration and dominates all other factors, so that the foam becomes too stiff and in some cases is

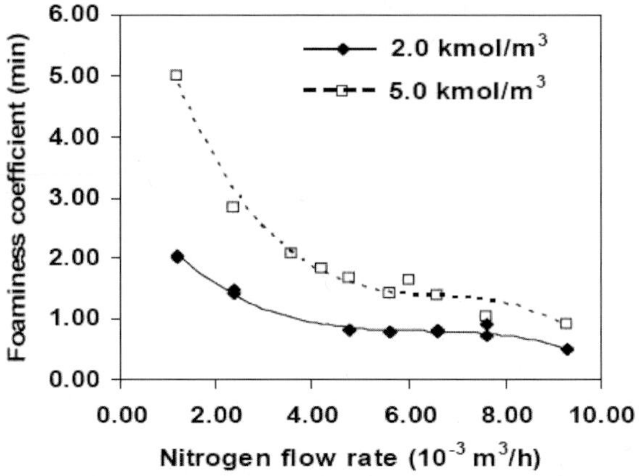

Fig. 8.15 Foaming tendency of MEA solution with given concentrations at various gas flow rates. CO_2 loading, 0.4 mol mol^{-1}; temperature, 40 °C (Thitakamol and Aroonwilas, 2008).

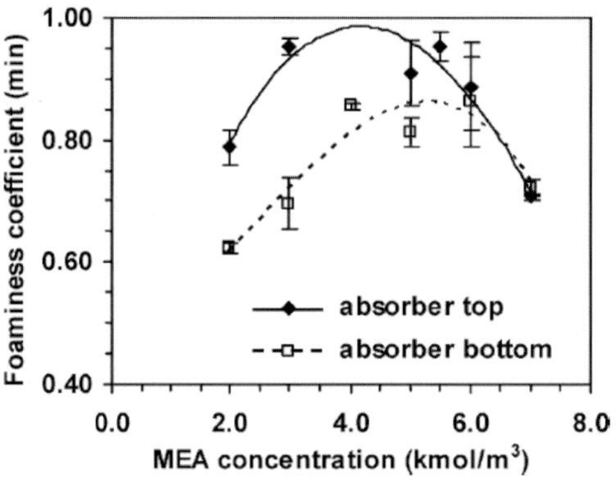

Fig. 8.16 Variation of foaming tendency of MEA solution with concentrations. CO_2 loading, 0.4 mol mol^{-1}; temperature, 40 °C (Thitakamol and Aroonwilas, 2008).

less stable. Hence disintegration of the foam bubbles is increased and the foaming tendency is reduced.

Impact of CO_2 Loading

An increase in CO_2 loading generally leads to an increase in the foaming tendency at temperatures between 40 and 90 °C (Fig. 8.17). With an increase in CO_2 loading, the wash solution density and viscosity also increase, and the surface tension decreases. A low surface tension and a high wash solution density lead to a lower foaming tendency, with which the surface forces are reduced and the buoyancy force of the gas rises within the liquid. The decreasing ten-

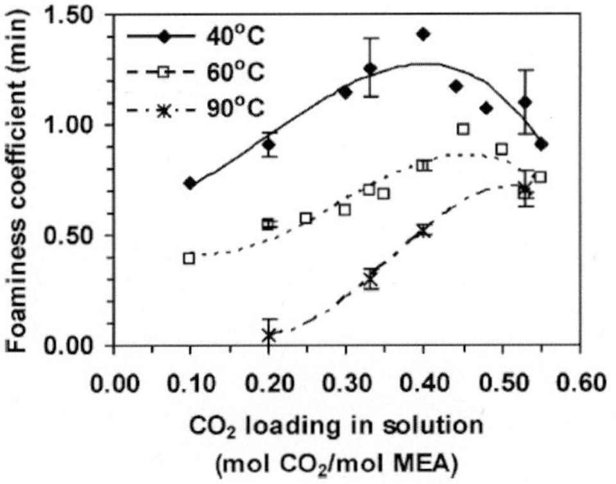

Fig. 8.17 Foaming tendency of 5 kmol m^{-3} MEA solution at various temperatures (Thitakamol and Aroonwilas, 2008).

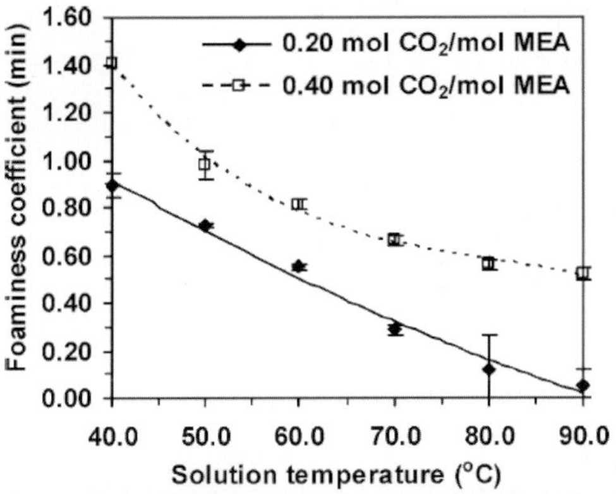

Fig. 8.18 Foaming tendency of 5 kmol m^{-3} MEA solution with temperature for CO_2 loadings of 0.2 and 0.4 mol mol^{-1} (Thitakamol and Aroonwilas, 2008).

dency for foam formation at 40 °C reflects the dominant effect of the surface viscosity, which increases continuously with increasing CO_2 loading, exactly as with the wash solution viscosity.

Impact of Temperature

Between 40 °C and 90 °C, the foaming tendency is reduced for a 5 kmol m^{-3} MEA wash solution, as shown in Fig. 8.18. With increase in temperature, the viscosity of the wash solution is reduced along with the stability of the foam that is formed.

Impact of Various Degradation Products

Most of the measures taken to reduce foaming tendencies in pilot plants aim to decrease and limit the amount of degradation products in the wash solution.

Table 8.7 Foaming tendency of a 5 kmol m^{-3} MEA solution with 10 000 ppm of various degradation products. CO_2 loading, 0.4 mol mol^{-1}; temperature, 60 °C (Thitakamol and Aroonwilas, 2008).

Degradation product	$\sum$ foaminess coefficient (min)
None	0.79
Ammonium thiosulfate	0.97
Glycolic acid	0.94
Sodium sulfite	0.92
Malonic acid	0.92
Oxalic acid	0.90
Sodium thiocyanate	0.90
Sodium chloride	0.85
Sodium thiosulfate	0.85
Bicine	0.83
Hydrochloric acid	0.83
Formic acid	0.83
Acetic acid	0.82
Sulfuric acid	0.77

For this purpose, reclaimers and active carbon filters are put in place to separate degradation products. Table 8.7 shows various degradation products and their inclination toward foam formation.

The great influence of degradation products on the foaminess coefficient can be traced back to the pre-existing anions such as carboxylate (COO^-), sulfonate (SO_3^-), and sulfate (SO_4^{2-}). These have a hydrophilic effect and therefore weaken the surface tension of the wash solution. This effect interferes with the mechanism of foam formation and promotes it.

Separation processes such as reclaimers and filtrations can equally be found in the reduction of the foaminess coefficient in CO_2 capture. Antifoaming solutions in comparison only act against the cause and therefore work only temporarily.

Impact of Corrosion Inhibitors

The inclination of amine solutions and their degradation products to corrode (i.e., various acids, compare with Table 8.7) has long been known. This problem is still present, so all components in direct contact with the wash solution in the plant are made of stainless steel to limit excessive corrosion. This gives rise to high costs and in fact provides no guarantee that corrosion will not appear.

For the prevention of excessive corrosion in plants with amine-based washes, corrosion inhibitors sidestep the corrosion problem and may instead bring forward the foaming tendency and thus create a new operational problem. Figure 8.19 shows three selected corrosion inhibitors in addition to a 5 kmol m^{-3} MEA solution in terms of the tendency toward foam formation relative to wash solutions without inhibitors.

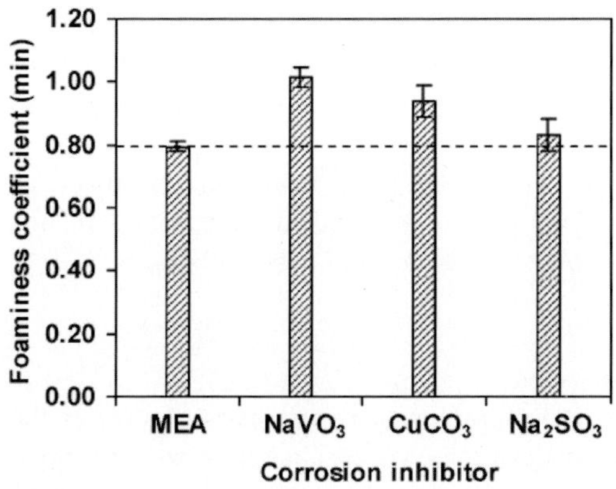

Fig. 8.19 Relative impact of corrosion inhibitors (1000 ppm) added to a 5 kmol m^{-3} MEA solution at 60 °C (Thitakamol and Aroonwilas, 2008).

Table 8.8 Foaming tendency of different amine solutions and blends at various mixing ratios with a total amine concentration of 4.0 kmol m^{-3}; temperature, 60 °C (Thitakamol and Aroonwilas, 2008).

Type of alkanolamine	$\sum$ foaminess coefficient (min)
MEA	0.85
DEA	No foam
MDEA	0.32
AMP	No foam
MEA + MDEA (1:2)	No foam
MEA + MDEA (1:1)	No foam
MEA + MDEA (2:1)	No foam
DEA + MDEA (1:2)	No foam
DEA + MDEA (1:1)	No foam
DEA + MDEA (2:1)	No foam
MEA + AMP (1:2)	No foam
MEA + AMP (1:1)	No foam
MEA + AMP (2:1)	0.13

As can be seen from Fig. 8.19, Na_2SO_3 (sodium sulfite) has almost no impact on foam tendency, whereas $NaVO_3$ (sodium metavanadate) and $CuCO_3$ (copper carbonate), as corrosion inhibitors, raise the foaming tendency.

Foaming Tendencies of Various Blends and Amines

It can be observed that not all amine solutions and blends lead to foam formation. This could also explain why some pilot plants have reported problems with foam whereas others clearly do not encounter any problems with foam. The foaming tendency of amine blends should not, as Table 8.8 shows, be taken trivially.

MEA and MDEA both show a foaming tendency that disappears with certain compound behaviors in blends. Interestingly, MEA combined with AMP shows a foaming tendency at a ratio of 2 : 1, although all other blends with AMP and also AMP alone show absolutely no foaming tendency.

It should be noted that Table 8.8 reports laboratory measurements made by Thitakamol and Aroonwilas (2008). This may be the reason for different results when used under real conditions in pilot plants or demonstration plants.

8.2.1.3 Foam Formation in CO_2 Absorption

The pilot plant in Boundary Dam (BD plant) and the test plant station at the University of Regina (UR plant) are reviewed jointly because they are both operated by the International Test Centre for CO_2 Capture (ITC), Regina. The wash solutions were tested at the test plant (UR plant) and then introduced to the pilot plant (BD plant). The test plant has a gas flow rate of up to 4800 $m^3\,d^{-1}$, corresponding to a CO_2 capture of 1 t d^{-1}.

Layout of the Plants

The layout of the scrubbing system is similar to that of the pilot plant in Esbjerg. Figure 8.20 shows the layout of the UR plant.

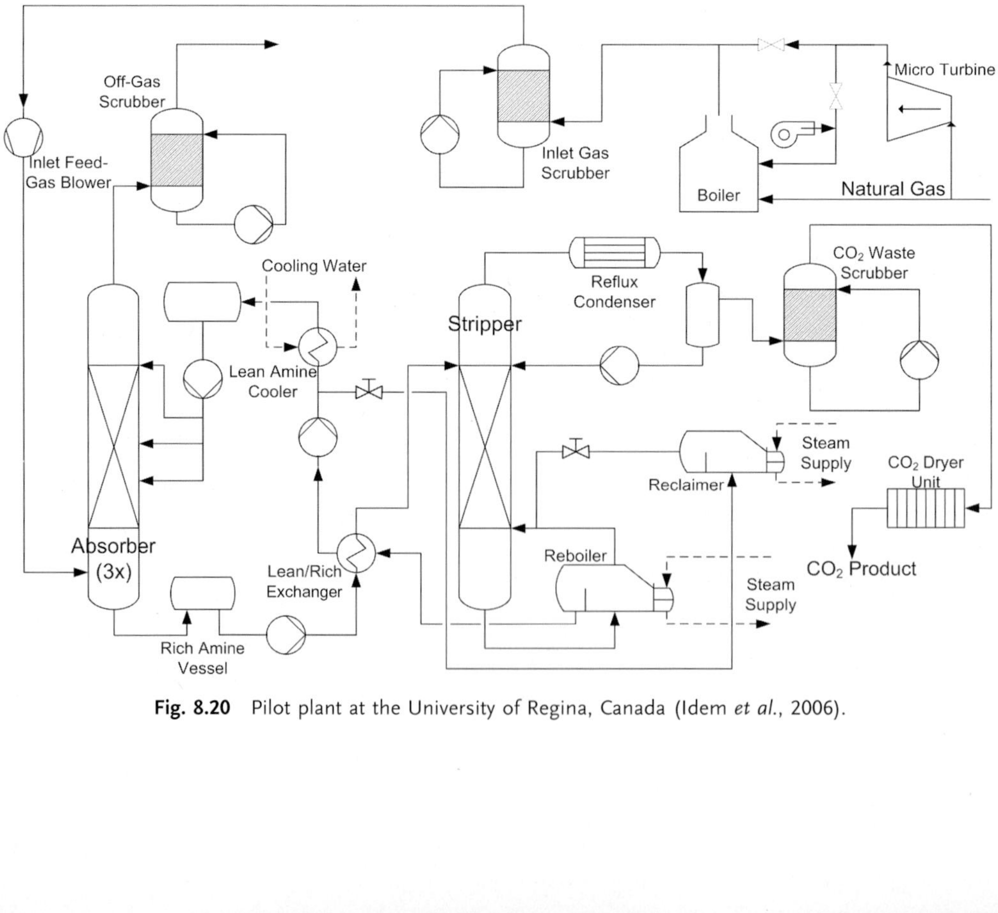

Fig. 8.20 Pilot plant at the University of Regina, Canada (Idem *et al.*, 2006).

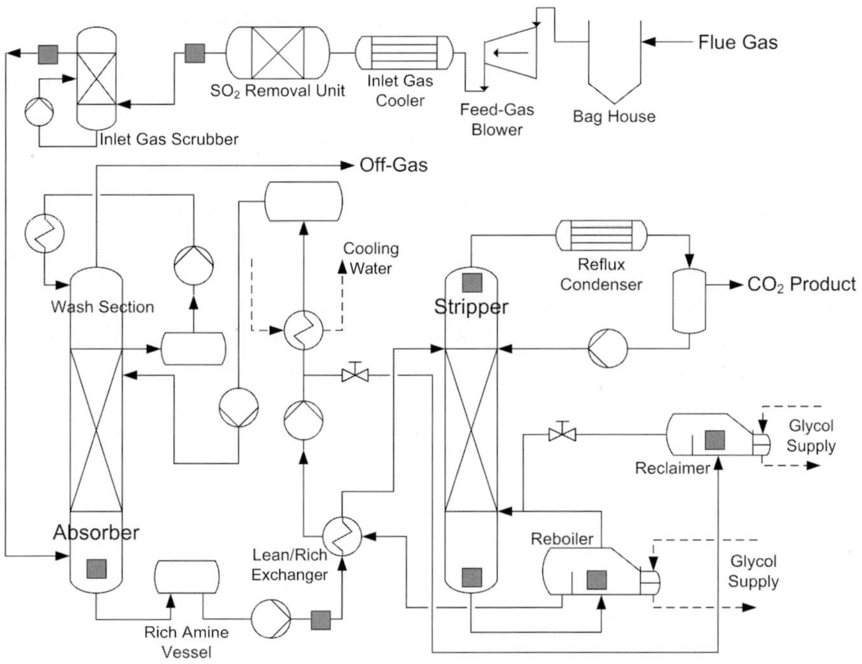

Fig. 8.21 Boundary Dam pilot plant. Shaded squares are corrosion testing points (Idem *et al.*, 2006).

The UR plant is equipped with a 250 kW gas-fired boiler and a 30 kW gas-fired turbine. Both units produce the flue gas and the steam required for the plant. By changing the connections of the gas turbine and the boiler, the CO_2 content in the flue gas flow can be varied. The flue gas runs through a prewasher (prescrubber) before it is sent to one of three absorbers. The primary function of the prewasher is to cool the flue gas, as it contains no flue ash and only small amounts of SO_x or NO_x due to the fuel used.

The three available absorbers are each divided into three sections, each of which can be loaded with wash solution. The washing section is located on the outside of the absorber. Each absorber has an overall height of 10 m and a diameter of 0.3 m. After the stripper, there is an additional purification unit installed in order to purify the captured CO_2 to pipeline quality.

The pilot plant in the Boundary Dam power station (BD plant) is operated with flue gas from a coal-fired boiler. The layout of the BD plant is shown in Fig. 8.21. The plant can separate a maximum of 4 t d^{-1} of CO_2. The flue gas is split off from a branch current at the power station.

In contrast to the UR plant, in the BD plant flue gas must be cleaned to remove flue ash and SO_2. For this there is a two-stage SO_2 removal unit, which uses an aqueous Na_2SO_3 solution and a fabric filter. Before entering the absorber column, the flue gas is quenched and additionally cooled to the desired feed temperature in a prescrubber. The CO_2 scrubber does not differ greatly in layout from that in the UR plant and the pilot plant in Esbjerg. The captured CO_2 cannot be brought to pipeline quality. As a heat source for the stripper, a glycol heat exchanger is used. The absorber column has a diameter of 0.46 m and the desorber column has a diameter of 0.41 m. In addition to the temperature and concentration measurements, the BD plant also has gauges to measure short- and long-term corrosion. These measuring points are represented in Fig. 8.21 by shaded squares.

Both layouts utilize MEA or MEA–MDEA solutions. In the case of MEA, a solution with 5 kmol m^{-3} of MEA is run, which equates to about 30 wt%. The MEA–MDEA blend is used with a concentration of 5 kmol m^{-3} and with a ratio of 4 : 1. In the pilot plant, additional corrosion inhibitors were added to the solutions, whereas in the test plant no inhibitors need to be added (no SO_2, dust, etc., present and therefore less degradation products which support corrosion). The CO_2 concentration in flue gas in the UR plant amounts to a maximum of 11% and in the BD plant to an average of 15%.

Test Results and Comparison

MEA was used as a reference wash solution in these test runs, although the MEA–MDEA solution was identified as suitable in the UR plant. Figure 8.22 shows a comparison of the two wash solutions in the UR plant.

The tests at the UR plant show that with the use of the MEA–MDEA solutions at the same specific heat output, more CO_2 will be captured. To investigate

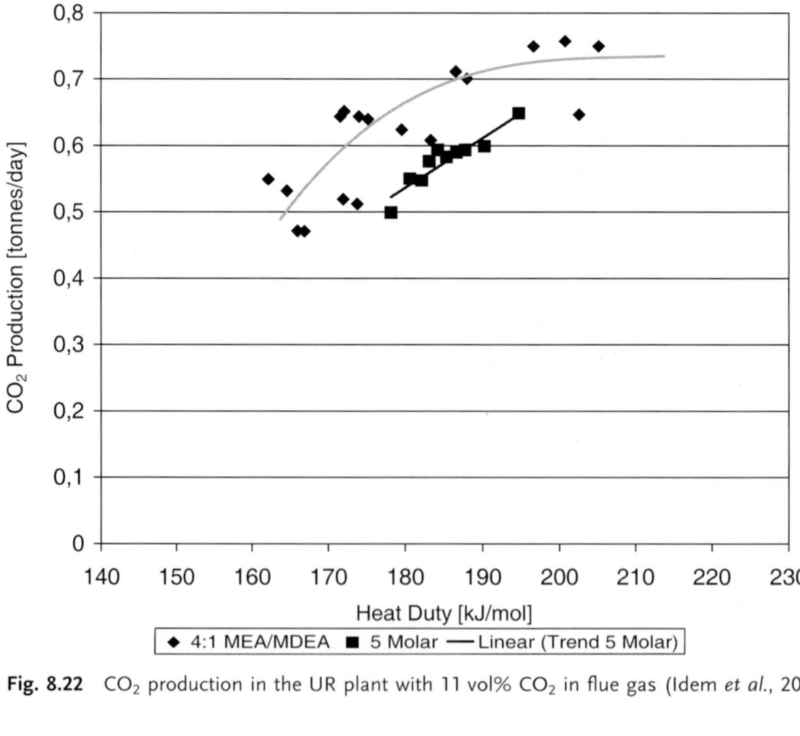

Fig. 8.22 CO_2 production in the UR plant with 11 vol% CO_2 in flue gas (Idem *et al.*, 2006).

this effect further, tests of lean and rich loading were carried out with both wash solutions.

Figures 8.23 and 8.24 show that in spite of the scatter of the measured values, the lean and rich loadings are hardly distinguishable from one another. The intake capacity of the wash solution is therefore identical and the effect in Fig. 8.22, because of this, can be reduced to a less expendable desorption energy. The performance of the MEA–MDEA solution confirms the lower regeneration energy required for MDEA through weak bonding of CO_2 to the amine. This process also works with the MEA–MDEA blend and affects the energy consumption of CO_2 capture positively in the UR plant.

Both solutions were implemented in connection with the tests in the UR plant in the pilot plant at Boundary Dam power station in order to investigate their behavior with flue gas from a coal fired boiler.

Figure 8.25 shows the CO_2 production of both scrubber liquors in direct comparison with one another in the BD plant, from a branch current of the flue gas from the power plant. In spite of the expectations for the test in the UR-plant, a poorer absorption efficiency was observed in the test runs with the MEA–MDEA blend. Although the scatter of the measured values cannot be neglected, it is easy to recognize the offset.

Investigations of lean and rich loadings because of the different flue gas compositions of the BD plant in comparison with the UR plant and the addition of corrosion inhibitors to the wash solution were undertaken in order to clarify the

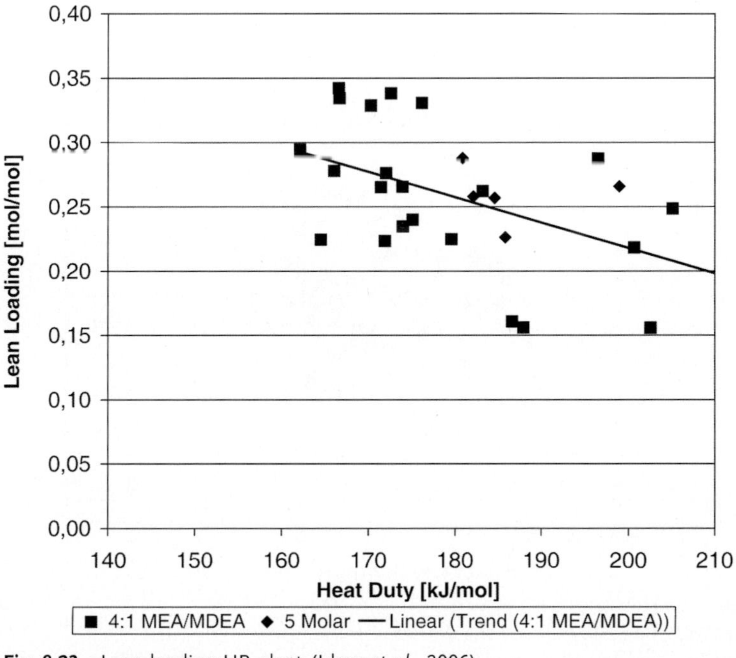

Fig. 8.23 Lean loading UR plant (Idem *et al.*, 2006).

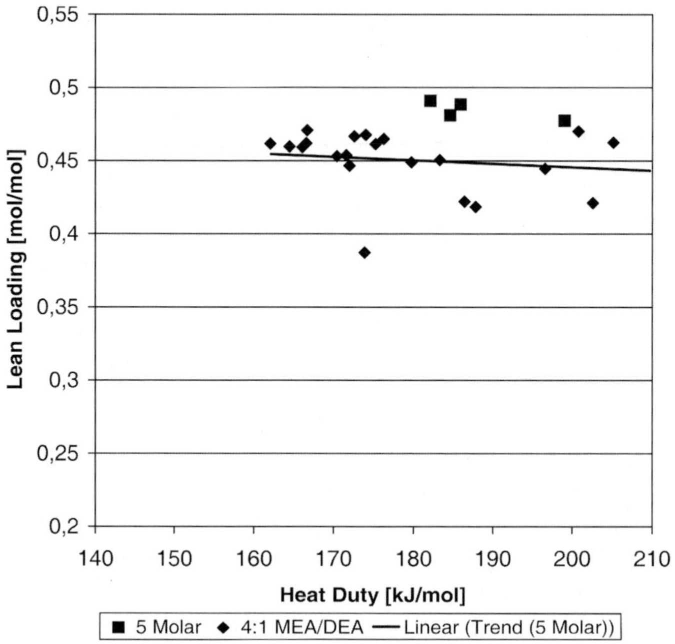

Fig. 8.24 Rich loading UR plant (Idem *et al.*, 2006).

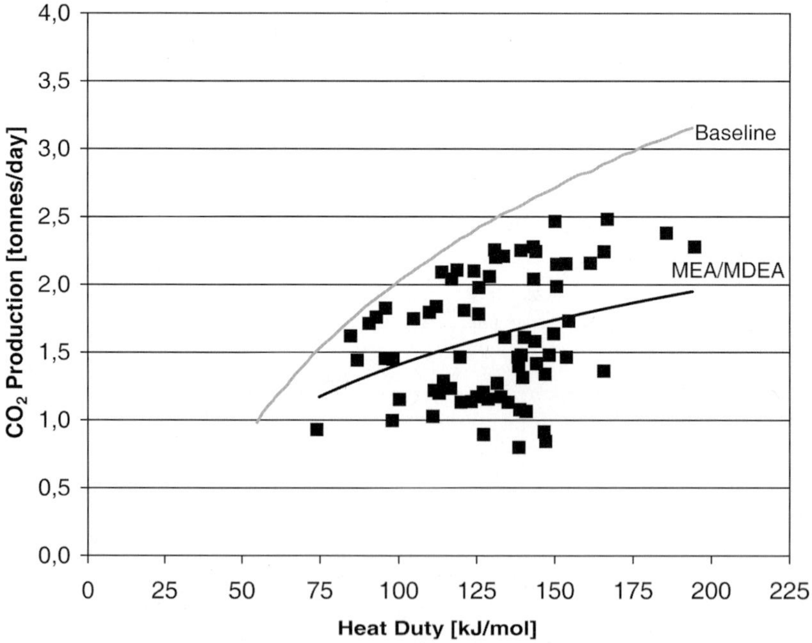

Fig. 8.25 CO_2 production in the BD plant with 15 vol% CO_2 in flue gas. Baseline, MEA; other line, MEA–MDEA blend (Idem *et al.*, 2006).

behavior of the blend. Figures 8.26 and 8.27 show the lean and rich loadings, respectively, for the tests in the BD plant.

Figures 8.26 and 8.27 give a first hint of an opposing tendency in the wash solutions in both plants with the loading of the input and output compositions. It should be remembered that the lean and rich loadings in the UR plant were identical. It can be recognized that MEA remains above the values of the blends for both lean and rich loadings. The difference between lean loadings lies on average at 0.05 mol mol^{-1}, whereas rich loadings differ from one another by 0.08 mol mol^{-1}. A higher intake capacity follows for MEA, at 0.03 mol mol^{-1}, in comparison with the MEA–MDEA blend. Figure 8.28 clarifies the difference between intake capacities.

The load difference (= intake capacity) is greater for MEA and therefore this solution captures more CO_2, as shown in Fig. 8.25. At the same time, the personnel at the plant assume that through the addition of inhibitors, the degradation of the MEA–MDEA blend is greater than that of the MEA wash solution. Early investigations support this conclusion. In 2004, in a CO_2 extraction plant that had been finalized in 2000 on the site of the Boundary Dam power station, Wilson *et al.* (2004) established that sulfate, oxalate, and thiocyanate are produced and accumulate to 0.5 wt% within 10 days of operation. Furthermore, more than 330 ppm of Fe ions were found in the wash solution within the same time period.

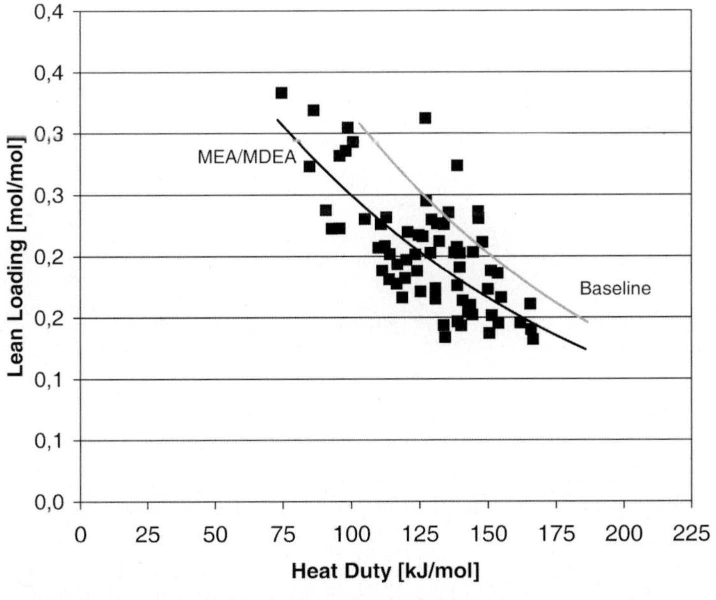

Fig. 8.26 Lean loading BD plant. Baseline, MEA; other line, MEA–MDEA blend (Idem *et al.*, 2006).

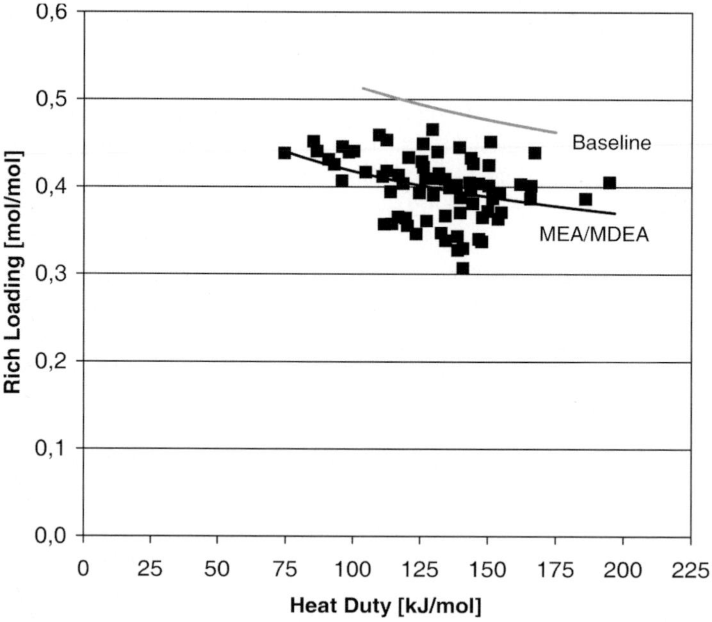

Fig. 8.27 Rich loading BD plant. Baseline, MEA; other line, MEA–MDEA blend (Idem *et al.*, 2006).

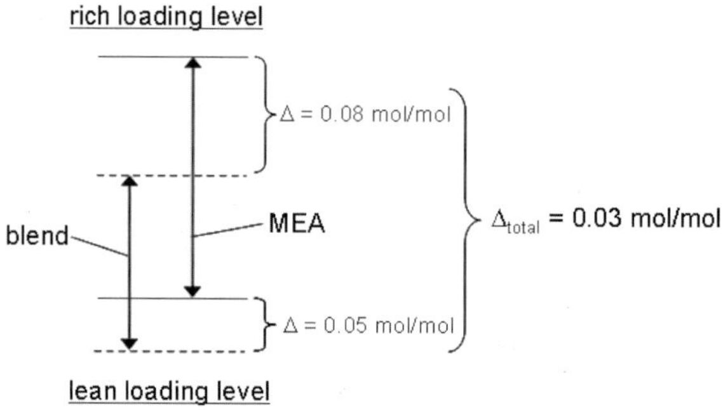

Fig. 8.28 CO_2 production in the UR plant with 11% CO_2 in flue gas (Idem *et al.*, 2006).

8.3 Conclusion

The research at the UR plant with the transfer of the results (preferred wash solutions) to the BD plant makes it very clear that pilot plants are an essential intermediate step towards the development of large-scale implementation of CO_2 capture in coal-fired power stations. It cannot be guaranteed that the techniques which led to good results in the laboratory can be applied under real circumstances in power stations generating equivalent results! The intermediate step from a pilot plant on the way to commercially usable CO_2 absorption is indispensable since not only is the number of influencing factors in power station operations large but also they differ considerably and cannot be disregarded. The large number of pilot plants and demonstration plants is therefore a consequence of the complexity of the process and the sheer number of influencing factors, as their consequences are just as multifaceted and unpredictable.

The commercial application of post-combustion plants for CO_2 capture from power plant flue gases in the range >600 MW_{el} is accordingly not yet a state-of-the-art technology. However, there are many pilot plants worldwide that will serve as the intermediate step on the path to industrial operation. Because of this, demonstration plants will enter into operation in the coming years, closing the gap up to the region of >600 MW_{el} (see the overview in Tables 8.3 and 8.4). The first industrial implementation of post-combustion technology is expected to occur, depending on operational experience in demonstration plants, in 2020–2022. It will certainly take until 2025 for CO_2 capture to become a state-of-the-art technology and for ready-to-use power stations to be built.

It is also interesting that amine-based wash solutions are still widely used for CO_2 capture in existing pilot plants and no other wash solutions. A consistent trend towards alternative wash solutions is not yet distinguishable. This is due to the fact that no known wash solution offers decided procedural advantages over the others, and rather there will always be a compromise made in the selection of wash solutions. Nevertheless, it will be interesting to keep track of operational experience that may be gained regarding amino salts and ammonia as wash solutions.

Symbols and Abbreviations

A	Area	m^2
Abs	Absorber	
aMDEA	Activated MDEA	
AMP	2-Amino-2-methyl-1-propanol	
APH	Air preheater	
Col	Column	
d	Diameter	m
DEA	Diethanolamine	
DeNOx	Denitrification unit	
ESP	Electrostatic precipitator	
F	Load factor	$Pa^{0.5}$
FGD	Flue gas desulfurization	
G	Gas stream	$m^3\ s^{-1}$
gas	Gaseous phase	
L/G	Liquid/gas-ratio	
MDEA	Methyldiethanolamine	

MEA	Monoethanolamine	
SCR	Selective catalytic reduction	
STP	Standard temperature and pressure	
TA-Luft	German Technical Instruction on Air Quality	
u	Velocity	m/s^{-1}
ρ	Density	$kg\ m^{-3}$

References

Alie, C.F. (2004) CO_2 capture with MEA: integrating the absorption process and steam cycle of an existing coal-fired power plant. University of Waterloo, Master of Applied Science thesis, p. 178.

Amann, M., Bertok, I., *et al.* (2005) A further emission control scenario for the Clean Air For Europe (CAFE) programme – CAFE Scenario Analysis Report No. 7. E. C. Environment, International Institute for Applied Systems Analysis (IIASA), Laxenburg.

Aroonwilas, A. and Veawab A. (2006) Integration of CO_2 capture unit using single- and blended-amines into supercritical coal-fired power plants: implications for emission and energy management. Presented at GHGT 8, Trondheim, pp. 6.

Benson, H.E., *et al.*, (1954) CO_2 Absorption – Employing Hot Potassium Cabonate Solutions. Chemical Engineering Progress **50**(7), 356–364.

Black, S. (2006) *Alstom: Chilled Ammonia Process for CO_2 Capture.* Position Paper, 29 November 2006. Alstom, Levallois-Perret.

Braine, B.H. (2007) *American Electric Power: AEP amd Climate Change.* American Electric Power, Columbus, OH.

Chakravarty, S., Gupta, A., *et al.* (2001) Advanced technology for the capture of carbon dioxide from flue gases. Presented at the First National Conference on Carbon Sequestration. Washington, DC.

Chapel, D.G. (1999) Recovery of CO_2 from flue gases: commercial trends. Presented at the Canadian Society of Chemical Engineers Annual Meeting. Saskatoon, Saskatchewan.

Chen, E.R., Rochelle, G.T., and Seibert, F. (2006) Pilot plant for CO_2 capture with aqueous piperazine/potassium carbonate. Presented at the 8th International Conference on Greenhouse Gas Control Technologies, GHGT 8, Trondheim.

Derks, P W.J., Kleingeld, T., van Aken, C., Hogendoorn, J.A., Versteeg, G.F. (2006) Kinetics of absorption of carbon dioxide in aqueous piperazine solutions. Chemical Engineering Science **6**, 6837–6854.

Engelking, W. (2007) Die Rauchgasreinigung im Zusammenspiel mit CO_2-Minderungsmaßnahmen. Conference: Kraftwerke sind mehr als nur Turbinen und Kessel, Gelsenkirchen.

Epp, B. (2011) Grundlegende Untersuchungen zum Absorptionsprozess bei der Decarbonisierung von Rauchgasen mit einem Membranabsorber. Doctoral thesis (submitted), Technische Universität Dortmund.

Epp, B., Fahlenkamp, H., *et al.* (2010) Degradation of solutions of monoethanolamine, diglycolamine and potassium glycinate in view of tail-end CO_2 absorption. Presented at the 10th International Conference on Greenhouse Gas Control Technologies, GHGT 10, Amsterdam.

Feldkamp, M. (2008) Simulation der Zwei-Phasen-Strömung in REA-Sprühturmwäschern fossil befeuerter Kraftwerke. Dr.-Ing. thesis, Technische Universität Dortmund.

Feraud, A., Marocco, L., *et al.* (2006) CASTOR study on technological requirements for flue gas clean-up prior to CO_2 capture. Presented at the 8th International Conference on Greenhouse Gas Control Technologies, GHGT 8, Trondheim.

Geuzebroek, F. (2007) Improvements and implementation of CO_2-capture technology. Presented at BCI-Kolloquium, Technische Universität Dortmund, Shell Global Solution.

Goff, G. S., and Rochelle, G. T. (2004) Monoethanolamine degradation: O_2 mass transfer effects under CO_2 capture conditions. *Ind. Eng. Chem. Res.*, **43**, 8.

Gosh, U. K., and Kentish, S. E. (2009) Absorption of carbon dioxide into aqueous potassium carbonate promoted by boric acid. Engery Procedia **01**, 1075–1081.

Göttlicher, G. (1999) Energetik der Kohlendioxidrückhaltung in Kraftwerken. *Fortschr.-Ber. VDI*, 187.

Idem, R., Wilson, M., Tontiwachwuthikul, P., Chakma, A., Veawab, A., Aroonwilas, A., and Gelowitz, D. (2006) Pilot plant studies of the CO_2 capture performance of aqueous MEA and mixed MEA/MDEA solvents at the University of Regina CO_2 capture technology development plant and the Boundary Dam CO_2 capture Demonstration plant. *Ind. Eng. Chem. Res.*, **45**(8): 2414–2420.

Kittel *et al.*, focused in [Quelle: Kittel, J. *et al.* (2009) Corrosion in MEA units for CO_2 capture: pilot plant studies. Energy Procedia **1**(1), 791–797] on the corrosion rates of different steels at various locations in a CO_2 capture pilot plant.

Knudsen, J., Jensen, J., Vilhelmsen, P., and Biede, O. (2007 a) First year operation experience with a 1 t/h CO_2 absorption pilot plant at Esbjerg coal-fired power plant. *VGB PowerTech* **87**(3), 57–61, 66.

Knudsen, J., Jensen, J., Vilhelmsen, P., and Biede, O. (2007 b) Second year operation experience with a 1 t/h CO_2 absorption pilot plant at Esbjerg coal-fired power station in Denmark. *Proc. Annu. Int. Pittsburgh Coal Conf.*, **24**(4), 462/1–462/12.

Knudsen, J., Jensen, J., Vilhelmsen, P., and Biede, O. (2009) Experience with CO_2 capture from coal flue gas in pilot-scale: testing of different amine solvents. *Energy Procedia*, **1**(1), 783–790.

Knudsen, J., Vilhelmsen, P.-J., Biede, O., and Jense, J. N. (2006) CASTOR 1 t/h CO_2 absorption pilot plant at the Elsam Kraft A/S Esbjerg power plant – first year operation experience. Presented at the 8th International Conference on Greenhouse Gas Control Technologies, GHGT 8, Trondheim.

Kohl, A. and Nielsen, R. (1997) Gas Purification. Gulf Publishing Company, Houston, Texas, 1997.

Kubacz, V. (2011) Vereinfachtes Modell zur Bewertung der Leistungsfähigkeit von Aminen und deren Mischungen bei der Entfernung von CO_2 aus Kraftwerksrauchgas. PhD thesis (submitted), Technische Universität Dortmund.

Lawal, A., Wang, M., *et al.* (2008) Dynamic modelling of CO_2 absorption for post combustion capture in coal-fired power plants. *Fuel*, **88**(12), 2455–2462.

Majchrowicz, M., Niederer, J. P. M., *et al.* (2006) Precipitation in amino acid salt CO_2 absorption systems. Presented at the 8th International Conference on Greenhouse Gas Control Technologies, GHGT 8, Trondheim, pp. 6.

McKetta, J. J. (1978) *Encyclopedia of Chemical Processing and Design, Calcination Equipment to Catalysis*, Vol. 6, Marcel Dekker Inc (June 1978).

Nsakala, Y. N., Marion, J., *et al.* (2001) Engineering feasibility of CO_2 capture on an existing US coal-fired power plant. Presented at the First National Conference on Carbon Sequestration, Washington, DC.

Park, S.-W., and Sung, D.-H. (2006) Carbonation Kinetics of Potassium Carbonate by Carbon Dioxide. Ind. Eng. Chem. Res. **12**(4), 522–530.

Rhudy, R. S. B. (2007) *Alstom: Chilled Ammonia Process Update.* Alstom, Levallois-Perret.

Rochelle, G. T., and Hilliard, M. D. (2005) CO_2 Capture by Absorption with Potassium Carbonate Fourth Quarterly Report 2005. Quarterly Progress Report University of Texas

Schubert, S. (2004) Untersuchungen zur Anwendung immobilisierter Aktivatoren bei der Absorption von CO_2 mit wässrigen Methyldiethanolamin-Lösungen. PhD thesis, Technische Universität Dortmund.

Stankewitz, C. (2011) Analyse der verbesserten energetischen Integration einer Postcombustion Carbon Capture Wäsche in den Kraftwerksprozess. PhD thesis (submitted), Technische Universität Dortmund.

Strazisar, B.R. *et al.* (2003) Degradation Pathways for Monoethanolamine in a CO_2 Capture Facility. Energy & Fuels **17**(4), 1034–1039

Stöver, B., Bergins, C., *et al.* (2010) Optimized post combustion carbon capturing on coal fired power plants. Presented at the 10th International Conference on Greenhouse Gas Control Technologies, GHGT 10, Amsterdam.

Suenson, M.M., Georgakis, C., *et al.* (1985) Steady-state and dynamic modeling of a gas absorber-stripper system. *Ind. Eng. Chem. Fundam.*, **24**(3), 288–295.

Thitakamol, B., Veaweb, A., and Aroonwilas, A. (2008) Foaming behavior in CO_2 absorption process using aqueous solutions of single and blended alkanolamines. *Ind. Eng. Chem. Res.*, **47**(1), 216–225.

Thitakamol, B., Veaweb, A., and Aroonwilas, A. (2009) Foaming in amine-based CO_2 capture process: experiment, modeling and simulation. *Energy Procedia*, **1**(1), 1381–1386.

Bartholomé, E., Biekert, E., Hellmann, H., Leg, H., Weigert, M., Ullmanns Encyclopädie der technischen Chemie (1975), 4. Auflage, Bd. 10, Erdgas Aufbereitung. Verlag Chemie, Weinheim.

Van Holst, J., Politiek, P.P., *et al.* (2006) CO_2 capture from flue gas using amino acid salt solution. Presented at the 8th International Conference on Greenhouse Gas Control Technologies, GHGT 8, Trondheim.

Wilson, M., Tontiwachwuthikul, P., Chakma, A., Idem, R., and Veawab, A. (2004) Test results from a CO_2 extraction pilot plant at boundary dam coal-fired power station. *Energy*, **29**(9–10), 1259–1267.

9
CO_2 Removal in Coal Power Plants via Pre-Combustion with Physical Absorption

Stéphane Walspurger, Eric van Dijk, and Ruud van den Brink

Abstract

Pre-combustion decarbonization of syngas streams benefits from high pressures and concentrations of CO_2, which constitutes a strong driving force to lower the energy required to capture CO_2 from power plants. A number of commercial technologies based on CO_2 absorption in a liquid phase may be implemented in coal gasification-based plants. Other commercially available CO_2 separation processes by regenerative physical adsorption on solid sorbents, by refrigeration. or by membrane-selective permeation need to be adapted for bulk CO_2 removal. Novel technologies, namely membrane reactors and sorption-enhanced reaction processes, allowing integrated fuel processing and CO_2 removal, are at the development stage for further scale-up and pilot-plant demonstration. Sorption-enhanced water gas shift (SEWGS) allows CO_2 removal at elevated temperature while producing essentially carbon-free hydrogen. Preliminary techno-economic studies showed the potential energetic benefit of performing the separation step at high temperature. The cyclic sorption/regeneration of CO_2 is carried out by pressure swings, similar to those used in pressure swing adsorption units. Bench-scale testing has demonstrated that high hydrogen recoveries and high CO_2 capture ratios with sufficient purity may be obtained using the technology. High temperatures and periodic exposure to intense steaming and carbonating conditions impose high stress on the sorbent material, which needs to preserve both mechanical integrity and high cyclic CO_2 capacity for extended periods. Alkali metal-promoted hydrotalcite-based and alumina-based materials are particularly attractive sorbents, showing a stable cyclic capacity for testing periods of thousands of cycles. Material research and development has led to the use of aluminum-rich materials that benefits from higher mechanical stabilities and improved carbonation reversibility. A definitive advantage of using SEWGS in coal-fired IGCC power plants could come from recent breakthroughs: (1) alkali-promoted hydrotalcite-based materials are WGS active, suppressing the need for extra WGS catalyst material and thus increasing the net volumetric CO_2 capacity and simplifying the regeneration procedure, (2) co-adsorption of H_2S and CO_2, which avoids the use of an additional sulfur removal section before the

Process Engineering for CCS Power Plants, First Edition. Edited by Detlef Stolten and V. Scherer

combustion of hydrogen for electricity generation, and (3) the hydrotalicte-based CO_2 sorbents have been shown to be stable for several thousand cycles.

Keywords: pre-combustion decarbonization, sorption enhanced water gas shift, carbon dioxide capture

9.1 Introduction

Significant anthropogenic CO_2 emission mitigation may be achieved by capturing CO_2 from large point sources such as natural gas- or coal-fired power plants and subsequent underground storage [1]. Current research on separation processes is focused on reducing the cost of capture resulting from decreased plant efficiency due to the capture. Available commercial processes for separating bulk CO_2 use either liquid alkaline solutions (generally alkanolamines) when the partial pressure of CO_2 is low, such as when the separation is carried out once the fuel has been burned (post-combustion), or solvents for physical absorption when the partial pressure is high enough, when the fuel is first converted into CO_2 and H_2 (pre-combustion). Implementation of post-combustion capture with existing technologies in existing coal-fired power plants leads to reduced plant efficiency and may require 25–50% additional fuel to maintain a constant power output [1, 2]. The lower figures may be achieved by smart heat integration, use of low-grade heat and abatement of regeneration energy requirement. In contrast, separation technologies for pressurized gas streams containing higher CO_2 concentrations typically present in integrated gasification combined cycle (IGCC) power plants are likely eventually to lead to lower capture penalties because they benefit from a larger concentration driving force for CO_2 separation [3]. In pre-combustion decarbonization, natural gas or coal is first converted to syngas (mostly $CO + H_2$), which is subsequently converted to CO_2 and H_2 by the water gas shift (WGS) catalytic process. Since these processes are usually carried out at relatively high pressure (typically 20–50 bar), the equipment required for CO_2 and acid gas removal is smaller and simpler. In addition, an advanced WGS reaction process recently has been suggested to achieve sufficiently high CO conversion by WGS while reducing the amount of pressurized steam needed [4]. Separation of CO_2 from hydrogen may then be achieved using commercial technologies or other processes based on physical absorption by solvents. In this section, first the commercially available absorption processes are presented while emphasizing operational windows, followed by a discussion on the characteristics and advantages of some more advanced separation processes in the development stage.

9.1.1 Liquid-Phase Absorption Commercial Technologies

Commercially available absorption processes involve scrubbing of syngas with a liquid solvent that selectively removes acid components such as H_2S and CO_2 [5, 6]. The acid components are recovered in a regenerator that is connected with the scrubber (Fig. 9.1).

CO_2 and H_2S may be absorbed by physical dissolution (physical solvents), chemical bonding (chemical solvents), or a combination of both processes by designing an optimal solvent system (hybrid physical–chemical solvents). The loading capacity of "physical solvents" depends primarily on Henry's law, which allows solvent regeneration by means of simple pressure reduction (flashing), stripping with an inert gas (stripping), or temperature swing with an intermediate reboiler using available steam (reboiling). Flash regeneration is considered the simplest and cheapest method, in which pressurized loaded solvent is depressurized in one or multiple stages. This is particularly attractive when bulk removal at high operating pressures and high product purities is desired. The optimization of the process design and operation requires a thorough knowledge of liquid–vapor equilibrium data covering all components and conditions encountered in the process, allowing the determination of the number of theoretical plates as a function of gas and liquid flow rates for a given desired gas removal rate and gas purity recovery [5]. The solvent circulation rate dictates the capital cost (size, circulation pumps, heat exchangers) of the whole system and may be reduced by chilling the solvent. The most commonly used processes are Rectisol (Lurgi), Selexol (UOP), Purisol (Lurgi), and Fluor Solvent (Fluor). Rectisol uses methanol at low temperature (–15 to –60 °C) to remove H_2S up to 0.1 ppm and CO_2 up to the ppm range, which makes it the designated process for acid gas recovery (AGR) within coal- and heavy oil-based synthesis of ammonia, methanol, hydrogen, substitute natural gas (SNG), and the Fischer–Tropsch

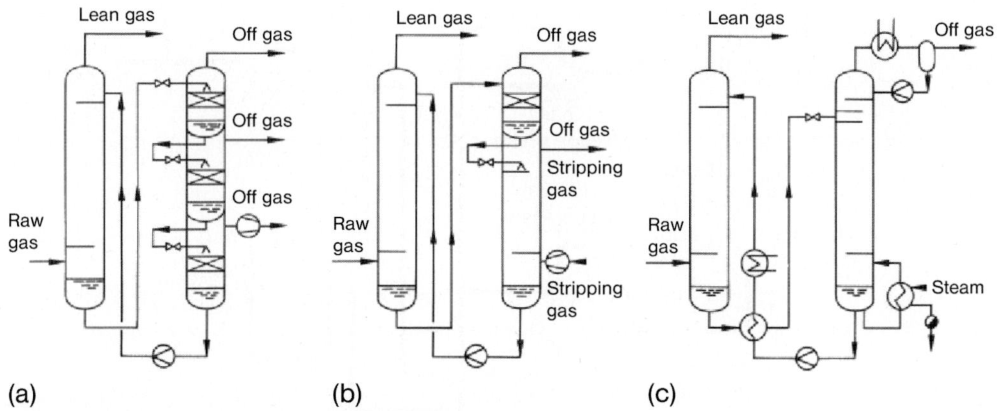

Fig. 9.1 Absorption with different regeneration methods: (a) flashing, (b) stripping and (c) reboiling.

process [7]. The low temperatures enhance H_2S and CO_2 capacities while preventing excessive methanol evaporation and maintaining the viscosity at acceptable levels, thus ensuring good heat- and mass-transfer properties. Selexol uses the dimethyl ether of poly(ethylene glycol) (DEMPG) [8] at typical operating temperatures ranging from 0 to 175 °C. Although a high capacity for H_2S and CO_2 may be reached at lower temperatures, the actual practical operating window of Selexol is strongly dependent on the viscosity. Selexol is currently mostly used for either combined CO_2 and H_2S removal or the selective removal of H_2S, and may be used for bulk CO_2 removal with a double-stage process, when the CO_2 concentration is high compared with that of H_2S (Fig. 9.2) [9]. Similar considerations are valid for the Purisol and Fluor processes, which use *N*-methyl-2-pyrrolidone (NMP) at 15–40 °C and propylene carbonate (PC) between –18 and 25 °C, respectively. Purisol is used in ammonia production from reformed natural gas for the removal of the high CO_2 content of gases to a low residual level, whereas the Fluor process is suitable for application in natural gas sweetening thanks to the high selectivity for CO_2 over CH_4.

For process selection for gas treatment, the level of partial pressure of CO_2 in the considered plant is an essential factor [10]. In the case of pre-combustion CO_2 capture from natural gas combined cycles, in which the CO_2 partial pressure is low, chemical solvents must be used. In IGCC, the partial pressure of CO_2 is certainly higher, leading to physical solvent selection. In contrast to physical solvents, the loading capacity of "chemical solvents" is dictated by the concentration of active component (alkanolamine) in the solution, which usually allows high capacities at low partial pressures – typically for natural gas-fired power plants – at the expense of significant heat addition through the reboiler

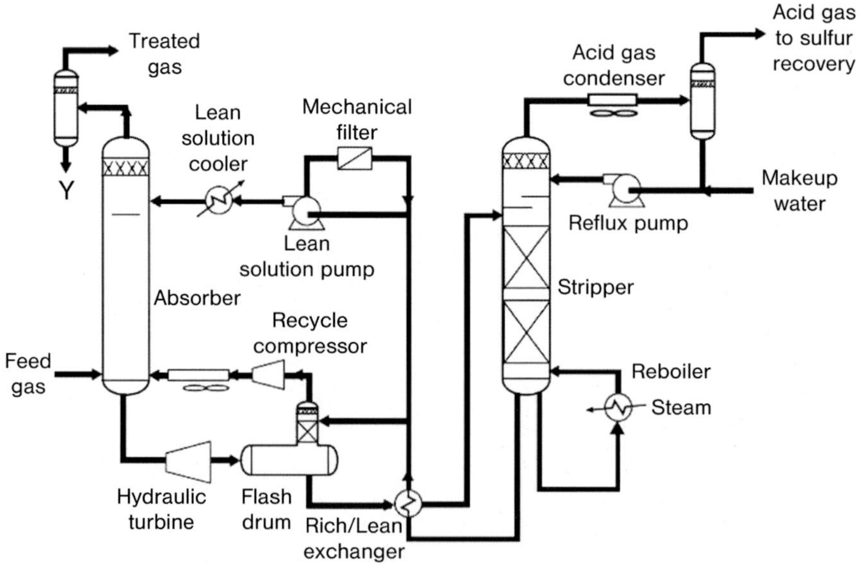

Fig. 9.2 Selexol process [8].

of the stripper column for the regeneration step. Alkanolamine-based systems are treated elsewhere in this book.

Hybrid physical–chemical absorption processes are also considered in the industry, such as the Sulfinol process (Shell), Amisol process (Lurgi), and Optisol (NATCO Group). These processes essentially combine physical and chemical absorption processes by using mixtures of solvents and amines. This allows the system to be tailored for as low regeneration heat requirements as possible at the expense of extra equipment in most cases.

The selection of the appropriate solvent system in a gas treatment plant should be considered with caution, since the solvent circulation rate governs the equipment size and thus capital and operating costs. As far as physical absorbers are concerned, the solvent circulation rate required is largely independent of the concentration of the component to be removed but proportional to the total quantity of crude gas. Physical absorbers such as Selexol are therefore usually the preferred option in IGCC plant design studies [11]. In contrast, for chemical absorbers the solvent circulation rate is dependent on the quantity of acid gas constituents to be removed and thus highly dependent on the concentration of these components, but practically independent of pressure. Other aspects must also be considered, such as solvent selectivity, resulting purity, utility consumption, crude gas impurities, and solvent corrosiveness.

In addition to physical solvents and aqueous solutions of amines, many processes have been developed for carbon dioxide removal utilizing alkali metal salts. Sodium and potassium carbonate are among the most popular salts in the industry. However, aqueous solutions of alkali metal carbonates have a number of problems in practice: (1) slow reaction rate with CO_2, which may be enhanced by addition of, for example, amines [12, 13]; (2) high heat requirements for regeneration of the solution; (3) other specific problems such as foaming, corrosion, and dispersion of alkali metal in the fuel stream as fine droplets, mist, or aerosol which may cause serious damage to the turbines. In contrast, low cost and minimal degradation are the main advantages of these processes.

The capture of CO_2 using physical solvents from coal gasification combined cycles causes a decrease in net electric efficiency. Damen *et al.* [3] gave a range of electric efficiencies of 32–35% for an IGCC with 85% CO_2 capture using Selexol. Existing IGCCs with unmitigated CO_2 emissions have an efficiency of 38–42%, depending on the type of gasifier. Dry entrained-flow gasifiers (so-called Shell gasifiers), which are used in European IGCCs in Buggenum (The Netherlands) and Puertollano (Spain), have a higher efficiency than slurry-fed gasifiers (Texaco or GE type).

The major part of the efficiency loss in a pre-combustion CO_2 capture section is caused by the WGS section, and more specifically by the steam needed to drive the shift equilibrium forward and also to prevent the reactors from overheating [4]. The H_2–CO_2 mixture coming out of the shift section still contains steam. The gas has to be cooled for the CO_2–H_2S physical solvent, and the hydrogen fuel coming from the acid gas removal has to be heated again for entry into the gas turbine combustion chamber.

9.1.2
Solid Physical Adsorption (Pressure Swing and Temperature Swing Adsorption)

Processes involving physical adsorption by solid sorbents have been employed for hydrogen purification by CO_2 removal, although this approach has not yet reached a commercial stage for bulk CO_2 recovery from flue gases. The technical feasibility is dictated by the adsorption step because there must be a sufficiently strong affinity to remove the undesired component selectively from the gas mixture. However, the stronger the affinity, the higher is the energy requirement for regeneration. The main advantages of physical adsorption on solid sorbents over chemical or physical absorption by liquids are its simple and energy-efficient operation and regeneration, which can be achieved by a pressure swing. Zeolitic molecular sieves are under consideration for flue gas, although both scale-up and the need to develop CO_2-specific adsorbents remain the main issues to be solved [14]. Furthermore, the gaseous stream has to be cooled and dried, which limits the attractiveness of such pressure swing adsorption (PSA) processes, or temperature swing adsorption (TSA) and electrical swing adsorption (ESA) processes, for pre-combustion applications. Activated carbons may have a larger CO_2 capacity than zeolite at high pressure, depending mostly on the pore structure [15]. However, the presence of water also affects the CO_2 adsorption capacity despite its hydrophobic properties [16], implying necessarily a drying step in most cases. Some metal–organic framework and hybrid organic–inorganic framework materials have shown promising performances for selective CO_2 adsorption, outperforming zeolites and activated carbons for high-pressure applications such as those encountered in pre-combustion [17]. However, there are several critical issues, such as material stability in the presence of air or water, for which only very limited information is available in the literature [18].

9.1.3
Low-Temperature Distillation

CO_2 can be separated from other gases by cryogenic separation processes, by a commercial purification process (licensed by Process System International) for CO_2 removal from streams that already have high CO_2 concentrations [5]. However, cryogenic separation of CO_2 suffers from the prohibitive amount of energy demanded by the refrigeration for dilute gas streams. Clodic *et al.* [19] reported CO_2 capture from diluted streams by cryogenic separation, relying on the frosting and defrosting CO_2 concept. However, the main technical challenge is to avoid blockages by condensable gases such as water, requiring thorough removal before cooling. Nevertheless, cryogenic separation has the advantage that it allows direct production of liquid CO_2, which would be needed for transport by ship. A novel process using dynamically operated packed beds has also been proposed more recently [20]. Cryogenics would optimally only be applied to high-pressure gases, such as in oxygen-fired combustion in which the input gas contains a high concentration of CO_2.

9.1.4 Combined CO_2 Separation and Water Gas Shift Reaction

Separation at high temperatures and high pressures is among the best options to reduce the energy penalty because cooling/reheating would be avoided between the WGS section and the gas turbine, and the steam would be preserved in the hot hydrogen stream, contributing to power generation [3, 21].

Interestingly, inorganic membranes allow the separation of gases at high temperatures. Porous inorganic membrane with carefully controlled pore structure may separate gases on the basis of their molecular size (small micropores) or on the basis of the difference in the mean path of gaseous molecules due to collision with pore walls implied by the Knudsen diffusion (for larger micropores). Since differences in molecular weight are usually small, the selectivity of this type of membrane is small at high temperatures and their long-term stability is a general issue [22], giving rise to a multi-stage plant which is not regarded to be feasible if economics determine the process. Silica, carbon, alumina, nitride, and hybrid-based membranes are among the most studied systems [23]. In addition to limited selectivities, these membranes have not shown sufficient stability towards humid conditions, reducing conditions, or oxidizing conditions [24]. In contrast, metallic "dense" membranes allow H_2 separation by dissociative chemisorption on the metallic surface, producing atomic hydrogen that diffuses through the metal lattice driven by the partial pressure difference between the permeate and retentate sides. Eventually atomic hydrogen recombines and H_2 diffuses away from the surface in the permeate side. Membranes of palladium and its alloys have been developed to separate hydrogen from gas mixtures in the past 50 years [25]. Recently they have attracted renewed attention because of their potential role in hydrogen production for fuel cells and also in CO_2 separation. Hydrogen membrane reactors may permit CO_2 capture with relatively lower cost and lower energy consumption than conventional pre-combustion CO_2 capture technologies [26]. An attractive approach is to integrate WGS or steam reforming (SR) reaction within a Pd-based membrane reactor since the *in situ* removal of H_2 allows high conversion of syngas or methane while CO_2 may be obtained in the retentate side. Therefore, a process development unit (PDU) has been built at the Energy Research Centre of The Netherlands (ECN) for testing the performance, durability, and robustness of H_2-selective membranes, thus providing an important step towards full-scale demonstration of the membrane technology and industrial application. First bench-scale tests have highlighted the effect of WGS or SR gas mixtures on Pd-based membrane H_2 permeation [27–29] and benchmarking of Pd-based membranes is ongoing within the European project CACHET-II and the Dutch funded project CATO-2.

Combination of CO_2 sorption and fuel processing step such as steam methane reforming and/or WGS reactions are innovative pre-combustion decar-

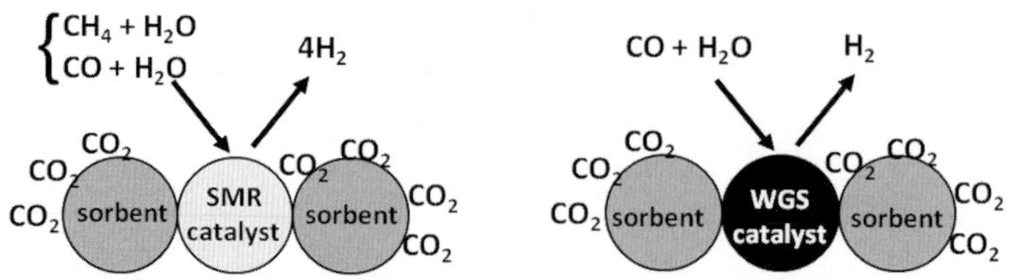

Scheme 9.1 Basic principle of sorption-enhanced steam methane reforming and WGS reaction processes.

bonization technologies [30]. Significant process development was carried out in the late 1990s at Air Products on sorption-enhanced steam methane reforming and sorption-enhanced reaction processes in general [31]. Basically, CH_4 and/or CO catalytic conversion, which is limited by thermodynamic equilibrium in a conventional reformer or WGS reactor, is enhanced in agreement with Le Chatelier's principle (Scheme 9.1).

Air Products pioneered the use of potassium-promoted hydrotalcite-based sorbent mixed with a steam methane reforming (SMR) catalyst to achieve high methane conversion at temperatures as low as 400 °C. Other systems based on calcium oxide [32], dolomite [33, 34], alkali metal-promoted alumina [35], alkali metal zirconates [36], and lithium orthosilicate [37] have also been used for sorption-enhanced SMR (SESMR). Calcium oxide and derivatives generally have a large CO_2 capacity but require high regeneration temperatures, which may be detrimental to the SMR catalyst. Moreover, the steam requirement for regeneration may be high if relatively low temperatures of desorption are targeted [38]. A general problem with CaO-based materials is the significant capacity drop as a function of time. Lithium zirconates also show good overall capacities but rather slow uptake kinetics, which limits their attractiveness, although recent material developments have allowed this problem to be overcome [39–41]. A study of the thermodynamics of the process conditions used in SESMR and sorption-enhanced water gas shift (SEWGS) revealed that it would be difficult to achieve high enough methane and CO conversion while operating SESMR at 400 °C (Fig. 9.3) [42].

Although laboratory-scale SESMR experiments carried out at 1 bar showed promising results, the effect of increased pressure on the required catalytic conversion is dramatic (see Fig. 9.3). The concentration of CO_2 that must be reached to keep the CH_4 conversion close to 100% at 17 bar SESMR operation is simply too low (sub-ppm level) for the sorbents considered. Operation at higher temperature would be beneficial but would also increase the stress on both the catalyst and sorbent materials. In contrast, for the SEWGS process, the CO conversion level is hardly affected by process pressure and it may operate with sorbents that allow up to 1000 ppm CO_2 slip through the bed. The SEWGS process, which combines CO_2 separation and WGS) operates at lower temperature than SMR with more robust catalytic material in a single unit. The process is carried out with multiple vessels to allow the continuous production of H_2

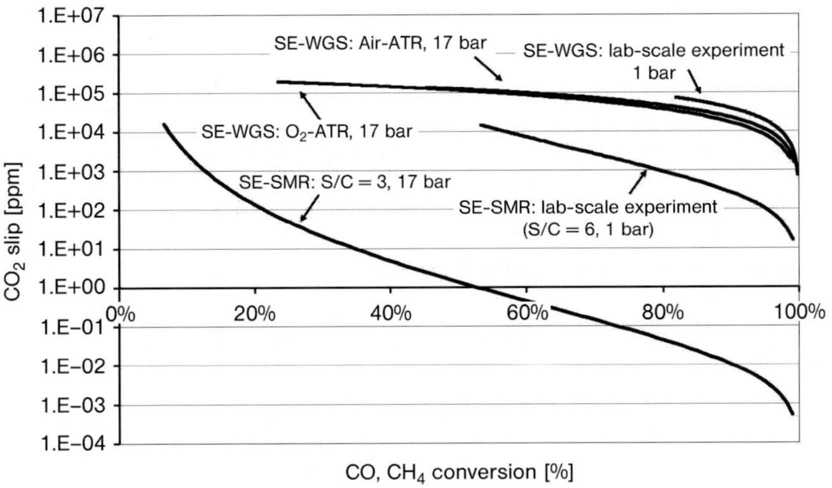

Fig. 9.3 CO_2 concentration at the outlet of a reactor at 400 °C loaded with sorbent and catalyst, assuming that equilibrium is reached in the gas phase, as a function of the conversion of CO or CH_4 for the SEWGS and SESMR systems on a laboratory scale (operational pressure 1 bar), and on the scale envisaged in the systems study (operational pressure 17 bar).

and continuous CO_2 removal. A partially pre-shifted syngas stream containing H_2, CO_2, CO, H_2O, CH_4, other inerts, and impurities (e.g., sulfur-containing compounds) may be generated by methane steam reforming or coal gasification and fed at high pressure and temperature (~30 bar absolute, 200–400 °C) and CO is catalytically converted while CO_2 is removed from the gas phase by sorption.

This catalytic sorption step effectively removes CO and CO_2 from the feed gas, producing a hot, pressurized, hydrogen-rich product stream that can be converted to electricity by combustion in a gas turbine. Once the sorbent is nearly saturated with CO_2, regeneration must be carried out. Mainly two options have been reported in the literature, regeneration by temperature swing and regeneration by pressure swing. Beaver and co-workers have shown that fuel cell-grade H_2 may be produced while capturing CO_2 using a low-temperature shift catalyst in a temperature swing reaction process between 200 and 400 °C [43, 44]. Although very attractive for high-grade hydrogen production, this option may be less suited for coal applications in which H_2S may severely deactivate the catalyst. In the other process configuration, specific pressure swing steps (also called equalization steps) are conducted before the actual regeneration, which results in a low-pressure CO_2-rich stream. Continuous operation is enabled by using multiple interconnected beds and a proper staged process cycle that allows high H_2 recovery, high CO_2 capture rate and suitable CO_2 purity to be achieved for further storage or use. Overall the process resembles the PSA used in H_2 purification units, for instance, the major difference being the higher process temperatures (350–550 °C).

The following sections first describe the process and present the challenges, then review recent advances in the R&D of the SEWGS process, and finally highlight the sorbent material – the heart of the process.

9.2 The Sorption-Enhanced Water Gas Shift Process

9.2.1 A Cyclic Process

A good understanding of each step of the SEWGS process cycle is of the utmost importance to design an optimal cycle that allows the highest possible H_2 recovery, a sufficient capture ratio, and a high CO_2 purity that may facilitate further compression work before transport and storage. The CO_2 rinse SEWGS cycle proposed by Allam *et al.* [31] is illustrated in Scheme 9.2 for eight interconnected SEWGS reactors.

- The feed step, or reaction–sorption step, where the syngas mixture (at ~30 bar and 400 °C) passes through the catalyst–sorbent packed bed. CO and steam react to yield H_2 and CO_2, which is removed by the sorbent. The hydrogen-rich product is delivered roughly at feed pressure and high temperature (450–550 °C, due to the exothermic WGS reaction and sorption process) and may further be mixed and cooled to an appropriate temperature for combustion and production of electricity in a gas turbine. After a given time on-stream, the feed gas stream is diverted to the next vessel and the current vessel is subjected to a rinse step, successive pressure decrease by equalization with other vessels, and finally regeneration.
- The rinse step, which consists in depleting up to half of the vessel from valuable feed gas by transferring it to another vessel, is a crucial step to achieve a high-purity CO_2 product and high H_2 recovery. As an alternative to the high-pressure CO_2 co-current rinse presented in Scheme 9.2, a countercurrent steam rinse may be used at the expense of IP steam, which would be valuable for electricity generation in the Hirn cycle.
- The next steps are co-current pressure equalizations between the vessel that needs to be depressurized and previously regenerated vessels that need to be pressurized in order to recover the remaining interstitial hydrogen after the rinse step and saving compression energy. The number of these equalization steps decides the number of vessels that need to be considered. At the end of the last equalization step, the CO_2 front must be at or very near to the product end of the vessel, and the interstitial gas phase has to consist of essentially CO_2 and steam.
- Recovery of CO_2 product is achieved in successive countercurrent blowdown (from the last equalization pressure to essentially ambient pressure) and

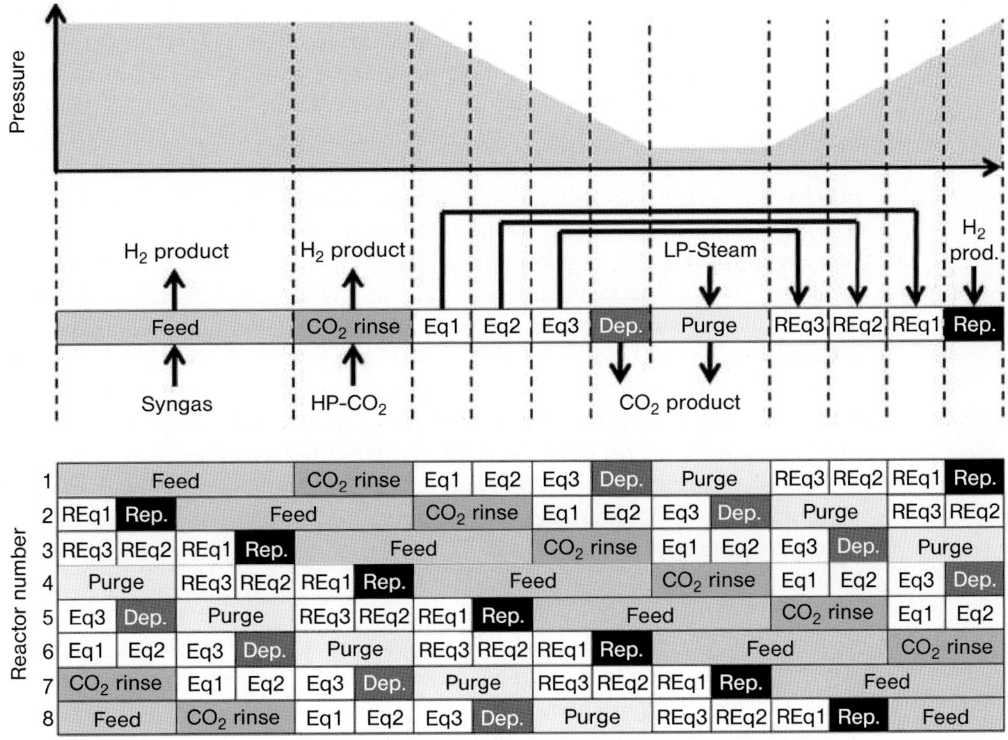

Scheme 9.2 Detailed cycle steps for SEWGS process using eight interconnected vessels.

purge with steam as stripping gas. The steam pressure must be high enough to overcome the pressure drop of the sorber/reactor and downstream piping/condenser, which is typically less than 0.7 bar. The effluent gas consists of CO_2 and steam at $\sim$1 bar and 350–450 °C. The final steps for cyclic process completion are re-pressurization of the vessel by equilibration with gas coming from other vessels undergoing pressure equalization and finally by receiving H_2 product from the rinsed reactor and feed gas for the final re-pressurization.

9.2.2
Process Potential

SEWGS was developed in the first instance for pre-combustion CO_2 capture in a natural gas combined cycle (NGCC). Two studies compared the performances of SEWGS and a methyldiethanolamine (MDEA) scrubbing system in the case of an NGCC, which also gives valuable information for the application of SEWGS in IGCC. In both cases, CO_2 is removed from syngas that has been pre-shifted in a high-temperature shift section and further compressed for transport and storage, while the hydrogen-rich product is combusted in a gas turbine. An advantage of SEWGS is the production of hot H_2-containing steam, which increases the efficiency of the gas turbine relative to the MDEA system, in which H_2 is relatively dry and at a low temperature (~40 °C). In Allam *et al.*'s study [31], syngas was generated from natural gas using an air-blown autothermal reformer (ATR), whereas Manzolini *et al.* [45] considered syngas generation by a more advanced gas heated reformer placed in front of an ATR (GHR–ATR) developed by Johnson Matthey. Various degrees of heat integration were considered in order to produce as much power as possible with the combined cycles. In the studies by Allam *et al.* [31] and Wright and co-workers [46,47], SEWGS allowed an increase in efficiency of 2–3% (the rinse step configuration used is important) compared with the MDEA system with overall lower costs for CO_2 avoided. Furthermore, in addition to a comparison of efficiencies, Manzolini *et al.* [48] reported the lower energy consumption offered by SEWGS for CO_2 removal [~2.7 MJ (LHV) kg^{-1} CO_2] under the conditions to limit the amount of steam used in the rinse and purge step. Further, exergy analysis convincingly showed the advantage of separating CO_2 at high temperature, significantly reducing heat transfer losses.

Recently, Gazzani *et al.* showed that integration of SEWGS in an IGCC plant leads to 2% lower efficiency loss compared with Selexol technology and a 10% lower cost of CO_2 avoided [49]. In the case of Selexol, the net electric efficiency of a 400 MWe dry coal gasifier (Shell type) combined cycle dropped from 47.7% without CO_2 capture to 36.5% when using a dual-stage WGS section and a Selexol unit for sulfur and CO_2 removal (see Table 9.1). Replacing the double WGS and Selexol unit with a single shift and a SEWGS unit, increased the net electric efficiency to 38.4%. Also, the amount of CO_2 avoided is higher in the SEWGS case: 98% versus 88% for Selexol. Gazzani *et al.* reported that the specific energy consumption for CO_2 capture is almost 30% lower for SEWGS than Selexol. A reason is that most of the other pre-combustion decarbonization technologies require much lower temperatures for CO_2 removal, which implies that steam in the raw syngas must be condensed, which will eventually result in

Table 9.1 Results of comparison of CO_2 capture using SEWGS with a state-of-the-art solvent for NGCC [48] and IGCC [49].

Parameter	System						
	NGCC	NGCC	NGCC–	IGCC	IGCC–	IGCC–	IGCC–
		MDEA	SEWGS		Selexol	SEWGS (clean)	SEWGS (sour)
Electric efficiency (LHV base) (%)	58.3	50.3	50.9	47.7	36.5	37.6	38.4
CO_2 avoided (%)	–	91.5	95.2	0	87.6	98.0	98.0

heat losses, and the H_2 needs to be reheated before it can enter the gas turbine. Although an improvement over conventional physical solvents, the steam that is used both in the adsorption–reaction step and in the desorption step of the SEWGS cyclic process still accounts for the major source of efficiency penalty. In the regeneration it facilitates CO_2 desorption and is condensed to yield a relatively pure CO_2 stream that is compressed for further transport and storage. Therefore, a major challenge for SEWGS optimization, and for separation-enhanced reaction processes in general, is to reduce the steam demand and thereby mitigate the impact of CO_2 removal on the overall electrical efficiency of the power plant [42].

9.2.3
Bench-Scale Results Towards Cycle Optimization

In order to develop the process cycles with minimum steam requirements, it is clear that a reliable and robust model needs to be developed to allow the prediction of the sorbent packed bed utilization and help in defining optimal cycle parameters as a function of gas composition and product specifications. In that perspective, testing of SEWGS under representative conditions is a definitive asset. The ECN has therefore commissioned two bench-scale demonstration units, namely SEWGS-1 [50], a single-column test rig, and SEWGS-7, a multi-column test rig to accelerate the development of this technology towards pilot-scale demonstration. The single-column unit (2 m in height × 38 mm i.d.) is able to withstand conditions up to 31 bar absolute and 550 °C and is used to measure breakthrough performance and also simple cycle experiments and elementary SEWGS cycle step effects on the sorbent performance. The experiments are also used to estimate the CO_2 working capacity of the adsorbent thanks to a set of six thermocouples that allow the adsorption fronts (exotherms) to be followed along the column. A multi-column test rig was also assembled to assess process performance under realistic industrial conditions and process cycles. The unit may handle flows of up to 0.2 kmol h^{-1} (about 1/150 000th of an anticipated industrial system). It consists of six tubular reactors (6 m × 38 mm i.d.) individually heat traced and insulated for operation at up to 31 bar absolute and 550 °C. The gas-phase superficial velocities that are used are similar to those expected industrially. A set of 48 automated high-pressure valves allows such a degree of interconnection that operation of all cyclic steps of the steam rinse and CO_2 rinse cycles can be tested. A pool of gas analyzers (mass spectrometer and IR detector) allows continuous on-line monitoring of the effluent streams.

A first series of experimental results were obtained on the single-column reactor which was filled with a mixture of pelletized sorbent material and commercial high-temperature shift catalyst [50]. The sorbent material used in this case was a potassium carbonate-promoted hydrotalcite-based material supplied by Sasol Germany. The material consisted of hydrotalcite with an Mg:Al ratio of 2.9, commercially available as MG70, which was heat treated and impregnated with 20 wt% K_2CO_3 and resulted in a material denoted K-MG70.

In the following discussion, material CO_2 uptake capacity is discussed and specified either as a total capacity (equilibrium capacity), a breakthrough capacity, which depends on material uptake kinetics and packed bed column dynamics, or a working capacity, which is the uptake capacity that is actually used during the feed step of the cyclic process and which strongly depends on regeneration procedure. The experimental breakthrough curve that is measured in the presence of the main syngas components CO_2, CO, H_2O, H_2, and N_2 is shown in Fig. 9.4 and demonstrates that quantitative conversion may be achieved during the sorption–reaction stage (feed step) at 400 °C and 28 bar, yielding decarbonized H_2. The WGS catalyst allows the CO conversion to reach about 55% after CO_2 breakthrough, which is the thermodynamic limit value (see Fig. 9.4).

CO_2 breakthrough corresponded to a CO_2 capacity of 1.4 mol kg^{-1}, which was very close to that obtained for a CO_2 and steam mixture (without syngas) in preliminary experiments. It may be deduced that a sufficient amount of catalyst was present in the reactor, and that the CO_2 capacity is not affected by the presence of syngas components. Furthermore, the stability of the sorbent and catalyst under extended cyclic conditions with a countercurrent steam rinse was assessed. Both the CO and CO_2 levels in the H_2-rich product during the feed step are plotted as a function of cycle number in Fig. 9.5.

Very low levels of CO_2 and CO appear in the H_2 product gas essentially from the start and increase slightly, reaching 0.15 and 0.02%, respectively, after 550 pressure swing cycles. Note that these numbers correspond to a CO_2 recovery

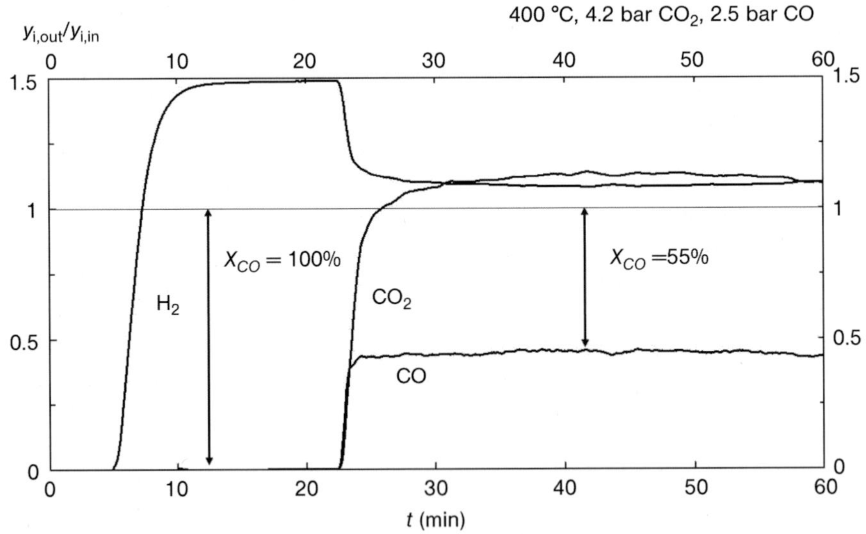

Fig. 9.4 Breakthrough experiment with syngas feed. Before simultaneous CO and CO_2 breakthrough, hydrogen is the only product present in the dry product gas, whereas after CO_2 breakthrough. CO hydrogen and CO_2 are close to the WGS equilibrium concentrations.

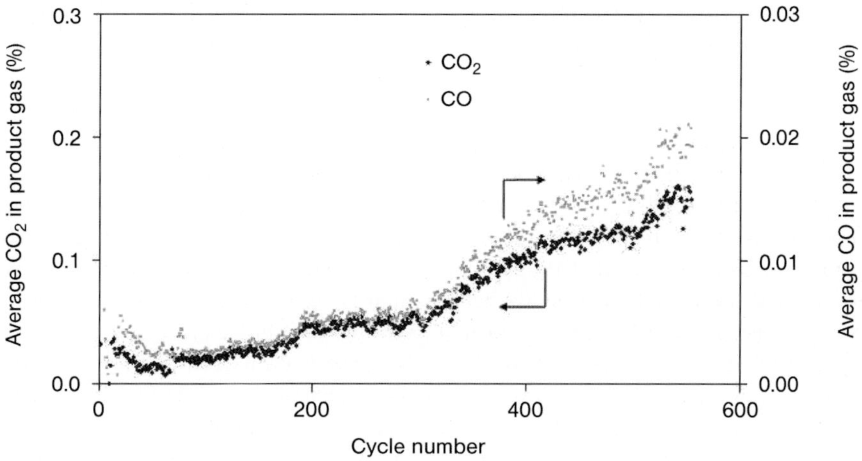

Fig. 9.5 CO and CO_2 concentration in Hydrogen product during extended cyclic experiment with syngas feed.

of 99.3% and a CO conversion of 99.2%, respectively. In a preliminary experiment with only CO_2, steam, and N_2, a similar observation was made: the CO_2 level reached 0.3% after 600 cycles and remained constant for the next 1000 cycles. The increased amount of CO_2 in the H_2 product over the first 500–600 cycles was correlated with continuous CO_2 uptake by the sorbent material. Long regenerative steps were carried out after cyclic operation and showed that the sorbent material stored a fairly large amount of CO_2 (11–12 mol kg^{-1}). The accumulation of CO_2 in the bulk phase of the sorbent is responsible for the CO_2 slip observed and calls for R&D on sorbent materials to improve the cyclic working CO_2 capacity by reducing the unused CO_2 capacity at the same time.

The SEWGS process performance was then assessed in the multi-column unit where all six reactors were loaded with a mixture of sorbent (K_2CO_3-promoted hydrotalcite-based material supplied by Sasol Germany) and commercial HTS catalyst. Comparable breakthrough capacities of around 1.3 mol kg^{-1} where obtained for all six reactors. The countercurrent steam rinse cycle was tested with 1.6 equiv. steam in the rinse step (excess steam) and with a feed gas containing 37% H_2, 12% CO_2, 4% CO, 34% N_2, and 13% H_2O over 75 cycles. This allowed a high recovery of H_2 in the product but the CO_2 purity of the product (around 80%) was low. The cycle working CO_2 capacity was reduced to 0.23 mol kg^{-1} because of the characteristics of the cycle with equalization steps and a re-press step, but this value may be optimized by adjusting the cyclic steps duration through modeling [51]. In summary, the first series of tests carried out on the multi-column unit showed excellent reproducibility of the results obtained on a single-column test rig and underlined the need (1) to develop robust and reliable process modeling to design optimal cyclic parameters and (2) to develop new sorbent materials with higher cyclic working capacities that are essential to reduce the size of the equipment.

Ongoing SEWGS modeling work is based primarily on the model used by Air Products and described by Kumar *et al.* [52], and on the models developed by other groups for separation-enhanced reaction processes. The model takes into account the dynamic behavior of the adsorption column, which depends on the adsorption kinetics, adsorption equilibrium, and fluid dynamics. More accurate and reliable simulations are obtained by the development of a nonisothermal, nonlinear, nonequilibrium, and nonisobaric model. The material balance, the energy balance, and the momentum equations are solved simultaneously by the finite difference method. Intraparticle mass transfer rates are usually conveniently and best described by using the linear driving force [53]. Langmuir [54, 55], Freundlich [55, 56], dual-site Langmuir [34], bi-Langmuirian [57], and Toth [54, 58] models for isotherm description have been used throughout the literature. Several kinetic models have also been proposed to describe adsorption and desorption processes. Overall, adsorption breakthrough is fairly well described by the available model curves. However, improvements are needed to describe sorption at high pressure in the presence of high pressure of steam which co-adsorbs on the sorbent materials and was found to have a beneficial effect on the CO_2 uptake properties of hydrotalcite-based materials [54]. In particular, CO_2 sorption isotherm measurements at high temperature in the presence of steam would bring further robustness and reliability to modeling work. In addition, long-term cyclic experiments have shown that the initially preferred sorbent material, K-MG70, required several hundred cycles to achieve a steady operational state, which was also shown by the huge amount of CO_2 released during extended regeneration steps. Model predictions are still significantly affected by these slow or irreversible absorption processes that lead to carbonate build-up during extended cyclic tests, especially for the desorption step prediction [55, 59].

Therefore, R&D on materials and on CO_2 sorption mechanisms at high temperatures is important to be able to make sound assumptions in modeling. Moreover, it gives leads to designing better materials with optimal performance that may strengthen the SEWGS technology's advantages to outperform the alternative pre-combustion decarbonization technologies. In the following section, some of the R&D on materials is considered and the further outlook for SEWGS technology application and potential is discussed.

9.3 Sorption Processes and Material Development for SEWGS

9.3.1 Acid–Base Interaction

CO_2 is an acidic gas and therefore sorption by a solid may be best interpreted in terms of acid–base interaction. Hydrotalcites are layered double hydroxides included in acid–base materials that are extensively used in the chemical industry for a wide range of applications [60]. Their chemical composition at low temperature may be described as $Mg_{1-x}Al_x(OH)_2(CO_3)_{x/2} \cdot nH_2O$, where x is usually between 0.3 and 0.4 in the case of naturally occurring hydrotalcites. When hydrotalcites are heated, a series of chemical rearrangements take place. The first consists of a dehydration step corresponding to water removal from the interlayer of the hydrotalcite structure at temperatures between 100 and 200 °C, resulting in roughly $Mg_{1-x}Al_x(OH)_2(CO_3)_{x/2}$. Then several dehydroxylation and decarbonation steps occur at temperatures above 300 °C [61, 62]. The solid that would be used for sorption experiment at 400 °C may be therefore best described as a mixed oxide, $Mg_{1-x}Al_xO_4$. Powder X-ray diffraction experiments have shown that this mixed oxide loses the long-range ordered layered structure, showing broad diffraction peaks corresponding to a periclase crystalline phase or salt rock phase consisting of Mg(Al)O phase. As such, the material may hardly adsorb any CO_2 at temperatures around 400 °C under a low partial pressure of CO_2 because of the very slow reaction between MgO and CO_2. However, it is now well known that the basic character and the CO_2 uptake capacity of Mg,Al mixed metal oxides may be enhanced by alkali metal promotion [63–67]. Studies of the effect of the amount of alkali metal promotion on hydrotalcite-based material on CO_2 adsorption capacities at high temperatures and ordinary pressures in the presence of steam have indicated an optimal loading of the alkali metal compound of about 20 wt% [68]. In a comparative study with alumina-based material, IR spectroscopy demonstrated that interaction between aluminum oxide centers and the alkali metal plays an essential role in CO_2 adsorption. In contrast, CO_2 sorption on potassium-promoted magnesia did not show significant reversible adsorption capacity at ordinary pressures and 400 °C [69].

9.3.2
Magnesium Carbonate Formation

Experimental research carried out on the SEWGS process using potassium-promoted hydrotalcite-based material for CO_2 adsorption at 350–450 °C and at a total pressure of 30 bar indicated that bulk carbonate phase could be generated in significant quantity in the presence of high partial pressures of both steam and CO_2. High capacities were detected by quantifying the amount of CO_2 recovered by extended regeneration of the sorbent bed after long-term cyclic operation [50]. Recently, the reaction pathway associated with this huge uptake has been identified by carrying out *ex situ* and *in situ* structural analysis of carbonated material produced at high pressures and temperatures. The design of a special gas cell in collaboration with the Swiss–Norwegian Beam Lines at the European Synchrotron Radiation Facility (ESRF, Grenoble, France) has allowed us to characterize the structural rearrangement of hydrotalcite-based material under temperature, pressure, and gas composition conditions relevant for SEWGS, that is, 5 bar of CO_2 and 5 bar of steam flowing through a sorbent material packed bed in a quartz capillary (total pressure 10 bar). A full cycle of CO_2 sorption–regeneration could be carried out while analyzing the material with high-resolution powder X-ray diffraction (XRD). Figure 9.6 shows the formation of a magne-

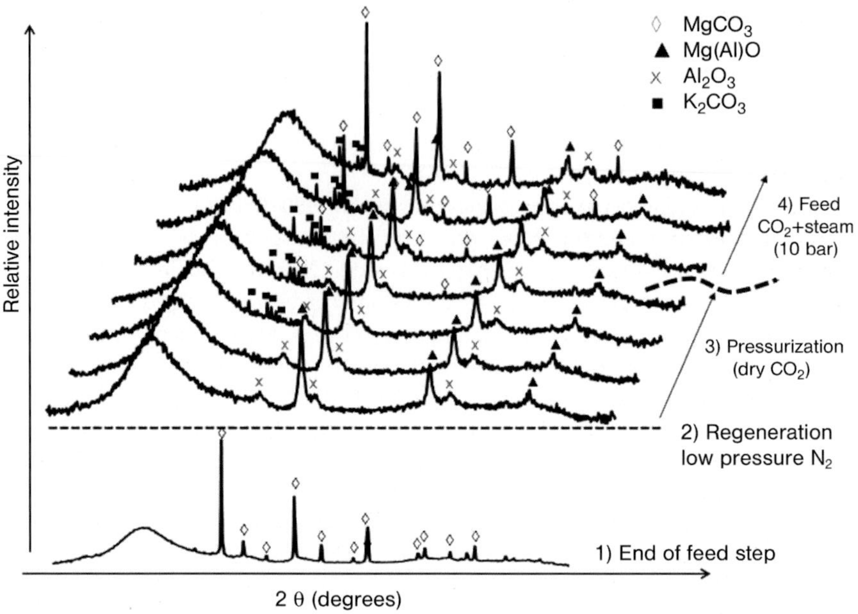

Fig. 9.6 *In situ* XRD analysis of 20 wt% K_2CO_3-promoted hydrotalcite-based material at 400 °C and under (1) 5 bar of steam, 5 bar of CO_2, (2) 1 bar dry N_2, (3) increased dry CO_2 pressure from 1 to 10 bar, and (4) increased steam partial pressure from 0 to 5 bar in 10 bar pressure balanced with CO_2.

sium carbonate crystalline phase by contacting K_2CO_3-promoted hydrotalcite-based material with the CO_2–steam mixture at 10 bar and 400 °C. The experiment was further extended by simulating a regeneration step which consisted of first decreasing the pressure from 10 to 1 bar and then switching the gas composition from 1 : 1 CO_2–H_2O to 100% dry N_2 while maintaining the temperature at 400 °C. As a result, the $MgCO_3$ crystalline phase completely disappeared while diffraction peaks of an MgO periclase crystalline phase and an ill-defined alumina phase became visible. The gas mixture was then switched from N_2 to CO_2 (dry) and the total pressure was increased to 10 bar. The XRD pattern recorded during the pressure increase showed no major structural changes. Only when steam was added to the CO_2 at a pressure of 10 bar did the magnesium carbonate crystalline phase progressively reappear. These observations are in good agreement with the thermodynamic data on magnesium oxide, magnesium hydroxide, and magnesium carbonate systems. The increase in steam pressure may induce a partial rehydroxylation of the MgO periclase phase which enhances the formation of magnesium carbonate [70, 71]. Obviously, K_2CO_3-promoted hydrotalcite-based material can reversibly form crystalline magnesium carbonate in the presence of steam, offering a total capacity of more than 10 mmol g^{-1}. However, the SEWGS experiments described in Section 9.2.3 clearly showed that only a small part of the total CO_2 capacity corresponding to magnesium carbonate actually participates in the pre-breakthrough region and actual cyclic working capacity.

A carbonation experiment was carried out at pressures of 20 bar CO_2 and 20 bar steam at 350 °C in an autoclave with K_2CO_3-promoted hydrotalcite-based material [72]. Thermogravimetric analysis was carried out on the resulting sample containing magnesium carbonate crystalline phase (identified by XRD) in order to obtain quantitative information about the extent of bulk $MgCO_3$ formation under these conditions (Fig. 9.7). The mass loss between 350 and 600 °C corresponded to 41.2 wt%, which is comparable to the theoretical loss (42.9 wt%) considering that fully carbonated K_2CO_3-promoted hydrotalcite (Mg : Al = 2.9, 11 wt% K_2CO_3) would contain $MgCO_3$ phase instead of MgO as a result of the carbonation reaction.

Assuming that steam does not form significant amounts of hydrated alumina or hydrated analogs of magnesium carbonate, this experiment shows that the equilibrium CO_2 capacity of 11 wt% K_2CO_3-promoted hydrotalcite (Mg : Al = 2.9) may be as high as 15.2 mol kg^{-1}. To corroborate further the extensive carbonation occurring in the material that could similarly happen during long cyclic experiments (in the case of incomplete regeneration), comparative SEM analyses of freshly calcined material and material treated in an autoclave under high partial pressures of CO_2 and steam were carried out (Fig. 9.8). The calcined material consisted of platelet-like particles, which are typical of double-layered material, at the surface of larger aggregates whereas the sample treated in the autoclave showed rhombohedric crystallites a few hundred nanometers in size that were densely packed at the surface of larger aggregates. Magnesium carbonate formation clearly modifies the overall morphology of the hydrotalcite-based material at the microscopic scale.

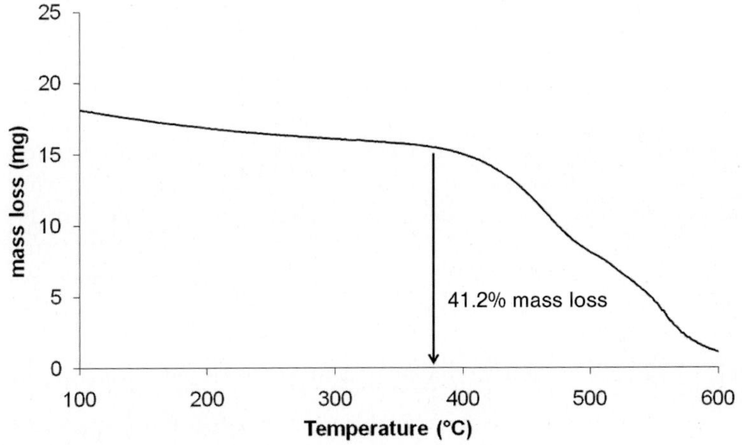

Fig. 9.7 Thermogravimetric analysis of 11 wt% K_2CO_3-promoted hydrotalcite-based material pretreated under hydrothermal conditions at 350 °C.

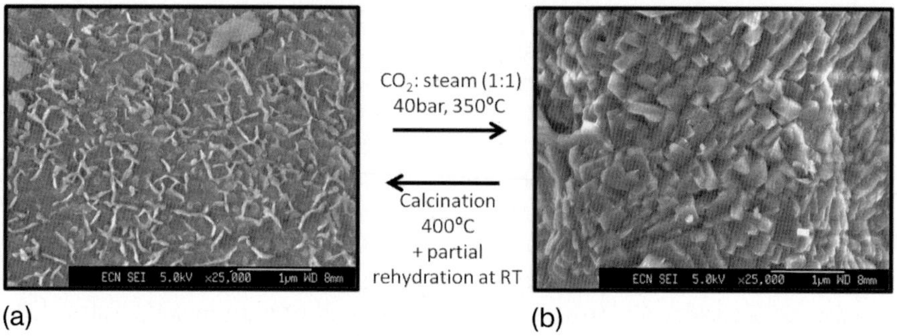

Fig. 9.8 SEM analysis of 11 wt% K_2CO_3-promoted hydrotalcite (Mg/Al = 2.9)-based material treated under 20 bar of CO_2 and 20 bar of steam at 350 °C (b) and calcined at 400 °C under 1 bar of N_2 and partially rehydrated in a laboratory atmosphere (a).

In addition to a large unused CO_2 capacity in such a dynamic adsorption process as SEWGS, the transformation of the material under the effect of successive carbonation and decarbonation, during an extensive regeneration period, for example, could alter significantly and detrimentally the mechanical strength of the material and lead eventually to pellet degradation. Therefore, an intense R&D effort has been made to develop novel sorbents with a lower amount of magnesium oxide and a higher aluminum content. Recent research work on alkali metal-promoted alumina-rich materials has shown that significant CO_2 capacity may be obtained by the formation/decomposition of potassium (hydr)oxyaluminocarbonate species [K–O(H)–Al–CO_3] [73, 74]. XRD experiments carried out under realistic conditions showed that a dawsonite-type [K–O(H)–Al–CO_3] phase [75, 76] may indeed be more suitable for dynamic sorption of CO_2 with a faster and more reversible process than bulk magnesium carbonate. Furthermore, the proper design of the method for the synthesis of alumina-rich material may lead both to improved mass transfer from the gas phase to the potassium-promoted species responsible for CO_2 uptake and also more robust materials which may withstand cyclic sorption/regeneration. For instance, within the CAESAR project (FP seventh program), a new sorbent with a smaller amount of Mg than in the K-MG70 material used in the long cyclic tests has shown a higher cyclic capacity (by 27%) together with a lower steam requirement for regeneration (36%). The formation of crystalline $MgCO_3$ was essentially suppressed, and mechanical stability was confirmed for over 1200 adsorption–desorption cycles. In the meantime, R&D work has widened the potential application of SEWGS to sour syngas feed and has highlighted the potential of alkali metal-promoted hydrotalcite- and alumina-based materials as WGS catalytically active sorbents.

9.3.3
Water Gas Shift Catalytic Activity and Opportunities for CO_2-H_2S Co-capture

The syngas generated in IGCC power plant contains significant amounts of H_2S, ranging from a few hundred parts per million to a few percent [6]. For the application of SEWGS technology within an IGCC, it is crucial to control the fate of H_2S. For example, if H_2S were to end up in the H_2 turbine fuel, subsequent H_2 desulfurization would become necessary, invoking an additional energy penalty and seriously lowering the benefit of applying the SEWGS technology. Alternatively, if H_2S seriously affected the CO_2 sorption behavior, upstream syngas desulfurization would be required, preferably at high temperature. Therefore, an in-depth understanding of the effect of H_2S on the SEWGS process is required.

Both CO_2 and H_2S are acidic gases, and recent studies [77, 78] have demonstrated that both CO_2 and H_2S reversibly adsorb on K_2CO_3-promoted hydrotalcite- and alumina-based materials, an example of which is given in Fig. 9.9, which shows the responses for N_2, CO_2, H_2O, and H_2S upon repeated periodic exposure to CO_2–H_2S–H_2O–N_2 adsorption and H_2O–Ar regeneration using an H_2S-free feed (symbols), and a feed containing 500 ppm H_2S (lines). Clearly, the presence of 500 ppm H_2S in the feed does not alter the CO_2 sorption behavior of these materials as the responses for CO_2, H_2O, and N_2 are identical with or without H_2S in the feed. Regeneration using steam is essential in maintaining simultaneous adsorption of CO_2 and H_2S [79], since dry conditions lead to irreversible uptake of H_2S [80]. The stability of these features has been observed for hundreds of adsorption–desorption cycles. Similar results have been obtained for alkali metal-promoted hydrotalcite-based materials. Recent testing in our SEWGS-1 single-column test rig [50] indicated that the sorbent simul-

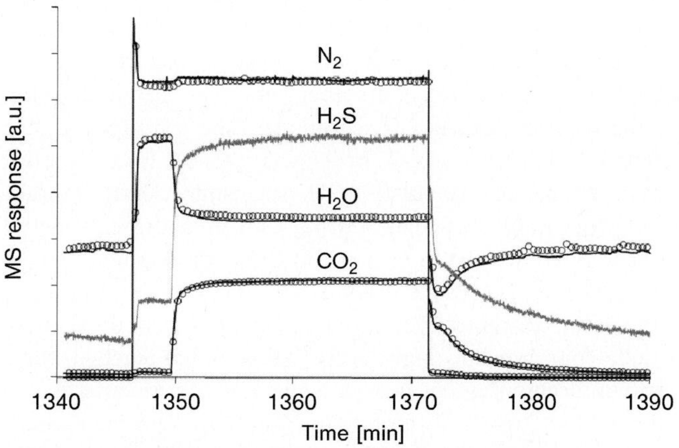

Fig. 9.9 Cyclic steady-state transient responses for N_2, CO_2, H_2O, and H_2S using K_2CO_3-promoted Al_2O_3 at 400 °C and 1.2 bar. Symbols, without H_2S; lines, with 500 ppm H_2S.

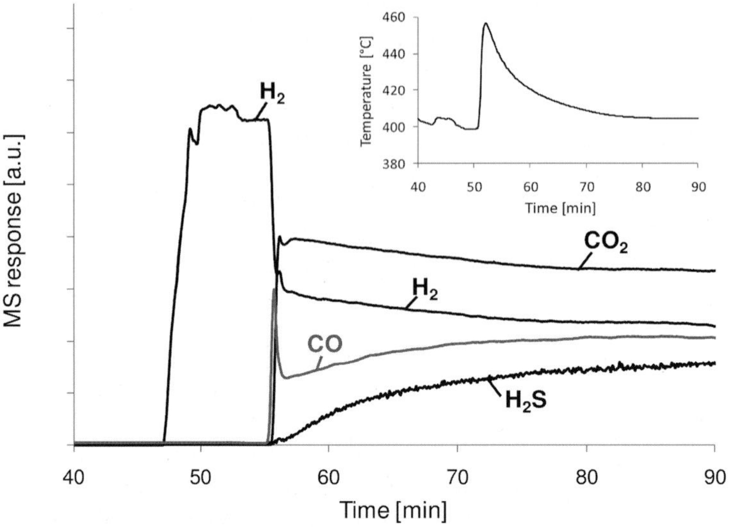

Fig. 9.10 Transient responses for CO, CO_2, H_2, and H_2S during a breakthrough experiment on K_2CO_3-promoted hydrotalcite-based material at 400 °C and 30 bar. Feed conditions: 200 ppm H_2S, 17% CO, 20% CO_2, 17% H_2, 40% H_2O in He. Inset: temperature in the middle of the column.

taneously decarbonizes and desulfurizes sour syngas under countercurrent adsorption–regeneration operation. The SEWGS technology may accordingly produce an H_2-rich product with low contents of CO_2 and H_2S and a CO_2+ H_2S-rich product.

In Fig. 9.10, a breakthrough experiment is illustrated for a syngas feed of 20% CO_2, 17% H_2, 40% H_2O, 17% CO, and 200 ppm H_2S. The column was filled with K_2CO_3-promoted hydrotalcite-based material only (no WGS catalyst). The operating temperature and pressure were 400 °C and 30 bar, respectively. Hydrogen did not noticeably interact with the sorbent and was readily detected at the column outlet. Remarkably, during CO_2 sorption in the column, the outlet product did not contain either CO_2 or CO, while an increased H_2 concentration was evidenced by mass balance calculation. These results clearly indicate that the sorbent has a sufficient WGS activity during CO_2 breakthrough in the presence of 200 ppm of H_2S in the feed to ensure full CO conversion.

Following breakthrough, the CO concentration decreased again and then slowly increased to the inlet concentration. Simultaneously, the high H_2 and CO_2 concentrations following breakthrough decreased and approached their feed values. This result indicates that the sorbent material has an overall low WGS activity once the sorbent reaches saturation with CO_2. Complementary steady-state activity measurements indeed confirmed a 2–4 orders of magnitude lower WGS activity compared with commercial FeCr- and CoMo-based WGS catalysts. The gradual change in the CO, H_2, and CO_2 concentrations towards their inlet values is actually related to the dynamic temperature profile within the column. The inset in Fig. 9.10 shows the temperature of the packed bed measured at half bed height. The full CO conversion prior to breakthrough and the simultaneous adsorption of CO_2 caused the temperature to rise to 457 °C. Once the breakthrough front had passed, the temperature slowly decreased to the feed temperature. Accordingly, directly after breakthrough the column remained hot and the steady-state WGS activity of the sorbent was high, resulting in a low response for CO and high responses for CO_2 and H_2. With decrease in column temperature, the WGS activity also decreased and the responses approached their inlet concentrations.

9.4
Conclusion and Outlook

The SEWGS process is an innovative route for pre-combustion decarbonization in coal-fired power plants which may combine syngas conversion, CO_2 removal, and desulfurization in a single unit at high temperature. Techno-economic analysis has shown the benefit of CO_2 separation at high temperature in comparison with state-of-the-art liquid scrubbing systems that operate mostly near ambient temperature. The SEWGS process concept has been demonstrated in a reactor packed with a mixture of a K_2CO_3-promoted hydrotalcite-based sorbent and a commercial iron–chromium shift catalyst for more than 2000 cycles. Carbon recoveries well above 90% could be obtained at the expense of steam used for regeneration. Further reduction of the steam demand can be achieved by process and cycle optimization and by allowing slightly higher CO_2 slip. R&D on sorbent materials has led to the introduction of novel high-capacity and robust alkali metal-promoted alumina-based and aluminum-rich hydrotalcite-based materials. Experiments using sour syngas have indicated that the sorbent material has sufficient WGS activity prior to CO_2 breakthrough to allow CO conversion without addition of WGS catalysts. Hence the volumetric CO_2 capacity of the column may be increased by substituting sorbent for catalyst, decreasing both the capital expenditure (smaller columns) and the operating costs (lower steam consumption). All these results suggest that the sour SEWGS technology is a promising technology for the simultaneous decarbonization and desulfurization of coal-derived syngas. Further R&D on sorbent material design and SEWGS cycle design are ongoing to (1) allow a decrease in the steam demand for both WGS and sorbent regeneration, which account for the main sources of energy losses for systems comprising SEWGS, and (2) increase further the cyclic capacity of the sorbent material.

Acknowledgments

Support from the European Commission (FP 6th CACHET and FP 7th CAESAR), CCP-2, and the CAPTECH program, which is financed by the Dutch Ministry of Economic Affairs, through the EOS (Energie Onderzoek Subsidie) program (see also www.CO2-captech.nl) is gratefully acknowledged. The authors gratefully acknowledge Daan Jansen, Ed van Selow, Paul Cobden, Michiel Carbo, Paul Verbraeken, Soledad van Eijk, and Coen Hoogland at ECN.

References

1 White, C. M., Strazisar, B. R., Granite, E. J., Hoffman, J. S., and Pennline, H. W. Critical review: separation and capture of CO_2 from large stationary sources and sequestration in geological formations – coalbeds and deep saline aquifers. *J. Air Waste Manage. Assoc.*, 2003, **53**, 643–715.

2 Zenz House, K., Harvey, C. F., Aziz, M. J., and Schrag, D. P. The energy penalty of post-combustion CO_2 capture and storage and its implications for retrofitting the U.S. installed base. *Energy Environ. Sci.*, 2009, **2**, 193–205.

3 Damen, K., van Troost, M., Faaij, A., and Turkenburg, W. A comparison of electricity and hydrogen production systems with CO_2 capture and storage. Part A: review and selection of promising conversion and capture technologies. *Prog. Energy Combust. Sci.*, 32006, **2**, 215–246.

4 Carbo, M. C., Boon, J., Jansen, D., van Dijk, H. A. J., Dijkstra, J. W., van den Brink, R. W., and Verkooijen, A. H. M. Steam demand reduction of water gas shift reaction in IGCC power plants with pre-combustion CO_2 capture. *Int. J. Greenhouse Gas Control*, 2009, **3**, 712–719.

5 Kohl, A. L. and Nielsen, R. B. (eds.). *Gas Purification*, 5th edn., Gulf Publishing, Houston, TX, 1997.

6 Higman, C. and Van Der Burgt, M. (eds.). *Gasification*, 1st edn., Gulf Professional Publishing, Boston, MA, 2003.

7 Lurgi GmbH. *The Rectisol Process, Lurgi's Leading Technology for Purification and Conditioning of Synthesis Gas*, Lurgi, Frankfurt, 2009.

8 UOP LLC. *Selexol Process*, UOP, Des Plaines, IL, 2002.

9 Chen, C. and Rubin, E. S. CO_2 control technology effects on IGCC plant performance and cost. *Energy Policy*, 2009, **37**, 915–924.

10 Hiller, H., Reimert, R., Marschner, F., Renner, H. J., Boll, W., Supp, E., Brejc, M., Liebner, W., Schaub, G., Hochgesand, G., Higman, C., Kalteier, P., Muller, W. D., Kriebel, M., Schlichting, H., Tanz, H., Stonner, H. M., Klein, H., Hilsebein, W., Gronemann, V., Zwiefelhofer, U., Albrecht, J., Cowper, C. J., and Driesen, H. E, 2006. Gas Production. *Ullmann's Encyclopedia of Industrial Chemistry.*

11 Kanniche, M., Gros-Bonnivard, R., Jaud, P., Valle-Marcos, J., Amann, J. M., and Bouallou, C. Pre-combustion, post-combustion and oxy-combustion in thermal power plant for CO_2 capture. *Appl. Thermal Eng.*, 2010, **30**, 53–62.

12 Cullinane, J. T. and Rochelle, G. T. Carbon dioxide absorption with aqueous potassium carbonate promoted by piperazine. *Chem. Eng. Sci.*, 2004, **59**, 3619–3630.

13 Oyenekan, B. A. and Rochelle, G. T. Rate modeling of CO_2 stripping from potassium carbonate promoted by piperazine. *Int. J. Greenhouse Gas Control*, 2009, **3**, 121–132.

14 Yang, S. I., Choi, D. Y., Jang, S. C., Kim, S. H., and Choi, D. K. Hydrogen separation by multi bed pressure swing adsorption of synthesis gas. *Adsorption*, 2008, **14**, 583–590.

15 Siriwardane, R.V., Shen, M.S., Fisher, E.P., and Poston, J.A. Adsorption of CO_2 on molecular sieves and activated carbon. *Energy Fuels*, 2001, **15**, 279–284.

16 Wang, Y.X., Zhou, Y.P., Liu, C.M., and Zhou, L. Comparative studies of CO_2 and CH_4 sorption on activated carbon in presence of water. *Colloids Surf. A: Physicochem. Eng. Aspects*, 2008, **322**, 14–18.

17 Wang, Q., Luo, J.Z., Zhong, Z.Y., and Borgna, A. CO_2 capture by solid adsorbents and their applications: current status and new trends. *Energy Environ. Sci.*, 2011, **4**, 42–55.

18 Keskin, S., van Heest, T.M., and Sholl, D.S. Can metal–organic framework materials play a useful role in large-scale carbon dioxide separations? *ChemSusChem*, 2010, **3**, 879–891.

19 Clodic, D., El Hitti, R., Younes, M., Bill, A., and Casier, F. CO_2 capture by anti-sublimation. Thermo-economic process evaluation. Presented at the 4th Annual Conference on Carbon Capture and Sequestration, Alexandria, VA, 2005.

20 Tuinier, M.J., van Sint Annaland, M., Kramer, G.J., and Kuipers, J.A.M. Cryogenic CO_2 capture using dynamically operated packed beds. *Chem. Eng. Sci.*, 2010, **65**, 114–119.

21 Woudstra, T. and Woudstra, N. Exergy analysis of hot-gas clean-up in IGCC systems. *J. Inst. Energy*, 1995, **68**, 157–166.

22 Bredesen, R., Jordal, K., and Bolland, A. High-temperature membranes in power generation with CO_2 capture. *Chem. Eng. Process.*, 2004, **43**, 1129–1158.

23 Prasad, R.M., Iwamoto, Y., Riedel, R., and Gurlo, A. Multilayer amorphous-Si–B–C–N/gamma-Al_2O_3/alpha-Al_2O_3 membranes for hydrogen purification. *Adv. Eng. Mater.*, 2010, **12**, 522–528.

24 Scholes, C.A., Smith, K.H., Kentish, S.E., and Stevens, G.W. CO_2 capture from pre-combustion processes – strategies for membrane gas separation. *Int. J. Greenhouse Gas Control*, 2010, **4**, 739–755.

25 Paglieri, S.N. and Way, J.D. Innovations in palladium membrane research. *Sep. Purif. Methods*, 2002, **31**, 1–169.

26 Dijkstra, J.W., Jansen, D., van den Brink, R.W., Peters, T.A., Stange, M., Bredesen, R., Goldbach, A., Xu, H.Y., Gottschalk, A., Tlatlik, S., and Doukelis, A. Development of hydrogen membrane reactors for CO_2 capture. In *Carbon Dioxide Capture for Storage in Deep Geologic Formations*, Vol. 3 (ed. Eide, L.I.), CPL Press and BP, London, 2009, p. 121–134.

27 Li, H., Dijkstra, J.W., Pieterse, J.A.Z., Boon, J., van den Brink, R.W., and Jansen, D. Towards full-scale demonstration of hydrogen-selective membranes for CO_2 capture: inhibition effect of WGS-components on the H_2 permeation through three Pd membranes of 44 cm long. *J. Membr. Sci.*, 2010, **363**, 204–211.

28 Li, H., Goldbach, A., Li, W.Z., and Xu, H.Y. On CH_4 decomposition during separation from H_2 mixtures with thin Pd membranes. *J. Membr. Sci.*, 2008, **324**, 95–101.

29 Li, H., Goldbach, A., Li, W.Z., and Xu, H.Y. CO_2 decomposition over Pd membrane surfaces. *J. Phys. Chem. B*, 2008, **112**, 12182–12184.

30 Harrison, D.P. Sorption-enhanced hydrogen production: a review. *Ind. Eng. Chem. Res.*, 2008, **47**, 6486–6501.

31 Allam, R.J., Chiang, R., Hufton, J.R., Middleton, P., Weist, E.L., and White, V. Development of the sorption enhanced water gas shift process. In *Carbon Dioxide Capture for Storage in Deep Geologic Formations* (ed. Thomas, D.C. and Benson, S.M.), Elsevier Science, Amsterdam, 2005, p. 227–256.

32 Balasubramanian, B., Ortiz, A.L., Kaytakoglu, S., and Harrison, D.P. Hydrogen from methane in a single-step process. *Chem. Eng. Sci.*, 1999, **54**, 3543–3552.

33 Silaban, A., Narcida, M., and Harrison, D.P. Characteristics of the reversible reaction between CO_2(g) and calcined dolomite. *Chem. Eng. Commun.*, 1996, **146**, 149–162.

34 Lee, K.B., Beaver, M.G., Caram, H.S., and Sircar, S. Novel thermal-swing sorption-enhanced reaction process concept for hydrogen production by low-temperature steam–methane reforming. *Ind. Eng. Chem. Res.*, 2007, **46**, 5003–5014.

35 Lee, K.B., Beaver, M.G., Caram, H.S., and Sircar, S. Chemisorption of carbon dioxide on sodium oxide promoted alumina. *AIChE J.*, 2007, **53**, 2824–2831.

36 Ochoa-Fernandez, E., Ronning, M., Yu, X.F., Grande, T., and Chen, D. Composi-

tional effects of nanocrystalline lithium zirconate on its CO_2 capture properties. *Ind. Eng. Chem. Res.*, 2008, **47**, 434–442.

37 Essaki, K., Kato, M., and Nakagawa, K. CO_2 removal at high temperature using packed bed of lithium silicate pellets. *J. Ceram. Soc. Jpn.*, 2006, **114**, 739–742.

38 Solieman, A.A.A., Dijkstra, J.W., Haije, W.G., Cobden, P.D., and van den Brink, R.W. Calcium oxide for CO_2 capture: operational window and efficiency penalty in sorption-enhanced steam methane reforming. *Int. J. Greenhouse Gas Control,* 2009, **3**, 393–400.

39 Escobedo Bretado, M., Delgado Vigil, M.D., Collins Martinez, V.H., and Lopez Ortiz, A. Thermodynamic analysis for the production of hydrogen through sorption enhanced water gas shift (SEWGS). *Int. J. Chem. Reactor Eng.*, 2008, **6**, A51.

40 Essaki, K., Nakagawa, K., and Kato, M. Acceleration effect of ternary carbonate on CO_2 absorption rate in lithium zirconate powder. *J. Ceram. Soc. Jpn.*, 2001, **109**, 829–833.

41 Ochoa-Fernandez, E., Zhao, T.J., Ronning, M., and Chen, D. Effects of steam addition on the properties of high temperature ceramic CO_2 acceptors. *J. Environ. Eng.*, 2009, **135**, 397–403.

42 Cobden, P.D., van Beurden, P., Reijers, H.T.J., Elzinga, G.D., Kluiters, S.C.A., Dijkstra, J.W., Jansen, D., and van den Brink, R.W. Sorption-enhanced hydrogen production for pre-combustion CO_2 capture: thermodynamic analysis and experimental results. *Int. J. Greenhouse Gas Control*, 2007, **1**, 170–179.

43 Lee, K.B., Beaver, M.G., Caram, H.S., and Sircar, S. Effect of reaction temperature on the performance of thermal swing sorption-enhanced reaction process for simultaneous production of fuel-cell-grade H_2 and compressed CO_2 from synthesis gas. *Ind. Eng. Chem. Res.*, 2008, **47**, 6759–6764.

44 Beaver, M.G., Caram, H.S., and Sircar, S. Sorption enhanced reaction process for direct production of fuel-cell grade hydrogen by low temperature catalytic steam–methane reforming. *J. Power Sources*, 2010, **195**, 1998–2002.

45 Manzolini, G., Macchi, E., Binotti, M., and Gazzani, M. Integration of SEWGS for carbon capture in natural gas combined cycle. Part A: thermodynamic performances. *Int. J. of Greenhouse Gas Control*, 2010, **5**, issue 2, March 2011, pp. 200–213.

46 Wright, A., White, V., Hufton, J., Selow, E., and Hinderink, P. Reduction in the cost of pre-combustion CO_2 capture through advancements in sorption-enhanced water gas shift. In *Energy Procedia*, 2009, *Proceedings of the 9th International Conference on Greenhouse Gas Technologies*, **1**, Washington, DC, 2009, pp. 707–714.

47 van Selow, E.R., Cobden, P.D., van den Brink, R.W., Hufton, J.R., and Wright, A. Performance of sorption-enhanced water gas shift as a pre-combustion CO_2 capture technology. In *Energy Procedia – Proceeding of the 9th Greenhouse Gas Control Technology Conference*, **1**, Washington, DC, 2009, pp. 689–696.

48 Manzolini, G., Macchi, E., Binotti, M., and Gazzani, M. Integration of SEWGS for carbon capture in natural gas combined cycle. Part B: reference case comparison. *Int. J. of Greenhouse Gas Control*, 2010.

49 Gazzani, M., Macchi, A., and Manzolini, G. CAESAR: SEWGS integration into an IGCC plant. Presented at the International Conference on Greenhouse Gas Technologies, Amsterdam, 2010.

50 van Selow, E.R., Cobden, P.D., Verbraeken, P.A., Hufton, J.R., and van den Brink, R.W. Carbon capture by sorption-enhanced water gas shift reaction process using hydrotalcite-based material. *Ind. Eng. Chem. Res.*, 2009, **48**, 4184–4193.

51 van Selow, E.R., Cobden, P.D., van den Brink, R.W., Wright, A., White, V., Hinderink, P., and Hufton, J. Pilot-scale development of the sorption enhanced water gas shift process. In *Carbon Dioxide Capture for Storage in Deep Geologic Formations*, Vol. 3 (ed. Eide, L.I.), CPL Press and BP, London, 2009, pp. 157–179.

52 Kumar, R., Fox, V.G., Hartzog, D.G., Larson, R.E., Chen, Y.C., Houghton,

P. A., and Naheiri, T. A versatile process simulator for adsorptive separations. *Chem. Eng. Sci.*, 1994, **49**, 3115–3125.

53 Hufton, J. R. and Sircar, S. Why does the linear driving force model for adsorption kinetics work? *Adsorption*, 2000, **6**, 137–147.

54 Ding, Y. and Alpay, E. Equilibria and kinetics of high temperature CO_2 adsorption on hydrotalcite adsorbent. *Chem. Eng. Sci.*, 2000, **55**, 3461–3474.

55 Reijers, H. T. J., Boon, J., Elzinga, G. D., Cobden, P. D., Haije, W. G., and van den Brink, R. W. Modeling study of the sorption-enhanced reaction process for CO_2 capture. I. Model development and validation. *Ind. Eng. Chem. Res.*, 2009, **48**, 6966–6974.

56 Soares, J. L., Casarin, G. L., Jose, H. J., and Moreira, R. F. P. M. Experimental and theoretical analysis for the CO_2 adsorption on hydrotalcite. *Adsorption*, 2005, **11**, 237–241.

57 Oliveira, E. L. G., Grande, C. A., and Rodrigues, A. E. CO_2 sorption on hydrotalcite and alkali-modified (K and Cs) hydrotalcites at high temperatures. *Sep. Purif. Technol.*, 2008, **62**, 137–147.

58 Du, H., Ebner, A. D., and Ritter, J. A. Pressure dependence of the nonequilibrium kinetic model that describes the adsorption and desorption behavior of CO_2 in K-promoted hydrotalcite like compounds. *Ind. Eng. Chem. Res.*, 2010, doi:10.1021/ie100965b.

59 Ebner, A. D., Reynolds, S. P., and Ritter, J. A. Understanding the adsorption and desorption behavior of CO_2 on a K-promoted hydrotalcite-like compound (HTlc) through nonequilibrium dynamic isotherms. *Ind. Eng. Chem. Res.*, 2006, **45**, 6387–6392.

60 Debecker, D. P., Gaigneaux, E. M., and Busca, G. Exploring, tuning, and exploiting the basicity of hydrotalcites for applications in heterogeneous catalysis. *Chem. Eur. J.*, 2009, **15**, 3920–3935.

61 Stanimirova, T., Hibino, T., and Balek, V. Thermal behavior of Mg–Al–CO_3 layered double hydroxide characterized by emanation thermal analysis. *J. Thermal Anal. Calorim.*, 2006, **84**, 473–478.

62 Millange, F., Walton, R. I., and O'Hare, D. Time-resolved *in situ* X-ray diffraction study of the liquid-phase reconstruction of Mg–Al–carbonate hydrotalcite-like compounds. *J. Mater. Chem.*, 2000, **10**, 1713–1720.

63 Wang, Y., Han, X. W., Ji, A., Shi, L. Y., and Hayashi, S. Basicity of potassium-salt modified hydrotalcite studied by H-1 MAS NMR using pyrrole as a probe molecule. *Micropor. Mesopor. Mater.*, 2005, **77**, 139–145.

64 Abello, S., Medina, F., Tichit, D., Perez-Ramirez, J., Rodriguez, X., Sueiras, J. E., Salagre, P., and Cesteros, Y. Study of alkaline-doping agents on the performance of reconstructed Mg–Al hydrotalcites in aldol condensations. *Appl. Catal. A*, 2005, **281**, 191–198.

65 Cross, H. E. and Brown, D. R. Entrained sodium in mixed metal oxide catalysts derived from layered double hydroxides. *Catal. Commun.*, 2010, **12**, 243–245.

66 Reijers, H. T. J., Valster-Schiermeier, S. E. A., Cobden, P. D., and van den Brink, R. W. Hydrotalcite as CO_2 sorbent for sorption-enhanced steam reforming of methane. *Ind. Eng. Chem. Res.*, 2006, **45**, 2522–2530.

67 Lee, J. M., Min, Y. J., Lee, K. B., Jeon, S. G., Na, J. G., and Ryu, H. J. Enhancement of CO_2 sorption uptake on hydrotalcite by impregnation with K_2CO_3. *Langmuir*, 2010, **26**, 18788–18797.

68 Yang, J. I. and Kim, J. N. Hydrotalcites for adsorption of CO_2 at high temperature. *Korean J. Chem. Eng.*, 2006, **23**, 77–80.

69 Walspurger, S., Boels, L., Cobden, P. D., Elzinga, G. D., Haije, W. G., and van den Brink, R. W. The crucial role of the K^+–aluminium oxide interaction in K^+-promoted alumina- and hydrotalcite-based materials for CO_2 sorption at high temperatures. *ChemSusChem*, 2008, **1**, 643–650.

70 Siriwardane, R. V. and Stevens, R. W. Novel regenerable magnesium hydroxide sorbents for CO_2 capture at warm gas temperatures. *Ind. Eng. Chem. Res.*, 2009, **48**, 2135–2141.

71 Bearat, H., McKelvy, M. J., Chizmeshya, A. V. G., Sharma, R., and Carpenter, R. W. Magnesium hydroxide dehydroxylation/

carbonation reaction processes: implications for carbon dioxide mineral sequestration. *J. Am. Ceram. Soc.*, 2002, **85**, 742–748.

72 Walspurger, S., Cobden, P. D., Safonova, O. V., Wu, Y. H., and Anthony, E. J. High CO_2 storage capacity in alkali-promoted hydrotalcite-based material: *in situ* detection of reversible formation of magnesium carbonate. *Chem. Eur. J.*, 2010, **16**, 12694–12700.

73 Wang, Y., Zhu, J. H., and Huang, W. Y. Synthesis and characterization of potassium-modified alumina superbases. *Phys. Chem. Chem. Phys.*, 2001, **3**, 2537–2543.

74 Walspurger, S., Cobden, P. D., Haije, W. G., Westerwaal, R., Elzinga, G. D., and Safonova, O. V. *In situ* XRD detection of reversible dawsonite formation on alkali promoted alumina: a cheap sorbent for CO_2 capture. *Eur. J. Inorg. Chem.*, **2010**, 2461–2464.

75 Stoica, G. and Perez-Ramirez, J. Reforming dawsonite by memory effect of AACH-derived aluminas. *Chem. Mater.*, 2007, **19**, 4783–4790.

76 Yalfani, M. S., Santiago, M., and Perez-Ramirez, J. *In situ* studies during thermal activation of dawsonite-type compounds to oxide catalysts. *J. Mater. Chem.*, 2007, **17**, 1222–1229.

77 Cobden, P. D., Walspurger, S., van Dijk, H. A. J., and van den Brink, R. W. Water gas shift process, Dutch Patent NL2009050709, 2009.

78 van Dijk, H. A. J., Walspurger, S., Cobden, P. D., van den Brink, R. W., and de Vos, F. G. Testing of hydrotalcite-based sorbents for CO_2 and H_2S capture for use in sorption enhanced water gas shift. *Int. J. of Greenhouse Gas Control*, 2010, in press.

79 Wakker, J. P., Gerritsen, A. W., and Moulijn J. A., High temperature H_2S and COS removal with MnO and FeO on γ-Al_2O_3 acceptors. *Ind. Eng. Chem. Res.*, 1993, **32**, 139–149.

80 Singh, R., Ram Reddy, M. K., Wilson, S., Joshi, K., Da Costa, J. C. D., and Webley, P., High temperature materials for CO_2 capture. *Energy Procedia*, 2009, **1**, 623–630.

Part 3
CO_2 Removal with Cryogenic Air Separation

Process Engineering for CCS Power Plants, First Edition. Edited by Detlef Stolten and V. Scherer

10
CO_2 Capture via the Oxyfuel Process with Cryogenic Air Separation

Alfons Kather and Mathias Klostermann

Abstract

A significant reduction in global CO_2 emissions may be achieved by the capture of CO_2 from coal-fired power plants, where the oxyfuel process is one promising option. Pure oxygen is used for the combustion of coal, resulting in a flue gas with high CO_2 concentration. Oxygen is obtained from a cryogenic air separation process, which is the only available state-of-the-art technology for large demands of oxygen. The separated nitrogen is substituted by an equivalent amount of recycled flue gas. This allows a boiler design similar to conventional air-operated boilers. The options for flue gas recycle are manifold, where each option comes along with different characteristics. Pollutant formation, flue gas composition, and flexibility are only a few of the issues influenced by flue gas recycle design. The CO_2 fraction of the non-recycled flue gas has to be increased and possible impurities have to be removed due to transport and storage issues. The overall net efficiency is expected to be lowered by 9–11%-points, due to the generation of oxygen and also the purification, liquefaction, and compression of CO_2. More efficient air separation unit designs and heat integration will contribute to efficiency improvements.

Keywords: oxyfuel, cryogenic air separation, purity, efficiency

10.1
Introduction

In the oxyfuel process, fossil fuel is combusted in an atmosphere of pure oxygen and recycled flue gas. By separating air into oxygen and nitrogen upstream of the combustor, CO_2 concentrations of up to 90 vol% (dry) are reachable in the flue gas, facilitating the capture and storage of a highly concentrated CO_2 stream. The underlying steam power process remains unmodified in its main features. A simplified schematic of the oxyfuel process is shown in Fig. 10.1.

Due to the large quantities of oxygen required in a state-of-the-art coal-fired power plant, the air separation is only feasible by means of cryogenic rectifica-

Process Engineering for CCS Power Plants, First Edition. Edited by Detlef Stolten and V. Scherer

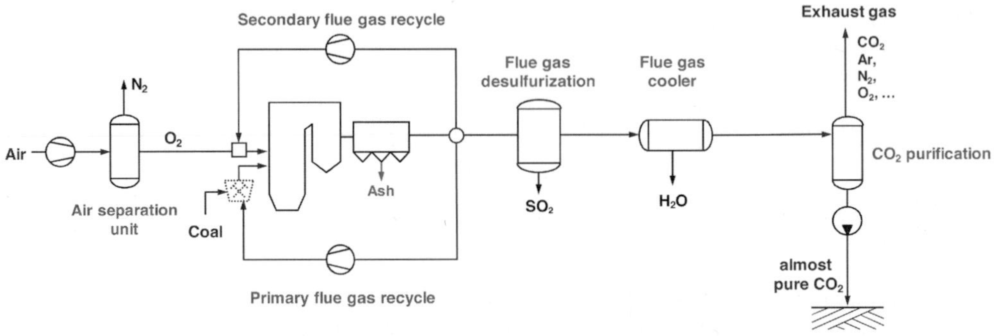

Fig. 10.1 Simplified schematic of oxyfuel process with cryogenic air separation and recirculation of untreated flue gas.

tion. This air separation process has been successfully applied throughout several decades on a commercial scale [1]. Additionally, the adaptation of this process to the specifics of the oxyfuel process offers considerable potential for reducing the efficiency decrease of the overall power plant process.

Since temperatures in the steam generator would increase significantly due to the missing nitrogen, cooled flue gas needs to be recycled to the combustor. The configuration and design of the flue gas recycle depend on coal composition, feasible limits of dust, sulfur oxides, and water vapor, and whether or not conventional combustion with air must still be possible.

The flue gas consists of CO_2, water vapor, oxygen, nitrogen, and argon in addition to pollutants such as dust, nitrogen oxides, sulfur oxides, and carbon monoxide. The pollutants must be removed to comply with the specifications of downstream process equipment and storage recommendations. Water vapor, dust, nitrogen oxides, and sulfur oxides can be almost entirely be removed with existing flue gas cleaning equipment, whereas low-boiling gas components (oxygen, nitrogen, argon) can only be eliminated by condensation of CO_2 at a temperature below 0 °C and elevated pressure. Rectification of the CO_2-rich flue gas needs to be applied if the oxygen concentration in the flue gas is subject to more severe limitation. The reachable CO_2 purity lies well above 99.5 vol%

The electrical net efficiency of a coal-fired oxyfuel power plant decreases by approximately 10% due to the auxiliary power demand for air separation, enhanced flue gas cleaning, and CO_2 compression. A reduction in the efficiency penalty by optimization of several components and sub-processes and also by reintegration of waste heat into the power plant process are possible.

10.2 Flue Gas Recycle

In the following, the calculation and possible configurations of the required flue gas recycle are discussed.

10.2.1 Calculation

The combustion of coal with oxygen would lead to a significant increase in temperature level within the furnace as the heat capacity of nitrogen is missing. The partial recycle of cool flue gas reduces the temperature to values known from the combustion of coal with air. This allows the application of conventional steam generator materials and only slightly modified designs, considerably reducing the involved technical risks.

A simplified method of estimating the required amount of recycled flue gas involves consideration of the adiabatic combustion temperature. By assuming that this quantity must be the same for the oxyfuel process as it is for the conventional combustion of coal with air, the necessary amount of flue gas for recirculation can be calculated. That rule of thumb is valid as long as flue gas for the recycle is branched downstream of the economizer.

The accurate amount of flue gas required is determined by two temperatures that are critical in the design of steam generators (see Fig. 10.2): first, the maximal feasible temperature of the water vapor in the membrane wall at the end of the combustion chamber; and second, the maximal feasible temperature of the flue gas at the entrance of the convective heating surfaces (also at the end of the combustion chamber).

The flue gas temperature is limited by the ash softening temperature of the coal to prevent slagging and fouling in the convective heating surfaces. The possible membrane wall temperature depends on the available materials. If this temperature is given and if the temperature of the water at the membrane wall inlet is known, the heat that is transferred on to the wall within the combustion

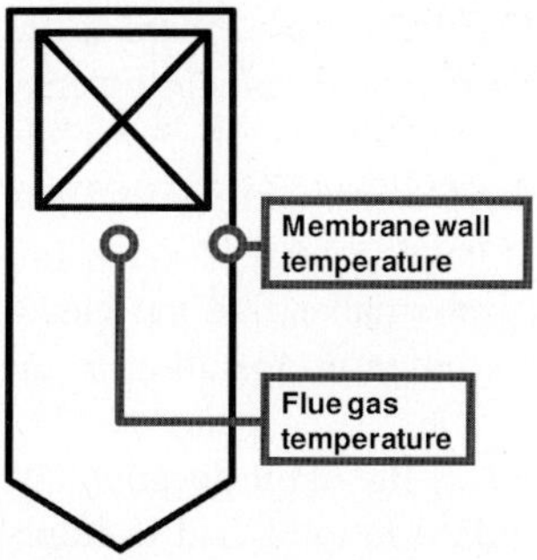

Fig. 10.2 Location of critical temperatures for determination of required flue gas recycle.

chamber can be determined. The heat transfer in the combustion chamber is dominated by radiation; therefore, an increase in recycled flue gas lowers the temperature in the combustion chamber and consequently reduces the amount of heat that is transferred on to the wall.

An increase of flue gas recycle beyond the determined amount is not advisable since the higher flue gas flow requires a larger cross-section and as the corresponding reduction in flue gas temperature requires larger convective heating surfaces.

The amount of recycle also depends on the temperature and composition of the flue gas. The lower the temperature and the higher the water content, the less is the amount of flue gas needed. Both the temperature and composition of the flue gas are determined by the configuration of the recycle system.

10.2.2 Configuration

The recycled flue gas is separated into a primary and a secondary stream. In hard coal-fired oxyfuel power plants, the primary stream is need for drying and transport of the pulverized coal. In pre-dried lignite-fired power plants, where drying of the coal is realized in an external unit, the primary stream must ensure proper transport of the coal to the burners only. The secondary stream controls the desired temperature level in the steam generator. There are multiple options for the location and for the design of the flue gas recycle (e.g., [2]). The choice of option significantly influences the composition and in particular the amount of pollutants in the flue gas. In the following, the two most differing variants are discussed in more detail.

10.2.2.1 Untreated Recycle

The most energy-efficient option is the recycling of untreated flue gas which is extracted downstream of the economizer at a temperature between 300 and 400 °C (see Fig. 10.1). The secondary recycle stream must be led to the burners by an adequate fan at temperatures of up to 400 °C. In this case, there is no need for a reheat of the recycled flue gas, which avoids an additional efficiency penalty.

This configuration allows for a short recycle path. The remaining flue gas shows only about one-quarter of the volume flow of a conventional power plant, allowing for small cross-sectional areas of the ducts and smaller flue gas cleaning equipment downstream of the recycle branch. A change to operation in air mode at higher loads is not feasible, however.

If the primary stream for the coal mill is extracted at the same location, its temperature needs to be controlled to ensure the required ratio of coal to transport gas and also the remaining water content of the coal at the outlet of the mill.

As there is no dust removal in the recycle stream, the dust content accumulates in the flue gas path including the steam generator. Assuming a constant flue gas velocity, an increase in erosion is expected. Adaptations might be required in particular in and around the mills, the burners, and the steam generator.

The missing removal of sulfur oxides and water leads to high concentrations of water vapor and SO_x in the flue gas. This can increase the risk of corrosion both in the steam generator and at the cold end of the boiler. Therefore, adequate modifications are necessary. For example, increasing the classifier outlet temperature of hard coal mills to approximately 160 °C can prevent low-temperature corrosion due to sulfuric acid condensation. Nitrogen oxides are not expected to accumulate, as the direct recycle into the combustion zone reduces most of the nitrogen oxide to nitrogen. NO_x concentrations are therefore expected to be insignificantly higher than in the oxyfuel mode with completely treated recycle (see Section 10.4.1).

10.2.2.2 **Treated Recycle**

An option with fewer risks and modifications is the recirculation of a completely treated flue gas. This option includes – in addition to the removal of dust – desulfurization and, if appropriate, drying of the entire flue gas stream before it is recycled. This recycle option is shown in Fig. 10.3. A more complex, longer, and larger flue gas duct system is necessary for this option. A heat transfer system for the reheating of the recycled flue gas is essential to reduce the efficiency loss due to heating of cold recycled flue gas by firing additional coal in the furnace. The temperature of the flue gas decreases when it passes the desulfurization unit and the cooler and part of heat contained in the flue gas is dissipated and discharged by cooling water. This leads to an energy disadvantage for this recycle option. Under operational conditions, the use of treated flue gas offers a number of advantages. No fundamental changes for mill internal drying and pulverization of hard coal are necessary. The formation of acidic condensate

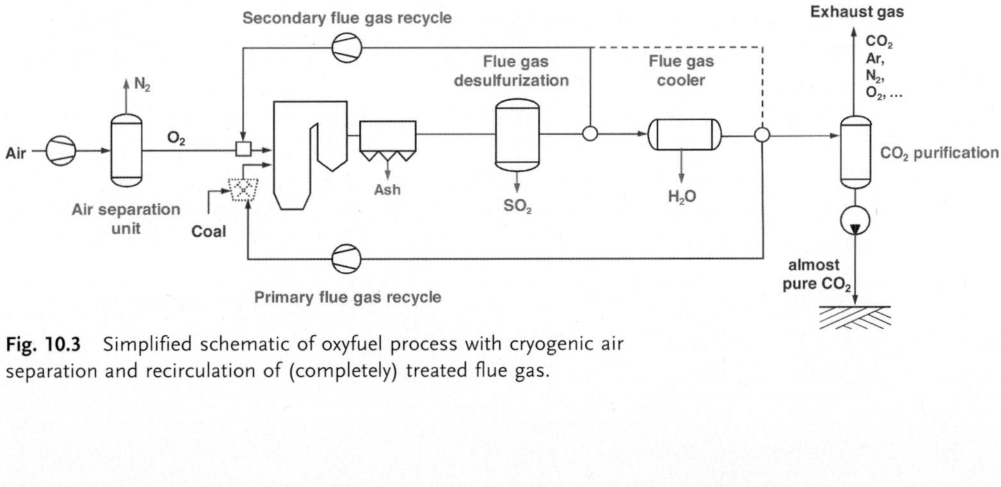

Fig. 10.3 Simplified schematic of oxyfuel process with cryogenic air separation and recirculation of (completely) treated flue gas.

in colder regions of the flue gas system is considerably reduced. The dimensions of the flue gas system permit the conventional operation (combustion with air) of the steam generator at high loads [3].

Between the two mentioned and fundamentally different recycle configurations, several variations are possible. For instance, the primary recycled flue gas could be treated completely to restrict changes or risks for the fuel supply and fuel processing unit. Dust removal might only be used for the secondary recirculation. Finally, the best choice of all flue gas recirculation options depends on the fuel used, the operation mode of the power plant, and the balance between an increase in efficiency and the operational flexibility of the plant.

10.3 Combustion

The combustion of coal under oxyfuel conditions only partly differs from the operation of a conventional power plant, if the principles for the design of the flue gas recycle system as explained above are considered. An excess of oxygen is still needed, but the change in combustion atmosphere has an effect on the formation of pollutants. The main aspects are explained in the following sections.

10.3.1 Oxygen Balance

One of the essential boundary conditions for the design of a coal-fired power plant which is based on the oxyfuel process is the determination of the necessary amount of oxygen for the maximal conversion of coal. The combustion of coal in a power plant is done with excess of oxygen to achieve a sufficient burnout of the coal and to reduce the risk of corrosion on the membrane walls of the combustion chamber, which is caused by an unbalanced distribution of oxygen and coal. Due to the constant concentration of oxygen in air, the air ratio is identical with the oxygen ratio for conventional air-operated coal-fired power plants without flue gas recirculation. The dimensionless oxygen ratio is calculated from the actual amount of oxygen supplied by the air separation unit plus the recycled oxygen in flue gas related to the stoichiometrically required amount of oxygen. The oxygen ratio can be reduced to 1.15 for the most modern coal-fired power plants, where the excess oxygen consequently amounts to 15%. Due to the explained negative effects, lower values of excess oxygen can be achieved only if the distribution of the coal to the burners is more homogeneous, for example, in an indirect firing system. For coal-fired power plants it is not likely that an oxygen ratio of less than 1.10 – as is possible in gas- or light oil-fired systems – can be realized. Therefore, a significant amount of oxygen remains in the flue gas of a coal-fired oxyfuel power plant.

For the oxyfuel process, the correct definition of the oxygen ratio $O_{2,ex}$ is particularly important since in principle different definitions are possible. A high amount of oxygen is introduced into the combustion process by the recycled flue gas. Following the recommendation in [4], the amounts of oxygen delivered by the air separation unit (ASU) $\dot{m}_{O_2,ASU}$, the oxygen mass flow that enters with air ingress $\dot{m}_{O_2,AI}$, and the residual mass fraction of oxygen in the recycled flue gas $\dot{m}_{O_2,Rec}$ are summarized in the equation

$$O_{2,ex} = \frac{\dot{m}_{O_2,ASU} + \dot{m}_{O_2,AI} + \dot{m}_{O_2,Rec}}{\dot{m}_{O_2,stoi}} \tag{10.1}$$

This overall balance is independent of the location of the mixing point of the oxygen produced by the air separation unit and the oxygen contained in the recycled flue gas. As can be seen from Fig. 10.4, the amount of oxygen supplied by the air separation unit is smaller than the necessary amount defined by the oxygen ratio. An oxygen balance without considering the oxygen contained in the recycled flue gas is unsuitable. For conventional air-operated power plants, the calculation of the oxygen ratio is easily possible as it is based on the residual oxygen content in the flue gas.

In the oxyfuel process, the calculation of the oxygen ratio is very important. If the oxygen ratio used is too high, the energy demand of the air separation unit increases due to the excessive amount of oxygen produced. Additionally, the energy demand for the CO_2 purification also increases due to the higher residual oxygen content in the flue gas. However, the calculation itself is not trivial due to the variation of the composition of the oxidant as a result of a varying flue gas recirculation. If the temperature in the furnace section is decreasing due to an increase in the amount of recycled flue gas, an incomplete combustion may result although the oxygen ratio is high enough. The oxygen ratio can be calculated, for instance, due to continuous measuring of the composition and the mass flow of the recycled flue gas.

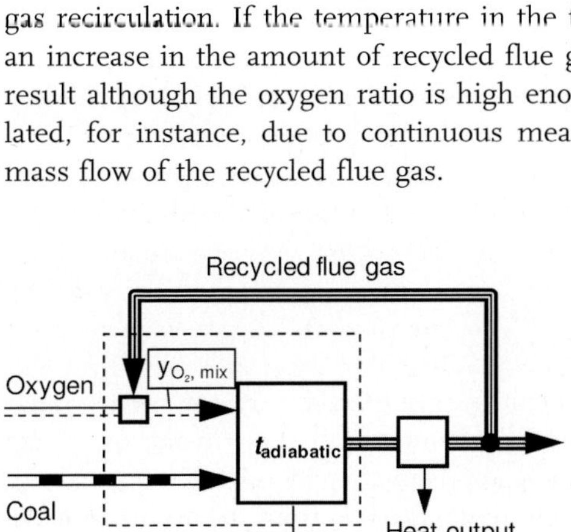

Fig. 10.4 Simplified balance of the oxyfuel process for oxygen ratio calculation.

10.3.2
Oxygen Concentration in the Oxidant Upstream of the Combustion

The oxyfuel process provides the possibility to feed oxygen supplied by an ASU and oxygen contained in recycled flue gas to the combustion independently. However, it has to be ensured that sufficient oxygen is available for combustion at any time. Although the overall requirements for oxygen in both the air-operated process and the oxyfuel process are similar due to the same oxygen ratio, the concentration of oxygen in the oxidant upstream of the combustion differs significantly. If oxygen supplied by the ASU and oxygen contained in recycled flue gas are completely mixed before entering the furnace, the oxygen concentration of the oxidant cannot be chosen arbitrarily. It depends on the amount of flue gas which is recycled. The composition of the oxidant is influenced by the temperature of the recycled flue gas, the fuel composition, and the thermal design of the steam generator. According to these effects, different oxygen concentrations upstream of the furnace result for the combustion of hard coal and for the combustion of pre-dried lignite. In general, the oxygen concentration in a fully mixed atmosphere decreases while the oxygen ratio is kept constant and the amount of recycled flue gas increases. Oxyfuel processes with cryogenic air separation are characterized by an oxygen concentration in the entire mixture of more than 21 vol% and thus by higher values compared with operation with air. For South African hard coal, an oxygen concentration in the oxidant of about 30 vol% is calculated; for pre-dried lignite with a total moisture in fuel of 12 wt% it is 24 vol% The higher oxygen concentration is caused by the higher heat capacity of recycled flue gas compared with nitrogen. Therefore, less flue gas has to be recycled and the overall flue gas molar flow decreases while the oxygen concentration in the oxidant increases. The combustion of pre-dried lignite needs more flue gas to be recycled because lower temperatures in the furnace section have to be maintained compared with the combustion of hard coal. These lower temperatures are required due to the lower ash softening point of lignite.

10.4
CO_2 Purification and Capture

The flue gas from an oxyfuel power plant does not contain CO_2 and water vapor only, as assumed in earlier times. The concentration of CO_2 in the dried flue gas reaches, under favorable conditions, approximately 85 vol% The increase in CO_2 concentration that results from the removal of N_2 from the air is desirable. Simultaneously, however, the concentrations of other gas components also increase, which influences the purity of the separated CO_2 stream and the concentrations of gaseous emissions to the atmosphere. The increase in the volumetric concentrations of most species in the dry flue gas of the oxyfuel process is determined by three factors:

- smaller specific mass flow of the flue gas compared with air operation
- higher flue gas density due to the higher CO_2 concentration
- higher water content of the raw flue gas.

This still does not include other effects that arise from the recirculation of individual gases and the resulting changes in chemical equilibrium and/or reaction kinetics.

Nitrogen, oxygen, and argon constitute the main impurities. Their concentrations must be reduced in order to operate the transportation and the geologic storage of the CO_2 as efficiently and steadily as possible. Nitrogen and argon are considered inert gases whereas oxygen can disturb the downstream process chain (e.g., pipeline, geology), especially when water is also present in relevant concentrations.

Nitrogen oxides, sulfur oxides, and carbon monoxide, which are also present in the flue gas, are at higher concentrations than under air operation and must therefore be reduced due to process or emission regulatory reasons. The following sections highlight the sources of the individual impurities and the possible options towards raising the CO_2 concentration by reducing their concentration.

10.4.1 Fuel

Complete and low-emission combustion of coal under oxyfuel conditions is possible. In fact, the operating conditions are even advantageous over using air as oxygen carrier provided that the conditions described in Section 10.2.1 are maintained.

The nitrogen that is bound in the coal contributes to the formation of nitrogen oxides. In comparison with the air-operated case, the NO_x concentration in the flue gas increases as a result of the altered flue gas composition and quantity. All studies that have been performed to date show, however, that in the oxyfuel process the NO_x emissions when scaled with the overall heat input are significantly reduced, independent of the composition of the recirculated flue gas [5, 6]. In the case of reinjection of NO_x-free flue gas (scenario "treated" in Fig. 10.5) the NO_x formation is approximately 25% lower than in the air case. This is due to the significantly reduced nitrogen fraction in the combustion chamber, which limits the formation of NO_x from the side of the reactants. In addition, the same amount of NO_x formed in the oxyfuel case leads to a higher concentration in the flue gas. Because the amount of flue gas is lower than in the air case, further formation of NO_x is suppressed. The recirculation of untreated flue gas, which is a method for NO_x reduction already known in conventional power plants (scenario "untreated" in Fig. 10.5), reduces the formation rate of NO_x to approximately one-third compared with the air case. The recirculated NO_x increases the concentration at the side of the products and suppresses the formation reaction. From these investigations, one can infer that by employing the recirculation of NO_x-rich flue gas, the resulting NO_x emissions are only in-

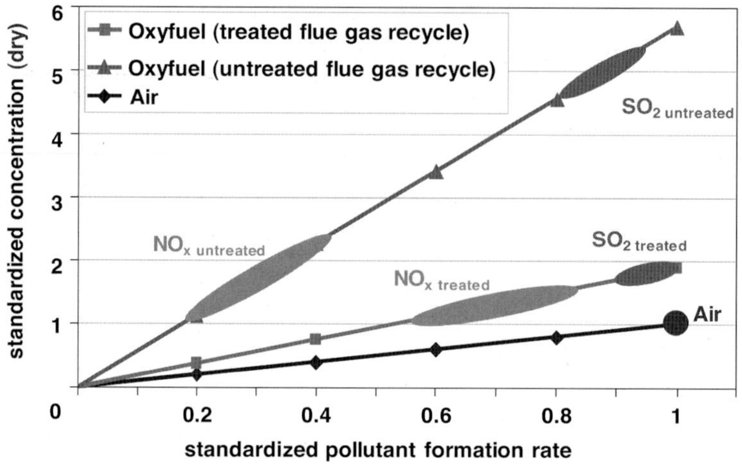

Fig. 10.5 Standardized concentrations and pollutant formation rates of NO_x and SO_2 under oxyfuel conditions in comparison with the air-fired case (set as 1). Results based on hard coal firing. Diagram according to [7].

significantly higher than achievable by the application of a separate SCR plant for the denitrification of the recirculated mass flow.

The sulfur content of the coal reacts during combustion with the excess oxygen to form mainly sulfur oxides. This process is independent of the nature of the oxygen carrier. As seen in Fig. 10.5, the formation of sulfur oxides is hardly influenced by the conditions of the oxyfuel process. The concentration of SO_2 in the dried flue gas increases almost as much as it is calculated from the changes in flue gas composition and amount. In the case of desulfurized recirculation, the concentration of SO_2 almost doubles whereas without desulfurization it can increase by a factor of about five. This occurs, in comparison with the air case, at a slightly lower formation rate. Several experiments have shown that the slightly lowered formation rate of SO_2 is caused by a reaction of SO_2 and ash components [7, 8]. Therefore, an untreated flue gas recycle with a higher dust load is able to capture more SO_2 and reduces the SO_2 concentration more significantly. The reduction also depends on the ash composition – hence the results shown in Fig. 10.5 have to be considered only qualitatively. An additional reduction is caused by the increased formation of SO_3, which commonly is not measured and promoted by higher oxygen and vapor concentrations in the flue gas [9].

High carbon monoxide concentrations are characteristic of a local lack of oxygen and/or low temperatures during coal combustion. If the firing system is designed to reach similar adiabatic temperatures as in the air mode and if oxygen, coal, and recycled flue gas are well mixed during combustion, it can be assumed that the formation of CO will decrease under oxyfuel conditions. However, the concentration in the dry flue gas will increase for the reasons explained above.

10.4.2
Oxygen Purity

Most cryogenic ASUs are designed to produce oxygen with a purity of 99.5 vol%. This oxygen predominantly contains argon and additionally traces of nitrogen. ASUs that are developed to reach a lower specific energy demand for oxygen production are characterized by a slightly increased fraction of impurities. Approximately 2.5–5 vol% of the gas stream produced is nitrogen and argon, where the argon fraction does not exceed ~3.8 vol%. Both components pass through the furnace. Argon is inert whereas nitrogen could promote the formation of nitrogen oxides. The CO_2 concentration in the flue gas is lowered by ~3% when using 95 instead of 99.5 vol% pure oxygen.

10.4.3
Residual Oxygen

The need for excess oxygen in coal combustion, as explained in Section 10.3.1, inevitably causes a significant fraction of residual oxygen in the flue gas. As shown in Fig. 10.6, for hard coal the oxygen concentration depends on the oxygen ratio and the amount of recycled flue gas. The higher the oxygen ratio at constant flue gas flow within the steam generator that is chosen, the higher is the oxygen concentration in the flue gas. The larger the recycled flue gas mass flow needs to be, the lower is the residual oxygen since the oxygen fraction is diluted by a larger flue gas flow. From Fig. 10.6, one can see that a residual oxygen fraction of ~4.7 vol% (dry, at STP) results from the combustion of South African hard coal in combination with 15% excess oxygen and a recycle rate of 67% relative to the overall flue gas mass flow. Hence the residual oxygen concentration is significantly higher than in conventional power plants, where only 2.8 vol% (dry, at STP) is measured under similar boundary conditions.

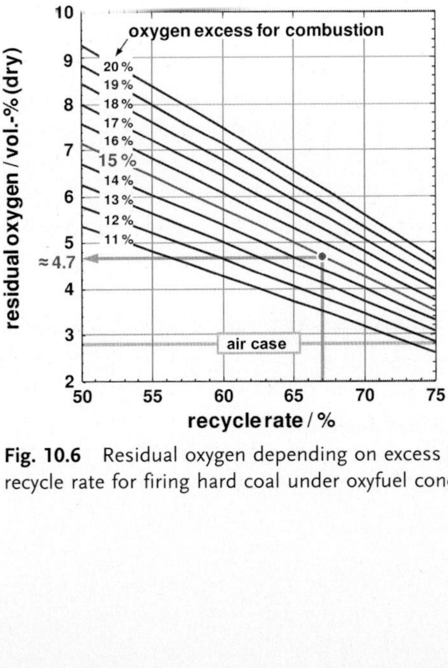

Fig. 10.6 Residual oxygen depending on excess oxygen and recycle rate for firing hard coal under oxyfuel conditions.

Firing pre-dried lignite causes a slightly lower residual oxygen content because of the larger amount of recycled flue gas when assuming the same excess oxygen as for hard coal.

10.4.4
Air Ingress

Large sections of the air and flue gas ducts of a conventional coal-fired power plant are operated below atmospheric pressure to prevent the slip of hot untreated flue gases to the atmosphere. This operation applied to the oxyfuel process would lead to unregulated air ingress, which significantly impacts the flue gas composition.

It can be assumed that new hard coal-fired power plants under construction today have an ingress of air of about 3% of the total flue gas mass flow at the outlet of the economizer [10]. In most cases, ingress of air significantly increases during years of plant operation. Air ingress of this extent is unfavorable for the oxyfuel process due to a decreasing CO_2 concentration in the flue gas. If air ingress occurs upstream of the furnace outlet, the incoming oxygen takes part in the combustion while the content of nitrogen in the resulting flue gas increases. If there is also air ingress downstream of the boiler, the concentrations of both oxygen and nitrogen in the flue gas increase. In particular, air ingress downstream of the boiler has to be avoided because the decrease in CO_2 concentration is severe.

As shown in Fig. 10.7, air ingress has to be limited to around 1% to achieve sufficient CO_2 concentrations when using oxygen of lower purity. An increase

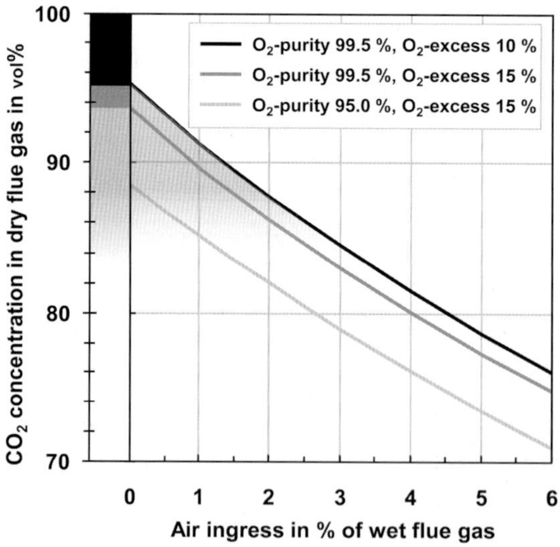

Fig. 10.7 Influence of air ingress upstream of the furnace on CO_2 concentration in dry flue gas. Calculation based on South African hard coal.

in air ingress over the years of power plant operation has to be limited as much as possible. It is also required that controlled feed air for cooling, sealing, and other purposes should be replaced, if possible, by CO_2-rich gas. Alternatively, other technical solutions have to be found.

10.4.5
CO_2 Purification and Capture

The operational requirements of the CO_2 transport chain and the geologic characteristics of the storage formation determine the minimal CO_2 fraction required and also the maximal allowable concentrations of impurities. It is envisaged that the CO_2 concentration must be raised to substantially above 90 vol%, whereas the chemically active gases, for example, NO_x, SO_x, water, and oxygen, must simultaneously be reduced. When a suitable CO_2 concentration is already present in the flue gas of the oxyfuel process, it is possible to achieve CO_2 capture rates of 90% or higher. An increase in CO_2 concentration towards food-grade quality is technically feasible but not always reasonable, because increasing the purity of the CO_2 leads to an additional energy penalty and, in most cases, also to a reduction in the capture rate of CO_2, raising disproportionately the specific extra cost for CO_2 separation [11].

The SO_2 can be removed by more than 99% with conventional wet desulfurization [12]. The retention of SO_3 in this cleaning stage is, however, limited. Both desulfurization and denitrification are promoted by the downstream flue gas compression. Under high pressure, the NO reacts with the residual oxygen to yield NO_2 and then contributes to the formation of highly acidic condensates in the intercoolers of the flue gas compressor. In addition to nitrous and nitric acids, the condensate also includes sulfurous and sulfuric acids. Under advantageous process operating conditions, the sulfur oxides and nitrogen oxides will be removed down to very low levels [13]. At high output concentrations of SO_x and NO_x, it is possible that damage to the construction materials may occur, due to the extremely low pH values of the condensate. This risk can be compensated through additional upstream measures, such as fixed-bed filters or scrubbers [14].

To increase the CO_2 fraction in the flue gas, processes are preferred in which the flue gas is compressed and subsequently cooled, so that a CO_2-rich condensate and a CO_2-poor rest gas result. The flue gas pressure must be increased to between 20 and 40 bar. The condensation can take place at temperatures close to the triple temperature of CO_2 (−56.6 °C). The process parameters are selected to match the available CO_2 fraction in the flue gas and the desired CO_2 purity level. The choice of parameters directly yields the maximum possible CO_2 capture rate, which cannot be selected independently.

At increasing pressures, the solubility of the accompanying gases in the condensing CO_2 also increases [15]. By applying downstream rectification of the CO_2 stream, it is possible to reduce the fractions of argon, nitrogen, and oxygen to such an extent that a CO_2 purity of more than 99.5 vol% can be achieved.

Rectification leads to partial re-evaporation of the CO_2 product, which then results in a slight reduction in the CO_2 capture rate.

The energy demand for the increase in concentration and the compression of the CO_2 depends on the CO_2 fraction in the flue gas, on the desired purity of the CO_2 stream, and on the required delivery pressure to the transportation network.

10.5 Efficiency

As mentioned above, the oxyfuel process is based on the steam power process, which is the predominant technology for coal-fired power plants worldwide. To calculate the efficiency losses, a conventional steam power plant as reference, which includes state-of-the-art components and developments in power plant technology, is needed. The following data are taken from the concept study "Reference Plant North-Rhine Westphalia," a German study of a USC hard coal-fired power plant published in 2004 [16]. Figure 10.8 shows a simplified scheme of the underlying reference power plant. Typical parameters of this type of power plant are as follows:

- steam parameters: 600 °C/620 °C/285 bar
- gross power output: 600 MW_{el}
- net efficiency: 45.5–46.5% (LHV), for typical cooling conditions in Germany.

The concept is very similar to currently deployed technology of various projects under construction in Germany. The simulation of the conventional reference process results in a process net efficiency of 45.6% (LHV). The auxiliary power demand is about 7.3% of the gross power output, resulting in a difference in gross

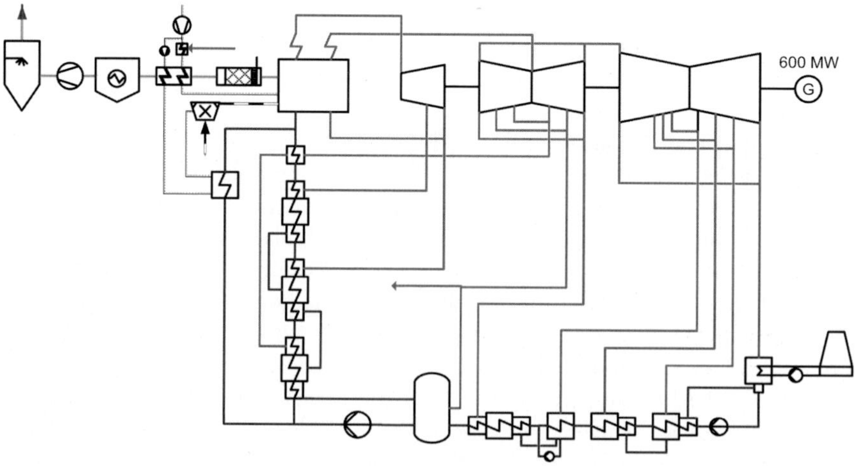

Fig. 10.8 Simplified flow sheet for a conventional reference power plant.

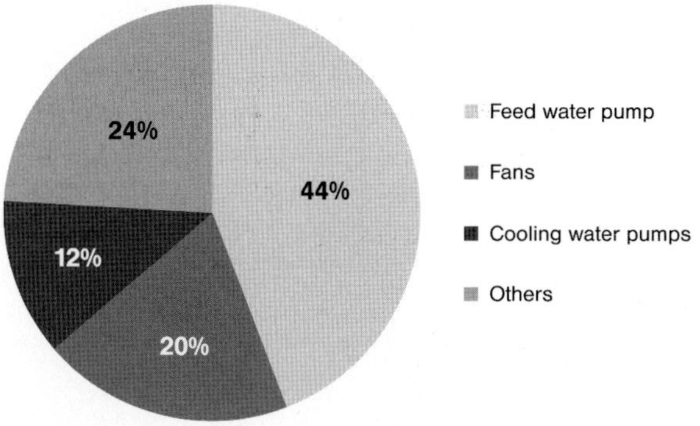

Fig. 10.9 Structure of the auxiliary power demand of the conventional hard coal-fired USC reference power plant with a gross power output of 600 MW_{el} and overall auxiliary power demand of 44 MW_{el}.

and net efficiency of ~3.6%-points As an electrically driven feed water pump is applied, the structure of the auxiliary power demand as shown in Fig. 10.9 is mainly caused by the demand of pumps followed by fans and other components.

10.5.1 Basic Design of the Oxyfuel Power Plant

A simplified process scheme of the oxyfuel power plant is shown in Fig. 10.10. The power plant is characterized by an inline configuration of ASU, boiler island, and gas processing unit (GPU) without any integration of waste heat between the ASU and GPU as heat sources and the water-steam cycle as the potential heat sink.

With a required heat input of the steam generator of ~1200 MW, hard coal has to be burned at a rate of 48 kg s^{-1} (172.8 t h^{-1}). Hence an oxygen supply capacity of ~8500 t d^{-1} is necessary, which is 2–3 times greater than today's largest single-train ASU capacity. The total amount of CO_2 being captured by the GPU is ~104 kg s^{-1} at a capture rate of 90%.

The concept branching the primary flue gas recycle directly downstream of the flue gas desulphurisation (FGD) is chosen to avoid the risk of formation of sulfurous and sulfuric acids within the coal mills. The secondary flue gas recycle is taken directly downstream of the hot electrostatic precipitator as it is presumed that owing to the higher temperature there is no corrosion due to sulfurous and sulfuric acids. A regenerative gas preheater is used for reheating the primary recycle to the temperatures required for mill internal drying. Commonly, the drying is controlled based on the classifier outlet temperature of the coal mills, which is regulated by the split fraction of the raw flue gas sent to the regenerative gas preheater.

To minimize the efficiency losses due to the quenching effect of the hot flue gases in a wet FGD, it is favorable to make use of the heat of the flue gases

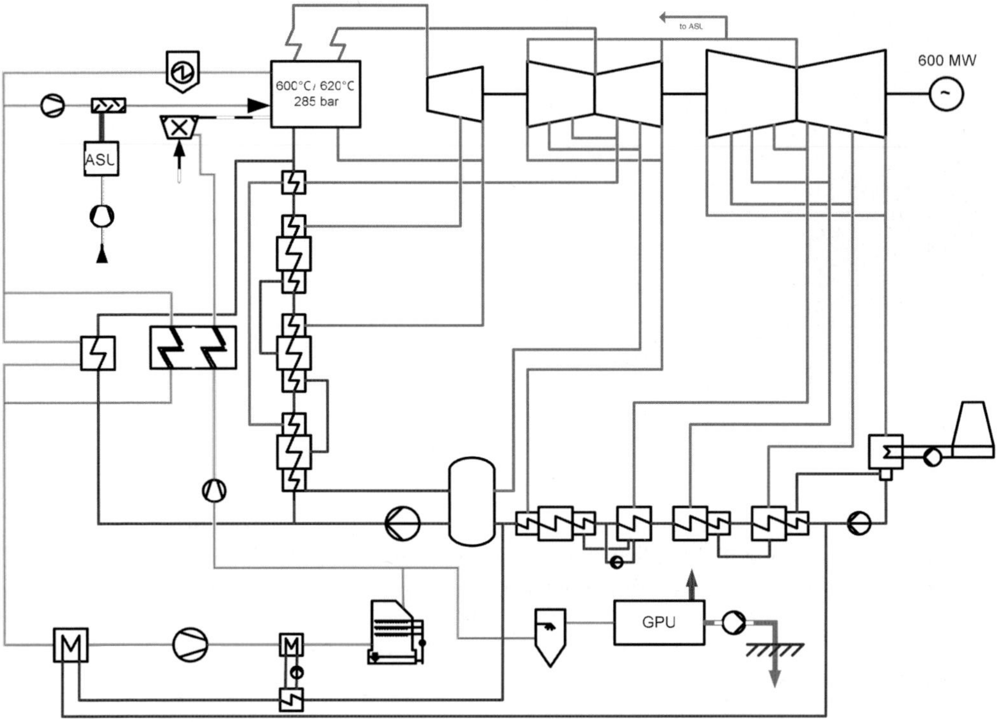

Fig. 10.10 Simplified flow sheet of the non-integrated oxyfuel power plant.

within the power plant process before entering the FGD. The corresponding available amount of heat and its temperature level match well the conditions of the feed water and condensate preheating system. The oxygen is preheated to 80 °C directly with the waste heat of the ASU and mixed with the secondary recycle upstream of the burners.

For the basic non-integrated design of the oxyfuel process, the calculated overall net efficiency loss is ~10.2%-points, assuming the application of a technically mature cryogenic process for the ASU (O_2: 99.5 mol%, 0.245 kWh_{el} kg O_2^{-1}) and a GPU process based on proven components (0.123 kWh_{el} kg CO_2^{-1}, p_{out}=10 MPa). The main reason for the efficiency losses is the considerably increased auxiliary power demand (about four times compared with the reference power plant). Figure 10.11 shows that the ASU and GPU cause around 75% of the auxiliary power demand whereas the absolute power demand of the 'conventional' components does not vary significantly.

As the ASU is the largest power consumer, energy savings for the production of oxygen are the most effective way to reduce the overall efficiency loss of the oxyfuel power plant. Two possible optimization levels of the ASU process to lower its energy demand are explained in the following.

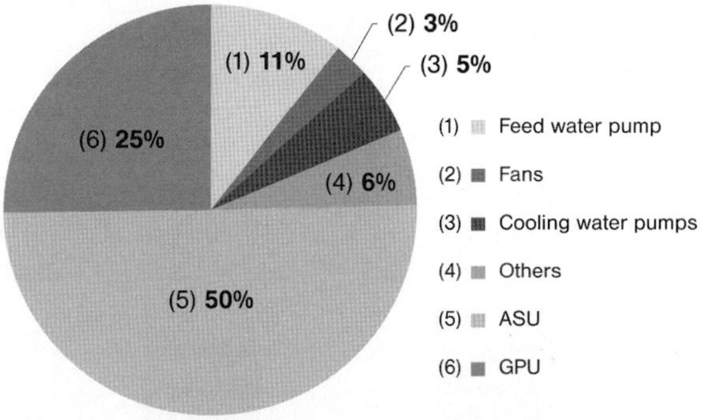

Fig. 10.11 Structure of the auxiliary power demand of the studied oxyfuel hard coal-fired USC power plant with a gross power output of 600 MW_{el} and overall auxiliary power demand of 175 MW_{el}. Power plant without optimized ASU and waste heat integration.

In contrast, there is only limited potential for energy savings within the GPU. The cryogenic processes of the GPU proposed by manufacturers rely on a high-level thermal integration to realize an efficient separation. Hence the energy demand is mainly affected by boundary conditions for the GPU:

- flue gas composition
- flue gas mass flow to be treated
- desired CO_2 capture rate and CO_2 purity
- design pressure of the CO_2 transport system.

The flue gas mass flow that has to be treated in the GPU at full load operation of the power plant is directly affected by the amount and the location of the air ingress and also by the composition of the coal used. In all calculations in this work, CO_2 capture rate, purity, and pipeline pressure are fixed boundary conditions.

10.5.2
Impact of Optimized ASU Designs

The oxygen production causes about 50% of the auxiliary power demand of a non-integrated oxyfuel power plant (cf., Fig. 10.11). This value corresponds to the application of a conventional ASU design, that is, a single condenser–reboiler and double column for high oxygen purity (>99.5 mol%.), which can be rated as a mature state-of-the-art process. Therefore, the potential for reducing the internal losses without major changes in the process design is very limited. However, even with this process there is the possibility to reduce the power demand by decreasing the oxygen purity, as shown below. With decreased requirements on oxygen purity, there is an additional optimization potential for ASU process design to lower the energy consumption considerably. In fact, lower oxygen purities (>95 mol%) are the most effective option for reducing the energy

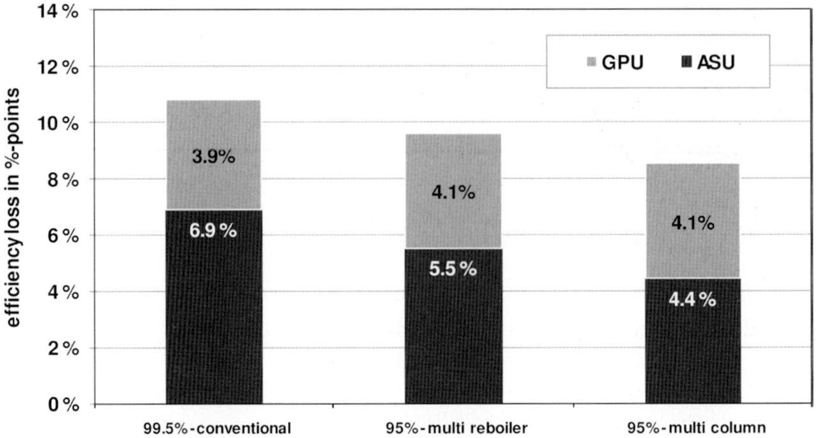

Fig. 10.12 Efficiency losses of different cryogenic ASU designs, including the influence on the GPU. Boundary conditions: capture rate = 90%, CO_2 purity > 96 mol%, 87% of oxygen supplied by ASU, South African hard coal, 2% air ingress. Specific power demands: 250, 200, 160 kWh t O_2^{-1}, pure.

demand of a cryogenic ASU. Therefore, different cryogenic ASU process designs with lower purities of gaseous oxygen are proposed by the manufacturers for oxyfuel power plants [17–19].

Double-column processes with an additional reboiler or processes with more than two rectification columns are proposed for lowering the specific energy demand of oxygen production. Figure 10.12 illustrates the impact of these processes compared with a conventional double-column process. The effects of increased level of impurities, that is, a lowered CO_2 concentration and an increase in flue gas mass flow, are included in these results. The efficiency loss due to oxygen production is reduced considerably. The moderate increase in the GPU energy demand only marginally affects the overall saving. The potential increase in net efficiency is more than 2%-points; however, the CO_2 purity is decreased by 1.4%-points to 96.0 mol% at constant CO_2 capture rate (not shown in Fig. 10.12).

If the air ingress is not strictly minimized and controlled during the lifetime of the power plant, it could become more difficult to meet CO_2 capture rate and purity requirements.

10.5.3
Reintegration of Waste Heat

The predominant part of the additional auxiliary power demand of the oxyfuel technology is caused by compressors. When the design of the compressor intercoolers leads to nearly isothermal conditions, the temperature level of the intercoolers waste heat is relatively low. If this waste heat is transferred to the water-steam cycle, only a small beneficial effect can be achieved concerning overall efficiency.

An intercooling concept closer to adiabatic compression conditions increases the specific energy demand for compression on the one hand. On the other hand, reintegrating the heat into the power plant process outweighs this demand and leads to an increase in the overall efficiency. The higher temperature level of the reintegrated waste heat results in a higher exergetic level of the recovered heat. Additionally, the compressor design and piping are less complex. One option is to use multistage axial compressors as used in gas turbines. However, material limits have to be taken into account.

10.5.4
Oxygen and Condensate Preheating by ASU Waste Heat

Nearly the entire driving power of the main air compressor (MAC) in the ASU is transformed into heat. With "isothermal" compression using several intercoolers, air temperatures of ~90 °C can be achieved. With an "adiabatic" compression without intercoolers, air temperatures up to 220 °C – depending on the pressure ratio and the inlet temperature of the air – are obtainable. Considering all possible heat sinks within the power plant, oxygen preheating is the most efficient option. However, the potential for efficiency improvements is limited, as significantly more waste heat is available from the compressed air than can be transferred by preheating the oxygen.

The oxyfuel base case (cf., Fig. 10.10) includes an "isothermal" compression, moderate oxygen preheating (80 °C; the excess waste heat is rejected to the cooling water). Without any oxygen preheating, the overall process efficiency would be relatively lowered by ~0.5% (0.24%-points) (cf. Fig. 10.13). Furthermore, this

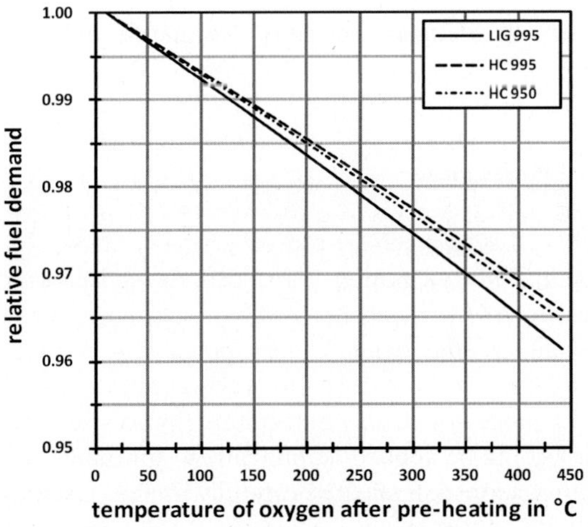

Fig. 10.13 Relative fuel demand reduction depending on the maximum oxygen temperature for different fuels (LIG: lignite; HC: hard coal) and oxygen purities (995: 99.5 mol%; 950: 95.0 mol%).

case seems to be suitable for possible retrofit cases because no thermodynamic interaction between the water–steam cycle and MAC is applied.

An optimized case with an "adiabatic" MAC without any intercoolers was investigated. Here the oxygen is heated by air to 200 °C. According to Fig. 10.11, this increase in oxygen temperature from 80 to 200 °C would lead to an efficiency increase of 1% relative (0.47%-points). The residual waste heat of the air is transferred to a condensate bypass economizer. The condensate slip stream is heated to 175 °C. Taking into account additionally the efficiency increase due to this and the efficiency decrease due to the higher power demand of the MAC, the overall net efficiency increases by 0.8 to 36.2%-points. An operational drawback of this alternative is the direct interaction between MAC/ASU and the water-steam cycle.

It should be noted that the energy savings shown in Fig. 10.11 are generated by the increase in the oxygen temperature without taking into account where the heat for this comes from and whether the use of this heat causes any decreases in the efficiency of the overall process. Therefore, this diagram represents the maximum potential energy saving and it can only be used if no energy disadvantages are caused by the oxygen preheating. As already mentioned, the use of waste heat from compressors with less intercooling cannot achieve the whole potential shown in Fig. 10.13, as the power demand of the MAC increases and thus negatively affects the efficiency of the overall power plant process. Using hot flue gas behind the economizer for oxygen preheating could utilize the whole potential shown in Fig. 10.11 but it has to be considered that preheating the oxygen with flue gas could cause major problems if leakages in the heat exchanger occur. Therefore, the marginally less effective option of feed water preheating is used for simulations (cf., Fig. 10.10).

These examples show that it is necessary to perform overall process calculations which take into account realistic boundary conditions to identify realistic savings.

10.5.5
Efficiency of an Optimized Oxyfuel Power Plant

The overall net efficiency loss is ~10.2%-points if proven cryogenic technologies for ASU and GPU are used and no integration is realized. Overall process evaluations show that a combination of multi-reboiler ASU, oxygen preheating, and MAC waste heat recovery reduces the net efficiency penalty to ~9.4%-points (Fig. 10.14). The efficiency loss can be reduced to ~8%-points by using further optimized cryogenic ASU processes (e.g., multi-column processes). An even higher level of heat recovery could be applied additionally to minimize the efficiency losses. The latter option has to be evaluated carefully as the operability or reliability can be negatively influenced.

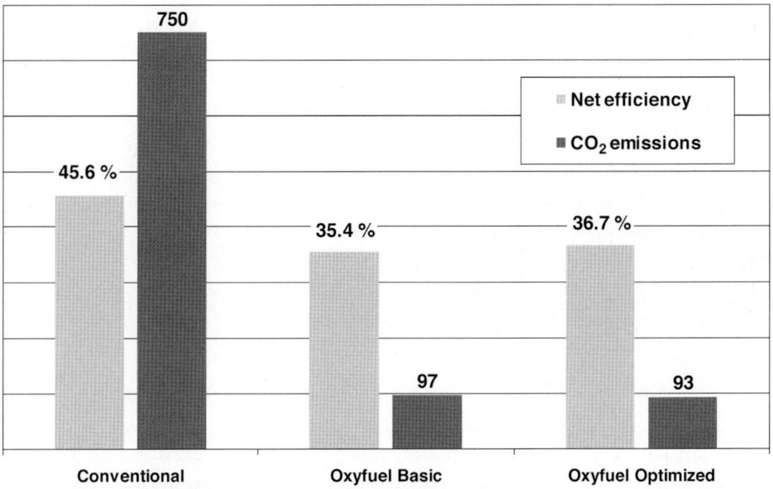

Fig. 10.14 Overview of the efficiencies and the specific CO_2 emissions (g kWh_{el}^{-1} net) of the described power plant processes.

10.6 Current Developments

The oxyfuel process is a subject of intensive research worldwide [20]. A large number of small-scale test rigs are operated by research institutes, manufacturers, and utilities [21]. Current research is focused on the corrosion behavior of different materials, for example, for steam generators, and on the removal of impurities which cause problems during compression of flue gas and condensation of CO_2 [22].

A first milestone towards the industrial deployment of the oxyfuel process was reached by Vattenfall. The 30 MW_{th} oxyfuel pilot plant was taken into operation in 2008 at the site of Schwarze Pumpe in Germany [23]. The plant has been operated successfully for several thousand hours in oxyfuel mode using pre-dried lignite and has demonstrated the feasibility of the oxyfuel process on the pilot scale. The whole process chain of an oxyfuel process is available, including a cryogenic ASU, steam generator, flue gas treatment, and CO_2 purification unit. The CO_2 capture and purification capacity is $\sim 9\,t\,h^{-1}$.

Based on the positive experiences with the oxyfuel pilot plant, Vattenfall decided to build a 250 MW_{el} demonstration plant at the site of Jänschwalde by 2015 [24]. This project is funded by the European Union as the most promising CCS demonstration project within Europe [25]. Vattenfall intends to establish a complete oxyfuel power plant with a net electric efficiency of $\sim$36%. Because of the large CO_2 stream that will be captured, the power plant will be operated in combination with a pipeline and geologic CO_2 storage.

Within the Callide A oxyfuel project, CS Energy is retrofitting a hard coal-fired power plant to oxyfuel. The gross power output is 30 MW_{el}. The captured CO_2 is planned to be stored in a geologic formation close to the site. Commissioning is planned for mid-2011 [26].

Several research projects are dealing with the so-called second generation of oxyfuel processes. These novel processes are characterized by completely different processes for oxygen production. Chemical looping combustion [27–29] and oxygen transfer membranes are promising, but are at a very early stage of development [30–32].

References

1 Hausen, H. and Linde, H. *Tieftemperaturtechnik – Erzeugung sehr tiefer Temperaturen, Gasverflüssigung und Zerlegung von Gasgemischen*, 2nd edn, Springer, Berlin, 1985.

2 Dillon, D. J., White, V., Allam, R. J., Wall, R. A., and Gibbins, J. *Oxy-Combustion Processes for CO_2 Capture from Power Plant*. IEA GHG R&D Programme, 2005.

3 Bergins, C. Design criteria for carbon capture ready power stations and technological options under development by power plant suppliers. Presented at the IZEC CCR Symposium, Tokyo, 2009.

4 FDBR. *Handbuch der Wärme und Strömungstechnik*, Vulkan Verlag, Essen, 2006.

5 Kather, A. and Scheffknecht, G. The oxycoal process with cryogenic oxygen supply. *Naturwissenschaften*, 2009, **96**(9), 993–1010.

6 Wegerich S. Experimentelle Untersuchungen zum Verbrennungsverhalten und zur NO_x-Entstehung bei der Verbrennung von Kohlenstaub in einer O_2/CO_2-Atmosphäre. Dissertation, TU Hamburg-Harburg, 2009.

7 Mieske, K. Schadstoffentstehung bei der Verbrennung im Oxyfuel Prozess. Dissertation, Hamburg University of Technology, 2009.

8 Mönckert, P., Dhungel, B., Kull, R., and Maier, J. Impact of combustion conditions on emission formation (SO_2, NO_x) and fly ash. Presented at the 3rd Workshop, IEAGHG International Oxy-Combustion Network, Yokohama, 2008.

9 Srivastava, R. K., Miller, C. A., *et al. Emissions of Sulfur Trioxide from Coal-Fired Power Plants*. Technical Report, Riley Power, Worcester, MA, 2002.

10 Kather, A., Hermsdorf, C., and Klostermann, M. Der kohlebefeuerte Oxyfuel-Prozess: Grundlagen zur Dampferzeugergestaltung und Möglichkeiten zur Verminderung der Verunreinigungen im CO_2. *VGB PowerTech*, 2007, **87**(4), 84–91.

11 Klostermann, M., Kather, A., Köpke, D., and Eggers, R. Energetische Betrachtungen zur Verflüssigung und Aufkonzentration des CO_2 aus Rauchgasen eines steinkohlegefeuerten Oxyfuel-Prozesses. *40. Kraftwerkstechnisches Kolloquium, Dresden, 2008, Tagungsband "Künftiges Brennstoff- und Technologieportfolio in der Kraftwerkstechnik"*, pp. 134–145.

12 Burchhardt, U., Lysk, S., Kosel, D., Griebe, S., Kass, H., and Preusche, R. In *Betriebserfahrungen aus zwei Jahren Testbetrieb der Oxyfuel-Forschungsanlage von Vattenfall* (eds. Beckmann, M. and Hurtado, A.), Kraftwerkstechnik – Sichere und nachhaltige Energieversorgung, 2010, TK Verlag Karl Thome-Kozmiensky, Neuruppin, Vol. 2, pp. 221–239.

12 White, V., Torrente-Murciano, L., Sturgeon, D., and Chadwick, D. Purification of oxyfuel-derived CO_2. *Energy Procedia*, 2009, **1**, 399–406.

14 Ritter, R., Holling, B., Altmann, H., and Biele, M. Konzepte und Ausblick für eine CO_2-Anlage eines Oxyfuel-Kraftwerkes am Beispiel Schwarze Pumpe. In *Verfahren und Anlagen der Hochtemperaturenergietechnik: Stand und Entwicklungs-*

perspektiven, Vol. 39. Kraftwerkstechnisches Kolloquium, TU Dresden, 2007, pp. 199–211.

15 Köpke, D., Eggers, R.: Experimentelle Untersuchungen des Phasengleichgewichts des Systems Kohlendioxid/Argon und Literaturauswertung zum System Kohlendioxid/Schwefeldioxid. *Chem.-Ing.-Tech.*, 2007, **79** (8), 1235–1239.

16 VGB PowerTech. *Konzeptstudie Referenzkraftwerk Nordrhein-Westfalen*. Report. VGB PowerTech, Essen, 2004.

17 Beysel, G. Enhanced cryogenic air separation: a proven process applied to oxyfuel – future prospects. Presented at the 1st Oxyfuel Combustion Conference, September 2009.

18 Tranier, J.-P., Perrin, N., Darde, A. ASU and CO_2 CPU for oxy-combustion. IEA GHG Oxy-Combustion Network, March 2008.

19 White, V. ASU and CO_2 processing units for oxyfuel CO_2 capture plants. Presented at the Workshop on Operating Flexibility of Power Plants with CCS, London, 2009.

20 IEA Greenhouse Gas R&D Programme, Oxyfuel Combustion Network, last accessed 28.2.2011. http://www.co2captureandstorage.info/networks/oxyfuel.htm.

21 Scheffknecht, G. Research perspective – review of the current understanding, identifying research gaps. Presented at the 1st IEA GHG Oxyfuel Combustion Conference, Cottbus, 2009.

22 Santos, S. and Yan, J. CO_2 processing unit – challenges in meeting the required CO_2 quality. Presented at the 1st IEA GHG Oxyfuel Combustion Conference, Cottbus, 2009.

23 http://www.vattenfall.de/de/die_pilotanlage.htm, Die CCS Pilotanlage, 28.2.2011.

24 Vattenfall Europe AG, http://www.vattenfall.de/de/ccs-demonstrationskraftwerk.htm. Das Demonstrationsprojekt, 28.2.2011.

25 ZEP. *EU Demonstration Progamme for CO_2 Capture and Storage*, http://www.zero-emissionplatform.eu, 2009, last accessed: 28.2.2011

26 CS Energy Limited http://www.callideoxyfuel.com/. Callide Oxyfuel Project, 28.2.2011.

27 Anthony, E.J. Solid looping cycles: a new technology for coal conversion. *Ind. Eng. Chem. Res.*, 2008, **47** (6), 1747–1754.

28 Lebas E. ENCAP SP4 – chemical looping combustion. Presented at the ENCAP–CASTOR Workshop, Billund, 16 March 2006.

29 Hongguang, J. and Masaru, I. A new type of coal gas fuelled chemical-looping combustion. *Fuel*, 2004, **83**, 2411–2417.

30 BMWi. *Leuchtturm COORETEC – Das zukunftsfähige, fossil befeuerte Kraftwerk*. Forschungsbericht Nr. 566. Projektträger Jülich, Forschungszentrum Jülich GmbH, (eds. Dr. Jochen Seier and Dr. Horst Markus), May 2007 (see also: http://bmwi.de/BMWi/Redaktion/PDF/Publikationen/leuchtturm-cooretec, property=pdf,bereich=b mwi,sprache=de,rwb=true.pdf).

31 WSA RWTH Aachen, Oxycoal AC http://www.oxycoal-ac.de/. 28.2.2011.

32 Shah, M., Christie, M., Degenstein, N., and Wilson, J. Oxy-combustion on oxygen transport membranes. Presented at the 1st Oxyfuel Combustion Conference, Cottbus 2009.

Part 4
Separation with Membranes

Process Engineering for CCS Power Plants, First Edition. Edited by Detlef Stolten and V. Scherer

11
Physics of Membrane Separation of CO_2

Matthias Wessling

Abstract

This chapter gives an overview of the physical phenomena that occur during gas separation using dense ceramic and polymeric membranes relevant for membrane processes and potentially contributing to CO_2 emission reduction. The ceramic membranes considered are perovskite membranes transporting oxygen over any other gas with infinite selectivity. The polymer membranes considered can be glassy or rubbery in nature. The chapter expands from the molecular visualization of transport towards membrane and module morphologies viable for processes potentially contributing to CO_2 emission reduction.

Keywords: gas separation, polymer membranes, ceramic membranes, diffusion, partitioning

11.1
Introduction

11.1.1
History

Gas separation using membranes has been industrial reality for more than 50 years. Initially, porous ceramic membranes were used to separate UF_6 isotopes for nuclear energy and weapon applications. The extremely low separation factors of these membranes required multiple sequential permeation and recompression processes in order to achieve the desired enrichment. This technology faded out and has now been replaced by specialized centrifugal processes.

During the last 25 years, polymer membranes have introduced into industrial applications and they represent a significant market for hydrogen recovery, nitrogen production from air, and sweetening of natural gas containing CO_2 and SO_2 [1]. Recently, ceramic membranes have gained significance again due to the urgent need for CO_2 capture technology. This section summarizes the major concepts governing the selective transport of molecules such as carbon dioxide,

Process Engineering for CCS Power Plants, First Edition. Edited by Detlef Stolten and V. Scherer

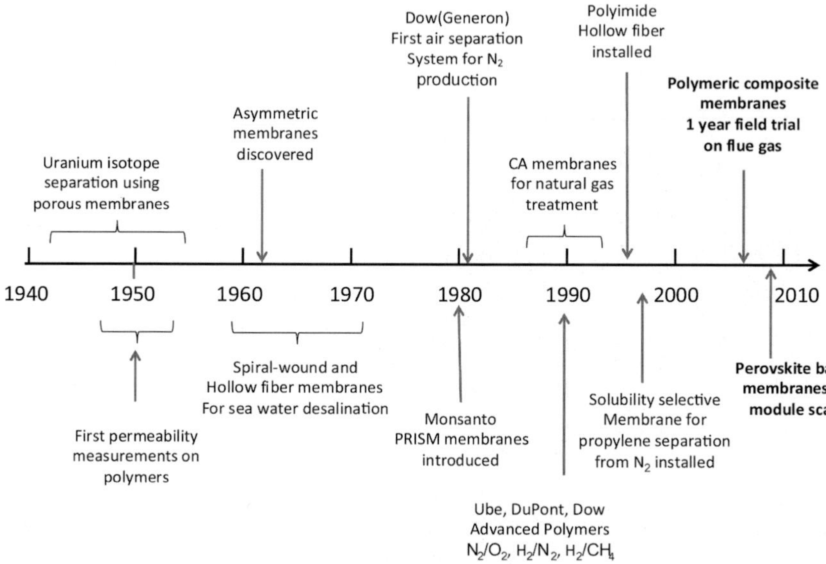

Fig. 11.1 Representation of major achievements in membrane gas separation processes. The end of the previous decade showed the emergence of technologies for CO_2 emission reduction in power plants.

water, nitrogen, and oxygen through the empty space (free volume or lattice defects) in dense polymeric and ceramic membrane materials.

Polymeric membranes have been developed by various companies over the past 30 years, and today most of them are integrated into gas production corporations such as Air Products, Praxair, and Linde. The historical milestones are illustrated in Fig. 11.1 [2]. Figure 11.1 also gives an indication of significant membrane developments in the field of carbon capture processes in flue gas and oxygen production for oxyfuel processes.

11.1.2
Gas Separation Processes for Carbon Capture

In the various schemes anticipated today to reduce CO_2 emissions, membrane separation processes are suggested for the following tasks:

- post-combustion CO_2 capture [3–5]
- hydrogen separation in water gas shift adjustment processes [6, 7]
- oxygen production for oxyfuel processes [7].

The first is most likely to be performed with polymeric membranes; the second may be accessible for both ceramic and polymeric membranes, and the last will be dominated by perovskite-based ceramic membranes. For all of these processes, field trials are currently in progress or have even been completed.

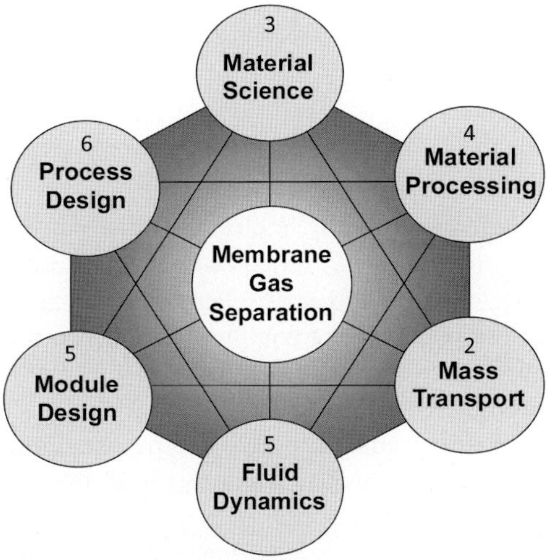

Fig. 11.2 Visualization of the integration of various scientific and design disciplines in order to develop viable membrane gas separation processes. The numbers indicate the sections of this chapter covering the specific topic.

11.1.3
Disciplines

Successful developments in membrane gas separation applications require the integration of various scientific and design disciplines. These disciplines are shown in Fig. 11.2 and range from (a) *materials science* and the molecular understanding and the molecular tailoring and design of materials, (b) cost-effective *materials processing* into large quantities of membrane materials, (c) the microscopic and macroscopic description of *mass transport* through membranes and adjacent fluid layers, (d) the characterization of feed and permeate side *fluid dynamics* conditions, (e) the design of a *membrane module* complying with optimized hydrodynamic conditions, and (f) *process design* with the integration of such modules into larger processes (pumps, heat exchangers, valves, piping) and eventually with other separation processes. Companies today integrate these disciplines well, whereas academic research often focuses on only one or two of these disciplines.

The following sections touch upon five of the six disciplines. Only process design is not explicitly treated. All the others are addressed for both dense polymer- and ceramic-based membrane materials and membrane structures.

11.2
Macroscopic Mass Transport

11.2.1
Permeation

In order to allow mass transport through a membrane, molecules must permeate through the empty space inside the membrane, as visualized in Fig. 11.3 [8]. For now, we do not consider the nature of the empty space. Permeation is generally considered to comprise first the establishment of equilibrium between the gas side boundary layer and the empty space in the surface layer of the membrane and subsequently diffusion through the empty space. Whereas the equilibrium is governed by thermodynamics, diffusion is a purely kinetic phenomenon. It is important to realize that the molecule that was previously in the gas phase is now a dissolved molecule without knowing that it ever was in a gas phase. For polymer materials, the transport model is called the "solution–diffusion model" (SDM) [9]. For dense ceramic materials, such as perovskites, this overall mechanism also holds, but the physical phenomena and their mathematical description are much more intricate.

The SDM assumes that the rate of mass transport is proportional to the concentration of the dissolved molecules and their molecular mobility (diffusivity) inside the empty space. The rate of mass transport of a component i per unit membrane area is quantified by its molar flux J_i (mole m^{-2} s^{-1}), which is a function of the driving force and the resistance R, which the membrane represents for the transport process: flux = driving force/resistance.

Tthe resistance of a membrane system may not be restricted to the resistance of the active layer of the membrane, but may include also contributions of the membrane support structure, of the boundary layer, and of the pressure drop in the feed and permeate channels. Some of these are treated in Section 11.5. If all other resistances are neglected, the remaining resistance of the active layer depends on its thickness and its material properties and their interaction with the permeating molecules. The use of the chemical potential μ as driving force

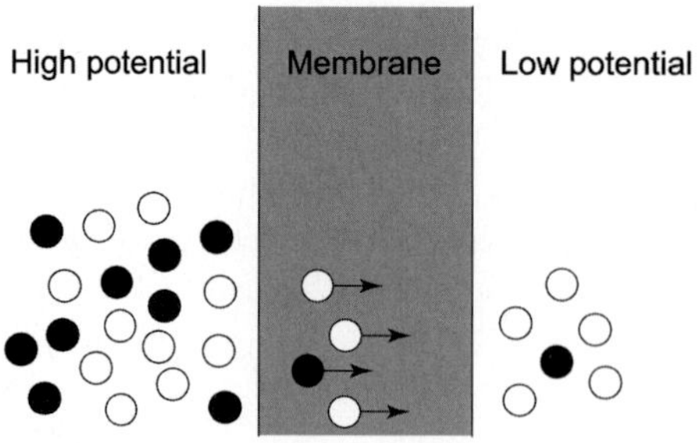

Fig. 11.3 Transport of gases from a feed phase with a high chemical potential (concentration) to a permeate phase with a low chemical potential (concentration).

avoids driving force discontinuities at phase boundaries. Inside the membrane, the local driving force then is grad μ, whereas the total driving force across the membrane is $\Delta\mu = \mu_F - \mu_P$, where the subscripts F and P represent feed and permeate, respectively. For ideal gases, $\Delta\mu_i$ is equal to the difference in the partial pressures of component i on the feed and the permeate sides.

Frequently, the reciprocal value $Q = 1/R$, which is called "permeance," is used for membrane characterization and Q is then further broken down as the ratio of the permeability P, which is an intrinsic material property, to the thickness μ of the active layer The resulting transport equation is then

$$J_i = \Delta\mu_i \frac{P_i}{\delta} \tag{11.1}$$

This equation holds for each of the gases permeating through the membrane.

Whereas the flux represents a quantitative productivity parameter, the membrane selectivity S_{ij} characterizes the quality of the separation achieved. Provided that the permeating molecules do not change the membrane structure (see Chapter 13), which also implies that two simultaneously permeating gases do not influence each other's permeation, the selectivity can simply be calculated as the ratio of permeabilities P_i and P_j for components i and j. The resulting term P_i/P_j is called "permselectivity."

A membrane housed in a module is visualized as shown in Fig. 11.4. The module splits the feed gas into a retentate and a permeate flow stream. In cases where a sweep gas is used to carry away the permeated species, the membrane module has four ports. Macroscopically, the mass balances must be fulfilled

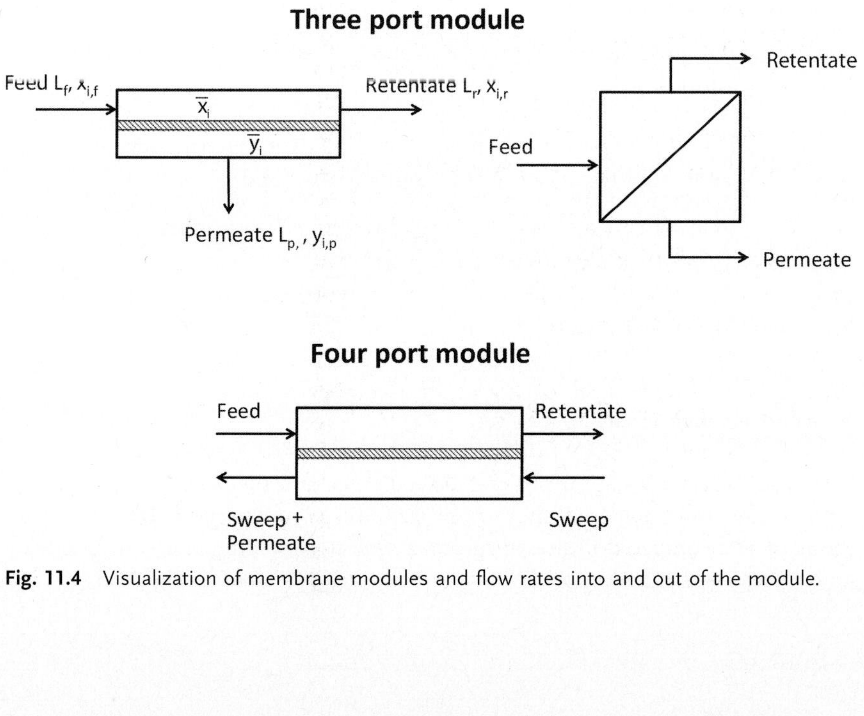

Fig. 11.4 Visualization of membrane modules and flow rates into and out of the module.

around the module for the overall flow rates [in units of volume at standard temperature and pressure (STP) per unit time] for the feed L_f, the retentate L_r, and the permeate L_p:

$$L_f = L_r + L_p \tag{11.2}$$

and also for each of the components i:

$$x_{i,f} L_f = x_{i,r} L_r + y_{i,p} L_p \tag{11.3}$$

where $x_{i,f}$ and $x_{i,r}$ are the mole fractions of the gas mixture components i in the feed and retentate and $y_{i,r}$ is the mole fraction in the permeate.

The compositions in the permeate $y_{i,r}$ and retentate $x_{i,r}$ are a strong function of a parameter stage cut Θ, which quantifies the split of the feed flow rate into the permeate and retentate flow rates:

$$\Theta = \frac{L_p}{L_f} \tag{11.4}$$

The separation performance of a module therefore depends on a variety of parameters related to the membrane properties and the process parameters.

For a given membrane module, at a given set of parameters, one can also quantify the separation performance by a separation factor α. This can be defined in different ways, a common one being based on the molar fractions in the feed and permeate:

$$\alpha_{ij} \equiv \frac{y_i/y_j}{x_i/x_j} \quad \text{and} \quad \alpha_i = \frac{y_i/(1-y_i)}{x_i/1-x_i} \tag{11.5}$$

Since the local mole fractions are a function of the stage cut, the separation factor depends on the process conditions. Only in the special case of negligible removal of a component from the feed, zero stage cut, and vacuum at the permeate side, is the module-related separation factor equal to the permselectivity, that is, the ratio of the permeabilities of the components i and j.

11.3 Permeation Through Materials

11.3.1 Transport Through Empty Space

Technical membranes are made of solid materials and their porosity can vary in amount and morphology such as pore size and pore structure. Depending on the pore size compared to the permeating molecules, membranes can permeate and separate gas mixtures to different extents, as illustrated in Fig. 11.5.

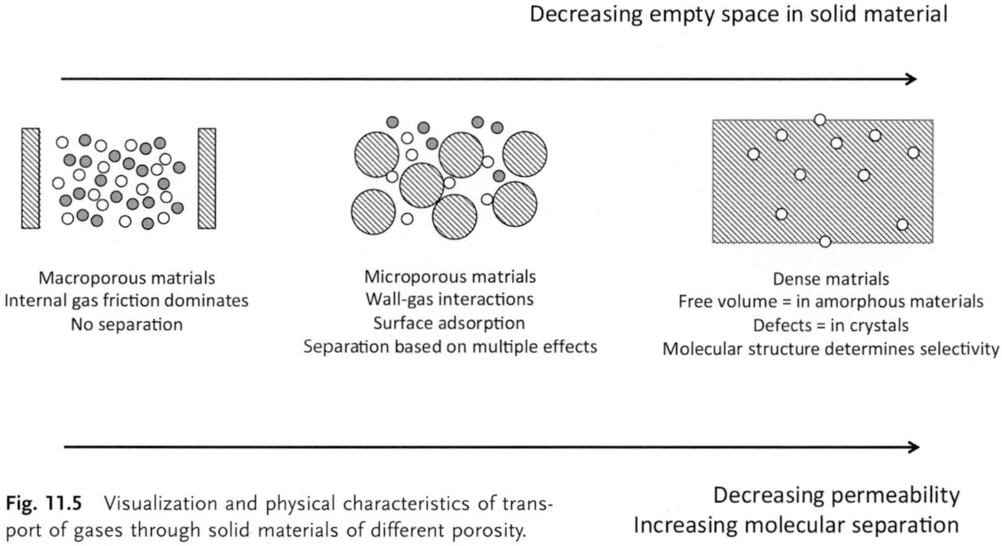

Fig. 11.5 Visualization and physical characteristics of transport of gases through solid materials of different porosity.

Macroporous materials do not separate gas mixtures. Significantly smaller pores can separate gas mixtures when the free path length of the diffusing molecule approaches the size of the pores. Then the collisions with the wall contribute to the dynamics of the diffusing gas mixture and separation occurs based on the ratio of the square root of the molar mass of the gases to be separated. This mechanism is called Knudsen diffusion. At smaller pore sizes, surface adsorption can also contribute to the transport with a significant lateral mobility of adsorbed gases along the internal solid membrane surface.

When the size of the pores approaches the order of 1 nm, the term pore is no longer appropriate. In polymeric membranes, the empty space is called the free volume. In crystalline materials, such as perovskites, oxygen is transported via lattice defects. It is important to realize that gas transport in the free volume of polymers or lattice defects of crystalline materials occurs through thermally caused stochastic cooperativity of the matrix: the dissolved molecule/atom can only move to a new position if a new vacant site opens in a crystalline or amorphous surrounding. This surrounding matrix needs to have many vacant sites with a sufficiently high thermal mobility in order to allow hopping of dissolved guest gas molecules at high rates.

11.3.2
Transport in Polymer Materials

The most widely known mass transport equation is Fick's first law:

$$J = -D\frac{\mathrm{d}c}{\mathrm{d}x} \tag{11.6}$$

The molar flux is proportional to its driving force, the concentration gradient with the proportionality factor being the Fickian diffusion coefficient. Strictly, Fick's law is only valid for dilute solutions. For concentrated solutions, the coefficient of diffusion becomes strongly dependent on the concentration.

If sorption equilibrium data are available, Fick's law can be used to quantify the permeability P, as shown in the case of a linear solubility function provided by Henry's law:

$$c_i = S_i p_i \tag{11.7}$$

where S_i is the solubility.

Substituting Henry's law into Fick's law and integrating it over the membrane thickness l with feed partial pressure $p_{i,\mathrm{f}}$ and permeate partial pressure $p_{i,\mathrm{p}}$ as boundary conditions gives

$$J_i = \frac{D_i S_i (p_{i,\mathrm{f}} - p_{i,\mathrm{p}})}{l} \tag{11.8}$$

where P_i defines the permeability of component i in a material under as the product of diffusivity D_i and solubility S_i:

$$P_i = D_i S_i \tag{11.9}$$

Displacement of a dissolved molecule inside a solvent occurs due to thermal motion of both the dissolved molecule and the solvent molecules. It is a purely stochastic process in which incidental cooperative motion of the solvent molecules generates empty space for the dissolved molecule to jump into (Fig. 11.6). Overall, a net flux of the dissolved molecules in the direction of lower concentration is observed. Locally, however, the dissolved molecule can move unrestricted in any of the three dimensions. The term "driving force" can be misleading: there is no force acting on the dissolved molecule. There is only a concentration difference, and in time statistically averaged over many diffusional jumps, dissolved molecules overall move by pure chance towards the lower concentration. This molecular picture of transport holds well for diffusion of dissolved gasses in a matrix of polymer chains.

Diffusional displacement in a polymer is, other than in a solvent, of a smaller extent since it requires the coordination of a large number of monomers in the

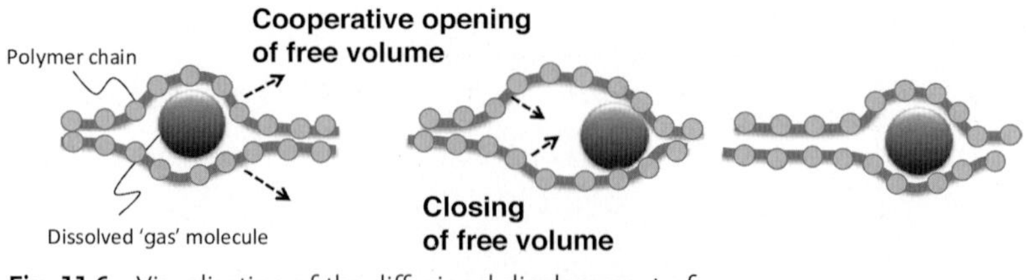

Fig. 11.6 Visualization of the diffusional displacement of a dissolved penetrant inside a polymer matrix through stochastic cooperative motion of the polymer chains.

polymer chain. The chain flexibility – often represented by the glass transition temperature of the polymer – governs the displacement jump of the penetrant molecule entrapped in a free volume pocket. Diffusional jumps in a stiff, glassy polymer therefore require much more energy than a jump in a rubbery polymer.

The mechanism of diffusional displacement has been investigated extensively through molecular dynamics simulations [10, 11] and today the influence of chemical features of the polymer and the penetrant molecule can be understood in detail.

The relationship between the gas-phase conditions and the concentration inside the membrane is often more complicated than described by Henry's law alone. Deviations from Henry's law are shown in Fig. 11.7. For highly swelling vapors in rubbery polymers, the sorption isotherm turns convex towards much higher concentrations [12]. For inert gases, and in some cases also for vapors, in glassy polymers the isotherm is concave. This isotherm shape has its origin in the non-equilibrium, frozen-in free volume of a glassy polymer. It can often be described by a dual-mode sorption approach where a Henry-type sorption adds on to a Langmuir-type sorption isotherm [13]. However, much more complex and realistic descriptions have also been developed over the decades [14].

Such nonlinear isotherms strongly influence the sorption behavior of both single gases and gas mixtures. However, the diffusional behavior is also strongly affected. A polymer that contains a large amount of penetrant has a much higher molecular mobility, and hence the diffusion coefficient of a penetrant will also be higher. This in fact means that a membrane swollen at the feed side shows a higher diffusion coefficient as compared with the much less swollen

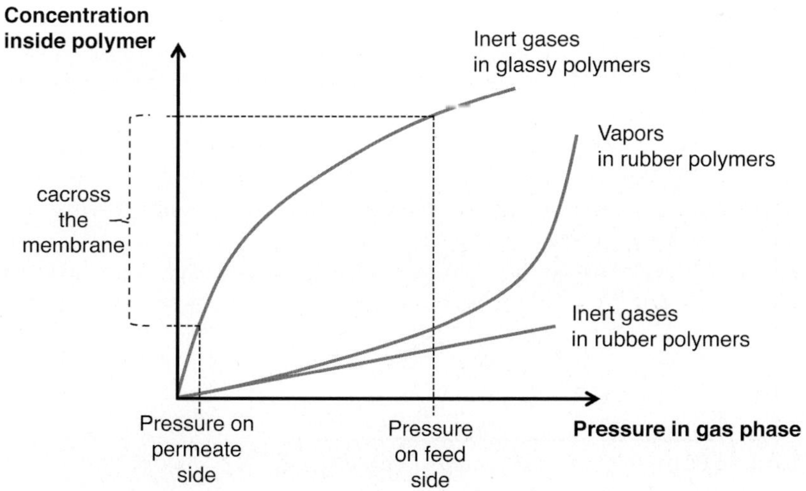

Fig. 11.7 Different sorption isotherms relating the gas-phase pressure to the concentration inside the membrane. For a glassy polymer, the concentration difference across a membrane is shown for a high feed pressure and a low permeate pressure.

permeate side. Owing to this concentration dependence, the local mobility of the penetrant molecule varies throughout the thickness of the membrane.

A comprehensive collection of all physical phenomena as a function of material chemistry is available [15]. Together with Crank and Park's book on diffusion in polymers [16], the reader has the most solid material at hand to understand the complex nature of permeation, diffusion, and solubility in polymers.

In the last decade, membrane gas separation has been dominated by polymer-based gas separation membranes. However, not many new commercial products have entered real applications, even though the number of newly synthesized polymers has increased continuously in a synthetic competition to "beat" a selectivity/permeability trade-off as shown in Fig. 11.8, also known as the Robeson plot [17]. This plot can be drawn for different gas pairs and is continuously expanding with new polymers shifting the upper bound to higher values. For gas pairs comprising non-condensable gases, there is a trade-off: highly permeable materials show lower selectivities. For condensable gases, one often observes both increasing permeability and selectivity, in particular for water as compared with N_2 or CO_2 [18].

In the field of dense ceramic membranes, synthetic activity is also increasing considerably; however, a Robeson plot type of summary does not yet exist. For instance, perovskite materials used for oxygen production generally show infinite selectivity, hence they cannot be benchmarked in a trade-off graph. However, a trade-off benchmarking the mechanical stability of a perovskite versus its permeability will most likely also give a distinct pattern, as high permeability/conductivity goes along with high dilution causing low mechanical strength.

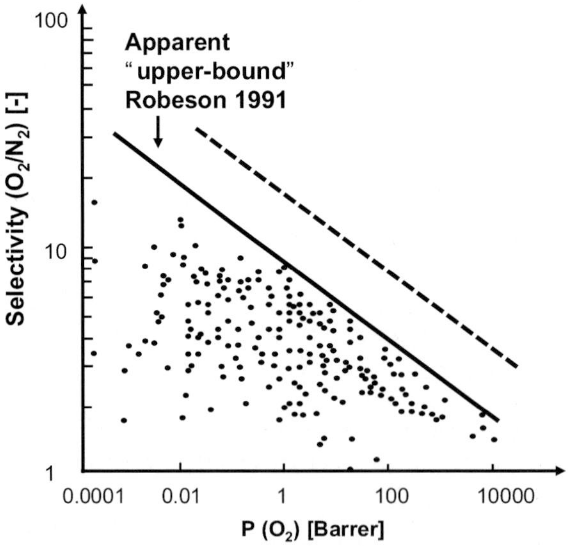

Fig. 11.8 Graphs for different polymers, represented by the data points, show a selectivity/permeability trade-off with an upper bound beyond which no materials have so far been identified.

11.3.3
Transport in Dense Ceramic Materials

Perovskite materials can be used to separate oxygen from gas mixtures. Perovskites are crystalline materials of the type ABO_3, where A represents large ions such as lanthanides, actinides, alkali and alkaline earth elements, Pb, and Bi, and B represents transition metal elements, Li, Mg, Ca, Al, Si, Ga, Ge, In, Sn, Sb, Pb, and Bi [19]. A large variety of different perovskites exist with this general molecular structure ABO_3. Not only can A and B be varied independently, but also both A and B may comprise different chemical elements. For instance, the widely applied perovskite $Ba_{1-x}Sr_xCo_{1-y}Fe_yO_{3-\delta}$ (BSCF) has $(1-x)$Ba and xSr as A and $(1-y)$Co and yFe as B. The oxygen deficiency in the lattice is indicated by $3-\delta$, representing the deviation from the stoichiometry of the oxygen sub-lattice (3). This non-stoichiometry depends strongly on the partial pressure of oxygen in the gas phase in contact with the material. Different materials can have different defect characteristics. It is known that $3-\delta$, for BSCF may have values as low as 2.2 at a low oxygen partial pressures of $p=10^{-3}$ atm and $T=1000$ K [20], indicating that in any of the octahedra two out of six oxygen atoms are missing. The transport of oxygen through the crystal lattice occurs via displacement of defects along the edges of the octahedra as illustrated in Fig. 11.9. Extensive modeling has been applied to understand the intricate transport phenomena [21].

The visualization of transport of oxygen through perovskite-based membranes (Fig. 11.10) is very similar to that of transport through polymeric membranes, with the difference that displacement of an O^{2-} must be compensated by a charge transport in the opposite direction. Hence suitable membrane materials are required to exhibit both electronic conductivity and ionic conductivity.

The simplest description of oxygen transport through such materials is via the Wagner equation:

$$j_{O_2}(\text{bulk})L = -\frac{RT}{4^2F^2}\int_{\ln p'_{O_2}}^{\ln p''_{O_2}} \frac{\sigma_{el}\sigma_{ion}}{\sigma_{el}+\sigma_{ion}} \mathrm{d}\ln p_{O_2} \qquad (11.10)$$

where R and T have their usual meanings, F is the Faraday constant, σ_{el} is the electronic conductivity, and σ_{ion} is the ionic conductivity. This equation is valid when the major resistance to the oxygen transport lies in the bulk diffusion.

Permeation occurs at extremely high temperatures in order to have sufficiently defect mobility. The partial oxygen pressure can vary significantly: while academic work is frequently carried out at atmospheric feed pressures, the permeate pressure is often very low, even below any technically feasible vacuum. For a real application, it is therefore important to know how the deficiency value μ depends on the oxygen partial pressure.

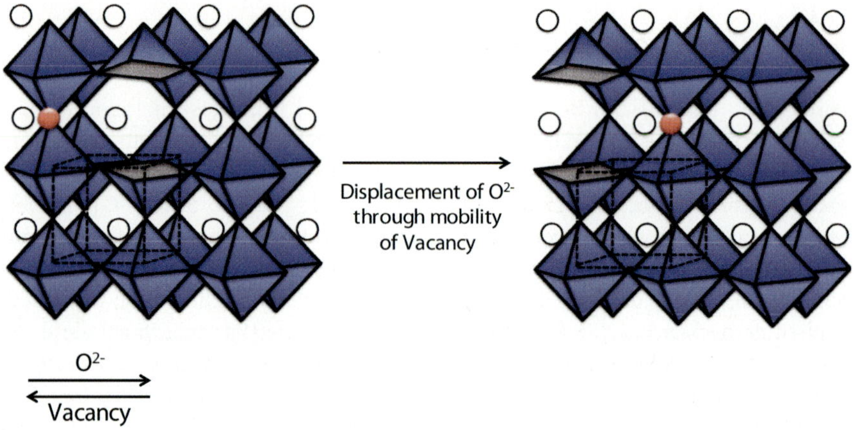

Fig. 11.9 Visualization of defect displacement with net motion of O^{2-} in the left direction. This displacement may be realized through at least two motions on the lattice along the proximity of lattice sites.

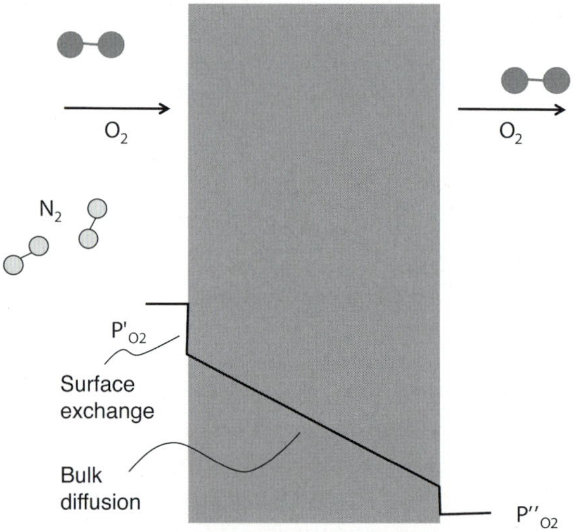

Fig. 11.10 Visualization of the infinitely selective transport of oxygen through a perovskite with an surface exchange reaction where O_2 has to take up four electrons to split up into two O^{2-} ions in order to partition into the crystalline lattice with subsequent diffusion through the bulk of the crystalline membrane.

11.4
Membrane Geometries and Morphologies

11.4.1
Membrane Geometries

Membranes are either shaped planar or tubular, as illustrated in Fig. 11.11. These geometries are packaged and arranged in membrane modules. The dimensions given for the different geometries are rough estimates. Flat sheet or planar membranes are used in so-called plate-and-frame and spiral-wound modules. Whether one uses hollow fiber, capillary, or tubular membranes depends very much on the application. Perovskite membranes, for instance, are difficult to prepare defect free as a hollow fiber. Hence one rather fabricates thick-walled, defect-free tubular perovskite membranes where the wall comprises multiple perovskite crystal grains which are well fused at their boundary. Hollow fiber membranes can be used easily when deposition is not an issue during operation; then the small diameter allows high packing densities of membrane area per module unit volume.

11.4.2
Membrane Morphologies

Polymer membranes are produced today as
- integrally skinned membranes comprising only one material with an active gas separating skin thickness of <100 nm (Fig. 11.12 a)
- composite membranes with a support membrane and a second material with a dense layer loss than 1 μ (Fig. 11.12 b).

The production process for integrally skinned membranes is industrially well established and highly automated, resulting in highly reproducible membrane

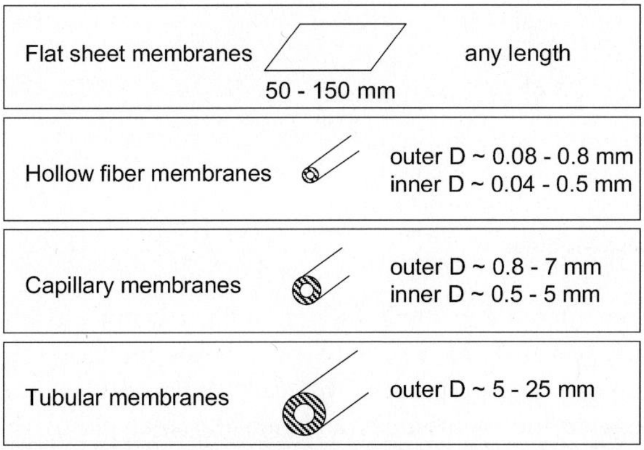

Fig. 11.11 Different membrane geometries and their typical dimensions.

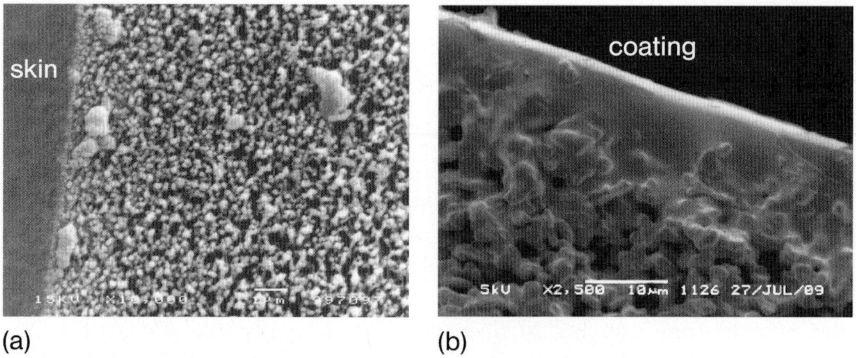

Fig. 11.12 Scanning electron micrographs of (a) an integrally skinned membrane prepared from polysulfone with a skin integrated with its support structure and (b) a composite membrane comprising a ceramic support and a polymeric PDMS coating on top of it.

qualities. The morphology of composite membranes is well known for seawater desalination in a flat sheet geometry; however, hollow fiber composite membranes are not yet available as commercial products.

For the perovskite membranes, flat or tubular geometries are currently used in various research settings. Asymmetric hollow fiber membranes have been reported, but non-selective transport may occur along grain boundaries that completely bridge the skin of the asymmetric perovskite membrane.

For both membrane types, it is in fact not important to have maximum permeability and maximum selectivity as often suggested by performance benchmarking plots such as the Robeson plot. Instead, the road to a viable technical membrane often depends on balancing properties such as material processability, longevity, and mechanical strength, especially under start-up and shut-down cycles. Information on such challenging issues is not readily available; the patent literature may give clues to the real practical challenges for specific applications.

11.5 Fluid Dynamics and Modules

11.5.1 Flow Distribution in Membrane Modules

Membranes are arranged into membrane modules. Generally, module performance – especially the overall selectivity of the module – is below the local values of the respective single membrane properties. This can have its origin in a lack of integrity of seals between the membranes and housing after gluing or potting. Once these challenges have been resolved, often through tedious trial-

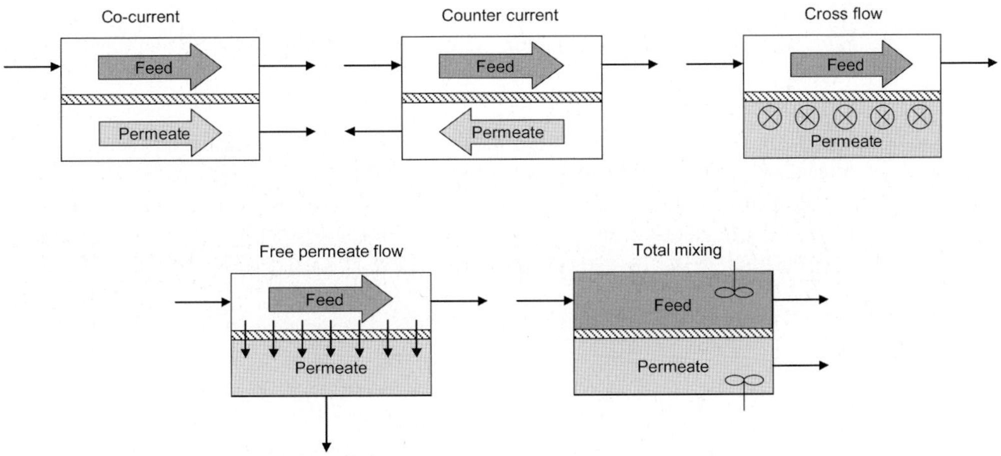

Fig. 11.13 Graphical representations of flow patterns in different modules showing co-current, counter-current, and cross-flow patterns and also free permeate flow and total mixing conditions.

and-error product development activities, one still observes worse performance of the membrane module as compared with the membrane properties measured under ideal laboratory conditions. This may be related to the flow patterns inside the membrane module.

Depending on the geometry of the membranes, different module constructions can be distinguished. A variety of the flow patterns are shown in Fig. 11.13. The details of the module construction affect the gas flow and its distribution significantly. The flow patterns inside the module determine the permeate and retentate concentrations leaving the module. In order to correlate, understand, and predict the average exit concentrations leaving the module, one needs simplified model representations of the flow distribution. Groundbreaking work on such module configurations and their effects on the separation performance of modules was described by Hwang and Kammermeyer [22]. Any further deviation between predicted performance and observed performance of a module may be related to mal-distribution of gas flow on the feed or the permeate side of a hollow fiber bundle, for instance [22]. A pressure drop in the connectors and flow channels of the module may create a significant loss of driving force, particularly on the permeate side in vacuum operation.

11.5.2
Concentration Polarization

Under laboratory conditions, one frequently succeeds in correctly measuring membrane properties. However, in a membrane module, in particular when gas mixtures are used, the flux of gases may be smaller than expected from single gas measurements. This can generally be observed with membranes having a high flux and separation factor. The phenomenon causing this deviation is called concentration polarization (Fig. 11.14).

The origin of concentration polarization is stagnant boundary layers with laminar flow conditions at the membrane surface. Owing to the rapid transport of one of the species, depletion in the laminar boundary layer occurs [23]. Hence, in addition to the membrane resistance, there evolves an additional resistance in the laminar boundary layer adjacent to the membrane surface. The resistance R is defined as the inverse of the mass transfer coefficient k:

$$R = \frac{1}{k} \tag{11.11}$$

In analogy with electrical circuits, membrane resistances ($1/k_m$) including potential surface resistances and those of adjacent boundary layers at the feed ($1/k_f$) and permeate ($1/k_p$) sides add up to the overall total resistance ($1/k_{ov}$). In fact, this overall resistance is the one that is measured in mixed gas and module experiments:

$$\frac{1}{k_{ov}} = \frac{1}{k_f} + \frac{1}{k_m} + \frac{1}{k_p} \tag{11.12}$$

Generally, in gas separation, concentration polarization is important for the feed side [18]. However, the resistances of the permeate side can become important when the permeate transport is obstructed [4]. This occurs particularly

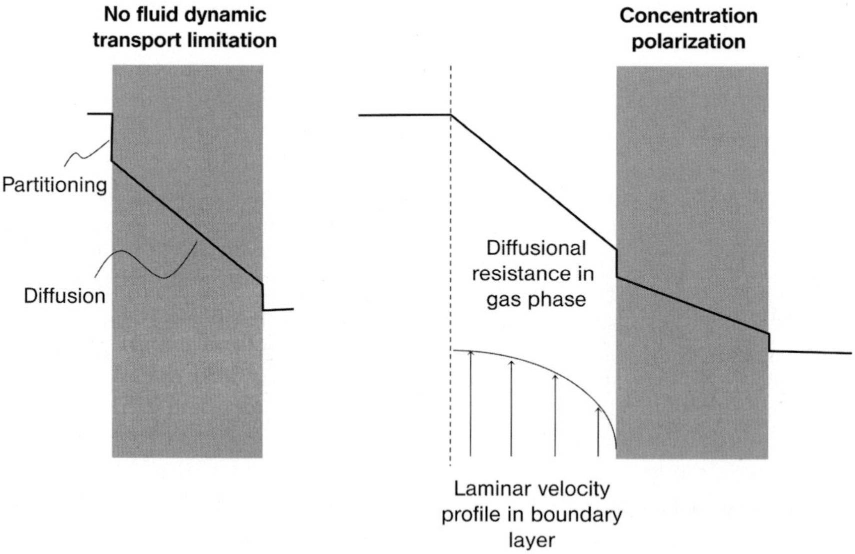

Fig. 11.14 Graphical representation of concentration profiles at dense gas separation membranes without and with transport limitations in the feed gas phase (concentration polarization).

when using a sweep gas or in counter-current operation. The resistances on the feed and permeate sides can be quantified experimentally for a give geometry by measuring the overall resistance for different membrane thicknesses [24], or the boundary layer resistances can be estimated through Sherwood relationships.

Whereas concentration polarization has frequently been investigated for liquid-phase flow along membranes, few data have been reported for gas separation processes. Often it is assumed that diffusion in the gas phase is much more rapid than that in dense membranes. When very high selectivities are observed, although even at moderate flux, concentration polarization can influence overall transport rates significantly. Quantitative relationships for the estimation of mass transport resistances/coefficients as a function of the process parameters for gases and vapors are needed.

11.6 Process Design

11.6.1 Selectivity and Driving Forces

Researchers motivated by trade-off curves such as the Robeson plot often aim for high permselectivity and high permeability. The advantage of the former requires certain process conditions to become effective. Equation 11.14 [25] shows that the gas purity obtained in the permeate is a function not only of the selectivity of the material, that is, the ratio of the permeabilities of the components i and j, but also of the pressure ratio ϕ between the feed and permeate and of the target gas volume fraction x_i on the feed side:

$$y_i = \frac{\phi}{2}\left[x_i + \frac{1}{\phi} + \frac{1}{a-1} - \sqrt{\left(x_i \frac{1}{\phi} + \frac{1}{a-1}\right)^2 - \frac{4ax_i}{\phi(a-1)}}\right] \tag{11.13}$$

This equation holds provided that the permeabilities are independent of the concentrations, and it is valid locally only. In a module, a reasonable approximation is to substitute the local concentration by an average one.

Figure 11.15 shows results of Equation 11.13 for three different permselectivities and two different feed gas volume fractions.

It appears that a certain minimum pressure ratio is needed for the selectivity to have a significant effect on the target gas volume fraction on the permeate side. This minimum depends on the order of magnitude of the target gas volume fraction on the feed side: if the target gas is a tracer as in Fig. 11.15 a, the minimum pressure ratio is of the same order of magnitude as the selectivity. If the target gas is a major component of the feed gas stream, as in air separation (Fig. 11.15 b) or in post-combustion CO_2 separation, high selectivities are effective at significantly lower pressure ratios.

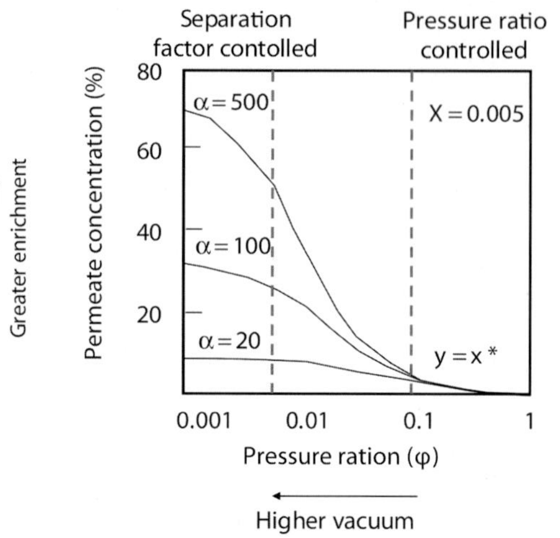

Fig. 11.15 Calculated permeate volume fractions y of a gas mixture for different feed volume fractions x as a function of pressure ratio ϕ and permselectivity α.

The effect of pressure-ratio is important and will be treated separately (see Chapter 14).

11.6.2
Membrane Process Configurations

Figure 11.16 shows the simplest process design of a membrane separation, a single-stage membrane process. However, it is often insufficient to achieve the required product purities. Hence one expands the single-stage process into a two-stage process, for instance. Generally, this requires an undesired recompression step as shown in Fig. 11.16 with an inter-stage compressor. Currently, many such process configurations are evaluated technically and economically in order to estimate their energy efficiency and to quantify the loss of power plant efficiency if such processes are incorporated into today's power plants. The details of the potential design options are treated comprehensively elsewhere in Chapter 14.

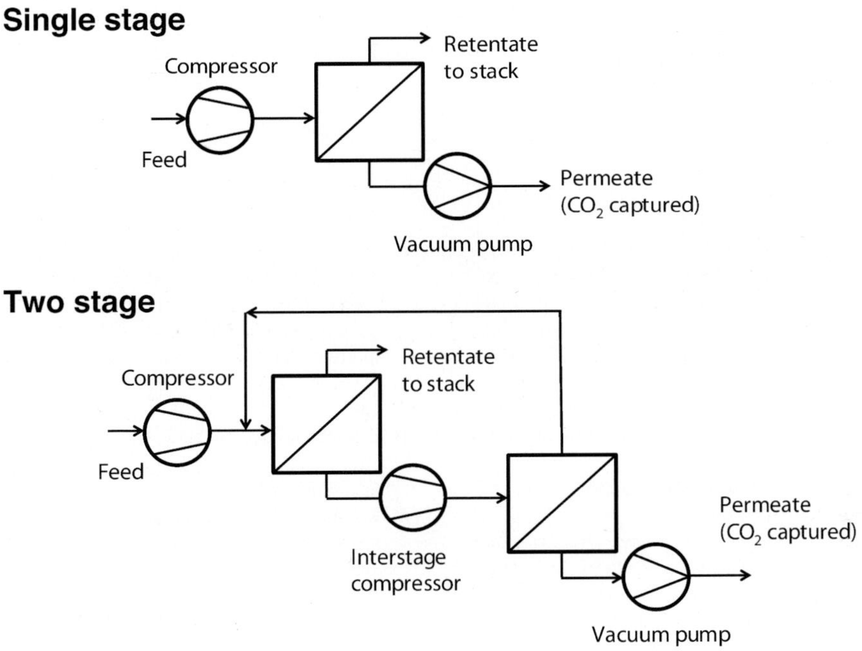

Fig. 11.16 Process flow diagrams of single-stage and two-stage processes for capturing CO_2 from flue gas.

11.7 Conclusion

This chapter summarizes the physical principles of membrane separation. It provides a general introduction for the other chapters dealing with membrane separation to build upon.

By the end of 2010, there were various players at the stage of piloting such technologies:

- MTR Inc. (Menlo Park, CA, USA) operates a post-combustion CO_2 capture pilot process, and also a process for hydrogen adjustment for water gas shift gases. In Europe, many players are active.
- KEMA, in The Netherlands, has demonstrated 1 year of operation of polymer membranes in a flue gas stack to recover water with membranes and modules developed at the University of Twente. KEMA coordinates a large EU consortium on CO_2 capture (NanoGlowa).
- Norway has significant polymer and ceramic research activities at NTNU Trondhem and Scintef.
- Germany has large consortia, with the Forschungszentrum Jülich, Fraunhofer Institut (IKTS) and RWTH Aachen playing central roles in coordination and piloting perovskite-based technologies.

The present decade will show which of today's approaches and products will prevail in reducing CO_2 emissions from power plants to acceptable levels with minimum loss of power plant efficiency.

References

1 Baker, R.W. Future directions of membrane gas separation technology. *Ind. Eng. Chem. Res.*, 2002, **41**(6), 1393–1411.

2 Baker R.W. *Membrane Technology and Applications*, 2nd edn, John Wiley & Sons, Inc., New York, 2004.

3 Merkel, T.C., Lin, H., Wei, X., and Baker, R.W. Power plant post-combustion carbon dioxide capture: an opportunity for membranes. *J. Membr. Sci.*, 2010, **359**(1–2), 126–139.

4 Sijbesma, H., Nymeijer, D.C., van Marwijk, R., Heijboer, R., Potreck, J., and Wessling, M. Flue gas dehydration using polymer membranes. *J. Membr. Sci.*, 2008, **313**(1–2), 263–276.

5 Bounaceur, R., Lape, N., Roizard, D., Vallieres, C., and Favre, E. Carbon dioxide recovery from post-combustion processes: can gas permeation membranes compete with absorption? *J. Membr. Sci.*, 2007, **294**(1–2), 50–59.

6 Criscuoli, A., Basile, A., Drioli, E., and Loiacono, O. An economic feasibility study for water gas shift membrane reactor. *J. Membr. Sci.*, 2001, **181**(1), 21–27.

7 Intergovernmental Panel on Climate Change. *IPCC Special Report. Carbon Dioxide Capture and Storage*, Cambridge University Press, Cambridge, 2005.

8 Mulder, M.H.V. *Basic Principles of Membrane Technology*, 2nd edn, Springer, Berlin, 2007.

9 Wijmans, J.G. and Baker, R.W. The solution-diffusion model. *J. Membr. Sci.*, 1995, **107**(1–2), 1–21.

10 Müller-Plathe, F. Permeation of polymers – a computational approach. *Acta Polym.*, 1994, **45**(4), 259–293.

11 van der Vegt, N.F.A., Briels, W.J., Wessling, M., and Strathmann, H. The sorption induced glass transition in amorphous glassy polymers. *J. Chem. Phys.*, 1999, **110**(22), 11061–11069.

12 Metz, S.J., van der Vegt, N.F.A, Mulder, M.H.V., and Wessling, M. Thermodynamics of water vapor sorption in poly(ethylene oxide)–poly(butylene terephthalate) block copolymers. *J. Chem. Phys. B*, 2003, **107**(49), 13629–13635.

13 Erb, A.J. and Paul, D.R. Gas sorption and transport in polysulfone. *J. Membr. Sci.*, 1981, **8**(1), 11–22.

14 Kirchheim, R. Sorption and partial molar volume of small molecules in glassy polymers. *Macromolecules*, 1992, **25**(25), 6952–6960.

15 Yampolskii, Y., Pinnau, I.., and Freeman, B. *Materials Science of Membranes for Gas and Vapor Separation*, John Wiley & Sons, Inc., New York, 2006.

16 Crank, J. and Park, G.S. *Diffusion in Polymers*, Academic Press, New York, 1968.

17 Robeson, L.M. The upper bound revisited. *J. Membr. Sci.*, 2008, **320**(1–2), 390–400.

18 Metz, S.J., van de Ven, W.J.C., Potreck, J., Mulder, M.H.V., and Wessling, M. Transport of water vapor and inert gas mixtures through highly selective and highly permeable polymer membranes, *J. Membr. Sci.*, 2005, **251**(1–2), 29–41.

19 Tejuca, L.G. *Properties and Applications of Perovskite-Type Oxides*, Marcel Dekker, New York, 1993.

20 Zhou, W., Ran, R., and Shao, Z. Progress in understanding and development of $Ba_{0.5}Sr_{0.5}Co_{0.8}Fe_{0.2}O_3$-based cathodes for intermediate-temperature solid-oxide fuel cells: a review. *J. Power Sources*, 2009, **192**(2), 231–246.

21 van Hassel, B.A., Kawada, T., Sakai, N., Yokokawa, H., Dokiya, M., and Bouwmeester, H.J.M. Oxygen permeation modelling of perovskites. *Solid State Ionics*, 1993, **66**(3–4), 295–305.

22 Hwang, S.T. and Kammermeyer, K. *Membranes in Separation. Techniques of Chemistry*, Vol. VII, Wiley-Interscience, New York, 1975.

23 Lemanski, J. and Lipscomb, G.G. Effect of shell-side flows on the performance of hollow-fiber gas separation modules. *J. Membr. Sci.*, 2001, **195**, 215–228.

24 Baker, R.W., Wijmans, J.G., Athayde, A.L., Daniels, R., Ly, J.H., and Le, M. The effect of concentration polarization on the separation of volatile organic compounds from water by pervaporation. *J. Membr. Sci.*, 1997, **137**(1–2), 159–172.

25 Melin, T. and Rautenbach, R. *Membranverfahren, Grundlagen der Modul- und Anlagenauslegung*, 3rd edn, 2007, pp. 473–475.

Materials for Membrane Separation of CO_2

12
Inorganic Membranes for CO_2 Separation

Wilhelm A. Meulenberg, Ingolf Voigt, Ralf Kriegel, Stefan Baumann, Mariya Ivanova, and Tim van Gestel

Abstract

This chapter gives an overview of ceramic materials and membranes for gas separation in fossil-fuelled power plants. These membranes can be used for the separation of oxygen from air or the separation of hydrogen from syngas. For all ceramic membrane materials and systems a detailed literature overview is given. Materials for oxygen separation membranes have to exhibit mixed ionic–electronic conductivity. The theory of oxygen tansport through the crystal lattice at high temperature, and also the variety of membrane materials, such as perovskites and fluorites, are discussed. Their structures and functional properties are presented and oxygen flux data are summarized. The manufacture and relevant properties of asymetric thin-film structures and enhanced surfaces are also discussed in terms of their influence on oxygen transport. Finally, different designs, for example, planar, tubular, and fiber concepts, are considered. For hydrogen separation from syngas, dense ceramic membranes with mixed proton–electron conductivity or microporous membranes are applicable. Similarly, for dense membranes, conductivity and hydrogen transport are discussed, together with an overview of material variety over the past two decades, including the classes cerates, zirconates, pyrochlores, niobates, and tungstates. The influence of different dopings and their influence on hydrogen transport as a function of temperature are discussed. Microporous membranes are split into two groups: amorphous sol–gel-derived membranes and crystalline zeolites. Both membrane types consist of SiO_2- or other oxide-based material systems. Hydrogen transport in micropores occurs through the free volumes formed by the amorphous networks or through well-defined pore channels of the zeolites in the nanometer range. Different transport mechanisms through a porous system/medium are considered as a function of the pore size. The manufacture of graded membrane structure is described along with the relevant properties of the microporous membranes and examples of other applications.

Process Engineering for CCS Power Plants, First Edition. Edited by Detlef Stolten and V. Scherer

Keywords: membrane technology, oxygen separation, hydrogen separation, mixed ionic–electronic conductors, microporous ceramic membranes, fossil-fuelled power plants

12.1
Introduction

Membrane technology can be integrated into many advanced system concepts for the production of liquid energy carriers and chemicals, for microfiltration, oxygen generation, low CO_2 emission power generation, hydrogen technology, and carbon dioxide capture. The separation of gases by inorganic membranes for energy applications is a growing field with respect to the high efficiency of membrane-based processes.

In a fossil-fuelled power plant, inorganic separation membranes can find application in, for example, the separation of oxygen from air (oxyfuel process) and the separation of hydrogen from syngas (pre-combustion process).

12.1.1
Oxygen Separation

The worldwide production of oxygen is approximately 100 million tonnes per year [1]. Nearly 50% is used in steel industry and around 25% for the production of ethylene oxide. Energy generation by the combustion of fossil fuels needs much more oxygen but mainly in the form of air. The CO_2 emissions of 29.4×10^9 t in 2008 [2] correspond to an oxygen consumption of 21.4×10^9 t.

The use of oxygen in energy production requires high oxygen production rates and a low specific energy demand. Therefore, different energy-efficient processes have been developed in recent years. Discontinuous processes called chemical looping combustion (CLC) [3, 4] and ceramic auto-thermal recovery (CAR) [5, 6] are based on oxygen storage materials (OSMs). Continuous O_2 production could be realized by membrane separation, but the cheap polymer membranes are not competitive for oxygen enrichment above 35 vol.%. Because of the small difference in the kinetic diameters of O_2 and N_2, porous membranes are not able to separate these molecules. The membrane separation process currently preferred for oxygen production uses mixed ionic–electronic conductors (MIECs) at high temperatures.

12.1.2
Hydrogen Separation

The main technological stages of the pre-combustion carbon capture approach are that the fossil fuel is partially oxidized (at 400–600 °C and a pressure of ~65 bar), resulting in a mixture of H_2+CO (synthesis gas, syngas) that is consecutively transformed [via the water gas shift reaction (WGSR)] to a mixture of

H_2+CO_2. The typical operating conditions are temperatures between 400 and 900 °C, pressures between 2 and 50 bar, a reducing gas environment with a high content of CO_2 and water, and the presence of, for example, H_2S, typically present at concentrations from 5 to 200 ppm. These harsh operating conditions make hydrogen separation a complex process, which is feasible, however, with the integration of a robust ceramic membrane. The demands for highly pure hydrogen that is required for a number of processes, and also carbon capture and storage (CCS)-based concepts, are at the focus of the modern society. Hence at the very core of the membrane separation technologies is the development of novel membrane materials with improved functional properties. In principle, two different membrane concepts are under consideration: dense membranes (mixed proton–electron conducting ceramic) and microporous ceramic membranes (amorphous and zeolitic).

The dense inorganic membranes have shown good prospects for integration into pre-combustion applications because of their high chemical and thermal stability (compared with the organic membranes), and practically infinite permselectivity, that is, pure product in the permeate stream (compared with the microporous membranes with Knudsen transport and separation based on the molecular sieving effect). Current microporous membrane systems, on the other hand, provide a much higher permeability and, therefore, lower demands on membrane area and material.

Therefore, research is focused on the different membrane concepts. They have in common that the research yields an optimized total solution in terms of economic and ecological feasibility. In all cases an asymmetric membrane design consisting of a thin functional membrane layer over a robust porous carrier is a favorable option because it offers a decreased membrane thickness that yields higher flux combined with improved mechanical stability.

12.2 Membranes for Gas Separation

12.2.1 Mixed Ionic–Electronic Conducting Membranes for Oxygen and Hydrogen Separation

12.2.1.1 Oxygen Transport Membranes (OTMs): Materials, Working Principle, and Membrane Development

The working principle of MIEC membranes, also called OTMs (oxygen transport membranes) or ITMs (ion transport membranes), is depicted schematically in Fig. 12.1. It is also related to the solid oxide fuel cell (SOFC) using an oxide ion-conducting solid electrolyte. In contrast to a pure ionic conductor in a SOFC, a MIEC membrane also conducts electronic charge carriers (holes or electrons) resulting in an inner short-circuit of the ion conductor. Therefore, the electrons or holes generated or consumed by surface reactions are transported

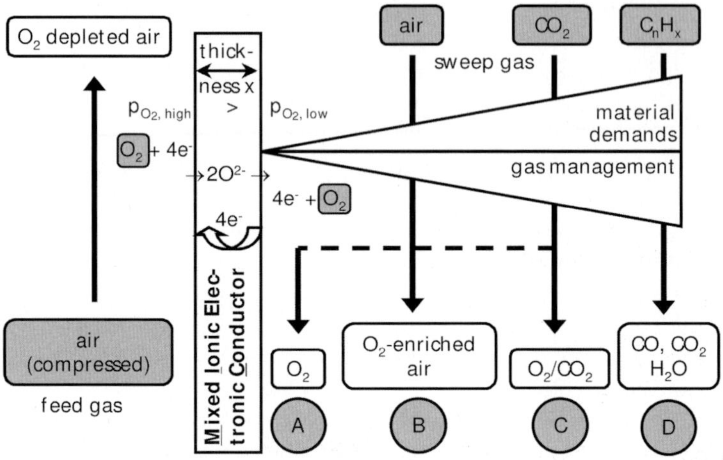

Fig. 12.1 Oxygen separation with MIEC membranes: functional principle and process variants.

through the membrane and electrostatic neutrality is obtained. Hence the driving force for the oxygen transport is generated by the different oxygen partial pressures of the different gas streams.

The first reports on oxygen permeation through dense oxide ceramics were published in the 1960s [7, 8]. In spite of the high temperatures, the oxygen permeation was very low due to the low electronic conductivity of the investigated oxides with a fluorite structure limiting the whole oxygen permeation [9] in ZrO_2, HfO_2, and ThO_2. It took 30 years to demonstrate oxygen permeation in the system $La_{1-x}Sr_xCo_{1-y}Fe_yO_{3-\delta}$ [10, 11] crystallizing in the perovskite structure. Nowadays material development of MIECs is focused on perovskites and partly on related structures, such as on brownmillerites [12], K_2NiF_4 [13], and substituted $Sr_4Fe_6O_{13}$ [14].

Perovskites are complex oxides with the general formula ABO_3, where A represents large ions such as lanthanides, actinides, alkali and alkaline earth elements, Pb, or Bi, and B represents transition metal elements, Li, Mg, Ca, Al, Si, Ga, Ge, In, Sn, Sb, Pb, or Bi. Hence synthetic perovskites show a wide variety of chemical composition and material properties, such as n–p conductivity, superconductivity, piezoelectricity, O^{2-} and H^+ ion conductivity, catalytic activity, and magnetism.

The perovskite structure depicted in Fig. 12.2 in a polyhedral representation is characterized by a cubic close packing of large A ions together with oxygen ions, and B cations occupy the octahedral cavities furthest from A ions. The stability of the structure can be estimated according to the Goldschmidt tolerance factor t [15]:

$$t = \frac{r_A + r_X}{\sqrt{2}(r_B + r_X)} \tag{12.1}$$

where r is the ionic radius.

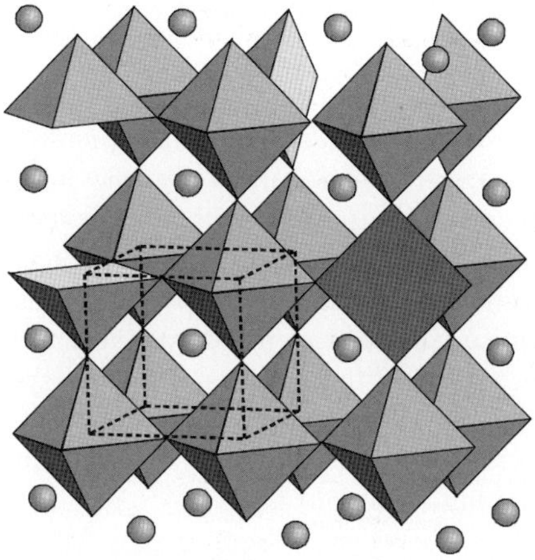

Fig. 12.2 Perovskite structure: A ions, spheres; $BO_{6/2}$ octahedra, with arbitrarily drawn oxygen vacancies ($BO_{5/2}$ pyramids).

A value of $t=1$ is obtained for the idealized case where all ions are in direct contact, but cubic perovskites are also found with t values down to 0.9. Structures resulting from combinations of hexagonal and cubic stacking sequences of AO_3 layers occur at t above 1, and rhombohedral, tetragonal, and orthorhombic distorted structures were observed below $t=0.9$.

The highest oxygen permeation is observed for alkaline earth cobaltites such as $Ba_{1-x}Sr_xCo_{1-y}Fe_yO_{3-\delta}$ [16–20], $BaCo_{1-y}Fe_yO_{3-\delta}$ [21, 22] and $SrCo_{1-y}Fe_yO_{3-\delta}$ [23, 24]. Cobalt and iron are often partly replaced by Ni, Cu, Zn, Cr, Fe, Al, Ga, In, Ce, Ti, Zr, Sn, V, and Nb [25] and also by Bi, Sc, La, and Y [26] and other elements. Frequently, the main target is enhancement of the oxygen permeation. The stabilization of the perovskite phase is also important because of the change in properties during the transition to brownmillerite phases [23, 24] and to hexagonal perovskites [27, 28]. Moreover, such phase transitions are mostly connected with significant volume changes [29] possessing the ability to destroy ceramic membranes [30].

Some B ions show a strong reducibility to the metal at low oxygen partial pressures. Further, the high affinity of alkaline earth elements to CO_2 and SO_2 is accompanied by the formation of carbonate and sulfate layers on the membrane surface, blocking the oxygen flux. For that reason, stability against chemical attack is also an important aspect of material development [31, 32]. Correspondingly, process routes C and D in Fig. 12.1 place higher demands on the MIEC membrane materials used.

Further, most MIEC materials showing high thermal expansion are additionally superimposed with a so-called chemical expansion [33], resulting in an overall nonlinear expansion behavior. The chemical expansion is caused by a reversible adjustment of the oxygen stoichiometry to temperature and oxygen partial pressure according to the following chemical reaction:

$$ABO_{3-\delta'} \underset{p_{O_2}\uparrow}{\overset{T\uparrow}{\rightleftharpoons}} ABO_{3-\delta''} + \frac{\delta' - \delta''}{2} O_2 \uparrow \qquad (12.2)$$

Therefore, an increase in temperature or a decrease in oxygen partial pressure results in an increase in concentration of oxygen vacancies and a decrease in reducible B ions due to an increase in their ionic radius. Further, the electrostatic repulsion between B ions is no longer compensated by an oxide ion corresponding to Fig. 12.2. Correspondingly, variations in temperature and oxygen partial pressure also result in variation of the lattice constants of the crystal structure [33], at least above a critical temperature indicating the freezing in of the system. High-flux materials such as BSCF5582 ($Ba_{0.5}Sr_{0.5}Co_{0.8}Fe_{0.2}O_{3-\delta}$) show large numbers of unoccupied oxygen sites, for example, up to one-sixth at room temperature and up to one-quarter under operating conditions [34].

Since an MIEC membrane is always located between different oxygen partial pressures under the operation conditions, there is also a gradient of oxygen vacancy concentration along the membrane thickness [35]. The corresponding different local expansions of the crystal lattice cause a chemically induced stress dominating the whole stress situation of a working membrane [36, 37]. Ceramic materials tolerate high compression stress, but low tensile stress caused by small oxygen partial pressure ratios is able to destroy a ceramic membrane, especially over long times. Therefore, a suitable combination of feed and sweep gas and absolute pressure differences can be used to minimize the tensile stress [36, 37], but the selection of MIEC materials with low chemical expansion coefficient and small changes in the oxygen stoichiometry under the operation conditions is also necessary.

The mechanical properties of MIECs depend distinctly on temperature and oxygen partial pressure. As an example, BSCF5582 shows a nonlinear variation of stiffness and fracture toughness with temperature with minima between 200 and 400 °C [38]. Much more critical seems to be the plastic deformation due to creeping at high temperatures described for different perovskites [39] and also for BSCF5582 [40]. This has to be kept in mind for the design of thin membrane components working in an absolute pressure gradient.

Because of the known stability problems of high-flux MIECs based on perovskites, further developments were carried out on fluorites, especially on ceria with its advantageous stability against CO_2 and SO_2. Significant enhancements of the electronic conductivity and also of the oxygen permeation were achieved by doping with Pr and Co [41], but the permeation level was located two orders of magnitude below that of BSCF5582. Higher oxygen permeation values around one order of magnitude below that of BSCF5582 were attained with dual-phase composites consisting of a stable ionic conductor together with an electronic conductor [32, 42, 43]. The current state of material development has been reviewed [44].

Oxygen transport inside a solid MIEC membrane can be described by the Wagner equation [45, 46]:

$$j_{O_2} = \frac{RT}{16F^2x} \int_{p_{O_2}(l)}^{p_{O_2}(h)} \frac{\sigma_e \sigma_i}{\sigma_e + \sigma_i} \, d\ln(p_{O_2}) \tag{12.3}$$

where j_{O_2} is the experimental O_2 flux, R the ideal gas constant, F the Faraday constant, x the membrane thickness, T the absolute temperature, $\sigma_{e,i}$ the electronic and ionic conductivity, p_{O_2} the O_2 partial pressure, (h) the high O_2 pressure side, and (l) the low O_2 pressure side.

The oxygen flux j_{O_2} is mainly determined by the ionic and electronic conductivity; the corresponding fraction inside Equation 12.3 is called the ambipolar conductivity. The majority of high-flux MIECs are p-type semiconductors characterized by a charge carrier mobility that increases with temperature. However, the concentration of holes decreases with increasing oxygen vacancy concentration caused by rising temperature and decreasing oxygen partial pressure according to Equation 12.2. Therefore, the electronic conductivity of high-flux MIECs often show a maximum in the intermediate temperature range [47].

Ionic conductivity is determined by the oxygen vacancy diffusion coefficient and the oxygen vacancy concentration depending mainly on material composition. An increase in temperature enhances the diffusivity and also the vacancy concentration; the latter is also increased by a decrease in oxygen partial pressure.

Therefore, the dependence on vacancy concentration differs for electronic and ionic conductivity. Accordingly, the ambipolar conductivity depends on the operating conditions in a complex manner, but in general a temperature dependence dominates. According to other transport processes, an Arrhenius approach can be used in most cases to calculate the variation of the ambipolar conductivity and the related oxygen permeation with temperature.

In addition to inherent material properties, there are further opportunities to enhance the oxygen permeation. The first results from the driving force determined by the ratio of the oxygen partial pressures. Obviously, very low oxygen partial pressures increase the oxygen permeation significantly. Therefore, the use of sweep gases that consume oxygen enhances the oxygen permeation significantly, but the material demands are very high, as mentioned before and illustrated by route D in Fig. 12.1. Sweep gases without oxygen consumption according to route C are less critically related to reducibility because of their higher oxygen concentration, but real industrial gases such as flue gas contain CO_2, NO_x, SO_2, and dust. Therefore, the majority of high-flux materials cannot be used with recirculated flue gas and stable materials have to be developed [32]. Oxygen enrichment of air according to route B in Fig. 12.1 increases the combustion efficiency [48] and CO_2 emissions decrease by saving of fuel. The driving force for this process is also the relatively low oxygen permeation. The driving force has to be realized by compression of feed air, consuming energy in competition with the energy saving mentioned. Therefore, high feed overpressures must be avoided, but this also limits the level of oxygen enrichment. The production of pure oxygen according to route A does not need a sweep gas and therefore the gas management is simpler. However, the whole driving force has to be realized by compression of air or extraction of oxygen with vacuum. The energy demand for compression of an excess of feed air is much higher than that for the extraction oxygen by vacuum because of the much lower gas volume. Additionally, a vacuum process using feed air at ambient pressure needs only one gas-tight oxygen supply and no external pressure vessel since the membrane itself serves as a pressure shell [49–51]. Otherwise, generation of overpressure is much simpler and cheaper than a vacuum process and recovery of the compression energy seems to be feasible for large industrial plants with high oxygen demand [52].

A second opportunity for a distinct enhancement of the oxygen flux for a given material under fixed operating conditions is a decrease in the separation layer thickness. Obviously, the mechanical stability required for the separation membrane, for example against different absolute pressures, can be achieved by a porous support coated with a dense MIEC separation layer. This concept, called an asymmetric membrane, is depicted in Fig. 12.3. Such membranes were developed by different groups and significant enhancements of oxygen permeation were described [53–59].

However, the use of asymmetric membranes for oxygen production on a larger scale has not been described in detail so far. The greatest issue seems to be leakage of the thin crystalline separation layers due to bad selectivity. In many cases such leakages were caused by the use of porous ceramic supports with an expansion behavior not well adjusted to the complex expansion behavior of the MIEC separation layer [59]. Therefore, the use of identical material compositions for the porous support and dense separation layer was recommended [60].

Figure 12.4 shows a scanning electron micrograph of a broken asymmetric BSCF5582 membrane multiply coated by dip coating at Fraunhofer Institut IKTS.

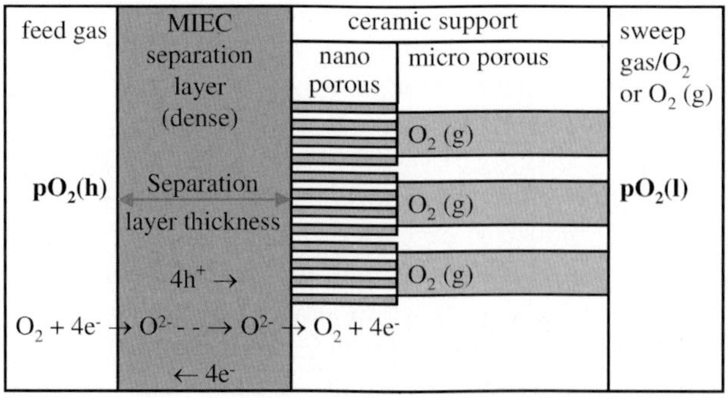

Fig. 12.3 Scheme of an asymmetric MIEC membrane.

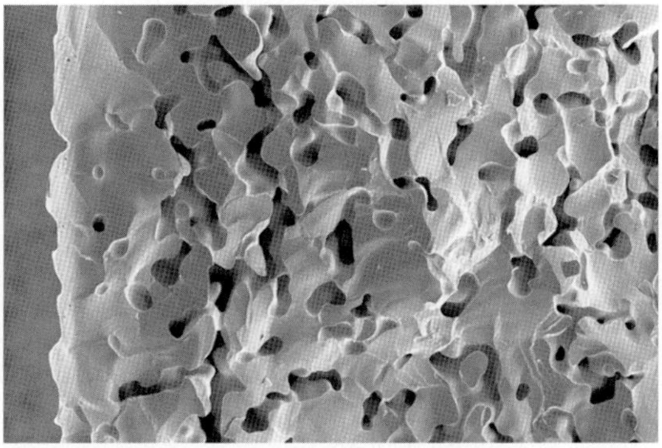

Fig. 12.4 SEM image of a cross break of an asymmetric BSCF5582 membrane, with leakage 0.01 ml (STP) cm^{-2} min^{-1} bar^{-1} [56].

The remaining leakage rate was determined as 0.01 ml(STP) cm^{-2} min^{-1} bar^{-1}. Therefore, product purity of better than 99 vol.% oxygen can be expected.

Oxygen permeation values of thin MIEC layers fall well below the expectations from the Wagner equation. This is caused by the surface exchange reactions limiting the oxygen transport for thin membranes. This means that no further increase in the oxygen permeation takes place below a critical membrane thickness L_c [61] given by the following equation:

$$L_c = \frac{D_V}{k_s} \tag{12.4}$$

In most cases, the critical thickness is determined by relaxation measurements on MIEC samples delivering the surface reaction constants k_s and the chemical diffusion coefficient D_{chem}. The latter has to be converted to the vacancy diffusion coefficient D_V using the thermodynamic factor. The critical length of BSCF5582 was determined as approximately 55 μm for the temperature range 550–725 °C [62]. It should be noted that the relaxation of high-flux materials is so fast that no reliable values for higher temperatures have been published up to now. Because of the decrease in the critical length with temperature, a smaller value can be expected for 850 °C.

The state of the art in oxygen production using asymmetric membranes is currently represented by Air Products and Chemicals (APC) (Allentown, PA, USA) [63, 64]. The developed planar quadratic cells are based on tape casting of a dense, a porous, and a punched tape, the last characterized by a channel structure for fast gas transport. Each cell consists of the outer dense layer supported by the porous layer. The central layer of a single cell contains the gas channels. About 60–70 cells are combined into a stack using a central tube with one end closed carrying the cells, which is also used for oxygen withdrawal. The conical open end is connected to the oxygen supply. Currently each stack produces 500 kg of O_2 per day; larger stacks producing 1 t per day are under development. Since 1988, the consortium around APC received generous financial support from the US Department of Energy. Unfortunately, no detailed information about the oxygen permeation rate and the materials used is available. However, the key technology for the production of the stacks described is the gas-tight joining of the different tapes. It should be kept in mind that these stacks operate with approximately 15 bar feed pressure and 0.5 bar oxygen pressure and the joining area per stack is huge. Therefore, small leakages would result in a decrease in the selectivity and in oxygen purity below 99 vol.% aimed for by APC.

There are further reactor and module concepts for oxygen separation using MIEC membranes. A comparison of planar systems with tubular systems and hollow fibers was made by Vente *et al.* [65], but the conclusion that single-channel tubes are to be preferred for oxygen production and syngas reaction seems to be oversimplified. The main arguments for that result were the better distribution of outer loads compared with planar systems and the great efforts in sealing and joining of hollow fibers.

A major advantage of hollow fibers is the high package density of membrane area per unit reactor volume together with the relatively low membrane thickness [66–68]. In principle, small reactors with high oxygen fluxes can be realized, as demonstrated by Tan *et al.* [69]. They used $La_{0.6}Sr_{0.4}Co_{0.2}Fe_{0.8}O_{3-\delta}$ hollow fibers for a demonstration unit producing 3 l(STP) min^{-1} of O_2 at 1070 °C; oxygen was extracted by a vacuum pump. However, the joining of hollow fibers was effected with a silicon polymer sealant that did not withstand temperatures higher than 350 °C. Therefore, the reactor design was not suitable for use with hot gases preheated in a heat exchanger and did not utilize the whole built-in membrane area. Moreover, no heat recovery was made and hot oxygen was not actively cooled. Therefore, the temperature of the sealed open fiber ends depends significantly on the operating temperature.

Several joining methods for BSCF5582 ceramic components were developed to overcome these limitations [70, 71]. Figure 12.5 shows a small hollow-fiber

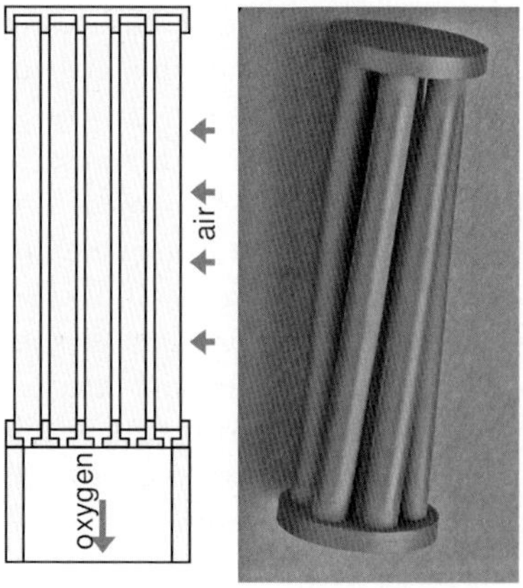

Fig. 12.5 Scheme of a hollow-fiber module and a corresponding ceramic joined module, made by Fraunhofer Institut IKTS/Vito.

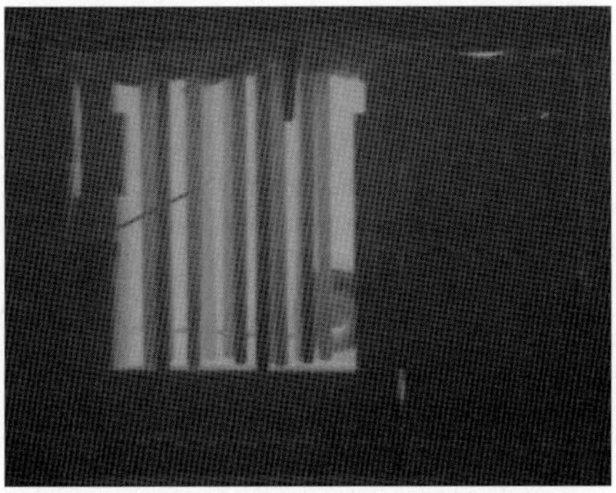

Fig. 12.6 Steel sheath tubes with BSCF5582 membrane tubes inside in a flue gas canal.

module prepared with BSCF5582 within the project MEM-BRAIN [72]. Hollow fibers were prepared by Vito (Mol, Belgium) and construction and joining were realized by Fraunhofer Institut IKTS (Hermsdorf, Germany). The pure ceramic joining withstands temperatures up to 900°C. Oxygen withdrawal will be realized by joining a BSCF5582 supply tube to the hollow fiber module.

The tubular concept preferred by some groups [52, 65, 73, 74] was developed within the project MEM-BRAIN [72] as a demonstration unit delivering up to 3 l(STP) min^{-1} of O_2 at 850 °C [75]. The portable device operating in vacuum extraction mode was constructed in spring 2009 and has been presented at different exhibitions and fairs. So far an operating time of 1700 h and 27 thermal cycles have been completed [76].

BSCF membrane tubes covered by steel sheath tubes were also tested in a flue gas canal under conditions of an industrial combustion plant as depicted in Fig. 12.6. Stable operation and production of oxygen were achieved. The BSCF membrane tubes tolerate heating and cooling rates of 5 K min^{-1} without and up to 15 K min^{-1} with steel sheath tubes. The oxygen permeation corresponds to laboratory results [77].

Overall, the technical feasibility of high-temperature oxygen separation with mixed conducting membranes has been proven. There seem to be no critical limitations for the industrial application of MIEC membranes for oxygen production.

12.2.1.2 Mixed Proton–Electron Conducting (MPEC) Membranes for Hydrogen Separation: Working Principle and Materials

In order to achieve H_2 transport through the bulk of a dense ceramic membrane, a partial pressure difference needs to be applied across the MPEC membrane that leads to bulk diffusion of protons simultaneously with the electron flow that is required to maintain the electrical balance. This gives rise to simultaneous proton and electron conduction occurring in the same direction and driven by the partial pressure gradient. From the application point of view, this simultaneous and one-directional conduction of the two charge carriers offers benefits because no external circuit is required, for example, to lead the current out of the device as in the case of fuel cells. The total conductivity σ_{tot} of a mixed proton–electron conductor consists of a proton conductivity term σ_{H^+} and an electronic conductivity term σ_{e^-}: $\sigma_{tot} = \sigma_{H^+} + \sigma_{e^-}$ [78]. The ability of the material to conduct one or another mobile species depends on their charge z, concentration c, and mobility μ, according to the equation

$$\sigma = zFc\mu \tag{12.5}$$

The correlation between the ion mobility and the diffusion coefficient D is given by the Einstein-Smoluchowski equation:

$$\mu = \frac{zD}{kT} \tag{12.6}$$

After combining Equations 12.5 and 12.6, for the conductivity of protons, for example, the following equation is obtained:

$$\sigma_{H^+} = \frac{z^2 F C_{H^+} D_{H^+}}{kT} \tag{12.7}$$

The equilibrium concentration of point defects increases exponentially with temperature, which leads to an exponential increase in conductivity as a function of temperature. The activation energy that is associated with the point defects conduction can be determined from the following dependence of Arrhenius type:

$$\sigma = \sigma_0 \exp\left(-\frac{E_{act}}{kT}\right) \tag{12.8}$$

where σ is the ionic conductivity, σ_0 is the so-called conductivity pre-exponential factor, E_{act} is the activation energy of ionic conduction, k is the Boltzmann constant, and T is the absolute temperature. The pre-exponential term σ_0 is a complex entity that accounts the influence of a number of parameters:

$$\sigma_0 = Ne^2a^2C(1 - C)\gamma k^{-1}\omega_0 \exp\left(\frac{\Delta S}{k}\right) \tag{12.9}$$

where N is the number of equivalent lattice sites per unit volume, e is the electronic charge, a is the hopping distance, C is the concentration of mobile ions, γ is a geometric factor, k is the Boltzmann constant, ω_0 is the vibrational frequency of the mobile ions in their lattice sites, and S is the entropy of hopping. The term $\omega_0 \exp(\Delta S/k)$ represents the frequency of effective jump attempts, ω_e [79].

The total conductivity of a mixed conducting material can be directly measured (by either a d.c. or an a.c. approach), and also the ionic partial conductivity (through e.m.f. measurements), whereas the electronic conductivity can only be calculated by subtracting the ionic term from the total conductivity term. The relation of the partial conductivity term (for example, the protonic term σ_{H^+}) to the total conductivity σ_{tot} gives the so-called transport number t of the corresponding charge carrier (for example, of the protons). The sum of the transport numbers of all charge carriers, t_i, is equal to unity:

$$t_{H^+} = \frac{\sigma_{H^+}}{\sigma_{tot}}; \quad 1 = \sum t_i \tag{12.10}$$

In order to create conduction of a mixed type in oxide ceramics, proton defects and free electrons have to be generated in sufficient concentrations. With respect to proton conduction, the presence of water or hydrogen-containing environments is a strictly required condition for the generation of proton defects. For oxides, the formation of proton defects can be written using Kröger-Vink notation according to the reaction

$$H_2O_{(v)} + V_O^{\bullet\bullet} + O_O^X \Leftrightarrow 2\,OH_O^{\bullet} \tag{12.11}$$

$$K_{OH^{\bullet}} = \frac{[OH_O^{\bullet}]^2}{([V_O^{\bullet\bullet}][O_O^X]p_{H_2O})} \tag{12.12}$$

As discussed in the literature [80–82], when water is present, proton defects are formed in the oxide material as a result of dissociative adsorption of water. Water from the gas phase disassociates to a hydroxide ion and a proton. The hydroxide ion is then adsorbed by an oxygen vacancy while the proton bonds to a lattice oxygen ion. Thus two positive proton defects $OH_O^{\bullet}$ are generated.

Electronic conduction in oxide ceramics may be generated when (i) the main elements of the compound have a variable valence, (ii) donor doping is used, and/or (iii) material is operated under reducing conditions (see Equation 12.13 for the third case):

$$O_O^X \rightarrow V_O^{\bullet\bullet} + 2e' + \frac{1}{2}O_2 \tag{12.13}$$

$$K_e = \frac{[V_O^{\bullet\bullet}][e']^2 p_{O_2}^{\frac{1}{2}}}{[O_O^X]} \tag{12.14}$$

The permeation of hydrogen through a dense ceramic membrane is proportionally dependent on the ambipolar conductivity:

$$\sigma_{amb} = \frac{\sigma_{ionic}\sigma_{electronic}}{\sigma_{ionic} + \sigma_{electronic}} \tag{12.15}$$

In the case of a mixed proton–electron conducting membrane, the proton flux density j_{H^+} can be expressed with the following equation:

$$j_{H^+} = \frac{-RT}{2F^2L}\int_{I}^{II} \sigma_{H^+} t_e \mathrm{d}\ln p_{H_2} \tag{12.16}$$

where R, T and, F have their usual meanings, L is the membrane thickness, σ_{H^+} is the proton conductivity, t_e is the transport number of electrons, and p_{H_2} is the partial pressure of hydrogen.

From this equation, it is clear that the hydrogen flux through the membrane depends directly on the protonic and electronic conductivity and on the magnitude of the applied partial pressure difference across the membrane, and it is inversely dependent on the membrane thickness. Furthermore, taking into consideration that the hydrogen transport through the dense ceramic membrane takes place in several steps [(i) adsorption of H_2 molecules; (ii) dissociation of H_2 into atomic H and proton defects formation; (iii) diffusion of protons across the membrane bulk; (iv) recombination of the protons to atomic and molecular hydrogen; and (v) desorption of the H_2 molecules from the membrane surface], it is obvious that both the catalytic properties of the membrane surface and the bulk diffusion contribute to the mass transport. At a membrane thickness L equal to the so-called characteristic thickness L_c, the hydrogen flux through the membrane is equally limited by the bulk diffusion and the surface exchange kinetics. Hence the development of stable membranes with reduced thickness, activated or optimized surface, and an improved mixed proton–electron conductivity is a priority of the R&D activities in this field.

A variety of material classes have been continuously studied in order to define those materials that combine high mixed proton–electron conductivity with improved chemical, phase, and thermal stability and high efficiency at a lower price. Among the numerous materials studied in a variety of structural classes (e.g., perovskites, fluorites, pyrochlores, brownmillerites, phosphates, fergusonites), the optimum levels of mixed proton–electron conductivity and H_2 flux were measured for representatives of the perovskite class. During recent decades, the research focus, however, was mainly on the defect chemistry and the proton conductivity of a number of doped and undoped oxide materials aiming to identify the most promising candidates for MPEC membrane applications [78, 80–92]. However, the ability of these materials to permeate hydrogen has received less attention. Of course, in the last 10 years, the number of studies on H_2-permeable materials has increased significantly. Recent publications on H_2 separating membranes give an overview of the structure, stability, and functional achievements of a number of materials [91, 93]. Table 12.1 presents selected literature examples of ceramic membrane materials for H_2 separation [93].

The proton conductivity in the perovskite type of oxide ceramics was first described for $SrCe_{0.95}Yb_{0.05}O_{3-\delta}$ (SCYb) [107]. Undoped $SrCeO_3$ does not exhibit proton conductivity. When Ce-ions are replaced by Yb, oxygen vacancies are generated within the crystal lattice that obviously play an important role in the formation and transport of proton defects. The proton conductivity of SCYb measured in an H_2 atmosphere at 900 °C was 7×10^{-3} S cm^{-1} [107]. This value is the

Table 12.1 Selected literature on ceramic membrane materials for H_2 separation according to [93].

Material	Membrane thickness (mm)	H_2 flux (ml cm^{-2} min^{-1})	*T* (°C)	Additional data	Ref.
$SrCe_{0.95}Tm_{0.05}O_3$	0.15	0.126	900	10% H_2–He (upstream) Air (downstream)	94
$SrCe_{0.95}Tm_{0.05}O_{3-\delta}$	1.6	0.039	900	10% H_2–He (upstream) 20% O_2–N_2 (downstream) No electrodes	95
	3	0.028	900	10% H_2–He (upstream) 20% O_2–N_2 (downstream) No electrodes	
$SrCe_{0.95}Yb_{0.05}O_{3-\delta}$	1.6	0.023	900	1% H_2–Ar (upstream) 20% O_2–Ar (downstream) Pt catalyst (electrode)	96
$SrCe_{0.95}Y_{0.05}O_{3-\delta}$	1.6	0.028 (extrapolated)	900	4% H_2–Ar (upstream) Ar (downstream) 2 Pt-electrodes (short-circuited)	97
$SrCe_{0.80}Y_{0.20}O_{3-\delta}$	1.6	0.048 (extrapolated)	900	4% H_2–N_2 (upstream) Ar (downstream) 2 Pt-electrodes (short-circuited)	98
$SrCe_{0.95}Tb_{0.05}O_3$	1 (effective area 1.8 cm^2)	1.4×10^{-2} 1.6×10^{-2}	900 900	0.001 atm CO (downstream) 0.1 atm CO (downstream)	99
$SrCe_{0.95}Y_{0.05}O_{3-\delta}$	0.05 0.62	0.102 7 times less	950	80% H_2–He (upstream) Ar (downstream)	100
$SrCe_{0.95}Yb_{0.05}O_{3-\delta}$	0.001 (layer)	13.44	677	H_2 (upstream) N_2–O_2 and He (downstream)	101
$SrCe_{0.95}Eu_{0.05}O_3$	1.72	3.085×10^{-3} 3.018×10^{-3} 2.683×10^{-3} 2.548×10^{-3}	850 850 850 850	Dry H_2 Wet H_2 (=0.028 atm) Wet H_2 (=0.051 atm) Wet H_2 (=0.086 atm) He (downstream)	102
$SrCe_{0.95}Sm_{0.05}O_3$	1.72 (thickness) and 24 (diameter)	3.125×10^{-3} 1.623×10^{-3}	850 850	Dry H_2 Wet H_2 (=0.028 atm) He (downstream)	103
$BaCe_{0.8}Y_{0.1}Ru_{0.1}O_3$	0.5	8.68×10^{-2}	800	H_2–Ar (upstream) Ar (downstream)	104
$BaCe_{0.95}Nd_{0.05}O_3$	0.7	2.6×10^{-2}	925	15% H_2 (upstream) Ar+Ne (downstream)	105
Nd_6WO_{12}	0.51	2.3×10^{-2}	1000	Wet 20% H_2–He (upstream) Ar (downstream)	106

highest measured in a hydrogen atmosphere for a variety of oxide ceramic materials.

$BaCe_{0.9}Nd_{0.1}O_{3-\delta}$ (BCNd) also exhibits proton conduction ($2{,}2\times10^{-2}$ S cm^{-1} measured at 900 °C in H_2) along with the oxygen ion conduction (nearly the same order of magnitude as the proton conductivity) at temperatures higher than 800 °C. The hydrogen flux through a $BaCe_{0.95}Nd_{0.05}O_{3-\delta}$ membrane with a thickness of 0.7 mm as a function of temperature was measured [105] and it was found to increase with increase in the hydrogen content and the stream concentration at the feed side. For a stream concentration of 15 vol.% the hydrogen permeation flux is 0.026 ml cm^{-2} min^{-1} at 925 °C.

Similar behavior has been confirmed for $BaCe_{1-x}Y_xO_3$ (BCY) in which the oxygen ion conductivity is comparable to the proton conductivity of the material. $SrCe_{0.95}Y_{0.05}O_{3-\delta}$ (SCY), for example, has electron conductivity that is too low for use as a membrane material and too high for use as an electrolyte in solid oxide fuel cells. Mixed conductivity of cerate-based oxide materials, however, can be further improved by aliovalent doping on the Ce^{3+} site, and also by adding a metallic phase (10–40 vol.%) to the proton-conducting ceramic powder [94], and consequently producing a mixed conducting composite (cermet).

As mentioned above, in the first studies of proton conduction in perovskites, the important role of doping in increased proton conduction was recognized. For this reason, a variety of doped materials have been developed in recent years aimed at increasing the ambipolar conductivity by doping. However, it is still an open issue to determine the effect on the proton conductivity of improving the electronic conductivity substantially, since the empirical observation is that enhancement of the electronic conductivity reduces the proton conductivity.

As an example of the effect of doping on the functional properties (conductivity, H_2 permeation), the influence of Ru on acceptor-doped $BaCeO_3$ can be considered [104]. No hydrogen flux was detected when the membrane did not contain Ru or when the Ru concentration was $x=0.05$. For $x=0.075$ and 1, the flux value measured was 0.0868 ml cm^{-2} min^{-1} at 800 °C. It can be concluded that the hydrogen flux through an acceptor-doped $BaCeO_3$ membrane can be improved by adding Ru as a dopant, which contributes mainly to the electronic conduction of the material.

The electronic conductivity of the material was correlated with the ionization potential of the doping element [108]. The electronic conductivity of $SrCeO_3$ (2.82×10^{-4} S cm^{-1} at 900 °C) and $SrCe_{0.95}Yb_{0.05}O_{3-\delta}$ (8.50×10^{-3} S cm^{-1} at 900 °C) appeared to be inversely related to the ionization potential of the lanthanide element in these ceramics. The ionization potentials of the doping ions in proton-conducting $SrCeO_3$ (36.8 eV) and $SrCe_{0.95}Yb_{0.05}O_{3-\delta}$ (25.0 eV) were chosen as reference values for the selection of Tb and Tm as doping elements having ionization potentials higher (39.8 eV for Tb) and lower (23.7 eV for Tm), respectively, than Ce and Yb. Solid solutions of $SrCe_{0.95}Tb_{0.05}O_{3-\delta}$ (SCTb) and $SrCe_{0.95}Tm_{0.05}O_{3-\delta}$ (SCTm) were prepared by the citrate route according to [95] and their electron conductivity was measured under different atmospheres by the four-point d.c. method. The results showed that the electronic conductivity increases with decreasing ionization potential. SCTb, with electron conductivity much lower than that of SCYb, can be used as an electrolyte for fuel cell, and SCTm, having higher electron conductivity, is a good membrane material. The hydrogen permeance of disc membranes with a thickness of 1–3 mm was measured with one membrane surface exposed to a 10% H_2–He mixture and the other to a 20% O_2–N_2 mixture. At 900 °C, the hydrogen flux increased from 0.028 to 0.039 ml cm^{-2} min^{-1} as the membrane thickness decreased. SCTm membranes could offer a higher hydrogen permeation flux if the thickness is further reduced.

The class of zirconate-based materials is also interesting for manufacturing hydrogen separation membranes, mainly due to the higher stability of zirconates under reducing atmospheres, particularly in CO_2 environments. Despite their better chemical stability, zirconates exhibit high grain boundary resistance that negatively influences their electrical performance. Furthermore, manufacturing membranes with high gas tightness requires extremely high sintering temperatures (e.g., ~1700 °C or even higher for zirconates). However, fewer experimental data exist on the hydrogen flux through zirconate-based membranes than for the cerate class of membranes.

Rather limited hydrogen flux was reported through an $SrZr_{1-x}Yb_xO_{3-\delta}$ membrane with no oxygen in the sweep gas due to its low electronic conductivity under non-oxidative atmospheres [96]. This result indicates that the electronic conductivity limits the hydrogen permeance of proton-conductive membranes operating in a non-galvanic mode. As was outlined above, both proton and electron conductivities are required and determine the hydrogen flux through the dense ceramic membrane. Hence an effective route for improving the hydrogen permeability of the membrane is to increase the electron conductivity of the material also. The measured hydrogen flux through a 1 mm thick $SrZr_{0.95}Y_{0.05}O_{3-\delta}$ disc at 700 °C and 200 kPa H_2 was lower than 10^{-8} mol cm^{-2} min^{-1}. Obviously the low hydrogen flux is due to the low conductivity of one of the required charge carriers (protons and electrons/electron holes). The lower hydrogen flux of Y-doped $SrZrO_3$ was found to be related to the lower proton conductivity [109].

Due to the pronounced instability of the classical proton conductors (cerates) under reducing conditions and the lower proton conductivity and H_2 flux (zirconates), in recent years increasing attention has been paid to rare earth tungstates with general formula Ln_6WO_{12} as promising membrane candidates for H_2 separation. For this reason, here we focus further on their prospects as H_2 transport membranes.

The proton conductivity in $La_{5.8}WO_{11.7}$ was first reported a decade ago [110], and it was found that the proton transport number of this material was between 0.7 and 0.9 when H_2 is present in the atmosphere. At 900 °C the proton conductivity was measured to be about 5×10^{-3} S cm^{-1}. Since then, this material has been the subject of more extensive studies.

With decrease in the ionic radius of the lanthanide elements, for example in the series La → Nd → Er, a tendency to a symmetry reduction from cubic to tetragonal to rhombohedral was reported [111]. Lanthanum tungstate (La_6WO_{12})

exhibits a defective fluorite structure although the crystal symmetry of this material has not yet been completely resolved.

It was reported [112] that the highest proton conductivity was found in Ca-doped La_6WO_{12} compared with the corresponding Nd- and Er-containing compounds. It is obvious that the symmetry decrease affects the proton conductivity of the rare earth tungstates. It was also demonstrated that when operated under oxidizing and reducing conditions and at elevated temperatures, these materials exhibit n- or p-type conductivity. The highest conductivity of about 5×10^{-3} S cm^{-1} was measured for the undoped La_6WO_{12} by means of the a.c. technique. Doping with 0.5% and 5% of Ca leads to a threefold decrease in the conductivity in the first case and by about one order of magnitude in the second case. A similarity to the Zr-doped La_6WO_{12} investigated by Shimura *et al.* [110] was also pointed out by Haugsrud [112] with respect to the decreased contribution of the ionic conductivity term under oxidizing and reducing conditions with increasing concentration of the acceptor doping.

Although it was suggested [112] that the chemical stability might be an issue for the practical application of the tungstates similarly to the Ba- and Sr-rich perovskites, the possibility of performing H_2 separation with Ln_6WO_{12}-based membranes is at the focus of estimation by calculations (based on the Wagner equation) and also by a few experimental studies [106, 110, 112–114]. For the theoretical estimation of the H_2 flux through a La_6WO_{12} membrane, the ambipolar conductivity that was determined experimentally in previous studies serves as a basis [110, 112, 113].

A theoretical estimation of the H_2 permeability and comparison among La_6WO_{12}, $SrCeO_3$, and Er_6WO_{12} was presented [114]. The flux was calculated for 10 m thick layers with a feed-side pressure of 10 atm H_2 and assuming that the flux occurs through the bulk diffusion mechanism. The predicted hydrogen permeation for La_6WO_{12} at 800 °C was about 2.0 ml$_n$ cm^{-2} min^{-1}, which is than for $SrCeO_3$ and Er_6WO_{12}, which yield a hydrogen flux at the same temperature of about 1.0 and <0.1 ml$_n$ cm^{-2} min^{-1} (ml$_n$ = Norm ml), respectively.

The hydrogen flux through a 51 m thick Nd_6WO_{12} membrane was measured as a function of temperature [106]. One side of the membrane was exposed to a mixture of moist 20% H_2–He (feed side) and at the sweep side Ar was supplied (both sides were kept at atmospheric pressure). The measured flux for an Nd_6WO_{12} membrane at 1000 °C was 0.023 ml cm^{-2} min^{-1}. At moderate temperatures (below 900 °C) the electronic conductivity seems to be the limiting factor, whereas at high temperatures (above 900 °C) p-type conductivity becomes predominant and the proton conductivity is the limiting step, likely due to the low concentration of protons in the oxide lattice.

Figure 12.7 shows initial permeability measurements performed on an La_6WO_{12} bulk sample with a thickness of 1 mm manufactured at Forschungszentrum Jülich (Jülich, Germany) and measured at Sintef (Oslo, Norway). The H_2 flux value obtained for the Nd_6WO_{12} bulk membrane tested [106] (thickness = 510 μm) is indicated by the open square symbol on the graph. As can be seen, the hydrogen content has a clear effect on the H_2 flux according to the

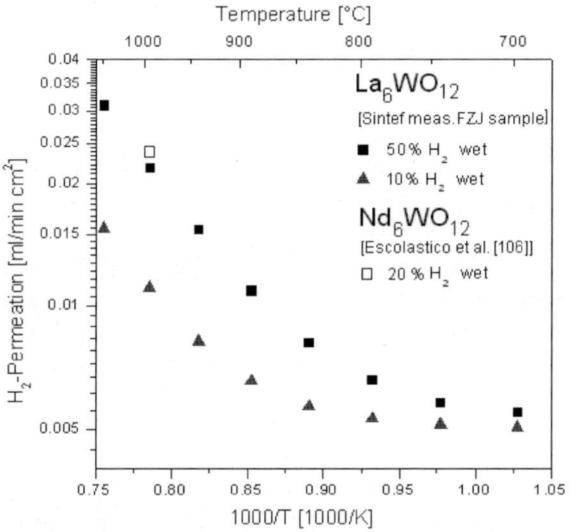

Fig. 12.7 Hydrogen permeation flux of an La_6WO_{12} membrane as a function of the temperature and H_2 content and flux at 1000 °C for Nd_6WO_{12} from [106].

Wagner equation. The thickness also has a pronounced effect, although it is difficult to compare these results.

Despite the limited number of studies on lanthanide tungstates as membranes for hydrogen separation, the attraction of this material class remains, especially after indications were given [106] that the short-term stability of tungstate materials in reducing atmospheres does not seem to be an issue. Structural stability tests carried out for Nd_6WO_{12} after exposure at dry and wet (2.5% H_2O) gas flows (90% CH_4–10% CO_2) at 700 °C for 72 h showed that the tested material remained unchanged regardless of the thermal treatment steps before the stability investigation.

Although the picture of the success of tungstates as H_2-separating membranes needs more studies to become complete, the most important targets at the present development stage are to improve the ambipolar conductivity (and especially the electronic conduction term) as the only material-related property, and to reduce significantly the membrane thickness (in the range of several microns). The electrical properties of the membrane material can be directly influenced via doping strategies that interact with and modify both the grain interior and the space-charged regions (grain boundary core and the doubly charged layers) in the solids [115]. The membrane thickness decrease is, of course, a technological issue connected with the manufacture of a robust substrate and the deposition of a defect-free functional layer. In conclusion, as a milestone target for H_2 flux through dense ceramic MPEC membranes is considered to be a value of $\sim 1\ \text{ml cm}^{-2}\ \text{min}^{-1}$ at 600 °C, this challenging field remains open for further activities.

12.2.2
Microporous Membranes for Hydrogen Separation

H_2–CO_2 separation in pre-combustion capture requires microporous membranes with a very narrow pore size distribution because the molecules are very small and of similar sizes (Table 12.2). H_2 is smaller than CO_2 but preferably H_2 will be separated from the mixture. According to the IUPAC notation, microporous materials are classified as having a pore diameter <2 nm [116]. In fact, pore sizes smaller than 0.5 nm are required where the conventional understanding of a pore can no longer be applied.

The transport mechanism with these small pore sizes is characterized by diffusion. Viscous flow no longer exists. Different diffusion mechanisms can be distinguished depending on the sizes of the molecules in relation to the pore size and also the operating conditions (Fig. 12.8). Knudsen diffusion is observed when the free path of the molecule is much greater than the diameter of the pore. In this case, selectivity is proportional to the square root of the molecular weight ratio of the gases and is independent of the temperature. The selectiv-

Table 12.2 Sizes according to Pauling [117] and kinetic diameters of the main gas molecules in post- and pre-combustion processes calculated from the minimum equilibrium diameter.

Molecule	Length (nm)	Width (nm)	Kinetic diameter (nm)
H_2O	0.39	0.315	0.265
H_2	0.31	0.24	0.289
CO_2	0.51	0.37	0.330
O_2	0.39	0.28	0.346
N_2	0.41	0.30	0.364
CO	0.42	0.37	0.376
NO	0.405	0.30	0.317
SO_2	0.528	0.40	0.360
H_2S	0.436	0.40	0.360

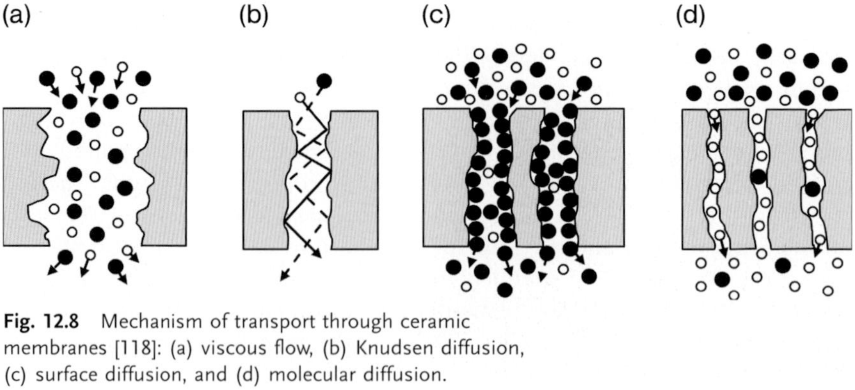

Fig. 12.8 Mechanism of transport through ceramic membranes [118]: (a) viscous flow, (b) Knudsen diffusion, (c) surface diffusion, and (d) molecular diffusion.

ities achieved for H_2–CO_2 and CO_2–N_2 are only 4.67 and 0.8, respectively, and are therefore too low for the desired separation.

Surface diffusion is usually reported as the transport mechanism for CO_2 separation membranes, for example, for the separation of CO_2 from H_2, N_2, or CH_4. The mechanism involves the surface diffusion of adsorbed CO_2 along the pore walls and CO_2 selectivity is obtained when the surface transport of CO_2 exceeds the contribution from Knudsen diffusion. The mechanism is usually reported for membranes with larger pore diameters in comparison with amorphous SiO_2 membranes ands zeolite membranes. In this case, the experimental test conditions are reported to play a major role. At lower temperatures, CO_2 adsorption is enhanced and the separation factor increases strongly. Further, the major effect of CO_2 adsorption on the separation factor is frequently confirmed by the difference in selectivity for a single gas and mixed gas operation. However, in comparison with the previously mentioned micropore diffusion type of membranes, the separation factors achieved are – particularly in experiments with smaller molecule gases – always much lower.

Molecular diffusion is often reported as a working mechanism in which the smaller gas molecule (He or H_2) permeates through the porous structure of the membrane, while the larger molecule is sieved out. Therefore, the size of the related gas molecules is an important issue for the development of gas separation membranes with a well-defined nanostructure. The transport of the smaller molecule tends to be an activated process, with the rate increasing exponentially with temperature. This type of mechanism is reported for amorphous SiO_2 membranes, for example, in experiments on the separation of H_2 or He from N_2 or CO_2, and in practice these membranes show high permeability for He and H_2. Usually, a pore diameter of 0.3–0.5 nm is reported for such materials, but in fact it is only a calculation from adsorption or permeation measurements.

Mainly two types of ceramic membranes have been reported for H_2–CO_2 separation, amorphous sol–gel membranes and crystalline zeolite membranes. Amorphous sol–gel membranes are well known from ceramic nanofiltration membranes, which are already commercially available (Fig. 12.9), and several

Fig. 12.9 Ceramic membranes from Inopor GmbH.

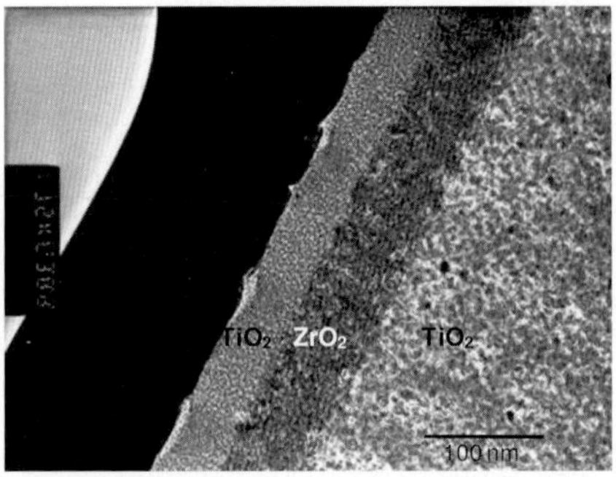

Fig. 12.10 Ceramic nanofiltration membrane made by the sol–gel technique.

thousand square meters are already in operation. These nanofiltration membranes consist of amorphous titania and have a thickness of only 50 nm. The small thickness is needed in order to obtain high fluxes but results in high quality requirements of the substrate to avoid defects and cracks. Therefore, a number of intermediate layers have to be applied to smooth the surface and to bridge the pores of the substrate. The last intermediate layer before membrane coating is a crystalline zirconia layer with a pore size of 3 nm and a thickness of 250 nm and a crystalline titania layer with a pore size of 5 nm and a thickness of 250 nm, respectively (Fig. 12.10).

The pore size of the nanofiltration membrane is 0.9 nm. In this range, the transport mechanism of gases is characterized by unselective viscous flow. A further reduction of the pore size is possible but unfortunately the amorphous structure of titania is not stable under pre-combustion conditions. At temperatures above 300 °C and in the presence of water vapor, amorphous titania is transformed into crystalline anatase and larger pores and defects are formed between the crystals.

An alternative to titania is silica, which has a much lower tendency to crystallize. That is why amorphous silica membranes can be synthesized not only by the sol–gel technique but also by chemical vapor deposition (CVD) [117, 119]. CVD methods for preparing an SiO_2 membrane are usually divided into two classes. In the first method, all precursors and reactants are provided on one side of the membrane, while the other side is usually vacuumed. In the second method, the reactants are supplied from the opposite side of the substrate. CVD membranes frequently have higher selectivity for H_2, but these membranes usually also have lower permeability. It should be noted, however, that literature data on the permeability of SiO_2 membranes show a large scattering, thus making a comparison of different types of membranes difficult.

After the first publication on highly selective silica membranes for gas separation in 1998 [120], a large number of papers on this topic have been published. As in the power plant hydrothermal conditions prevail, it is necessary to use membranes that do not degrade in contact with water vapor and there is evidence that current SiO_2 gas separation membranes suffer from serious degradation. For this reason, a number of researchers have focused on developing alternative SiO_2 materials with improved stability. A variety of modified materials have been developed with dopants/additives incorporated such as metals (Ni, Co), metal oxides (Al_2O_3, ZrO_2, TiO_2, Nb_2O_3) and organic groups (e.g., [121, 122]).

Limitations on the stability of silica membranes under pre-combustion conditions are based not only on the nature of the silica but also on the normally used intermediate γ-Al_2O_3 which is transformed into boehmite (γ-AlOOH). Therefore, replacement of the γ-Al_2O_3 layer is needed with the same quality to obtain defect-free thin silica membranes. Recently, Forschungszentrum Jülich succeeded in the preparation of highly permeable and selective silica membranes on mesoporous ZrO_2 (8 mol% Y_2O_3) which is schown in Fig. 12.11. The investigation of the stability of these membranes is still in progress.

Another interesting approach reported by many researchers is the preparation of crystalline zeolite membranes, because some of these materials can be made with a pore diameter in the lower microporous region. In addition, these crystalline materials have the required thermal stability and also some compositions show the required chemical stability. Zeolites are tectosilicates with SiO_4 tetrahedra forming rings which are permeable to small gas molecules and the number of SiO_4 tetrahedra of a ring determines the pore size of the zeolite. Six-ring zeolites such as sodalite (SOD) have a pore size of 0.25 nm; eight-ring zeolites such as Linde type A (LTA) have a pore size of 0.4 nm (Fig. 12.12).

LTA zeolite membranes are already commercially available in form of the NaA type. They are used for the dehydration of organic solvents via vapor permeation, especially for the dewatering of bioethanol (Fig. 12.13) [123, 124]. Unfortunately, the NaA zeolite membrane has a poor H_2–CO_2 selectivity, for two reasons: the pore size is larger than the sizes of both molecules, and the presence of intercrystalline mesopores. In terms of pore size, SOD zeolite should be better suited for H_2 separation. Khajavi *et al.* [125] succeeded in the preparation of aluminum-containing SOD zeolite membranes (H-SOD) and demonstrated

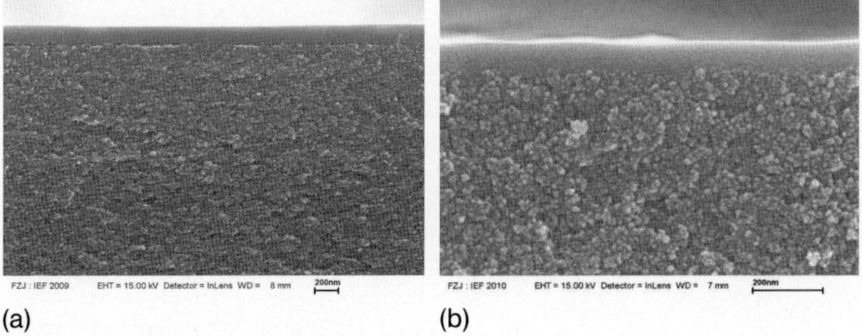

(a) (b)

Fig. 12.11 Amorphous SiO_2 membranes on top of a mesoporous γ-Al_2O_3 (a) and zirconia (b) layer.

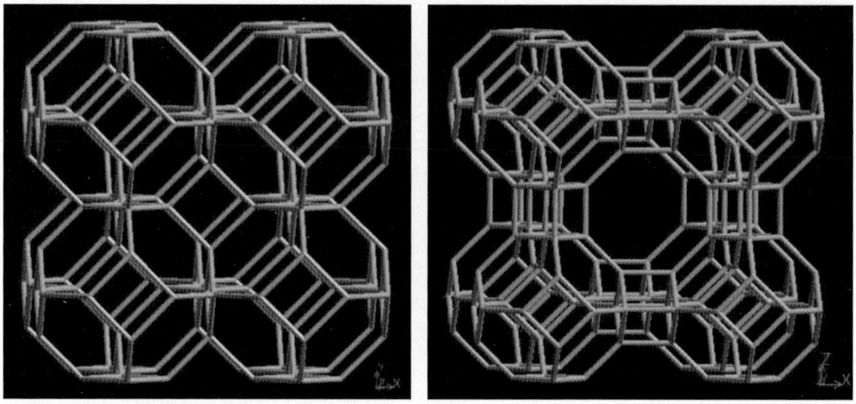

(a) (b)

Fig. 12.12 Wire frame models of the six-ring zeolite sodalite (SOD) (a) and of the eight-ring zeolite Linde type A (LTA) (b).

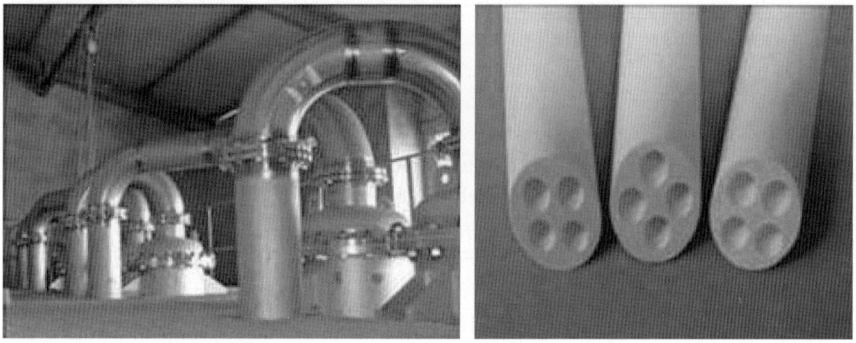

(a) (b)

Fig. 12.13 Industrial plant for the dehydration of bioethanol (a) containing 120 m^2 NaA zeolite membranes, for example, inside four-channel Al_2O_3 substrates (b).

their use for dehydration. However, H-SOD is not stable under pre-combustion conditions. Münzer *et al.* [126] prepared aluminum-free SOD zeolite membranes and showed that the removal of the template during synthesis results in destruction of the membrane layer. Recently, Richter *et al.* [127] reported the successful stabilization of the SOD zeolite structure and the preparation of thin supported layers (Fig. 12.14). The H_2–CO_2 selectivity of these membranes is slightly above Knudsen behavior, indicating that defects are still present. Therefore, studies focused on the improvement of the synthesis are needed in order to avoid defects.

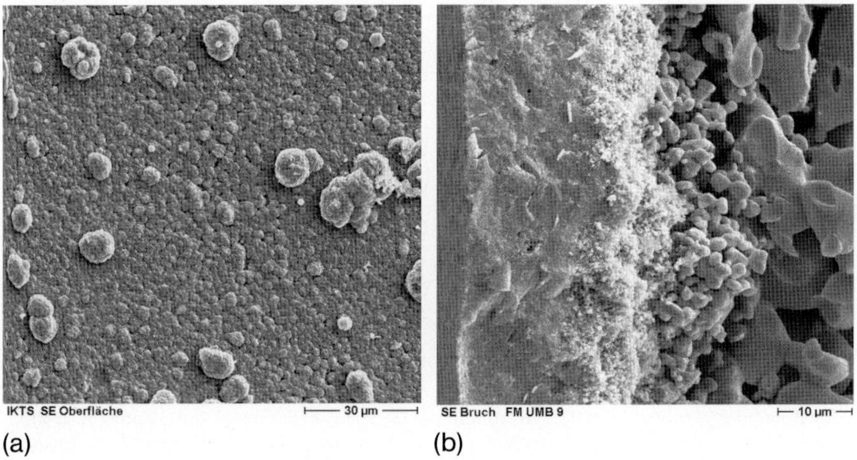

Fig. 12.14 Surface (a) and cross-section (b) of a SOD zeolite membrane prepared on top of a porous ceramic substrate.

12.3 Conclusion and Outlook

Membranes for separating the technically relevant gases in power plants have different development status, but they are all not available for implementation in a power plant. Nevertheless, membranes are already used for separation tasks in other fields (e.g., in the chemical industry). Owing to their low efficiency penalty, membranes are an interesting alternative to conventional separation technologies. Strategies for the development of novel, application-robust membranes rely on progress in materials science-based improvements and also technology-driven optimization of the process and application conditions. Research and technology activities are consequently directed towards functional layers and porous structures in the nanometer range, the development of mixed conducting oxides by means of theoretical materials design approaches, and the design of components under operating conditions based on their physico-chemical and mechanical properties. Decisive results based on a parallel investigation of (i) new materials and the design and processing of components and equipment, (ii) integration into power plants and the related process engineering, and (iii) energy systems analysis will pave the way to efficient, affordable CCS including hydrogen generation from fossil fuels by means of new gasification processes for fossil and biomass energy carriers.

References

1 Emsley, J. Oxygen. In *Nature's Building Blocks: an A–Z Guide to the Elements*, Oxford University Press, Oxford, 2001, pp. 297–304.

2 IEA. *CO_2 Emissions from Fuel Combustion. Highlights (2010 Edition)*, www.iea.org/co2highlights/co2highlights.pdf, last accessed 18. Feb. 2011.

3 Ishida, M. and Jin, H. Chemical looping combustion power generation plant system. US Patent 5 447 024, 1994.

4 Hossain, M.M. and de Lasa, H.I. Chemical looping combustion (CLC) for inherent CO_2 separations – a review. *Chem. Eng. Sci.*, 2008, **63**, 4433–4451.

5 Lin, Y.-S., Maclean, C., and Zeng, X. Elevated temperature adsorption process. European Patent, EP0913184B1, 1997.

6 Ullmann, H., Guth, U., Vashook, V.V., Burckhardt, W., Götz, R., and Bülow, M. Oxid-keramiken mit hohem Sauerstoff-transport. *Keram. Z.*, 2005, **57**(2), 72–78.

7 Möbius, H.H. and Hartung, H. Die Prüfung der Sauerstoffdurchlässigkeit feuerfester Materialien bei hohen Temperaturen durch potentiometrische Gasanalyse mit Festelektrolytzellen. *Silikattechnik*, 1965, **16**, 276–280.

8 Smith, A.W., Meszaros, F.W., and Amata, C.D. Permeability of zirconia, hafnia and thoria to oxygen. *J. Am. Ceram. Soc.*, 1966, **49**(5), 240–244.

9 Reetz, T. Sauerstoffpermeabilität oxidionenleitender Festelektrolyte. *Z. Phys. Chem.*, 1972, **249**, 369–375.

10 Teraoka, Y., Zhang, H.-M., Furukawa, S., and Yamazoe, N. Oxygen permeation through perovskite-type oxides. *Chem. Lett.*, 1985, **11**, 1743–1746.

11 Teraoka, Y., Zhang, H.M., Okamoto, K., and Yamazoe, N. Mixed ionic–electronic conductivity of $La_{1-x}Sr_xCo_{1-y}Fe_yO_{3-\delta}$ perovskite-type oxides. *Mater. Res. Bull.*, 1988, **23**(1), 51–58.

12 Anderson, M.T., Vaughey, J.T., and Poeppelmeier, K.R. Structural similarities among oxygen-deficient perovskites. *Chem. Mater.*, 1993, **5**, 151–165.

13 Bochkov, D.M., Kharton, V.V., Kovalevsky, A.V., Viskup, A.P., and Naumovich, E.N. Oxygen permeability of $La_2Cu(Co)O_{4+\delta}$ solid solutions. *Solid State Ionics*, 1999, **120**(1–4), 281–288.

14 Ma, B., Balachandran, U., Hodges, J.P., Jorgensen, J.D., Miller, D.J., and Richardson, J.W. Synthesis, conductivity and oxygen diffusivity of $Sr_2Fe_3O_x$. *Mater. Lett.*, 1998, **35**, 303–308.

15 Goldschmidt, V.M. Die Gesetze der Krystallochemie. *Naturwissenschaften*, 1926, **14**, 477–485.

16 Shao, Z.P., Yang, W.S., Cong, Y., Dong, H., Tong, J.H., and Xiong, G.X. Investigation of the permeation behaviour and stability of a $Ba_{0.5}Sr_{0.5}Co_{0.8}Fe_{0.2}O_{3-\delta}$ oxygen membrane. *J. Membr. Sci.*, 2000, **172**, 177–188.

17 Wang, H., Yang, W., Tablet, C., and Caro, J. Oxygen diffusion in oxide crystals – tracing new routes to identify the rate limiting step of oxygen permeation through perovskite membranes. *Diffus. Fundam.*, 2005, **2**, 46.1–46.15.

18 Lu, H., Tong, J., Cong, Y., and Yang, W. Partial oxidation of methane in $Ba_{0.5}Sr_{0.5}Co_{0.8}Fe_{0.2}O_{3-\delta}$ membrane reactor at high pressures. *Catal. Today*, 2005, **104**, 154–159.

19 Vente, J.F., Haije, W.G., and Rak, Z.S. Performance of functional perovskite membranes for oxygen production. *J. Membr. Sci.*, 2006, **276**, 178–184.

20 Vente, J.F., McIntosh, S., Haije, W.G., and Bouwmeester, H.J.M. Properties and performance of $Ba_xSr_{1-x}Co_{1-y}Fe_yO_{3-\delta}$ materials for oxygen transport membranes. J. *Solid State Electrochem.*, 2006, **10**, 581–588.

21 Tong, J., Yang, W., Zhu, B., and Cai, R. Investigation of ideal zirconium-doped perovskite-type ceramic membrane materials for oxygen separation. *J. Membr. Sci.*, 2002, **203**, 175–189.

22 Tong, J., Yang, W., Cai, R., Zhu, B., Xiong, G., and Lin, L. Investigation on the structure stability and oxygen permeability of titanium-doped perovskite-type oxides of $BaTi_{0.2}Co_xFe_{0.8-x}O_{3-\delta}$

(x=0.2–0.6). *Sep. Purif. Technol.*, 2003, **32**, 289–299.

23 Kruidhof, H., Bouwmeester, H. J. M., van Doorn, R. H. E., and Burggraaf, A. J. Influence of order–disorder transitions on oxygen permeability through selected nonstoichiometric perovskite-type oxides. *Solid State Ionics*, 1993, **63–65**, 816–822.

24 Qiu, L., Lee, T. H., Liu, L.-M., Yang, Y. L., and Jacobson, A. J. Oxygen permeation studies of $SrCo_{0.8}Fe_{0.2}O_{3-\delta}$. *Solid State Ionics*, 1995, **76**(3–4), 321–329.

25 Nagai, T., Ito, W., and Sakon, T. Relationship between cation substitution and stability of perovskite structure in $SrCoO_{3-\delta}$-based mixed conductors. *Solid State Ionics*, 2007, **177**, 3433–3444.

26 Zeng, P., Shao, Z., Liu, S., and Xu, X. P. Influence of M cations on structural, thermal and electrical properties of new oxygen selective membranes based on $SrCo_{0.95}M_{0.05}O_{3-\delta}$ perovskite. *Sep. Purif. Technol.*, 2009, **67**, 304–311.

27 Svarcova, S., Wiik, K., Tolchard, J., Bouwmeester, H. J. M., and Grande, V. Structural instability of cubic perovskite $Ba_xSr_{1-x}Co_{1-y}Fe_yO_{3-\delta}$. *Solid State Ionics*, 2008, **178**, 1787–1791.

28 Arnold, M., Gesing, T. M., Martynczuk, J., and Feldhoff, A. Correlation of the formation and the decomposition of the BSCF perovskite at intermediate temperatures. *Chem. Mater.*, 2008, **20**, 5851–5858.

29 Kriegel, R. and Preuss, N. Dilatometric determination of phase transition temperatures and oxidation temperatures on the compounds $SrMnO_{3-y}$ and Sr_2MnO_{4-y}. *Thermochim. Acta*, 1996, **285**, 91–98.

30 Kriegel, R., Schulz, M., and Kaps, Ch. *Materialentwicklungen für Membranschichten zur Sauerstoffseparation aus Luft, Tagungsband 1, Thüringer Grenz- und Oberflächentage*, Innovent, Jena, 2005, pp. 102–107.

31 Zeng, Q., Zuo, Y., Fan, C., and Chen, C. CO_2-tolerant oxygen separation membranes targeting CO_2 capture application. *J. Membr. Sci.*, 2009, **335**(1–2), 140–144.

32 Schulz, M., Kriegel, R., and Kämpfer, A. Assessment of CO_2-stability and oxygen flux of oxygen permeable membranes. *J. Membr. Sci.*, submitted.

33 Atkinson, A. and Ramos, T. M. G. M. Chemically-induced stresses in ceramic oxygen ion-conducting membranes. *Solid State Ionics*, 2000, **129**, 259–269.

34 Kriegel, R., Kircheisen, R., and Töpfer, J. Oxygen stoichiometry and expansion behaviour of $Ba_{0.5}Sr_{0.5}Co_{0.8}Fe_{0.2}O_{3-\delta}$. *Solid State Ionics*, 2010, **181**, 64–70.

35 Müller, F., Kaps, Ch., Kriegel, R., Voigt, I., and Sommer, E. Optimierte Membran-Werkstoffe für die Sauerstoffanreicherung – ein Beitrag zum emissionsarmen fossil befeuerten Kraftwerk. In *Schriftenreihe Werkstoffwissenschaften*, Bd. 18 (ed. L. Spiess *et al.*), Dr. H.-J. Köster, Berlin, 2004, pp. 77–82.

36 Schulz, M., Kriegel, R., and Kaps, Ch. Simulation of the complex loads of ceramic membranes used for oxygen separation. Presented at "A Global Roadmap for Ceramics", 1st International Conference on Ceramics, Toronto, 2006, CD-ROM.

37 Schulz, M. FEM-Simulation der Sauerstoffpermeation durch MIEC Membranen im Vakuumbetrieb und Vergleich mit experimentellen Daten. Presented at DKG-Jahrestagung, Hermsdorf, 22–24 March 2010.

38 Huang, B. X., Malzbender, J., Steinbrech, R. W., and Singheiser, L. Discussion of the complex thermo-mechanical behaviour of $Ba_{0.5}Sr_{0.5}Co_{0.8}Fe_{0.2}O_{3-\delta}$. *J. Membr. Sci.*, 2010, **359**(1–2), 80–85.

39 Routbort, J. L., Goretta, K. C., Cook, R. E., and Wolfenstine, J. Deformation of perovskite electronic ceramics – a review. *Solid State Ionics*, 2000, **129**, 53–62.

40 Yi, J., Bouwmeester, H. J. M., Lein, H. L., and Grande, T. High-temperature creep behaviour of perovskite oxide $Ba_{0.5}Sr_{0.5}Co_{0.8}Fe_{0.2}O_{3-\delta}$. Presented at the 10th International Conference on Inorganic Membranes, Tokyo, 2008, 2c02.

41 Fagg D. P., Shaula, A. L., Kharton, V. V., and Frade J. R. High oxygen permeability in fluorite-type $Ce_{0.8}Pr_{0.2}O_{2-\delta}$ via the use of sintering aids. *J. Membr. Sci.*, 2007, **299**, 1–7.

42 Chen, C. S., Kruidhof, H., Bouwmeester, H. J. M., Verweij, H., and Burggraaf, A. J.

Oxygen permeation through oxygen ion oxide–noble metal dual phase composites. *Solid State Ionics*, 1996, **86–88**, 569–572.

43 Kim, J. and Lin, Y. S. Synthesis and oxygen permeation properties of ceramic–metal dual-phase membranes. *J. Membr. Sci.*, 2000, **167**, 123–133.

44 Sunarso, J., Baumann, S., Serra, J. M., Meulenberg, W. A., Liu, S., Lin, Y. S., and Diniz da Costa, J. C. Mixed ionic–electronic conducting (MIEC) ceramic-based membranes for oxygen separation. *J. Membr. Sci.*, 2008, **320**, 13–41.

45 Wagner, C. Beitrag zur Theorie des Anlaufvorgangs. *Z. Phys. Chem. B*, 1933, **21**, 25–41.

46 Wagner, C. Elementarvorgänge bei der Bildung von Metalloxyd aus Metall und Sauerstoff sowie bei verwandten Reaktionen. *Angew. Chem.*, 1938, **49**, 735–746.

47 Jung, J.-I., Scott T. Misture, S. T. and Edwards, D. D. The electronic conductivity of $Ba_{0.5}Sr_{0.5}Co_xFe_{1-x}O_{3-\delta}$ (BSCF: $x = 0$–1.0) under different oxygen partial pressures. *J. Electroceram.*, 2010, **24**, 261–269.

48 Jeschar, R., Specht, E., and Bittner, H.-G. Brennstoff- und Kosteneinsparung durch Sauerstoffanreicherung der Verbrennungsluft in Industrieöfen. *Stahl Eisen*, 1988, **108**(18), 835–840.

49 Meulenberg, W. A., Baumann, S., Blum, L., and Riensche, E. Membran-Kraftwerk und Verfahren zum Betreiben eines solchen. German Patent DE102007056841A1.

50 Kriegel, R., Burckhardt, W., Voigt, I., Schulz, M., and Sommer, E. Benefits of the vacuum process for oxygen production from air using $Ba_{0.5}Sr_{0.5}Co_{0.8}Fe_{0.2}O_{3-\delta}$ membranes. Presented at the 10th ICIM, Tokyo, 18–22 August 2008.

51 Kriegel, R. Energieeffiziente Sauerstoff-Separation mit mischleitenden Membranen. In *Thüringer Werkstofftag 2010, Schriftenreihe Werkstofftechnik Aktuell*, Bd. 2 (ed. Spiess, L., *et al.*), Universitätsverlag Ilmenau, Ilmenau, 2010, pp. 15–20.

52 Engels, S., Beggel, F., Modigell, M., and Stadler, S. Simulation of a membrane unit for oxyfuel power plants under consideration of realistic BSCF membrane properties, *J. Membr. Sci.*, 2010, **359**(1–2), 93–101.

53 Muydinov, R. Y., Popova, M. N., and Kaul, A. R. Development and characterization of thin-film oxygen membranes based on $Ba_{0.5}Sr_{0.5}Co_{0.8}Fe_{0.2}O_{3-\delta}$. *Dokl. Chem.*, 2005, **402**, 88–90.

54 Chen, Z., Shao, Z., Ran, R., Zhou, W., Zeng, P., and Liu, S. A dense oxygen separation membrane with a layered morphologic structure, *J. Membr. Sci.*, 2007, **300**(1–2), 182–190.

55 Büchler, O., Serra, J. M., Meulenberg, W. A., Sebold, D., and Buchkremer, H. P. Preparation and properties of thin $La_{1-x}Sr_xCo_{1-y}Fe_yO_{3-\delta}$ perovskitic membranes supported on tailored ceramic substrates. *Solid State Ionics*, 2007, **178**, 91–99.

56 Böer, J., Kiesel, L., and Pippardt, U. Keramische asymmetrische Membranen aus $Ba_{0,5}Sr_{0,5}Co_{0,8}Fe_{0,2}O_{3-\delta}$ zur Sauerstoffseparation. Presented at DKG-Jahrestagung 2010, Hermsdorf, 22–24 March 2010.

57 Fontaine, M.-L., Smith, J. B., Larring, Y., and Bredesen, R. On the preparation of asymmetric $CaTi_{0.9}Fe_{0.1}O_{3-\delta}$ membranes by tape-casting and co-sintering process. *J. Membr. Sci.*, 2009, **326**, 310–315.

58 Chen, Z., Shao, Z., Ran, R., Zhou, W., Zeng, P., and Liu, S. A dense oxygen separation membrane with a layered morphologic structure. *J. Membr. Sci.*, 2007, **300**, 182.

59 Hong, L., Chen, X., and Cao, Z. Preparation of a perovskite $La_{0.2}Sr_{0.8}CoO_{3-\delta}$ membrane on a porous MgO substrate. *J. Eur. Ceram. Soc.*, 2001, **21**, 2207–2215.

60 Kriegel, R., Voigt, I., and Burckhardt, W. Oxygen separation with mixed conducting ceramic membranes – application potential and state of the art. In *Proceedings of the 10th ECerS Conference* (ed. Heinrich, J. G. and Aneziris, C.), Göller Verlag, Baden-Baden, 2007, pp. 2161–2168.

61 Bouwmeester, H. J. M. and Burggraaf, A. J. Dense ceramic membranes for oxygen separation. In *Fundamentals of Inorganic Membrane Science and Technology*, 1996, Ch. 10, pp. 435–528.

62 Bucher, E., Egger, A., Ried, P., Sitte, W., and Holtappels, P. Oxygen non-stoichiometry and exchange kinetics of $Ba_{0.5}Sr_{0.5}Co_{0.8}Fe_{0.2}O_{3-\delta}$. *Solid State Ionics*, 2008, **179**, 1032–1035.

63 Dyer, P.N., Minford, E., Russek, L.E., Carolan, M.F., Taylor, D.M., Bright, J.D., Cutler, R.A., Prouse, D.W., Richards, R.E., and Wilson, M.A. Planar solid-state membrane module. European Patent EP0732138B1, 1995.

64 Carolan, M.F., van Stein, E.E., and Armstrong, P.A. Development of planar ceramic membranes for tonnage oxygen production in advanced energy applications. Presented at the Jülicher Werkstoff-Kolloquium, Jülich, 2009.

65 Vente, J., Haije, W.G., IJpelaan, R., and Rusting, F.T. On the full-scale module design of an air separation unit using mixed ionic electronic conducting membranes. *J. Membr. Sci.*, 2006, **278**, 66–71.

66 Zhang, C., Xu, Z., Chang, X., Zhang, Z., and Jin, W. Preparation and characterization of mixed-conducting thin tubular membrane. *J. Membr. Sci.*, 2007, **299**, 261–267.

67 Wang, H., Tablet, C., and Caro, J. Oxygen production at low temperature using dense perovskite hollow fiber membranes. *J. Membr. Sci.*, 2008, **322**, 214–217.

68 Leo, A., Liu, S., da Costa, J.C.D, and Shao, Z. Oxygen permeation through perovskite membranes and the improvement of oxygen flux by surface modification. *Sci. Technol. Adv. Mater.*, 2006, **7**, 819–825.

69 Tan, X., Wang, Z., Meng, B., Meng, X., and Li, K. Pilot-scale production of oxygen from air using perovskite hollow fibre membranes. *J. Membr. Sci.*, 2010, **352**, 189–196.

70 Kriegel, R. Aktuelle Entwicklungstrends von Funktionskeramiken und fügetechnische Anforderungen. Presented at the DVS-Kolloquium "Fügen von Metall, Keramik und Glas," Berlin, 24 February 2009.

71 Kriegel, R., Kircheisen. R., and Ritter, K. Verfahren zur hochtemperaturfesten Verbindung von sauerstoff-permeablen Oxidkeramiken auf der Basis substituierter Erdalkalicobaltate durch dotierungsunterstütztes diffusives Reaktionssintern. German Patent DE102009050019, 2009.

72 Czyperek, M., Zapp, P., Bouwmeester, H.J.M., Modigell, M., Peinemann, K.-V., Voigt, I., Meulenberg, W.A., Singheiser, L., and Stöver, D. MEM-BRAIN gas separation membranes for zero-emission fossil power plants. *Energy Procedia*, 2008, **1**, 303–310.

73 Xu, N., Li, S., Jin, W., and Shi, J. Experimental and modeling study on tubular dense membranes for oxygen permeation. *AIChE J.*, 1999, **45**(12), 2519–2526.

74 Leo, A., Liu, S., and da Costa, J.C.D. Development of mixed conducting membranes for clean coal energy delivery. *Int. J. Greenhouse Gas Control*, 2009, **3**(4), 357–367.

75 Kriegel, R. and Schulz, M. Konstruktion, Aufbau und Betrieb des ersten Stand-Alone-Demonstrators zur Sauerstoffseparation mit keramischen Membranen. Presented at the 13th Sitzung des DKG-Fachausschusses 6 "Werkstoffanwendung", Nürnberg, 5 November 2009.

76 Kriegel, R. Einsatz keramischer BSCF-Membranen in einem transportablen Sauerstoff-Erzeuger. In *DKG Handbuch Technische Keramische Werkstoffe, Loseblattwerk* (ed. Kriegesmann, J.), HvB-Verlag, Ellerau, 119, Delivery complement, Nov. 2010, Chapter 8.10.1.1, pp. 1–48.

77 Voigt, I., Kriegel, R., Adler, W., and Sommer, E. Vacuum driven oxygen separation with BSCF membranes. Presented at the 11th ICIM, Washington, DC, 19–22 July 2010.

78 Norby, T. and Larring, Y. Concentration and transport of protons in oxides. *Curr. Opin. Solid State Mater. Sci.*, 1997, **2**(5), 593–599.

79 Almond, D.P., Duncan, G.K., and West, A.R. The determination of hopping rates and carrier concentrations in ionic conductors by a new analysis of ac conductivity. *Solid State Ionics*, 1983, **8**, 159–164.

80 Kreuer, K.D. Proton conducting oxides. *Annu. Rev. Mater. Res.*, 2003, **33**, 333–359.

81 Kreuer, K. D. Proton conductivity: materials and applications. *Chem. Mater.*, 1996, **8**, 610–641.

82 Kreuer, K. D. On the complexity of proton conduction phenomena. *Solid State Ionics*, 2000, **136–137**, 149–160.

83 Fontaine, M.-L., Norby, T., Larring, Y., Grande, T., and Bredesen R. Oxygen and hydrogen separation membranes based on dense ceramic conductors. In *Membrane Science and Technology*, Vol. 13 (ed. Mallada, R. and Menéndez, M.), Elsevier, Amsterdam, 2008, Ch. 10.

84 Stotz, S., Wagner, C. Die Lösigkeit von wasserdampf und Wasserstoff in festen Oxiden. *Ber. Bunsenges. Phys. Chem.*, 1966, **70**(8), 781–788.

85 Bonanos, N. Oxide-based protonic conductors: point defects and transport properties. *Solid State Ionics*, 2001, **145**, 265–274.

86 Iwahara, H. Proton conducting ceramics and applications. *Solid State Ionics*, 1996, **86–88**, 9–15.

87 Nowick, A. S. and Du, Y. High-temperature protonic conductors with perovskite-related structures. *Solid State Ionics*, 1995, **77**, 137–146.

88 Norby, T. Solid-state protonic conductors: principles, properties, progress and prospects. *Solid State Ionics*, 1999, **125**, 1–11.

89 Phair, J. W. and Badwal, S. P. S., Materials for separation membranes in hydrogen and oxygen production and future power generation. *Sci. Technol. Adv. Mater.*, 2006, **7**, 792–805.

90 Norby, T. and Haugsrud, R. Dense ceramic membranes for hydrogen separation. In *Nonporous Inorganic Membranes* (ed. Sammells, A. F. and Mundschau, M. V.), Wiley-VCH Verlag GmbH, Weinheim, 2006, pp. 1–48.

91 Meulenberg, W., Ivanova, M., van Gestel, T., Bram, M., Buchkremer, H.-P., Stöver, D., and Alfaro-Serra, J. M. State of the art of ceramic membranes for hydrogen separation. In *Hydrogen and Fuel Cells* (ed. Stolten, D.), Wiley-VCH Verlag GmbH, Weinheim, 2010, pp. 321–349.

92 Li, L. and Iglesia, E. Modeling and analysis of hydrogen permeation in mixed proton–electronic conductors. *Chem. Eng. Sci.*, 2003, **58**, 1977–1988.

93 Meulenberg, W. A., Ivanova, M., Serra, J. M., and Roitsch, St. Proton conducting ceramic membranes: materials and properties. In *Advanced Membrane Science and Technology for Sustainable Energy and Environmental Applications, Part 2 – Membranes for Coal and Gas Power Plants and the Petrochemical Industry* (ed. Basile, A. and Nunes, S.), Woodhead Publishing, Cambridge, 2011, to be published Sept. 2011.

94 Cheng, S., Gupta, V., and Lin Y. S. Synthesis and hydrogen permeation properties of asymmetric proton-conducting ceramic membranes. *Solid State Ionics*, 2005, **176**, 2653–2662.

95 Qi, X. and Lin, Y. S. Electrical conductivity and hydrogen permeation through mixed proton–electron conducting strontium cerate membranes. *Solid State Ionics*, 2000, **130**, 149–156.

96 Hamakawa, S., Hibino, T., and Iwahara, H. Electrochemical H_2 permeation in a proton–hole mixed conductor and its application to a membrane reactor. *J. Electrochem. Soc.*, 1994, **141**(7), 1720–1725.

97 Guan, J., Dorris, S. E., Balachandran, U., and Liu, M. Transport properties of $SrCe_{0.95}Y_{0.05}O_{3-\delta}$ and its application for hydrogen separation. *Solid State Ionics*, 1998, **110**, 303–310.

98 Balachandran, U., Guan, J., Dorris, S. E., Bose, A. C., and Stiegel, G. J. Presented at the 5th International Conference on Inorganic Materials, Nagoya, 22–26 June 1998, A-410.

99 Wei, X., Kniep, J., and Lin, Y. S. Hydrogen permeation through terbium doped strontium cerate membranes enabled by presence of reducing gas in the downstream. *J. Membr. Sci.*, 2009, **345**, 201–206.

100 Zhan, S., Zhu, X., Ji, B., Wang, W., Zhang, X., Wang, J., Yang, W., and Lin, L. Preparation and hydrogen permeation of $SrCe_{0.95}Y_{0.05}O_{3-\delta}$ asymmetrical membranes. *J. Membr. Sci.*, 2009, **340**, 241–248.

101 Hamakawa, S., Li, L., Li, A., and Iglesia, E. Synthesis and hydrogen permeation properties of membranes based on

dense $SrCe_{0.95}Yb_{0.05}O_{3-\alpha}$ thin films. *Solid State Ionics*, 2002, **48**, 71–81.

102 Song, S.J., Wachsman, E.D., Rhodes, J., Yoon, H.S., Lee, K.H., Zhang, G., Dorris, S.E., and Balachandran, U. Hydrogen permeability and effect of microstructure on mixed protonic–electronic conducting Eu-doped strontium cerate. *J. Mater. Sci.*, 2005, **40**, 4061–4066.

103 Song, S.J., Wachsman, E.D., Rhodes, J., Dorris, S.E., and Balachandran, U. Hydrogen permeability of $SrCe_{1-x}M_xO_{3-\delta}$ (x=0.05, M=Eu, Sm). *Solid State Ionics*, 2004, **167**, 99–105.

104 Matsumoto, H., Shimura, T., Higuchi, T., Tanaka, H., Katahira, K., Otake, T., Kudo, T., Yashiro, K., Kaimai, A., Kawada, T., and Mizusaki, J. Protonic–electronic mixed conduction and hydrogen permeation in $BaCe_{0.9-x}Y_{0.1}Ru_xO_{3-\delta}$. *J. Electrochem. Soc.*, 2005, **152**(3), A488–A492.

105 Cai, M., Liu, S., Efimov, K., Caro, J., Feldhoff, A., and Wang, H. Preparation and hydrogen permeation of $BaCe_{0.95}Nd_{0.05}O_{3-\delta}$ membranes. *J. Membr. Sci.*, 2009, **343**, 90–96.

106 Escolástico, S., Vert, V.B., and Serra, J.M. Preparation and characterization of nanocrystalline mixed proton–electronic conducting materials based on the system Ln_6WO_{12}. *Chem. Mater.*, 2009, **21**, 3079–3089.

107 Iwahara, H., Esaka, T., Uchida, H., and Maeda, N. Proton conduction in sintered oxides and its application to steam electrolysis for hydrogen production. *Solid State Ionics*, 1981, **3/4**, 359.

108 Lin, Y.S. Microporous and dense inorganic membranes: current status and prospective. *Sep. Purif. Technol.*, 2001, **25**, 39–55.

109 Hibino, T., Mizutani, K., Yajima, T., and Iwahara H. Characterization of proton in Y-doped $SrZrO_3$ polycrystal by IR spectroscopy. *Solid State Ionics*, 1992, **58**(1–2), 85–88.

110 Shimura, T., Fujimoto, S., and Iwahara, H. Proton conduction in non-perovskite-type oxides at elevated temperatures. *Solid State Ionics*, 2001, **143**, 117–123.

111 McCarthy, G.J., Fisher, R.D., Johnson, G.G., Jr., and Gooden, C.E. Crystal chemistry and compound formation in the systems rare earth sesquioxide–WO_3. In *Solid State Chemistry. Proceedings of the 5th Materials Research Symposium*, NBS Special Publication 364, National Bureau of Standards, Gaithersburg, MD, 1972, p. 397.

112 Haugsrud, R. Defects and transport properties in Ln_6WO_{12} (Ln=La, Nd, Gd, Er). *Solid State Ionics*, 2007, **178**, 555–560.

113 Haugsrud, R., and Kjølseth, Ch. Effects of protons and acceptor substitution on the electrical conductivity of La_6WO_{12}. *J. Phys. Chem. Solids*, 2008, **69**, 1758–1765.

114 Norby, T., and Haugsrud, R. High temperature proton conducting materials for H_2-separation. Invited lecture at FZJ Workshop, 16 November 2007.

115 Maier, J., Ionic conduction in space charge regions. *Prog. Solid State Chem.*, 1995, **23**, 171–263.

116 IUPAC. www.iupac.org/goldbook/M03853.pdf. Compendium of Chemical Terminology, 2nd Ed. (1997).

117 Pauling, L. *Nature of the Chemical Bond*, Cornell University Press, Ithaca, NY, 1960.

118 Marano, J.J. and Ciferino, J.P. Integration of gas separation membranes with IGCC. *Energy Procedia*, 2009, **1**, 361–368.

119 Lee, D. and Oyama, S. Gas permeation characteristics of a hydrogen selective supported silica membrane. *J. Membr. Sci.*, 2002, **210**, 291–306.

120 de Vos, R.M. and Verweij, H. High Selectivity, High Flux Silica Membranes for Gas Separation. *Science*, 1998, **279**, 1710–1711 (1998).

121 de Vos, R.M., Maier, W.F., and Verweij, H. Hydrophobic silica membranes for gas separation. *J. Membr. Sci.*, 1999, **158**, 277–288.

122 Battersby, S., Tasaki, T., Smart, S., Ladewig, B., Liu, S., Duke, M.C., Rudolph, V., and Diniz da Costa, J.C. Performance of cobalt silica membranes in gas mixture separation. *J. Membr. Sci.*, 2009, **329**, 91–98.

123 Richter, H., Voigt, I., and Kühnert, J.-T. *Desalination*, 2006, **199**, 92–93.

124 Weyd, M., Richter, H., Kühnert, J.-T., Voigt, I., Tusel, E., and Brüschke, H. E. A. *Chem.-Ing.-Tech.*, 2010, **82**, 1257–1260.

125 Khajavi, S., Kapteijn, F., and Jansen, J. C. *J. Membr. Sci.*, 2007, **299**, 63–72.

126 Münzer, S., Caro, J., and Behrens, P. *Micropor. Mesopor. Mater.*, 2008, **110**, 3–10.

127 Richter, H., Voigt, I., Weyd, M., and Puhlfürss, P. Presented at the 10th Jülicher Werkstoffsymposium, Jülich, 15–16 June 2009.

13
Polymer Membranes for CO_2 Separation

*Sander R. Reijerkerk, Kitty Nijmeijer, Jens Potreck,
Katja Simons, and Matthias Wessling*

Abstract

Polymer membranes have the potential to be a viable separation medium for carbon capture processes. Currently, two distinctly different classes of materials are under pilot plant trial. This chapter reflects extensively on the transport of CO_2 through two representatives of these classes for post-combustion capture applications: a hydrophilic rubbery poly(ethylene oxide)-based polymer (PEBAX®) and a sulfonated glassy aromatic poly(ether ether ketone) (S-PEEK). The understanding of gas and water vapor transport phenomena through these membrane materials is particularly interesting because both materials are different in their chemistry and physical state: PEBAX shows molecular transport of gases and vapors through the soft rubbery, poly(ethylene oxide)-based phase, whereas S-PEEK shows gas and vapor transport through an amorphous glassy phase containing ionic groups that will preferentially hydrate over the apolar matrix. Few data have been reported on multi-component mass transport through these polymers. This chapter addresses the transport of both binary (water vapor–N_2 or water vapor–CO_2) as well as ternary (water vapor–CO_2–N_2) gas mixtures. The results show that the presence of N_2 when comparing the binary and ternary systems does not change the transport behavior of CO_2 through both membrane types, as is the case for the effect of the presence of CO_2 on the N_2 permeability in the rubbery material PEBAX. The presence of CO_2 does, however, increase the transport of N_2 through the glassy polymer membrane S-PEEK due to the occurrence of plasticization and relaxation phenomena, resulting in considerably lower selectivities for ternary systems than compared binary systems in the case of such glassy polymer membranes.

Keywords: polymer membranes, flue gas treatment, carbon dioxide capture, gas/vapor transport, rubbery poly(ethylene oxide), glassy poly(ether ether ketone)

Process Engineering for CCS Power Plants, First Edition. Edited by Detlef Stolten and V. Scherer

13.1 Introduction

Recent years have shown growing interest in the application of membrane technology as a viable technological solution for carbon capture and storage (CCS) in post-combustion to recover carbon dioxide directly from flue gas, but also in pre-combustion applications and oxyfuel processes. The use of membrane technology has inherent advantages: no regeneration (one-step process), small footprint, mechanical simplicity (no rotating equipment), and especially a high energy efficiency as it does not require phase transitions no suffer from undesirable liquid losses due to evaporation [1]. The enormous scale of CO_2 capture from flue gases, however, requires large membrane areas and/or highly permeable membranes with sufficiently high CO_2/light gas (e.g., He, H_2, N_2, O_2, CH_4) selectivities to remove the CO_2. Figure 13.1 presents the principle of polymer membranes for gas separation.

Powell and Qiao [2] provided an excellent review on novel materials for gas separation membranes, and Baker and Lokhandwala evaluated a number of industrial applications of polymer membranes for gas separation and CO_2 separation in particular [3–5].

Polymeric membranes for gas separation consist of a thin, dense polymer layer on a porous support, prepared either as an asymmetric membrane in a single step (Fig. 13.2a and b) or as composite membrane using dip coating (Fig. 13.2c).

The porous support layer provides the mechanical stability, whereas the dense top layer of the membrane is the actual separating layer and allows the selective transport of one component over the other. The dense separating layer usually has a thickness of less than 200 nm to guarantee sufficiently high fluxes. The chemical and physical nature of the dense polymeric separating layer determines the gas and vapor transport properties of the membrane.

Recently, we demonstrated that polymeric membranes can operate for an extended period in an end of pipe flue gas dehydration application with direct exposure to the flue gas [7]. We investigated whether removal of water from the nearly saturated flue gas is an option to (a) eliminate a superheating step in the flue gas prior to the entry of the stack and (b) re-use the recovered water as make-up water to the steam cycle. This concept has been evaluated up to pilot-

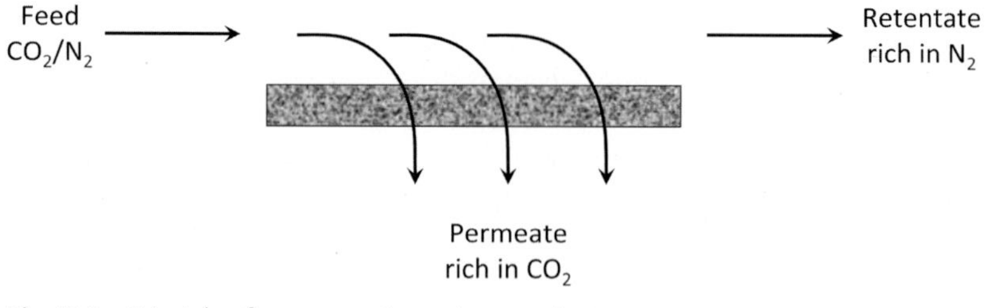

Fig. 13.1 Principle of gas separation using membranes.

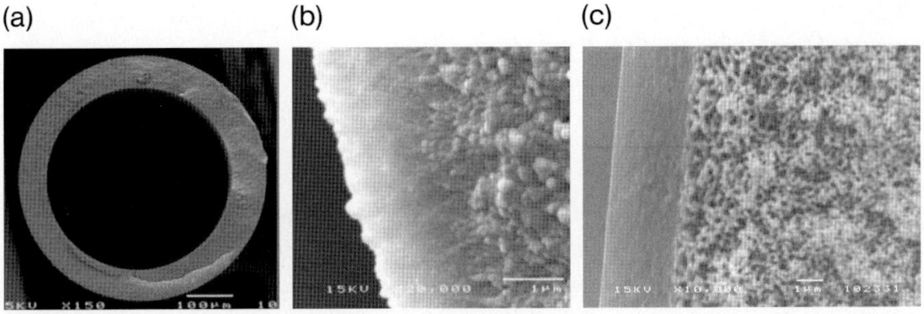

Fig. 13.2 Scanning electron micrograph images of the cross section of (a) an asymmetric hollow-fiber membrane and a zoom in at (b) the selective layer [6] and (c) a composite hollow fiber membrane [7].

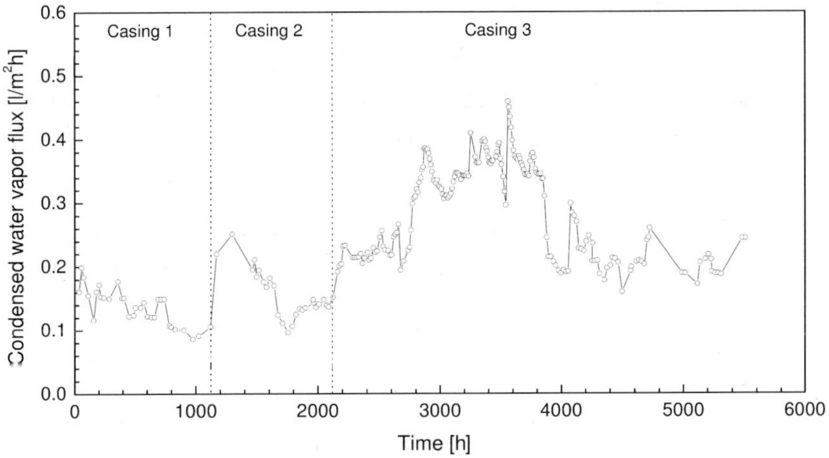

Fig. 13.3 Water vapor flux as a function of the operation time during a field test with the membranes directly and continuously positioned in the real flue gas stream. Data are continuous averages over five data points [7].

plant scale (Fig. 13.3 [7]) and long-term tests (5300 h) proved that although the flux showed variations in time, mainly due to variations in the process temperature, an average water vapor removal rate of 0.2–0.46 $l\,m^{-2}\,h^{-1}$ could be obtained continuously under real flue gas conditions. Due to the deposition of fly ash, dust, and gypsum crystals on the membrane surface, the flux slightly decreased in time but the selectivity (>100 000) and permeability remained sufficiently high for flue gas dehydration. This shows the industrial potential of membrane technology for flue gas treatment.

13.2
Polymer Membranes for CO_2 Capture

13.2.1
Membrane Materials

Especially the state of the polymer, that is, if the polymer is in its glassy or rubbery state, significantly determines its transport performance. In the glassy state, the mobility of the polymer chains is restricted and the polymer segments cannot rotate freely. The polymer chains are fixed. In contrast, in the rubbery state, the polymer segments can rotate freely along the polymer backbone and the polymer chains have a high degree of mobility. The transition between the glassy and the rubbery state is characterized by the glass transition temperature (T_g) of the polymer. Many physical and mechanical properties change substantially at the glass transition temperature.

Permeation through dense polymer layers preferentially occurs through the free volume of the polymer. In general, rubbery polymers have higher permeabilities but lower selectivities than glassy polymers. The free volume is defined as that part of the specific volume of a polymer that is not occupied by the polymer chains. The specific volume of a polymer depends strongly on whether the polymer is in its glassy or rubbery state. Figure 13.4 shows the specific volume of a polymer as a function of temperature [8].

The total polymer volume is composed of the polymer's occupied van der Waals volume and the free volume. The occupied van der Waals volume is only slightly dependent on temperature, but the total specific volume decreases significantly with decrease in temperature. Consequently, the free volume of a polymer also depends strongly on temperature. Above the glass transition temperature of the polymer, the polymer chains have the ability to adapt instanta-

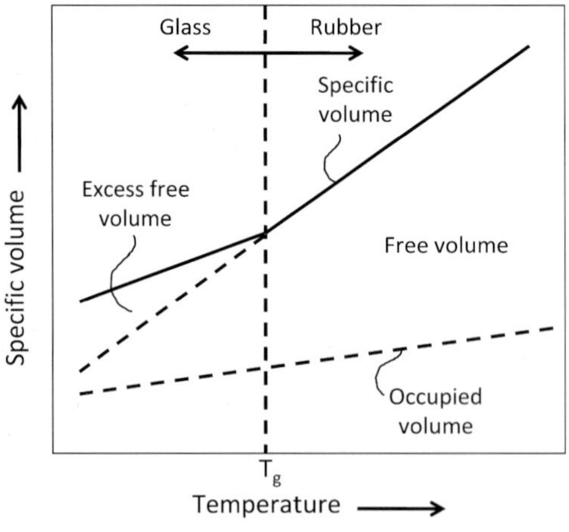

Fig. 13.4 Specific volume of a polymer as a function of temperature [8].

neously to changes. Cooling a polymer in its rubbery state from above the glass transition temperature results in a decrease in the specific volume with decreasing temperature. At the glass transition, a sudden change in physical and mechanical properties is observed and the molecular motions of the polymer chains are frozen in, as is the molecular volume of the polymer.

The difference between the specific volume as determined from the extrapolation of the specific volume of the polymer in its rubbery state and the actual specific volume of the polymer in its glassy state is referred to as the unrelaxed or non-equilibrium excess free volume. Glassy polymers are in a non-equilibrium state in terms of their free volume and especially this non-equilibrium excess free volume of glassy polymers has a strong effect on the performance of the polymer in terms of gas transport and gas separation.

Especially this non-equilibrium excess free volume in glassy polymers has severe consequences for the permeation behavior of membranes prepared from such glassy polymers. Sorption of plasticizing species such as CO_2 may induce relaxational and plasticization phenomena related to swelling stresses due to swelling of the polymer network by the plasticizing penetrant [9, 10]. On a molecular level, this causes an enhancement of the local segmental mobility of the polymer chains, whereas on a macroscopic scale, physical and mechanical properties are significantly affected, resulting in, for example, a decrease in glass transition temperature and moduli. In membrane operations, plasticization is usually considered negative as it causes a decrease in membrane separating ability [11–14].

An increase in the local segmental mobility of the polymer chains induces an enhancement of the transport rate of all components, but especially accelerates the transport of the slower component. The presence of one component influences the permeation behavior of the other components. Plasticization typically results in a decrease in membrane selectivity. Consequently, data obtained for pure, single gas permeation do not allow the prediction of the membrane performance under mixed gas conditions. Mixed gas permeation data for glassy polymers usually result in significant deviations from ideal transport behavior. Mixed gas membrane performance data are highly dependent on the feed gas composition and conditions [6]. Plasticization is directly related to the concentration of the plasticizing agent, introducing a certain degree of dilation of the polymer matrix, which can be controlled by the partial pressure of the component and/or the temperature [9, 15].

In addition to CO_2 [9], H_2O is also known to be able to induce strong relaxation phenomena in glassy polymers [10]. Although plasticization phenomena in rubbery polymers can also play a role, their effect is usually less dominant [15].

Figure 13.5 shows a typical example of the behavior of a glassy and a rubbery polymer membrane upon exposure to (high-pressure) CO_2 and the impact of plasticization on the separation performance of these membranes. The glassy polymer membrane shows a strong decrease in CO_2/CH_4 selectivity at a CO_2 fugacity of only 4–6 bar, whereas the corresponding values for the rubbery polymer membrane remain almost constant over the full pressure range.

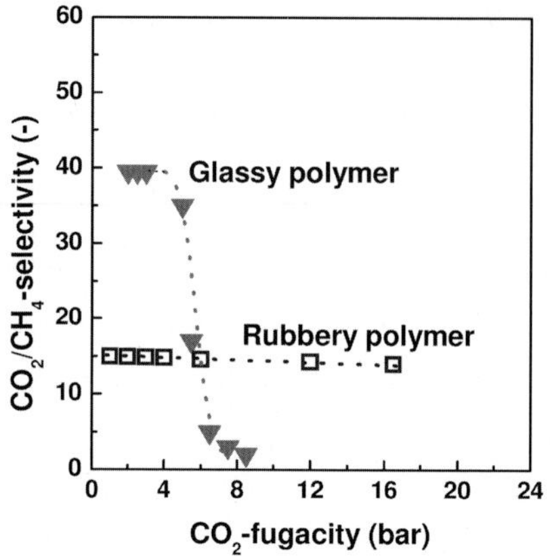

Fig. 13.5 CO_2/CH_4 selectivity as a function of the CO_2 fugacity for the glassy polymer 6FDA-polyimide [16] and the rubbery polymer XL-PEO [17].

In addition, film thickness plays a role in the plasticization behavior of glassy polymer membranes for gas and vapor separation. Usually, relaxation times in thin films are an order of magnitude faster than those in thicker films [18].

Numerous attempts to increase the plasticization resistance of glassy polymer membranes have been made. Especially cross-linking of the polymer (e.g., [18–21]), thermal annealing (e.g., [19, 22]), and polymer blending [23] increase the plasticization resistance of the membranes while maintaining reasonable permeances.

In contrast to glassy polymers, rubbery materials are only to a limited extent susceptible to plasticization and relaxation phenomena. Lin and Freeman [24] presented an extensive materials selection guide to provide materials that combine high CO_2 permeabilities with reasonable CO_2/N_2 selectivities. Especially materials containing polar groups, more specifically ethylene oxide (EO) units, were identified as suitable materials for CO_2 capture. The polar ether oxygen linkage favors the solubility of the quadrupolar CO_2 over the other light gases and the rubbery nature of the flexible EO linkages allows high diffusivity, overall resulting in high CO_2 permeabilities, combined with reasonably high CO_2/light gas selectivities. Freeman and co-workers designed a cross-linked poly(ethylene oxide) (PEO)-based polymer with an excellent CO_2 permeation performance and high resistance against impurities [17, 25–31].

As pure PEO lacks mechanical stability, another approach is the use of block copolymers combining soft PEO blocks that allow gas permeation and hard, rigid blocks of, for example, polyamide that provide the mechanical stability. Metz *et al.* [32] studied the gas permeation properties of a PEO–PBT block copolymer membrane, whereas Metz *et al.* [32] and Potreck *et al.* [33] investigated the gas and vapor separation performance of the commercially available polymer PEBAX®. Disadvantages of such commercially available block copolymers are the poor phase separation behavior of the soft and hard blocks and the presence of a semi-crystalline soft phase, impermeable to CO_2. Reijerkerk and coworkers [15, 34–38] developed a PEO-based block copolymer system with improved phase separation and very low soft phase crystallinity using soft segments with a random distribution of ethylene oxide and propylene oxide units, which prevents regular packing of the polymer chains. More than a fivefold increase in CO_2 permeability was obtained while maintaining sufficiently high CO_2/N_2 selectivities of ~ 40.

Powell and Qiao [2] prepared a comprehensive review on polymer materials for membranes for CO_2 post-combustion capture applications. It discusses possible design strategies, synthesis, fabrication, and the role of novel materials.

Polymer membranes should exhibit a number of critical properties in order to be effective in CCS [2]: (1) have high CO_2 permeability; (2) have high CO_2/N_2 selectivity; (3) be thermally and chemically robust; (4) be resistant to plasticization; (5) be resistant to aging; (7) be cost effective; and (8) membrane module manufacture should be inexpensive.

Major classes of polymers for gas separation membranes for CCS as summarized by Powell and Qiao are polyacetylenes, polyaniline, poly(arylene ether)s, polyarylates, polycarbonates, poly(ether imide)s, poly(ethylene oxide), polyimides, poly(phenylene oxide)s., polypyrrolones and polysulfones. The review by Powell and Qiao gives numerous examples of polymers for all these different classes and summarizes molecular structure, conditions (temperature and pressure), and permeabilities and selectivities.

Transport through polymer membranes can be described by the solution–diffusion model, in which the gas first dissolves in the polymer membranes, subsequently diffuses through the membrane from free volume pocket to free volume pocket, and finally the gas molecule desorbs again at the low-pressure side. Attempts to improve the transport properties of existing polymers materials consequently focus on two routes: diffusivity related (to increase the diffusivity of CO_2 or to decrease that of the other gases through the membrane), or solubility related (to increase the solubility of CO_2 in the membrane or to decrease that of the other gases). An increase in the diffusivity of CO_2 implies an increase in the free volume available for gas transport through the dense membrane.

Overall, several major strategies can be distinguished, as summarized in the review [2]. The main strategy usually considered is the incorporation of bulky side groups in the polymer backbone. The presence of such bulky groups prevents regular packing of the polymer chains and as such limits the occurrence of crystallization and creates additional free volume for permeation. Usually, the increase in CO_2 permeability obtained also leads to a decrease in selectivity, although the increase in permeability is more significant. *Para* substitution of phenyl rings, compared with meta substitution, usually gives higher permeabilities for all gases, at the expense of the selectivity. Closely packed polymers usually have a better size sieving ability, resulting in higher selectivities but lower permeabilities.

The solubility ratio is in principle easier to influence, as CO_2 has a considerably higher polarizability than, for example, nitrogen. This offers ways to improve the solubility of CO_2 over the other gases by incorporating functional groups in the polymer backbone that interact selectively with CO_2, thus increasing its solubility. An example of such a strongly interacting polymer is the class of ethylene oxide-based polymers.

Based on the large volume of research already available on this class of polymers, Powell and Qiao [2] identified especially polyimides as an interesting class of glassy polymer materials for CO_2 capture. Nevertheless, especially polymers with high solubilities for CO_2, which ensures high permeabilities combined with good CO_2/N_2 selectivities, are of interest. Examples are poly(ethylene oxide)-based materials. Finally, polymers with a high fractional free volume will be of interest.

In general, a highly selective membrane gives lower permeabilities, whereas highly permeable membranes usually have lower selectivities. Robeson [39] recognized this permeability–selectivity trade-off, characterized by an upper bound relationship in a plot of the logarithm of the separation factor versus the logarithm of the permeability of the faster permeating component. In 1991, Robeson summarized all data available for a specific separation using single component permeation data at 1 bar and 35 °C in such a so-called Robeson plot [39]. Later, in 2008, Robeson revisited the plots of 1991 and added the most recent data [40]. Figure 13.6 shows the Robeson plot for CO_2–N_2 separations sum-

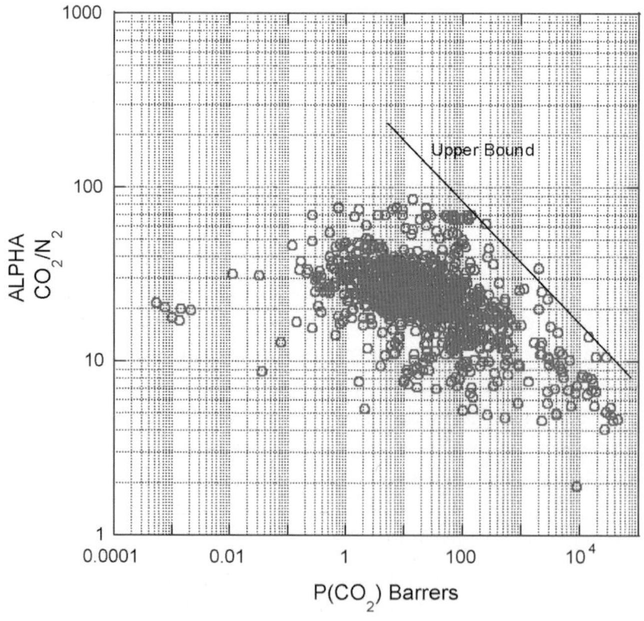

Fig. 13.6 Robeson plot or upper bound correlation for CO_2–N_2 separation [40].

marizing all single gas membrane permeation data (at 1 bar and 35 °C) available for this separation.

13.2.2
Membrane Process Design

The US-based company Membrane Technology and Research (MTR) has developed a highly permeable and CO_2-selective membrane, the MTR Polaris membrane (which is a polar, hydrophilic, solubility-selective material), for the separation of CO_2 from post-combustion flue gas streams. In addition to to membrane development, MTR performed a process design and economic analysis study for a membrane-based post-combustion capture process, starting from the boundary conditions set by the US Department of Energy [41]:

- capture performance: $\geq$90% CO_2
- capture costs: <\$ 30 t^{-1} CO_2
- power plant energy losses: <20%

A process scheme of the proposed process is presented in Fig. 13.7 [42].

The feed entering the membrane unit is slightly compressed using a blower to a pressure over 1.1 bar. Compression to higher pressures is undesirable due to the high energy consumption associated with the compression of a flue gas stream containing only 10–15% CO_2. The proposed driving force for permeation is achieved using a vacuum at the permeate side of the membrane unit. This results in a permeate pressure of 0.11 bar. The permeate stream of this membrane unit is enriched in CO_2 and contains a significant amount of water vapor, as the membrane is also permeable to water vapor owing to its hydrophilic nature. After removal of the water and subsequent compression, the CO_2 can be used for sequestration purposes.

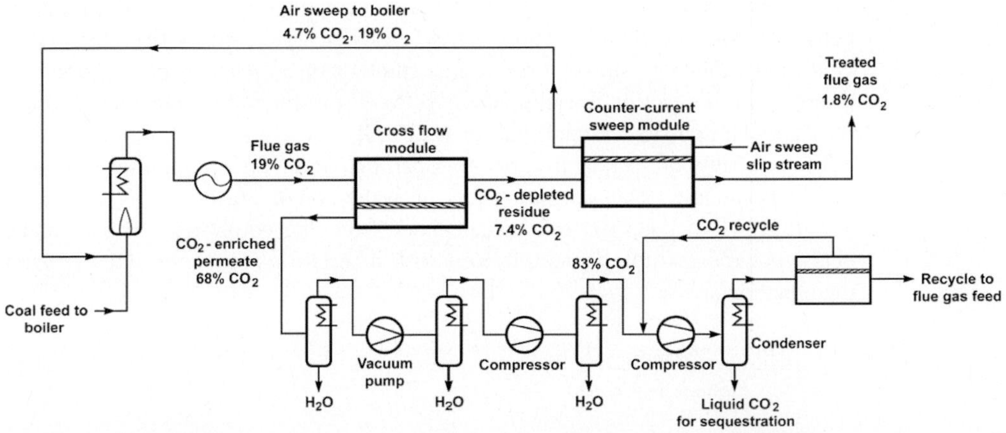

Fig. 13.7 Process design for a CO_2 post-combustion process as proposed by MTR [42].

The retentate, which is depleted in CO_2, is subsequently used as feed for a second membrane unit. In this unit, air is used as a sweep gas stream to provide an additional driving force for CO_2 permeation. This air sweep, which contains a certain volume percentage of CO_2, is subsequently recirculated to fire the boiler. As such it increases the CO_2 concentration to 19 vol.% (compared with only 10–15 vol.% in a standard boiler) in the flue gas exiting the boiler. The retentate flue gas exiting the second membrane module contains only 1.8 vol.% of CO_2 and is released to the atmosphere.

Overall, the process captures 90% of the CO_2 at the expense of only 12% of the power plant's energy output.

13.3
Theoretical Gas and Vapor Transport Through Dense Polymer Membranes

13.3.1
Sorption and Diffusion

The solubility of a gas in a polymer matrix is defined as the concentration of dissolved gas or vapor in the polymer, normalized to the applied equilibrium vapor pressure. In the case of an ideal system, for example, the sorption of light gases (e.g., He, H_2, N_2, O_2, CH_4) in rubbers, solubility is independent of concentration [43] and the sorption isotherms follow Henry's law: the concentration inside the polymer is proportional to the applied pressure. For non-ideal sorption behavior, the solubility coefficient is dependent on the solute concentration and the sorption isotherms are either concave or convex towards the pressure axis, especially at high gas or vapor concentrations [44]. Sorption isotherms which show a concave behavior are typical of glassy polymers [43], and can be described by the Dual Mode sorption model [45–47] or the more extended energy site distribution model [48, 49]. Systems that show convex sorption isotherms are usual for the sorption of strongly interacting gases and vapors in polymers [43], and their sorption behavior can be described by the Flory–Huggins model [50–53]. The Flory–Huggins model can be used to fit sorption isotherms and to describe the interaction between a solute and a polymer through the Flory–Huggins interaction parameter χ [54].

The solubility of a vapor in a polymer can be obtained from equilibrium sorption experiments. The concentration of water vapor inside a polymer film [cm^3 (STP) cm^{-3} polymer) can be calculated from the equilibrium mass uptake of the polymer sample in equilibrium with a certain water vapor activity using the equation

$$c_i = \frac{(m_\infty - m_{\text{polymer, dry}})\, V_{H_2O}}{V_{\text{polymer, dry}}\, M_{w,\, H_2O}} \tag{13.1}$$

where m_∞ (g) is the equilibrium mass of the polymer sample under the conditions of interest, $m_{\text{polymer, dry}}$ (g) is the dry weight of the polymer, V_{H_2O} (22 414 cm^3) is the volume of 1 mol of H_2O at standard temperature (273.15 K) and standard pressure (1.01325 bar), $V_{\text{polymer, dry}}$ (cm^3) is the volume of the dry polymer, and M_{w, H_2O} (18 g mol^{-1}) is the molecular mass of water.

Kinetic sorption data can be used to extract diffusion coefficients of penetrants in polymers [33] and for thin films, when diffusion from the edges can be neglected, Crank [55] deduced that the diffusion coefficient at short time scales can be described by

$$\frac{M_t}{M_\infty} = \frac{4}{l_p}\sqrt{\frac{Dt}{\pi}} \tag{13.2}$$

where M_t (g) is the mass of sorbed water vapor at time t (s), M_∞ (g) is the mass sorbed at $t=\infty$, l_p (m) is the polymer film thickness, and D (m^2 s^{-1}) is the Fickian diffusion coefficient. This equation can be used to extract the Fickian diffusion coefficient for a gas or a vapor in a polymer matrix from the initial slope of a plot of M_t/M_∞ versus $t^{1/2}$.

At longer time scales, the following equation allows the calculation of the Fickian diffusion coefficient from the slope of a plot of ln $(1-M_t/M_\infty)$ versus t:

$$\ln\left(1 - \frac{M_t}{M_\infty}\right) = \ln\left(\frac{8}{\pi^2}\right) - \frac{\pi^2 Dt}{l_p^2} \tag{13.3}$$

where M_t (g) is the mass of sorbed water at time t (s), M_∞ (g) is the mass penetrated at $t=\infty$, l_p (m) is the polymer film thickness, and D (m^2 s^{-1}) is the Fickian diffusion coefficient.

The sorption kinetics of highly sorbing gases and vapors (e.g., water vapor) into glassy polymers are complex and, in addition to to Fickian sorption behavior, additional mass uptake due to complex non-Fickian relaxation phenomena related to the non-equilibrium excess free volume may be observed [9, 56–59]. Fickian transport behavior can be characterized as a rapid, elastic, and reversible process, whereas non-Fickian transport and relaxational phenomena occur on much longer time scales. As a consequence, Berens and Hopfenberg [60] considered the overall non-Fickian sorption behavior of a penetrant in a polymer matrix $[M(t)_{\text{total}}]$ as the sum of Fickian sorption $[M(t)_F]$ and relaxational phenomena $[M(t)_R]$:

$$M(t)_{\text{total}} = M(t)_F + M(t)_R \tag{13.4}$$

Relaxational phenomena, the second term of the approach proposed by Berens and Hopfenberg [60], can be described as a series of relaxational regimes, each having its specific relaxation time constant τ_R. Consequently, non-Fickian diffusion of gases and vapors in glassy polymers can be described as by

$$\frac{M(t)}{M_\infty} = M_{F,\infty}\left[1 - \frac{8}{\pi^2}\sum_{m=0}^{\infty}\frac{1}{(2m+1)^2}\right]\exp\left[-\frac{D(2m+1)^2\pi^2 t}{L^2}\right] + \sum_{i=1}^{\infty} M_{R_i}\left[1 - \left(-\frac{1}{\tau_{R_i}}\right)\right] \tag{13.5}$$

where $M_{F,\infty}$ (g) and M_{R_i} (g) represent the infinite sorbed mass of the Fickian part and the relaxation part of sorption, respectively, and τ_{R_i} (s) is the characteristic time constant for relaxation. The relative contribution of Fickian and relaxational diffusion can only be determined when Fickian diffusion is much faster than the rate of relaxation [9, 61].

If a polymer sorbs significant amounts of vapor, the assumed frame of constant polymer dimensions is no longer true. This swelling effect can be taken into account and the Fickian diffusion coefficient can be corrected for this effect using the equation

$$D_c = \frac{D}{(1-\Phi_w)^{\frac{5}{3}}\partial \ln a/\partial \ln \Phi_w} \tag{13.6}$$

where D_c is the Fickian diffusion coefficient corrected for the swelling of the polymer upon water vapor sorption ($m^2\,s^{-1}$), a is the water vapor activity of the vapor phase, and Φ_w is the volume fraction of the penetrant obtained from the weight uptake due to water sorption and the partial molar volume of the penetrant at each specific water vapor activity and measurement temperature. In this case, the swelling effect is expressed in the volume fraction of water vapor in the swollen polymer [33]. This thermodynamic correction assures the use of the actual driving force for water vapor sorption in the polymer based on activities and the actual volume fraction of water in the polymer matrix.

When absorbed in a polymer, solvent molecules have a tendency to cluster, stemming from hydrogen bonding between water molecules [62]. To provide a measure for clustering, Zimm and Lundberg defined the clustering function [62]:

$$\frac{G_{ww}}{V_w} = -\Phi_p\left(\frac{\partial(a/\Phi_w)}{\partial a}\right) - 1 \tag{13.7}$$

where G_{ww} is the water–water cluster integral, V_w is the molar volume of the water vapor penetrant ($cm^3\,mol^{-1}$), a is the vapor activity ($p_{H_2O}/p_{H_2O,\,saturated}$), and Φ_p and Φ_w are the polymer and water volume fractions, respectively, as determined from the equilibrium mass in sorption runs. When $G_{ww}/V_w < -1$, the solution is ideal and water molecules are statistically distributed in the polymer matrix. When $G_{ww}/V_w > 0$, water molecules tend to cluster and form regions with an excess amount of water molecules relative to the average value based on Φ_w. When $G_{ww}/V_w = 0$, the excluding effect of the central water molecule is just sufficient to prevent clustering.

13.3.2
Permeation

According to Fick's law, gas diffusion through non-porous, dense polymer membranes can be described by [8]

$$J = -D\frac{\mathrm{d}c}{\mathrm{d}x} \tag{13.8}$$

where J is the gas flux through the membrane [cm^3 (STP) cm^{-2} s^{-1}], D is the diffusion coefficient (cm^2 s^{-1}) and $\mathrm{d}c/\mathrm{d}x$ is the driving force, or the concentration gradient over the membrane. Assuming steady-state conditions, Equation 13.8 can be integrated and results in the following equation:

$$J_i = \frac{D_i(c_{i,\mathrm{f}} - p_{i,\mathrm{p}})}{l} \tag{13.9}$$

where D_i is the diffusion coefficient (cm^2 s^{-1}) of component i, $c_{i,\mathrm{f}}$ and $c_{i,\mathrm{p}}$ are the concentrations (mol cm^{-3}) of component i on the feed and permeate side of the membrane, respectively, and l is the thickness (cm) of the membrane.

According to Henry's law, the concentration of component $i\,(c_i)$ can be linearly related to its partial pressure (p_i):

$$c_i = S_i\,p_i \tag{13.10}$$

where S_i is the solubility coefficient of component i [cm^3 (STP) cm^{-3} cmHg^{-1}]. Substitution of Equation 13.10 in Equation 13.9, and taking into account that the product of solubility (S) and diffusivity (D) equals the permeability (P), this results in the following flux equation:

$$J_i = \frac{D_i S_i(p_{i,\mathrm{f}} - p_{i,\mathrm{p}})}{l} \tag{13.11}$$

where D_i is the average effective diffusion coefficient (cm^2 s^{-1}) and S_i is the solubility coefficient [cm^3 (STP) cm^{-3} cmHg^{-1}] of component i, $p_{i,\mathrm{f}}$ and $p_{i,\mathrm{p}}$ are the partial pressure of component i at the feed and at the permeate side, respectively, and l is the thickness (cm) of the membrane.

The transport of gases and vapors through non-porous, dense polymer membranes is generally described with the solution–diffusion model [43, 63]. This model describes the transport as a three-step process. First, a component absorbs in the membrane at the feed side. Next, it diffuses through the membrane, and finally it desorbs at the permeate side. Starting from Fick's first law, one can define the permeability coefficient of component i (P_i) according to the equation

$$P_i = D_i S_i \tag{13.12}$$

Table 13.1 Physicochemical properties of CO2 and N2 [8].

Gas	Kinetic diameter (Å)	Critical temperature (K)
CO_2	3.30	304.2
N_2	3.64	126.1

The permeability coefficient is usually expressed in units of barrer, where 1 barrer equals 1×10^{-10} cm^3 (STP) cm cm^{-2} s^{-1} cmHg^{-1} or 7.5×10^{-18} m^3 (STP) m m^{-2} s^{-1} Pa^{-1}. The separation of gases and/or vapors is obtained due to differences in diffusivity and/or solubility of the components in the dense membrane layer and the selectivity of a membrane for component A over component B is given by the ratio of their respective permeability coefficients, P_A/P_B, according to the equation

$$\alpha = \frac{P_A}{P_B} = \frac{D_A}{D_B}\frac{S_A}{S_B} \tag{13.13}$$

where D_A/D_B is the diffusivity selectivity and S_A/S_B is the solubility selectivity. The membrane selectivity can be expressed either as an ideal selectivity (based on the measured permeability of pure, single gases) or as a real selectivity (based on the measured permeability using multicomponent gas mixtures).

Gas diffusivity increases with decreasing penetrant size (kinetic diameter), increasing polymer chain flexibility, and decreasing polymer–penetrant interactions. Penetrant solubility increases with increasing condensability of the penetrant (which is reflected in an increase in critical temperature and boiling point of the penetrant) and increasing polymer–penetrant interactions [8]. Table 13.1 summarizes the physicochemical properties of the main components of flue gas. The data clearly show that both the kinetic diameter (related to the diffusivity) and the critical temperature (related to the solubility) favor the permeation of CO_2 over N_2.

13.4 Gas and Vapor Transport Through Dense Polymer Membranes for Flue Gas Treatment

13.4.1 Introduction

A question not addressed in our previous publications is the extent to which these polymeric membranes can also be used for post-combustion CO_2 capture from flue gas and the extent to which the presence of water vapor influences the CO_2 separating performance of such membranes.

The work presented here addresses these issues and is dedicated to the characterization and understanding of molecular transport properties of two distinctively different polymeric membrane materials: the segmented block copolymer PEBAX 1074 and the sulfonated polymer S-PEEK. The understanding of gas and water vapor transport phenomena through these membrane materials is particularly interesting because the two materials are different in their chemistry and physical state:

- PEBAX shows molecular transport of gases and vapors through a soft rubbery, poly(ethylene oxide)-based phase.
- S-PEEK shows gas and vapor transport through an amorphous glassy phase containing ionic groups that will preferentially hydrate over the apolar matrix.

These two polymeric systems may vary strongly in relation to their polymer–gas interactions, relaxation phenomena, and separation performance of the membranes. The focus is on the transport of single components and binary and ternary mixtures as the presence of one component in the feed mixture may also influence the transport behavior of the other components, more specifically the transport of CO_2, N_2, and water vapor, through polymeric membranes for post-combustion CCS applications.

PEBAX is a flexible, rubbery material known as PEBAX 1074 (Arkema, Amsterdam, The Netherlands). It is a hydrophilic, commercially available block copolymer consisting of 45 wt% hard polyamide (PA) blocks (PA12) and 55 wt% soft poly(ethylene oxide) (PEO) blocks. The dissolution of gas and water vapor in and the subsequent diffusion through the polymer primarily occur in the (soft) amorphous PEO phase. The hard crystalline PA blocks provide mechanical and thermal stability to the membrane. Figure 13.8a shows the chemical structure of PEBAX.

The second material is a rigid, glassy polymer known as sulfonated poly(ether ether ketone) (S-PEEK). S-PEEK (Fig. 13.8b) can be prepared by sulfonation of PEEK according to the procedure described by Komkova *et al.* [64]. For the experiments described here, S-PEEK with a degree of sulfonation of 70% was used ($x=0.7$).

Thin films of both materials have been prepared [10, 33] and the transport behaviors of both materials for binary and ternary gas mixtures have been compared. The water vapor sorption in both materials at 50 °C has been determined

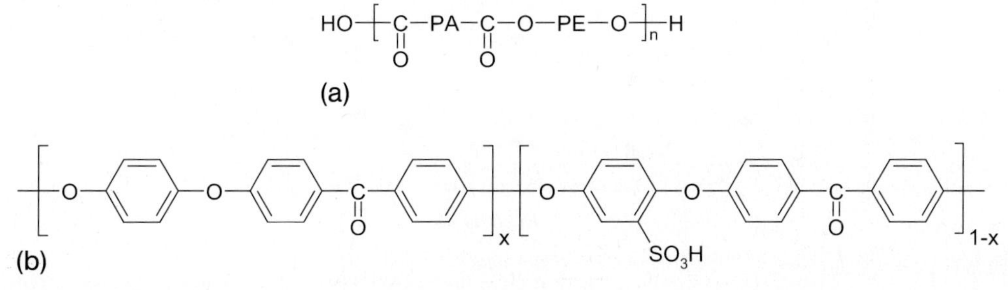

Fig. 13.8 Chemical structures of (a) PEBAX and (b) S-PEEK.

using a gravimetric sorption method [37] and the mixed water vapor–gas permeation behavior of both types of polymer films has been evaluated [7, 33, 37, 65].

13.4.2
Water Vapor Sorption

13.4.2.1 Sorption Isotherms

Water vapor sorption experiments were performed on PEBAX 1074 and S-PEEK to analyze the uptake of water vapor at 50 °C. Figure 13.9 shows the concentration of water vapor inside the polymer as a function of the water vapor activity for both polymers during sorption.

The concentration of water vapor inside PEBAX 1074 increases exponentially with increase in water vapor activity. The sorption isotherm represents Flory–Huggins-type sorption (convex increase towards the x-axis). This behavior is typical of rubbery polymers such as PEBAX 1074 [43], and especially at higher water vapor activities ($a > 0.6$) the amount of water vapor absorbed into the polymer is significant.

For S-PEEK, the water vapor concentration in the polymer films also increases with increasing water vapor activity. However, the sorption isotherm shows a concave increase for water vapor activities $a < 0.5$ (often described by the simple Dual Mode sorption model [45, 46] or the more extended energy site distribution model [48, 49]). For activities $a > 0.5$, the sorption isotherm has an inflection point turning to convex with an exponential increase, which is often described by the Flory–Huggins model.

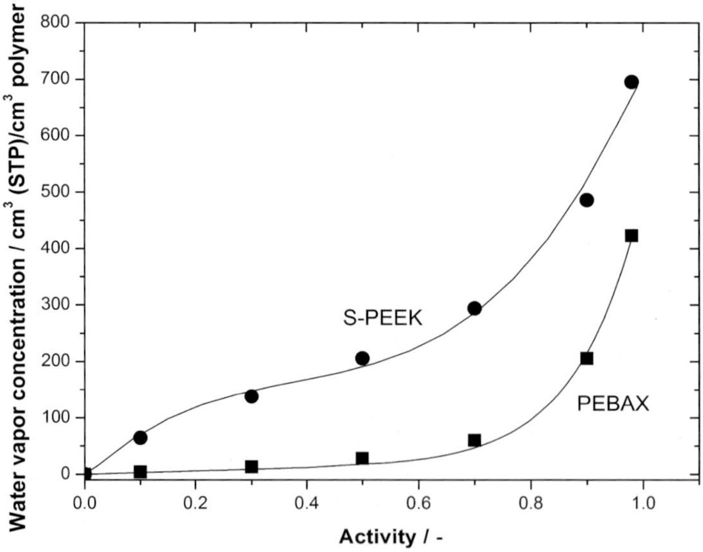

Fig. 13.9 Concentration of water in the polymer membranes as a function of water vapor activity at 50 °C for PEBAX 1074 and S-PEEK.

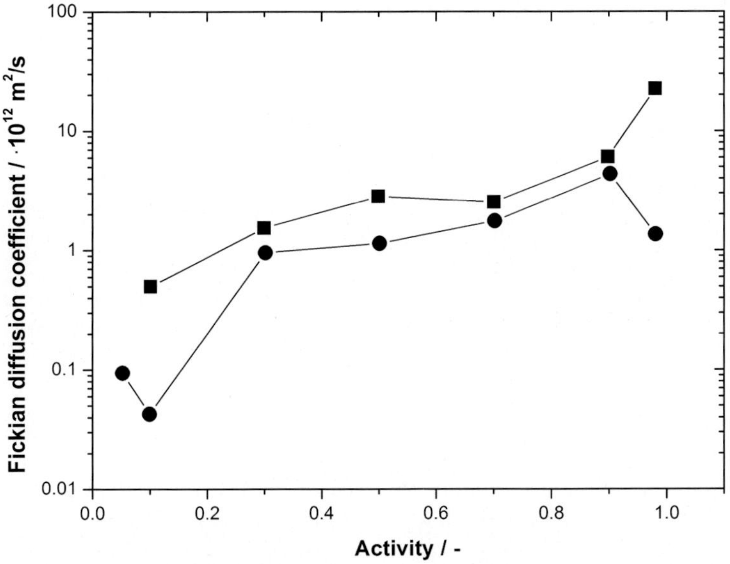

Fig. 13.10 Fickian diffusion coefficient corrected for swelling as a function of water vapor activity at 50 °C for (■) PEBAX 1074 and (●) S-PEEK.

13.4.2.2 **Fickian Diffusion**

The experimentally obtained kinetic sorption data were analyzed and the diffusion coefficient of water vapor in both polymeric materials was determined according to Equations 13.2–13.6. The resulting Fickian diffusion coefficients (corrected for swelling) are presented in Fig. 13.10.

For PEBAX 1074, the diffusion coefficient increases with increase in temperature and increase water vapor activity over the full range of activities. The strong increase in Fickian diffusion coefficient at especially higher water vapor activities is due to the exponential increase in water vapor uptake with increasing activity.

For S-PEEK, the Hopfenberg-Berens model has been used to fit the sorption data and allows the extraction of the Fickian diffusion coefficient from the experimental results. Also for S-PEEK the Fickian diffusion coefficient increases with increase in water vapor concentration. As discussed earlier, the sorption isotherms show two different sorption mechanisms: Dual Mode sorption for lower water vapor activities ($a < 0.5$) and Flory–Huggins sorption for higher water vapor activities ($a > 0.5$). According to Yampolskii *et al.* [43], an increase in diffusion coefficient with increase in water vapor concentration can be expected for both Dual Mode and Flory–Huggins sorption behavior.

13.4.2.3 Clustering Analysis of Sorption Isotherms

The state of the absorbed water molecules in both hydrophilic polymers can be evaluated using the cluster integral as described by Zimm and Lundberg [62]. Figure 13.11 presents the cluster integral as a function of the water vapor activity for PEBAX 1074 and S-PEEK.

The water molecules have a tendency to form clusters if the value of the cluster integral (G_{ww}/V_w) is >0. It is obvious from Fig. 13.11 that sorbed water molecules in PEBAX 1074 have a significant tendency to form water clusters in the polymer at each water vapor activity (clustering occurs when $G_{ww}/V_w>0$). This effect of water clustering in polymers is attributed to self-hydrogen bonding of water molecules [62]. One may expect cluster formation when penetrant–penetrant interactions are stronger than penetrant–polymer interactions [66]. At high water vapor activities ($a>0.6$), the cluster tendency decreases slightly as an additional water molecule that absorbs in the highly water swollen membrane will increasingly experience a more water-like environment due to the exponential increase in water vapor concentration observed at these high activities (Fig. 13.9) [33]. This decreases slightly the tendency to form clusters. A detailed thermodynamic analysis of water vapor sorption in PEBAX 1074 shows that enthalpically penetrant–penetrant interactions are indeed more favorable than polymer–penetrant interactions in PEBAX 1074 [67]. At higher water vapor activities, the thermodynamic data show that a newly sorbed water vapor molecule experiences a water-like environment in the polymer PEBAX 1074. The results of the cluster analysis show the same observation.

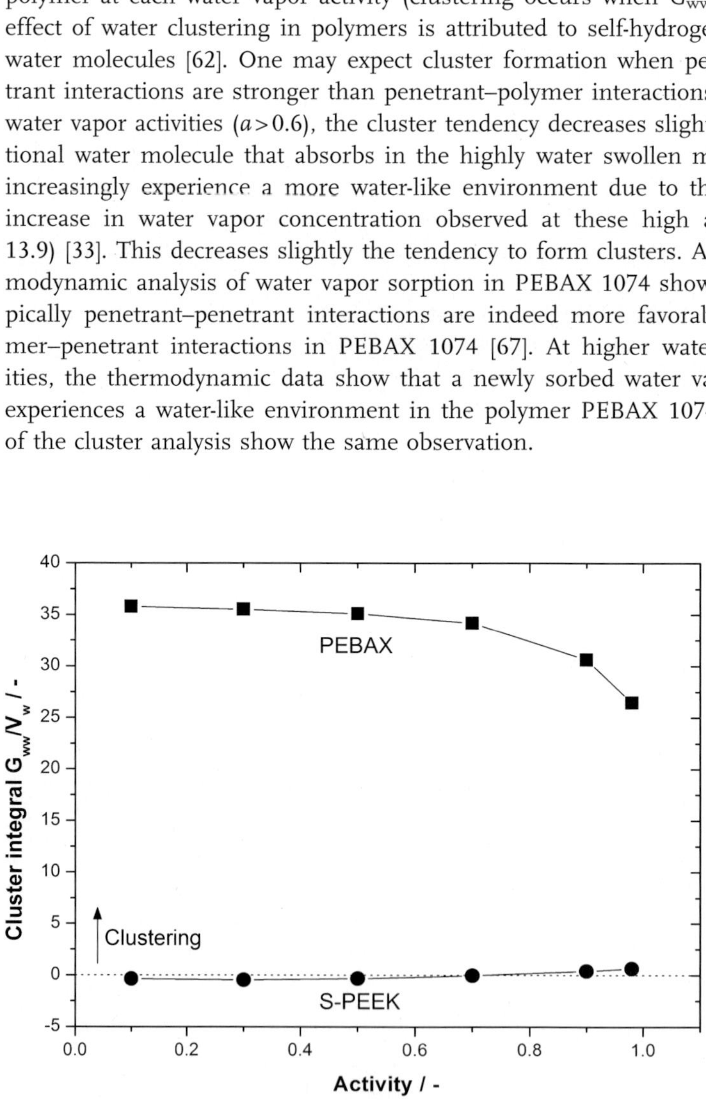

Fig. 13.11 Cluster integral as a function of water vapor activity at 50 °C for PEBAX 1074 and S-PEEK.

S-PEEK shows distinctly different behavior to PEBAX 1074. The cluster integral increases slightly with increase in water vapor activity and is nearly zero, indicating that the water molecules do not tend to cluster in the S-PEEK films investigated. Thus, the water molecules remain isolated at almost all water vapor concentrations, and as such cluster formation does not influence the sorption kinetics of water vapor molecules in the S-PEEK.

13.4.3
Water Vapor–Gas Transport

13.4.3.1 Binary Water Vapor–Gas Transport

Figure 13.12 shows the pure gas permeability for CO_2 and N_2 for both polymeric materials as a function of water vapor activity as determined from binary water vapor–gas permeation experiments.

Figure 13.12 shows that a distinct difference exists between the gas permeabilities observed for PEBAX 1074 and those for S-PEEK. S-PEEK, a rigid, glassy polymer, has very low gas permeabilities, especially for N_2. The presence of water vapor in the feed gas stream has a large influence on the gas permeability of S-PEEK due to the highly hydrophilic nature of the sulfonic acid group the N_2 permeability increases with increase in water vapor activity. The absorbed water swells the polymer and lowers the glass transition temperature [10], enhancing diffusion, resulting in higher gas permeabilities as the water vapor ac-

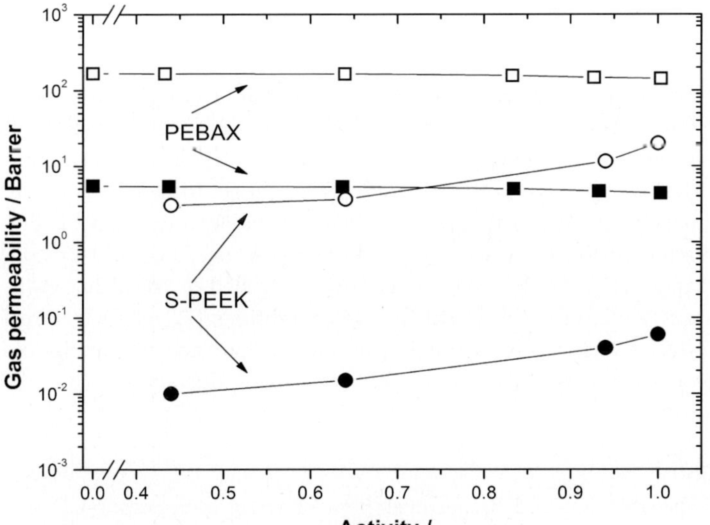

Fig. 13.12 CO_2 permeability (open symbols) and N_2 permeability (closed symbols) as a function of water vapor activity for PEBAX 1074 (squares) and S-PEEK (circles) as determined from binary water vapor–CO_2 and water vapor–N_2 gas mixtures.

tivity in the feed stream increases (and thus the glass transition temperature decreases). A high ideal CO_2/N_2 gas selectivity of $\sim$300 is achieved over the entire activity range investigated.

PEBAX 1074, on the other hand, is a flexible, rubbery polymer. Owing to its flexibility, gas diffusion coefficients are high, resulting in significantly higher overall gas permeabilities for both CO_2 and N_2, which would reduce membrane area and capital costs compared with S-PEEK. The disadvantage of such a high flexibility is a low CO_2/N_2 diffusivity selectivity ($\sim$1). However, the reasonable CO_2/N_2 selectivity ($\sim$30) is derived from the polar ether oxygen linkages, which favor the solubility of the quadrupolar CO_2 over the nonpolar N_2, resulting in high solubility selectivities [24]. The presence of water vapor has a completely different effect on PEBAX 1074 compared with S-PEEK. In this case, the gas permeability decreases as the water vapor activity in the feed stream increases.

A similar decrease has been observed by other researchers in polyether-based block copolymers for water vapor–N_2 [33, 37] and water vapor–CO_2 gas mixtures [37], and is believed to be related to reduced gas diffusion coefficients, but changes in gas solubility can also contribute to the observed behavior.

13.4.3.2 **Ternary Water Vapor–Gas Transport**

Although the use of binary water vapor–pure gas measurements shows that the presence of water vapor in the feed gas stream can have a profound effect on the gas permeability of polymeric membranes (especially in the case of rigid, glassy polymer membranes), it still lacks resemblance to the real-life situation and only an ideal CO_2/N_2 selectivity is obtained. To assess the viability of these membrane materials for application in real flue gas streams, more in-depth measurements are required. This is especially relevant as it is well known that the presence of CO_2 has the ability to influence (increase) the permeability of other, especially non-condensable, gases such as N_2 and O_2. This effect, commonly referred to as "plasticization," generally has a larger influence on rigid, glassy membranes compared with their flexible, rubbery counterparts. Therefore, the membrane performance was assessed using a ternary gas mixture containing water vapor, CO_2, and N_2. The gas mixture consisted of 10 vol.% CO_2 in N_2 (90 vol.%) to resemble closely typical flue gas conditions [7]. As such, a real mixed gas CO_2/N_2 selectivity as a function of water vapor activity can be obtained. The gas permeability and selectivity obtained using this specific gas mixture are discussed below.

Gas Permeability

Fig. 13.13 presents the CO_2 and N_2 permeability of ternary water vapor–CO_2–N_2 gas mixtures as a function of the water vapor activity. The gas permeabilities as measured using binary water vapor–CO_2 and water vapor–N_2 gas mixtures as discussed above (Fig. 13.12) have been added for comparison.

Using these ternary mixtures, the influence of the presence of CO_2 on the N_2 permeability (and vice versa) can be studied. Figure 13.13a shows that the CO_2

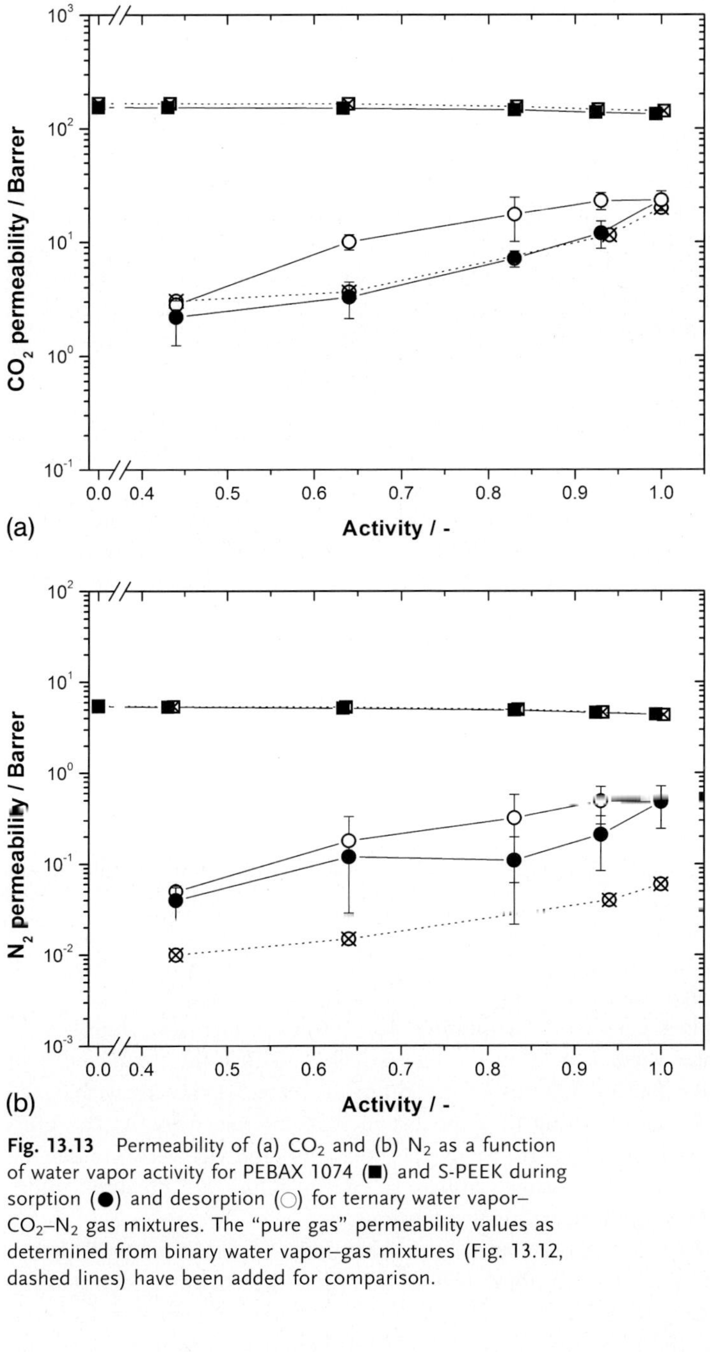

Fig. 13.13 Permeability of (a) CO_2 and (b) N_2 as a function of water vapor activity for PEBAX 1074 (■) and S-PEEK during sorption (●) and desorption (○) for ternary water vapor–CO_2–N_2 gas mixtures. The "pure gas" permeability values as determined from binary water vapor–gas mixtures (Fig. 13.12, dashed lines) have been added for comparison.

permeability is not influenced by the presence of N_2 in the gas mixture and CO_2 permeabilities similar to those using binary mixtures of water vapor–CO_2 are obtained for both the rubbery PEBAX 1074 and the glassy S-PEEK. This indicates that the presence of N_2 does not change the permeation behavior of CO_2 in these materials.

Significantly different behavior can be observed for the influence of the presence of CO_2 on the N_2 permeability (Fig. 13.13 b). In the case of the rubbery PEBAX 1074, no differences in N_2 permeability are observed between binary and ternary experimental conditions. This indicates that at the low feed pressures (and thus low partial pressures of CO_2) typically encountered in post-combustion CO_2 capture processes, the simultaneous presence of CO_2 and N_2 is not expected to influence the membrane performance compared with pure gas conditions. This is opposite to the behavior generally observed seen in glassy polymers, where already low partial pressures of the plasticizing component might induce a significant decrease in membrane performance under realistic mixed gas conditions.

This decrease in membrane performance (due to the plasticizing component) is exactly what is observed for the S-PEEK membrane material. Already at a very low CO_2 partial pressure (0.25 bar) it induces plasticization and relaxation phenomena within the membrane, resulting in an increase in the N_2 permeability over the full activity range of about one order of magnitude. The actual membrane separation performance is thus influenced by the simultaneous presence of the components that need to be separated. This influence is discussed in more detail below.

CO_2/N_2 Gas Selectivity

Figure 13.14 presents the real CO_2/N_2 selectivity as a function of water vapor activity for both materials. The ideal selectivity as determined from separate binary water vapor–CO_2 and water vapor–N_2 mixtures has been added for comparison.

Owing to the specific interactions of the ether oxygen linkages with CO_2 and the high flexibility of the polyether chains, a high CO_2 permeability combined with a reasonably high selectivity of CO_2 over N_2 is obtained at 50 °C for the PEBAX 1074 block copolymer membrane. The selectivity increases slightly with increase in water vapor activity in the feed stream, because the permeability of N_2 is decreased relatively more as a function of the water vapor activity than the permeability of CO_2, resulting in a modest increase in selectivity. As the presence of CO_2 has no influence on the N_2 permeability (and vice versa) under the operating conditions investigated, similar selectivity values are obtained under both binary (ideal) and ternary (real) conditions.

On the other hand, the plasticization that has been observed for S-PEEK has a large influence on the observed real CO_2/N_2 gas selectivity. Under realistic conditions, a CO_2/N_2 selectivity of ~50–100 is obtained, which is a factor of 3–6 lower than the ideal selectivity of ~300. Although this decrease is significant, the real selectivity of S-PEEK remains higher than that of PEBAX 1074 over the full activity range.

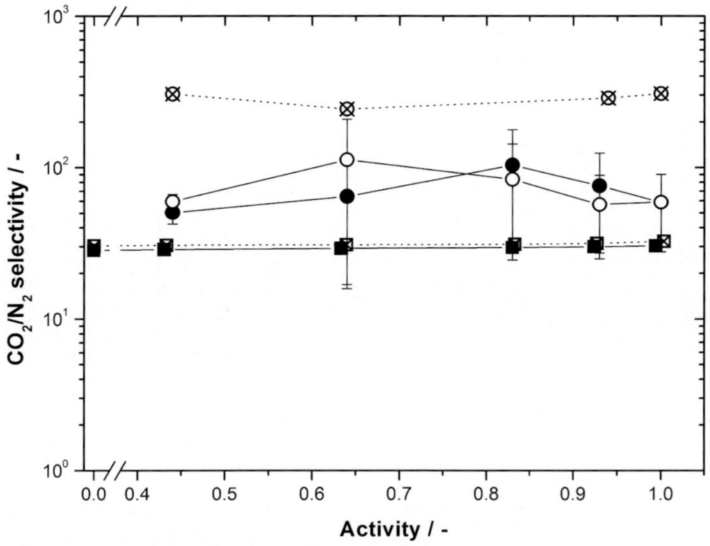

Fig. 13.14 Real CO_2/N_2 selectivity as a function of water vapor activity for PEBAX 1074 (■) and S-PEEK during sorption (●) and desorption (○) for ternary water vapor–CO_2–N_2 gas mixtures. The ideal selectivity as determined from separate binary water vapor–CO_2 and water vapor–N_2 mixtures (dashed lines) has been added for comparison.

H_2O Permeability and Selectivity

Figure 13.15 presents the water vapor permeability selectivity as a function of water vapor activity for both materials using the ternary water vapor–CO_2–N_2 gas mixture. A comparison with the water vapor permeability using binary water vapor–pure gas mixtures (not shown) has demonstrated that the composition of the gas mixture has no influence on the overall water vapor permeability.

The water vapor permeability in both membrane materials is high as water vapor has both a high solubility (due to its high critical temperature, $T_c = 647$ K) and a high diffusivity (due to its small kinetic diameter, $d_k = 2.65$ Å) in the membrane materials. The water vapor permeability increases with increase in water vapor activity. This increase is dominated by an increase in water vapor solubility. At high water vapor activity ($a > 0.6$), the water vapor permeability of S-PEEK is higher than that of PEBAX 1074 as the increase in the water vapor solubility of S-PEEK is more significant than that of PEBAX 1074 (see Fig. 13.9).

The high water vapor permeability has an added advantage as it not only makes these membranes suitable for the capture of CO_2 from flue gas, but also raises the possibility of combining the separation of CO_2 and water vapor in a single step.

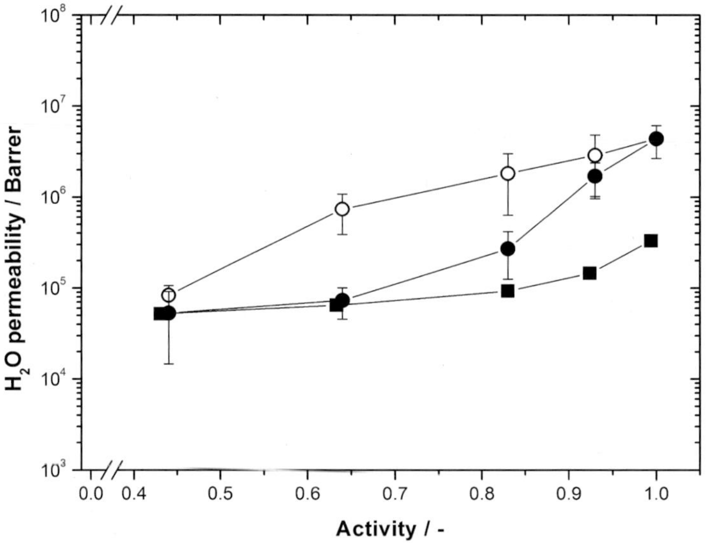

Fig. 13.15 Water vapor permeability as a function of water vapor activity for PEBAX 1074 (■) and S-PEEK during sorption (●) and desorption (○) for ternary water vapor–CO_2–N_2 gas mixtures.

Figure 13.16 presents the water vapor/gas selectivity for water vapor over CO_2 and water vapor over N_2 as a function of the water vapor activity as determined from ternary water vapor/gas permeation experiments.

For both materials, the water vapor/gas selectivity increases as a function of the water vapor activity. This increase is mainly the result of an increase in water vapor permeability (Fig. 13.15) due to the strong sorption of water vapor. The water vapor/N_2 selectivity is approximately 1–2 orders of magnitude higher than the water vapor/CO_2 selectivity. This is mainly the result of a significantly higher CO_2 permeability due to its specific interaction with the ether oxygen linkages (in the case of PEBAX 1074) or the sulfonic acid groups (in the case of S-PEEK). Furthermore, most likely, and especially at higher water vapor activities, the higher solubility of CO_2 in water vapor also plays a role at higher activities as in that case the gas will also experience a more water-like environment.

13.5 Conclusion

This chapter discusses the transport of CO_2 through two distinctively different polymeric membranes, PEBAX and S-PEEK, for post-combustion capture applications. Both materials are different in their chemistry and physical state: PEBAX shows molecular transport of gases and vapors through a soft, rubbery, poly(ethylene oxide)-based phase, whereas S-PEEK shows gas and vapor trans-

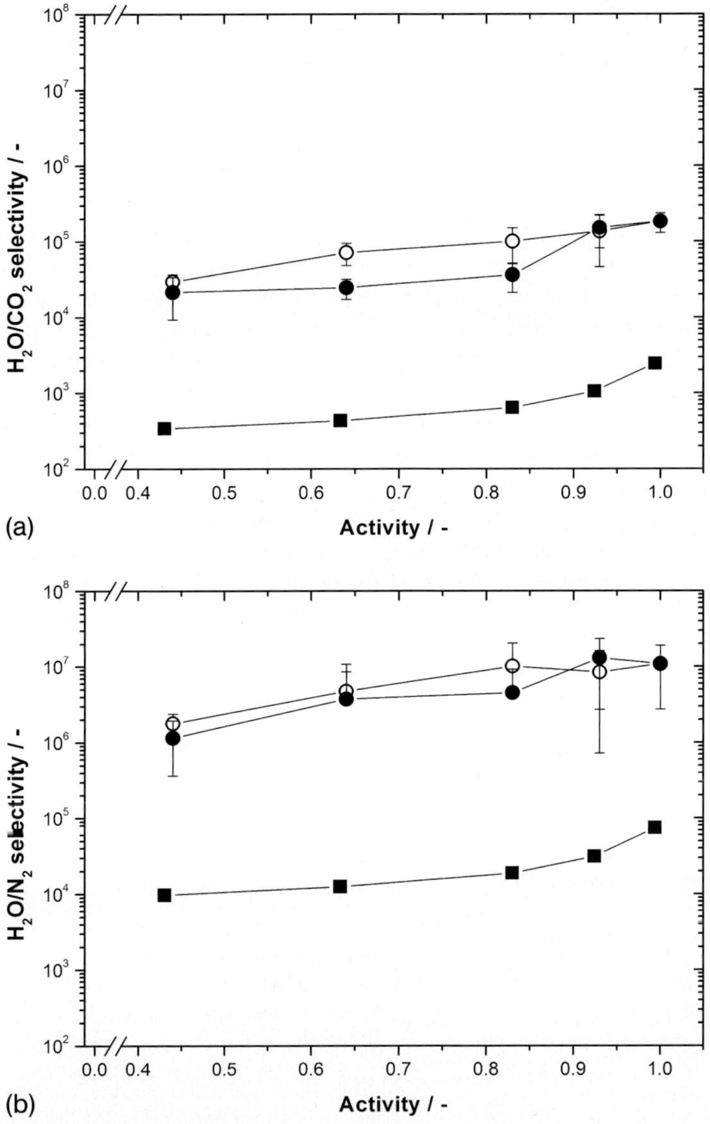

Fig. 13.16 Water vapor/gas selectivity for (a) H_2O–CO_2 and (b) H_2O–N_2 as a function of water vapor activity for PEBAX 1074 (■) and S-PEEK during sorption (●) and desorption (○) for ternary water vapor–CO_2–N_2 gas mixtures.

port through an amorphous glassy phase containing ionic groups that will preferentially hydrate over the apolar matrix. In both types of membranes the water vapor sorption increases with increase in water vapor activity, showing Flory–Huggins-type sorption for the rubbery polymer PEBAX, whereas the glassy polymer S-PEEK shows dual mode sorption at lower water vapor activities and Flory–Huggins-type sorption at higher water vapor activities. The kinetic water vapor diffusion coefficients corrected for the extensive swelling of the material upon water vapor sorption are almost equal for both polymer types and increase with increase in water vapor activity. Water vapor molecules in PEBAX show the occurrence of clustering attributed to self-hydrogen bonding of water molecules and penetrant–penetrant interactions are expected to be stronger than penetrant–polymer interactions. The results show that the presence of N_2 when comparing the binary and ternary systems does not change the transport behavior of CO_2 through both membrane types, as is the case for the effect of the presence of CO_2 on the N_2 permeability in the rubbery material PEBAX. The presence of CO_2 does, however, increase the transport of N_2 through the glassy polymer membrane S-PEEK due to the occurrence of plasticization and relaxation phenomena, resulting in considerably lower selectivities for ternary systems compared with binary systems in the case of such glassy polymer membranes. Nevertheless, the CO_2/N_2 selectivities obtained are sufficiently high for an economically viable membrane process for post-combustion CO_2 capture [42, 68].

References

1 Intergovernmental Panel on Climate Change (IPCC). *Third Assessment Report: Climate Change 2001*, Cambridge University Press, Cambridge, 2001.

2 2. Powell, C.E. and Qiao, G.G. Polymeric CO_2N_2 gas separation membranes for the capture of carbon dioxide from power plant flue gases. *J. Membr. Sci.*, 2006, **279**(12), 149.

3 Baker, R.W. *Membrane Technology and Applications*, 2nd edn, John Wiley & Sons, Ltd, Chichester, 2000.

4 Baker, R.W. Future directions of membrane gas separation technology. *Ind. Eng. Chem. Res.*, 2002, **41**(6), 1393–1411.

5 Baker, R.W. and Lokhandwala, K. Natural gas processing with membranes: an overview. *Ind. Eng. Chem. Res.*, 2008, **47**(7), 2109–2121.

6 Visser, T., Koops, G.H., and Wessling, M. On the subtle balance between competitive sorption and plasticization effects in asymmetric hollow fiber gas separation membranes. *J. Membr. Sci.*, 2005, **252**(12), 265–277.

7 Sijbesma, H., Nymeijer, K., van Marwijk, R., Heijboer, R., Potreck, J., and Wessling, M. Flue gas dehydration using polymer membranes. *J. Membr. Sci.*, 2008, **313** (12), 263–276.

8 Mulder, M.H.V. *Basic Principles of Membrane Technology*, 2nd edn, Kluwer, Dordrecht, 1996.

9 Visser, T. and Wessling, M. When do sorption-induced relaxations in glassy polymers set in? *Macromolecules*, 2007, **40**(14), 4992–5000.

10 Potreck, J., Uyar, F., Sijbesma, H., Nijmeijer, K., Stamatialis, D., and Wessling, M. Sorption induced relaxations during water diffusion in S-PEEK. *Phys. Chem. Chem. Phys.*, 2009, **11**(2), 298–308.

11 Bos, A., Pünt, I.G.M., Wessling, M., and Strathmann, H. CO_2-induced plasticization phenomena in glassy polymers. *J. Membr. Sci.*, 1999, **155**(1), 67–78.

12 Sanders, E.S. Penetrant-induced plasticization and gas permeation in glassy polymers. *J. Membr. Sci.*, 1988, **37**(1), 63–80.

13 Wessling, M., Schoeman, S., van der Boomgaard, T., and Smolders, C.A. Plasticization of gas separation membranes.*Gas Sep. Purif.*, 1991, **5**(4), 222–228.

14 Chiou, J.S., Barlow, J.W., and Paul, D.R. Plasticization of glassy polymers by CO_2. *J. Appl. Polym. Sci.*, 1985, **30**(6), 2633–2642.

15 Reijerkerk, S.R., Arun, A., Gaymans, R.J., Nijmeijer, K., and Wessling, M. Tuning of mass transport properties of multi-block copolymers for CO_2 capture applications. *J. Membr. Sci.*, 2010, **359**(12), 5463.

16 Staudt-Bickel, C., Koros, W.J. Improvement of CO_2CH4 separation characteristics of polyimides by chemical crosslinking. *J. Membr. Sci.*, 1999, **155**(1), 145–154.

17 Lin, H., Van Wagner, E., Raharjo, R., Freeman, B.D., and Roman, I. High-performance polymer membranes for natural-gas sweetening. *Adv. Mater.*, 2006, **18**(1), 39–44.

18 Wind, J.D., Sirard, S.M., Paul, D.R., Green, P.F., Johnston, K.P., and Koros, W.J. Relaxation dynamics of CO_2 diffusion, sorption, and polymer swelling for plasticized polyimide membranes. *Macromolecules*, 2003, **36**(17), 6442–6448.

19 Wind, J.D., Paul, D.R., and Koros, W.J. Natural gas permeation in polyimide membranes. *J. Membr. Sci.*, 2004, **228**(2), 227–236.

20 Wind, J.D., Staudt-Bickel, C., Paul, D.R., and Koros, W.J. The effects of cross-linking chemistry on CO_2 plasticization of polyimide gas separation membranes. *Ind. Eng. Chem. Res.*, 2002, **41**(24), 6139–6148.

21 Hillock, A.M.W. and Koros, W.J. Cross-linkable polyimide membrane for natural gas purification and carbon dioxide plasticization reduction. *Macromolecules*, 2007, **40**(3), 583–587.

22 Wind, J.D., Sirard, S.M., Paul, D.R., Green, P.F., Johnston, K.P., and Koros, W.J. Carbon dioxide-induced plasticization of polyimide membranes: pseudo-equilibrium relationships of diffusion, sorption, and swelling. *Macromolecules*, 2003, **36**(17), 6433–6441.

23 Visser, T., Masetto, N., and Wessling, M. Materials dependence of mixed gas plasticization behavior in asymmetric membranes. *J. Membr. Sci.*, 2007, **306**(12), 16–28.

24 Lin, H. and Freeman, B.D. Materials selection guidelines for membranes that remove CO_2 from gas mixtures. *J. Mol. Struct.*, 2005, **739**(13), 57–74.

25 Lin, H. and Freeman, B.D. Gas and vapor solubility in cross-linked poly(ethylene glycol diacrylate). *Macromolecules*, 2005, **38**(20), 8394–8407.

26 Lin, H. and Freeman, B.D. Gas permeation and diffusion in cross-linked poly(ethylene glycol diacrylate). *Macromolecules*, 2006, **39**(10), 3568–3580.

27 Lin, H., Kai, T., Freeman, B.D., Kalakkunnath, S., and Kalika, D.S. The effect of cross-linking on gas permeability in cross-linked poly(ethylene glycol diacrylate). *Macromolecules*, 2005, **38**(20), 8381–8393.

28 Lin, H., Wagner, E.V., Swinnea, J.S., Freeman, B.D., Pas, S.J., Hill, A.J., Kalakkunnath, S., and Kalika, D.S. Transport and structural characteristics of crosslinked poly(ethylene oxide) rubbers. *J. Membr. Sci.*, 2006, **276**(12), 145–161.

29 Lin, H., Van Wagner, E., Freeman, B.D., Toy, L.G., and Gupta, R.P. Plasticization-enhanced hydrogen purification using polymeric membranes. *Science*, 2006, **311**(5761), 639–642.

30 Kelman, S., Lin, H., Sanders, E.S., and Freeman, B.D. $CO_2C_2H_6$ separation using solubility selective membranes. *J. Membr. Sci.*, 2007, **305**(12), 57–68.

31 Lin, H., Freeman, B.D., Kalakkunnath, S., and Kalika, D.S. Effect of copolymer composition, temperature, and carbon dioxide fugacity on pure- and mixed-gas permeability in poly(ethylene glycol)-based materials: free volume interpretation. *J. Membr. Sci.*, 2007, **291**(12), 131–139.

32 Metz, S.J., Mulder, M.H.V., and Wessling, M. Gas-permeation properties of poly(ethylene oxide)poly(butylene terephthalate) block copolymers. *Macromolecules*, 2004, **37**(12), 4590–4597.

33 Potreck, J., Nijmeijer, K., Kosinski, T., and Wessling, M. (2009,) Mixed water vaporgas transport through the rubbery

polymer PEBAX® 1074. *J. Membr. Sci.*, **338**(12), 11–16.

34 IJzer, A.C., Arun, A., Reijerkerk, S.R., Nijmeijer, K., Wessling, M., and Gaymans, R.J. Synthesis and properties of hydrophilic segmented block copolymers based on poly(ethylene oxide)-ran-poly(propylene oxide). *J. Appl. Polym. Sci.*, 2010, **117**(3), 1394–1404.

35 Reijerkerk, S.R., IJzer, A.C., Nijmeijer, K., Arun, A., Gaymans, R.J., and Wessling, M. Subambient temperature CO_2 and light gas permeation through segmented block copolymers with tailored soft phase. *ACS Appl. Mater. Interfaces*, 2010, **2**(2), 551–560.

36 Reijerkerk, S.R., Knoef, M.H., Nijmeijer, K., and Wessling, M. Poly(ethylene glycol) and poly(dimethyl siloxane): combining their advantages into efficient CO_2 gas separation membranes. *J. Membr. Sci.*, 2010, **352**(12), 126–135.

37 Reijerkerk, S.R., Jordana, R., Nijmeijer, K., and Wessling, M. Highly hydrophilic, rubbery membranes for CO_2 capture and dehydration of flue gas. *Int. J. Greenhouse Gas Control*, 2010, **5**(1) 2011, 26–36.

38 Reijerkerk, S.R., Nijmeijer, K., Ribeiro C.P. Jr, Freeman, B.D., and Wessling, M. On the effects of plasticization in CO_2light gas separation using polymeric solubility selective membranes. *J. Membr. Sci.*, 2011, **367**(12), 33–44.

39 Robeson, L.M. Correlation of separation factor versus permeability for polymeric membranes. *J. Membr. Sci.*, 1991, **62**(2), 165–185.

40 Robeson, L.M. The upper bound revisited. *J. Membr. Sci.*, 2008, **320**(12), 390–400.

41 US Department of Energy. *Carbon Sequestration Technology Roadmap and Program Plan 2007*, US Department of Energy, Washington, DC, 2007.

42 Merkel, T.C., Lin, H., Wei, X., and Baker, R. Power plant post-combustion carbon dioxide capture: an opportunity for membranes. *J. Membr. Sci.*, 2010, **359**(12), 126–139.

43 Yampolskii, Y., Pinnau, I., and Freeman, B.D. *Materials Science of Membranes for Gas and Vapor Separation*, 1st edn, John Wiley & Sons, Ltd, Chichester, 2006.

44 Metz, S.J., van der Vegt, N.F.A., Mulder, M.H.V., and Wessling, M. Thermodynamics of water vapor sorption in poly(ethylene oxide)poly(butylene terephthalate) block copolymers. *J. Phys. Chem. B*, 2003, **107**(49), 13629–13635.

45 Barrer, R.M. (1984,) Diffusivities in glassy polymers for the dual mode sorption model. *J. Membr. Sci.*, **18**, 2535.

46 Horas, J.A. and Nieto, F. Generalization of dual mode transport theory for glassy polymers. *J. Polym. Sci. Part B: Polym. Phys.*, 1994, **32**(11), 1889–1898.

47 Paul, D.R. Gas sorption and transport in glassy polymers. *Ber. Bunsenges. Phys. Chem.*, 1979, **83**(4), 294–302.

48 Kirchheim, R. Sorption and partial molar volume of small molecules in glassy polymers. *Macromolecules*, 1992, **25**(25), 6952–6960.

49 Kirchheim, R. Partial molar volume of small molecules in glassy polymers. *J. Polym. Sci. Part B: Polym. Phys.*, 1993, **31**(10), 1373–1382.

50 Lee, D.K., Kim, Y.W., Lee, K.J., Min, B.R., and Kim, J.H. Thermodynamic model of gas permeability in polymer membranes. *J. Polym. Sci. Part B: Polym. Phys.*, 2007, **45**(6), 661665.

51 Ubbink, J., Giardiello, M.I., and Limbach, H.J. Sorption of water by bidisperse mixtures of carbohydrates in glassy and rubbery states. *Biomacromolecules*, 2007, **8**(9), 2862–2873.

52 Serad, G.E., Freeman, B.D., Stewart, M.E., and Hill, A.J. Gas and vapor sorption and diffusion in poly(ethylene terephthalate). *Polymer*, 2001, **42**(16), 6929–6943.

53 Flory, P.J. *Principles of Polymer Chemistry*, Cornell University Press, Ithaca, NY, 1963.

54 Price, G.J., Haddon, D.A., Bainbridge, A., and Buley, J.M. Vapour sorption studies of polymer-solution thermodynamics using a piezoelectric quartz crystal microbalance. *Polym. Int.*, 2006, **55**(7), 816–824.

55 Crank, J. The Mathematics of Diffusion, 2nd edn, Oxford University Press, London, 1975.

56 Del Nobile, M. A., Mensitieri, G., Netti, P. A., and Nicolais, L. Anomalous diffusion in poly-ether-ether-ketone. *Chem. Eng. Sci.*, 1994, **49**(5), 633–644.

57 Grzywna, Z. J. and Stolarczyk, J. Diffusion in glassy polymer membranes submitted to induced stresses. *Polimery*, 2001, **46**(5), 351–358.

58 Joshi, S. and Astarita, G. Diffusion-relaxation coupling in polymers which show two-stage sorption phenomena. *Polymer*, 1979, **20**(4), 455–458.

59 Sun, Y. M. Sorption/desorption properties of water vapour in poly(2-hydroxyethyl methacrylate): 2. Two-stage sorption models. *Polymer*, 1996, **37**(17), 39213928.

60 Berens, A. R. and Hopfenberg, H. B. Diffusion and relaxation in glassy polymer powders: 2. Separation of diffusion and relaxation parameters. *Polymer*, 1978, **19**(5), 489–496.

61 Vrentas, J. S., Jarzebski, C. M., and Duda, J. L. Deborah number for diffusion in polymersolvent systems. *AIChE J.*, 1975, **21**(5), 894–901.

62 Zimm, B. H. and Lundberg, J. L. Sorption of vapors by high polymers. *J. Phys. Chem.*, 1956, **60**(4), 425–428.

63 Wijmans, J. G. and Baker, R. W. The solution-diffusion model: a review. *J. Membr. Sci.*, 1995, **107**(12), 121.

64 Komkova, E. N., Wessling, M., Krol, J., Strathmann, H., and Berezina, N. P. Influence of the nature of polymer matrix and the degree of sulfonation on the properties of membranes. *Polym. Sci. Ser. A*, 2001, **43**(3), 300–307.

65 Metz, S. J., van de Ven, W. J.C., Potreck, J., Mulder, M. H.V., and Wessling, M. Transport of water vapor and inert gas mixtures through highly selective and highly permeable polymer membranes. *J. Membr. Sci.*, 2005, **251**(12), 2941.

66 Barrie, J. A. and Platt, B. The diffusion and clustering of water vapour in polymers. *Polymer*, 1963, **4**, 303–313.

67 Potreck, J., Nijmeijer, K., van der Vegt, N. F.A., and Wessling, M. Thermodynamics of water vapor sorption in hydrophilic polymers. 2010, to be submitted for publication.

68 Shelley, S. Capturing CO_2: membrane-based systems move forward. *Chem. Eng. Progr.*, 2009, **105**(4), 42–47.

Processes for Membrane Separation of CO_2

14
CO_2 Separation via the Post-Combustion Process with Membranes in Coal Power Plants

Peter Michael Follmann, Christoph Bayer, Matthias Wessling, and Thomas Melin

Abstract

This chapter provides an integral view on membrane-based post-combustion CO_2 capture, beginning with a review of the process boundary conditions. Subsequently, suitable membrane materials for CO_2 capture are identified. In addition to the membrane material, driving force generation constitutes another critical issue in terms of technical realization. The review on driving force generation includes a detailed discussion of trade-offs associated with membrane gas separation processes and a summary of published studies on this subject. The chapter closes with a brief discussion of process economics.

Keywords: coal-fired power plant, membranes, post-combustion CO_2 capture, driving force generation

14.1
Introduction

Chemical absorption is currently considered the best available technology for post-combustion CO_2 capture. Recent publications, however, indicate that the energy requirement for solvent recovery may cause efficiency losses of up to 16% [1]. Moreover, the chemical degradation of the solvent leads to the continuous need for replacement and thus additional material and disposal costs. Due to important modifications of the steam cycle, the integration of chemical absorption units into existing power plants appears to be a challenging task [1–3].

In contrast to absorption, membrane processes operate without additional chemicals. Moreover, they may offer higher energy efficiencies, greater operational flexibility, and simplicity of operation and maintenance [4, 5]. The important number of publications and large-scale research projects such as NanoGLOWA [6] and Membrain [7] reflect the great interest in membrane technology for CO_2 capture.

Process Engineering for CCS Power Plants, First Edition. Edited by Detlef Stolten and V. Scherer

Membranes generally feature lower selectivities towards CO_2 than chemical absorption processes. Hence, the specification of the process boundary conditions – particularly downstream of the membrane separation process – is very important and is reviewed first. Subsequently, potential membrane materials for post-combustion CO_2 capture are briefly examined. In addition to the membrane material, the driving force across the membrane constitutes another critical issue for CO_2 capture. Apart from the examination of trade-offs in membrane gas separation technology, special attention is devoted to driving force generation. Finally, the economics of membrane-based CO_2 capture processes are addressed.

14.2
Process Boundary Conditions

The definition of the respective process boundary conditions is required for the design and optimization of any separation process. Boundary conditions usually comprise compositions, temperatures, pressures, and flow rates of all streams connected to the separation unit.

Figure 14.1 indicates that the flue gas stream establishes the inlet conditions of any CO_2 capture unit. Whereas composition and temperature lead to specific requirements regarding chemical and thermal resistance of the process equipment, the flow rate determines the dimensions of the CO_2 separation process. CO_2 transport and CO_2 storage set constraints on the CO_2-rich product stream. Technical, safety, environmental, and legal aspects may further restrict the concentration of co-captured species such as O_2, SO_x, and NO_x.

14.2.1
Upstream Conditions: the Coal-Fired Power Plant

Figure 14.2 represents schematically a state-of-the-art flue gas purification system including NO_x reduction by selective catalytic reduction (SCR), dust removal by electrostatic precipitators (ESPs), and a wet flue gas desulfurization unit (FGD). The scrubbed flue gas passes through a mist eliminator prior to its release into the atmosphere by the stack or cooling tower [9].

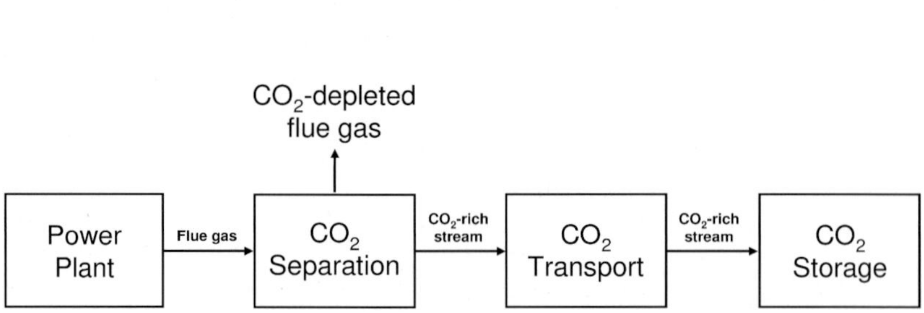

Fig. 14.1 Schematic overview of post-combustion CO_2 capture. Adapted from [8].

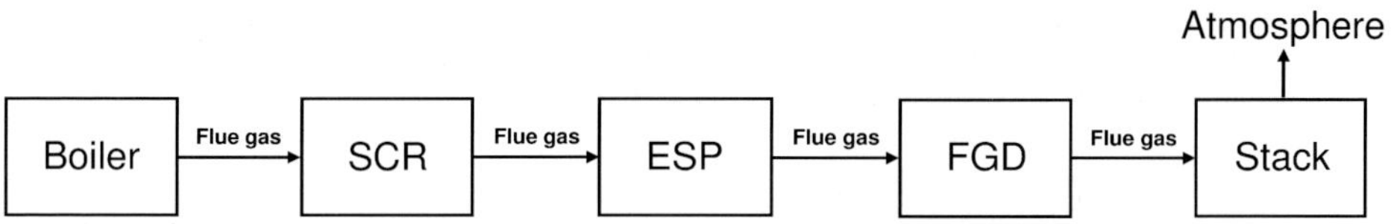

Fig. 14.2 Schematic representation of a state-of-the-art flue gas purification system.

Table 14.1 Molar flue gas compositions at FGD outlet.

Ref.	N_2 (mol%)	CO_2 (mol%)	O_2 (mol%)	H_2O (mol%)
12	72.0	13.0	3.0	12.0
13	71.6	13.3	3.8	11.3
14	71.8	13.6	3.4	11.2

To minimize the required modifications of the power plant, the optimal position of a CO_2 capture process is between the FGD unit and the stack. Moreover, particulate matter and also SO_x and NO_x concentrations have been reduced to a large extent [10]. Hence, the flue gas condition subsequent to the FGD unit represents the inlet condition of the CO_2 capture facility. The flue gas stream leaves the desulfurization unit saturated with water at about 50 °C and 1.013 bar. The volume flow depends on the electrical output of the power plant and is case specific. Table 14.1 summarizes the flue gas compositions used in different engineering studies. In addition to the gaseous constituents, the flue gas still contains particles. Pakala and Bhown [11] reported a particle load of 2–5 mg m^{-3} and particle diameters of 0.2–0.3 μm.

14.2.2
CO_2 Pipelining

Among the variety of options for CO_2 transmission from power plants to storage sites, only pipelines and ships provide reasonable economics for large-scale applications [15]. The topology of Germany and most other European countries and also the location of most power plants clearly favor pipeline transmission in conjunction with carbon capture and storage (CCS).

CO_2 pipelines have been operated in the United States since the late 1970s and have reached a cumulated length of about 2300 km and a cumulated throughput of almost 44 Mt a^{-1} of CO_2 [16]. CO_2 pipelining is a commercially available, mature technology [17, 18].

Since CO_2 pipelining in the liquid phase has no commercial application and transmission in the gaseous state is uneconomic for large volumes, commercial pipelines are operated in the dense phase [19–21]. The literature, however, indicates that operating conditions are very case specific [18, 20, 22]. The injection pressure, for instance, depends considerably on the required delivery pressure, the pipeline length, and the number of booster stations.

Table 14.2 Summary of requirements for CO_2 transmission by pipeline. Adapted from [8].

Parameter	Ref.			
	18	24	25	26
CO_2 (mol%)	>95	>95.5	97–99	96
N_2 (mol%)	<4	<4	<3	0.6
Ar (mol%)	–	<4	–	–
O_2 (ppm)	<10	0.1–4	–	100
Total air gases (mol%)	–	<4	–	–
H_2O (ppm)	<600	<500	<600	20

Noncondensable gases such as N_2, O_2, and Ar have an adverse effect on pipeline flow properties, operating pressures, and compression requirements [20, 23]. Even minor quantities of these gases lead to a significant increase in the lower pipeline operating pressure and thus in compression requirements. Therefore, the concentration of noncondensable gases should be as low as possible to ensure economic CO_2 transmission. Since general quality standards for CO_2 transmission do not exist so far [24], we need to rely on published data to estimate the concentration limits of noncondensable gases. Table 14.2 summarizes CO_2 specifications for pipeline transport. The data are related to existing pipelines that transmit CO_2 for enhanced oil recovery (EOR) purposes.

In addition to noncondensable gases, CO_2 originating from coal-fired power plants may be contaminated with SO_x and NO_x. According to Race *et al.* [27], no experience exists regarding the transmission of CO_2 streams with such contaminants. In addition to technical aspects, SO_x and NO_x levels may be relevant regarding health and safety. However, so far no regulation deals with the restrictions of toxic substances accompanying the CO_2 [24].

The elevated vapor concentration in power plant flue gases (compare Table 14.1) may result in the co-capture of water in conjunction with CO_2. Because carbon and low-alloy steels are typical construction materials for CO_2 pipelines, corrosion and hydrate formation are major concerns if moisture is present [22, 28]. Hence the precipitation of free water must be avoided to prevent corrosion and hydrate formation. Mohitpour *et al.* [28], for example, stated that industry generally accepts 300–500 ppm of residual water content for carbon steel pipelines. The operating data in Table 14.2 seem to reflect that statement.

14.2.3
CO_2 Storage

The overall objective associated with storing CO_2 is its long-term isolation from the atmosphere until a significant decline in atmospheric levels has taken place [29]. Advantages and drawbacks of CO_2 sinks are widely discussed in the literature [18, 30, 31]. There is strong evidence that subsurface sinks are particularly interesting for CO_2 storage. Among subsurface sinks, oil and gas reservoirs and also deep saline aquifers have been identified as the most attractive storage options [32]. May and Turkovic [33] recommended geologic storage only at supercritical conditions to guarantee the effective use of storage volumes.

CO_2 purity requirements for geologic storage may be different from the specifications derived for pipeline transmission. In addition to technical considerations, legal issues and public acceptance may also influence the purity requirements for geologic CO_2 storage.

As Table 14.2 indicates, the CO_2 purity requirements for EOR are case specific. Nevertheless, the concentration of lighter impurities such as N_2 and Ar should not exceed about 4 mol% to maintain the oil-extraction properties of the CO_2. Since O_2 may react with the reservoir oil, resulting in an even more difficult extraction, the residual O_2 content must be very low.

Unlike oil reservoirs, there is little information available about purity requirements for CO_2 storage in depleted gas reservoirs. Hence it is currently impossible to quantify the concentration limits for impurities co-captured with CO_2.

Deep aquifers contain fossil high-salinity connate water that is not of interest for industrial and agricultural use or for human consumption. Light components co-injected with CO_2 do not dissolve in the formation water as well as CO_2 does [34]. They remain mostly in a separate phase occupying pore volume that is not available for CO_2 storage [35].

In summary, the literature provides data mainly on EOR applications that feature very strict CO_2 quality standards. Owing to a lack of data, the same standards may be assumed for CO_2 storage in natural gas reservoirs. Deep saline aquifers are currently the subject of extensive investigations, but the impact of O_2, NO_x, and SO_x on the reservoir conditions and the integrity of the cap rock layer are not yet completely understood.

14.2.4
Summary of Process Boundary Conditions

It was the objective of this section to review the up- and downstream process boundary conditions. Probably the best position to integrate a membrane unit within the flue gas purification process is after the FGD unit. The respective flue gas conditions define the requirements with which the equipment used in the capture process should comply.

The transmission of the captured CO_2 from the power plant to an appropriate sink is most likely to be realized by pipeline. The composition of the pipelined CO_2 may depend on technical and also safety and legal aspects. Due to the present uncertainty, particularly regarding co-captured contaminates, purity requirements are not exactly specified as yet.

Concerning CO_2 storage, there is strong evidence that subsurface sinks are the preferred solution. Among potential subsurface sinks, depleted oil and gas reservoirs and also deep aquifers have attracted the greatest attention. The literature, however, currently addresses only CO_2 purity requirements for EOR, and provides very little tangible information for CO_2 storage in natural gas fields and deep saline aquifers. The impact of co-captured species such as O_2, SO_2, and NO is currently not completely understood. In analogy with CO_2 transmission, the purity requirements for CO_2 storage are not completely known at present.

Owing to the limited selectivity of membranes, the insufficient specification of downstream boundary conditions renders it difficult to design and optimize a membrane-based CO_2 capture processes. Slight variations of the allowed content of co-captured species may result in considerable changes in CO_2 recovery and process economics.

14.3 Membranes and Membrane Modeling

14.3.1 Introduction to Membrane Gas Separation

This section provides a brief introduction to membrane-based gas separation. It mostly contains the definition of relevant terms used in this context. More detailed information is available elsewhere [36–40].

For technical applications, membranes are arranged in modules. As displayed in Fig. 14.3, they act as filters to separate one or more gases from a feed mixture to generate a specific gas-rich permeate. In addition, a sweep stream may be introduced at the permeate side [36].

The performance of a membrane depends generally on three properties: permeability, membrane thickness, and selectivity [36]. The permeability P_i denotes the rate at which a species i traverses a membrane. It is a function of the physical and chemical membrane structure, the nature of the permeating species (size, shape, polarity), and its interaction with the membrane [4]. The permeability is an intrinsic material property that does not take into account the thickness of the membrane. To limit the mass transfer resistance induced by the membrane, thin membranes are desired. Thus, the permeance Q (ratio of permeability and thickness) rather than the permeability P has to be considered as the relevant membrane property for industrial applications. It can be mostly attributed to manufacturing issues that materials featuring a high intrinsic per-

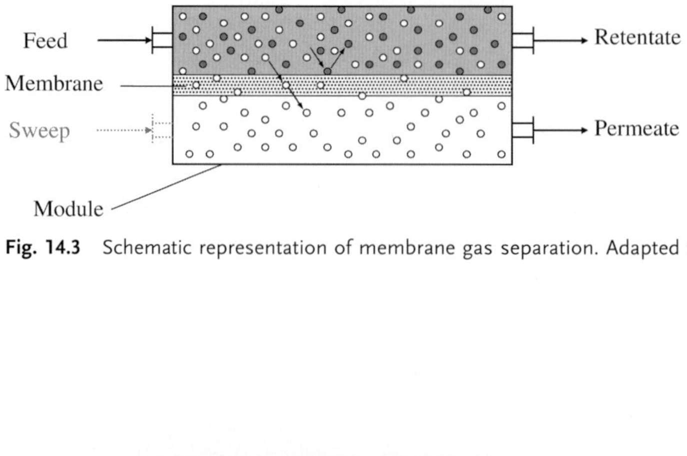

Fig. 14.3 Schematic representation of membrane gas separation. Adapted from [36].

meability and hence preferred properties do not necessarily exhibit a high permeance, and vice versa.

The other important property of the membrane is the ability to discriminate among different species, for example, the gases i and j. A measure of this property is the ideal selectivity $\alpha_{i/j}$, which is defined as the ratio of the respective pure gas permeabilities P or permeances Q of the species i and j [39–40]:

$$\alpha_{i/j} = \frac{P_i}{P_j} = \frac{Q_i}{Q_j} \quad (14.1)$$

Ideal selectivities are commonly reported in the literature and usually used to perform process calculations. If species interact strongly with the membrane, ideal selectivities often lead to an overestimation of the membrane and consequently of the process performance [40]. However, such an overestimation may be considered as a best case scenario. If a membrane process does not even show benefits over competing technologies under such optimal assumptions, its potential for application and commercialization is low.

In summary, both permeance and selectivity are important membrane properties. Whereas a low permeance can be balanced with additional membrane area, insufficient selectivity may result in multistage membrane processes. Such multistage membrane processes are supposedly not competitive with standard unit operations such as absorption [36]. Therefore, from an economic point of view, it appears that selectivity is in general more important than permeance.

14.3.2 Membrane Requirements

A variety of CO_2-selective membrane materials have been reported in the literature [11, 12]. According to Powell and Qiao [11], membranes applicable for CO_2 capture from coal-fired power plants have to comply with the following criteria:

- high CO_2 selectivity and CO_2 permeance
- thermal, chemical, and mechanical resistance towards flue gas conditions
- cost-effective membrane and module production.

A detailed quantification of membrane requirements has been derived in a joint effort among power plant operators, module producers, and membrane developers in the course of the NanoGLOWA project [43].

14.3.3 Membrane Materials

Figure 14.4 represents a general classification of membrane materials [36]. For technical applications, only solid synthetic membranes are relevant. Solid membranes can be further categorized with respect to the material of which they are made. State-of-the-art organic gas separation membranes consist of a thin,

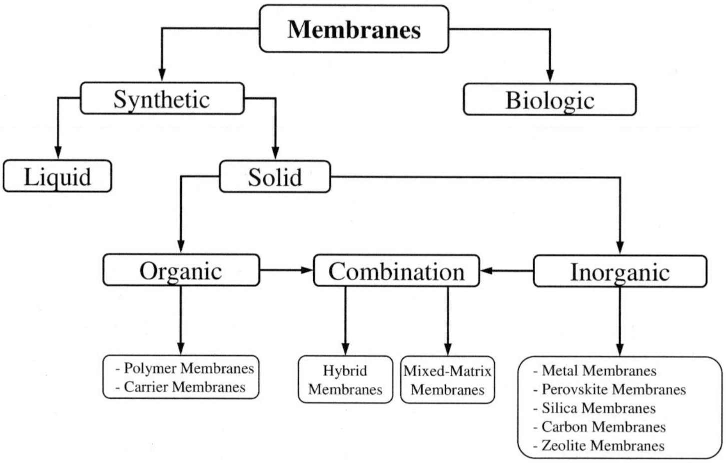

Fig. 14.4 General classification of membrane materials. Adapted from [36].

dense polymer film and a porous support. Detailed information about the morphology and structure of polymer membranes is available elsewhere [36, 38]. The class of inorganic materials includes dense metals but also porous materials such as carbon, zeolite, and silica membranes [36]. In addition to purely organic and inorganic membranes, combinations of both materials as in mixed-matrix or hybrid membranes have been reported. The selectivity and permeability of membranes are predominantly related to the prevalent gas transport mechanisms and hence the material itself. The literature provides extensive descriptions and reviews of these mechanisms [36, 37, 44–47].

Polymer membranes generally feature an inverse permeability–selectivity behavior, that is, the selectivity to different gas pairs increases as their permeability decreases [48, 49]. Rubbery polymers are known to exhibit high permeabilities and low selectivities, whereas glassy polymers show low permeabilities and high selectivities. Commercial applications are mostly based on glassy polymers, because they show superior mechanical properties and more favorable permeability–selectivity trade-offs relative to their rubbery counterparts [36, 39, 50]. Common glassy polymers include polysulfones (PSF), polyimides (PI), polyamides (PA), polycarbonates (PC), poly(phenylene oxide)s (PPO), and cellulose derivatives. Polydimethylsiloxane (PDMS) is the only rubbery polymer applied in commercial separation systems so far [36, 39, 50]. About 8–10 polymer materials have at present at least a 90% market share for gas separation applications, whereas other polymers have failed in industrial practice so far [40]. In terms of CO_2 separation from power plant flue gases, most commercially available membrane materials show either a low CO_2 selectivity or low CO_2 permeability. Moreover, many of them suffer from CO_2-induced plasticization or react sensitively to water. Among the commercialized polymers, PPO seems to show the least trade-offs.

Among rubbery polymers, PEO-based block polymers attracted considerable attention in the recent past [42]. The high polarity of PEO results in a high CO_2/N_2 gas separation factor and also a high CO_2 permeance [51]. Due to the rubbery nature of PEO-based block copolymers, water vapor and plasticization are expected to have only a light impact.

Facilitated transport membranes and in particular fixed-site carrier membranes feature very interesting properties in the separation of CO_2 and N_2 [42]. However, wetting requirements at the low-pressure side, saturation effects, and a high uncertainty with respect to carrier degradation are significant shortcomings. Carrier membranes still need to prove their suitability for CO_2 capture from coal-fired power plants.

Two major groups of inorganic membranes exist: dense and porous inorganic membranes [36]. Since dense inorganic membranes allow only certain gases (generally either H_2 or O_2) to pass [36], they are not applicable for CO_2 capture. Different types of materials, including alumina, silica, carbon, and zeolite, are suitable for membrane manufacture. In comparison with polymers, most inorganic membranes feature a better separation performance in terms of permeability and/or selectivity. Their higher chemical, thermal, and mechanical durability make them in principle very attractive for CO_2 capture from flue gases. Nevertheless, inorganic membranes are much more difficult to handle, explaining their high unit price [52]. Moreover, their inherent brittleness and their low volume-to-surface area ratios are still serious issues that limit their commercial applicability [42]. Accordingly, Nunes and Peinemann [39] stated that the market for inorganic gas separation membranes is small and will not increase notably in the future [39, 42].

The development of mixed organic–inorganic membranes is driven by the objective of merging the processability of polymer membranes with the gas separation performance of inorganic materials [36]. Two main categories of mixed organic–inorganic membranes exist:

- Mixed-matrix membranes (MMMs) are characterized by a heterogeneous gas separation layer that consists of a dispersed phase of discrete inorganic particles in a continuous polymeric phase [36].
- In hybrid membranes, the organic and inorganic phases intermingle at the molecular level due to covalent bonds [53].

Because of weak interfacial adhesion between the hosting organic and the discrete inorganic phase, the gas separation performance of MMMs must be considered moderate [39, 54]. Although reasonable permeabilities have been achieved, the required thickness of MMMs results in low fluxes. The brittleness, cost, and difficulties in production are currently the major barriers to their large-scale application [36]. A variety of hybrid combinations have been reported in the literature, but it seems that only moderate improvements in comparison with the base polymer have been achieved. The thermal, chemical, and mechanical properties of hybrid membranes are hardly been investigated so far. Hybrid membranes are not used on an industrial scale for gas separation applications [42].

The above discussion reveals that polymeric membranes appear currently the best option for membrane-based post-combustion CO_2 capture in terms of manufacturing, handling, and cost issues. Owing to the low flue gas temperature (see Section 14.2.1), the additional expense for inorganic or mixed membranes seems to be not justifiable for post-combustion CO_2 capture.

14.3.4 Membrane Models

Mass transfer in dense polymeric membranes is mostly described by the widely accepted solution diffusion mechanism. It consists of three consecutive steps:

- sorption of the gas molecules from the feed phase
- molecular diffusion through the membrane
- desorption of the gas molecules at the low-pressure side [36–37].

Hence separation is not only diffusion dependent, but also relies on the physical–chemical interaction between the various gas species and the polymer. The relationship between permeability, diffusivity, and solubility can be described according to the equation

$$P_i = D_i S_i \tag{14.2}$$

where P_i is the permeability, D_i is the diffusivity, and S_i is the solubility of component i in the polymer. The gas transport through dense polymeric membranes is governed by the equation

$$J_i = \frac{D_i S_i}{\delta}(p_{i_F} - p_{i_P}) = Q_i(p_{i_F} - p_{i_P}) \tag{14.3}$$

where J_i represents the flux density of component i across the membrane p_{i_F}, p_{i_P} and are the partial pressures of component i on either side of the membrane, δ is the membrane thickness and Q_i is the permeance of component i. Thus, the driving force for mass transfer relies on a partial pressure difference across the membrane, which can be generated by a difference in total pressures and/or different concentrations of the species i.

The majority of engineering studies [55–61] on membrane-based post-combustion CO_2 capture have considered rather simple models for CO_2 separation. Nonideal effects such as concentration polarization, pressure drop, and the flux of coupled gas species are mostly neglected.

14.4 Membrane Processes

14.4.1 Driving Force Generation

The driving force for mass transfer in gas permeation processes depends on the partial pressure difference of the respective component across the membrane [36]. Implementation may be achieved in three different ways:

- by increasing the partial pressure of the respective component on the upstream side by compression of the feed stream (Fig. 14.5 a)
- by lowering the partial pressure on the downstream side (permeate) by applying vacuum (Fig. 14.5 b) and/or sweeping with an inert gas (Fig. 14.5 c)
- Combinations of all the above options.

Merkel *et al.* [62] demonstrated recently that the pressure ratio is the limiting factor for CO_2 enrichment in the permeate stream. Consequently, highly selective membranes do not offer a significant advantage, unless the pressure ratio across the membrane and hence the driving force become sufficiently large. Therefore, not only the development and improvement of highly selective membranes but also the generation of the required driving force are crucial issues, particularly with respect to technical implementation in power plants. A variety of engineering and economic studies have explored the driving force generation for post-combustion capture processes [55–64]. Table 14.3 provides a brief summary of the respective process conditions.

Analysis of Table 14.3 indicates that relatively high pressure ratios are required across the membrane, regardless of the studied concept of driving force generation. For efficiency, the driving force is often generated by suction or a combination of different concepts. So far, engineering studies have usually focused on the required membrane properties and also on energetic and economic process optimization. Results were mostly generated on a theoretical basis, and aspects regarding technical realization have not been examined in detail

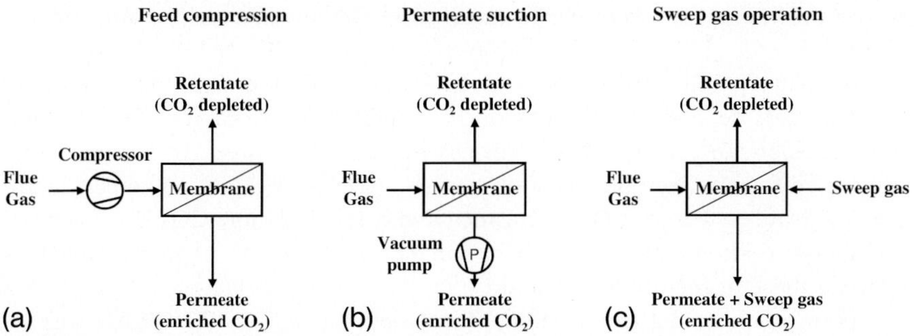

Fig. 14.5 Different concepts of driving force generation for membrane-based post-combustion CO_2 capture.

Table 14.3 Literature regarding driving force generation.

Ref.	Process configuration	Driving force generation/pressure limits
55	Multistage process with/without recycle stream	Compression + vacuum: compression 10 bar/vacuum 30 mbar
56	Single-stage process	Vacuum: 30 mbar
58	Single-stage process	Compression + vacuum: compression 20 bar/vacuum 2 mbar
59	Single-stage process	Compression + vacuum: compression 1.5 bar/vacuum 5 mbar
60	Two-stage process with recycle	Compression + vacuum: compression 1.5 bar/vacuum 80 mbar
61	Single stage process Two-stage process	Compression: 54 bar
62	Two-stage process	Compression + vacuum + sweep: compression 3 bar/vacuum 200 mbar
63	Single-stage process Two-stage process with recycle	Compression + vacuum: compression 20.8 bar/vacuum 30 mbar
64	Two-stage process	Compression + vacuum + sweep: compression 2.5–5 bar/vacuum 25–125 mbar

yet. Such aspects may include the selection of a specific type of machine, which tolerates the flue gas conditions and achieves the required target pressure and volume flow at the same time.

14.4.2 Trade-Offs in Membrane Gas Separation Technology

The general trade-offs associated with membrane gas separation processes and the trade-offs among feed compression and suction at the permeate side will be illustrated on the basis of a single-stage membrane process. It is assumed in this section that compression and suction machines operate in isothermal mode, representing the lower bound of energy consumption. Potential means to recover energy, for example by expansion of the retentate stream, are neglected.

The recovery and enrichment and also the energy requirement of a membrane process depend considerably on the separation performance of the membrane, notably on permeance and selectivity. Therefore, three generic polymeric solution diffusion membranes are selected, (see Table 14.4). Whereas membrane A exhibits a high CO_2 permeance and a relatively low CO_2-N_2 selectivity, membrane C shows the inverse behavior. The performance of membrane B is between those of membranes A and C.

A permeation model (see Section 14.3.4) implemented in FORTRAN was interfaced with the process design software Aspen Plus. A summary of the relevant model equations is available elsewhere [36]. Most studies regarding post-

Table 14.4 Permeance and selectivity of generic membranes.

Membrane	CO_2 permeance ($m_N^3\ m^{-2}\ h^{-1}\ bar^{-1}$)	H_2O permeance ($m_N^3\ m^{-2}\ h^{-1}\ bar^{-1}$)	CO_2/N_2 selectivity	CO_2/O_2 selectivity
A	4.10	$\gg CO_2$	20	4.5
B	1.25	$\gg CO_2$	45	15
C	0.07	$\gg CO_2$	85	28

combustion CO_2 capture rely on model assumptions that are very similar to or identical with the assumptions in this work [55–61, 63].

According to Melin and Rautenbach [36], a set of three dimensionless parameters is sufficient to characterize the separation performance of a single-stage membrane process: the membrane selectivity α (Equation 14.4), the ratio of the feed pressure p_{feed} and the permeate pressure p_{perm} (Equation 14.5) and the permeation number π (Equation 14.6):

$$\alpha = \frac{Q_{CO_2}}{Q_{N_2}} \tag{14.4}$$

$$\phi = \frac{p_{feed}}{p_{perm}} \tag{14.5}$$

$$\pi = \frac{Q_{CO_2} A_{mem} p_{perm}}{\dot{V}_{flue}} \tag{14.6}$$

where Q_{CO_2} is the CO_2 permeance of the membrane, A is the membrane area, and $\dot{V}_{flue}$ corresponds to the flue gas stream. To demonstrate the impact of the membrane selectivity, the permeation number π is kept constant for the different membranes given in Table 14.4 while the pressure ratio ϕ changes.

Figures 14.6–14.9 show the CO_2 recovery, the CO_2 enrichment, the required energy, and the specific CO_2 flux as a function of the pressure ratio ϕ. The driving force is generated by

- increasing the flue gas pressure (solid lines) at 1 bar permeate pressure, or
- lowering the permeate pressure (dashed lines) at 1 bar flue gas pressure.

The analysis of flue gas compression (solid lines in Fig. 14.6–14.9) leads to the following conclusions:

- A trade-off among CO_2 recovery and CO_2 enrichment exists. At a similar pressure ratio, highly selective membranes lead to higher enrichments in the permeate stream. The higher enrichment is at the expense of a lower recovery (Fig. 14.6 and 14.7).
- To take advantage of the superior separation performance of highly selective membranes, high pressure ratios are mandatory (Fig. 14.6 and 14.7).
- The specific energy requirement per unit of CO_2 captured initially decreases with higher pressure ratios (Fig. 14.8). This initial drop is a result of the in-

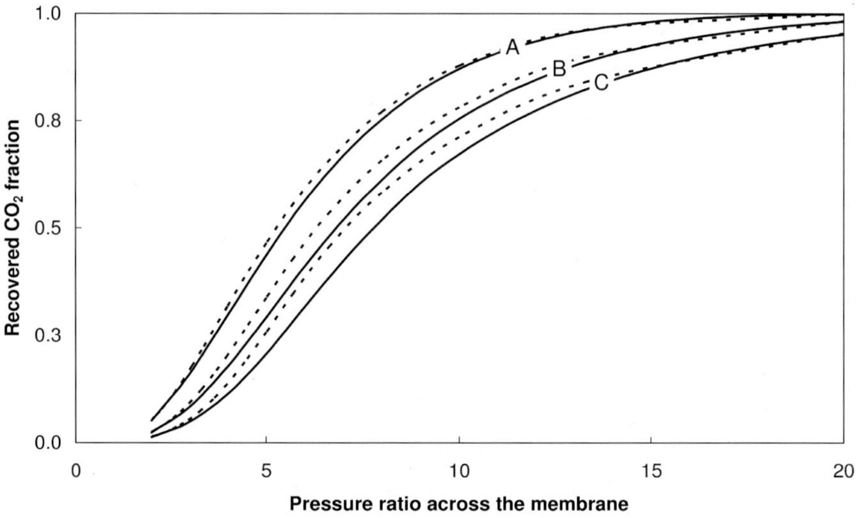

Fig. 14.6 Recovered CO_2 fraction as a function of the pressure ratio for membranes A, B, and C (Table 14.4). Solid lines, compression; dashed lines, suction.

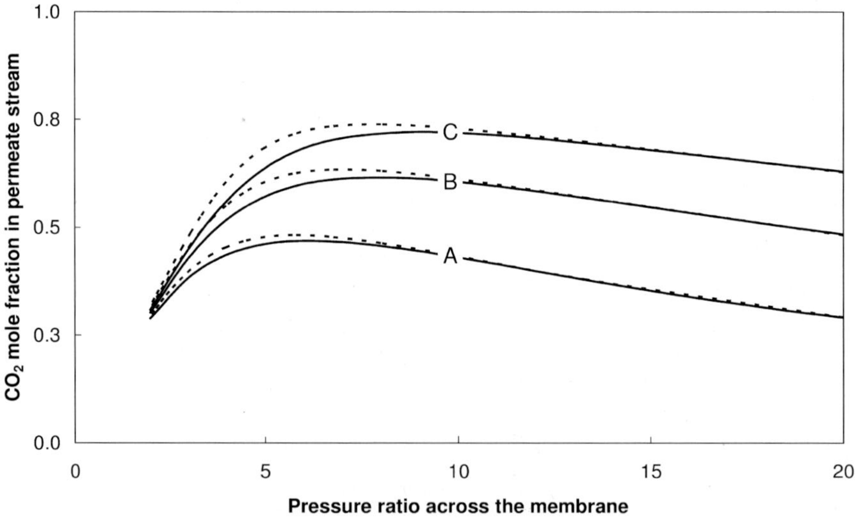

Fig. 14.7 CO_2 mole fraction in the permeate stream as a function of the pressure ratio for membranes A, B, and C (Table 14.4). Solid lines, compression; dashed lines, suction.

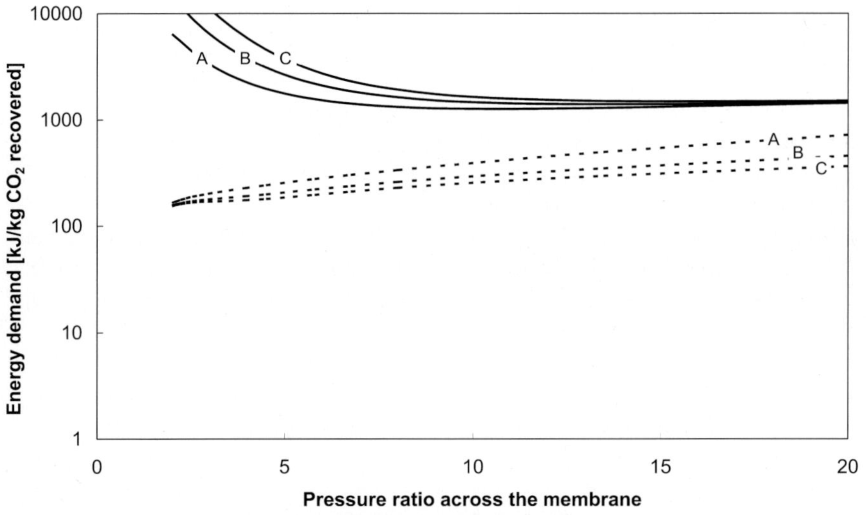

Fig. 14.8 Specific energy consumption per kg CO_2 recovered as a function of the pressure ratio for membranes A, B, and C (Table 14.4). Solid lines; compression, dashed lines, suction.

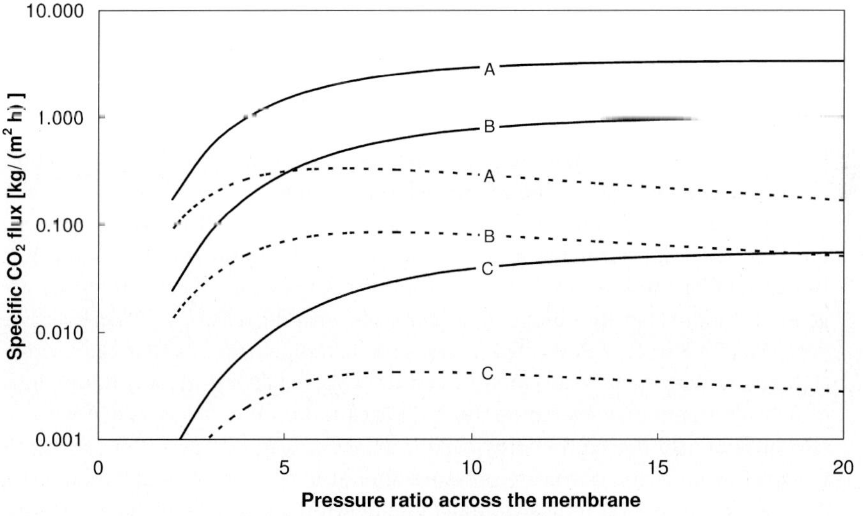

Fig. 14.9 Specific CO_2 flux as a function of the pressure ratio for membranes A, B, and C (Table 14.4). Solid lines, compression; dashed lines, suction.

creasing CO_2 recovery (Fig. 14.6). However, this effect diminishes at elevated pressure ratios.
- Owing to the considerably higher CO_2 recovery at low pressure ratios, less selective membranes cause lower energy consumptions per unit of CO_2-captured (Fig. 14.6 and 14.8). Nevertheless, the lower specific energy consumption does not take into account the lower quality (CO_2 enrichment) of the permeate stream (Fig. 14.7).
- The higher permeance of less selective membranes results in a more efficient use of the installed membrane area (higher specific flux) and reduces the required membrane area compared with more selective membranes (Fig. 14.9).

Results for suction at the permeate side (dashed lines in Fig. 14.6–14.9) indicate:
- The separation performance of such a membrane process is very similar to feed compression. The slight deviations result from the presence of vapor, which is mostly removed by condensation if the driving force is based on gas compression.
- In contrast to feed compression, the specific energy requirement increases with increasing pressure ratio. The compression of the permeate stream from a lower pressure level back to ambient pressure requires more energy (higher pressure ratio).
- Owing to the lower mass flow on the permeate side, the theoretical energy requirement for vacuum operation is lower relative to feed compression. However, with vacuum operation the feed pressure remains at 1 bar. In consequence, the larger volume flows on the feed side cause a considerably larger membrane area.

In summary, the combination of the results for feed compression and suction at the permeate side lead to the following conclusions:
- The driving force generation by compression features small membrane areas but requires a substantial amount of energy (Fig. 14.8 and 14.9, solid lines). In contrast, suction seems to be very efficient with respect to driving force generation, but the membrane area increases with decreasing suction pressure (Fig. 14.8 and 14.9, dashed lines). The discussion reveals that each concept encounters specific advantages and drawbacks. Therefore, a combination of both concepts may overcome the high area requirements associated with suction and also the high energy consumption of feed compression. Owing to practical aspects, the permeate pressure should not be below $\sim$200 mbar [62, 65]. As Table 14.3 indicates, most engineering studies have chosen a combined concept of feed compression and moderate suction at the permeate side.
- A CO_2 concentration of 95% and higher cannot be obtained in a single-stage process. The reduction of the membrane area and hence a lower permeation number π (Equation. 14.6) may yield the targeted CO_2 permeate fraction. However, this comes at the expense of vanishing CO_2 recovery, resulting in extremely high specific energy consumptions.

- Highly selective membranes may accomplish 95% CO_2 in conjunction with reasonable CO_2 recoveries, although this requires elevated pressure ratios. Furthermore, more selective membranes usually feature a lower CO_2 permeance, increasing the required membrane area. Consequently, two-stage processes appear to be the only reasonable option to achieve high recoveries in conjunction with the desired CO_2 enrichment. Table 14.3 reveals that most studies on membrane gas separation from power plants have focused on two-stage concepts.

14.4.3 Two-Stage Membrane Processes

Figure 14.10 depicts a simplified flow sheet of a typical two-stage membrane cascade with retentate recycling. Retentate recycling to the feed of the previous stage offers a simple means to enhance the overall recovery of more permeable species [66]. Rautenbach and Albrecht [67] reported that two-stage cascades with and without recycling showed only minor differences in terms of economics. However, two-stage cascades with recycling yield higher recoveries.

Figures 14.11 and 14.12 represent the power plant penalty and the required membrane area of the membrane process given in Fig. 14.10 as function of the overall CO_2 recovery. The driving force generation in both stages relies on a combination of feed compression (90% isentropic efficiency) and suction at the permeate side (75% isentropic efficiency). While flue gas compression is limited to 4 bar, the feed pressure of the second stage does not exceed 5 bar. A permeate pressure of 200 mbar is assumed in both stages. Energy from the N_2-rich retentate stream of the first membrane stage is recovered by an expansion unit (75% isentropic efficiency). A multistage compressor boosts the CO_2

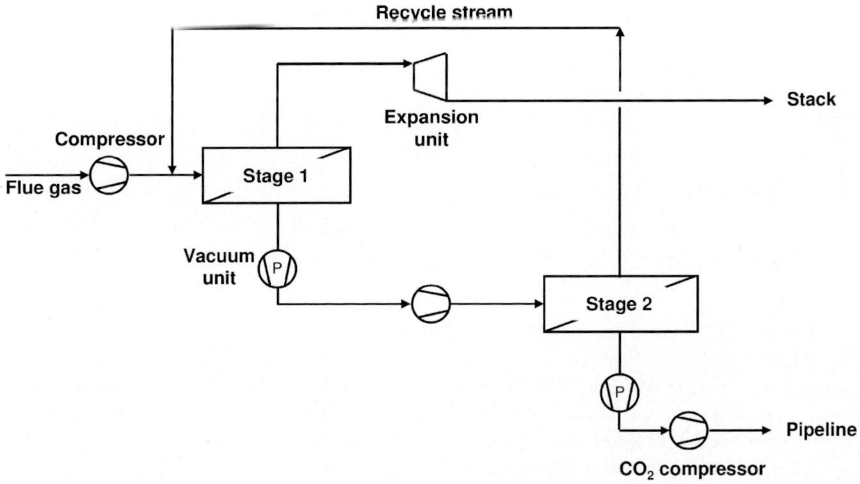

Fig. 14.10 Typical two-stage membrane cascade with retentate recycling.

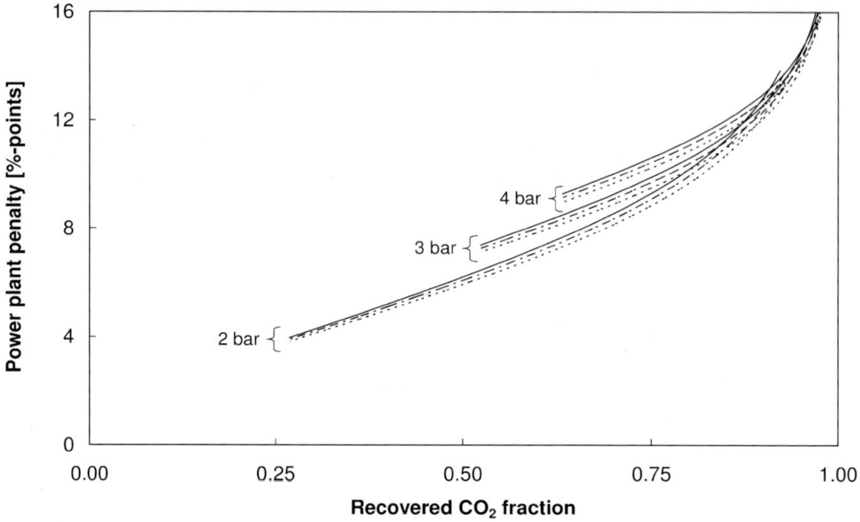

Fig. 14.11 Power plant penalty of a two-stage membrane process with retentate recycling as a function of the overall CO_2 recovery for different flue gas pressures (curly brackets) and different feed pressures of the second stage (dotted line, 3 bar; dashed line, 4 bar; solid line, 5 bar); 0.2 bar permeate pressure in both stages; membrane B in both stages.

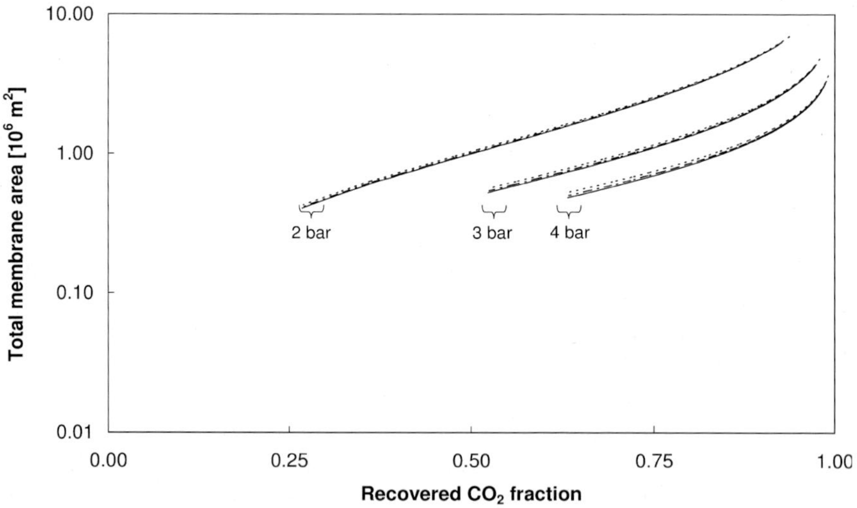

Fig. 14.12 Total membrane area of a two-stage membrane process with retentate recycling as a function of the overall CO_2 recovery for different flue gas pressures (curly brackets) and different feed pressures of the second stage (dotted line, 3 bar; dashed line, 4 bar; solid line, 5 bar); 0.2 bar permeate pressure in both stages; membrane B in both stages [8].

(95.5 mol% CO_2, 4.5 mol% co-captured species) up to pipeline pressure. Membrane B (Table 14.4) is considered in both membrane stages. The base power plant (supercritical coal-fired power plant without CO_2 capture) exhibits a thermal duty of 1210 MW and a net thermal efficiency of 45% (LHV based). The flue gas composition is in agreement with Table 14.1 and the flue gas stream accounts for 560 kg s^{-1} at the FGD outlet.

Figures 14.11 and 14.12 indicate the following:

- Two-stage membrane processes with retentate recycling may achieve up to 97% CO_2 recovery in conjunction with 95.5% CO_2 purity.
- The flue gas pressure (first membrane stage) has a significant impact on the power plant penalty and the total membrane area, whereas the feed pressure of the second membrane stage is less important. Thus, the first membrane stage dominates in terms of energy consumption and membrane area.
- A trade-off among energy consumption and membrane area exists. The optimal CO_2 recovery results from an economic evaluation of the CO_2 capture process.

14.5 Economics of Membrane Processes for CO_2 Capture

The specification of a separation problem requires the derivation of not only the product composition but also the respective product recovery. With respect to CCS, the optimal CO_2 recovery depends predominantly on the associated CO_2 avoidance cost, which includes the costs of CO_2 separation, transmission, and storage. The avoidance cost corresponds to the required carbon credit that matches the additional capital and operating expense of the capture facility and of the power plant. Since the capture cost dominates the CO_2 avoidance cost, expenses for CO_2 transport and storage are not taken into account in most studies.

The economic evaluation of the CO_2 capture process for different sets of technical/operating parameters permits the derivation of the optimal CO_2 avoidance cost. Such a techno-economic analysis offers the opportunity to identify the trade-offs among operation expenses (energy consumption, utilities), capital investment, and product recovery. Abu-Zahra *et al.* [13] applied this methodology to chemical absorption processes and demonstrated that ~90% CO_2 recovery yields the lowest avoidance cost. While lower recoveries do not fully exploit the scaling effects, that is, the specific costs decrease to some extent with increasing capture, higher capture levels caused additional energy consumption and consequently higher cost.

A variety of literature sources have provided cost estimates for post-combustion CO_2 capture, particularly for amine-based processes. At 90% CO_2 recovery, cost estimates per ton of CO_2 range from about € 40 [13] to € 60 [68]. Oexmann and Kather [69] and Hamilton *et al.* [70] calculated about € 52 and US$ 52, respectively. Follmann *et al.* [8], Zhao *et al.* [55], Ho *et al.* [60], and Yang *et al.* [63]

performed cost assessments for membrane-based capture processes. At 90% CO_2 recovery, the calculated costs show about the same range as the cost of amine absorption processes. It is important to note that technical parameters (e.g., isentropic efficiency of compressors or pumps, flue gas composition) and economic assumptions (e.g., membrane costs, membrane lifetime, investment costs of the power plant, interest rate, book life) may differ considerably among studies. Consequently, it is difficult to compare results and to determine the most favorable capture concept in terms of process economics.

14.6 Summary and Conclusions

This chapter has presented an integral view on membrane-based post-combustion CO_2 capture from coal-fired power plants. The review of the process boundary conditions revealed that flue gas composition and flow are easily accessible data. In contrast, the literature currently provides little tangible information about CO_2 purity requirements for pipelining and storage.

Owing to the moderate flue gas conditions at the FGD outlet, polymeric membranes appear most suitable for post-combustion CO_2 capture. Owing to the ambient pressure of the flue gas stream in conjunction with the rather low CO_2 concentrations, driving force generation becomes a difficult task. The combination of flue gas compression and suction at the permeate side currently appears to be the most practical option and is considered in the majority of technical studies. The examination of the trade-offs associated with membrane gas separation technology revealed that two-stage membrane processes are mandatory to achieve 95% CO_2 purity at the permeate side in combination with reasonable CO_2 recoveries. Therefore, most economic evaluations have concentrated on two-stage membrane processes, particularly with retentate recycling. Since technical parameters, such as isentropic efficiency, and economic assumptions, such as membrane costs or interest rate, differ considerably among the studies, it is difficult to compare results.

Acknowledgment

This work is part of the NanoGLOWA project (NMP3-CT-2007-026735), which has been supported financially by the EU Commission within the thematic priority NMP of the Sixth Framework Program.

References

1 Cifre, P. G., *et al.* Integration of a chemical process model in a power plant modelling tool for the simulation of an amine based CO_2 scrubber. *Fuel*, 2009, **88**, 2481–2488.
2 Korkmaz, Ö., *et al.* Analysis of retrofitting coal-fired power plants with carbon dioxide. *Energy Procedia*, 2009, **1**, 1289–1295.
3 Sanpasertparnich, T., *et al.* Integration of post-combustion capture and storage into a pulverized coal fired power plant. *Int. J. of Greenhouse Gas Control*, 2010, **4**, 499–510
4 Kohl, A. and Nielsen, R. *Gas Purification*, 4th edn., Gulf Publishing, Houston, TX, 1997.
1 Kaldis, S. P., *et al.* Application of polymer membrane technology in coal combustion processes. *Chem. Eng. Commun.*, 2007, **194**, 22.
6 NanoGLOWA. *NanoGLOWA Project*, www.nanoglowa.com (last accessed 27 December 2010).
7 Helmholtz Allianz MEM-BRAIN. *Das Forschungsprojekt MEM-BRAIN*, http://www.fz-juelich.de/ief/ief-1/membrain, last accessed 2 March 2011.
8 Follmann, P. M., *et al.* Membrane gas separation processes for post-combustion CO_2 capture. In *Membrane Engineering for the Treatment of Gases* (ed. Drioli, E. and Barbieri, G.), RSC Publishing, Cambridge, 2011, to be published
9 Umweltbundesamt. *Merkblatt für beste verfügbare Techniken für Großfeuerungsanlagen*, July 2006, www.bvt.umweltbundesamt.de/sevilla/kurzue.htm, last accessed 2 March 2011.
10 *NanoGLOWA Deliverable D9.2.2, Refined Set of Feasible System Solutions*, June 2008.
11 Pakala, N. R. and Bhown, A. S. Effect of particulate matter on mass transfer through microporous hollow fiber membranes. *J. Membr. Sci.*, 1996, **111**, 71–79.
12 Hendriks, Ch. *Carbon Dioxide Removal from Coal-Fired Power Plants*, Kluwer, Dordrecht, 1994.
13 Abu-Zahra, M., *et al.* CO_2 capture from power plants. Part I. A parametric study of the technical performance based on mono-ethanolamine. *Int. J. Greenhouse Gas Control*, 2007, **1**, 37–46.
14 Sijbesma, H., *et al.* Flue gas dehydration using polymer membranes. *J. Membr. Sci.*, 2008, **313**, 263–276.
15 Svensson, R., *et al.* Transportation systems for CO_2 – application to carbon capture and storage. *Energy Convers. Manage.*, 2004, **45**, 2342–2353.
16 Gale, J. and Davidson, J. Transmission of CO_2 – safety and economic considerations. Presented at the Sixth International Conference on Greenhouse Gas Control Technologies, Kyoto, 1–4 September 2002.
17 Doctor, R. D., *et al.* Transporting carbon dioxide recovered from fossil-energy cycles. In *Fifth International Conference on Greenhouse Gas Control Technologies*, Cairns, 13–16 August, 2000, pp. 567–571. Eds.: William, D. J., Durie, R. A., McMullan, P., Paulson, C. A. J., and Smith, A. Y., CSIRO Publishing, Australia.
18 Metz, B., *et al.* *IPCC Special Report on Carbon Dioxide Capture and Storage*, Cambridge University Press, Cambridge, 2005.
19 Fayed, A. S. CO_2 injection for enhanced oil recovery benefits from improved technology. *Oil Gas J.*, 1983 (3 January), 92–96.
20 Farris, C. B. Unusual design factors for supercritical CO_2 pipelines. *Energy Prog.*, 1983, **3**(3), 150–158.
21 Recht, D. L. CO_2 line design needs. *Pipeline Gas J.*, 1986 (January), 41–44.
22 Wallace, C. B. Drying supercritical CO_2 demands care. *Oil Gas J.*, 1985 (24 June), 98–104.
23 Li, H. and Yan, J. Impacts of impurities in CO_2-fluids on CO_2 transport process. In *Proceedings of ASME Turbo Expo 2006: Power for Land, Sea and Air*, 8–11 May 2006, Barcelona, GT2006-90954.
24 de Visser, E., *et al.* Dynamis CO_2 quality recommendations. *Int. J. Greenhouse Gas Control*, 2008, **2**, 478–484.
25 Johnson, E. and Walter, F. B. Gas processing needs for EOR. *Hydrocarbon Process.*, 1985 (October), 62–66.

26 Liljedahl, G., *et al.* Alstom's oxygen-fired CFB technology development status for CO_2 mitigation. Presented at the 31st International Technical Conference on Coal Utilization and Fuel Systems, Clearwater, FL, 21–25 May 2006.

27 Race, J.M., *et al.* Challenges for offshore transport of anthropogenic carbon dioxide. Presented at the 26th International Conference on Offshore Mechanics and Arctic Engineering, San Diego, CA, 2007, 10–15 June.

28 Mohitpour, M., *et al. Pipeline Design and Construction: a Practical Approach*, 2nd edn., American Society of Mechanical Engineers, New York, 2003.

29 Holloway, S. Underground sequestration of carbon dioxide – a viable greenhouse gas mitigation option. *Energy*, 2005, **30**, 2318–2333.

30 Voormeij, D.A. and Simandl, G. Geological, ocean and mineral CO_2 sequestration options: a technical review. *Geosci. Can.*, 2004, **31**, 11–22.

31 Bachu, S. CO_2 storage in geological media: role, means, status and barriers to deployment. *Prog. Energy Combust. Sci.*, 2008, **34**, 254–273.

32 May, F., *et al.* Möglichkeiten zur untertägigen Speicherung von CO_2 in Deutschland – eine Bestandsaufnahme. *Glückauf Forschungshefte*, 2003, **64**(4), 138–146.

33 May, F. and Turkovic, R. Minderung von Treibhausgas-Emissionen durch CO_2 Speicherung in tiefen Aquiferen. *Z. Angew. Geol.*, 2003, (1), 65–71.

34 Bachu, S. and Bennion, D.B. Chromatographic partitioning of impurities contained in a CO_2 stream injected into a deep saline aquifer. Part 1. Effects on gas composition and in situ conditions. *Int. J. Greenhouse Gas Control*, 2009, **3**, 458–467.

35 IEAGHG. *Impact of Impurities on CO_2 Capture, Transport and Storage*, Report No. PH4/32, IEA Greenhouse Gas R&D Programme, Cheltenham. August 2004.

36 Melin, T. and Rautenbach, R. *Membranverfahren: Grundlagen der Modul- und Anlagenauslegung*, 3rd edn., Springer, Berlin, 2007.

37 Mulder, J. *Basic Principles of Membrane Technology*, 2nd edn., Kluwer, Dordrecht, 1996.

38 Ohlrogge, K. and Ebert, K. *Membranen: Grundlagen, Verfahren und Industrielle Anwendungen*, Wiley-VCH Verlag GmbH, Weinheim, 2006.

39 Nunes, S.P. and Peinemann, K.-V. *Membrane Technology in the Chemical Industry*, Wiley-VCH Verlag GmbH, Weinheim, 2001.

40 Baker, R.W. Future directions of membrane gas separation technology. *Ind. Eng. Chem. Res.*, 2002, **41**, 1393–1411.

41 Powell, C.E. and Qiao, G.G. Polymeric CO_2/N_2 gas separation membranes for the capture of carbon dioxide from power plant flue gases. *J. Membr. Sci.*, 2006, **279**, 1–49.

42 Ritter, J.A. and Ebner, A.D. *Chemical Industry Vision 2020 Technology Partnership: CO_2 Separation Technology: R&D Needs for the Chemical and Petrochemical Industries*, 2007, www.chemicalvision2020.org/library.html (last accessed 28 May 2008).

43 *NanoGLOWA Deliverable D6.2.2, Updated Set of Requirements*, June 2010.

44 Hsieh, H.P. *Inorganic Membranes for Separation and Reaction*, 1st edn., Elsevier, Amsterdam, 1996.

45 Wijmans, J.G. and Baker, R.W.,The solution diffusion model: a review. *J. Membr. Sci.*, 1997, **107**, 1–21.

46 Barrer, R.M. *Diffusion in Polymers*, Academic Press, New York, 1968.

47 Behave, R.R. *Inorganic Membranes – Synthesis, Characteristics and Applications*, Van Nostrand Reinhold, New York, 1991.

48 Robeson, L.M. Correlation of separation factor versus permeability for polymeric membranes. *J. Membr. Sci.*, 1991, **62**, 165–185.

49 Robeson, L.M. The upper bound revisited. *J. Membr. Sci.*, 2008, **320**, 390–400.

50 Baker, R.W. *Membrane Technology and Applications*, 2nd ed., John Wiley & Sons, Ltd, Chichester, 2004.

51 Reijerkerk, S.R., *et al.* Tuning of mass transport properties of multi-block copolymers for CO_2 capture applications. *J. Membr. Sci.*, 2010, **359**, 54–63.

52 Zhang, L., *et al.* Supported inorganic membranes: promises and challenges. *JOM*, 2009, **61**(4), 61–71.

53 Shekhawat, D., *et al.* *A R eview of CO_2 Selective Membranes – a Topical Report*, DOE/NETL-2003/1200, National Energy Technology Laboratory, United States Department of Energy, Washington, DC.

54 Anson, M., *et al.* ABS copolymer-activated carbon mixed matrix membranes for CO_2/CH_4 separation. *J. Membr. Sci.*, 2004, **243**, 19–28.

55 Zhao, L., *et al.* Concepts and investment cost analyses of multi-stage membrane systems used in post-combustion processes. *Energy Procedia*, 2009, **1**, 269–278.

56 Zhao, L., *et al.* A parametric study of CO_2/N_2 gas separation membrane processes for post-combustion capture. *J. Membr. Sci.*, 2008, **325**, 284–294.

57 Favre, E. Carbon dioxide recovery from post-combustion processes: can gas permeation membranes compete with absorption? *J. Membr. Sci.*, 2007, **294**, 50–59.

58 Bounaceur, R., *et al.* Membrane processes for post-combustion carbon dioxide capture: a parametric study. *Energy*, 2006, **31**, 2556–2570.

59 Kotowicz, J., *et al.* The influence of membrane CO_2 separation on the efficiency of coal-fired power plants. *Energy*, 2010, **35**, 841–850.

60 Ho, M.T., *et al.* Reducing the cost of CO_2 capture from flue gases using membrane technology. *Ind. Eng. Chem. Res.*, 2008, **47**, 1562–1568.

61 Van der Sluijs, J.P., *et al.* Feasibility of polymer membranes for carbon dioxide recovery from flue gases. *Energy Convers. Manage.*, 1992, **33**, 429–436.

62 Merkel, T.C., *et al.* Power plant post-combustion carbon dioxide capture: an opportunity for membranes. *J. Membr. Sci.*, 2010, **359**, 126–139.

63 Yang, D., *et al.* Potential of two stage membrane systems with recycle stream for CO_2 capture from post-combustion gas. *Energy Fuels*, 2009, **23**, 4755–4762.

64 Hussain, A. and Hägg, M.-B. A feasibility study of CO_2 capture from flue gas by a facilitated transport membrane. *J. Membr. Sci.*, 2010, **359**, 140–148.

65 Follmann, P.M., *et al.* Driving force generation for membrane based post-combustion CO_2 capture from coal fired power plants. In preparation.

66 Spillman, R.W. Economics of gas separation membranes. *Chem. Eng. Prog.*, 1989, **85**, 41–62.

67 Rautenbach, R. and Albrecht, R. *Membrane Processes*, John Wiley & Sons, Ltd, Chichester, 1989, p. 196.

68 Romeo, L.M., *et al.* Designing a supercritical steam cycle to integrate the energy requirements of CO_2 amine scrubbing. *Int. J. Greenhouse Gas Control*, 2008, **2**, 563–570.

69 Oexmann, J. and Kather, A. Post-combustion CO_2 capture in coal-fired power plants: comparison of integrated chemical absorption processes with piperazine promoted potassium carbonate and MEA. *Energy Procedia*, 2009, **1**, 799–806.

70 Hamilton, M.R., *et al.* Cost and U.S. public policy for new coal power plants with carbon capture and sequestration. *Energy Procedia*, 2009, **1**, 4487–4494.

15
CO_2 Separation via the Oxyfuel Process with O_2-Transport Membranes in Coal Power Plants

Franz Beggel, Nicolas Nauels, and Michael Modigell

Abstract

The membrane-based oxyfuel power plant process is considered to be the most energy-efficient carbon capture and storage technology, but its efficiency depends strongly on the process integration of the membrane unit. Two different methods of membrane integration are presented in detail and the results for process parameters are discussed. Several coal- and gas-fired process concepts and approaches towards the realization of membrane-based oxyfuel technology are also reviewed.

Keywords: oxyfuel, oxycoal, membrane, hard coal membrane, process integration, carbon capture, oxygen production

15.1
Introduction

Three technical lines for CO_2 capture have been investigated in recent decades: post-combustion capture, pre-combustion capture, and oxyfuel combustion [1, 2]. In the oxyfuel combustion process, the fuel is combusted with a mixture of oxygen and recirculated flue gas. In the absence of air, and therefore N_2, this combustion leads to a flue gas that consists mainly of CO_2 and H_2O, which can easily be purified by condensation to a gas stream with high CO_2 contents (>90%). This stream can be compressed for transportation and storage. Compared with the currently available post-combustion process, in which the CO_2 is separated from the flue gas at the very end of the process, the energy penalty of the oxyfuel technology is considered to be lower.

Nevertheless, there is still a high energy demand for the production of pure oxygen required by the process. The state-of-the-art production process for oxygen is cryogenic air separation, in which total efficiency losses of 8–12% have been estimated for a Greenfield oxyfuel power plant [3, 4], mainly caused by the cryogenic air separation unit (ASU).

Process Engineering for CCS Power Plants, First Edition. Edited by Detlef Stolten and V. Scherer

Another possibility for providing the oxygen for the combustion is the use of ceramic membranes known as mixed ion electron-conducting (MIEC) membranes [5–8]. At high temperatures, these membranes are able to separate oxygen from air with infinite selectivity. The required operating temperature is achieved by efficient integration of the membrane into the power plant process. The OXYCOAL-AC research project is concerned with the process development of a coal-fired oxyfuel power plant [8]. Two different ways of process integration are presented in the article. Furthermore, other membrane-based oxyfuel process concepts also with natural gas fuel are reviewed and an overview of membrane integration with different kinds of module design is given.

15.2
MIEC Membrane Operating Concepts

Membranes for oxygen separation in oxyfuel processes can be operated with a four-end [6, 9, 10] or three-end design [11, 12]. In the case of the four-end concept, a sweep stream is applied on the low-pressure side in order to remove the permeating oxygen (Fig. 15.1 a). Thereby, the concentration of oxygen on the permeate side of the membrane is reduced and the driving potential is increased (cf. Equation 15.1). In the three-end concept, the driving potential has to be sustained by an increased feed pressure or by applying vacuum on the permeate side (Fig. 15.1 b).

Membrane materials for the two concepts have to fulfill different requirements regarding chemical stability. In the three-end concept, the membrane material is only exposed to air, allowing many different materials to be employed that can be optimized for a high flux of oxygen without many constraints. Membrane material development and performance under idealized laboratory-scale conditions have been reviewed recently [13, 14]. Since in the four-

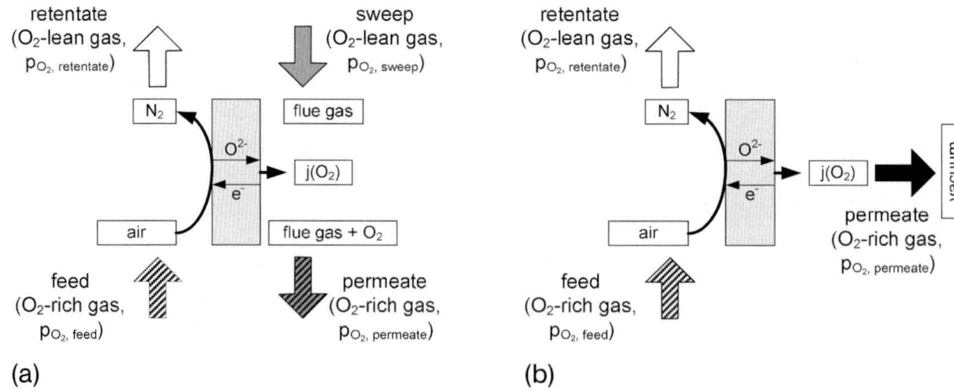

Fig. 15.1 Oxygen separation by MIEC membranes: (a) four-end operation; (b) three-end operation. Adapted from [6].

end concept the membrane has direct contact with flue gas, the material has to resist all components of the flue gas and still has to provide an acceptable flux of oxygen. This problem is currently being addressed in various research projects such as Mem-Oxycoal [15] and MEM-BRAIN [16]. Typical MIEC materials, BSCF ($Ba_{0.5}Sr_{0.5}Co_{0.8}F_{0.2}O_{3-\delta}$) and La_2NiO_4, have been investigated experimentally with regard to chemical stability of the material [17, 18]. Both material types were exposed to an artificial flue gas consisting of a mixture of CO_2 and SO_2 on the sweep side. The results showed a collapsing flux due to degradation of the membrane surface on the exposed side. It can be concluded that neither of the investigated membrane materials is suitable for integration within the flue gas recirculation as is necessary in the four-end concept. For the three-end operation, BSCF was proposed to be applied, since the oxygen flux is 20 times higher than that of La_2NiO_4. The heat required to reach the operating temperature is supplied by the sweep gas in the four-end concept or the air in the three-end concept. The specific O_2 permeation, j_{O_2}, of the membrane can be calculated by the Wagner equation:

$$j_{O_2} = C_{\text{Wagner}} \frac{T_{\text{Mem}}}{d_{\text{Mem}}} e^{-\frac{k_{\text{Wagner}}}{T_{\text{Mem}}}} \ln \Pi_{O_2} \tag{15.1}$$

where j_{O_2} is the specific flux of oxygen (mol s^{-1} m^{-2}) and C_{Wagner} and K_{Wagner} are intrinsic membrane material constants which have to be determined experimentally (measurements with perovskites such as BSCF showed values of 1.004×10^{-6} mol m^{-1} s^{-1} K^{-1} and 6201 K, respectively), T is the operating temperature, and d is the active thickness of the membrane in meters. The different types of membrane modules are presented in Section 15.5.

The mean O_2 partial pressure ratio Π_{O_2} (Equations 15.2 and 15.3) depends on the process variant and the chosen boundary conditions. It is influenced by

- the O_2 separation ratio ξ, defining the ratio of the separated O_2 flow to the entering O_2 flow of the air, influencing the oxygen partial pressure in the retentate
- the applied feed pressure
- the mean O_2 partial pressure on the sweep and permeate sides.

$$\text{Four-end: } \Pi_{O_2} = \frac{p(O_{2,\text{feed}}) + p(O_{2,\text{retentate}})}{p(O_{2,\text{sweep}}) + p(O_{2,\text{permeate}})} \tag{15.2}$$

$$\text{Three-end: } \Pi_{O_2} = \frac{p(O_{2,\text{feed}}) + p(O_{2,\text{retentate}})}{2p(O_{2,\text{permeate}})} \tag{15.3}$$

15.3
Hard Coal Membrane-Based Oxyfuel Process

15.3.1
Recirculation

Flue gas is recirculated to the boiler to replace the absent N_2 of the air as a heat sink. To reach similar adiabatic temperatures in the combustion chamber to those in air-fired boilers, about 60% of the flue gas has to be recirculated [19]. To compensate for the pressure loss within the boiler and the recirculation, a recirculation fan is installed. The recirculation temperature is between 160 and 400 °C [4] for the three-end process and up to 850 °C for the four-end process [6]. The latter demands the development of a high-temperature recirculation fan and proper optimization of the process since this fan is one of the main energy consumers of the oxyfuel plant [20].

15.3.2
Combustion

The combustion of fuel in oxyfuel boilers differs from that in air-fired facilities in having a lower flame temperature and reduced NO_x and SO_x emissions. This results from the different composition of the generated flue gas. In contrast to air firing, the main components are CO_2 and H_2 instead of N_2, leading to different flue gas properties [19, 21]:

- Density ($MW_{N_2} < MW_{CO_2}$).
- Heat capacity ($c_{p_{CO_2}} > c_{p_{N_2}}$).
- Diffusivity of O_2 in CO_2 is lower than in N_2.
- Radiative properties (oxyfuel flue gas consists of, in addition to CO_2, a high fraction of H_2O. Both species have a high emittance).

To compare oxyfuel and air combustion, it is common to keep the adiabatic temperature similar, resulting in O_2 contents at the burner of around 30% referred to wet recycled flue gas. Owing to the higher gas radiation, it was found for the boiler of the reference power plant NRW that an oxygen content of 24% is sufficient to keep the heat transfer by radiation in the furnace equal to conventional combustion with air [22].

15.3.3
Compression

The CO_2 purification and compression unit (CPU) is one of the main energy consumers of the oxyfuel process. The energy consumption is between 100 and 135 kWh t^{-1} CO_2 [23] and depends on the CO_2 content of the flue gas entering the CPU, and also on the required separation rate and purity for transport. The composition of the flue gas is influenced by

- *The excess ratio of oxygen during combustion:* for oxyfuel combustion, an excess ratio of O_2 of 10–15%, similar to air, should be chosen [4].
- *The air leakage:* old boilers have air leakage rates of 3–10%, resulting in high unwanted oxygen contents in the flue gas. For oxyfuel combustion, this value has to be drastically minimized. In today's boilers, leakages of less than 3% can be realized [24].
- *The quality of the separated oxygen:* state-of-the-art cryogenic ASUs produce oxygen with a purity of 99.5% [25]. Nevertheless, O_2 purities of 97% are favored as they do not require the difficult Ar–N_2 separation with a higher energy demand and up to twice the number of separation stages in the column [5]. Oxygen purity from air separation by MIEC-membranes is 100% due to their 100% selectivity.
- *The efficiency of the installed denitrification (DeNOx) and desulfurization units* defines the quality of the raw gas entering the CPU. It has been stated [26] that SO_x and NO_x can be removed during compression and additional desulfurization and DeNOx are not required.

15.3.4 Three-End Integration

By integrating the membrane without contact with the flue gas, it is possible to use highly permeable MIEC materials, such as BSCF. In Fig. 15.2, the flow sheet of a three-end membrane-based oxyfuel power plant is shown. The plant is divided into two sections: the flue gas path that can be seen on the right-hand side and the oxygen separation unit depicted on the left-hand side.

15.3.4.1 Flue Gas Path

The flue gas exits the boiler at temperatures comparable to the conventional hard coal power plants at $\sim$370 °C [27]. Subsequently, the dust is removed by a high-temperature [6] or a conventional electrical precipitator [4], depending on the temperature necessary for the primary recycling. The flue gas passes through a heat exchanger to heat up the secondary recycling stream, which is used for coal drying, and the primary recycling stream, if lower recycling temperatures are anticipated [7, 28]. The remaining heat is integrated into the high-pressure (HP) and low-pressure (LP) preheating sections of the water–steam cycle. The acid dew point of oxyfuel flue gas is 14–21 °C higher than in air-fired cases since the SO_x concentration is 3.5 times higher [29], referred to a primary wet recycling without prior desulfurization. Hence the flue gas can be cooled only to $\sim$160 °C to avoid sulfuric acid condensation. Subsequently, the gas is quenched to reach the entry temperature of a conventional desulfurization unit. After water condensation, the dry secondary recycling stream exits and the remaining part enters the CPU. The secondary recycling is heated and is used to dry the coal and also to convey the crushed coal from the mill to the burners.

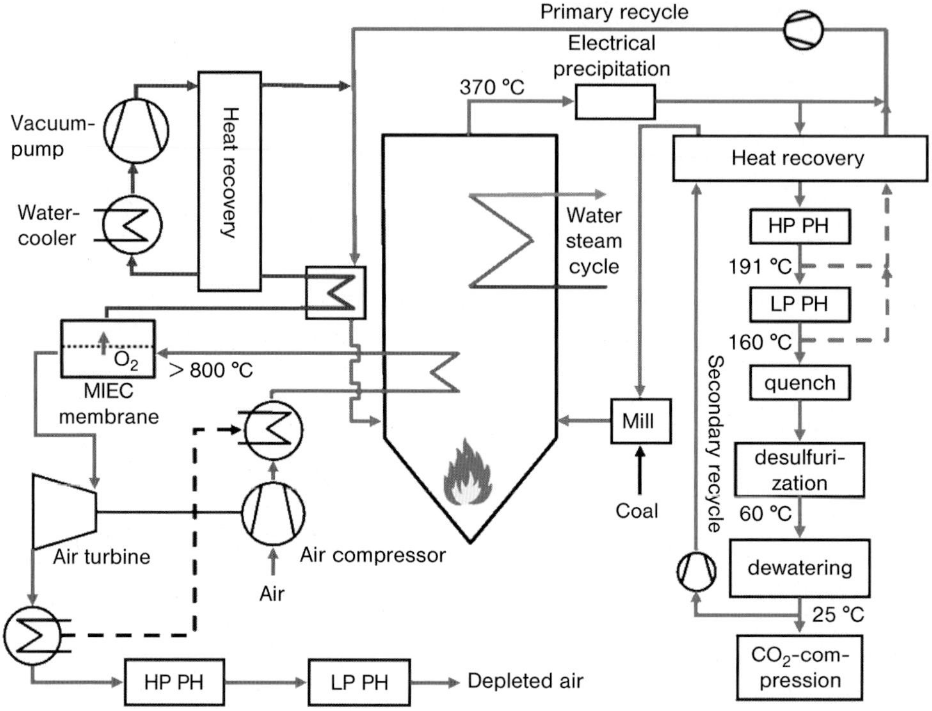

Fig. 15.2 Three-end integration of a MIEC membrane.

15.3.4.2 **Oxygen Separation Unit**

The ambient air is compressed by an air compressor and is heated to the operating temperature of the membrane (about 800–850 °C). For this purpose, expanded air after the air turbine can be used if the temperature is higher than the compression outlet temperature. However, the main heating energy must be provided by the boiler. Therefore, high-temperature gas/gas heat exchangers must be installed within the boiler. These types of heat exchangers are currently being investigated owing to their promising application in externally fired cycles [30]. Incoloy has been proposed as a adequate construction material [31]. Other concepts include heating by combustion of natural gas [28].

In the membrane unit, the oxygen is separated and the hot, pressurized, depleted air is subsequently expanded in a turbine to power the air compression. After recuperation, the remaining heat of the air is integrated into the HP and LP preheating section of the water–steam cycle.

The separated oxygen leaves the membrane unit and must be cooled before entering the vacuum pump. Before entering the flue gas primary recycling, it is reheated by a recuperator.

To achieve an oxygen partial pressure ratio over the membrane as driving force, three different operating conditions can be used:

- atmospheric air feed and vacuum on the permeate side [32]

- pressurized air feed and atmospheric permeate
- combination of pressurized air feed and vacuum on the permeate side [6, 7, 16, 28].

The mean oxygen partial pressure ratio is between 3 and 12.

15.3.5
Four-End Integration

The four-end integration of an MIEC membrane makes use of the flue gas as a sweep gas stream. Hence it is easier to achieve high oxygen partial pressure ratios but only membranes stable to flue gas can be applied. Figure 15.3 shows a simplified scheme of the process.

The flue gas exits the boiler at ~850 °C, since the heat needed to reach the operating temperature of the membrane must be supplied by the recirculating flue gas. Owing to the sensitivity of the membrane materials to dust and sulfur, a hot gas cleaning step is installed. To remove the SO_x, hot gas desulfurization with limestone as adsorbent is applied. The solid products of the desulfurization and the dust are separated in a hot gas filtration unit, consisting of ceramic filter candles. Part of the flue gas serves as sweep gas for the MIEC membrane unit, resulting in a Π_{O_2} of about 26. The remaining flue gas passes through an economizer and, subsequently, the remaining heat is integrated into the HP and LP preheating section of the water–steam cycle. The flue gas can be cooled

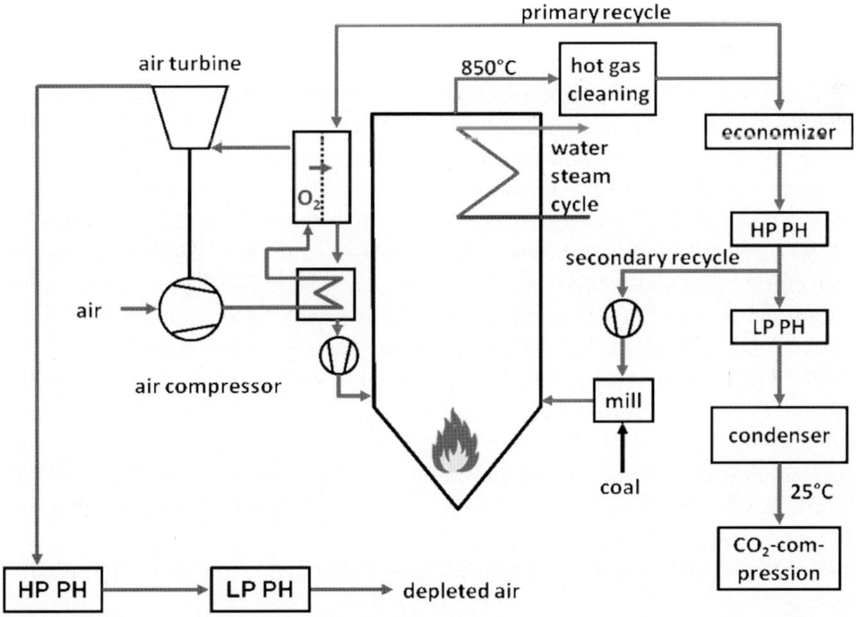

Fig. 15.3 Four-end integration of an MIEC membrane.

to lower temperatures than in the three-end case, since the SO_x concentration is reduced by the hot gas desulfurization, which results in lower acid dew points. After passing through the condenser, the flue gas enters the compression unit. The secondary recycling stream exits after the HP preheating heat exchanger to dry and to convey the coal to the burners.

The air is compressed and heated by a heat exchanger to 750 °C to minimize the thermal stress and to maintain a mean temperature of 825 °C inside the MIEC membrane unit. Unlike in the three-end concept, the oxygen separation ratio is not simply selectable. The minimum ratio is 81% [6]. Lower ratios would require additional heat integration or supply of external heat to heat the air since a certain flow ratio between the air and the flue gas has to be maintained. It has been proposed to heat the air by combustion of natural gas prior to the membrane unit, which increases the flexibility of the process [33]. The depleted air is subsequently expanded in a turbine, which is used to power the air compressor. The remaining heat is integrated into the HP and LP preheating section of the water–steam cycle.

15.3.6
Efficiency

The efficiencies of the described processes are given in Table 15.1. The CO_2 compression unit is the main energy consumer for all integration concepts (36–65% of the total auxiliary energy demand). For the three-end integration, the vacuum pump is the second main energy consumer. Ciferno [28] described a similar vacuum system to that reported by Stadler *et al.* [6], but no details of the energy consumption were given. The efficiencies of the air compressor and the turbine have a large influence on the overall efficiency. Stadler *et al.* [6] assumed state-of-the-art turbo machinery for calculation of the overall efficiency ($\eta_{\text{poly,compressor}} = 90\%$, $\eta_{\text{poly,turbine}} = 91\%$) and a surplus of electricity can be achieved. The stated efficiency drop of the processes was between 9.89 and 3.4%.

For the four-end integration case presented in Table 15.1, a mean total pressure ratio of 20 and a mean oxygen partial pressure ratio of 26 at a degree of separation of 90% were chosen. These process conditions provide an oxygen concentration of about 16.5% for the combustion. The high degree of separation allows the thermodynamic integration to be as simple as possible, but does not provide the highest net efficiency. Reducing the separation ratio to the lower limit of 81% as mentioned before leads to a net efficiency of 40.7% and therefore to a lower efficiency drop compared with the three-end integration in the same study, but for this case heat integration is more complex. The preferred combination of parameters will be the result of a thorough technical and economic analysis. Nevertheless, it can be seen from the presented case that in the four-end integration the hot gas recirculation fan contributes 35.1% to the efficiency loss, which is comparable to the power demand of the CO_2 compression unit [6].

Table 15.1 Auxiliary power demand of the described 3 and four-end membrane-based oxyfuel processes.

Parameter	Three-end [6]		Three-end [28]		Three-end [34]		Four-end [6]	
	Absolute (MW)	Relative (%)	Absolute (MW)	Relative (%)	Absolute (MW)	Relative (%)	Absolute (MW)	Relative (–)
Auxiliary loads								
CO_2 compression	47.5	49.5	58.4	64.4			47.6	36.3
Vacuum pump	17.0	17.7	n.a.	n.a.			–	–
Hot gas recirculation fan	–	–	–	–			46.0	35.1
Water pumps	14.5	15.1	6.39	7.1			20.4	15.5
Blowers	5.9	6.1	12.04	13.3			5.5	4.2
Flue gas cleaning	1.0	1.0	4.64	5.1			1.7	1.3
Others	10	10.4	9.16	10.1			10.0	7.6
Total	*95.9*	*100*	*90.6*	*100*			*131.2*	*100*
Thermal heat input	1210		1875		1283		1210	
Gross output								
Steam turbine	–470.3		–687.9				–585.5	
Membrane compressor/turbine	–109.4		43.9				–21.6	
Net efficiency		39.98		29.51		36.1	39.9	
Efficiency drop[a] (%)		5.92		9.89		3.4		6

a) Compared with the air-fired reference power plant.

The influence of the oxygen separation ratio on the reduction of the efficiency is explained exemplarily by means of Fig. 15.4 and 15.5. Low oxygen separation ratios are characterized by high flow rates of air. This results in an increased heat demand for air preheating. Consequently, the amount of water within the water–steam cycle has to be reduced if equal steam parameters are to be achieved. The gross output of the steam turbines decreases. In addition, the heat of the expanded depleted air cannot be fully integrated into the HP and LP preheating section of the water–steam cycle and a large amount of non-utilizable exhaust heat leaves the process. Conversely, the net output of the turbo machinery benefits from the higher air flow rates. Additionally, the power demand of the vacuum pump is lower if similar oxygen partial pressures ratios are applied. Since the mean oxygen partial pressure on the feed side is higher if low oxygen separation rates are applied, the required permeate pressure for achieving the same oxygen partial pressure ratio is higher compared with higher separation ratios.

The presented effects have an influence on the efficiency of the power plant. The minimum efficiency drop depends on the feed pressure and oxygen separation ratio. In Fig. 15.5, the optimum for different oxygen partial pressure ratios is presented. The optimum is between 45 and 55%. With increasing oxygen par-

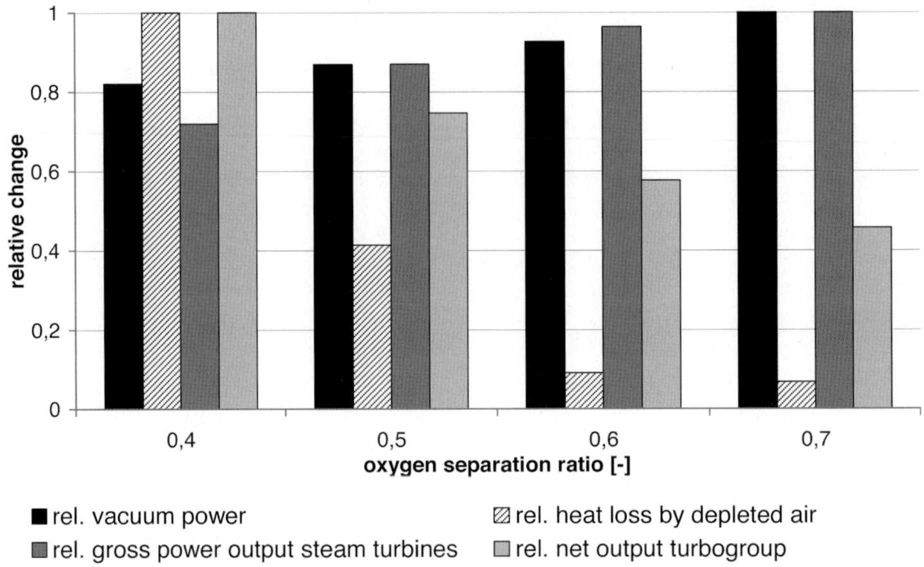

Fig. 15.4 Influence of the oxygen separation ratio on important power plant parameters.

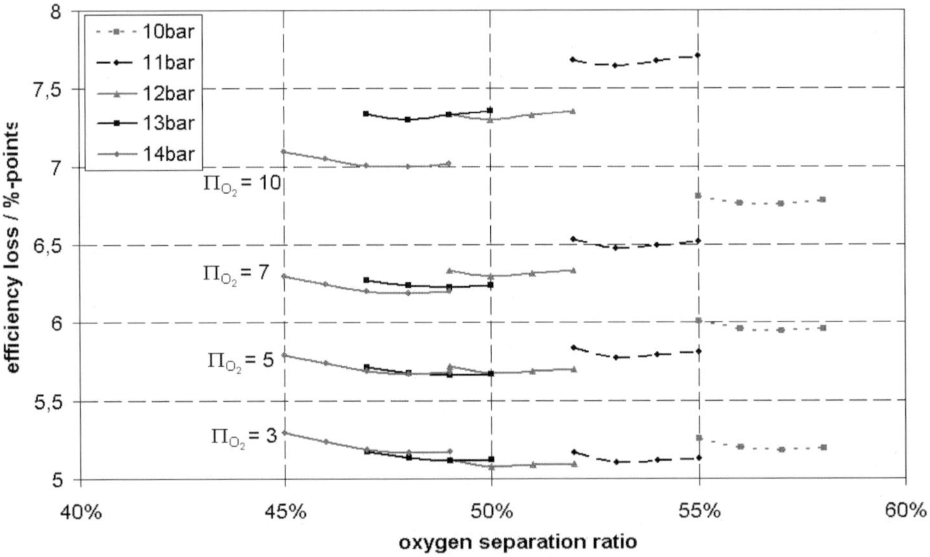

Fig. 15.5 Influence of the oxygen separation ratio and the partial pressure ratio on the efficiency of the membrane-based power plant.

tial pressure ratio, higher feed pressures are favorable since the energy demand of the vacuum pump has an exponential slope towards low vacuum pressures below 200 mbar. Higher oxygen partial pressure ratios have a direct influence on the membrane area, which is discussed in the next section.

15.3.7
Membrane Area

Membrane units for the three- and four-end processes described by Stadler *et al.* [6] were simulated by the same group [13] to determine the required membrane area. The results showed a dependence of the efficiency on the membrane area (see Fig. 15.6). Generally, the four-end concept is characterized by lower membrane area and lower efficiency drop compared with the three-end oxyfuel process. The specific area demand for the simulated power plant is 0.3–0.7 $m^2\ kW_{el}^{-1}$ for the four-end and 0.5–1.5 $m^2\ kW_{el}^{-1}$ for the three-end process.

The membrane area and the related specific permeation rate of the membrane depend on the temperature, the oxygen partial pressure ratio, and the thickness of the membrane (cf. Equation 15.1). The curve for the three-end process shows the optimum net efficiency for the different oxygen partial pressure ratios presented in Fig. 15.5 at a constant temperature of 825 °C. With higher oxygen partial pressure ratios, the membrane area demand decreases, but the efficiency loss increases due to a higher energy demand for the vacuum pump to reach these ratios. Another measure for decreasing the specific membrane area is to apply higher temperatures at the membrane unit, but then the effi-

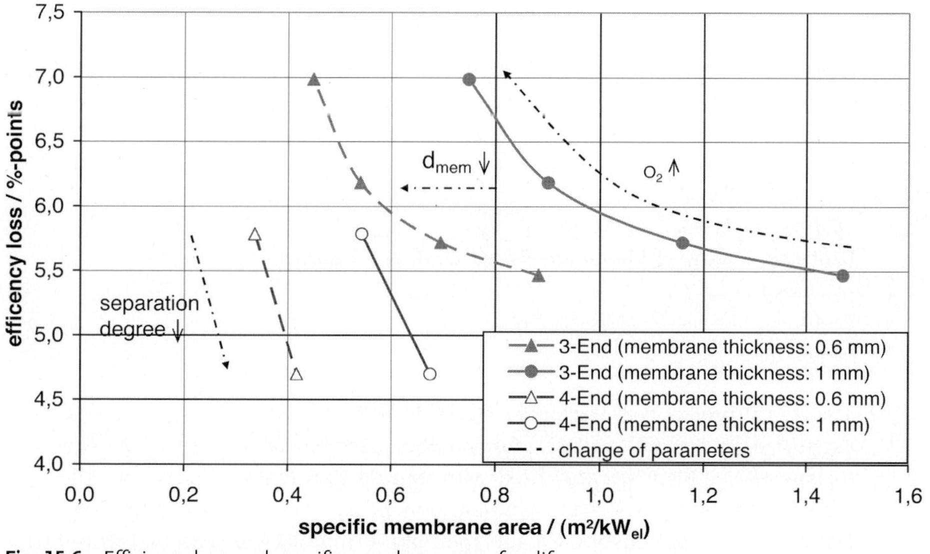

Fig. 15.6 Efficiency loss and specific membrane area for different process conditions. For the membrane, a BSCF membrane was assumed. Adapted from [13].

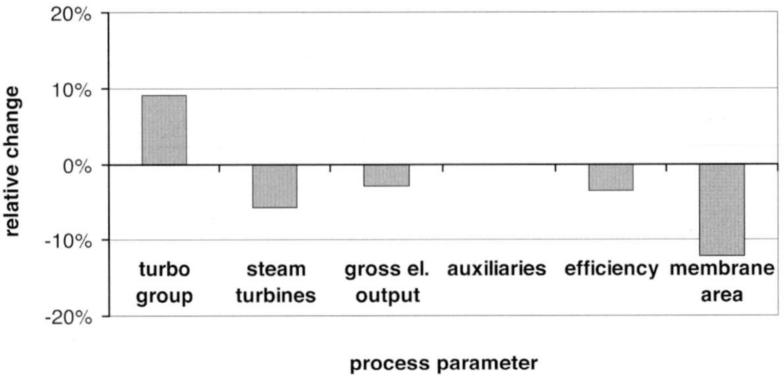

Fig. 15.7 Influence on the process of a 25 °C higher air inlet temperature at the membrane unit.

ciency of the oxyfuel process would be lower (see Fig. 15.7). This results from the higher heat demand for air preheating. The effect is comparable to the influence of the oxygen separation ratio. For the efficiency calculation of the four-end process, the mean oxygen partial pressure ratio was set constant at a value of 26, but the oxygen separation ratio was varied from 90 to 81% as described in Section 15.3.5. The change of this parameter results in a lower air inlet temperature and consequently in a lower mean temperature of the membrane, but has a positive effect on the efficiency of the oxyfuel process. Decreasing the membrane thickness would reduce the membrane area demand since the permeation ratio is inversely proportional to d_{mem}.

In absolute terms, the membrane area for a 1210 MW_{th} oxyfuel power plant that is based on the reference power plant NRW [27] with an oxygen demand of 8400 t d^{-1} ranges between 350 000 and 700 000 m^2 for currently available tubular membrane materials [13]. The materials have permeation rates for the given process parameters between 0.5 and 1.5 Nml min^{-1} cm^{-2}.

15.4
Literature Review of Membrane-Based Oxyfuel Processes

15.4.1
AZEP

The AZEP project was started by Norsk Hydro and Alstom Power in 2001. It is supported by the EU and several industry partners to develop a CO_2-free gas turbine power plant process fired with natural gas. Figure 15.8 shows the flow scheme of this process. The main component is a mixed conducting membrane (MCM) reactor which besides the membrane module consists of a combustor and a high-temperature heat exchanger.

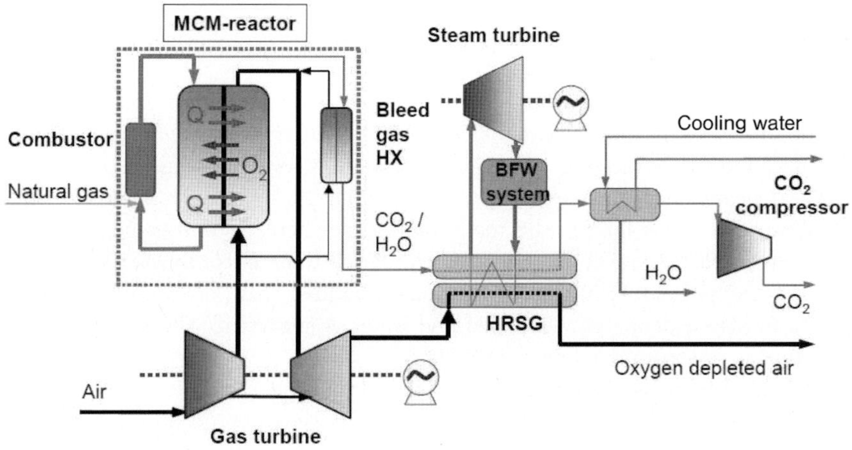

Fig. 15.8 Scheme of the AZEP process [36].

In a compressor, ambient air is compressed to 20 bar and thereby reaches temperatures of up to 450 °C. About 90% of this compressed air is fed to the membrane module, where an exchange of oxygen to the combustion gas and heat transfer to the air takes place. The temperature of the air, which is heated in the lower part of the MCM reactor, is about 900–1100 °C.

About 40–50% of the oxygen in the airstream is separated with the membrane and transported from the exhaust gas to the combustor. At the entrance of the natural gas-fired combustor, the oxygen concentration is about 10% and the temperature can exceed 1200 °C. About 90% of the hot exhaust gas is led into the MCM reactor and heats the leaving air in the upper zone of the MCM reactor to more than 1200 °C. The heat of the remaining 10% of the hot exhaust gas is used in a bleed gas heat exchanger to heat the compressed bleed air before mixing it with the depleted air stream.

The hot, compressed airstream is expanded in the gas turbine for electricity generation. The remaining heat of the airstream and the exhaust gas is used in a heat recovery steam generator to produce steam at different pressure levels for a steam turbine process. The last step is to separate the water from the exhaust stream, which consists of water and CO_2. Finally, the CO_2 is compressed for disposal [35].

Compared with a similar standard gas and steam power plant without CO_2 separation, the thermal efficiency is reduced from 57.9 to 49.6% [36]. This high loss results mainly from the reduced turbine entry temperature due to the temperature limitation of the membrane material. The thermal efficiency can be increased to 53.4% by a gas co-firing before the turbine, but only 85% of the CO_2 can be separated. The efficiency of a comparable MEA CO_2 separation from flue gas with the same separation rate is much lower at 49%.

15.4.2
ZEITMOP

Another gas-fired power plant concept is the Zero Emission Ion Transport Membrane Oxygen Power Cycle (ZEITMOP), which was developed independently of the AZEP concept. The process scheme is shown in Fig. 15.9.

Air is heated in a heat exchanger to 800–900 °C after compression (1). Subsequently, the air is lead to the membrane unit where O_2 is transferred to the recycled exhaust gas. The oxygen-enriched exhaust gas is combusted with compressed methane at 1300–1600 °C (7) and expanded in a low-pressure turbine (5 and 6). The exhaust gas is cooled and water is separated. The water-free carbon dioxide is compressed and partly recycled.

Compared with the AZEP cycle, one can obtain higher turbine entry temperatures because the combustion takes place after the membrane unit. There is no heat exchange at high temperatures because the exhaust gas is expanded directly. This plant needs gas turbines for carbon dioxide, which are not yet ready for the market but should be available soon [37]. Simulations with Aspen Plus provided efficiencies of 45.2% with a turbine entry temperature of 1300 °C, 46.3% at 1400 °C, and 46.2% at 1500 °C [38, 39]. In this process, not only the membrane material is a limiting factor but also the material of the turbine, since entry temperatures of commercial gas turbines are limited to 1200–1300 °C.

A process modification links the combustion and the membrane unit [40, 41]. This leads to a decrease in membrane area because the oxygen on the permeate side is directly consumed for combustion and the partial pressure of oxygen drops immediately on the permeate side. Also, the oxygen permeability is higher due to a higher temperature. At these temperatures, fluorite membrane structures could come into operation since perovskite membranes are stable in

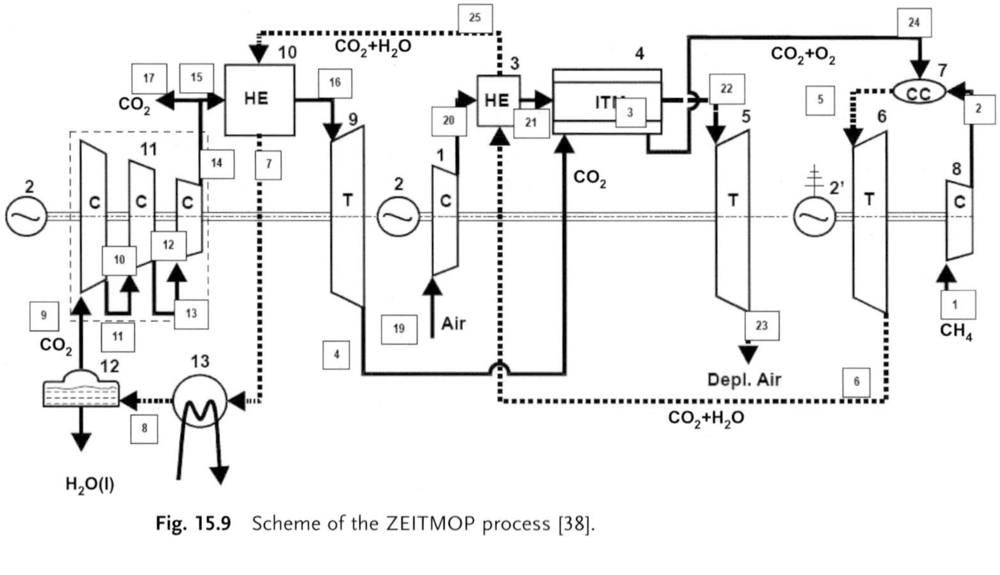

Fig. 15.9 Scheme of the ZEITMOP process [38].

the temperature range 800–1000 °C [41]. The efficiencies for the combined plant are up to 46% at 1300 °C, 48% at 1400 °C, and 51% at 1500 °C [41].

15.4.3 Milano

A coal-fired membrane-based oxyfuel power plant process was presented by Romano [42]. The Milano cycle is an ultra-supercritical (USC) steam plant with fluidized bed combustion. The process is shown in Fig. 15.10. The process simulation was made with a computer-code which was developed at the Politecnico di Milano.

Air is compressed and heated through the exhaust gas in a heat exchanger. Subsequently, the air reaches the membrane unit and the oxygen permeates through it to the sweep exhaust gas. The oxygen-enriched exhaust gas is cooled

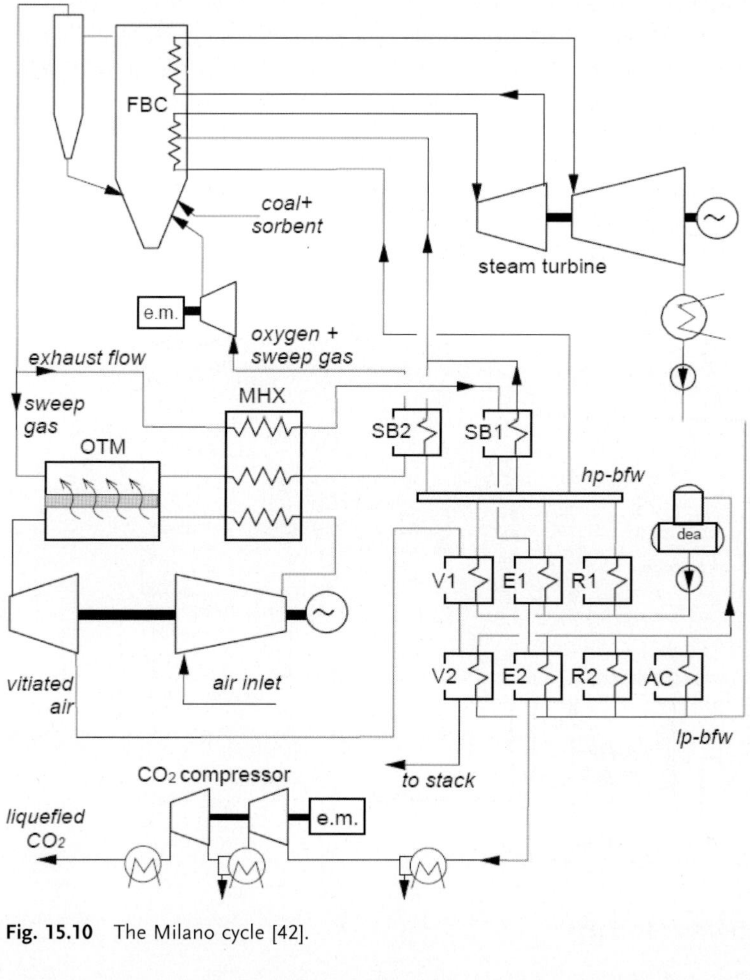

Fig. 15.10 The Milano cycle [42].

and fed to the burner. The heat of the exhaust gas is used for the Rankine process. Part of the flue gas is recycled and used as sweep gas. The other part is used for preheating and goes to the stack. The water–steam cycle has a normal reheat cycle with steam parameters of 250 bar/600 °C and 60 bar/610 °C, and the combustion takes place at 860 °C. With these parameters, the efficiency is 41.34% ($p_{\text{combustion}} = 1.15$ bar) or 41.89% ($p_{\text{combustion}} = 10$ bar). The steam turbine mainly produces electricity and the gas turbine mainly drives the compressor.

15.4.4 EFCC Oxyfuel

Another power plant concept is the EFCC (Externally Fired Combined Cycle), which is also adapted to oxyfuel power plants (Fig. 15.11). This process was also developed at the Politecnico di Milano. In contrast to the Milano process, the electricity is produced in equal parts by the gas and steam turbine.

The heat exchange from the exhaust gases to the depleted air takes place directly after the combustion. Consequently, the temperature at the entrance of the gas turbine is higher. The expanded air and the flue gas streams provide sufficient heat for a Rankine process. Branched-off exhaust gas is desulfurized

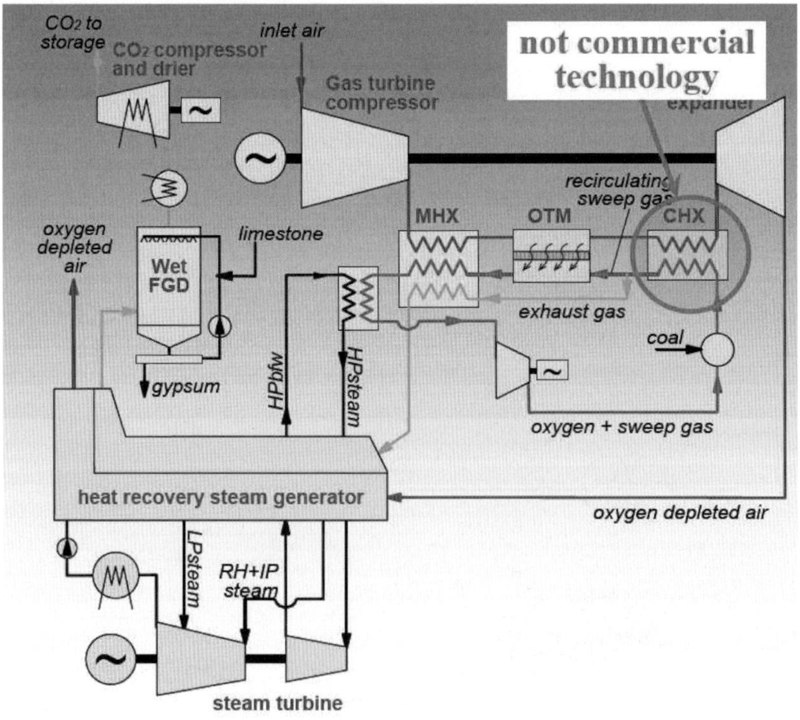

Fig. 15.11 The EFCC cycle [42].

and compressed. Owing to the elevated inlet temperatures for the gas turbine, this process reaches a thermal efficiency of 44.94% at 1050 °C and 49.25% at 1350 °C.

The idea of the EFCC process without oxyfuel combustion has long existed, but has not been realized in practice due to the high-temperature heat exchanger. It requires high mechanical stability due to the pressure difference of the working medium and exhaust gas and also it has to be thermally stable and resistant to erosion and corrosion [43]. These requirements are basically realizable with ceramics but such heat exchangers are not the state of the art and are still the subject of basic research [44].

15.5 Towards Realization – Module Design

Since process calculations indicate that membrane-based oxyfuel technology is the most promising carbon capture and storage measure, various efforts have been started towards the realization of membrane technology for oxyfuel combustion. Since most of the process components are adaptable from state-of-the-art power plant processes, current research is focused on the membrane material and how to integrate it into a fossil fuel-fired power plant.

Figure 15.12 shows the most important issues for the application of MIEC membranes in a power plant process. All these requirements interact with or even contradict each other. For instance, the best reported oxygen fluxes were obtained with BSCF membranes [45] but this material shows poor chemical stability in the presence of CO_2, which makes it unacceptable for operation in the four-end mode [17]. High oxygen fluxes can be obtained by elevating the temperature or manufacturing thinner membranes, which has amongst others an influence on the thermal stability or the mechanical strength. All these issues have to be taken into account when designing a membrane unit that can come into operation in an oxyfuel power plant process. Some research and development activities in this field are presented in this section.

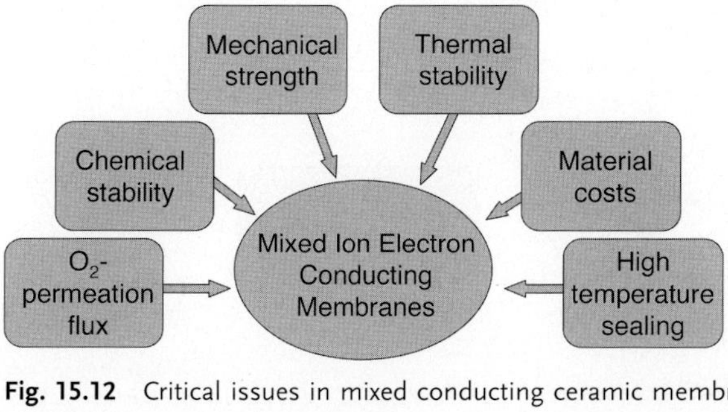

Fig. 15.12 Critical issues in mixed conducting ceramic membranes. Adapted from [46].

15.5.1
Air Products ITM Technology

In collaboration with the US Department of Energy (DOE), Air Products is developing their ion transport membrane (ITM) technology with the aim of reducing capital and operating costs for oxygen production. Compared with cryogenic air separation, Air Products expects savings in both the power demand and capital costs of more than 30% [47].

In the oxygen production process, air is compressed to 10–34 bar and then heated via direct combustion of H_2 fuel gas (Fig. 15.13). This feed stream enters the membrane unit, where oxygen is separated from the air stream. The depleted air is reheated by additional combustion and then enters a turbine expander for electricity production. Thermal energy of the expanded retentate stream is partly recovered in a steam generator.

The membrane unit itself (Fig. 15.14) consists of a pressure vessel in which a number of planar ion transport membranes are installed and arranged in series. Each of these modules consists of several wafers having planar parallel surfaces as shown in Fig. 15.14b. Oxygen can diffuse through the outside dense, thin, active membrane layer that is fixed on a porous support and enters the interior of the wafer, which consists of a channeled support to let the oxygen flow to the central collection tube (Fig. 15.15). Via this manifold, the oxygen is delivered to the outside of the pressure vessel [12].

One stacked module reaches an oxygen production capacity of $0.5\ t\ d^{-1}$. A pilot plant in Maryland is already delivering $5\ t\ d^{-1}$ of oxygen [47]. In the next phase, scale-up to $150\ t\ d^{-1}$ is planned.

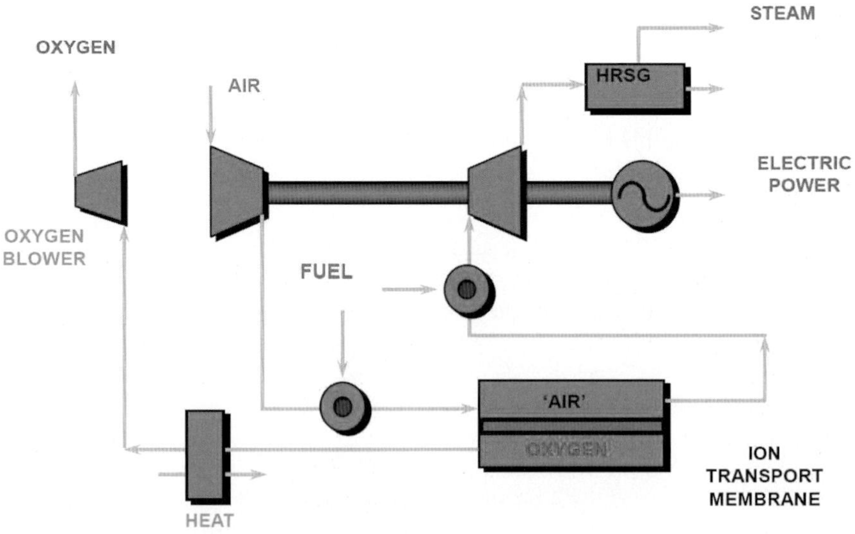

Fig. 15.13 Scheme of the Air Products process for ITM oxygen production [48].

Fig. 15.14 Air Products membrane unit (a) and single 0.5 t d^{-1} membrane module (b) [48].

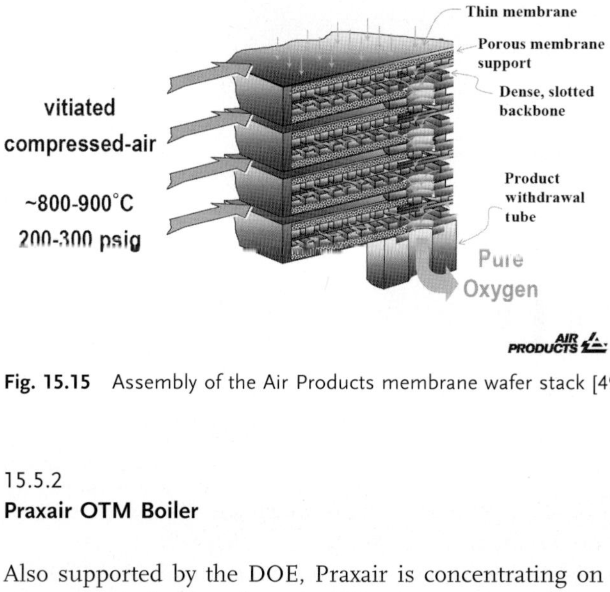

Fig. 15.15 Assembly of the Air Products membrane wafer stack [49].

15.5.2
Praxair OTM Boiler

Also supported by the DOE, Praxair is concentrating on the development of an "oxygen transfer membrane" (OTM) system. By utilizing an oxygen-consuming reaction on the membrane permeate side, the oxygen partial pressure ratio is enhanced (Fig. 15.16). The OTM system delivers oxygen to a power cycle where gasified coal is used for electricity generation [50]. In conceptual designs, the tubular membranes are integrated directly in the boiler (Fig. 15.17). The combustion reaction on the fuel side of the membranes creates a very low oxygen partial pressure compared with the air side of the membrane. This chemical potential difference drives oxygen through the membrane without the need for additional air compression. As an additional advantage, the membrane does not have to withstand high mechanical stress caused by large pressure differences.

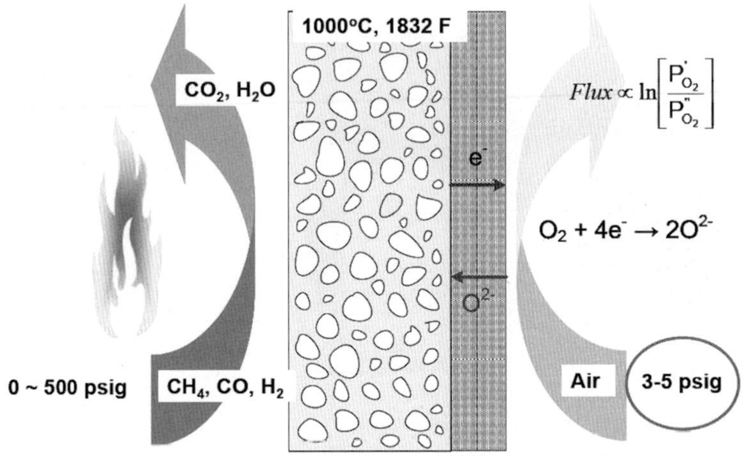

Fig. 15.16 Principle of the Praxair OTM system [55].

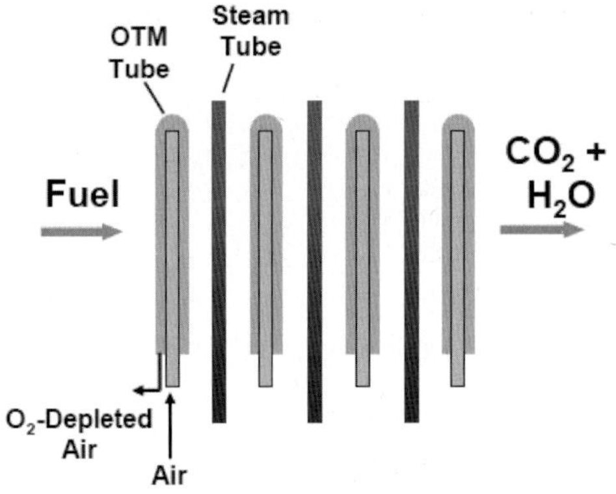

Fig. 15.17 Principle of the OTM boiler [56].

Test results have indicated the potential for a 70% decrease in auxiliary power consumption for supplying O_2 on a kWh t^{-1} O_2 basis [51, 52].

The steam generator pipes are located in the combustion chamber in addition to the membrane tubes. Due to the direct heat release to the steam generator pipes during combustion, no recycled exhaust gas is necessary. For the membrane, a tubular design is chosen. The tubes consist of a zirconia porous support and a dual-phase zirconia and metal oxide separation layer at the inner side of the tubes [53, 54].

Laboratory-scale OTM reactors have been built for natural gas and for hard coal combustion. Laboratory tests demonstrate that the natural gas-fired reactor shows good combustion and a stable membrane at a temperature of 1020 °C [57]. During coal combustion, problems in attaining the required temperature have emerged [53]. Current work is concerned with increasing the permeability,

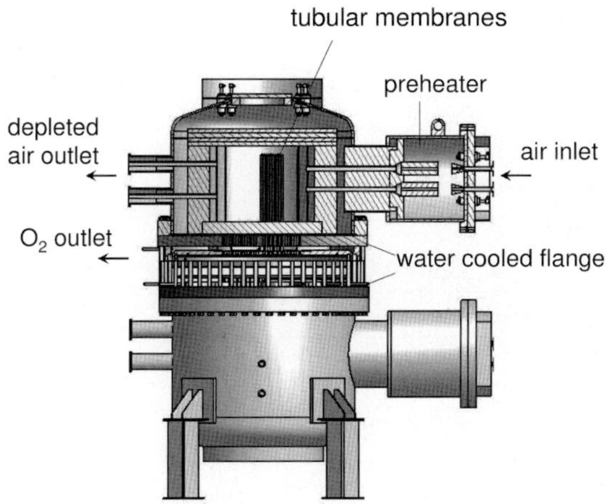

Fig. 15.18 Pilot-scale module of the OXYCOAL-AC project.

up-scaling, and expansion to fossil fuels with an SO_2-resistant membrane material [56, 58].

15.5.3
OXYCOAL-AC Pilot Module

For the OXYCOAL-AC project, a pilot-scale module is currently under construction. The working principle is shown in Fig. 15.18. The symmetrical assembly consists of a pressure vessel with closed-ended tubular membranes enclosed in a flange between the two parts of the vessel.

Compressed and preheated feed air at up to 20 bar and 900 °C flows around the membrane tubes, forcing the oxygen to permeate through the membrane. The oxygen is collected in the flange and conveyed by means of a vacuum pump. To avoid high-temperature sealing, the flange is cooled by water. Under optimized process conditions, which means high feed air pressure and low permeate pressure, the capacity exceeds 0.5 t d^{-1} of oxygen.

15.5.4
Ceramic Autothermal Recovery (CAR)

Another method of transferring oxygen to an oxyfuel combustion system is the ceramic autothermal recovery (CAR) process. This process makes use of the oxygen storage capacity of perovskites at high temperatures. Among others, the process under investigation by BOC (now Linde group), Western Research Institute (WRI), and Alstom Power Plant Laboratories.

The working principle is shown in Fig. 15.19 and 15.20b. Two separate beds of ceramic oxygen carriers are cycled to deliver O_2 for oxyfuel combustion. The perovskite sorbent is alternately exposed to feed air and subsequent regenera-

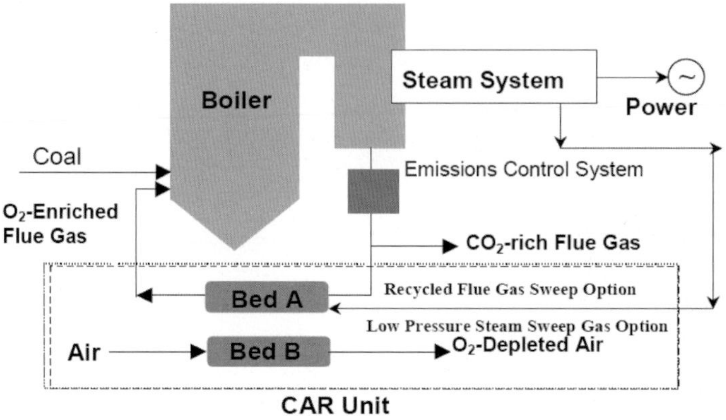

Fig. 15.19 Schematic of a CAR unit integrated with a coal-fired boiler [59, 60].

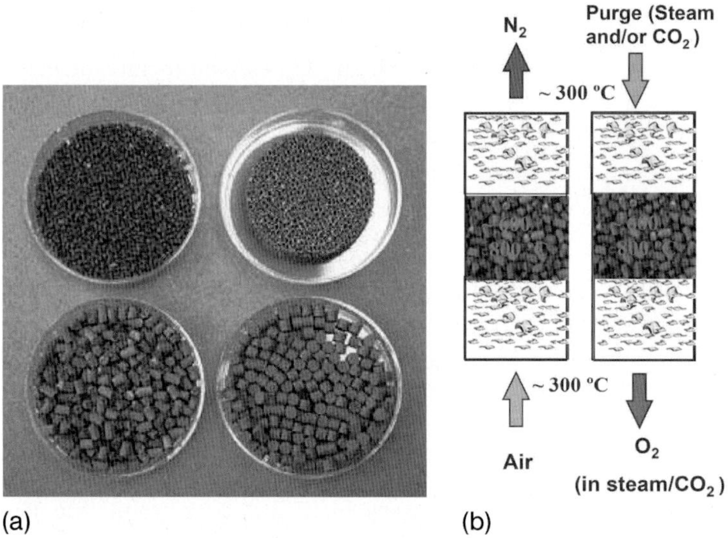

Fig. 15.20 Perovskite pellets (a) and process scheme (b) [61].

tion or purge gas flows with a small oxygen partial pressure, such as CO_2 or steam. Owing to the partial pressure swing, oxygen is absorbed and desorbed in the packed bed, and a continuous gas flow can be achieved by using a couple of continuously cycled beds.

Internal regenerative heat transfer is used to maintain temperature; the perovskite zone in each bed is sandwiched between two zones of heat transfer material. As carrier materials for the oxygen, pelletized perovskite LaSrCoFe (LSCF) materials as shown in Fig. 15.20a are used. Cycling times lie in the range 30–60 s, and within this time the carrier is able to gain an oxygen capacity of up to 0.5 wt%.

Thermo-economic parametric analyses have been performed to determine the specific oxygen production cost compared with a cryogenic ASU. It has been found that for an O_2 storage capacity of 1 wt%, the oxygen production costs are about 40% lower than for cryogenic production. Considering the so far best observed case of 0.5 wt% oxygen in the carrier material, it has been concluded that the cost advantage is too small to justify the effort of commercializing this technology at present.

A pilot-scale plant producing 0.7 t d^{-1} of oxygen was installed adjacent to WRI's pilot-scale coal combustion test facility (CTF) in August 2006 and was successfully demonstrated. Owing to high heat losses and a resulting unequal temperature distribution in the packed beds, only half of the aimed for capacity could be reached. Furthermore, it was found that the investigated LSCF perovskite materials show sulfate and carbonate formation on their surfaces in the presence of SO_2 and CO, respectively. These surface formations block the oxygen exchange reaction and therefore hinder the sorption of oxygen in the material. It was decided not to proceed to the next process phase owing to the overall cost potential compared with cryogenic ASU, being only nominal at current levels of performance, and owing to the high risks related to both material and process scale-up [61].

15.6 Conclusion

Increased interest in membrane-based oxyfuel processes has arisen during the past decade. Process calculations indicate this technology to be one of the most promising carbon capture measures, since it is also able to deliver a high-purity CO_2 stream. Especially flue gas-stable membranes offer the high benefit of oxyfuel technology compared with post- or pre-combustion capture. Also, without contact with flue gas, membranes show good potential for oxygen production for an oxyfuel process. There are currently still some problems to be solved prior to application of this technology. Most of them involve the integration of a large membrane area into the power plant process, but also the development and improvement of the membrane itself is the subject of many research projects. Finally, the cost aspects have to be taken into account before commercialization of this technology.

References

1 Working Group III of the Intergovernmental Panel on Climate Change. *IPCC Special Report on Carbon Dioxide Capture and Storage*, Cambridge University Press, Cambridge, 2005.

2 Buhre, B.J.P., Elliot, L.K., Sheng, C.D., Gupta, R.P., and Wall, T.F. Oxyfuel combustion technology for coal-fired power generation. *Prog. Energy Combust. Sci.*, 2005, **31**, 283–307.

3 Farzan, H., Vecci, S., McDonald, D., and K. McCauley, Pranda, P., Varagani, R., Gautier, F., Tranier, J.P. *State of the art of oxy-coal combustion technology for CO_2 control from coal-fired boilers*. Presented at the 32nd International Technical Conference on Coal Utilization and Fuel Systems, Clearwater, FL, USA, 2007.

4 Kather, A. and Scheffknecht, G. The oxy-coal process with cryogenic oxygen supply. *Naturwissenschaften*, 2009, **96**, 993–1010.

5 Allam, R.J. Improved oxygen production technologies. *Energy Procedia*, 2009, **1**, 461–470.

6 Stadler, H., Beggel, F., Habermehl, M., Persigehl, B., Kneer, R., Modigell, M., and Jeschke, P. (2011) Oxyfuel coal combustion by efficient integration of oxygen transport membranes. *Int. J. of Greenhouse Gas Control*, **5**, 7–15.

7 Seltzer, A., Fan, Z., and Hack, H. Oxy-fuel coal combustion power plant system optimization. Presented at the 7th Annual COAL-GEN Conference, Milwaukee, WI, 2007.

8 Kneer, R., Toporov, D., Förster, M., Christ, D., Broeckmann, C., Pfaff, E., Zwick, M., Engels, S., and Modigell, M. OXYCOAL-AC: towards an integrated coal-fired power plant process with ion transport membrane-based oxygen supply. *Energy Environ. Sci.*, 2010, **3**, 198–207.

9 Wilson, J., Christie, M., Degenstein, N., Shah, M., Li, J., Venkateswaran, V., Eddings, E,. and Adams, J. Presented at the 34th International Technical Conference on Coal Utilization and Fuel Systems, Clearwater, FL, 2009.

10 Renz, U. European Patent EP1717420A1, 2005.

11 den Exter, M.J. In *Inorganic Membranes for Energy and Enviromnental Applications* (ed. Bose, A.C.), Springer, Berlin, 2009, pp. 27–51.

12 Air Products and Chemicals. European Patent EP1676811A2, 2005.

13 Engels, S., Beggel, F., Modigell, M., and Stadler, H. Simulation of a membrane unit for oxyfuel power plants under consideration of realistic BSCF membrane properties. *J. Membr. Sci.*, 2010, **359**, 93–101.

14 Leo, A., Liu, S., and da Costa, J.C.D. Development of mixed conducting membranes for clean coal energy delivery. *Int. J. Greenhouse Gas Control*, 2009, **3**, 357–367.

15 Martin, M. MEM-OXYCOAL – oxygen permeation membranes for coal-fired power plants. Presented at the 4th International Conference on Clean Coal Technologies (CCT), Dresden, Germany, 2009.

16 Czyperek, M., Zapp, P., Bouwmeester, H.J.M., Modigell, M., Ebert, K., Voigt, I., Meulenberg, W.A., Singheiser, L., and Stöver, D. Gas separation membranes for zero-emission fossil power plants: MEM-BRAIN. *J. Membr. Sci.*, 2010, **359**, 149–159.

17 Engels, S., Markus, T., Modigell, M., and Singheiser, L. (2011) Oxygen permeation and stability investigations on MIEC Membrane Materials under Operating Conditions for Power Plant Processes. *J. Membr. Sci.*, **370**, 58–69.

18 Arnold, M., Wang, H., and Feldhoff, A. Influence of CO_2 on the oxygen permeation performance and the microstructure of perovskite-type $(Ba_{0.5}Sr_{0.5})(Co_{0.8}Fe_{0.2})O_{3-\delta}$ membranes. *J. Membr. Sci.*, 2007, **293**, 44–52.

19 Wall, T., Liu, Y., Spero, C., Elliott, L., Khare, S., Rathnam, R., Zeenathal, F., Moghtaderi, B., Buhre, B., Sheng, C., *et al.* An overview on oxyfuel coal combustion – state of the art research and technology development. *Chem. Eng. Res. Des.*, 2009, **87**, 1003–1016.

20 Haag, J.-C. and Jeschke, P. Presented at the 13th International Symposium on Transport Phenomena and Dynamics of Rotating Machinery, Honolulu, 2010.

21 Toporov, D., Förster, M., and Kneer, R. *Combustion of Pulverized Fuel in a CO_2 Atmosphere*, Wiley-VCH Verlag GmbH, Weinheim, 2010.

22 Erfurth, J., Toporov, D., Förster, M., and Kneer, R. Numerical simulation of a 1200MW pulverised fuel oxy-firing furnace. Presented at the 4th International Conference on Clean Coal Technologies (CCT), Dresden, Germany, 2009.

23 Aspelund, A. and Jordal, K. Gas conditioning – the interface between CO_2 capture and transport. *Int. J. Greenhouse Gas Control*, 2007, **1**, 343–354.

24 Tigges, K.D., Klauke, F., Bergins, C., Busekrus, K., Niesbach, J., Ehmann, M., Kuhr, C., Hoffmeister, F., Vollmer, B., Buddenberg, T., *et al.* Conversion of existing coal-fired power plants to oxyfuel combustion: case study with experimental results and CFD simulations. *Energy Procedia*, 2009, **1**, 549–556.

25 Agrawal, R. and Herron, D.M. In *Encyclopedia of Separation Science* (ed. Ian, D.W.), Academic Press, Oxford, 2000, pp. 1895–1910.

26 White, V., Torrente-Muricano, L., Sturgeon, D., and Chadwick, D. Purication of oxyfuel-derived CO_2. *Int. J. Greenhouse Gas Control*, 2009, **4**, 137–142.

27 VGB PowerTech. *Konzeptstudie Referenzkraftwerk Nordrhein-Westfalen (RKW NRW)*, VGB PowerTech, Essen, 2004.

28 Ciferno, J. *Pulverized Coal Oxycombustion Power Plants*, Office of Systems, Analysis and Planning, National Energy Technical Laboratory, Pittsburgh, PA, 2008.

29 Stanger, R. and Wall, T. Sulphur impacts during pulverised coal combustion in oxy-fuel technology for carbon capture and storage. *Prog. Energy Combust. Sci.*, 2011, **37**, 69–88.

30 Aquaro, D. and Pieve, M. High temperature heat exchangers for power plants: performance of advanced metallic recuperators. *Appl. Thermal Eng.*, 2007, **27**, 389–400.

31 Seltzer, A. *Furnace and Heat Recovery Area Design and Analysis for Conceptual Design of Supercritical O_2-Based PC Boiler*, Foster Wheeler North America, Livingston, NJ, 2006.

32 Forschungszentrum Jülich. *IEF-3 Report 2009*, Forschungszentrum Jülich, Jülich, 2009.

33 Carbo, M.C., Jansen, D., Hendriks, C., de Visser, E., Ruijg, G.J., and Davison, J. Opportunities for CO_2 capture through oxygen conducting membranes at medium-scale oxyfuel coal boilers. *Energy Procedia*, 2009, **1**, 487–494.

34 Seltzer, A. and Robertson, A. *Economic Analysis for Conceptual Design of Supercritical O_2-Based PC Boiler*, Foster Wheel North America, Livingston, NJ, 2006.

35 Griffin, T., Sundkvist, S.G., Asen, K., and Bruun, T. *Advanced Zero Emissions Gas Turbine Power Plant*, ALSTOM Power Technology Center, Baden-Dättwil, 2003.

36 Sundkvist, S.G., Julsrud, S., Vigeland, B., Naas, T., Budd, M., Leistner, H., and Winkler, D. Development and testing of AZEP reactor components. *Int. J. Greenhouse Gas Control*, 2007, **1**, 180–187.

37 Foy, K. and Yantovski, E. History and state-of-the-art of fuel fired zero emissions power cycles. *Int. J. Thermodyn.*, 2006, **9**, 37–63.

38 Yantovski, E., Gorski, J., Smyth, B., and Elshof, J. Zero-emission fuel-fired power plants with ion transport membrane. *Energy*, 2004, **29**, 2077–2088.

39 Yantovski, E., Gorski, J., and Warchol. Toward the optimization of ZEITMOP cycle. Presented at ECOS 2006, International Conference on Efficiency, Cost, Optimization, Simulation, Crete, 2006.

40 Foy, K. and McGovern, J. Analysis of the effects of combining air separation with combustion in a zero emissions (ZEITMOP) cycle. *Energy Convers. Manage.*, 2007, **48**, 3046–3052.

41 Yantovski, E. A Detailed simulation of the ZEITMOP cycle with combined air separation and combustion. Presented at ECOS 2007, International Conference on Efficiency, Cost, Optimization, Simulation, Padova, 2007.

42 Romano, M. Presented at the Fourth Annual Conference on Carbon Capture and Sequestration, Alexandria, VA, 2005.

43 Edelmann, H., and Steinmüller, F. EFCC – ein zukünftiges Konzept für Kohle-Kombi-Kraftwerke. *VGB Kraftwerkstech.*, 1997, **77**, 538–543.

44 COORETEC. *Research and Development Concept for Zero-Emission Fossil-Fuelled Power Plants*, COORETEC, Forschungszentrum Jülich, Jülich, 2003.

45 Sunarso, J., Liu, S., Lin, A., and Diniz da Costa, J. Oxygen permeation performance of $BaBiO_{3-\delta}$ ceramic membranes. *J. Membr. Sci.*, 2009, **344**, 281–287.

46 Hashim, S.M., Mohamed, A.R., and Bhatia, S. Current status of ceramic-based membranes for oxygen separation from air. *Adv. Colloid Interface Sci.*, 2010, **160**, 88–100.

47 Air Products and EPRI further develop ion transport membranes. *Membr. Technol.*, 2009, (7), 2–3. Ed.: S. Atkinson.

48 White, V. and Fogash, K. Oxygen Supply for Oxycoal CO_2-Capture. Presented at the 1st Oxyfuel Combustion Conference, Cottbus, 2009.

49 Armstrong, P.A., Benett, D.L., Foster, E.P.T., and Stein, V.E. *Ceramic Membrane Development for Oxygen Supply to Gasification Applications*, Air Products and Chemicals, Allentown, PA, 2002.

50 Shah, M.M., Jamal, A., Dmevich, R.F., Van Hassel, B.A., Christie, G.M., Kobayashi, H., and Bool, L.E. United States, patent no. 014672, 2008.

51 NETL. *DOE/NETL Advanced Carbon Dioxide Capture R&D Program: Technology Update.* National Energy Technology Laboratory, Pittsburgh, PA, 2010.

52 Praxair awarded DOE grant for ceramic membrane development. *Membr. Technol.*, 2010, (1), 16–11, 16. Ed.: S. Atkinson.

53 Christie, M. Oxygen Transport Membrane Based Oxycombustion for CO_2-Capture from Coal Power Plants. Presented at the 2010 NETL CO_2 Capture Technology Meeting, Pittsburgh, PA, 2010.

54 Nagabhushana, N., Lane, J.A., Christie, G.M., and Van Hassel, B.A. US Patent 7 556 676, 2009.

55 Christie, M., Wilson, J., Degenstein, N., and Shah, M.. Oxygen transport membrane based oxycombustion for CO_2 capture from coal power plants. Presented at the Annual NETL CO_2 Capture Technology for Existing Plants. R&D Meeting, Pittsburgh, PA, 2009.

56 Shah, M. Oxy-fuel combustion using OTM for CO_2 capture from coal power plants. Presented at the 2nd Workshop on International Oxy-Combustion Research Network,Windsor, CT, 2007.

57 Li, J., van Hassel, B.A., and Christie, G.M. Progress towards the Demonstration Of An 'Advanced' OTM Oxy-Fuel Boiler. Presented at the 31st International Technical Conference on Coal Utilization and Fuel Systems, Clearwater, FL, 2006.

58 Christie, M., Van Hassel, B., Li, J., Shah, M. OTM Based Oxycombustion for CO_2-Capture from Coal Power Plants, in Proceedings of the 32nd *International Technical Conference on Coal Utilization and Fuel Systems*, Vol. 1, 2007, pp. 41–46.

59 Acharya, D., Krishnamurthy, K.R., Leison, M., MacAdam, S., Sethi, V.K., Anheden, M., Jordal, K., and Yan, J. Development of a oxycombustion high temperature oxygen generation process and its application to oxycombustion powerplants with carbon dioxide capture. Presented at the Pittsburgh Coal Conference, Pittsburgh, PA, 2005.

60 NETL. *Carbon Capture Technology Research and Breakthrough Concepts, Project Overview*, National Energy Technology Laboratory, Pittsburgh, PA, 2008.

61 DOE/NETL. *Pilot-Scale Demonstration of a Novel, Low-Cost Oxygen Supply Process and its Integration with Oxy-Fuel Coal-Fired Boilers*, National Energy Technology Laboratory, Pittsburgh, PA, 2008.

16
CO_2 Separation via Pre-Combustion Utilizing Membranes in Coal Power Plants

Viktor Scherer and Johannes Franz

Abstract

This chapter gives a detailed overview of existing concepts for the use of membranes for pre-combustion capture of CO_2 in integrated gasification combined cycles (IGCC) and a short introduction to the concepts of natural gas (reforming) combined cycles (NGCC). The chapter summarizes which type of membranes (dense proton-conducting membranes, porous membranes, etc.) are feasible for pre-combustion capture and how they can be incorporated into the power plant cycle based on their operational and stability characteristics. The resulting influence on the power plants performance is evaluated. A comparison with state-of-the-art absorption technologies is given.

Keywords: pre-combustion, carbon capture, membranes, integrated gasification combined cycles, natural gas (reforming) combined cycles

16.1
Introduction

Membrane technologies (either H_2- or CO_2-selective membranes) show high potential for pre-combustion carbon capture, because the high gasification (reforming) pressure in integrated gasification combined cycle (IGCC) and natural gas (reforming) combined cycle (NGCC) power plants ensures high driving forces across separation membranes and allows compact-sized modules. In addition, the availability of sweep gases and the possibility of integrating them into an IGCC/NGCC power plant increases driving forces further and also can reduce recompression energy for either the gas turbine fuel (H_2) or the separated CO_2. Unlike absorption technologies, membrane technologies do not require chemicals which are consumed during operation.

Membranes can operate over a wide range of temperatures and pressure levels (depending on the underlying physical principle of separation) but show different thermal and chemical stabilities. On the one hand, this allows integration of the membranes at different locations in an IGCC or NGCC power plant. On

Process Engineering for CCS Power Plants, First Edition. Edited by Detlef Stolten and V. Scherer

the other hand, this sets the boundary conditions to the power plant process in terms of gas quality to be delivered to the membrane. Membranes are especially suitable for a modular design and hence have a good capability for an up-scaling approach.

Pre-combustion capture by membranes has to outperform existing technologies such as chemical and physical absorption processes in order to be competitive. Therefore, the cost of membrane integration and the efficiency losses accompanied by CO_2 separation are key factors for an evaluation of the potential of membranes for pre-combustion capture of CO_2. This chapter concentrates on the evaluation of efficiency losses.

16.2
Process Conditions, Membrane Characteristics, Classification Numbers, Permeation Laws, and Water Gas Shift

To evaluate how and where membrane reactors can be integrated into an IGCC, it is important to know about the process conditions (temperatures, pressures, gas composition, mass flows) in these power plants. These process conditions set the limits for the membrane selection based on the membrane operational characteristics and their chemical and thermal stability. As the focus of this book is on carbon capture in coal-fired power plants, the IGCC power plant is discussed in detail in this contribution. However, to give a complete picture of pre-combustion carbon capture, a brief introduction to NGCC processes is also given, although not in the same depth as for IGCCs.

16.2.1
IGCC Power Plants Without Carbon Capture

Figure 16.1 shows a simplified IGCC process without CO_2 capture. The coal is gasified at elevated temperatures (950–2000 °C in the reaction zone) and pressures (10–80 bar). Oxygen and steam serve as gasification agents. In order to achieve high coal conversion rates, it is necessary to use almost pure oxygen produced by an air separation unit (ASU). Air-blown gasifiers are limited to conversion ratios of 92% whereas oxygen-blown gasifiers can reach ratios above 99% [1]. The raw syngas stream produced consists mainly of H_2 and CO and temperature levels vary between 430 °C (solid-bed gasifiers) and 1300–1600 °C (entrained flow gasifiers). The common choice (see, for example, the two European IGCCs in Buggenum [2, 3] and Puertollano [2, 4, 5]) is entrained flow gasifiers as the required residence time of the coal is low. Hence reactor volumes are relatively low, which leads to limited investment costs. Downstream of the gasifier the raw syngas is cooled by a wet quench, radiative coolers, or recirculated cold syngas. In addition to H_2 and CO, the syngas also contains dust and gaseous pollutants (HF, COS, H_2S). Dust can be removed by ceramic filters, cyclones, or venturi scrubbers. Venturi scrubbers also remove water-soluble pollu-

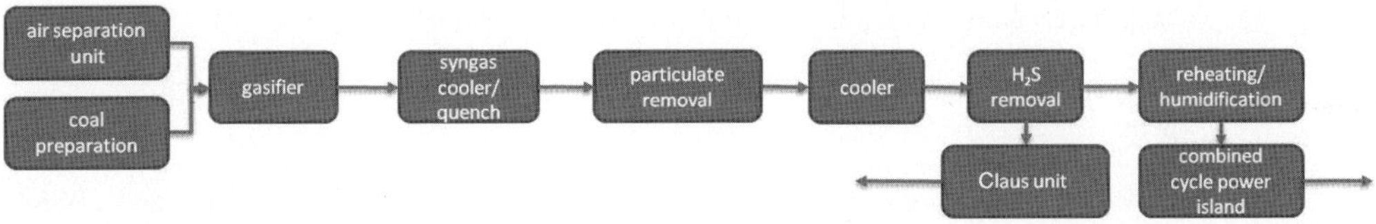

Fig. 16.1 Simplified IGCC process without CO_2 capture.

tants such as HCl and NH_3. COS is converted together with H_2O to H_2S and CO_2 in a catalytic COS hydrolysis unit. A chemical scrubbing unit based on amine solutions [methyldiethanolamine (MDEA), monoethanolamine (MEA), etc.] is the common choice for H_2S removal. As these systems work at low temperatures, further cooling steps are necessary. Because of the accompanying exergy losses of cooling, hot acid gas removal systems are under investigation [6–8]. However, these systems are not state of the art. After desulfurization, the syngas is reheated (to increase power plant efficiency) and humidified (to achieve low thermal NO_x in the gas turbine combustor) and then fed to the combined cycle.

The Puertollano IGCC in Spain – the most modern IGCC in Europe, commissioned in 1997 – achieves a net efficiency of 45%. With new gas turbine technology at hand, better heat integration and hot acid gas removal third-generation IGCC power plants are expected to reach net efficiencies as high as 55% without CO_2 capture and should be available between 2015 and 2020 [9]. Further information on IGCC power plants without CO_2 capture can be found elsewhere [10, 11].

16.2.2
Possible Points of Membrane Integration in IGCC and NGCC Power Plants

As mentioned before, IGCC and NGCC power plants with pre-combustion CO_2 capture offer multiple integration variants for membranes. The membrane characteristics have to match the process conditions at the integration point. If the syngas has to be cooled/heated or compressed/expanded just to integrate the membrane systems, further efficiency losses will result. The most feasible locations for the integration of membranes into IGCC power plants are depicted in Fig. 16.2–16.5. Syngas process conditions for different locations (indicated by letters in Fig. 16.2–16.5) are given in Table 16.1.

Figure 16.2 shows the standard parts of the IGCC which remain basically unaltered in all membrane integration setups. Downstream of the particulate removal, in the process area (B) between the dotted lines in Fig. 16.2, different CO_2 capture routes can be integrated. All the different routes consist of a water gas shift (WGS) reactor and a membrane/membrane reactor for gas separation and/or H_2S removal. The resulting streams of H_2S (A3), fuel gas (A4) and CO_2 (A5) downstream of process area B are then fed to the Claus unit, the combined cycle, and the CO_2 processing unit, respectively. Compression to pipeline pres-

Table 16.1 Boundary conditions for the different integration points for an IGCC.

Integration point	H_2 (vol.%)	CO_2 (vol.%)	H_2O (vol.%)	T (°C)	Sour
A	Low-intermediate (32%)	Low (2%)	Low-high (3%)	400–1000	Yes
C	Intermediate-high (37%)	Intermediate-high (25%)	High (34%)	300–500	Yes
D	High (55%)	↑ High (40%)	↑ Low (1%)	↓ <100	No
E	High (55%)	= High (40%)	= Low (1%)	↓ >200	No
F	High (45%)	↓ High (30%)	↓ High (20%)	↑ >200	No
G	Intermediate-high (35%)	High (23%)	High (33%)	300–500	No
H	High (52%)	↑ High (35%)	↑ Intermediate (10%)	↓ >200	No
I	High (35%)	High (23%)	High (20%)	300–500	No
J	High (52%)	↑ High (35%)	↑ High (10%)	↓ >200	No
K	High (35%)	High (23%)	High (20%)	300–500	Yes
L	High (52%)	↑ High (35%)	↑ High (10%)	↓ >200	Yes
M	High (52%)	= High (35%)	= High (10%)	= >200	No

sure is achieved in the CO_2 processing unit. If a significant amount of fuel species such as H_2 and CO are present in the CO_2 stream, further steps such as post-combustion or cryogenic recovery of the fuel species might be necessary.

As mentioned before, the gasification unit produces a syngas with a high CO and H_2 content. The level of CO_2 is relatively low (just about 1–15%; see Table 16.1). Without further treatment, the maximum CO_2 separation is about 14% for a wet coal feed [12]. For dry coal feed, the possible CO_2 separation ratio is even lower. Therefore, for all pre-combustion concepts a WGS is mandatory. The exothermic WGS reaction shown in Equation 16.1 converts CO to CO_2 and simultaneously preserves most of the energy by the formation of H_2.

$$CO + H_2O_{vap} \rightleftharpoons CO_2 + H_2 - 44.477\ \text{kJ mol}^{-1} \tag{16.1}$$

The reaction occurs sufficiently fast at elevated temperatures between 950 and 1000 °C in syngases; at lower temperatures a catalyst is necessary. Two facts have to be kept in mind for further understanding of the WGS system layout. First, according to the principle of chemical thermodynamics, the exothermic WGS reaction is favored by low temperatures and by a stoichiometric excess of water vapor. Second, by the continuous removal of one of the reaction products the equilibrium can be shifted further to the right-hand side of Equation 16.1. Because of the importance of the WGS reaction on power plant performance, more details on WGS are presented in Section 16.2.5.

Downstream of an entrained flow gasification, the temperature level matches the necessary 950–1000 °C during the cooling process. Hence the WGS reaction

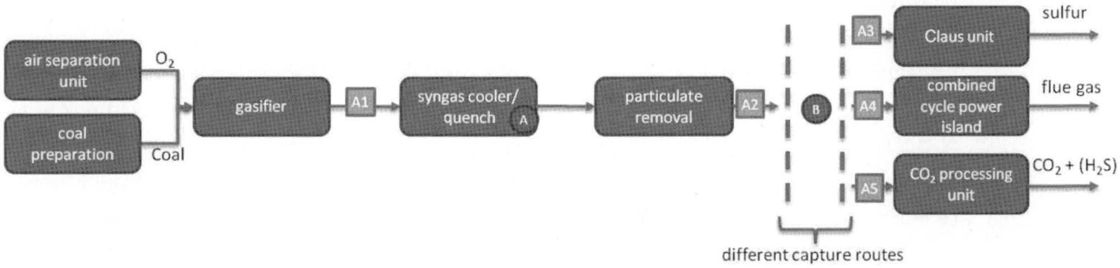

Fig. 16.2 IGCC with pre-combustion capture, main part.

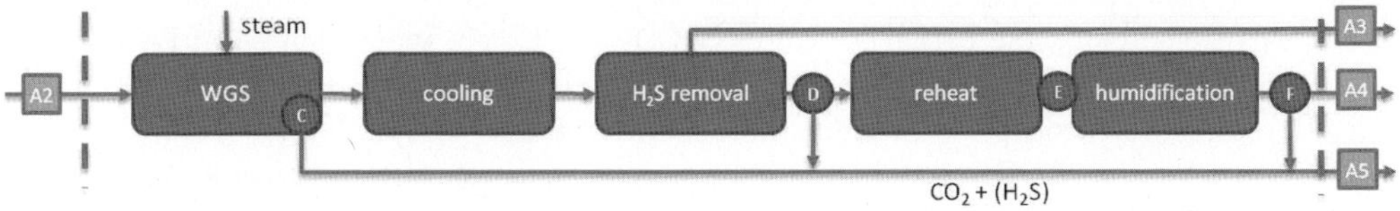

Fig. 16.3 IGCC pre-combustion route 1, sour WGS.

could be performed without the use of a catalyst. However, the high temperature will lead to low CO conversion ratios in chemical equilibrium, because of the exothermic nature of the WGS reaction. By the introduction of either an H_2- or CO_2-selective membrane at this point (A), one of the reaction products could be removed and the reaction will be promoted. For entrained flow gasifiers with an exit temperature of 1300–1600 °C this would mean that first a cooling step to 950–1000 °C has to be performed, for example, by cold gas/wet quenching or radiative coolers. The conditions at point A are severe, because of the presence of high temperatures, particulates, and poisonous species (H_2S, HF, COS, trace metals, etc.). State-of-the-art membranes may not be suited to the conditions encountered.

Figure 16.3 shows a more conservative capture approach. After syngas cooling and particulate removal, a sour WGS is performed. Because of the sour conditions, a sulfur-tolerant WGS catalyst (e.g., CoMo/Al oxide) has to be used, frequently applied in a two-stage setup with inter-cooling [12].

Depending on the water level upstream of the WGS unit, steam has to be added to the syngas. Prior to the H_2S removal, the syngas has to be cooled, because H_2S removal systems such as amine or physical solutions work at low temperatures. Downstream of the H_2S removal, the syngas has to be reheated for better power plant efficiency. In order to keep NO_x formation during combustion low, a certain amount of water is necessary in the syngas. Hence a humidification step is also incorporated before the syngas is fed to the gas turbine. CO_2- or H_2-selective membranes could be integrated at point C, D, E, or F. The integration during the WGS (C) can be performed between the two WGS steps or in a so-called WGS membrane reactor (WGSMR), where separation and reaction occur simultaneously. Both concepts are beneficial for the WGS reaction: The steam demand can be reduced and CO conversion ratios increased. However, coke formation, a side product of the WGS reaction, and temperature gradients because of the heat production during the WGS reaction could lead to operational problems. The membrane would have to be sulfur tolerant and impermeable to water vapor. Otherwise, a membrane would be disadvantageous as it decreases the H_2O/CO ratio for the WGS reaction (see Section 2.5). Less severe conditions occur at integration points D, E, and F. Points D, E, and F show different temperature levels and water contents so that different membrane types can be integrated. At points D and E, coke formation might lead to problems, as coke formation is hindered by a high water content, which is low at points D and E. Humidification leads to an increased H_2O content in the syngas stream and, hence, automatically lowers the content of the other species. Therefore, the partial pressures of CO_2 and H_2 are reduced at point F.

A slightly altered integration design is depicted in Fig. 16.4. H_2S removal is performed upstream of the WGS reaction (sweet WGS). Higher steam demands for the WGS reaction might be necessary, since the water content is significantly reduced before the H_2S removal (chemical scrubbing which requires low temperatures). Similarly to the sour WGS, an interstage or WGSMR integration (G) is possible. In this setup, sulfur-tolerant membranes are not required. A membrane integrated at point H would be exposed to moderately high temperatures and water contents. Coke formation is not problematic under the conditions at point H.

Sweet WGS concepts with low-temperature H_2S removal cannot utilize the high water contents downstream of the gasification, hence requiring more steam addition for the WGS reaction. Therefore, sweet WGS concepts usually achieve lower efficiencies than sour WGS concepts. On the other hand, sweet WGS concepts are sometimes preferable, because the overall WGS conversion might be higher (see Section 16.2.5). As more heating value is shifted from CO to H_2, a higher recovery of H_2 is possible, which can be routed to the gas turbine.

A hot/warm H_2S removal is depicted in Fig. 16.5 for a sour and a sweet WGS setup. A hot H_2S removal increases the overall plant efficiency because no water is removed from the syngas prior to the H_2S removal and exergy losses due to syngas cooling are avoided. Points I and J show the same demands towards the membrane materials as C and F, respectively. Sulfur tolerance would be mandatory for membranes integrated at points J and K. For all integration points membranes would be required which can operate at an intermediate to high temperature level. Another integration concept is a co-sequestration of H_2S with the separated CO_2, so that the H_2S removal step could be

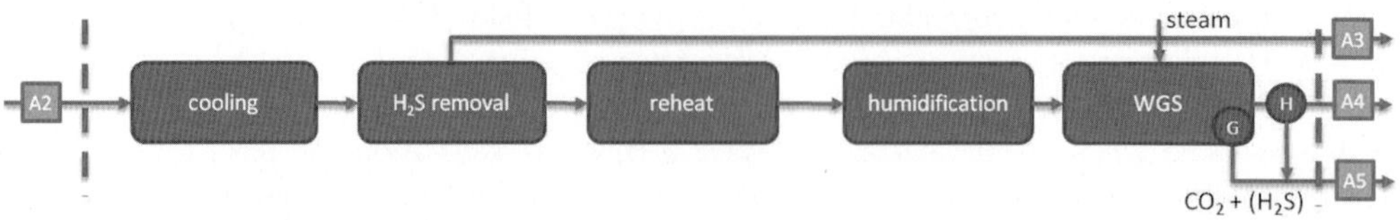

Fig. 16.4 IGCC pre-combustion route 2, sweet WGS.

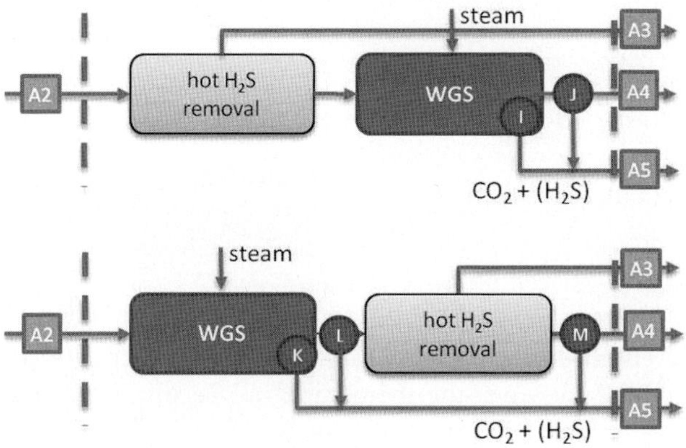

Fig. 16.5 IGCC pre-combustion routes 3 and 4, hot H_2S removal.

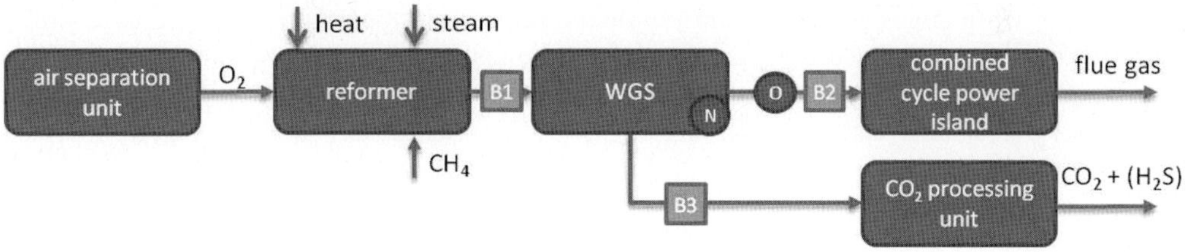

Fig. 16.6 NGCC pre-combustion route 1, separate reforming and WGS.

avoided. The membranes would have to allow co-permeation (CO_2-selective membrane) or co-retention (H_2 selective membrane) of H_2S with the CO_2 stream.

Figures 16.6 and 16.7 show two routes for membrane integration in NGCC power plants. The first possibility is to perform a separate reforming and WGS step (Fig. 16.6). The relevant reactions for reforming are methane–steam reforming (Equation 16.2) and partial oxidation of methane (Equation 16.3).

$$CH_4 + H_2O_{vap} \rightleftharpoons CO + 3H_2 + 206\ \text{kJ mol}^{-1} \tag{16.2}$$

$$CH_4 + 0.5O_2 \rightleftharpoons CO + 2H_2 - 509\ \text{kJ mol}^{-1} \tag{16.3}$$

The methane–steam reforming reaction (Equation 16.2) is highly endothermic. The necessary heat can be supplied by different setups. In a conventional steam–methane reforming (SMR) the heat is supplied by an external source such as combustion of natural gas and a subsequent heat exchanger. Another approach is to mix H_2O, CH_4, and O_2, as done in autothermal reformers (ATRs). Thereby, the partial oxidation reaction (Equation 16.3) provides the heat for the reforming step. An SMR and ATR step in series is called gas heating reforming (GHR) and in parallel heat integrated reforming (HIR), whereby the ATR step provides the heat for the SMR unit [13]. Depending on the reforming

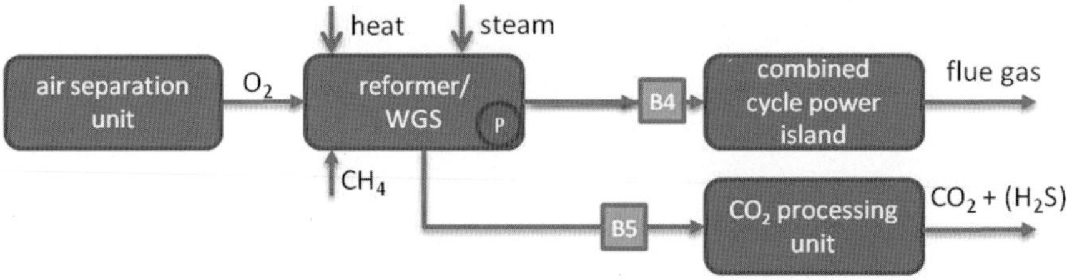

Fig. 16.7 NGCC pre-combustion route 2, integrated WGS and reforming.

concept, H_2O and O_2 levels may vary. Downstream of the reforming a WGS step follows. As for the IGCC concepts, the membrane can be integrated into the WGS interstage or as a WGSMR (N), or be applied downstream of the WGS (O).

Membranes offer the possibility to reduce the number of necessary process units. As Fig. 16.7 shows, a membrane could even integrate reformer and WGS into one methane–steam membrane reformer (MSMR) [13, 14], an option which is not applicable in IGCCs due to the harsh conditions in gasifiers (presence of particulates, poisonous species, very high temperatures). In an MSMR reaction, reactions 16.1 and 16.2 are occurring simultaneously and can be summarized as

$$CH_4 + 2\,H_2O \rightleftharpoons CO_2 + 4\,H_2 + 165\text{ kJ mol}^{-1} \qquad (16.4)$$

Typical conditions at the points of integration are listed in Table 16.1. The data are based on simulation data (modified Puertollano process, hard coal), but are given only qualitatively since the specific process setups and coals applied may vary. Arrows up and down indicate if the content has increased or decreased with respect to the unit before. Further, the heat integration can be varied so that other temperature levels are possible.

16.2.3 Membrane Classification

This section provides a brief classification of different membrane materials suitable for pre-combustion capture. Membranes can be differentiated by the membrane material, the main permeating species, and the transport mechanism. H_2- and CO_2-selective membranes are relevant for application in IGCC power plants.

Dense membranes are the first important H_2-selective membrane class. Dense metallic membranes are H_2 selective and are generally based on palladium or palladium alloys. The transport mechanism is based on the ionization of H_2 and the transport of the ions through the membrane lattice. The temperature range of palladium-based membranes is 300–500 °C [15, 16]. Below 300 °C , palladium membranes can undergo a phase change, which causes hydrogen embitterment. CO, NH_3, and chlorine can cause surface fouling [15]. A further problem with Pd membranes is the relatively short lifespan in syngas environments [15]. As for metallic membranes in dense ceramics based on rare earth mixed oxides, H_2 is ionized into two protons, which are transported through the membrane. The temperature required for the proper operation of this membrane class is 600–700 °C [16]. Cermets are metallic–ceramic composites and fall into the class of dense membranes, with an operating range of 600–900 °C. Non-porous silica-based and silica–alumina-based membranes transport H_2 by a solution–diffusion mechanism and are applicable from 150 to 550 °C. The same transport mechanism takes place in glassy polymeric H_2-selective membranes. The temperature range for this membrane type is below 110 °C for commercially available and up to 400 °C for advanced polybenzimidazole (PBI) membranes [16]. The presence of water and H_2S can cause plasticization and subsequent mechanical failure [15].

Porous inorganic membranes constitute another H_2 selective membrane class. A wide range of materials can be used: silica, zeolites, alumina, nitrides, and oxides [15]. The pores of the membrane have to be sufficiently small that the smaller species such as H_2 permeate while CO_2 is retained. The highest permeabilities are reported for silica membranes. The presence of water vapor may lead to problems and a decline of permeance [15]. The temperature range of inorganic porous membranes is 150–700 °C for ceramics, 150–450 °C for zeolites, and 150–550 °C for carbon-based membranes [16].

The most important class of CO_2-selective membranes is formed by rubbery polymers [16], which operate at temperatures below 110 °C. Lower temperatures favor the selectivity towards CO_2/H_2 for this membrane type.

Table 16.2 summarizes the different membrane materials with their field of operation and links them to the possible points of integration identified in Section 16.2.2. The data are based on [15, 16].

16.2.4
Classification Numbers and Permeation Laws

For the evaluation of membranes and carbon capture performance. the following classification numbers are convenient.

The shift rate (*Shr*) classifies the amount of converted CO in the WGS unit:

$$Shr = 1 - \frac{\dot{n}_{\mathrm{CO,\,shifted}}}{\dot{n}_{\mathrm{CO,\,preshift}}} \tag{16.5}$$

The separation rate (*Sep*) defines the ratio of captured CO_2 to the carbon present in the coal feed:

$$Sep = \frac{\dot{n}_{\mathrm{CO_2,\,pipeline}}}{\dot{n}_{\mathrm{carbon\ coal}}} \tag{16.6}$$

Table 16.2 Pre-combustion capture membrane operation fields.

Membrane type – H_2 selective	T range (°C)	Permeability – H_2 (mol s^{-1} m^{-1} $Pa^{-\frac{1}{2}}$)[a)]	Selectivity (H_2/CO_2)	Problems/critical conditions	Integration points
Metallic					
Pd	320–450	9×10^{-9}	>1000	(<300 °C) phase changes, possible problems with CO, NH_3, Cl, sulfur compounds	C, G, I, K, N, O, P
Pd alloys	230–450	6×10^{-5}–2×10^{-12}	>1000		C, E, F, G, H, I, J, K, L, M, N, O, P
Dense ceramics					
Silica-based	150–550		>1000		C, E, F, G, H, I, J, K, L, M, N, O, P
Doped rare-earth oxides	600–900		>1000		P
Porous inorganic		[GPU]			
Ceramic-based	150–700	20–6000	2–<30	High water contents may lead to performance decline (silica)	C, E, F, G, H, I, J, K, L, M, N, O, P
Zeolite-based	150–450	300	1–<90		
Carbon-based	150–550	50–80	10–<60	High water contents may lead to performance decline	
Polymeric		[barrer]			
Commercial	<110	0.1–45	0.1–<7	Problems for some polymers with H_2O, H_2S → plasticization	D
Advanced PBI	<400		40–58		C, E, F, G, H, I, J, K, L, M
Membrane type – CO_2 selective		[barrer]	[CO_2/H_2]		
Polymeric	<50	0.01–18200	1–<15		D
Facilitated transport	20–110	6–1500	7–144	Increasing CO_2 partial pressure will lead to a situation with all carriers fixed → decrease in permeability and selectivity Degradation because of NH_3 and H_2S possible	D

a) GPU, 10^{-6} cm^3 (STP) cm cm^{-2} s $cmHg^{-1}$; barrer, 10^{-10} cm^3 (STP) cm cm^{-2} s $cmHg^{-1}$.
Source: data based on [15, 16].

For transport and storage, the purity (*Pur*) of the separated CO_2 is an important measure:

$$Pur = \frac{\dot{n}_{CO_2,\text{pipeline}}}{\dot{n}_{\text{total, pipeline}}} \tag{16.7}$$

In CO_2-selective membranes, in addition to CO_2, H_2 can also be permeable. Hence the H_2 slip (*Slp*) is another important classification number:

$$Slp = \frac{\dot{n}_{H_2,\text{permeate}}}{\dot{n}_{H_2,\text{feed}}} \tag{16.8}$$

Finally, the H_2 recovery (*Rec*) defines the amount of H_2 that is recovered in an H_2-selective membrane and can be routed to the combined cycle:

$$Rec = \frac{\dot{n}_{H_2,\text{permeate}}}{\dot{n}_{H_2,\text{permeate}} + \dot{n}_{H_2,\text{retentate}}} \tag{16.9}$$

Polymeric membranes separate CO_2 by a combined solution–diffusion process, which can be described by Equation (16.10):

$$\mathrm{d}\dot{n}_i = \mathrm{d}AQ_i(p_{i,\mathrm{f}} - p_{i,\mathrm{p}}) \tag{16.10}$$

The permeation $\mathrm{d}\dot{n}_i$ of species i over the area increment $\mathrm{d}A$ is proportional to the permeability Q_i and the difference in the partial pressure p on the feed side (f) and permeate side (p). The selectivity of species i to species j can be defined as

$$a_{i,j} = \frac{Q_i}{Q_j} \tag{16.11}$$

Facilitated transport membranes enhance the CO_2 permeation by the reversible combination of CO_2 and H_2O to HCO_3^-, resulting in an additional transport term for CO_2 in the form of HCO_3^- [17]:

$$\mathrm{d}\dot{n}_{CO_2} = \mathrm{d}A\left(\frac{D_{CO_2}}{l}\right)(c_{CO_2,\mathrm{f}} - c_{CO_2,\mathrm{p}}) + \mathrm{d}A\left(\frac{D_{HCO_3^-,\mathrm{f}}}{l}\right)(c_{HCO_3^-,\mathrm{f}} - c_{HCO_3^-,\mathrm{p}}) \tag{16.12}$$

where D defines the diffusivity and c represents the concentration of a species.

The transportation law generally applied for proton-conducting membranes is proportional to the square root of the partial pressure of H_2 in feed and permeate:

$$\mathrm{d}\dot{n}_{H_2} = \mathrm{d}AQ_{H_2}\left(\sqrt{p_{H_2,\mathrm{f}}} - \sqrt{p_{H_2,\mathrm{p}}}\right) \tag{16.13}$$

16.2.5
Influence of the Water Gas Shift

The WGS unit influences the efficiency of a power plant. For a standard WGS design, the efficiency loss for an IGCC with CO_2 separation due to WGS (without having encountered any measures for separation or CO_2 compression) is in the order of 3–4%-points [17, 18]. Hence it is of critical importance to understand the details of the WGS to be able to improve power plant efficiency.

The loss of efficiency caused by the WGS is to a major extent governed by the steam demands necessary for sufficiently high CO conversion ratios. Furthermore, due to the temperature level, the heat produced during the WGS reaction can typically only be reintegrated into the bottoming cycle (steam cycle) of the IGCC and not into the topping cycle (gas turbine), which would be thermodynamically favorable.

For standard WGS units, the three most important catalysts available are the following [12]:

Sweet shift:

1. High-temperature shift:
   - temperature range: 300–530 °C
   - catalyst: Ni/Cr oxide, Fe/Cr oxide
   - H_2S tolerance: <0.03 wt%
   - minimum H_2O/CO ratio: 2–2.2
2. Low-temperature shift:
   - temperature range: 180–270 °C
   - catalyst: Cu/Zn oxide
   - H_2S tolerance: <0.1 ppm
   - minimum H_2O/CO ratio: 2–2.2

Sour shift:

3. Raw gas shift:
   - temperature range: 230–500 °C
   - catalyst: COMo/Al oxide
   - H_2S necessary for reaction
   - minimum H_2O/CO ratio: 1.8–2.

Achievable CO conversion ratios are strongly dependent on the H_2O/CO ratio and the working temperature of the catalyst. The lower temperature range of the sweet shift favors CO conversion. The maximum CO conversion can be achieved by isothermal reactors; however, these are complicated, costly, and difficult to maintain. Therefore, a staged setup is preferable [19]. Equilibrium calculations for the two-stage setup in Fig. 16.8 are shown in Fig. 16.9. The positive effect of a two-stage setup is easily derived from these results. The first catalyst converts the bulk amount of CO, and the second stage brings conversion to its maximum. Furthermore, it follows that a certain water content is necessary in order to keep the maximum temperature specifications of the catalysts. The equilibrium calculations for sour and sweet shift suggest that conversion ratios

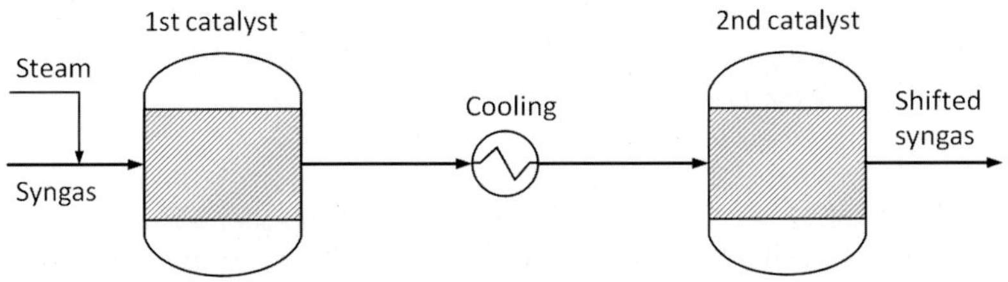

Fig. 16.8 Staged WGS setup.

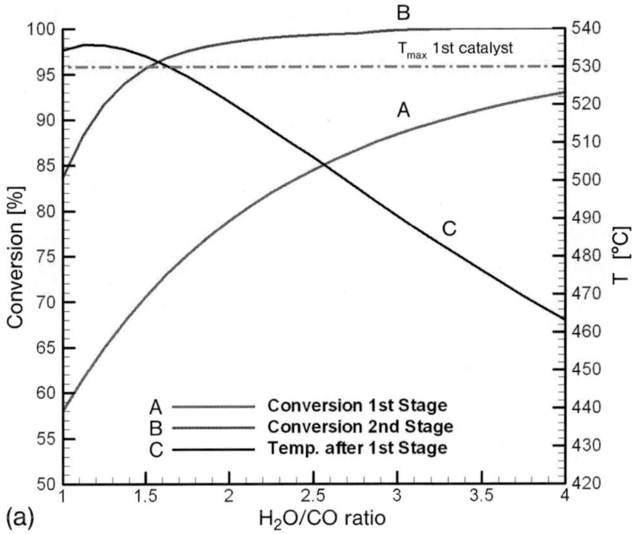

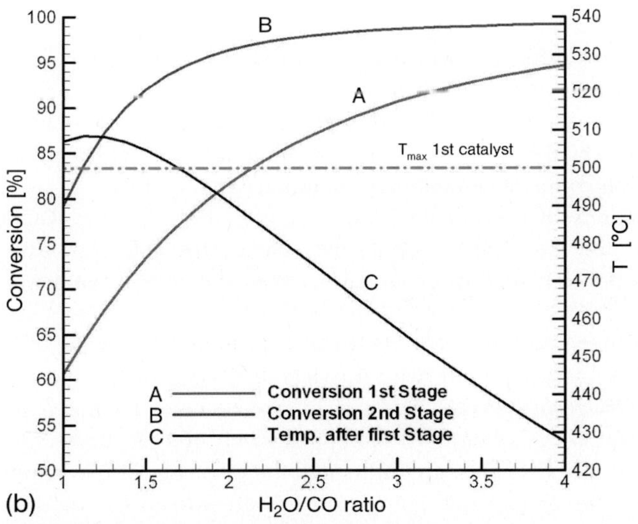

Fig. 16.9 Influence of H_2O/CO ratio on temperature and conversion for (a) sweet and (b) sour shift.

above 99% are possible. The minimum H_2O/CO ratio of 1.8 for the sour shift corresponds to a conversion of roughly 95% and for the sweet WGS the conversion ratio is 98% for the H_2O/CO minimum ratio of 2.

In principle, the parasitic efficiency impact of the WGS unit can be minimized by an isothermal shift or by a multiple stage shift. Carbo and co-workers analyzed an advanced WGS process with four stages and identified efficiency losses of only 1.4%-points for a dry-fed IGCC [19, 20]. As mentioned above, membranes offer another viable opportunity to reduce this impact.

16.3 Pre-Combustion Concepts with Scrubbing Technologies

Physical scrubbing processes for CO_2 removal in IGCC power plants are the most mature technologies and frequently described in the literature. They serve as reference technologies for other concepts such as membranes or adsorption (PSA) processes. The most common options investigated are processes based on methanol and Selexol. Physical absorption of CO_2 is favored by low temperatures [21]. Hence hot H_2S removal is usually not an option for scrubbing processes. Note that there have been some investigations aimed at the development of high-temperature physical solvents [7, 22, 23]. Subsequent to the absorption, a stripping step has to follow in order to recover the valuable physical solvent and the captured CO_2. Stripping is achieved by pressure release in flash drums and/or thermal regeneration of the solvent. The pressure release for the regeneration of the solvent leads to high CO_2 compression energies. Processes based on methanol require additional cooling energy, because CO_2 can be captured most efficiently at temperatures of about $-30\,°C$.

Descamps *et al.* stated that the comparison of different publications available on absorption processes is difficult because of a lack of detail regarding the underlying mass and energy fluxes and thermodynamics of the models applied [24]. Hence we strongly recommend only comparing directly physical scrubbing and membrane results which stem from the same source to assure at least that the studies are based on comparable boundary conditions and models.

For realistic estimations of the effect of CO_2 capture, the reference IGCC should be optimized for the highest efficiencies. Hence the reference case should be a setup with steam production by raw syngas cooling and not a wet quench.

Based on the latest literature references, the most realistic results for the efficiency losses caused by carbon capture with physical scrubbing units reported are in the range 10–12%-points [25, 26]. The carbon capture ratios for this is in the region of 90% and CO_2 is compressed to approximately 100 bar. Beyond liquefaction ($p \approx 74$ bar), the CO_2 pressure influences the efficiency of the power plant only slightly. On the other hand, going to lower pressures may improve the efficiency significantly. The purity achieved by the physical scrubbing processes is usually very high and may reach levels above 99% [25]. Kunze and

Spliethoff also pointed out that the plant configuration, ASU integration level, fuel composition and CO_2 purity requirements have a significant effect on the reported efficiency values and therefore should be understood as only approximate values [25].

Results for CO_2 scrubbing in IGCCs for Selexol are available, for example [21, 27–29], and also for methanol processes [24, 25, 30]. NGCC processes with conventional scrubbing techniques have been reported [28, 31, 32].

16.4 Pre-Combustion Concepts with CO_2-Selective Membranes

Polymeric or facilitated transport membranes are the most important class of CO_2-selective membranes. Polymeric membranes are cheap, there is extensive working experience in industry, and they can be produced on a large scale [18]. Their low thermal stability limits the integration of CO_2-selective polymeric membranes to the low-temperature end of IGCC processes. For porous ceramic membranes it is very challenging to retain the smaller H_2 molecule while the larger CO_2 molecule can permeate. This can only be achieved by either surface diffusion or capillary condensation. For the few CO_2-selective ceramic membranes reported, extensive surface modifications had to be undertaken to maximize the surface diffusion [15]. Therefore, CO_2-selective membrane concepts usually concentrate on polymeric membranes. The application of CO_2-selective membranes is normally limited to IGCC processes. For NGCCs, the syngas contains much less CO_2 than in an IGCC, hence the driving force for CO_2-selective membranes in NGCCs is low, which makes the application of CO_2-selective membranes less attractive.

16.4.1 Single-Stage Concepts

A single-stage setup of a CO_2-selective membrane is depicted in Fig. 16.10. The feed coming from the WGS enters the membrane at a high pressure, whereby, the permeate side (CO_2) of the membrane has to be operated at low pressure in order to realize sufficient driving forces (see Equations 16.10 and 16.12). Depending on the selectivities, other species besides CO_2, such as H_2 or N_2, might permeate across the membrane surface area and lead to a decrease of purity of the CO_2 to be sequestered. Because H_2 remains at a high pressure, usually no recompression before the gas turbine combustion chamber is necessary. In contrast to H_2-selective membranes, incomplete CO conversion does not reduce the IGCC efficiency, but it lowers the CO_2 capture ratio. Lower conversion ratios even increase the power plant's efficiency since more CO is routed to the gas turbine. The setup in Fig. 16.10 is called a three-end module, because only three ends of the membrane are connected to other process units.

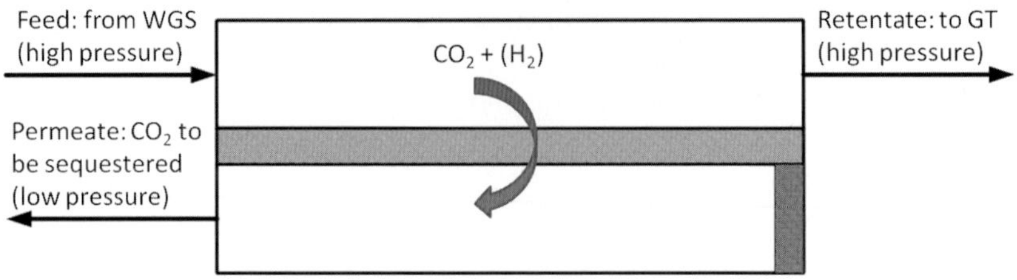

Fig. 16.10 Single-stage CO_2-selective membrane. GT, gas turbine.

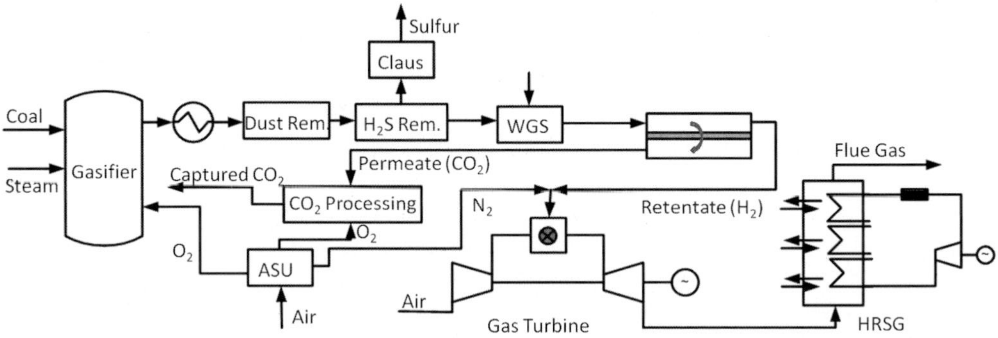

Fig. 16.11 CO_2-selective membrane integration for a sour WGS IGCC process.

The only possible way to realize a four-end module would be the use of low-temperature water vapor as a sweep gas with subsequent condensation. Although reducing the CO_2 compression energy, this may seriously influence the steam turbine power output and lead to higher investment costs. All other sweep gases reduce the CO_2 purity and, thereby, lead to additional process steps. Therefore, four-end modules are not the common option for CO_2-selective membranes in IGCC pre-combustion capture.

Investigations on the integration of CO_2-selective single-stage membrane concepts into IGCC power plants were performed by Grainger and Hägg [17] and Franz and Scherer [18]. A sour WGS setup is depicted in Fig. 16.11.

The WGS unit is integrated downstream of gasification, syngas cooling, and particulate removal. The shifted syngas is then cooled to the operating temperature of the H_2S removal unit and subsequently fed to the membrane. The separated H_2 is routed to the combined cycle, while CO_2 is compressed to pipeline pressure.

The efficiency of the IGCC process is decreased by

- compression of CO_2 to pipeline pressure
- pressure losses because of additional process units
- co-permeation of H_2 (slip)
- lost decompression energy of separated CO_2 for the gas turbine
- the WGS unit.

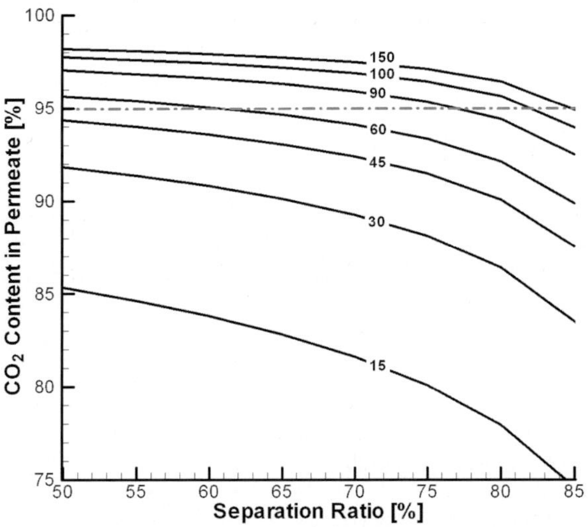

Fig. 16.12 Variation of separation ratio (abscissa) and selectivity (curve numbers).

For a single-stage setup, the loss because of H_2 slip is directly linked to two variables: the amount of CO_2 separated and the membrane selectivity. If more CO_2 is to be separated, the membrane area has to be increased. Hence more H_2 co-permeates across the membrane. This effect can be compensated by enhanced selectivity of the membrane. Figure 16.12 shows the results of variation of the selectivity and separation ratio for a typical shifted syngas with a dry composition of 57% H_2, 36.5% CO_2, 4% CO, 0.6% Ar, and 1.9% N_2. While the separation ratio increases, the content of CO_2 in the permeate drops. Because of the high H_2 content in the feed stream, the main species which co-permeates is H_2. Since H_2 is a valuable fuel gas, the membranes applied have to ensure that this amount is as low as possible. Hence certain minimum standards can be defined for CO_2-selective membranes for pre-combustion capture. For the syngas in Fig. 16.12, a selectivity of 150 is required in order to achieve 85% CO_2 separation with 5% H_2 slip.

The commercially available reference membrane investigated by Franz and Scherer [18] (Pebax/PG) has a CO_2/H_2 selectivity of 15, which is far below the required standard. Therefore, process simulations were carried out with a selectivity of 150. In order to recover some of the energy of the co-permeated H_2 stream, a post-combustion can be integrated. With post-combustion, the heating value of the slipped H_2 can be at least partly recovered and integrated into the steam cycle. The efficiency drop with post-combustion can be reduced by 0.8%-points for the setup investigated [18]. Pure O_2 coming from the ASU has to be used for the post-combustion process instead of air in order to prevent a decrease in the CO_2 purity. The products of post-combustion are CO_2 and H_2O, and the H_2O can be condensed. Thereby, a post-combustion step can even increase the overall CO_2 purity. On the other hand, post-combustion requires additional units, hence increasing the investment costs. Chiesa *et al.* [33] described another possibility for the recovery of fuel species such as H_2 and CO. By a cryogenic separation step the H_2 could be separated, rerouted to the gas turbine, and, hence, used in the topping cycle of the IGCC. Grainger and Hägg [17] evaluated a facilitated transport membrane with a selectivity of 100. A CO_2 separation degree of 85% with 95% purity was achieved. The differences between the simulations of Grainger and Hägg [17] and Franz and Scherer [18] can be explained by variations in feed gas composition, different inlet and outlet pressures, and selectivities towards N_2 and CO. Table 16.3 lists the results and boundary conditions reported [17, 18].

Table 16.3 Simulation results and boundary conditions for single-stage CO_2-selective membrane concepts (BC, base case; SwWGS, sweet WGS; SoWGS, sour WGS).

Parameter	Grainger and Hägg [17]			Franz and Scherer [18]	
	BC	SwWGS	SoWGS	BC	SoWGS
Capture method	–	PFTM [a)]	PFTM [a)]	–	PM [b)]
Gas turbine power (MW)	182	152	155	235.3	212.8
Steam turbine power (MW)	133	112	116	145.9	124.6
Gross power output (MW)	315	265	271	381.2	337.4
Auxiliary demand (MW)	35	35	35	41.2	42.1
CO_2 compression (MW)	0	20	20	–	21.6
Net power (MW)	280	210	216	340.0	273.8
Thermal input [MW (LHV)]	668	668	668	760.3	n.s. [c)]
Gross efficiency [% (LHV)]	47.2	39.6	40.6	50.1	n.s.
Net efficiency [% (LHV)]	41.9	31.4	32.3	44.7	36.0
Efficiency drop (%-points)	–	10.5	9.6	–	8.7
CO_2 compression pressure (bar)	–	110	110	–	200.0
CO_2 purity (%)	–	95	95	–	99.5
CO_2 separation (%)	–	85	85	–	85
Membrane area (m^2)	–	710000	610000	–	n.s.
Post-combustion	–	No	No	–	Yes
Selectivity:					
CO_2/H_2	–	100	100	–	151
CO_2/N_2	–	125	125	–	457.9
CO_2/CO	–	25	25	–	457.9
CO_2 emissions (kg kWh^{-1})	n.s.	n.s.	n.s.	n.s.	0.13

a) Polymeric-facilitated transport membrane.
b) Polymeric membrane.
c) n.s.: not stated.

16.4.2
Staged and Cascaded CO_2-Selective Membrane Concepts

A possible method to reduce the fuel slip and recompression energy is a combined staged and cascaded concept as reported by Grainger and Hägg [17], depicted in Fig. 16.13. As the separation of CO_2 takes place at different pressure levels, only part of the CO_2 has to be recompressed from the lowest pressure to pipeline pressure. In order to reduce the H_2 slip, Grainger and Hägg [17] introduced a recompression to 11 bar after the fourth stage. By this means, a larger amount of the H_2 can be retained in the retentate of membrane stage 5.

For the realization of low H_2 slip and high separation ratios with low selectivity, a cascaded membrane concept is a solution. One possible setup, proposed by Franz and Scherer [18]. is shown in Fig. 16.14. The permeate stream of the first stage is recompressed in an intercooled compression unit and subsequently cooled to the membrane working temperature. The enriched feed stream of the second stage has a higher CO_2 purity, resulting in an increased driving force of the CO_2 across the membrane. Thereby, higher purities are achieved in the permeate stream of the second stage. The depleted retentate stream of the second stage is recycled to the feed of membrane 1. With this setup, the drawbacks of low selectivity can be overcome by the investment of electrical energy and ad-

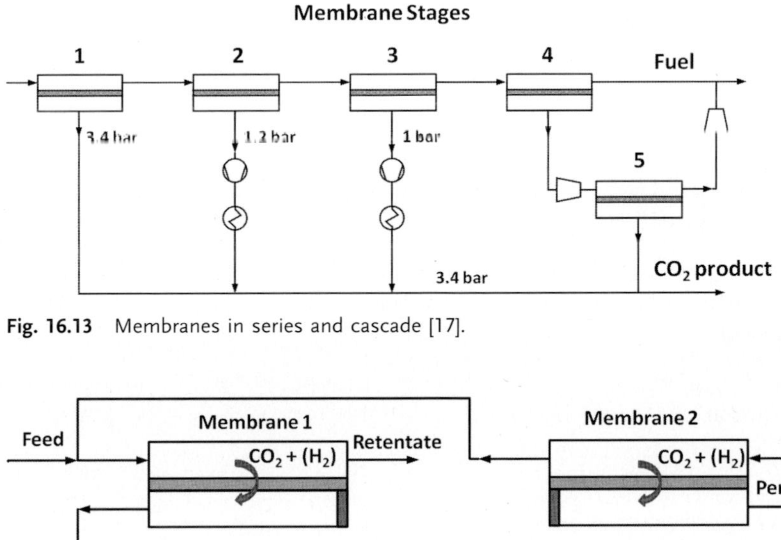

Fig. 16.13 Membranes in series and cascade [17].

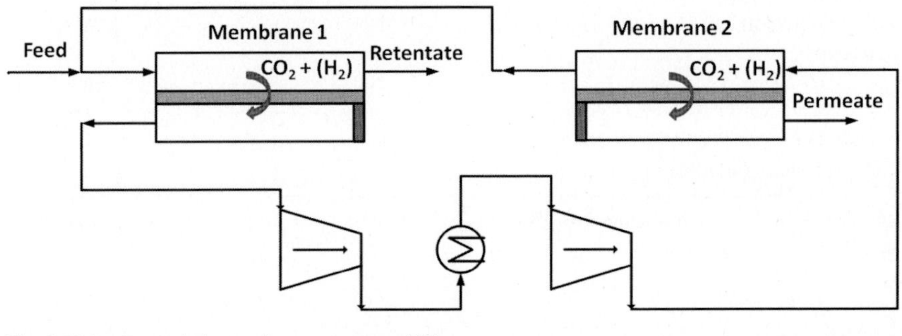

Fig. 16.14 Cascaded membrane concept [18].

ditional process units. Franz and Scherer's results show that efficiency losses below 10% and separation ratios of 85% can be achieved with a reduced membrane selectivity of 60 [18]. Lower selectivities would lead to increased recompression energies and higher efficiency losses. The results of the staged and cascade concepts are reported in Table 16.4.

Although they eliminate some of the drawbacks of CO_2-selective membranes, it has to be kept in mind that staged and cascaded membrane concepts lead to higher investment costs and increased system complexity, because of the additional process units. Furthermore, cascaded concepts require compression units, which reduce the net power output and thereby the efficiency of the power plant. Hence it is usually more favorable to use single-stage membrane setups if the membrane characteristics allow sufficiently high purities and separation ratios and low hydrogen slippage.

Table 16.4 Simulation results and boundary conditions for staged and cascaded CO_2-selective membrane concepts (BC, base case; SwWGS, sweet WGS; SoWGS, sour WGS).

Parameter	Grainger and Hägg [17]		Franz and Scherer [18]	
	BC	SoWGS	BC	SoWGS
Capture method	–	PFTM [a)]	–	PM [b)]
Gas turbine power (MW)	182	158	235.3	218.1
Steam turbine power (MW)	133	117	145.9	126.6
Gross power output (MW)	315	275	381.2	344.7
Auxiliary demand (MW)	35	35	41.2	57.8
CO_2 compression (MW)	0	21	–	20.3
Net power (MW)	280	219	340.0	266.6
Thermal input [MW (LHV)]	668	668	760.3	n.s. [c)]
Gross efficiency [% (LHV)]	47.2	41.1	50.1	n.s.
Net efficiency [% (LHV)]	41.9	32.8	44.7	35.0
Efficiency drop (%-points)	–	9.1	–	9.7
CO_2 compression pressure (bar)	–	110	–	200.0
CO_2 purity (%)	–	97	–	99.5
CO_2 separation (%)	–	85	–	85
Membrane area (m^2)	–	655000	–	n.s.
Post-combustion	–	No	–	Yes
Selectivity:				
CO_2/H_2	–	100	–	60
CO_2/N_2	–		–	182
CO_2/CO	–		–	182
CO_2 emissions (kg kWh^{-1})				0.14

a) Polymeric facilitated transport membrane.
b) Polymeric membrane.
c) n.s.: not stated.

16.4.3
CO_2-Selective Water Gas Shift Membrane Reactor (WGSMR-CO_2)

As mentioned earlier, the WGS accounts for up to 4%-points of the parasitic efficiency losses accompanied by pre-combustion CO_2 capture. Therefore, optimization of the WGS offers great potential for the improvement of the power plant's efficiency. Although high-temperature CO_2 separation membranes – a basic need for the realization of WGS membrane reactors – are far from realization [14], the basic potential of CO_2-selective membrane reactors will be highlighted here. Figure 16.15 shows a CO_2-selective membrane reactor. As the feed gas stream flows through the reactor, CO is catalytically converted and CO_2 is separated at the same time. Hence the performance of the WGS reaction is enhanced.

Because gasifiers always produce syngases with higher H_2 than CO_2 content, H_2-selective membrane reactors show better CO conversion ratios, but CO_2-selective reactors do not perform much worse. An IGCC process with CO_2-selective membrane reactors was investigated by Carbo *et al.* [34], depicted schematically in Fig. 16.16. In this setup, low-pressure steam is used as a sweep gas for the membrane reactor in order to increase the driving force of CO_2 and to reduce CO_2 recompression energy. The assumptions for a CO_2-selective membrane able to operate at WGS conditions were derived from H_2-selective Pd/Ag membranes.

The efficiency penalty calculated by Carbo *et al.* was 9.5%-points, which is not better than the results shown in Tables 16.3 and 16.4 for conventional staged WGS concepts. The results of the simulations of Carbo *et al.* are reported in Table 16.5. The direct comparison with a conventional physical scrubbing unit also simulated by Carbo *et al.* shows that the WGSMR-CO_2 setup is not able to achieve lower efficiency losses. The reported efficiency losses were 8.5%-points for the Selexol case and 9.5% for the membrane reactor setup. In both instances a capture ratio of 90% was achieved. The low-pressure steam sweep gas overcompensates the beneficial effects of a membrane reactor as it accounts for approximately 3%-points of the encountered efficiency losses. Therefore, a setup without a sweep gas stream might be favorable for a CO_2-selective WGS membrane reactor concept, in order to facilitate this technology. Carbo *et al.* pointed out that an advantage of a CO_2-selective membrane setup is their operational

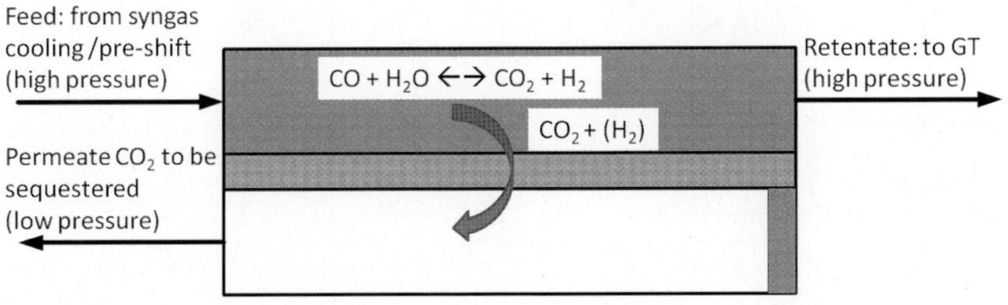

Fig. 16.15 CO_2-selective WGSMR.

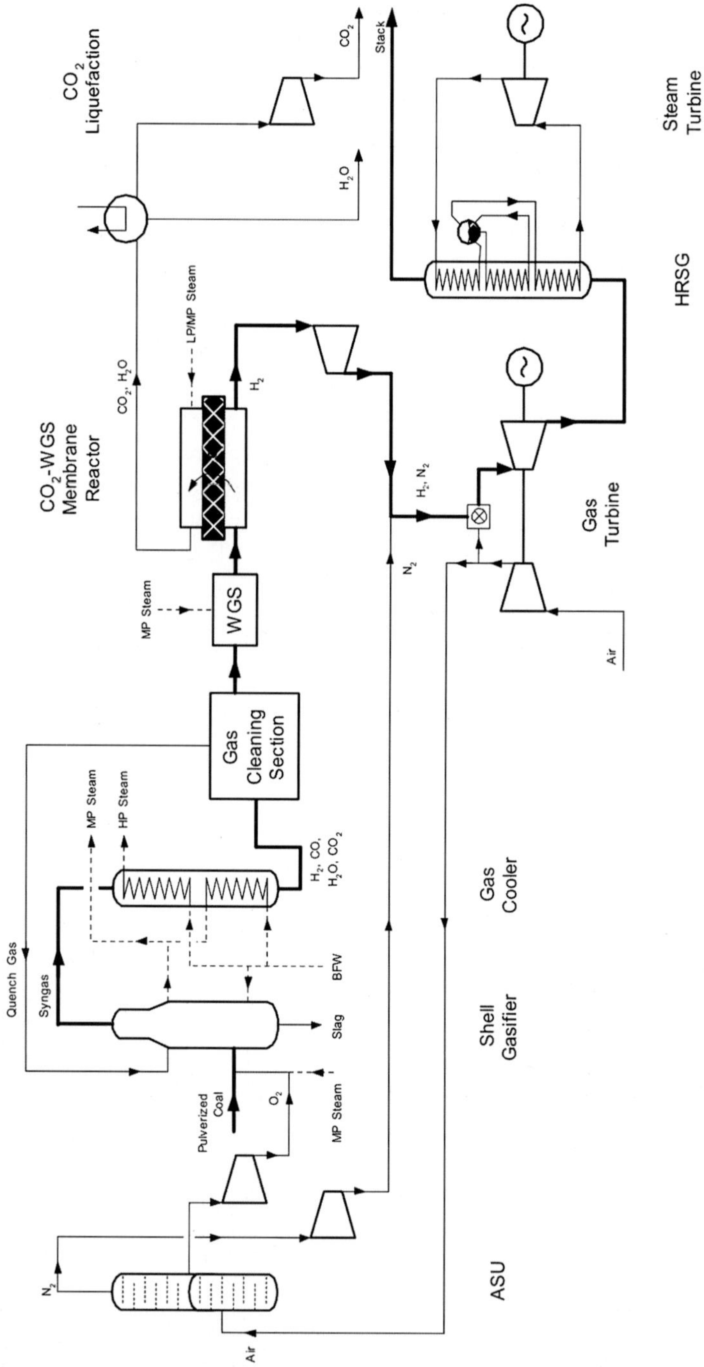

Fig. 16.16 IGCC process with a CO_2-selective WGSMR [34].

Table 16.5 CO_2 simulation results and boundary conditions for CO_2-selective membrane reactor concept according to Carbo *et al.* [34] (BC, base case; PS, physical scrubbing; MR, membrane reactor).

Parameter	BC	PS	MR
Membrane type	–	–	PM [a)]
Net power (MW)	500	413.5	403.3
Net efficiency [% (LHV)]	49	39.4	40.5
Efficiency drop (%-points)	–	8.5	9.5
CO_2 compression pressure (bar)	–	110	110
CO_2 purity (%)	–	n.s.	n.s. [b)]
CO_2 separation (%)	–	90	90
Membrane area (m^2)	–	–	30 000
Post-combustion	–	No	
Selectivity:		–	
CO_2/H_2	–	–	Infinity
CO_2/N_2	–	–	Infinity
CO_2/CO	–	–	Infinity

a) Polymeric membrane.
b) n.s.: not stated.

flexibility in terms of carbon capture ratios. A decrease in CO conversion leads to less separated CO_2 but has no negative effect on the IGCC efficiency. Thereby, utilities are able to respond to CO_2 certificate market price fluctuations by an adapted IGCC operating mode [34].

16.5 Pre-Combustion Concepts with H_2-Selective Membranes

Most of the literature on pre-combustion CO_2 capture with membranes is concerned with evaluating H_2-selective membranes or, more specifically, H_2-selective membrane reactors. Most analyses investigated dense metallic membranes, and ceramic or polymeric membranes formed a smaller subclass of these studies. Dense ceramic and metallic membranes can achieve very high selectivities. Supposing that a crack-free surface can be manufactured, even an infinitely selective membrane is possible.

16.5.1 H_2-Selective Membrane Units

H_2-selective membranes can be operated as four-end modules (see Fig. 16.17). The application of sweep gases, for example, excess N_2 from the ASU, will lower the partial pressure of H_2 on the permeate side and thereby increase the driving force across the membrane. By the use of sweep gases it is possible to de-

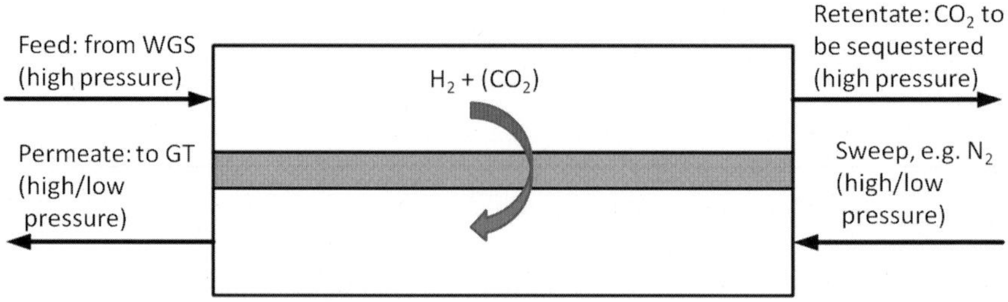

Fig. 16.17 Single-stage H_2-selective membrane.

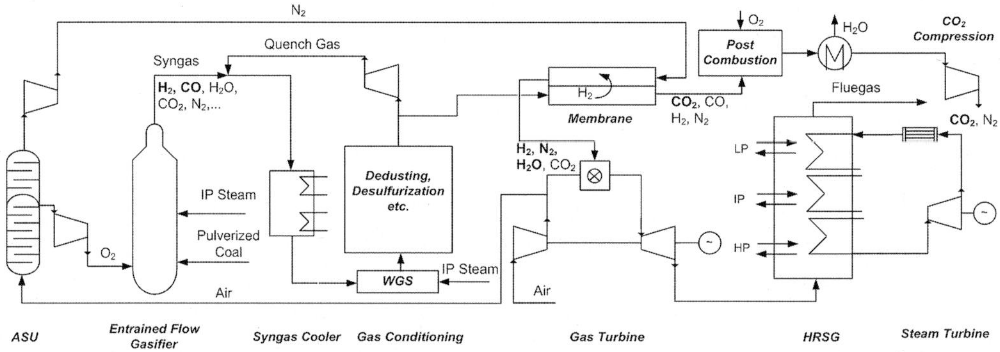

Fig. 16.18 Separated sour WGS and H_2-selective membrane [35].

sign processes in which both CO_2 and H_2 remain at high pressure. Hence the compression energy of CO_2 can be reduced, while at the same time recompression of the fuel is not needed.

Concepts with a conventional WGS and a downstream membrane unit are usually similar to those described in Section 16.4. Again, sour and sweet shift concepts are possible. Figure 16.18 shows a schematic process outline of pre-combustion CO_2 capture in an IGCC by separated sour WGS and an H_2-selective membrane. The results [35] show that sour shift concepts are preferable with respect to efficiency losses. On the other hand, the water content in a sweet shift design upstream of the membrane may be higher and can, thereby, hinder coke formation (see Section 16.2).

Krishnan *et al.* [36] investigated the integration of H_2-selective polymeric membranes based on polybenzimidazole (PBI) able to operate at elevated temperatures (200–400 °C). Two membrane application scenarios were analyzed. In both concepts the membrane directly follows downstream of a sour WGS. In the first scenario, it was assumed that steam and H_2 are permeable through the membrane, whereas CO, H_2S, and CO_2 are mainly retained. The pressure of the sweep gas was set to 17 bar, followed by recompression of the permeate to 31 bar to meet turbine design specifications. In the first scenario, the H_2S is co-sequestered with the retained CO_2. The second scenario assumed that in addition to H_2 and H_2O, H_2S is also permeating. Therefore, an H_2S removal step is integrated downstream of the membrane separation. The hydrogen recovery level in both scenarios was about 91%, and 90% of the CO_2 was separated. Whether the fuel species CO and H_2 in the CO_2 stream were recovered further remains unclear in the paper. In addition, the authors did not state the membrane selectivity assumed. Krishnan *et al.* also investigated a reference IGCC and conventional CO_2 separation by a physical scrubbing unit. The efficiency loss in the conventional capture case was reported as 8.9%-points based on the higher heating value. Both membrane scenarios showed higher efficiency losses of 10.2%-points (co-sequestration of H_2S) and 11.6%-points (subsequent H_2S removal). Especially for the scenario with co-sequestration one would expect smaller efficiency losses than for a physical scrubbing unit. An explanation for this may be the need for recompression of the permeate to meet the gas turbine specifications and the loss of the unrecovered fuel species in the investigated membrane integration designs. On the other hand, the study shows that polymeric H_2-selective membranes might play a role in pre-combustion CO_2 capture and advanced materials such as PBI show interesting characteristics.

Low temperature H_2-selective polymeric membranes (40 °C integration level) were investigated by Franz and Scherer [18]. Again a four-end module was considered. The membrane was integrated after sour WGS and subsequent H_2S removal. In order to reduce the associated efficiency losses, the aim of the simulations was to obtain an H_2 recovery above 99% while capturing 85% of the CO_2. Note that for a certain membrane material it is problematic in a four-end module to increase the H_2 recovery for two reasons. As the membrane area has to be increased to recover larger amounts of H_2 more and more CO_2 co-permeates, this decreases the CO_2 separation ratio. Furthermore, as the sweep gas is usually at high pressure, back-diffusion of N_2 to the feed side might decrease the purity of the retentate (e.g., the purity of CO_2). The base case membrane with an H_2/CO_2 selectivity of 6 (165 for H_2/N_2) proved to be insufficient for the proposed recovery ratio. Therefore, a selectivity variation was performed by decreasing the permeabilities of CO_2 and N_2 by the same factor. A selectivity of 50 towards H_2/CO_2 and above 1400 for H_2/N_2 was shown to be sufficient for the separation and recovery ratio. A post-combustion step was incorporated to regain the heating value of the retained fuel species (mainly CO) and to increase the purity level. The efficiency loss was calculated to be 9.07%-points. The purity of the separated CO_2 stream was 91%.

Kaldis *et al.* [37] performed simulations for H_2-selective high-temperature ceramic and low-temperature polymeric membranes. The setup considered in their study was a three-end module. It was not stated where or if an H_2S removal step was incorporated into the power plant. Furthermore, selectivities were not reported. Because of low purities in a single-stage layout, staging was investigated. Up to 57% of the CO_2 could be recovered at a purity of 88% with three polymeric stages. The ceramic membrane considered performed much worse, with a maximum purity of 66% for 7% CO_2 separation. The efficiency loss for the three-stage polymeric membrane was 9.6%-points. The simulation did not include compression of the CO_2 to pipeline pressure.

Detailed high-temperature ceramic membrane simulations have been reported [35]. Special attention was paid to the selectivities necessary for very high H_2 recoveries. A sour and a sweet WGS setup were investigated. The aim was to recover 99.5% of the H_2 in the feed gas while maintaining a purity of at least 95%. It was assumed that CO, CO_2, and N_2 permeate equally fast. Therefore, back-diffusion of nitrogen into the retentate stream is an obstacle for the CO_2 purity. Post-combustion of retained fuel species was incorporated into the simulations.

Figure 16.19 shows the results of the selectivity variations performed. For a purity of 95%, a selectivity of 440 proved to be necessary, which is demanding for microporous ceramic membranes. Very high hydrogen recovery rates result in high demands with respect to membrane selectivity and increased membrane area. Concepts with lower hydrogen recovery rates might be economically favorable and lead to selectivities feasible in the mid-term future. On the other hand, high hydrogen recoveries increase the process efficiency as more fuel can be routed to the topping gas turbine cycle and decrease investment costs for the post-combustion unit and heat integration into the steam cycle. Cascaded concepts might reduce the demands for membrane selectivity, but lead to higher investment costs and increased efficiency losses. Therefore, cascaded concepts should only be considered if the membrane characteristics cannot meet the process requirements. For a selectivity of 500, efficiency losses of 9.07%-points (sour WGS) and 9.43%-points (sweet WGS) were calculated. The results and assumptions of the investigations described are presented in Table 16.6.

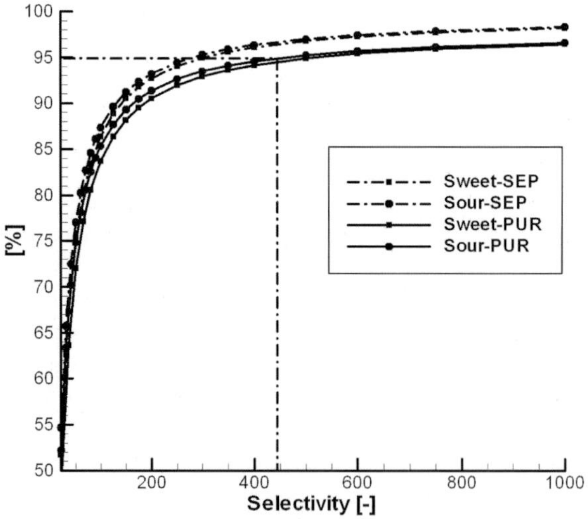

Fig. 16.19 Results for purity (PUR) and separation degree (SEP) of sour and sweet WGS simulations with an H_2-selective membrane [35].

Table 16.6 H_2-selective membrane results and simulation boundary conditions (BC, base case; M, membrane; M-H_2SCo, co-sequestration of H_2S; M-H_2Sre, removal of H_2S; M-MS, multistage; M-So, sour shift; M-sw, sweet shift).

Parameter	Krishnan *et al.* [36]				Franz and Scherer [18]		Kaldis *et al.* [37]			Franz and Scherer [35]		
	BC	Selexol	M-H_2SCo	M-H_2SCo	BC	M	BC	M-MS	M-MS	BC	M-So	M-Sw
Membrane type			PBI	PBI		Polymeric		Polymeric	Ceramic		Ceramic	Ceramic
Gas turbine power (MW)	464.0	464.0	464.0	464.0	235.3	194.2	317.5	251.6	240.4	250.0	250.0	250.0
Steam turbine power (MW)	302.9	277.0	273.1	271.8	145.9	126.4	138.7	113.1	111.1	150.8	130.6	131.9
Additional power prod. (MW)	6.6	6.4	–	–	381.2	–	–	–	–	–	–	–
Gross power output (MW)	773.5	747.4	737.1	735.8	41.2	320.5	456.2	364.7	351.5	400.8	380.6	381.9
Auxiliary demand (MW)	144.2	181.4	181.8	199.0	–	43.8	n.s.	n.s.	n.s.	34.9	41.2	46.0
CO_2 compression (MW)	–	50.4	12.6	15.6	–	8.4	–	–		–	9.7	9.4
Net power (MW)	629.3	515.7	542.8	521.2	340.0	271.2	n.s.	n.s.	n.s.	365.9	329.7	326.5
Thermal input (MW)	1605.7	1700.2	1869.8	1889.9	760.3	n.s.[a]	n.s.	n.s.	n.s.	776.7	866.9	866.3
Gross efficiency [% (LHV)]	48.2 (HHV)	44.0 (HHV)	39.4 (HHV)	38.9 (HHV	50.1	n.s.	48.4	38.8	37.3	51.6	43.9	44.1
Net efficiency [% (LHV)]	39.2 (HHV)	30.3 (HHV)	29 (HHV)	27.6 (HHV	44.7	35.6				47.1	38.0	37.7
Efficiency drop (%-points)	–	8.9	10.2	11.6	–	9.1	–	9.6	11.1	–	9.1	9.4
CO_2 compression pressure (bar)	–	152.0	152.0	152.0	–	200.0	–	–	–	–	200.0	200.0
CO_2 purity (%)	–	n.s.	n.s.	n.s.	–	91.0	–	88	66	–	95.3	94.9
CO_2 separation (%)	–	89.7	89.9	89.9	–	85.0	–	57	7	–	97.0	96.8
Membrane area (m^2)	–	–	n.s.	n.s.	–	n.s.	–	1230[b]	235[b]	–	82 291	111 515
Post-combustion	–	–	n.s.	n.s.	–	Yes	–	No	No	–	Yes	Yes
Selectivity:							–	n.s.	n.s.	–		
H_2/CO_2	–	–	n.s.	n.s.	–	50.0	–	n.s.	n.s.	–	500.0	500.0
H_2/N_2	–	–	n.s.	n.s.	–	1403.6	–	n.s.	n.s.	–	500.0	500.0
H_2/CO	–	–	n.s.	n.s	–	1403.6	–	n.s.	n.s.	–	500.0	500.0
CO_2 emissions (kg kWh^{-1})	0.776	0.103	0.105	0.111	n.s.	0.130	–	n.s.	n.s.	0.710	0.030	0.030

a) n.s.: not stated.
b) Membrane area is in $m^2\ mol^{-1}\ s^{-1}$.

16.5.2
H_2-Selective Water Gas Shift Membrane Reactors (WGSMR-H_2)

A WGSMR-H_2 promotes the WGS reaction (Equation 16.1) by the selective removal of H_2 from the reaction side. A schematic WGSMR-H_2 is depicted in Fig. 16.20. As the feed stream rich in CO, H_2, and H_2O is moving through the membrane channels, the WGS reaction shifts CO to H_2, while H_2 is permeating across the membrane material at the same time. Thus, the WGS reaction is promoted. The total number of patents on membrane reactors was 180 in 2009 [38], which is at least an indication of its potential.

Typically, upstream of the membrane reactor a traditional WGS as a pre-shift is included. A pre-shift converts the bulk amount of CO to CO_2 and therefore reduces the necessary membrane reactor size. Furthermore, a pre-shift reduces thermal gradients encountered in a membrane reactor.

An early study on H_2-selective WGS membrane reactors for integration into an IGCC was performed by Bracht *et al.* [39]. The design of the reactor assumed is a microporous silica membrane with an H_2/CO_2 selectivity of 15, where the membrane tubes are filled with either Fe–Cr or Pt/ZrO_2 catalysts. Bracht *et al.* stated that for the low H_2O/CO ratios (1.3) applied in their simulations, accompanying experiments did not show any undesirable by-products such as CH_4 or carbon. The design targets were set to 90% CO conversion and 80% CO_2 recovery. The reactor was operated as a four-end module with 31 bar feed and 21 bar sweep pressure, with N_2 coming from the ASU functioning as the sweep gas. CO_2 compression to pipeline pressure seemed not to be included in the calculations. The heating value of unpermeated fuel gas species was recovered in a post-combustion step. Two membrane reactor concepts were compared with a conventional CO_2 removal based on Rectisol scrubbing. For a membrane reactor with conventional H_2S removal, 3.9%-points was calculated. The results for a setup with a hot H_2S removal showed efficiency losses of 3.3%-points. In contrast, the efficiency losses of the conventional physical scrubbing system were 6.2%-points.

Another investigation on ceramic H_2-selective WGS membrane reactor application was performed by Franz and Scherer [35]. The membrane reactor was placed upstream of a conventional H_2S removal unit. It was assumed that because of the low driving forces, the H_2S is retained in the CO_2 stream. The

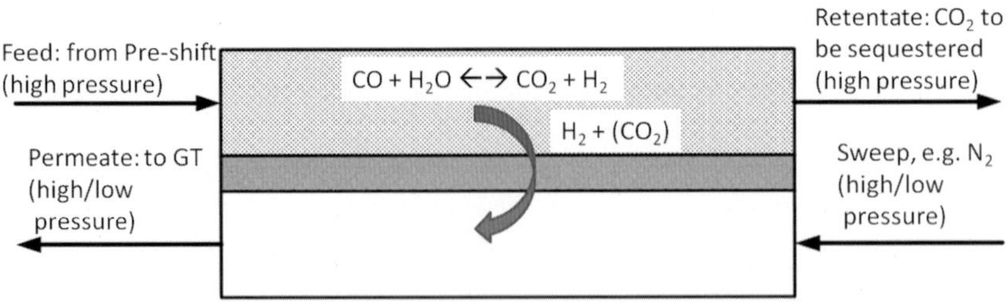

Fig. 16.20 WGSMR-H_2.

CO_2-enriched retentate was then fed to the H_2S removal and subsequently to post-combustion and compression. Upstream of the membrane reactor, a pre-shift was performed. First calculations showed that with the available N_2 stream and a pressure of 25 bar on the feed side and 23 bar on the sweep side, the driving forces were not sufficient for high shifting and hydrogen recovery ratios. Therefore, the gasification pressure was varied in order to increase the feed pressure. Although increasing the compression energy for O_2 and coal fluidization on the one hand, this measure decreases CO_2 compression energy to pipeline pressure on the other. Based on a gasification pressure of 52 bar and a membrane selectivity of 500, almost complete CO conversion and H_2 recovery could be achieved. The WGS membrane reactor concept outperformed a separated WGS and membrane unit by 2.3% efficiency loss due to a reduced steam demand and higher CO conversion ratios. In total, an efficiency loss of 6.7%-points for 96.6% CO_2 separation and 95%-points purity was calculated.

Amelio *et al.* [40] analyzed the integration of Pd–Ag membrane reactors. In the layout proposed, the gasification was followed by syngas cooling and H_2S removal. The following membrane reactor was operated in a four-end mode with low-pressure steam acting as the sweep gas. Upstream of the membrane the syngas was expanded to 11.5 bar and heated. It is not clear why the authors proposes a setup where the syngas is expanded, thereby lowering the CO_2 retentate pressure. The sweep gas was operated at about 1 bar, so after condensation of the water in the permeate stream, subsequent recompression of the separated H_2 stream was necessary. The separated CO_2 was compressed to 80 bar. The authors paid much attention to the impact of flow forms, namely co- and counter-current flow, on the separation and conversion behavior of the WGS membrane reactor. As expected, the results for the counter-current flow proved to be better. In a co-current setup, the driving forces decrease as the H_2 partial pressures on the feed and sweep sides converge. Additional driving forces can only be generated by further H_2 production and a thereby increased H_2 partial pressure on the feed side. Especially for ideal selective metallic membranes the analogy of membrane separation to heat exchangers holds. In a counter-current mode, it is possible to recover almost 100% of the hydrogen in the feed side, given the suitable sweep conditions. For high H_2 recovery ratios, the amount of the sweep stream is always lower for a counter-current layout. Therefore, not surprisingly, the authors chose a counter-current setup in the end, which could shift almost all CO and recover nearly 100% of the hydrogen produced. Furthermore, variation of the membrane thickness was studied. Decreasing the membrane thickness has two beneficial effects: first, the hydrogen permeability is improved, thereby reducing the membrane area, and second, the membrane material required per square meter is also decreased. The efficiency loss calculated was 8.02%-points for 91.5% CO_2 separation. Why the CO_2 separation is below 100% was not stated. It is not clear whether a post-combustion was integrated into the concept. The main reasons for the efficiency loss can be ascribed to the sweep gas production, recompression of the fuel gas stream, CO_2 compression and maybe unrecovered fuel species.

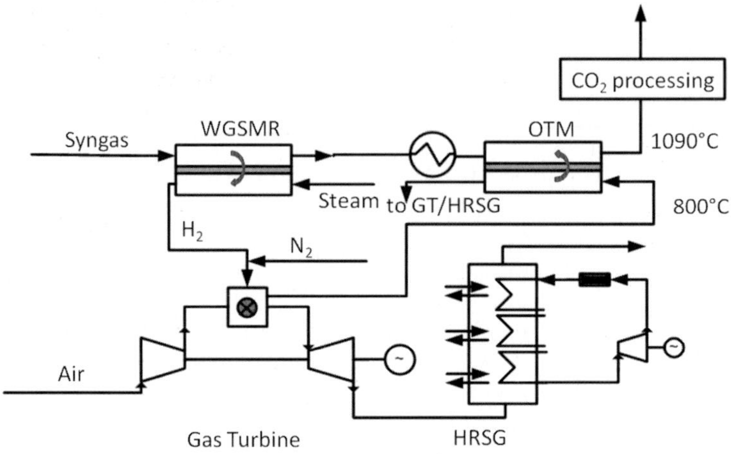

Fig. 16.21 OTM for post-combustion of retained fuel gas species.

Rezvani *et al.* [27] compared conventional removal by a physical scrubbing unit, a membrane reactor setup and a chemical looping process. The membrane reactor layout incorporates an oxygen transport membrane (OTM) based on perovskites for an advanced post-combustion of retained fuel species. The setup of an OTM for post-combustion is depicted schematically in Fig. 16.21. OTMs are fabricated from ionic and mixed conducting ceramic oxides which transport oxygen ions at elevated temperatures [27]. The gas stream entering the feed side of the OTM comes from the gas turbine burners at a temperature of about 800 °C. On the permeate side the CO_2 stream after the WGS-H_2 membrane reactors is fed to the OTM after heating from 400 to 600 °C, providing the right working temperature for the OTM membrane. Because of the partial pressure differences between the permeate and feed sides, O_2 permeates through the membrane lattice and reacts with the retained fuel species, thereby producing additional heat. The O_2-depleted retentate is recycled to the gas turbine, and the heat of the CO_2 stream is used to produce steam. After subsequent condensation of H_2O, the CO_2 stream is compressed. An efficiency loss of 9%-points was reported, which is about 0.8%-points lower than the efficiency losses for the conventional CO_2 capture case in this study. The CO_2 separation ratio was 93%. Other studies with metallic H_2-selective WGS setups and a subsequent post-combustion showed 100% CO_2 capture [33, 34]. Therefore, it is not clear where the CO_2 is lost in this setup, except for cracks in the WGS-membrane reactor. The purity of the separated CO_2 stream also was not stated.

Chiesa *et al.* [33] investigated a WGS-MR based on Pd–Cu with two different configurations for the treatment of unrecovered fuel species. Both concepts are based on a wet quench downstream of the gasifier. The wet quench humidifies the syngas stream, which is then fed to a high-temperature pre-shift followed by the H_2-selective WGS membrane reactor. Chiesa *et al.* used compressed nitrogen as a sweep gas, so that recompression of the permeated H_2 was unnecessary. In order to keep the membrane areas low, the rate of H_2 permeated to H_2+CO in the feed was capped at 85% in this simulation. In one configuration, a catalytic post-combustion step with pure O_2 was proposed for the recovery of the heating value of the unseparated fuel species. Because in this concept no H_2S removal step is incorporated, the following heat exchangers must be able to operate at high pressures in an extremely acidic environment [33]. Furthermore, the amount of heat to be integrated into the steam turbine cycle is huge. The steam turbine power output would rise by about 50% compared with the reference case without CO_2 capture. The other concept investigated was a cryogenic separation of the retained fuel gas species in the CO_2 stream. The separated fuel species can be recycled to the gas turbine and, thereby, used in the topping cycle. In the post-combustion concept H_2S is co-combusted in the post-combustion unit, whereas in the cryogenic concept it is co-sequestered with the CO_2. The efficiency loss of the post-combustion design was 4.57%-points and therefore is outperformed by the cryogenic recovery set-up with 4.34%-points. This can be explained by the different power balance: whereas the post-combustion concept utilizes the fuel energy of the unrecovered species in the bottom cycle, the cryogenic concept reroutes them to the topping cycle. For coals with a low sulfur content the difference should even be higher, because less heating value of H_2S can be used in the post-combustion concept. On the other hand, the post-combustion layout achieves an ultimate CO_2 separation ratio of 100%, whereas the cryogenic separation reduces the separation ratios to 92%. Chiesa *et al.* also made calculations for a conventional physical scrubbing unit, which achieved an efficiency loss of about 6%-points. In all setups the CO_2 was compressed to 150 bar. Note that the efficiency losses in these simulations were calculated by a comparison with a reference IGCC using a wet quench for syngas cooling. Wet quench syngas cooling reduces the efficiency of IGCC power plants. Compared with an IGCC with radiative and convective syngas coolers, the efficiency losses would be higher.

More studies on H_2-selective membrane integration into IGCC power plants have been reported [34], and also studies concerning hydrogen production and co-production of electricity from coal [41, 42]. Results of experimental and theoretical work on WGS membrane reactors have also been published [43–51].

Table 16.7 summarizes the results for WGSMR integration described in this chapter.

16.5.3 Hydrogen-Selective Membranes for NGCC Applications

Dijkstra *et al.* described a possible process layout [13], depicted schematically in Fig. 16.22. As pointed out in Section 16.2.2, methane can be reformed by different process setups, such as SMR, ATR, GHR, and HIR. Downstream of the reformer the syngas is cooled and fed to a pre-shifting unit in order to increase the H_2 partial pressure. Subsequently, the pre-shifted syngas is fed to a WGSMR based on Pd alloy membranes operated at 350–400 °C. The amount of

Table 16.7 WGSMR-H_2 results (BC, base case).

Parameter	Bracht *et al.* [39]				Franz and Scherer [35]	
	BC	Rectisol	WGSMR	WGSMR-H	BC	WGSMR
Membrane type			Ceramic	Ceramic		Ceramic
Gas turbine power (MW)					250.0	250.0
Steam turbine power (MW)					150.8	135.7
Additional power prod. (MW)					–	
Gross power output (MW)					400.8	385.7
Auxiliary demand (MW)					34.9	39.1
CO_2 compression (MW)					–	5.3
Net power (MW)					365.9	341.3
Thermal input (MW)					776.7	846.1
Gross efficiency [% (LHV)]					51.6	45.6
Net efficiency [% (LHV)]	46.7	40.5	42.8	43.4	47.1	40.3
Efficiency drop (%-points)		6.2	3.9	3.3	–	6.8
CO_2 compression pressure (bar)		1.3	1.0		–	200.0
CO_2 purity (%)					–	95.2
CO_2 separation (%)		88.0	80.0		–	96.7
Membrane area (m^2)			2176	2176	–	90 000
Post-combustion			Yes	Yes	–	Yes
Selectivity:						
H_2/CO_2					–	500.0
H_2/N_2					–	500.0
H_2/CO					–	500.0
CO_2 emissions (kg kWh^{-1})					0.710	0.02

a) n.s.: not stated.

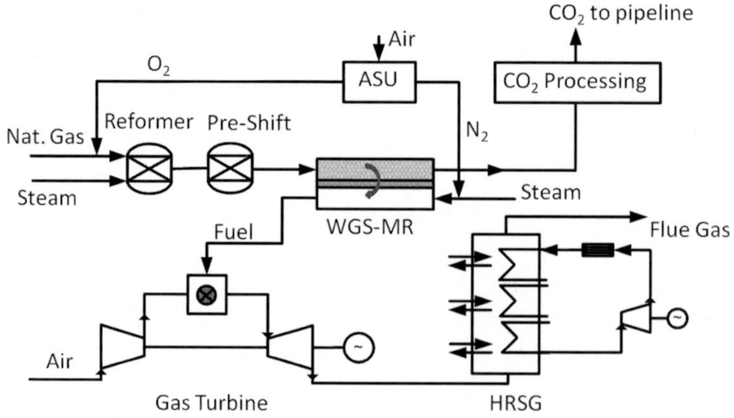

Fig. 16.22 WGSMR-H_2 NGCC application [13].

Table 16.7 (Cont.)

Amelio *et al.* [40]			Rezvani *et al.* [27]			Chiesa *et al.* [52]			
BC	**Selexol**	**WGSMR**	**BC**	**Selexol**	**WGSMR**	**BC**	**Selexol**	**WGSMR-PC**	**WGSMR-CR**
		Pd–Ag			Pd–Ag			Pd–Cu	Pd–Cu
242	238.3	271.86	335.2	304.5	278.5	294.3	293.7	315.5	300.2
204.7	195.4	160.6	201.8	152.9	178.4	178.6	179.3	266.3	181.0
11.2	11.2	11.5				9.5	9.8	0.0	0.0
457.9	444.9	444.0	537	457.4	456.9	482.4	482.8	581.8	481.2
53.9	57.6	59.0	43.2	45.8	45.1	92.0	92.3	156.5	106.8
0.0	21.3	31.7		24.8	16		28.6	3.8	7.8
404.0	366.0	353.3	493.8	386.8	395.8	390.1	361.9	421.5	366.6
879.4	931.0	931.4	n.s.[a)]	n.s.	n.s.	908.2	983.7	1098.1	949.4
52.1	47.8	47.7	n.s.	n.s.	n.s.	53.1	49.1	53.0	50.7
45.9	39.3	37.9	45.4	35.6	36.4	43.0	36.8	38.4	38.6
–	6.6	8.0	–	9.8	9	–	6.2	4.6	4.3
–	80.0	80.0	–	110	110	–	150.0	150.0	150.0
–			–	n.s.	n.s.	–	n.s.	n.s.	n.s.
–	91.5	91.5	–	91	93	–	92.0	100.0	92.2
–			–	–	1850	–	–	33202.0	34 337.0
–		n.s.	–		Yes	–	–	Yes	Cryo
–		Inf.	–	–	Inf.	–	–	Inf.	Inf.
–		Inf.	–	–	Inf.	–	–	Inf.	Inf.
–		Inf.	–	–	Inf.	–	–	Inf.	Inf.
0.709	0.071	0.072	n.s.	n.s.	n.s.	0.752	0.070	0.000	0.066

O_2 necessary for reforming processes is lower than that for the gasification of coal. Therefore, unlike in an IGCC, there is not enough N_2 to act solely as a sweep gas. Therefore, steam is mixed with the N_2 to increase the driving force across the membrane. The permeated H_2 leaves the membrane unit at a pressure of 22 bar and can be fed to the gas turbine without recompression. Dijkstra *et al.* investigated the influence of the different reforming concepts on the power plant's efficiency.

A post-combustion step of the fuel species in the retentate is included, followed by drying and compression of the CO_2 stream, the compression pressure not being further specified. The results showed that an SMR unit suffers from too low CH_4 conversion ratios, so the concept was abandoned. Since the membranes show infinite selectivity, the CO_2 separation ratio was 100% for the investigated ATR, GHR, and HIR concepts in combination with a WGSMR. The results were compared with those for a combined cycle power plant without carbon capture and a combined cycle with autothermal reforming and physical scrubbing (Selexol). All membrane concepts outperformed the conventional physical scrubbing process with respect to efficiency losses: the reported efficiency losses were 15.6%-points for ATR-Selexol, 12.2%-points for ATR-

WGSMR, 10.5%-points for GHR-WGSMR and 11.2%-points for HIR-WGSMR. The lowest necessary membrane area was achieved by the GHR concept.

H_2-selective membranes can also be used to perform the methane–steam reforming, the WGS, and the separation of H_2 in one process unit (MSMR) (Fig. 16.23). The MSMR is fed with a mixture of CH_4 and H_2O. The equilibrium of both the methane–steam reforming and the WGS reaction is shifted to the right by the parallel removal of H_2. Because the overall reaction (Equation 16.4) is endothermic, an external heat supply is necessary. The selective removal of H_2 allows high conversion ratios to be achieved at relatively low temperatures (500–600 °C) compared with conventional methane–steam reforming without a membrane [13].

The integration of an MSMR into an NGCC process was also investigated by Dijkstra *et al.* [13]. A simplified process layout is depicted in Fig. 16.24. Analogously to the WGS reaction setup, a pre-reforming step is applied to increase the H_2 partial pressure and, thereby, lower the necessary membrane area. In this setup, the pre-reforming is a conventional methane–steam reforming step. The heat supply for the MSMR is achieved by the combustion of natural gas with air. Again, a mixture of N_2 and steam acts as a sweep gas; this time, the amount of N_2 coming from the ASU is much smaller, because the oxygen in

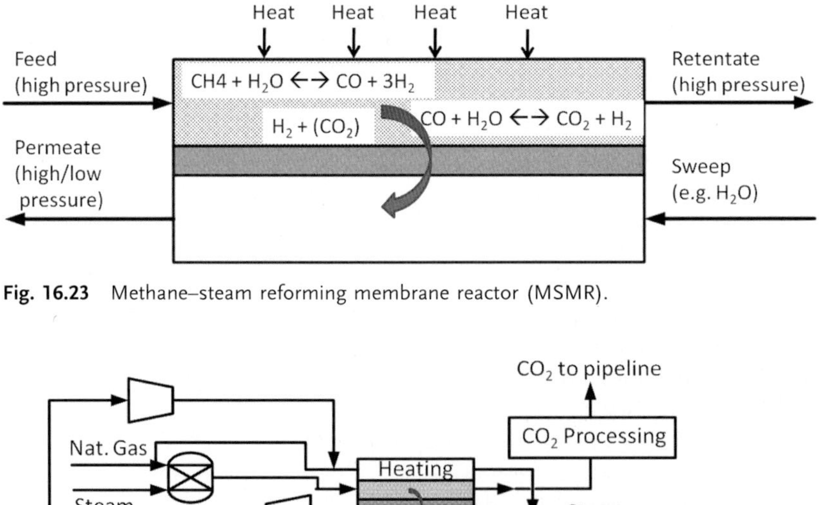

Fig. 16.23 Methane–steam reforming membrane reactor (MSMR).

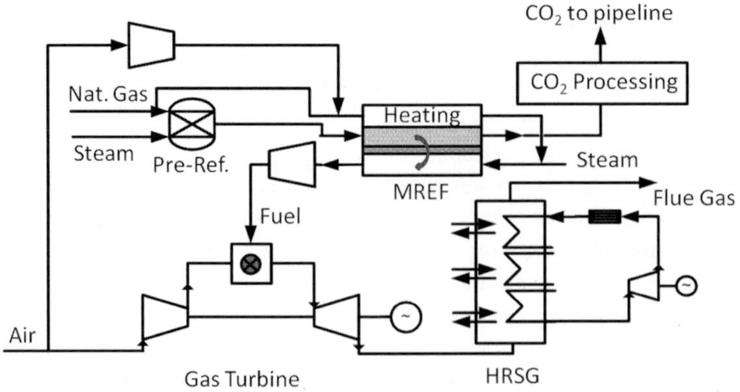

Fig. 16.24 MSMR NGCC application.

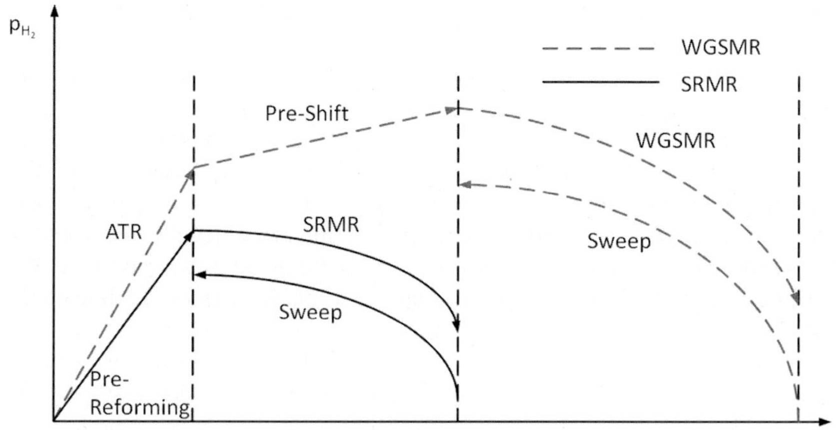

Fig. 16.25 Qualitative mapping of the H_2 partial pressure for WGSMR and MSMR [13].

this setup is only used for the post-combustion of unpermeated fuel species and not for the reforming step itself as in Fig. 16.23.

Figure 16.25 explains why the permeate stream in this setup has to be recompressed: unlike in the WGSMR setup, the partial pressure at the inlet of the MSMR is low. The assumed natural gas pressure for both setups was 40 bar. Therefore, a low sweep pressure has to be applied in order to achieve sufficiently high hydrogen recovery rates. With a sweep pressure of 22 bar, which would be required for direct routing to the gas turbine, the hydrogen recovery rate would only reach 50%. In order to achieve a recovery of 95–98%, the sweep pressure has to be lowered to about 10 bar [13]. Hence recompression is necessary, which reduces the efficiency of the MSMR setup.

Dijkstra *et al.* calculated an efficiency loss for this process setup of 10.3%-points. Hence the MSMR shows slightly better efficiency than the GHR-WGSMR process with 10.5%-points. On the other hand, the MSMR achieves only 90% CO_2 separation, because of the natural gas combusted for the heat supply of the membrane reactor. Furthermore, the membrane area required for the MSMR process is 7.4 times higher than that for the best concept described above.

A similar concept was investigated by Jordal *et al.* [53] for Pd-based and microporous ceramic membranes, except that in this study the heating for the MSMR was achieved by gas turbine exhaust gas with a supplementary firing with H_2 coming from the MSMR. Thereby, 100% CO_2 capture could be achieved with infinitely selective membranes. Furthermore, it was assumed that the retentate can be expanded efficiently in a steam turbine prior to compression, thereby regaining the pressure energy of the H_2O upstream of the condensation. As the natural gas pressure in these simulations was assumed to be 48 bar, a sweep pressure of 18 bar could be applied. Hence no recompression of the permeated hydrogen was necessary upstream of the gas turbine combustor operated at 15 bar. Different reforming temperature levels were evaluated. Furthermore, the efficiencies with and without the application of post-combustion were reported. For the microporous membranes, the selectivities of H_2 towards CO_2, CO, and H_2O were set to 50 and 100 towards CH_4. CO_2 capture was simulated to be 100% for infinitely selective Pd-based membranes, and 100% and 82.3% for microporous ceramic membranes using hydrogen for supplementary firing. By the application of a post-combustion step, up to 3.5%-points efficiency for Pd-based and up to 2.0%-points efficiency for ceramic membranes could be regained. In simulations by Jordal *et al.* no base case was simulated, but a comparison with the base case of Dijkstra *et al.* shows efficiency losses of 9.9%-points for the Pd-based and 9.4%-points for the ceramic based MSMRs.

More studies on MSMRs and WGSMRs have been published [54, 55] and experimental and theoretical work has been reported [56, 57, 58, 59]. The production of hydrogen has been investigated [60].The results and boundary conditions of the simulations described in this section are given in Table 16.8.

16.6 Conclusion

For the single-stage CO_2-selective concepts, efficiency losses of 8.7–10.5%-points were reported for CO_2 separation ratios of 90% compared with 10 –12%-points for physical scrubbing. A major problem for the integration of CO_2-selective membranes is the slip of H_2 to the permeate side of the membrane. The co-permeated H_2 is not available for the topping gas turbine cycle and therefore has to be recovered in a subsequent post-combustion step, which is thermodynamically less favorable. The H_2 slip is directly linked to the CO_2 separation ratios and the selectivity of the membrane. For lower selectivity requirements, a cascade concept such as that reported by Franz and Scherer [18] may be the solution. The parasitic efficiency loss for the cascaded concept was 9.7%-points.

The efficiency losses reported suggest that CO_2-selective membranes may be competitive with physical scrubbing as they achieve efficiency losses below 10%-points for separation ratios between 85 and 90%. On the other hand, the membrane area requirement reported by Grainger and Hägg [17] is at least 610 000 m^2. Therefore, an economical evaluation has to show if membranes can also compete in terms of CO_2 avoidance costs.

H_2-selective membranes show a very important inherent advantage compared with conventional scrubbing techniques and CO_2-selective membranes: by the use of a sweep gas, the retentate (CO_2) and also the permeate (H_2) remain at a high pressure, so compression energies can be significantly reduced. Hence a four-end mode is always preferable for H_2-selective membranes. Furthermore, the high working temperature of ceramic and metallic H_2-selective materials allows integration together with hot/warm H_2S removal. In order to utilize the valuable H_2 in the topping gas turbine cycle, as much H_2 as possible should be

Table 16.8 Results and boundary conditions of NGCC pre-combustion calculations (BC, base case).

Parameter	Dijkstra *et al.* [13]						Jordal *et al.* [53]	
	BC	ATR-Selexol	MSMR	ATR-WGSMR	GHR-WGSMR	HIR-WGSMR	MSMR	MSMR
Membrane type			Pd alloy	Pd alloy	Pd alloy	Pd alloy	Pd-based	Ceramic
Net efficiency [% (LHV)]	57.3	41.7	47.0	45.1	46.8	46.1	47.4	47.9
Efficiency drop (%-points)	–	15.6	10.3	12.2	10.5	11.2	9.4 [b]	9.9 [b]
CO_2 compression pressure (bar)	–	n.s. [a]	n.s.	n.s.	n.s.	n.s.	n.s.	n.s.
CO_2 purity (%)	–	n.s.	n.s.	n.s.	n.s.	n.s.	n.s.	n.s.
CO_2 separation (%)	–	84.2	90.0	99.9	99.9	99.9	100	81.5
Membrane area (m^2)	–	–	28 000	4560	3775	4560	n.s.	n.s.
Post-combustion	–	–	Yes	Yes	Yes	Yes	Yes	Yes
Selectivity:							100 000	50
H_2/CO_2	–	–	Inf.	Inf.	Inf.	Inf.	10 000	50
H_2/N_2	–	–	Inf.	Inf.	Inf.	Inf.	100 000	50
H_2/CO	–	–	Inf.	Inf.	Inf.	Inf.	10 000	100

a) n.s.: not stated.
b) Compared to the base case of Dijkstra [13].

recovered. On the other hand, this leads to high membrane areas and the demands necessary for membrane selectivity are fairly high. The results for a separated WGS unit and membrane separation showed efficiency losses between 9.1 and 11.1%-points. For a low efficiency loss of 9.1%-points, high H_2 recovery ratios and the utilization of fuel species in the retentate by a post-combustion step are mandatory. Because CO_2 is the retained species, high CO_2 separation ratios (>95%) are possible, if the H_2/CO_2 selectivity is sufficiently high. For infinitively selective membranes, even zero CO_2 emission is achievable. Note that the CO_2-selective membrane setups reported captured only 85% of the CO_2 for a separated WGS and CO_2 separation; for higher capture ratios, the H_2 slip would be too high and the power plant's efficiency would decrease. The H_2-selective CO_2 capture potential is therefore higher with a similar range of efficiency losses and usually preferable.

Although the reported efficiency losses with separated WGS and H_2 separation are competitive, the results do not show essential advantages. Much better results have been reported for WGS membrane reactors (WGSMRs). The reported efficiency gains were between 0.8 and 2.9%-points. Hence for IGCC processes it can be concluded that H_2-selective WGSMRs show the greatest potential for pre-combustion capture and the results indicate that they will be more efficient than conventional scrubbing techniques.

For NGCC applications, H_2-selective WGSMRs and MSMRs outperform the combination of a autothermal reformer (ATR) and a conventional Selexol unit by up to 5.0%-points, so this is another promising field for membrane application.

References

1 Kloster, R. Thermodynamische Analyse und Optimierung von Gas-/Dampfturbinenkraftwerken mit integrierter Kohlevergasung. Dissertation, Fortschritt-Berichte VDI, Nr. 409, Reihe 6, Energietechnik, 1998.

2 Minchener, A. Coal gasification for advanced power generation. *Fuel*, 2005, **84**, 2222–2235.

3 Eurlings, J.T.G.M. Process performance of the SCGP at Buggenum IGCC. Presented at the Gasification Technologies Conference, 1999.

4 Gómez, F.S., Schellberg, W. and Günster W. Puertollano IGCC plant: operating experience and further technology development for commercial generation of clean energy from coal. In *Russia Power Conference*, Moscow, Russia, 2004, pp. 1–13.

5 Geosits, R.F. *IGCC – the Challenges of Integration*. ASME Paper GT2005-68997, ASME, New York, 2005, pp. 1–5.

6 Pennline, H.W. *Carbon Dioxide Capture and Separation Techniques for Advanced Power Generation Point Sources*. Report DOE/NETL-IR-2006-202, US Department of Energy, Washington, DC, 2006, pp. 1–10.

7 Pennline, H., Luebke, D., Jones, K., Myers, C., Morsi, B., Heintz, Y., *et al.* Progress in carbon dioxide capture and separation research for gasification-based power generation point sources. *Fuel Process. Technol.*, 2008, **89**, 897–907.

8 Korens, N., Simbeck, D.R., and Wilhelm, D.J. *Process Screening Analysis of Alternative Gas Treating and Sulfur Removal for Gasification*. Final Report by SFA Pacific for US Department of En-

ergy, National Energy Technology Laboratory. SFA
Pacific, Mountain View, CA, 2002.

9 Ogriseck, S. IGCC – Vergasungsbasierte Erzeugung von Strom. Presentation at TU-Darmstadt, 2009, www.exergie.info (accessed 29 September 2010).

10 Haupt, G., Zimmermann, G., and Ullrich, N. Modern IGCC stations as highly effective means for CO_2 reduction. Proceedings of the 5th International Conference on Greenhouse Gas Technologies, 2000, Cairns, Qld., 815–820. CSIRO Publishing.

11 Emun, F., Gadalla, M., Majozi, T., and Boer, D. Integrated gasification combined cycle (IGCC) process simulation and optimization. *Comput. Chem. Eng.*, 2010, **34**, 331–338.

12 Göttlicher, G. Energetik der Kohlendioxidrückhaltung in Kraftwerken. Dissertation, University of Essen, 1999.

13 Dijkstra, J.W., Raju, G., Peppink, G., and Jansen, D. Techno-economic evaluation of membrane technology for pre-combustion decarbonisation: water gas shift versus reforming. Presented at GHGT10 in Amsterdam, *Energy Procedia*, 2010, www.ghgt.info, in press.

14 Bredesen, R., Jordal, K., and Bolland, O. High-temperature membranes in power generation with CO_2 capture. *Chem. Eng. Process.*, 2004, **43**, 1129–1158.

15 Scholes, C.A., Smith, K.H., Kentish, S.E., and Stevens, G.W.CO_2 capture from pre-combustion processes – strategies for membrane gas separation. *Int. J. Greenhouse Gas Control*, 2010, **4**, 739–755.

16 Marano, J.J.and Ciferino, J.P. Integration of gas separation membranes with IGCC: identifying the right membrane for the right job. *Energy Procedia*, 2009, **1**, 361–368.

17 Grainger, D. and Hägg, M. Techno-economic evaluation of a PVAm CO_2-selective membrane in an IGCC power plant with CO_2 capture. *Fuel*, 2008, **87**, 14–24.

18 Franz, J. and Scherer, V. An evaluation of CO_2 and H_2 selective polymeric membranes for CO_2 separation in IGCC processes. *J. Membr. Sci.*, 2010, **359**, 173–183.

19 Carbo, M.C., Jansen, D., Boon, J., Dijkstra, J.W., van den Brink, R.W., and Verkooijen, A.H.M.Staged water gas shift configuration: key to efficiency penalty reduction during pre-combustion decarbonisation in IGCC. *Energy Procedia*, 2009, **1**, 661–668.

20 Carbo, M.C., Boon, J., Jansen, D., van Dijk, H.A.J., Dijkstra, J.W., van den Brink, R.W., *et al.* Steam demand reduction of water-gas shift reaction in IGCC power plants with pre-combustion CO_2 capture. *Int. J. Greenhouse Gas Control*, 2009, **3**, 712–719.

21 Chiesa, P. and Consonni, S. Shift reactors and physical absorption for low-CO_2 emission IGCCs. *J. Eng. Gas Turbines Power*, 1999, **121**, 295–305.

22 Heintz, Y.J., Sehabiague, L., Morsi, B.I., Jones, K.L., and Pennline, H.W. Novel physical solvents for selective CO_2 capture from fuel gas streams at elevated pressures and temperatures. *Energy Fuels*, 2008, **22**, 3824–3837.

23 Heintz, Y.J., Sehabiague, L., Morsi, B.I., Jones, K.L., Luebke, D.R., and Pennline, H.W. Hydrogen sulfide and carbon dioxide removal from dry fuel gas streams using an ionic liquid as a physical solvent. *Energy Fuels*, 2009, **23**, 4822–4830.

24 Descamps, C., Bouallou, C., and Kanniche, M. Efficiency of an integrated gasification combined cycle (IGCC) power plant including CO_2 removal. *Energy*, 2008, **33**, 874–881.

25 Kunze, C. and Spliethoff, H. Modelling of an IGCC plant with carbon capture for 2020. *Fuel Process. Technol.*, 2010, **91**, 934–941.

26 Renzenbrink, W., Ewers, J., Keller, D., Wolf, K.J., and Apel, W. RWE's 450 MW IGCC/CCS project – status and outlook. *Energy Procedia*, 2009, **1**, 615–622.

27 Rezvani, S., Huang, Y., McIlveen-Wright, D., Hewitt, N., and Mondol, J.D. Comparative assessment of coal fired IGCC systems with CO_2 capture using physical absorption, membrane reactors and chemical looping. *Fuel*, 2009, **88**, 2463–2472.

28 Kanniche, M., Gros-Bonnivard, R., Jaud, P., Valle-Marcos, J., Amann, J.-M., and Bouallou, C. Pre-combustion, post-combustion and oxy-combustion in thermal power plant for CO_2 capture. *Appl. Thermal Eng.*, 2010, **30**, 53–62.
29 Chen, C. A technical and economic assessment of CO_2 capture technology for IGCC power plants. Dissertation. Carnegie Melton University, Pittsburgh, 2005.
30 Gräbner, M., Morstein, O.V., Rappold, D., Günster, W., Beysel, G., and Meyer, B. Constructability study on a German reference IGCC power plant with and without CO_2-capture for hard coal and lignite. *Energy Convers. Manage.*, 2010, **51**, 2179–2187.
31 Romano, M.C., Chiesa, P., and Lozza, G. Pre-combustion CO_2 capture from natural gas power plants, with ATR and MDEA processes. *Int. J. Greenhouse Gas Control*, 2010, **4**, 785–797.
32 Amann, J.-M., Kanniche, M., and Bouallou, C. Reforming natural gas for CO_2 pre-combustion capture in combined cycle power plant. *Clean Technol. Environ. Policy*, 2008, **11**, 67–76.
33 Chiesa, P., Kreutz, T.G., and Lozza, G.G.CO_2 sequestration from IGCC power plants by means of metallic membranes. *J. Eng. Gas Turbines Power*, 2007, **129**, 123.
34 Carbo, M.C., Jansen, D., and Verkooijen, A.H.M. Advanced membrane reactors for fuel decarbonisation in IGCC: H_2 or CO_2 separation? In *Fifth Annual Conference on Carbon Capture and Sequestration*, 8–11 May 2006, Alexandria VA, USA, pp. 1–15.
35 Franz, J. and Scherer, V. Impact of ceramic membranes for CO_2 separation on IGCC power plant performance. Presented at GHGT10 in Amsterdam, *Energy Procedia*, 2010, www.ghgt.info, **2**, 1–8.
36 Krishnan, G., Steele, D., O'Brien, K., Callahan, R., Berchtold, K., and Figueroa, J. Simulation of a process to capture CO_2 from IGCC syngas using a high temperature PBI membrane. *Energy Procedia*, 2009, **1**, 4079–4088.
37 Kaldis, S.P., Skordas, G., and Sakellaropoulos, G.P.Energy and capital cost analysis of CO_2 capture in coal IGCC processes via gas separation membranes. *Fuel Process. Technol.*, 2004, **85**, 337–346.
38 Gallucci, F., Basile, A., Iulianelli, A., and Kuipers, H.J.A.M.A review on patents for hydrogen production using membrane reactors. *Recent Patents Chem. Eng.*, 2010, **2**, 207–222.
39 Bracht, M., Alderlisten, P.T., Kloster, R., Haupt, G., Xue, E., Ross, J.R.H., *et al.* Water gas shift membrane reactor for CO_2 control in IGCC systems: techno-economic feasibility study. *Energy Convers. Manage.*, 1997, **38**, 159–164.
40 Amelio, M., Morrone, P., Gallucci, F., and Basile, A. Integrated gasification gas combined cycle plant with membrane reactors: technological and economical analysis. *Energy Convers. Manage.*, 2007, **48**, 2680–2693.
41 De Lorenzo, L., Kreutz, T.G., Chiesa, P., and Williams, R.H. Carbon-Free hydrogen and electricity from coal: options for syngas cooling in systems using a hydrogen separation membrane reactor. *J. Eng. Gas Turbines Power*, 2008, **130**, 031401.
42 Chiesa, P., Consonni, S., Kreutz, T.G., and Williams, R.H.Co-production of hydrogen, electricity and CO_2 from coal using commercially-ready technology. In *Second Annual Conference on Carbon Sequestration*, Alexandria VA, USA, 5–8 May 2003, pp. 1–23 or: PARTA: Performance and Emissions International Journal of Hydrogen Energy, Vol. 30, 747–767, 2005.
43 Pinacci, P., Broglia, M., Valli, C., Capannelli, G., and Comite, A. Evaluation of the water gas shift reaction in a palladium membrane reactor. *Catal. Today*, 2010, 1–8, Vol. 156, Issues 3–4, 165–172
44 Diniz da Costa, J.C., Reed, G.P., and Thambimuthu, K. High temperature gas separation membranes in coal gasification. *Energy Procedia*, 2009, **1**, 295–302.
45 Jansen, D., Dijkstra, J.W., van den Brink, R.W., Peters, T.A., Stange, M., Bredesen, R., *et al.*, Hydrogen membrane reactors for CO_2 capture. *Energy Procedia*, 2009, **1**, 253–260.

46 Li, H., Dijkstra, J.W., Pieterse, J.A.Z., Boon, J., van den Brink, R.W., and Jansen, D., Towards full-scale demonstration of hydrogen-selective membranes for CO_2 capture: inhibition effect of WGS-components on the H_2 permeation through three Pd membranes of 44 cm long. 2010, Journal of Membrane Science **363**, 204–211.

47 Borio, D.O., Pedernera, M.N., Adrover, E., and Lo, E. Theoretical study of a membrane reactor for the water gas shift reaction under nonisothermal conditions, *AIChE J.*, 2009, **55**, 3206–3213.

48 Gosiewski, K., Warmuzinski, K., and Tanczyk, M. Mathematical simulation of WGS membrane reactor for gas from coal gasification. Catal. Today, 2010, **156**, 229–236.

49 Boutikos, P. and Nikolakis, V. A simulation study of the effect of operating and design parameters on the performance of a water gas shift membrane reactor. *J. Membr. Sci.*, 2010, **350**(5), 378–386.

50 Battersby, S., Ladewig, B.P., Duke, M., Rudolph, V., and Diniz da Costa, J.C. Membrane reactor modelling, validation and simulation for the WGS reaction using metal doped silica membranes. *Asia-Pacific J. Chem. Eng.*, **2009**, 83–92.

51 Tang, Z., Kim, S.-J., Reddy, G.K., Dong, J., and Smirniotis, P. Modified zeolite membrane reactor for high temperature water gas shift reaction. *J. Membr. Sci.*, 2010, **354**, 114–122.

52 Chiesa, P., Kreutz, T.G., and Lozza, G.G. CO_2 sequestration from IGCC power plants by means of metallic membranes. *J. Eng. Gas Turbines Power*, 2007, **129**, 123.

53 Jordal, K., Bredesen, R., Kvamsdal, H.M., and Bolland, O. Integration of H_2-separating membrane technology in gas turbine processes for CO_2 capture. *Energy*, 2004, **29**, 1269–1278.

54 Najmi, B. and Soltanieh, M. Process integration of membrane reactor for steam methane reforming for hydrogen separation with CO_2 capture in power production by natural gas combined cycle. *Energy Procedia*, 2009, **1**, 279–286.

55 Dijkstra, J.W., Pieterse, J.A.Z., Li, H., Boon, J., Delft, Y.C.V., Raju, G., *et al.* Development of membrane reactor technology for power production with pre-combustion CO_2 capture. Presented at GHGT10 in Amsterdam, *Energy Procedia*, 2010, www.ghgt.info, in press.

56 Li, H., Dijkstra, J.W., Pieterse, J.A.Z., Boon, J., van den Brink, R.W., and Jansen, D. WGS-mixture separation and WGS reaction test in a bench-scale multi-tubular membrane reactor. Presented at GHGT10 in Amsterdam, *Energy Procedia*, 2010, www.ghgt.info, in press.

57 Jansen, D., Dijkstra, J.W., van den Brink, R.W., Peters, T.A., Stange, M., Bredesen, R., *et al.* Hydrogen membrane reactors for CO_2 capture. *Energy Procedia*, 2009, **1**, 253–260.

58 Johannessen, E. and Jordal, K. Study of a H_2 separating membrane reactor for methane steam reforming at conditions relevant for power processes with CO_2 capture. *Energy Convers. Manage.*, 2005, **46**, 1059–1071.

59 Fernandes, F. and Soares, A., Jr. Methane steam reforming modeling in a palladium membrane reactor. *Fuel*, 2006, **85**, 569–573.

60 Manzolini, G. and Vigan, F. Co-production of hydrogen and electricity from autothermal reforming of natural gas by means of Pd–Ag membranes. *Energy Procedia*, 2009, **1**, 319–326.

Part 5
Chemical Looping for CO_2 Separation

Process Engineering for CCS Power Plants, First Edition. Edited by Detlef Stolten and V. Scherer

17
Chemical Looping Materials for CO_2 Separation

Anders Lyngfelt and Tobias Mattisson

Abstract

Chemical looping combustion (CLC) is a combustion technology with inherent separation of the greenhouse gas CO_2. Two interconnected fluidized beds, a fuel reactor and an air reactor, are used. The fuel is oxidized by the metal oxide in the fuel reactor, and the metal oxide is regenerated in the air reactor. The outlet gas from the fuel reactor consists of CO_2 and H_2O, easily separated by condensation. Oxides of Ni, Co, Fe, Cu, and Mn are used as oxygen-carrier materials. More than 900 materials have been investigated and some have been used in actual operation in chemical looping combustors in the size range 0.3–140 kW. The total time of operational experience is more than 4000 h. The results indicate that almost complete conversion of the fuel can be obtained and 100% CO_2 capture is possible. Most work so far has been focused on gaseous fuels, but direct application to solid fuels is also rapidly advancing. Moreover, chemical looping technologies to produce hydrogen with inherent CO_2 capture are being developed. This chapter presents an overview of the current status of the technology with focus on materials.

Keywords: chemical looping combustion, oxygen carrier, carbon dioxide capture, separation

17.1
Introduction

Chemical looping combustion has emerged as an attractive option for carbon dioxide capture because CO_2 is inherently separated from the other flue gas components, namely N_2 and unused O_2, and therefore no energy is expended for the gas separation and no gas separation equipment is needed. The CLC system is composed of two reactors, an air and a fuel reactor (Fig. 17.1). The fuel is introduced into the fuel reactor, which contains a metal oxide, Me_xO_y. The fuel and the metal oxide react according to

Process Engineering for CCS Power Plants, First Edition. Edited by Detlef Stolten and V. Scherer

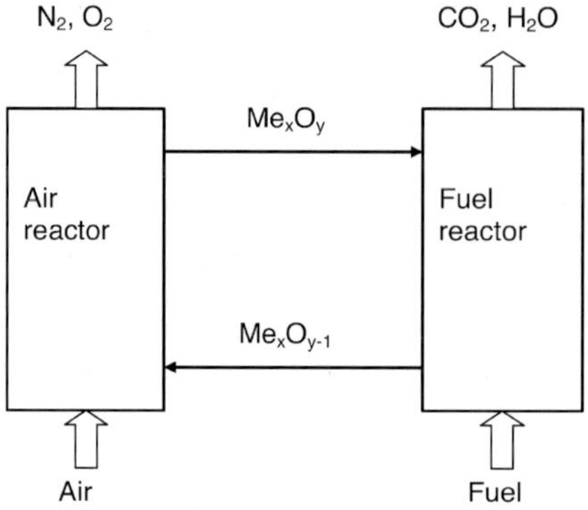

Fig. 17.1 Chemical looping combustion. Me_xO_y/Me_xO_{y-1} denotes recirculating oxygen carrier material.

$$(2n+m)\mathrm{Me}_x\mathrm{O}_y + \mathrm{C}_n\mathrm{H}_{2m} \rightarrow (2n+m)\mathrm{Me}_x\mathrm{O}_{y-1} + m\mathrm{H_2O} + n\mathrm{CO_2} \tag{17.1}$$

The exit gas stream from the fuel reactor contains CO_2 and H_2O, and a stream of CO_2 is obtained when H_2O is condensed. The reduced metal oxide, Me_xO_{y-1}, is transferred to the air reactor, where it is oxidized:

$$\mathrm{Me}_x\mathrm{O}_{y-1} + \tfrac{1}{2}\mathrm{O}_2 \rightarrow \mathrm{Me}_x\mathrm{O}_y \tag{17.2}$$

The air which oxidizes the metal oxide produces a flue gas containing only N_2 and some unused O_2. Depending on the metal oxide and fuel used, reaction 17.1 is often endothermic, whereas reaction 17.2 is exothermic. The total amount of heat evolved from reactions 17.1 and 17.2 is the same as for normal combustion, where the oxygen is in direct contact with the fuel. The advantage of chemical looping combustion over normal combustion is that CO_2 is not diluted with N_2, but obtained in a separate stream without the need for any active separation of gases.

The metal oxides used for the oxygen transfer are called oxygen carriers. The reactor system is made up of two interconnected fluidized beds, with the oxygen carrier in the form of particles being circulated between the two beds (Fig. 17.2). Clearly, the need to develop suitable oxygen-carrier materials for the process, and comprehensive testing of these under realistic conditions, are one of the cornerstones in the development of this technology. Another issue is, of course, to adapt the comprehensive experience from circulating fluidized-bed boilers to this application.

CLC was first introduced by Lewis and Gilliland *et al.* in 1954 as way to produce pure CO_2 from fossil fuels, using two interconnected fluidized beds [1]. Much later, Ishida and Jin proposed the use of chemical looping combustion for climate mitigation and also started laboratory research on oxygen-carrier materials [2]. Ishida *et al.* introduced the name of the process, chemical looping combustion [3]. In 2001, a design based on the circulating fluidized-bed prin-

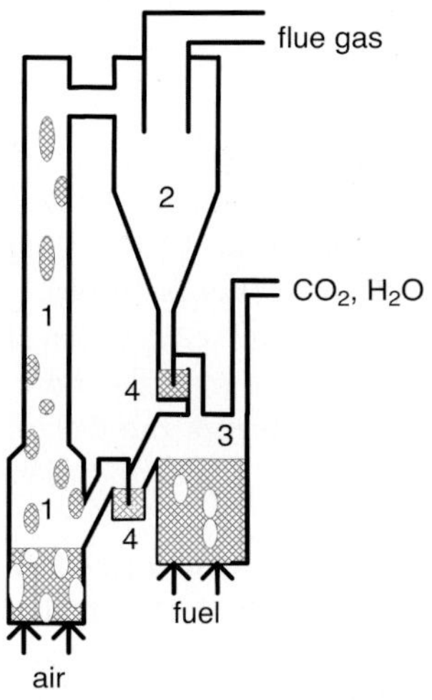

Fig. 17.2 CLC process: example with two interconnected fluidized reactors. 1, Air reactor and riser; 2, cyclone; 3, fuel reactor; 4, loop seals.

ciple was presented (Fig. 17.2), investigating the critical design parameters of a system such as the solids inventory and recirculation rate of oxygen carriers between the reactors and identifying the relationship between these and the oxygen carrier properties [4].

The meaning of CLC as given above is consistent with the first use of the name given by Ishida *et al.*, and is today well established, although a few other names have also been proposed, such as unmixed combustion [5], sorbent energy transfer system (SETS) [6], and mediator recirculation integrating technology (MERIT) [7]. Also, a number of related chemical looping processes have been proposed, where the oxygen transfer is used for purposes other than full combustion, for example, chemical looping reforming and chemical looping gasification processes. Anthony [8] introduced the term solid looping cycles, which includes both oxygen looping, that is, chemical looping, and the transfer of carbon dioxide, that is, . calcium looping. The term high-temperature solid looping is also in use for these processes. Fan *et al.* [9] defined chemical looping as decomposing a reaction "into multiple subreactions in a reaction scheme using chemical intermediates that are reacted and regenerated through the progress of the subreactions." This definition is logical but has the disadvantage that a very large number of chemical processes use intermediates that are regenerated, and would consequently be chemical looping processes. Examples from CO_2 capture include post-combustion capture processes using amines such as monoethanolamine (MEA) and the chilled ammonia process. In this chapter, chemical looping is used solely in connection with the transfer of oxygen.

The desired properties of the chemical looping materials will depend on the application, and an overview of important applications is presented in Sections 17.2–17.5.

17.2
Chemical Looping Combustion of Solid Fuels

CLC research was initially focused on gaseous fuels, but in recent years important work has been dedicated to adapting the process to solid fuels. The first to study solid fuels for CLC were Lewis and co-workers [1, 10] in the 1950s, using copper and iron oxides, and much later, in 2000–2006, new studies involving the same oxides were reported [11–14]. In 2008, Leion and co-workers published a number of papers using different fuels and oxygen carriers in a small laboratory fluidized bed (e.g., [15–17]), and since then there have been a number of publications on laboratory work with solid fuels, and also from actual operation (see Section 17.7).

There are also more complex systems involving hydrogen production (see Section 17.5). However, the most straightforward option is to use the CLC–circulating fluidized-bed concept outlined in Fig. 17.2, but adapting the fuel reactor system for direct addition of solid fuels, and testing in a 10 kW unit suggests that direct application to solid fuels should be feasible [18, 19]. When using solid fuels, the reaction between the oxygen carrier and the char remaining after release of volatiles is not direct, but involves an intermediate gasification step. This is determinant for the fuel reactor design, and several key issues have been identified in relation to fuel reactor performance, namely solid fuel conversion, gas conversion, and CO_2 capture [20].

17.3
Chemical Looping with Oxygen Uncoupling (CLOU)

CLOU is closely related to CLC, but differs in the spontaneous release of oxygen in the fuel reactor. Thus, instead of the fuel gas reacting directly with the oxide, the oxidation of the fuel involves two distinct steps, first the release of gaseous oxygen and then the combustion of the fuel by the oxygen released.

Although this mechanism is clearly useful for any fuel, the advantages are very obvious for solid fuels. In other proposed schemes for using CLC of solid fuels there is a need for an intermediate gasification step of the char with steam or carbon dioxide to form reactive gaseous compounds, which then react with the oxygen carrier particles. The gasification of char with H_2O and CO_2 is inherently slow, resulting in slow overall rates of reaction. This slow gasification is avoided in CLOU, since there no intermediate gasification step is needed and the char reacts directly with gas-phase oxygen.

The process demands an oxygen carrier which has the ability to react with the oxygen in the combustion air in the air reactor but which decomposes to a reduced metal oxide and gas-phase oxygen in the fuel reactor. Three metal oxide systems with suitable thermodynamic properties have been identified, and a thermal analysis has shown that Mn_2O_3/Mn_3O_4, CuO/Cu_2O, and Co_3O_4/CoO have suitable thermodynamic properties [21]. However, the last system has the disadvantage of an overall endothermic reaction in the fuel reactor, and also high costs and risks with respect to health and safety. The equilibrium oxygen partial pressure is 5% for Mn_2O_3/Mn_3O_4 around 800 °C and for CuO/Cu_2O around 950 °C. In a combustion process, most of the oxygen in the combustion air needs to be consumed in order to avoid large flows and thermal losses with exiting flue gas. This means that we want to reduce the O_2 concentration to typically 5% or lower in the air reactor. Thus, for the Mn system, the temperature in the air reactor would need to be below about 800 °C. It is not unlikely that the reactions at these temperatures are too slow.

However, batch laboratory fluidized-bed tests with CuO and solid fuel have been successful [22–24]. Thus, the conversion rate of petroleum coke is a factor of ~50 higher when using CLOU in comparison with the reaction rate of the same fuel with an iron-based oxygen carrier in normal CLC.

As will be discussed below, an additional option could be combined manganese oxides, having lower oxygen partial pressures than the pure manganese system.

17.4
Chemical Looping Reforming

The chemical looping technology can also be adapted for the production of hydrogen with inherent CO_2 capture. In this section, hydrogen production from gaseous fuels is discussed, and hydrogen production from solid fuels is discussed in Section 17.5. Chemical looping processes for hydrogen production include autothermal chemical looping reforming (CLR-a), chemical looping steam reforming (CLR-s), and chemical looping with water splitting.

CLR-a uses chemical looping for partial oxidation to form a syngas that after a water gas shift reaction can be separated into CO_2 and H_2. With nickel-based oxygen carriers, a methane free syngas can be produced (e.g., [25–27]). Several groups have explored the possibility of using oxygen storage materials for the production of syngas (e.g., [28–30]).

CLR-s, or chemical looping combustion with steam reforming, is a marriage between conventional steam reforming and CLC [31]. Just as in commercial steam reforming, the reactions take place inside tubes using suitable catalysts and working at elevated pressure. However, the steam reforming tubes are placed inside the fuel reactor in a CLC unit, or in a fluidized-bed heat exchanger. Hence the reformer tubes are not heated by direct firing but by the oxygen carrier particles in the normal CLC process. The feed gas to the fuel reactor is the offgas from the steam reforming, which is a gas mixture of CH_4, CO_2, CO, and H_2. The principle is illustrated in Fig. 17.3.

The process has a number of advantages: (i) only one gaseous component, namely H_2, needs to be separated, in contrast to CLR-a, where two essentially pure streams of CO_2 and H_2 are needed; (ii) the chemical looping can be at atmospheric pressure, whereas the reforming can be at high pressure; and (iii)

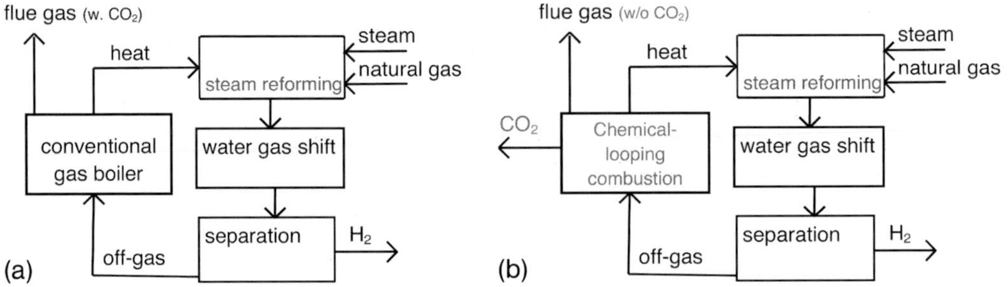

Fig. 17.3 (a) Conventional steam reforming and (b) steam reforming with CO_2 capture by CLC.

compared with the normal gas boiler used in conventional steam reforming, the temperature around the tubes is considerably lower and more uniform. The lower temperature means that a greater fraction of the combustion heat is used for steam reforming, with the consequence that the reforming efficiency is increased. This may well be the only CO_2 capture technology that results in increased efficiency, that is, if the power of CO_2 compression is not included.

Chemical looping with water-splitting, also known as one-step decarbonization, uses three reactors. The process requires an oxygen carrier which is reduced in steps through different oxidation states, for example, $Fe_2O_3 > Fe_3O_4 > FeO > Fe$. In the fuel reactor, the fuel and oxygen carrier need to move in countercurrent; in the top Fe_2O_3 is reduced to Fe_3O_4 while accomplishing complete combustion of the fuel, and in the bottom Fe_3O_4 is further reduced to Fe/FeO. In a second reactor, the Fe/FeO is oxidized by steam giving hydrogen, so-called water splitting. Finally, the material is led to the air reactor where it is oxidized back to Fe_2O_3. Note that two changes in oxidation state are needed: Fe_2O_3 to Fe_3O_4 is needed to oxidize the fuel fully, and Fe/FeO to Fe_3O_4 is needed for water splitting. The process elegantly avoids any gas separation in the hydrogen production, but at the price of added complexity of the reactor system.

The three concepts were compared in the Cachet project for the purpose of producing power with decarbonized hydrogen [32]. From the results, it is possible to deduce that the CLR-s process would produce CO_2-free hydrogen with an energy efficiency of around 80%, and at a cost of € 38 MWh^{-1}. This would be similar to the price of crude oil today (December 2010), namely around US$88 per barrel.

17.5
Chemical Looping Gasification of Solid Fuels

In many cases, the work with chemical looping of solid fuels is directed towards gasification and hydrogen production, rather than complete combustion, similar to the chemical looping reforming and water-splitting processes proposed for gaseous fuels. A chemical looping process for the production of syngas using solid fuels and two interconnected fluidized beds was patented as early as in 1950 [33]. A number of processes share the common approach of using lime to enhance fuel conversion to H_2 by *in situ* CO_2 removal, including the CO_2 acceptor process, the HyPr-Ring process, the Alstom Hybrid Combustion–Gasification process and the GE Fuel-Flexible process [9]. With respect to water splitting, it should be mentioned that going back 80–90 years, the main process for hydrogen production was the so-called steam–iron process, in which iron oxide was reduced by coal to iron, and the iron was then reacted with steam to form hydrogen [9]. Today, a number of related processes are concerned with the direct production of hydrogen through water splitting using Fe/FeO, for example, the Syngas Chemical Looping (SCL) process and the Coal Direct Chemical Looping process [9].

17.6 Oxygen Carrier Development

For the fluidized-bed systems outlined above, important criteria for a good oxygen carrier are the following:

- high reactivity with fuel and oxygen, and ability to convert the fuel fully to CO_2 and H_2O
- low fragmentation and attrition, and also a low tendency for agglomeration
- low production cost and preferably environmentally sound.

A number of reviews related to chemical looping combustion have been published previously. A review by Hossain and Lasa [34] is mostly focused on materials, and includes 107 references, of which approximately 65 are related to experimental work with oxygen carriers. A review by Lyngfelt *et al.* [35] includes 93 references to experimental studies of oxygen carriers. A later review by Fan [36] includes 100 references on CLC. A recent overview by Lyngfelt [37], with focus on the actual operation with oxygen carriers in chemical looping combustors, will be summarized in Section 17.7. Moreover, a comprehensive book overviewing chemical looping systems has recently been published by Fan [9]. Comprehensive material overviews are also given in PhD theses (e.g., [38, 39]).

Taking into account the comprehensive reviews available the material developments up to around 2007–2008, the following overview will be divided in two phases. The first phase will give a more general overview of all the work reported previously, and the second phase covers more recent developments and trends in development.

17.6.1 Oxygen Carrier Development – Overview of First Phase

A comprehensive table presented by Lyngfelt *et al.* [35] summarizes experimental work on oxygen carrier particles up to 2007. Major contributors are the Tokyo Institute of Technology, Chalmers University of Technology in Sweden, CSIC-ECB in Spain, and the Korea Institute of Energy Research. All of the early oxygen carrier development was carried out by the Japanese group led by Ishida. In total, the table includes more than 600 tested materials, of which the majority were tested by CSIC-ECB (50%) and Chalmers University of Technology (25%), with additional important contributions from Japan, South Korea, North America, and Europe. A large part of the CSIC/Chalmers work was performed in common EU projects, and thus well coordinated. The most studied oxides were based on Ni, followed by Fe, Cu, Mn, and Co in order of number of materials studied.

Most of the active metal oxides studied were combined with an inert material, such as Al_2O_3. There are some studies of unsupported materials, such as iron ore [40]. Although such materials may have low costs, reactivity experiments simulating CLC performed on natural ores or unsupported metal oxides suggested rapid degeneration or low reactivity of these materials [2, 41–43]. The use of an inert material is believed to increase the porosity and reactivity of the particles, help to maintain the structure, and possibly also increase the ionic conductivity of the particles. Even though the proportion of free oxygen in a particle decreases with the addition of an inert material, the reactivity with the fuel and oxygen can still be higher [2].

The ability of the oxygen carrier to convert a fuel gas completely to CO_2 and H_2O has been investigated thermodynamically and it was found to be feasible to use the metal oxide systems NiO/Ni, Mn_3O_4/MnO, Fe_2O_3/Fe_3O_4, Cu_2O/Cu, and CoO/Co as oxygen carriers [41]. For CoO/Co, the thermodynamics are not so favorable, with a maximum 93% conversion at 1000 °C, and moreover this oxygen carrier is expensive and has health and safety risks.

The four oxides of copper, iron, manganese, and nickel have advantages and disadvantages, as can be seen in Table 17.1. Note, for instance, that the most reactive materials, namely nickel oxides, are unfortunately also the most expensive. For NiO there are also health, safety, and environmental aspects to be considered. Furthermore, NiO also differs from the other oxides in having a thermodynamic restriction – it cannot convert fuels completely to CO_2 and H_2O. The maximum conversion is 99–99.5%, depending on temperature. All the oxides have a more or less exothermic reaction in both reactors if the fuel is H_2 or CO, but with methane the reaction is endothermic for all the oxides except CuO. This is clearly an advantage for CuO, since it reduces the particle circulation needed to maintain the fuel reactor temperature. On the other hand, Cu has the disadvantage of a low melting temperature. As noted previously, CuO–Cu_2O can be used with release of gaseous oxygen in the CLOU process. Both CuO and Cu_2O have considerably higher melting temperatures than Cu.

Table 17.1 Qualitative estimation of pros and cons for the active oxides.

Item	Fe_2O_3/Fe_3O_4	Mn_3O_4/MnO	CuO/Cu	NiO/Ni	Comments
R_0	0.03	0.07	0.20	0.21	Oxygen ratio
Reactivity towards methane	☹	← decreasing	increasing →	☺	
Cost	☺	← decreasing	increasing →	☹	
Health, safety, and environmental risks	☺	☺	← decreasing risk	☹	
Thermodynamics				☹	<99.5% conversion for NiO
Reaction enthalpy with CH_4			☺		CuO exothermic with CH_4
Melting point			☹		1085 °C for Cu
CLOU			☺		CuO/Cu_2O

The oxygen carrier needs to react at a sufficient rate. As the amount of oxygen carrier needed in the reactors is directly related to the reactivity of the oxygen carrier, a fast rate would mean less material and hence smaller reactor sizes and lower material production costs.

The oxygen carriers must also be able to transfer a sufficient amount of oxygen to the fuel. The oxygen transfer capacity, R_0, that is, the maximum mass fraction of oxygen that can be transferred, is also given in Table 17.1. The R_0 data in Table 17.1 are for pure systems, and will therefore be reduced if diluted with support materials. For the case where the reaction in the fuel reactor is endothermic, the circulation needed is normally not set by the oxygen transfer capacity, but by the heat balance, in order to reach a sufficient fuel reactor temperature [41].

There are a few studies, by Adanez *et al.* [44] and Johansson and co-workers [45–47], which compared a large number of different oxygen carriers. These generally indicate a large difference in reactivity between the different oxides following the ranking shown in Table 17.1. However, most of the reactivity comparisons are focused on CH_4. In general, the reactivity is significantly higher using H_2 and CO compared with CH_4, and may be fairly high for cheaper materials showing low reactivity with CH_4.

Reactivity may increase significantly with reaction temperature, especially in the case of low-reactivity materials and CH_4. On the other hand, the reactivity of iron and manganese materials with CO and H_2 may be high also at lower temperatures, and the same applies to nickel materials with CH_4.

Clearly, there are important advantages and disadvantages with all of the different oxide systems and the choice of oxygen carrier will be dependent on the application. Generally, oxygen carriers with low cost have an advantage for ash-containing solid fuels, whereas more expensive and reactive oxygen carriers could be suitable for gaseous fuels.

17.6.2
Oxygen Carrier Development – Second Phase

As a continuation of the table published by Lyngfelt *et al.* [35] covering 600 materials studied up to 2007, an overview of materials tested in more recent years is given in Table 17.2. This table includes more than 300 materials. An important difference between the phases 1 and 2 is the much larger fraction of materials tested in fluidized conditions. Most of the studies used similar materials to those in the first phase. Thus, half of the materials tested were nickel-based, followed by copper, iron, and combined oxides, at around 30–40 materials each. Perovskites, manganese, $CaSO_4$, and cobalt include around 10 materials or fewer each. As can be seen, the table includes some interesting new materials: (i) combined metal oxides, (ii) $CaSO_4$–CaS, and (iii) low-cost materials such as ores.

Combined metal oxides, where two or more oxides are combined not only physically, but also chemically, constituting new oxides, include, for example, $Cu_{0.95}Fe_{1.05}AlO_4$, $Co_{0.5}Ni_{0.5}FeAlO_4$, $CoFeAlO_4$, $CuFeGaO_4$, and $NiFeAlO_4$ [48]. Some of these materials have a perovskite structure, for example $La_{1-x}Sr_xFe_{1-y}Co_yO_{3-\delta}$ and $Sr(Mn_{1-x}Ni_x)O_3$ [49, 50]. Also of interest are combined Mn oxides with partial CLOU properties, that is, with the ability to release some oxygen. These include Mn combined with Ca, Mg, Ni, and Fe [51, 52]. Many of these combined materials are promising, but few have yet been successfully tested in actual operation. An exception is $CaMn_{0.875}Ti_{0.125}O_3$ [53] (see Section 17.7).

$CaSO_4$–CaS has been studied as a low-cost oxygen carrier for solid fuels [54–61]. It has a uniquely high oxygen transfer capacity of 47%, but it has a thermodynamic constraint and cannot convert CO and H_2 more than 98–99%. There is also a risk of sulfur being lost, converting the oxygen carrier to CaO [55, 59]. Loss of sulfur is difficult to predict as it takes place in the shifts between oxidizing and reducing conditions and will be very dependent on the process conditions, including temperature, fuel sulfur content, and extent of fuel conversion.

Low-cost materials have been studied mainly for use with solid fuels; this includes natural ores such as ilmenite ($FeTiO_3$), iron ore, and manganese ore, and also industrial waste materials such as iron oxide scales (e.g., [15, 62–69]). Many of the studies used ilmenite, being a cheap ore, having a reasonably high reactivity towards syngas, and showing good fluidization behavior. Laboratory testing show that iron oxide scales generally have higher reactivity than ilmenite [65].

An interesting new aspect is also the development of CLOU, already discussed in Section 17.3.

As noted, a large majority of the work on gaseous fuels was focused on nickel materials. Although this work may not have revealed any dramatic new insights, it provided important experience and demonstrated the reliability of these materials. Another important aspect is the need to look at industrially relevant produc-

Table 17.2 Overview of materials investigated in phase 2[a)].

Study	Ref.	Oxygen carrier (Me_xO_y/support)	O.C.	Preparation method	Reduction agent	T_{red} (°C)	D_p (mm)	Apparatus	Notes[b)]
Abad *et al.*, 2007	73	Fe_2O_3/Al_2O_3, $NiO/NiAl_2O_4$, CuO/Al_2O_3	3	Freeze granulation, impregnation	CH_4 or CO or H_2 with H_2O/CO_2	500–950	0.15–0.2	TGA	d
Abad *et al.*, 2007	74	Fe_2O_3/Al_2O_3, $NiO/NiAl_2O_4$, CuO/Al_2O_3	3	Freeze granulation, impregnation	$H_2/H_2O/N_2$, $CO/CO_2/N_2$, $H_2/CO/CO_2/H_2O$, $H_2/CO/CO_2$	450–1000	0.15–0.2	TGA	d
Erri and Varma, 2007	75	$NiO/NiAl_2O_4$, $NiO/Ni_{1-x}Mg_xAl_2O_4$	2	Solution combustion	$CH_4/H_2O/Ar$	800–1200	0.21–0.43	TGA	c
Erri and Varma, 2007	76	$NiO/NiAl_2O_4$, $NiO/Ni_{1-x}Mg_xAl_2O_4$	7	Solution combustion	$CH_4/H_2O/Ar$, $H_2/CO/CO_2$	800–1200	0.21–0.43	TGA	c
He *et al.*, 2007	77	Fe_2O_3/Al_2O_3	1	Impregnation	CH_4	900	0.1–0.3	TGA, FxB	
Hossain *et al.*, 2007	78	Ni/Al_2O_3, $Co\text{–}Ni/Al_2O_3$	3	Impregnation	CH_4	680	0.06–0.1	FzB	
Hossain and de Lasa, 2007	79	Ni/Al_2O_3, $Co\text{–}Ni/Al_2O_3$, Co/Al_2O_3	6	Impregnation	CH_4, H_2/Ar[c]	650, 0–750[c]	0.07	FzB	c, d
Yang *et al.*, 2007	80	Fe_2O_3	1	n.a.	Lignite char	430–900	0.2–0.45	FxB	
Zafar *et al.*, 2007	81	$Mn_3O_4/Mg\text{–}ZrO_2$	1	Freeze granulation	$CH_4/H_2O/N_2$	800–950	0.09–0.125, 0.125–0.18, 0.18–0.25	TGA	d
Berguerand and Lynfelt, 2008	62	Ilmenite	1	Natural mineral	Petroleum coke	950	0.14	CFzB	
Berguerand and Lynffelt, 2008	82	Ilmenite	1	Natural mineral	South African coal	950	0.14	CFzB	
Chuang *et al.*, 2008	83	CuO/Al_2O_3	17	Mechanical mixing, impregnation, coprecipitations	$CO/CO_2/N_2$	800–900	0.5–0.71	FzB	

Table 17.2 (continued)

Study	Ref.	Oxygen carrier (Me_xO_y/support)	O.C.	Preparation method	Reduction agent	T_{red} (°C)	D_p (mm)	Apparatus	Notes[b)]
Gao *et al.*, 2008	84	$NiO/NiAl_2O_4$	1	Impregnation	Bituminous coal	800–960	0.30–0.45	FzB	
Gayan *et al.*, 2008	85	NiO/Al_2O_3, $NiO/MgAl_2O_4$, $NiO/CaAl_2O_4$	16	Impregnation	$CH_4/H_2O/N_2$, CH_4/N_2	950	0.1–0.3, 0.1–0.5	TGA, FzB	
Johansson *et al.*, 2008	86	$NiO/NiAl_2O_4$, $NiO/MgAl_2O_4$	2	Freeze granulation	CH_4/H_2O	950	0.125–0.180	FzB	
Ksepko *et al.*, 2008	49	$SrNiO_3$, $Sr(Mn_{0.2}Ni_{0.8})O_3$, $Sr(Mn_{0.5}Ni_{0.})O_3$, $Sr(Mn_{0.8}Ni_{0.2})O_3$, $SrMnO_3$	5	Mechanical mixing	H_2/N_2	800	3	TGA	a
Leion *et al.*, 2008	15	Ilmenite	1	Natural mineral	CH_4, H_2/CO	950	0.125–0.180	FzB	
Leion *et al.*, 2008	16	Ilmenite, $Fe_2O_3/MgAl_2O_4$	2	Natural mineral, freeze granulation	Coal, petroleum coke	950	0.125–0.180	FzB	
Leion *et al.*, 2008	17	Ilmenite	1	Natural mineral	Coal, petroleum coke, wood	970	0.125–0.180	FzB	
Leion *et al.*, 2008	23	CuO/ZrO_2	1	Freeze granulation	German lignite	850–985	0.125–0.180	FzB	b
Linderholm *et al.*, 2008	87	$NiO/NiAl_2O_4$	1	Spin flash drying	Natural gas	660–950	0.132	CFzB	
Mattisson *et al.*, 2008	88	$NiO/NiAl_2O_4$	1	Freeze granulation	CH_4/H_2O, $CH_4/H_2O/CO_2/N_2$	750–950	0.09–0.125, 0.125–0.180, 0.180–0.250	TGA, FzB	
Rydén *et al.*, 2008	89	$NiO/MgAl_2O_4$, NiO/Al_2O_3	3	Freeze granulation, impregnation	Natural gas, natural gas/H_2O	848–964	0.09–0.212, 0.09–0.25	CFzB	
Rydén *et al.*, 2008	50	$La_{0.5}Sr_{0.5}Fe_{0.5}Co_{0.5}O_{3-\delta}$, $LaFeO_{3-\delta}$, $La_{0.8}Sr_{0.2}FeO_{3-\delta}$, $La_{0.5}Sr_{0.5}FeO_{3-\delta}$, $NiO/MgAl_2O_4$, $Fe_2O_3/MgAl_2O_4$, $Mn_3O_4/Mg–ZrO_2$	7	Glycine nitrate combustion + spray drying, freeze granulation	CH_4/Ar	900	0.09–0.15, 0.15–0.25	FxB	

Table 17.2 (continued)

Study	Ref.	Oxygen carrier (Me_xO_y/support)	O.C.	Preparation method	Reduction agent	T_{red} (°C)	D_p (mm)	Apparatus	Notes[b)]
Ryu *et al.*, 2008	90	NiO/bentonite, $NiO/NiAl_2O_4$, $Co_xO_y/CoAl_2O_4$	3	n.a.	Natural gas	900	0.106–0.212	FzB	
Sedor *et al.*, 2008	91	NiO/Al_2O_3	1	Impregnation	H_2, CH_4	600–680	n.a.	FxB, FzB	c, d
Sedor *et al.*, 2008	92	NiO/Al_2O_3	4	Impregnation	H_2, CH_4	680	0.095	FxB, FzB	c
Shen *et al.*, 2008	54	$CaSO_4$	1	Analytical grade	CO/N_2, H_2/N_2	800–1050	0.05–0.1	TGA	
Song *et al.*, 2008	55	$CaSO_4$	1	Natural mineral	$H_2/CO/CO_2$	950	0.15–0.2	FzB	
Song *et al.*, 2008	56	$CaSO_4$	1	Natural mineral	CH_4	800–1000	0.2–0.4, 0.4–0.6, 0.6–0.8	FxB	
Song *et al.*, 2008	57	$CaSO_4$	1	Natural mineral	$H_2/CO/CO_2$	890–950	0.15–0.2	FzB	d
Tian *et al.*, 2008	93	CuO/bentonite (1), CuO/barium-hexaaluminate (2)	2	Mechanical mixing, sol–gel	$H_2/CO/CO_2/He$	700–900	0.07–0.15 (1), >0.84 (1) As prep. (2)	TGA, Fxb	d
Tian *et al.*, 2008	58	$CaSO_4$	1	Analytical grade	$CO/CO_2/N_2$	HR	0.009	TGA	
Zhao *et al.*, 2008	94	$NiO/NiAl_2O_4$	1	Sol–gel	CH_4/N_2	900	0.125–0.18	TGA	
Adanez *et al.*, 2009	95	NiO/Al_2O_3	1	Impregnation	CH_4/N_2	800–880	0.1–0.3	CFzB	
Adanez *et al.*, 2009	96	NiO/Al_2O_3	1	Impregnation	$C_xH_y/N_2/H_2O$, $CH_4/C_xH_y/N_2$, $CH_4/C_xH_y/H_2O/N_2$	700–950	0.1–0.3	TGA, FzB, CFzB	
Adanez *et al.*, 2009	97	NiO/Al_2O_3	1	Impregnation	CH_4/N_2, $CH_4/C_2H_6/N_2$, CH_4/C_3H_8, N_2, H_2O, $CH_4/N_2/H_2S$	800–880	0.1–0.3	CFzB	
Berguerand and Lyngfelt, 2009	63	Ilmenite	1	Natural mineral	Petroleum coke	950–1000	0.14	CFzB	
Berguerand and Lyngfelt, 2009	98	Ilmenite	1	Natural mineral	Petroleum coke	950–1000	0.14	CFzB	
Bolhar-Nordenkampf *et al.*, 2009	99	$NiO/NiAl_2O_4$, $NiO/NiAl_2O_4/MgO$	2	Spray drying	CH_4	780–950	0.12	CFzB	

Table 17.2 (continued)

Study	Ref.	Oxygen carrier (Me_xO_y/support)	O.C.	Preparation method	Reduction agent	T_{red} (°C)	D_p (mm)	Apparatus	Notes [b)]
Chandel *et al.*, 2009	100	$NiO/NiAl_2O_4$ (1), $Cu_{0.95}Fe_{1.05}AlO_4$ (2), $CuO/Cu_{0.95}Fe_{1.05}AlO_4$	3	Precipitation (1), n.a. (2)	CH_4, CH_4/H_2O	800–900	0.1–0.3	FzB	
Chuang *et al.*, 2009	101	CuO/Al_2O_3	1	Coprecipitation	CO/N_2	300–750	0.355–0.5	FzB	d
Dahl *et al.*, 2009	102	CuO/Al_2O_3	1	Impregnation	CH_4/N_2	700	2	TGA	
Dong *et al.*, 2009	103	Fe_2O_3	1	n.a.	CO/N_2, $H_2/CO/CH_4/CO_2/N_2$	800–880	0.1–0.2	FzB	
Dueso *et al.*, 2009	104	NiO/Al_2O_3	1	Impregnation	$CO/H_2/CO_2/H_2O/N_2$ $CO/CO_2/N_2$, H_2/N_2	800–880, 950	0.1–0.3	TGA, FzB, CFzB	
Forero et el., 2009	105	CuO/Al_2O_3	1	Impregnation	$CO/H_2/CO_2/H_2O/N_2$ $CO/CO_2/N_2$, H_2/N_2	800, 880	0.3–0.5	TGA, CFzB	
Garcia-Labiano *et al.*, 2009	106	NiO/Al_2O_3	1	Impregnation	$CH_4/N_2/H_2S$	830–870	0.1–0.3	CFzB	
Gayan *et al.*, 2009	107	NiO/Al_2O_3	9	Impregnation, deposition–precipitation	$CH_4/H_2O/N_2$, CH_4/N_2	950	0.1–0.3	TGA, FzB	
Hossain *et al.*, 2009	108	Ni/La–γ-Al_2O_3	2	Impregnation	H_2, CH_4	650	0.1	FxB, FzB	c
Hoteit *et al.*, 2009	109	$NiO/NiAl_2O_4$ (1), $Cu_{0.95}Fe_{1.05}AlO_4$ (2)	2	Precipitation (1), n.a. (2)	CH_4, CH_4/H_2O	800–900	0.171, 0.234	FzB	
Erri and Varma, 2009	110	$NiO/MgAl_2O_4$	1	Solution combustion	H_2/Ar	HR	0.125–0.15	TGA	d
Hoteit *et al.*, 2009	111	$NiAl_{0.44}O_{1.67}$ (1), $Cu_{0.95}Fe_{1.05}AlO_4$ (2)	2	Precipitation (1), n.a. (2)	Biogas	850–900	0.171 (1), 0.150 (2)	FzB	
Jerndal *et al.*, 2009	70	$NiO/NiAl_2O_4$, $NiO/NiAl_2O_4/MgO$, $NiO/MgAl_2O_4$	24	Spray drying	CH_4	950	0.125–0.180	FzB	
Kolbitsch *et al.*, 2009	112	Ilmenite, $NiO/NiAl_2O_4$	2	Natural mineral (1), spray drying (2)	H_2, CO, CH_4	850–950	~0.2 (mean) (1)	CFzB	
Kolbitsch *et al.*, 2009	113	$NiO/NiAl_2O_4$	1	Spray drying	H_2, CH_4	780–950	0.09–0.21	CFzB	
Kolbitsch *et al.*, 2009	114	$NiO/NiAl_2O_4$, $NiO/NiAl_2O_4/MgO$	2	Spray drying	NG	780–950	0.9–0.12	CFzB	

Table 17.2 (continued)

Study	Ref.	Oxygen carrier (Me_xO_y/support)	O.C.	Preparation method	Reduction agent	T_{red} (°C)	D_p (mm)	Apparatus	Notes[b)]
Kuusik *et al.*, 2009	115	$NiO/NiAl_2O_4$, $NiO/NiAl_2O_4/MgO$,	$NiO/MgAl_2O_4$	5	Spray drying	CH_4/O_2	1000–1185 0.106–0.212	FzB	
Lambert *et al.*, 2009	116	NiO/YSA, Mn–Fe mixed oxides	16	Coprecipitation Hydrothermal synthesis	$CH_4/CO_2/N_2$	900	0.03–0.04	TGA	
Leion *et al.*, 2009	64	Fe_2O_3-based, Mn_3O_4-based	19	Natural ores, industrial products/ wastes	CH_4, CO_2/H_2	950	0.125–0.18	FzB	
Leion *et al.*, 2009	117	$NiO/NiAl_2O_4$	1	Spin flash drying	CH_4, CO_2/H_2, coal, petroleum coke	970	0.09–0.125, 0.125–0.180	FzB	
Leion *et al.*, 2009	65	Iron ore, oxide scales	2	Natural ore, industrial products/ waste	Petroleum coke, coal, biomass, CO/H_2	900–1000	0.09–0.125, 0.125–0.180	FzB	
Leion *et al.*, 2009	118	Ilmenite	1	Natural ore	$Coal/N_2$, $coal/H_2O/N_2$ $coal/CO_2/N_2$ $coal/H_2O/CO_2/N_2$	970	0.125–0.18	FzB	
Leion *et al.*, 2009	119	$CaMn_xTi_{1-x}O_3$	8	Spray pyrolysis, freeze granulation	$H_2/CO_2/Ar$, $CH_4/CO_2/Ar$, CH_4, petroleum coke	800–1000	Powder, 0.090–0.125, 0.125–0.180, 1.4–4	TGA, FzB	b
Leion *et al.*, 2009	24	CuO/ZrO_2	1	Freeze granulation	Petroleum coke, coal, biomass	950	0.09–0.125	FzB	b
Linderholm *et al.*, 2009	72	$NiO/NiAl_2O_4$, $NiO/NiAl_2O_4/MgO$	2	Spray drying	NG	800–900	0.9–0.12	CFzB	
Mattisson *et al.*, 2009	21	CuO/ZrO_2, CuO/Al_2O_3	2	Freeze granulation	CH_4, petroleum coke	950, 985	0.125–0.18	FzB	
Mattisson *et al.*, 2009	22	CuO/ZrO_2	1	Freeze granulation	Petroleum coke	885–985	0.125–0.18	FzB	b

Table 17.2 (continued)

Study	Ref.	Oxygen carrier (Me_xO_y/support)	O.C.	Preparation method	Reduction agent	T_{red} (°C)	D_p (mm)	Apparatus	Notes [b)]
Muller *et al.*, 2009	120	CuO/Al_2O_3	1	Coprecipitation	CO/N_2, lignite, bituminous coal, petroleum coke, wood chips	800, 850	0.355–0.5	FzB	
Pröll *et al.*, 2009	66	Ilmenite, ilmenite + olivine	1	Natural mineral	NG, H_2/CO	813–986	~0.2	CFzB	
Pröll *et al.*, 2009	121	$NiO/NiAl_2O_4$	1	Spray drying	Natural gas, propane	800–950	0.135	CFzB	
Rydén *et al.*, 2009	122	$NiO/Mg\text{-}ZrO_2$	1	Freeze granulation	CH_4, CH_4/H_2O, natural gas, natural gas/H_2O	670–950	0.9–0.212	FzB, CFzB	
Rydén *et al.*, 2009	123	Red iron oxide, brown iron oxide, oxide scales, Ni-olivine,	NiO/ $MgAl_2O_4$	5	Waste materials from steel industry	CH_4	850–950 0.125–0.180	FzB	e
Rubel *et al.*, 2009	124	NiO, Fe_2O_3, CoO, CuO, Ni catalyst, Fe catalyst	8	Fine powders, commercial catalysts	H_2, $CO/CO_2/Ar$, coal char	800–900	n.a.	TGA	
Ryu *et al.*, 2009	125	NiO/bentonite, $NiO/NiAl_2O_4$, $Co_xO_y/CoAl_2O_4$, OCN-650	4	Mechanical mixing, dissolution, coprecipitation/impregnation, spray drying	$H_2/CO/CO_2/N_2$	900	0.106–0.212	FzB	
Shen *et al.*, 2009	126	$NiO/NiAl_2O_4$	1	Impregnation	Bituminous coal	870–970	0.2–0.4	CFzB	
Shen *et al.*, 2009	127	Fe_2O_3	1	Mixing powder	Pine sawdust	740–920	0.3–0.6	CFzB	
Shen *et al.*, 2009	128	NiO/Al_2O_3	1	Coprecipitation	Bituminous coal	920–960	0.2–0.4	CFzB	
Song *et al.*, 2009	59	$CaSO_4$	1	Natural mineral	CH_4	950	0.15–0.2	FxB	
Solunke and Veser, 2009	129	Ni/BHA, Cu/BHA	2	Sol–gel	$H_2/CO/CO_2/He$, $H_2/CO/CO_2/H_2S/He$	900	0.01–few mm	TGA	
Shulman *et al.*, 2009	130	$NiO/NiAl_2O_4$, $NiO/NiAl_2O_4/MgO$	2	Spray drying	CH_4	850, 950	0.125–0.180	FzB	

Table 17.2 (continued)

Study	Ref.	Oxygen carrier (Me_xO_y/support)	O.C.	Preparation method	Reduction agent	T_{red} (°C)	D_p (mm)	Apparatus	Notes[b)]
Shulman *et al.*, 2009	51	Mn_3O_4/Fe_2O_3, Mn_3O_4/NiO, Mn_3O_4/SiO_2	11	Freeze granulation	CH_4	810–900	0.125–0.180	FzB	b
Siriwardane *et al.*, 2009	131	CuO (1), NiO (2), Mn_2O_3 (3), Fe_2O_3 (4), Co_3O_4 (5)	5	Analytical grade	Illinois #6 coal	900–1000, HR	0.005 (1), 0.044 (2–4), 0.07 (5)	TGA, FxB	
Tian and Guo, 2009	60	$CaSO_4$	1	Analytical grade	$CO/CO_2/N_2$	HR	0.009	TGA	d
Tian *et al.*, 2009	132	Cu_2O/bentonite, NiO/bentonite, Mn_2O_3/bentonite, Fe_2O_3/bentonite	4	Mechanical mixing	$CO/CO_2/H_2/He$, $CO/CO_2/H_2/He/H_2S$	800–900	0.074–0.088, 0.250–0.595	TGA	
Xiang *et al.*, 2009	133	Fe_3O_4	1	n.a.	Shenhua coal	HR	<0.075	TGA	
Xiao *et al.*, 2009	61	$CaSO_4$	1	Natural ore	CO/N_2	880–950	0.15–0.2	FxB	d
Zhang *et al.*, 2009	134	Fe_2O_3	1	n.a.	HCl/N_2	700–900	n.a.	FxB	
Abad *et al.*, 2010	135	CuO/Al_2O_3	1	Impregnation	CH_4/H_2O, CH_4	600–800	0.1–0.3, 0.2–0.5	TGA, CFzB	g
Adanez *et al.*, 2010	67	Ilmenite	1	Natural mineral	H_2/H_2O, CO/CO_2, CH_4/H_2O	900	0.15–0.3	TGA	
Dueso *et al.*, 2010	136	NiO/Al_2O_3	1	Impregnation	CH_4/N_2, H_2/N_2	950	0.1–0.3	TGA, FzB	
Berguerand and Lyngfelt, 2010	137	Ilmenite	1	Natural mineral	Petroleum coke, bituminous coal	950–1030	0.14 (mean)	FzB	
Gayán *et al.*, 2010	138	CuO/Al_2O_3	1	Impregnation	$C_xH_y/N_2/H_2O$, $CH_4/C_xH_y/N_2$, $CH_4/C_xH_y/H_2O/N_2$	700–880	0.3–0.5	TGA, FzB, CFzB	
Hossain and de Lasa, 2010	139	$Co–Ni/Al_2O_3$	1	Impregnation	H_2, CH_4	550–650, R	0.095	FxB, FzB	c, d
Iliuta *et al.*, 2010	140	NiO/FCC	1	Impregnation	H_2/Ar, CH_4/Ar, CO/Ar	600–900	0.14	FxB, FzB	d
Jerndal *et al.*, 2010	71	$NiO/NiAl_2O_4$	17	Spray drying	CH_4	950	0.125–0.180	FzB	

Table 17.2 (continued)

Study	Ref.	Oxygen carrier (Me_xO_y/support)	O.C.	Preparation method	Reduction agent	T_{red} (°C)	D_p (mm)	Apparatus	Notes[b)]
Noorman *et al.*, 2010	141	CuO/Al_2O_3	1	n.a.	CH_4/N_2	650	1.1	FxB	
Kolbitsch *et al.*, 2010	142	Ilmenite, $NiO/NiAl_2O_4$	2	Natural mineral (1), spray drying (2)	H_2, CO, CH_4	850–950	~0.2 (1), 0.09–0.212 (2)	CFzB	
Xiao *et al.*, 2010	68	Fe_2O_3	1	Natural ore	Bituminous coal	970	0.09–0.125	PFxB	f
Lambert *et al.*, 2010	48	$NiO/NiAl_2O_4$, $CuAl_2O_4$, $Cu_{0.5}Ni_{0.5}FeAlO_4$, $Co_{0.5}Ni_{0.5}FeAlO_4$, $CoFeAlO_4$, $CuFeGaO_4$, $NiFeAlO_4$, NiO/CuO/$Cu_{0.95}Fe_{1.05}AlO_4$	17	Coprecipitation, Impregnation	CH_4	900	0.03–0.04	TGA	
Rifflart *et al.*, 2010	143	$NiO/NiAl_2O_4$	1	Precipitation	CH_4	750–850	~125	CFzB	
Bidwe *et al.*, 2010	69	Ilmenite	1	Natural ore	CO, H2, CH_4, coal	900–950	n.a.	FzB, CFzB	
Shulman *et al.*, 2010	52	Mg_2MnO_4, Mg_2MnO_4 + CaO/TiO_2	12	Freeze granulation	CH_4	810–950	125–180	FzB	
Rydén *et al.*, 2010	53	$CaMn_{0.875}Ti_{0.125}O_3$	1	Spray pyrolysis	CH_4	770–950	90–212	CFzB	

a) Abbreviations and symbols: O.C. = number of oxygen carriers studied in the paper. Oxygen carriers can vary by being prepared by different authors, using different manufacturing techniques, different raw materials, different ratios of active to inert material, and using different calcination temperatures (or times). D_p = particle diameter. Here a span of particle sizes is often reported. If only one number is presented, it is usually some type of average or mean. TGA, thermogravimetric analyzer; FxB, fixed bed; FzB, batch fluidized bed; PFxB, pressurized batch fixed-bed reactor; CFzB, circulating fluidized beds, i.e., chemical looping combustor; HR, heating ramp; n/a, not applicable.

b) Notes: a, cylindrical form, diameter × height; b, investigation of CLOU (chemical looping with oxygen uncoupling; c, temperature-programmed reduction; d, study on kinetics; e, mixed oxides, i.e., two or more particles mixed in reactor; f, effect of pressure; g, modeling of CFzB.

tion technology. When new material development is started, the first stage involves building an understanding, and may use very pure but expensive materials and also production technologies that are unrealistic for large-scale production. Therefore, work was initiated to verify that material production could be commercialized at reasonable cost. Hence particles were produced using technology suitable for large-scale production, such as spray-drying, and using raw materials available commercially in large amounts and at reasonable cost [70, 71]. These materials have also been evaluated in long-term testing for 1000 h [72] (see below).

17.7 Reactor Design and Operational Experience in Chemical Looping Combustors

With respect to reactor design, a number of studies have utilized cold-flow modeling, in order to identify stable and suitable operating conditions for various designs [144–149].

Although laboratory experiments give very valuable information, they cannot be used to prove the workability of the particles in the actual process. Therefore, it is crucial to test the particles in actual chemical looping combustors under realistic conditions and over sufficient time periods. A review of operational experiences from 12 CLC units of sizes 0.3–140 kW including 41 references and 29 oxygen-carrier materials was given by Lyngfelt [37]. The units are presented in Table 17.3, including two additional units [69, 143]. This work is summarized below.

Table 17.3 Testing in chemical looping combustors.

Location	Unit	Oxides tested	Time (h)	Fuel	Year
Chalmers	10 kW	NiO, Fe_2O_3	1350	Natural gas	2004
KIER	50 kW	NiO, CoO	28	Natural gas	2004
CSIC	10 kW	CuO, NiO	140	Natural gas	2006
Chalmers	0.3 kW	NiO, Mn_3O_4, Fe_2O_3, ilmenite	730	Natural gas, syngas	2006
Chalmers	10 kW–SF	Ilmenite	90	Coal, petroleum coke	2008
CSIC	0.5 kW	CuO, NiO	660	Natural gas	2009
KAIST	1 kW	NiO + Fe_2O_3	?	CH_4	2009
Vienna UT	140 kW	Ilmenite, NiO	390	Natural gas, CO, H_2	2009
Alstom	15 kW	NiO	100	Natural gas	2009
Nanjing	10 kW–SF	NiO, Fe_2O_3	230	Coal, biomass	2009
KIER	50 kW	NiO, CoO	300	Natural gas, syngas	2010
Nanjing	1 kW–SF	Fe_2O_3 (ore)	10	Coal	2010
IFP-Lyon	10 kW	NiO	?	CH_4	2010
Stuttgart	10 kW	Ilmenite	?	Syngas	2010

SF: solid fuel

In 2004, Lyngfelt and co-workers presented results from a 10 kW prototype unit [150, 151]. Here, an oxygen carrier based on nickel oxide was operated for 100 h with natural gas as fuel. A fuel conversion efficiency of 99.5% was achieved, and no carbon dioxide escaped to the air reactor, hence all carbon dioxide was captured in the process. Moreover, the attrition of particles was found to be low [152]. Later testing in this unit verified long-term operation for 1000 h and very low material loss [72]. Ryu *et al.* also found a high conversion of 98% in a 50 kW combustor operating with methane as fuel; a nickel oxide oxygen carrier was tested for 3.5 h and a cobalt oxide carrier was tested for 25 h [153]. Adanez *et al.* operated a 10 kW CLC unit for 120 h using a copper oxide-based oxygen carrier of two particle sizes [154]. Complete methane conversion was achieved and no deactivation of the particles was observed. Son and Kim presented results with a mixed oxide system of $NiO–Fe_2O_3$/bentonite in a circulating fluidized-bed reactor using methane at a thermal power of about 1 kW [155]. Almost complete conversion to CO_2 and H_2O was achieved. Sixteen different oxygen carriers based on Ni, Mn, and Fe have been used in a 300 W CLC reactor with syngas and/or natural gas (e.g., [27, 89, 122, 156–160]). Operation with a combined manganese oxide and perovskite, $CaMn_{0.875}Ti_{0.125}O_3$, showed both spontaneous release of oxygen when the fuel reactor was fluidized with nitrogen and complete conversion of methane at lower fuel flows [53].

Berguerand and Lyngfelt [62, 63, 82, 98, 137] used bituminous coal and petroleum coke in a 10 kW CLC combustor designed for solid fuels. Incomplete combustion of the gas was noted, mainly caused by the top feeding of fuel through a chute, meaning that the volatiles escaped the system without having contact with the oxygen carrier. Significant testing of nickel and copper materials were performed in a 500 W unit by Adanez and co-workers (e.g., [95–97]). The work included studies of ethane and propane, indicating similar reactivities to methane, and the negative effect of sulfur on nickel materials. The 140 kW unit in Vienna provided operational experience including a wide range of operating conditions, loads of 20–145 kW, temperatures of 780–960 °C, and air ratios of 0.8–1.25 (e.g., [99, 112–114]). The lower air ratios were used for syngas production. The unit showed a fuel conversion of up to 95% for CH_4 and 99% for H_2 and CO.

Shen and co-workers operated with coal and biomass using nickel- and iron-based oxygen carriers in a 10 kW unit (e.g., [126–128]). The unit demonstrated the first use of biomass in CLC, using both iron oxide and a nickel-based oxygen carrier. Alstom Power operated four nickel-based materials with natural gas, reaching conversions of up to 99% in a compact, highly integrated, 15 kW design [161]. Ryu *et al.* designed a new 50 kW chemical looping combustor which was operated with nickel and cobalt oxide particles [162]. With a nickel-based oxygen carrier, 99.2% CO_2 was attained in the exit from the fuel reactor. A 1 kW unit with fuel feeding from below was operated by Wu *et al.* using coal and an iron ore as bed material [163]. The ratio of carbon-containing gases from the fuel reactor was typically 1% CH_4, 4% CO, and 95% CO_2. The data indicate that reasonably high conversion of the gas from the fuel reactor should be possible using bottom bed feeding of coal.

Rifflart *et al.* operated a 10 kW system with three interconnected fluidized beds using nickel-based oxygen carriers and achieved gas conversions to CO_2 of higher than 99% [143]. Bidwe *et al.* operated another 10 kW unit with diluted syngas at low power using ilmenite, reaching 90% conversion [69]. Interestingly, the unit, which has a two-stage fuel reactor, was also operated with coal in semi-batch mode, that is, without circulation. Initially, before the reactivity of ilmenite was lost, the gas conversion was in the order of 90%.

In total, more than 4000 h of operation using a number of different oxygen carriers and fuels have been accomplished. This clearly demonstrates both that the process works and that there are suitable materials for this new combustion technology.

17.8
Reactivity and Solids Inventory

Adanez's group calculated solids inventories based on kinetic data for Ni-, Fe-, and Cu-based oxygen carriers using CH_4, CO, and H_2 as fuel [164, 165]. The minimum solids inventories depended on the fuel gas used, and followed the order $CH_4 > CO > H_2$. The minimum solids inventories ranged from 30 to 1000 kg MW_{th}^{-1} for the three investigated carriers, using CH_4 as fuel [164]. For the more reactive CO and H_2, the range narrowed to 25–90 kg MW_{th}^{-1}. It should be emphasized that these calculated inventories do not consider mass transfer resistances related to the fluidization, as these are highly case dependent. Hence considerably larger solids inventories would be needed to achieve full conversion in reality [135, 166].

17.9
Conclusion

In total, more than 900 oxygen-carrier materials have been investigated in the laboratory. Moreover, more than 30 materials have been tested in actual operation in CLC units totaling 4000 h. There are wide differences between the materials with respect to, for instance, reactivity, price, fluidization properties, oxygen transfer capacity, toxicity, and thermodynamic constraints. Moreover, different applications, such as gaseous fuels, solid fuels, reforming, and gasification, will have different emphases on the important properties. For example, ash-rich solid fuels will put more emphasis on low cost and reactivity with syngas, whereas natural gas will have high emphasis on reactivity with methane. This is also reflected in the main focus on expensive nickel-based materials for gaseous fuels and low-cost ilmenite ore for solid fuels. For these two applications, these two oxygen carriers have been studied most intensively and found to be reliable. However, it is too early to select any final winners, as there are still a number of very interesting materials waiting to be tested.

References

1 Lewis, W.K. and Gilliland, E.R. Production of pure carbon dioxide. US Patent 2 665 972, 1954.

2 Ishida, M. and Jin, H. A novel combustor based on chemical-looping reactions and its reaction kinetics. *J. Chem. Eng. Jpn.*, 1994, **27**(3), 296–301.

3 Ishida, M., Zheng, D., and Akehata, T. Evaluation of a chemical-looping-combustion power-generation system by graphic exergy analysis. *Energy*, 1987, **12**(2), 147–54.

4 Lyngfelt, A., Leckner B., and Mattisson, T. A fluidized-bed combustion process with inherent CO_2 separation; application of chemical-looping combustion. *Chem. Eng. Sci.*, 2001, **56**, 3101–3113.

5 Lyon, R.K. and Cole, J.A. Unmixed combustion: an alternative to fire. *Combust. Flame*, 2000, **121**, 249–261.

6 Copeland, J.R., Alptekin, G., Cesaria, M., and Gershkovich, Y. Sorbent Energy Transfer System (SETS) for CO_2 Separation With High Efficiency, in 27th International Conference on Coal Utilization & Fuel Systems. 2002: Clearwater, USA.

7 Hatanaka, T., Matsuda S., and Hatano, H. A new-concept gas-solid combustion system "merit" for high combustion efficiency and low emissions. In *Proceedings of Intersociety Energy Conversion Engineering Conference*, 1997, Hondulu, Hawaii, American Institute of Chemical Engineers, New York, pp. 944–948.

8 Anthony, E. Solid looping cycles: a new technology for coal conversion. *Ind. Eng. Chem. Res.*, 2008, **47**(6), 1747–1754.

9 Fan, L.S. Chemical looping systems for fossil energy conversions. John Wiley and Sons 2010, Hoboken, New Jersey.

10 Lewis, W., Gilliland, E., and Sweeney, M. Gasification of carbon. *Chem. Eng. Prog.*, 1951, **47**(5), 251–256.

11 Lyon, R.K.and Cole, J.A. Unmixed combustion: an alternative to fire. *Combust. Flame*, 2000, **121**, 249–261.

12 Cao, Y., *et al.* Reduction of solid oxygen carrier (CuO) by solid fuel (coal) in chemical looping combustion. *Prepr. Symp. Am. Chem. Soc. Div. Fuel Chem.*, 2005, **50**(1), 99–102.

13 Gupta, P., *et al.* Chemical looping combustion of coal to produce hydrogen. In *Proceedings of the 30th International Technical Conference on Coal Utilization and Fuel Systems*, Clearwater, FL, Ed. by Sakkestad, B.A., Coal Technology, Association, Gaithenburg, MD, 2005, pp. 349–352.

14 Dennis, J.S., Scott, S.A., and Hayhurst, A.N. *In situ* gasification of coal using steam with chemical looping: a technique for isolating CO_2 from burning a solid fuel. *J. Energy Inst.*, 2006, **79**(3), 187–190.

15 Leion, H., *et al.* The use of ilmenite as an oxygen carrier in chemical-looping combustion. *Chem. Eng. Res. Des.*, 2008, **86**, 1017–1026.

16 Leion, H., Mattisson, T., and Lyngfelt, A. Solid fuels in chemical-looping combustion. *Int. J. Greenhouse Gas Control*, 2008, **2**, 180–193.

17 Leion, H., Mattisson, T., and Lyngfelt, A. CO_2 capture from direct combustion of solid fuels with chemical-looping combustion. Presented at the Clearwater Coal Conference – the 33rd International Technical Conference on Coal Utilization and Fuel Systems, Clearwater, FL, 2008.

18 Berguerand, N. and Lyngfelt, A. Design and operation of a 10 kW_{th} chemical-looping combustor for solid fuels – testing with South African coal. Fuel, 2008, **87**, 2731–2726.

19 Berguerand, N. and Lyngfelt, A. The use of petroleum coke as fuel in a 10 kW_{th} chemical-looping combustor. Int. J. of Greenhouse Gas Control, 2008, **2**, 169–179.

20 Berguerand, N. Design and operation of a 10 kW_{th} chemical-looping combustor for solid fuels. PhD thesis, Chalmers University of Technology, Göteborg, 2009.

21 Mattisson, T., Lyngfelt, A., and Leion, H. Chemical-looping oxygen uncoupling for combustion of solid fuels. *Int. J. Greenhouse Gas Control*, 2009, **3**, 11–19.

22 Mattisson, T., Leion, H., and Lyngfelt, A. Chemical-looping oxygen uncoupling

using CuO/ZrO_2 with petroleum coke. *Fuel*, 2009, **88**, 683–690.

23 Leion, H., Mattisson, T., and Lyngfelt, A. Combustion of a German lignite using chemical-looping with oxygen uncoupling (CLOU). Presented at the Clearwater Coal Conference – the 33rd International Technical Conference on Coal Utilization and Fuel Systems, Clearwater, FL, 2008.

24 Leion, H., Mattisson, T., and Lyngfelt, A. Using chemical-looping with oxygen uncoupling (CLOU) for combustion six different of solid fuels. *Energy Procedia*, 2009, **1**, 447–453.

25 Rydén, M. and Lyngfelt, A. Hydrogen and power production with integrated carbon dioxide capture by chemical-looping reforming. Presented at the 7th International Conference on Greenhouse Gas Control Technologies, Vancouver, 2004.

26 Rydén, M., Lyngfelt, A., and Mattisson, T. Two novel approaches for hydrogen production; chemical-looping reforming and steam reforming with carbon dioxide capture by chemical-looping combustion. Presented at the 16th World Hydrogen Energy Conference, Lyon, June 2006.

27 Rydén, M., Lyngfelt, A., and Mattisson, T. Synthesis gas generation by chemical-looping reforming in a continuously operating laboratory reactor. *Fuel*, 2006, **85**(12–13), 1631–1641.

28 Fathi, M., *et al.* Partial oxidation of methane to synthesis gas: elimination of gas phase oxygen. *Catal. Today*, 2000, **63**, 489–497.

29 Stobbe, E. R., de Boer, B. A., and Geus, J. W. The reduction and oxidation behavior of manganese oxides. *Catal. Today*, 1999, **47**(1–4), 161–167.

30 Gavalas, G., Kim, K., and Pantu, O. Methane partial oxidation on $Pt/CeO_2–ZrO_2$ in absence of gaseous oxygen. *Appl. Catal. A*, 2000, **193**, 203–214.

31 Rydén, M. and Lyngfelt, A. Using steam reforming to produce hydrogen with carbon dioxide capture by chemical-looping combustion. *Int. J. Hydrogen Energy*, 2006, **31**(10), 1271–1283.

32 Beavis, R. Introduction to CACHET. In *Carbon Dioxide Capture for Storage in Deep Geological Formations, Vol. 3, Advances in* CO_2 *Capture and Storage Technology (2004–2009)*. Ed. by Eide, L. I. CPL Press, Newbury, Berkshire, 2009, pp. 67–74.

33 Process for the production of carbon monoxide and hydrogen. British Patent 636 206, 1950.

34 Hossain, M. and de Lasa, H. Chemical-looping combustion (CLC) for inherent CO_2 separations – a review. *Chem. Eng. Sci.*, 2008, **63**, 4433–4451.

35 Lyngfelt, A., Johansson, M., and Mattisson, T. Chemical-looping combustion – status of development. Presented at the 9th International Conference on Circulating Fluidized Beds, Hamburg, 2008.

36 Fan, H., Haibin, L., Zengli, Z. Advancements in development of chemical looping combustion (CLC) – a review. Int. J. Chem. Eng. 2009, Article ID 710515, doi 101155/2009/710515.

37 Lyngfelt, A. Oxygen carriers for chemical-looping combustion – 4000 h of operational experience. *Oil Gas Sci. Technol. Rev. IFP Energies Nouv.*, 2010, accepted for publication, doi 102516/ogst/2010038.

38 Johansson, M. Screening of oxygen-carrier particles based on iron-, manganese-, copper- and nickel oxides for use in chemical looping technologies. PhD thesis, Chalmers University of Technology, Göteborg, 2007.

39 Jerndal, E. Investigation of nickel- and iron-based oxygen carriers for chemical-looping combustion. PhD thesis, Chalmers University of Technology, Göteborg, 2010.

40 Mattisson, T., Lyngfelt, A., and Cho, P. The use of iron oxide as an oxygen carrier in chemical-looping combustion of methane with inherent separation of CO_2. *Fuel*, 2001, **80**, 1953–1962.

41 Mattisson, T. and Lyngfelt, A. Capture of CO_2 using chemical-looping combustion. Paper presented at the First Biemial Meeting of the Scandinavian–Nordic Section of Combustion Institute, Göteborg, April 18–20, 2001, pp. 163–168, www. entek.chalmers.se/~anly/co2/co2publ.htm.

42 de Diego, L. F., *et al.* Development of Cu-based oxygen carriers for chemical-looping combustion. *Fuel*, 2004, **83**(13), 1749–1757.

43 Lee, J.-B., *et al.* Redox characteristics of various kinds of oxygen carriers for hydrogen fueled chemical-looping combustion. *J. Ind. Eng. Chem. (Seoul)*, 2005, **11**(1), 96–102.

44 Adanez, J., *et al.* Selection of oxygen carriers for chemical-looping combustion. *Energy Fuels*, 2004, **18**(2), 371–377.

45 Johansson, M. Selection of oxygen-carriers particles based on iron-, manganese-, copper- and nickel-oxides for use in chemical-looping technologies. PHD Thesis, Department of Chemical and Biological Engineering, Environmental Inorganic Chemistry, Chalmers University of Technology: Göteborg, 2005.

46 Johansson, M., Mattisson, T., and Lyngfelt, A. Comparison of oxygen carriers for chemical-looping combustion. *Thermal Sci.* 2008, **10**, 93–107

47 Johansson, M., Mattisson, T., and Lyngfelt, A. Comparison of oxygen carriers for chemical-looping combustion of methane-rich fuels. Presented at the 19th FBC Conference, Vienna, 21–24 May 2006.

48 Lambert, A., Briault, P., and Comte, E. Spinel mixed oxides as oxygen carriers for chemical looping combustion. *Energy Procedia*, 2010, in press. www.ghgt.info/index.php/Content-GHGT10/conference-proceedings.html.

49 Ksepko, E., Talik, E., and Figa, J. Preparation and characterization of $Sr(Mn_{1-x}Ni_x)O_3$ solid solution in relation to their use in chemical looping oxygen transfer. Presented at the 25th Annual International Pittsburgh Coal Conference, Pittsburgh, PA, 2008.

50 Rydén, M., *et al.* Novel oxygen-carrier materials for chemical-looping combustion and chemical-looping reforming; $La_xSr_{1-x}Fe_yCo_{1-y}O_{3-\delta}$ perovskites and mixed-metal oxides of NiO, Fe_2O_3 and Mn_3O_4. *Int. J, Greenhouse Gas Control*, 2008, **2**, 21–36.

51 Shulman, A., *et al.* Manganese/Iron, manganese/nickel, and manganese/silicon oxides used in chemical-looping with oxygen uncoupling (CLOU) for combustion with methane. *Energy Fuels*, 2009, **24**, 5269–5275.

52 Shulman, A., Cleverstam, E., Mattisson, T., and Lyngfelt, A. Chemical-looping with oxygen uncoupling using Mn/Mg-based oxygen carriers for methane combustion. *Fuel*, 2011, **90**, 941–950

53 Rydén, M., Lyngfelt, A., and Mattisson, T. $CaMn_{0.875}Ti_{0.125}O_3$ as oxygen carrier for chemical-looping combustion with oxygen uncoupling (CLOU) – experiments in continuously operating fluidized bed reactor system. *Int. J. Greenhouse Gas Control*, 2011, **15**, 356–366.

54 Shen, L., *et al.* A mechanistic investigation of a calcium-based oxygen carrier for chemical looping combustion. *Combust. Flame*, 2008, **154**, 489–506.

55 Song, Q., *et al.* Multicycle study on chemical-looping combustion of simulated coal gas with a $CaSO_4$ oxygen carrier in a fluidized bed reactor. *Energy Fuels*, 2008, **22**, 3661–3672.

56 Song, Q., *et al.* Chemical-looping combustion of methane with $CaSO_4$ oxygen carrier in a fixed bed reactor. *Energy Convers. Manage.*, 2008, **49**, 3178–3187.

57 Song, Q., *et al.* Effect of temperature on reduction of $CaSO_4$ oxygen carrier in chemical-looping combustion of simulated coal gas in a fluidized bed reactor. *Ind. Eng. Chem. Res.*, 2008, **47**, 8148–8159.

58 Tian, H., Guo, Q., and Chang, J. Investigation into decomposition behavior of $CaSO_4$ in chemical-looping combustion. *Energy Fuels*, 2008, **22**, 3915–3921.

59 Song, Q., *et al.* Reactivity of a $CaSO_4$-oxygen carrier in chemical-looping combustion of methane in a fixed bed reactor. *Korean J. Chem. Eng.*, 2009, **26**, 592–602.

60 Tian, H. and Guo, Q. Investigation into the behavior of reductive decomposition of calcium sulfate by carbon monoxide in chemical-looping combustion. *Ind. Eng. Chem. Res.*, 2009, **48**, 5624–5632.

61 Xiao, R., *et al.* Reduction kinetics of a $CaSO_4$ based oxygen carrier for chemical-looping combustion. in *Proceedings of the 20th International Conference on Fluidized Bed Combustion*, Ed. by Yue, G., Zang, H., Zhao, C., Luo, Z. Springer Berlin 2009, pp. 519–526.

62 Berguerand, N. and Lyngfelt, A. The use of petroleum coke in a 10 kW_{th} chemical-looping combustor. *Int. J. Greenhouse Gas Control*, 2008, **2**, 169–179.

63 Berguerand, N. and Lyngfelt, A. Chemical-looping combustion of petroleum coke using ilmenite in a 10 kW_{th} unit – high temperature operation. *Energy Fuels*, 2009, **23**, 5257–5268

64 Leion, H., Mattisson, T., and Lyngfelt, A. Use of ores and industrial products as oxygen carriers in chemical looping combustion. *Energy Fuels*, 2009, **23**, 3207–2315.

65 Leion, H., *et al.* Solid fuels in chemical-looping combustion using oxide scale and unprocessed iron ore as oxygen carriers. *Fuel*, 2009, **88**, 1945–1954.

66 Pröll, T., *et al.* Natural minerals as oxygen carriers for chemical looping combustion in a dual circulating fluidized bed system. *Energy Procedia*, 2009, **1**, 27–34.

67 Adanez, J., *et al.* Ilmenite activation during consecutive redox cycles in chemical-looping combustion. *Energy Fuels*, 2010, **24**, 1402–1413.

68 Xiao, R., *et al.* Pressurized chemical-looping combustion of chinese bituminous coal: cyclic performance and characterization of iron ore-based oxygen carrier. *Energy Fuels*, 2010, **24**, 1449–1463.

69 Bidwe, A., Mayer, F., Hawthorne, C., Charitos, A., Schuster, A., and Scheffknecht, G. Use of ilmenite as an oxygen carrier in chemical looping combustion-batch and continuous dual fluidized bed investigation. *Energy Procedia*, 2010, in press.

70 Jerndal, E., *et al.* NiO particles with Ca and Mg based additives produced by spray-drying as oxygen carriers for chemical-looping combustion. *Energy Procedia*, 2009, **1**, 479–486.

71 Jerndal, E., *et al.* Investigation of NiO/$NiAl_2O_4$ oxygen carriers for chemical-looping combustion produced by spray-drying. *Int. J. Greenhouse Gas Control*, 2010, **4**, 23–25.

72 Linderholm, C., Mattisson, T., and Lyngfelt, A. Long-term integrity testing of spray-dried particles in a 10-kW chemical-looping combustor using natural gas as fuel. *Fuel*, 2009, **88**, 2083–2096.

73 Abad, A., *et al.* Mapping of the range of operational conditions for Cu-, Fe-, and Ni-based oxygen carriers in chemical-looping combustion. *Chem. Eng. Sci.*, 2007, **62**(1–2), 533–549.

74 Abad, A., *et al.* Reduction kinetics of Cu-, Ni-, and Fe-based oxygen carriers using syngas ($CO + H_2$) for chemical-looping combustion. *Energy Fuels*, 2007, **21**(4), 1843–1853.

75 Erri, P. and Varma, A. Solution combustion synthesized oxygen carriers for chemical looping combustion. *Chem. Eng. Sci.*, 2007, **62**(18–20), 5682–5687.

76 Erri, P. and Varma, A. Spinel-supported oxygen carriers for inherent CO_2 separation during power generation. *Ind. Eng. Chem. Res.*, 2007, **46**, 8597–8601.

77 He, F., Wang, H., and Dai, Y. Application of Fe_2O_3/Al_2O_3 composite particles as oxygen carrier of chemical looping combustion. *J. Nat. Gas Chem.*, 2007, **16**(2), 155–161.

78 Hossain, M.M., Sedor, K.E., and de Lasa, H.I. Co–Ni/Al_2O_3 oxygen carrier for fluidized bed chemical-looping combustion: desorption kinetics and metal–support interaction. *Chem. Eng. Sci.*, 2007, **62**(18–20), 5464–5472.

79 Hossain, M.M. and de Lasa, H.I. Reactivity and stability of Co–Ni/Al_2O_3 oxygen carrier in multicycle CLC. *AIChE J.*, 2007, **53**(7), 1817–1829.

80 Yang, J., Cai, N., and Li, Z. Reduction of iron oxide as an oxygen carrier by coal pyrolysis and steam char gasification intermediate products. *Energy Fuels*, 2007, **21**, 3360–3368.

81 Zafar, Q., *et al.* Reduction and oxidation kinetics of Mn_3O_4/Mg–ZrO_2 oxygen carrier particles for chemical-looping combustion. *Chem. Eng. Sci.*, 2007, **62**(23), 6556–6567.

82 Berguerand, N. and Lyngfelt, A. Design and operation of a 10 kW chemical-looping combustor for solid fuels – testing with South African coal. *Fuel*, 2008, **87**, 2713–2726.

83 Chuang, S., *et al.* Development and performance of Cu-based oxygen carriers for chemical-looping combustion. *Combust. Flame*, 2008, **154**, 109–121.

84 Gao, Z., *et al.* Chemical-looping combustion with Ni-based oxygen carrier. *Ind. Eng. Chem. Res.*, 2008, **47**, 9279–9287.

85 Gayan, P., *et al.* Effect of support on reactivity and selectivity of Ni-based oxygen carriers for chemical-looping combustion. *Fuel*, 2008, 2641–2650.

86 Johansson, M., *et al.* Using continuous and pulse experiments to compare two promising nickel-based oxygen-carriers for use in chemical-looping technologies. *Fuel*, 2008, **87**, 988–1001.

87 Linderholm, C., *et al.* 160 hours of chemical-looping combustion in a 10 kW reactor system with a NiO-based oxygen carrier. *Int. J. Greenhouse Gas Control*, 2008, **2**, 520–530.

88 Mattisson, T., *et al.* The reaction of $NiO/NiAl_2O_4$ particles with alternating methane and oxygen. *Can. J. Chem. Eng.*, 2008, **86**, 756–767.

89 Rydén, M., Lyngfelt, A., and Mattisson, T. Chemical-looping combustion and chemical-looping reforming in a circulating fluidized-bed reactor using Ni-based oxygen carriers. *Energy Fuels*, 2008, **22**, 2585–2597.

90 Ryu, H., *et al.* Natural gas combustion characteristics in a chemical-looping combustor with three different oxygen carrier particles. *J. Chem. Eng. Jpn.*, 2008, **41**, 716–720.

91 Sedor, K. E., Hossain, M. M., and de Lasa, H. I. Reduction kinetics of a fluidizable nickel-alumina oxygen carrier for chemical-looping combustion. *Can. J. Chem. Eng.*, 2008, **86**, 323–333.

92 Sedor, K. E., Hossain, M. M., and de Lasa, H. I. Reactivity and stability of Ni/Al_2O_3 oxygen carrier for chemical-looping combustion (CLC). *Chem. Eng. Sci.*, 2008, **63**, 2994–3007.

93 Tian, H., *et al.* Chemical-looping combustion of coal-derived synthesis gas over copper oxide oxygen carrier. *Energy Fuels*, 2008, **22**, 3744–3755.

94 Zhao, H., *et al.* $NiO/NiAl_2O_4$ oxygen carriers prepared by sol–gel for chemical-looping combustion fueled by gas. *J. Fuel Chem. Technol.*, 2008, **36**, 261–266.

95 Adanez, J., *et al.* Methane combustion in a 500 W_{th} chemical-looping combustion system using an impregnated Ni-based oxygen carrier. *Energy Fuels*, 2009, **23**, 130–142.

96 Adanez, J., *et al.* Effect of fuel gas composition in chemical-looping combustion with Ni-based oxygen carriers. 2. Fate of light hydrocarbons. *Ind. Eng. Chem. Res.*, 2009, **48**, 2509–2518.

97 Adanez, J., *et al.* Effect of gas impurities on the behavior of Ni-based oxygen carriers on chemical-looping combustion. *Energy Procedia*, 2009, **1**, 11–18.

98 Berguerand, N. and Lyngfelt, A. Operation in a 10 kW_{th} chemical-looping combustor for solid fuel – testing with a Mexican petroleum coke. *Energy Procedia*, **1**, 407–414.

99 Bolhar-Nordenkampf, J., Pröll, T., and Kolbitsch, P. Performance of a NiO-based oxygen carrier for chemical looping combustion and reforming in a 120 kW unit. *Energy Procedia*, 2009, **1**, 19–25.

100 Chandel, M., Hoteit, A., and Delebarre, A. Experimental investigation of some metal oxides for chemical looping combustion in a fluidized bed reactor. *Fuel*, 2009, **88**, 898–908.

101 Chuang, S., Dennis, J., and Hayhurst, A. Kinetics of oxidation of a reduced form of the Cu-based oxygen-carrier for use in chemical-looping combustion. In *Proceedings of the 20th International Conference on Fluidized Bed Combustion*, 2009. Ed. by Yue, G., Zang, H., Zhao, C., Luo, Z., Springer, Berlin, pp. 512–518.

102 Dahl, I., *et al.* On the development of novel reactor concepts for chemical looping combustion. *Energy Procedia*, 2009, **1**, 1513–1519.

103 Dong, C., *et al.* The use of iron oxide as an oxygen carrier in chemical looping combustion taking CO and biomass gas as fuels. Presented at the 1st

International Conference on Sustainable Power Generation and Supply, SUPERGEN '09, 2009.

104 Dueso, C., *et al.* Syngas combustion in a chemical-looping combustion system using an impregnated Ni-based oxygen carrier. *Fuel*, 2009, **88**, 2357–2364.

105 Forero, C. R., *et al.* Syngas combustion in a 500 W_{th} chemical-looping combustion system using an impregnated Cu-based oxygen carrier. *Fuel Process. Technol.*, 2009, **90**, 1471–1479.

106 Garcia-Labiano, F., *et al.* Effect of fuel gas composition in chemical-looping combustion with Ni-based oxygen carriers. 1. Fate of sulfur. *Ind. Eng. Chem. Res.*, 2009, **48**, 2499–2508.

107 Gayan, P., *et al.* NiO/Al_2O_3 oxygen carriers for chemical-looping combustion prepared by impregnation and deposition–precipitation methods. *Fuel*, 2009, **88**, 1016–1023.

108 Hossain, M. M., *et al.* Nickel on lanthanum-modified gamma-Al_2O_3 oxygen carrier for CLC: reactivity and stability. *Catal. Today*, 2009, **143**, 179–186.

109 Hoteit, A., Chandel, M., and Delebarre, A. Nickel- and copper-based oxygen carriers for chemical looping combustion. *Chem. Eng. Technol.*, 2009, **32**, 443–449.

110 Erri, P. and Varma, A. Diffusional effects in nickel oxide reduction kinetics. *Ind. Eng. Chem. Res.*, 2009, **48**, 4–6.

111 Hoteit, A., *et al.* Biogas combustion in a chemical looping fluidized bed reactor. *Int. J. Greenhouse Gas Control*, 2009, **3**, 561–567.

112 Kolbitsch, P., *et al.* Operating experience with chemical looping combustion in a 120 kW dual circulating fluidized bed (DCFB) unit. *Energy Procedia*, 2009, **1**, 1465–1472.

113 Kolbitsch, P., *et al.* Characterization of chemical looping pilot plant performance via experimental determination of solids conversion. *Energy Fuels*, 2009, **23**, 1450–1455.

114 Kolbitsch, P., *et al.* Comparison of two Ni-based oxygen carriers for chemical looping combustion of natural gas in 140 kW continuous looping operation. *Ind. Eng. Chem. Res.*, 2009, **48**, 5542–5547.

115 Kuusik, R., *et al.* High temperature behavior of NiO-based oxygen carriers for chemical looping combustion. *Energy Procedia*, 2009, **1**, 3885–3892.

116 Lambert, A., *et al.* Synthesis and characterization of bimetallic Fe/Mn oxides for chemical looping combustion. *Energy Procedia*, 2009, **1**, 375–381.

117 Leion, H., Lyngfelt, A., and Mattisson, T. Solid fuels in chemical-looping combustion using a NiO-based oxygen carrier. *Chem. Eng. Res. Des.*, 2009, **87**, 1543–1550.

118 Leion, H., Lyngfelt, A., and Mattisson, T. Effects of steam and CO_2 in the fluidizing gas when using bituminous coal in chemical-looping combustion. In *Proceedings of the 20th International Conference on Fluidized Bed Combustion*, 2009. Ed. by Yue, G., Zang, H., Zhao, C., Luo, Z. Springer, Berlin, pp. 608–611.

119 Leion, H., *et al.* Use of $CaMn_{0.875}Ti_{0.125}O_3$ as oxygen carrier in chemical-looping with oxygen uncoupling. *Energy Fuels*, 2009, **23**, 5276–5283.

120 Muller, C., *et al.* Experimental investigation of two modified chemical-looping combustion cycles using syngas from cylinders and the gasification of solid fuels. In *Proceedings of the 20th International Conference on Fluidized Bed Combustion*, 2009. Ed. by Yue, G., Zang, H., Zhao, C., Luo, Z. Springer, Berlin, pp. 590–595.

121 Pröll, T., *et al.* A novel dual circulating fluidized bed system for chemical looping processes. *AIChE J.*, 2009, **55**, 3255–3266..

122 Rydén, M., *et al.* NiO supported on Mg–ZrO_2 as oxygen carrier for chemical-looping combustion and chemical-looping reforming. *Energy Environ. Sci.*, 2009, **2**, 970–981.

123 Rydén, M., *et al.* Waste products from the steel industry with NiO as additive as oxygen carrier for chemical-looping combustion. *Int. J. Greenhouse Gas Control*, 2009, **3**, 693–703.

124 Rubel, A., *et al.* Oxygen carriers for chemical looping combustion of solid fuels. *Fuel*, 2009, **88**, 876–884.

125 Ryu, H., *et al.* Syngas combustion characteristics of four oxygen carrier particles for chemical-looping combustion in a batch fludized bed reactor. *Korean J. Chem. Eng.*, 2009, **26**, 523–527.

126 Shen, L., *et al.* Experiments on chemical looping combustion of coal with a NiO based oxygen carrier. *Combust. Flame*, 2009, **156**, 721–728.

127 Shen, L., *et al.* Chemical-looping combustion of biomass in a 10 kW reactor with iron oxide as an oxygen carrier. *Energy Fuels*, 2009, **23**, 2498–2505.

128 Shen, L., *et al.* Reactivity deterioration of NiO/Al_2O_3 oxygen carrier for chemical-looping combustion of coal in a 10 kW reactor. *Combust. Flame*, 2009, **156**, 1377–1385.

129 Solunke, R. and Veser, G. Nanocomposite oxygen carriers for chemical-looping combustion of sulfur-contaminated synthesis gas. *Energy Fuels*, 2009, **23**, 4787–4796.

130 Shulman, A., *et al.* High reactivity and mechanical durability of $NiO/NiAl_2O_4$ and $NiO/NiAl_2O_4/MgAl_2O_4$ oxygen carrier particles used for more than 1000 h in a 10 kW CLC reactor. *Ind. Eng. Chem. Res.*, 2009, **48**, 7400–7405.

131 Siriwardane, R., *et al.* Chemical-looping combustion of coal with metal oxide oxygen carriers. *Energy Fuels*, 2009, **23**, 3885–3892.

132 Tian, H., *et al.* Effect of hydrogen sulfide on chemical looping combustion of coal-derived synthesis gas over bentonite-supported metal-oxide oxygen carrier. *Ind. Eng. Chem. Res.*, 2009, **48**, 8418–8430.

133 Xiang, W., *et al.* Investigation of coal fueled chemical looping combustion using Fe_3O_4 as oxygen carrier. In *Proceedings of the 20th International Conference on Fluidized Bed Combustion*, 2009, pp. 527–532.

134 Zhang, J., *et al.* The interaction between HCl and Fe_2O_3 during the chemical looping combustion of MSW. Presented at the 1st International Conference on Sustainable Power Generation and Supply, SUPERGEN '09, 2009.

135 Abad, A., *et al.* Modeling of the chemical-looping combustion of methane using a Cu-based oxygen carrier. *Combust. Flame*, 2010, **157**, 602–615.

136 Dueso, C., *et al.* Reactivity of a NiO/Al_2O_3 oxygen carrier prepared by impregnation for chemical-looping combustion. *Fuel*, 2010, **89**, 3399–3409.

137 Berguerand, N. and Lyngfelt, A. Batch testing of solid fuels with ilmenite in a 10 kW_{th} chemical-looping combustor. *Fuel*, 2010, **89**, 1749–1762.

138 Gayán, P., *et al.* Effect of gas composition in chemical-looping combustion with copper-based oxygen carriers: fate of light hydrocarbons. *Int. J. Greenhouse Gas Control*, 2010, **4**, 13–22.

139 Hossain, M.M. and de Lasa, H.I. Reduction and oxidation kinetics of $Co–Ni/Al_2O_3$ oxygen carrier involved in a chemical-looping combustion cycles. *Chem. Eng. Sci.*, 2010, **65**, 98–106.

140 Iliuta, I., *et al.* Chemical-looping combustion process: kinetics and mathematical modeling. *AIChE J.*, 2010, **56**, 1063–1079.

141 Noorman, S., van Sint Annaland, M., and Kuipers, J.A.M. Experimental validation of packed bed chemical-looping combustion. *Chem. Eng. Sci.*, 2010, **65**, 92–97.

142 Kolbitsch, P., *et al.* Operating experience with chemical looping combustion in a 120 kW dual circulating fluidized bed (DCFB). *Int. J. Greenhouse Gas Control*, 2010, **4**, 180–185.

143 Rifflart, S., Hoteit, A., Yazdanpanah, M., Pelletant, W., and Surla, K. Construction and operation of a 10 kW CLC unit with circulation configuration enabling independent solid flow control. *Energy Procedia*, 2010, in press.

144 Johansson, E., *et al.* A circulating fluidized bed combustor system with inherent CO_2 separation – application of chemical looping combustion. Presented at the 7th International Conference on Circulating Fluidized Beds, Niagara Falls, Canada, 2002.

145 Johansson, E., *et al.* Gas leakage measurements in a cold model of an

interconnected fluidized bed for chemical-looping combustion. *Powder Technol.*, 2003, **134**, 210–217.

146 Kronberger, B., *et al.* A two-compartment fluidized bed reactor for CO_2-capture by chemical-looping combustion. *Chem. Eng. Technol.*, 2004, **27**, 1318–1326.

147 Kronberger, B., *et al.* Design and hydrodynamic testing of a 10-kW prototype for continuous chemical-looping combustion. *Ind. Eng. Chem. Res.*, 2005, **44**, 546–556.

148 Xu, M.E., N., Jim Lim, C., and Ryu, H.-J. Mapping of the operating conditions for an interconnected fluidized bed reactor for CO_2 separation by chemical looping combustion. *Chem. Eng. Technol.*, 2009, **32**(3), 404–409.

149 Bischia, A.L., ø., Saanum, I., Bakken, J., Seljeskog, M., Bysveen, M., Morin, J.-X., and Bolland, O. Design study of a 150 kW_{th} double loop circulating fluidized bed reactor system for chemical looping combustion with focus on industrial applicability and pressurization. *Int. J. Greenhouse Gas Control*, 2010, in press.

150 Lyngfelt, A., *et al.* The GRACE project. Development of oxygen carrier particles for chemical-looping combustion. Design and operation of a 10 kW chemical-looping combustor. Presented at the 7th International Conference on Greenhouse Gas Control Technologies Vancouver, 2004.

151 Lyngfelt, A. and Thunman, H. Construction and 100 h of operational experience of a 10-kW chemical-looping combustor. In *Carbon Dioxide Capture for Storage in Deep Geologic Formations – Results from the CO_2 Capture Project*, 2005, Vol. 1, pp. 625–645.

152 Johansson, M., Mattisson, T., and Lyngfelt, A. Use of $NiO/NiAl_2O_4$ particles in a 10 kW chemical-looping combustor. *Ind. Eng. Chem. Res.*, 2006, **45**(17), 5911–5919.

153 Ryu, H.-J., Jin, G.-T., and Yi, C.-K. Demonstration of inherent CO_2 separation and no NO_x emission in a 50 kW chemical-looping combustor: continuous reduction and oxidation experiment. Presented at the 7th International Conference on Greenhouse Gas Control Technologies Vancouver, 2004.

154 Adanez, J., *et al.* Chemical looping combustion in a 10 kW_{th} prototype using a CuO/Al2O3 oxygen carrier: effect of operating conditions on methane combustion. *Ind. Eng. Chem. Res.*, 2006, **45**(17), 6075–6080.

155 Son, S.R. and Kim, S.D. Chemical-looping combustion with NiO and Fe_2O_3 in a thermobalance and circulating fluidized bed reactor with double loops. *Ind. Eng. Chem. Res.*, 2006, **45**(8), 2689–2696.

156 Rydén, M., Lyngfelt, A., and Mattisson, T. $CaMn_{0.875}Ti_{0.125}O_3$ as oxygen carrier for chemical-looping combustion with oxygen uncoupling (CLOU) – experiments in continuously operating fluidized bed reactor system. *Int. J. Greenhouse Gas Control*, 2010, in press.

157 Johansson, E., *et al.* A 300 W laboratory reactor system for chemical-looping combustion with particle circulation. *Fuel*, 2006, **85**(10–11), 1428–1438.

158 Abad, A., *et al.* Chemical-looping combustion in a 300 W continuously operating reactor system using a manganese-based oxygen carrier. *Fuel*, 2006, **85**(9), 1174–1185.

159 Abad, A., *et al.* The use of iron oxide as oxygen carrier in a chemical-looping reactor. *Fuel*, 2007, **86**(7–8), 1021–1035.

160 Johansson, E., *et al.* Combustion of syngas and natural gas in a 300 W chemical-looping combustor. *Chem. Eng. Res. Des.*, 2006, **84**(A9), 819–827.

161 Mattisson, T., Adanez, J., Proell, T., Kuusik, R., Beal, C., Assink, J., Snijkers, F., and Lyngfelt, A. Chemical-looping combustion CO_2 ready gas power. *Energy Procedia*, 2009, **1**, 1557–1564.

162 Ryu, H.-J., Jo, S-H., Park, Y.C., Bae, D-H., and Kim, S. Long-term operation experience in a 50 kW_{th} chemical looping combustor using natural gas and syngas as fuels, Presented at the 1st International Conference on Chemical Looping, Lyon, 2010.

163 Wu, J., Shen, L., Hao, J., and Gu, H. Chemical looping combustion of coal in a 1 kW_{th} reactor. Presented at the

1st International Conference on Chemical Looping, Lyon, 2010.
164 Abad, A., *et al.* Mapping of the range of operational conditions for Cu-, Fe-, and Ni-based oxygen carriers in chemical-looping combustion. Chem. Eng. Sci., 2006, in press.
165 Garcia-Labiano, F., *et al.* Reduction and oxidation kinetics of a copper-based oxygen carrier prepared by impregnation for chemical-looping combustion. *Ind. Eng. Chem. Res.*, 2004, **43**(26), 8168–8177.
166 Mattisson, T., Jerndal, E., Linderholm, C., and Lyngfelt, A. Reactivity of a spray-dried $NiO/NiAl_2O_4$ oxygen carrier for chemical-looping combustion. 2010, submitted for publication.

18 Chemical Looping in Power Plants

Bernd Epple and Jochen Ströhle

Abstract

Chemical and carbonate looping are two efficient and environmentally sound processes for CO_2 capture from power plants. The chemical looping process is a promising option for new power plants and retrofit for existing circulating fluidized bed (CFB)-based power plants. It implies the potentially highest electrical efficiency that is around 53% for natural gas firing integrated into a combined cycle power plant, and up to 42% for coal firing. The carbonate looping process is particularly suited for the retrofit of existing plants. The electrical efficiency of a coal-fired power plant integrating post-combustion CO_2 capture by carbonate looping with an oxyfuel calciner lies in the range 39–40%. The process efficiency can even be improved by integrating an indirectly heated calciner. The efficiency of both chemical and carbonate looping processes is significantly higher than for oxyfuel combustion or monoethanolamine scrubbing, but significant research work is needed to reach commercial application.

Keywords: carbon dioxide capture, chemical looping, carbonate looping, power plant

18.1 Introduction

Fossil fuels will remain the dominant energy source for power generation worldwide for the next few decades. However, fossil fuel-fired power plants are the largest emitters of the greenhouse gas CO_2, hence solutions for the efficient and environmentally friendly use of fossil fuels are needed. Capture and storage of CO_2 from power plants are one option for a significant reduction of CO_2 emissions in the field of power generation. In recent years, three processes for CO_2 capture have been developed and brought close to large-scale demonstration, namely pre-combustion CO_2 capture based on the IGCC process, oxyfuel combustion, and amine-based post-combustion CO_2 capture. However, all these processes have the disadvantage of enormous energetic efficiency losses that

Process Engineering for CCS Power Plants, First Edition. Edited by Detlef Stolten and V. Scherer

have been estimated as 8–14% [1–7], significantly increasing fuel consumption and leading to a higher cost of power generation.

As an alternative, chemical and carbonate looping processes promise significantly lower efficiency losses and costs than the processes mentioned above. However, they are still in the early stages of development. Chemical looping represents an oxyfuel process, where the oxygen for combustion is supplied by a metal oxide, thus avoiding the use of an energy-intensive air separation unit. The carbonate looping process involves post-combustion absorption of CO_2 by lime (CaO). Both processes are characterized by a cyclic exchange of solids between two fluidized bed reactors. Fluidized bed systems have the advantage of very good mixing and a high solids surface area, leading to high heat and mass transfer rates at an almost uniform temperature in the bed.

This chapter presents a review of the state of research regarding the application of chemical and carbonate looping processes to fossil fuel-fired power plants.

18.2 Chemical Looping Combustion

Chemical looping is a relatively new carbon capture and storage (CCS) technology for the conversion of fossil fuels. The fuel conversion process is separated into two steps by means of a solid catalyst, the so-called oxygen carrier, in form of a metal oxide, Me_xO_y. In the air reactor, the oxygen carrier is oxidized by air. The oxygen carrier is then transferred to the fuel reactor supplying the oxygen for the conversion of the fuel. The reduced oxygen carrier is subsequently fed back to the air reactor, closing the cycle. A schematic of the process is shown in Fig. 18.1.

The global reactions in the two reactors for the conversion of a general hydrocarbon fuel, C_nH_{2m}, are given in Equations 18.1 and 18.2.

$$\text{Fuel reactor: } (2n+m)M_yO_x + C_nH_{2m} \rightarrow (2n+m)M_yO_{x-1} + mH_2O + nCO_2 \quad (18.1)$$

$$\text{Air reactor: } (2n+m)M_yO_{x-1} + (n+\tfrac{1}{2}m)O_2 \rightarrow (2n+m)M_yO_x \quad (18.2)$$

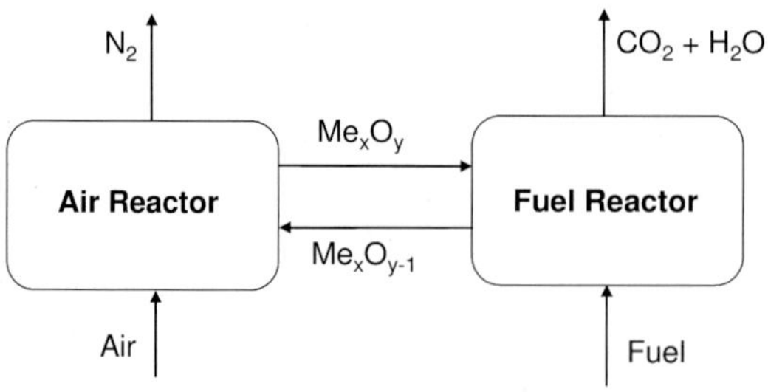

Fig. 18.1 Schematic illustration of chemical looping combustion.

$$\text{Net reaction: } C_nH_{2m} + (n + \tfrac{1}{2}m)O_2 \rightarrow mH_2O + nCO_2 \qquad (18.3)$$

Equation 18.2 is an exothermic reaction, whereas Equation 18.1 can be either exothermic or endothermic. In summation, the same energy is released as in conventional fuel conversion according to Equation 18.3. However, the fuel is never in contact with the N_2 in the air, so that a stream rich in CO_2 and H_2O leaves the fuel reactor. After drying, this stream can be processed for subsequent storage of CO_2.

Chemical looping can be used for different purposes, for example, H_2 production or power generation. The application to power generation implies complete fuel conversion, hence the process is named chemical looping combustion (CLC). Various fuels can be converted in chemical looping processes. Recent developments in the application of gaseous and solids fuels are summarized below.

18.2.1 Gaseous Fuels

A typical reactor system for CLC of gaseous fuel is illustrated in Fig. 18.2 [8]. It consists of a riser as air reactor and a bubbling bed as fuel reactor. The oxygen carrier particles entrained with the air flow in the air reactor are separated from the gas flow by a cyclone and fall by gravity via a loop seal into the fuel reactor.

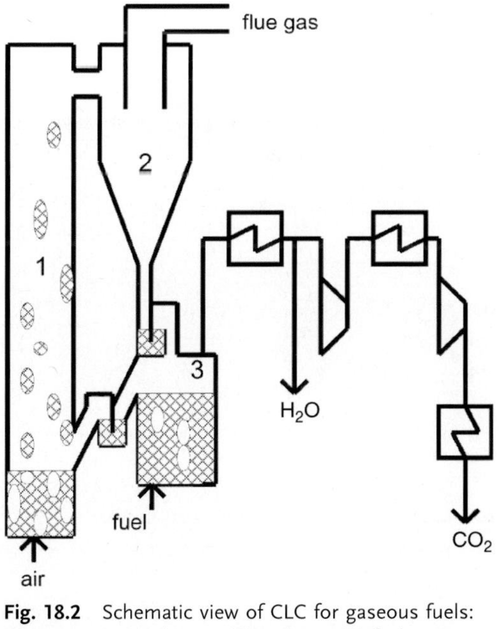

Fig. 18.2 Schematic view of CLC for gaseous fuels: (1) air reactor and riser; (2) cyclone; (3) fuel reactor. Taken from Johansson *et al.* [8].

The flow of solids back to air reactor is realized by an overflow weir with subsequent loop seal.

CLC of gaseous fuels has been tested on the bench and laboratory scale ranging from 300 W [9–11] to 10 kW [8, 12] and up to 120 kW [13]. The 120 kW test rig uses a circulating fluidized bed as fuel reactor. Various oxygen carrier materials have been investigated in numerous studies [8–14]. The oxygen carrier has to meet certain criteria regarding thermodynamic properties, namely the temperature range of oxidation/reduction, high O_2 carrying capacity, high reactivity towards O_2 and fuel, high mechanical stability, low tendency towards agglomeration, and also environmental constraints such as toxicity. Typical oxygen carrier particles consist of a highly reactive Ni-, Fe-, Mn-, or Cu-based metal oxide layer on a mechanically and thermally stable support material such as. Al_2O_3. A detailed review of oxygen carrier materials is available elsewhere (see Chapter 17).

A CLC plant for gaseous fuels can principally be operated in either the atmospheric or pressurized mode. A pressurized system offers the possibility of using a highly efficient combined cycle for power generation. Compressed air and natural gas are then fed to the reactor system, and the hot flue gases from the air reactor are expanded in a gas turbine. The heat in the off-gas from the gas turbine and in the CO_2-rich stream from the fuel reactor can be used for steam production in a heat recovery steam generator. A flow diagram of such a combined cycle is shown in Fig. 18.3 [15].

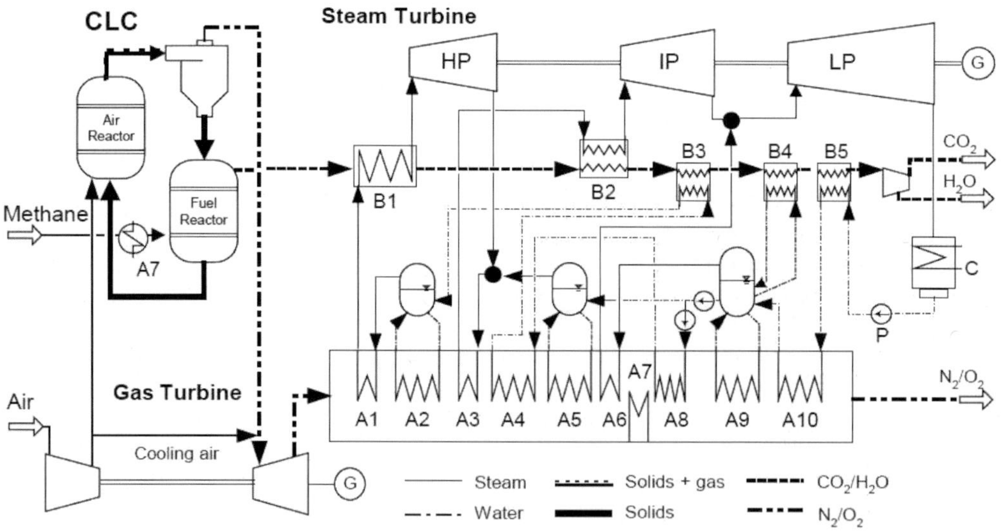

Fig. 18.3 Combined cycle with CLC. A1, A3, A6 = superheater; A2, A5, A9 = evaporator; A4, A7, A8, A10 = preheater; B1, B2 = superheater; B = economizer; B4 = supplemental evaporator; B5 = preheater; P = pump; C = condenser [15].

The efficiency of the process depends strongly on the gas turbine inlet temperature, which is equal to the temperature in the air reactor. Wolf *et al.* [15] calculated a net electrical efficiency of 53% at 1200 °C and 48–49% at 1000 °C for an 800 MW_{th} CLC plant using Ni- and Fe-based oxygen carriers, respectively. If the air reactor has to be operated at 900 °C due to material constraints, top-firing by natural gas will be an option to increase the turbine inlet temperature and thus permit acceptable efficiency. The efficiency penalty compared with a conventional combined cycle results mainly from CO_2 compression and is in the range 2–3%. The principle layout of an 800 MW_{th} CLC reactor is illustrated in Fig. 18.4 [15].

Naqvi *et al.* [16] calculated a net efficiency of 52.2% of a 364 MW_{el} CLC combined cycle at full load, and reducing the load to 60% resulted in a decrease in efficiency of 2.6%. Zhang *et al.* [17] proposed a methanol-based CLC system with a reduction reaction of iron oxides, in which the thermal energy for the endothermic reaction is supplied by the intercooling heat of the compressor of the gas turbine, and an oxidation reaction of iron oxides, where the compressed air is heated by the iron oxides. The thermal efficiency of this novel cycle is expected to be 56.8% with 90% CO_2 recovery. The economics of the CLC process for a 450 MW_{el} coal-fired power plant have been analyzed within the EU-ENCAP (2004–2009) project, resulting in CO_2 mitigation cost of less than € 10 t^{-1} CO_2.

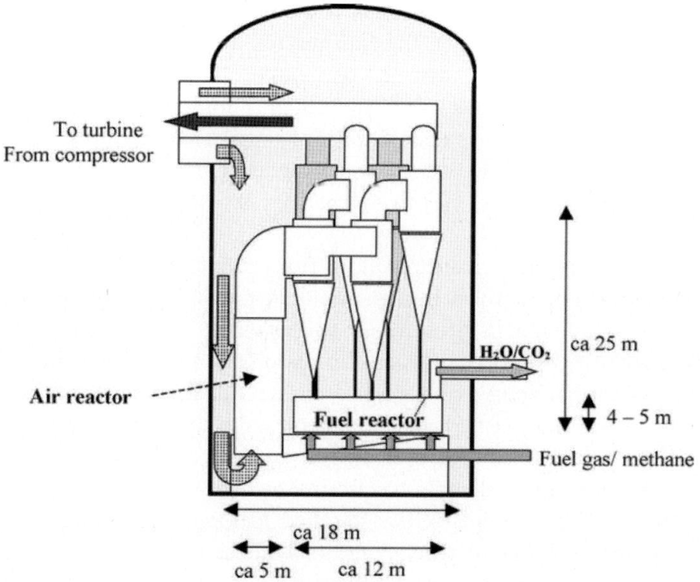

Fig. 18.4 Example of a ~800 MW CLC reactor system. Taken from Wolf *et al.* [15].

18.2.2
Solid Fuels

The application of solid fuels to CLC is not straightforward, since the contact efficiency of a pure solid–solid reaction is too slow to reach acceptable conversion rates. In principle, three options have been proposed to face this challenge: (1) the coal is gasified in a pre-reactor, and the syngas is converted in the fuel reactor; (2) coal gasification and metal oxide reduction occur simultaneously in a single reactor; (3) molecular oxygen is released in the fuel reactor and burns the coal. These three options are discussed below.

18.2.2.1 Gasification in a Pre-reactor

The CLC reactor system for syngas will look similar to that shown in Fig. 18.2, where syngas from coal gasification is used as a fuel. Several Fe-, Mn-, and Cu-based oxygen carriers have been successfully tested for conversion of syngas in various laboratory-scale test rigs [18]. Coal gasification takes place in a separate reactor. However, energy needs to be supplied to maintain the endothermic gasification reaction, for example, by partial oxidation of coal. The use of air as oxidizer would dilute the CO_2 stream with N_2 and is therefore not possible, so that pure oxygen is needed, leading to additional energy losses and to increased complexity of the plant.

18.2.2.2 Gasification and Metal Oxide Reduction in a Single Reactor

The coal is devolatilized and gasified in the fuel reactor, and the gaseous products, mainly CO, H_2, and CH_4, react with the oxygen carrier, forming H_2O and CO_2. These two gases again serve as gasification agents for unconverted char. The net reaction in the fuel reactor is endothermic. The heat required to sustain this reaction is delivered by the flow of hot oxygen carrier material coming from the air reactor. Hence the air reactor has to be operated at a significantly higher temperature than the fuel reactor.

In theory, there is no energy penalty resulting from this CLC process, apart from the supply of steam or recycled flue gas as fluidization agent in the fuel reactor. However, the simultaneous presence of two different solids in the reactor system implies some technical challenges. On the one hand, the coal should remain in the fuel reactor as long as possible to ensure full conversion of fuel, and the residence time is limited by the size and pressure drop in the reactor. On the other hand, a continuous stream of oxygen carrier must be transferred to the air reactor. The separation of coal/char from the solids stream leaving the fuel reactor occurs in a separate device, the so-called "carbon stripper" [19, 20]. Leakage of unburnt coal into the air reactor inevitably leads to the formation of CO_2 that is emitted to the environment. Furthermore, the ash resulting from coal combustion must be separated from the oxygen carrier and extracted from the process in an adequate manner in order to prevent the accumulation of ash in the cycling solids.

Furthermore, the gasification products formed close to the exit of the fuel reactor do not have the possibility of reacting completely with the oxygen carrier due to the low solids loading in the lean region, so that they will leave the reactor partly unburnt. For the complete conversion of these unburnt materials, pure oxygen may be injected into a separate device, the so-called "oxygen polishing unit," leading to additional energy losses due to the supply of pure oxygen.

In recent years, several laboratory tests have been performed in the frame of research projects. The natural ore ilmenite was found to be particularly suited for the application of coal since it is available at low cost, and it has high mechanical strength and high reactivity towards the oxidation of CO and H_2. However, the reactivity towards CH_4 is relatively low. The reaction rate increases exponentially with temperature and linearly with the water steam content of the fluidizing gas [21]. Furthermore, oxide scale and unprocessed iron ore may also be used as low-cost oxygen carriers [22].

The conversion of South African coal has already been successfully tested in a 10 kW_{th} test facility [19]. This facility consists of an air reactor, a riser, two cyclones, and a fuel reactor, as illustrated in Fig. 18.5. The fuel reactor is designed

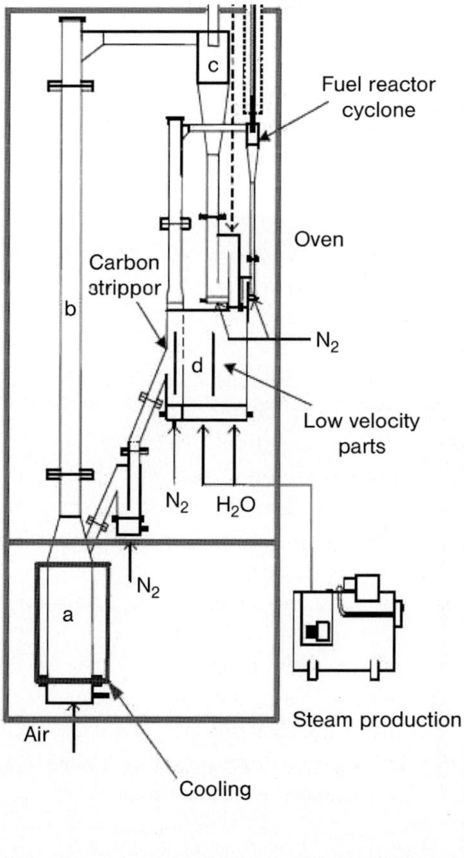

Fig. 18.5 10 kW_{th} CLC reactor system for solid fuels. Taken from Berguerand and Lyngfelt [19].

as a bubbling bed with barriers, allowing a relatively long residence time of fuel in the reactor.

A concept based on a moving bed combined with a circulating fluidized bed (CFB) has been proposed [23]. The status of CLC describing further alternative concepts was published in 2008 [24].

The key component of CLC is reactors based on CFB technology. Especially for the scale-up to full-scale power plants, CFB technology will be the preferred option. Therefore, a test plant based on CFB technology with a capacity of 1 MW_{th} has recently been erected at the Technische Universität (TU) Darmstadt in order to perform experimental investigations regarding the technical implementation of the CLC of coal within the framework of an EU-funded project [25]. The plant is designed as a system of two interconnected CFB reactors. The air reactor has an inner diameter of 0.6 m and is 8 m high, whereas the fuel reactor has an inner diameter of 0.4 m and is 11 m high. All reactor parts are refractory lined to minimize heat losses, enabling autothermal operation. A CAD view of the reactor system is shown in Fig. 18.6. Basic simulations of the 1 MW_{th} reactor system suggest that a CO_2 capture efficiency of 90% can be reached when the carbon stripper efficiency exceeds ~85% [20].

Fillman *et al.* [26] evaluated the effect of various parameters on the efficiency of a 455 MW_{el} CLC plant for coal. A schematic diagram of a CLC reactor system

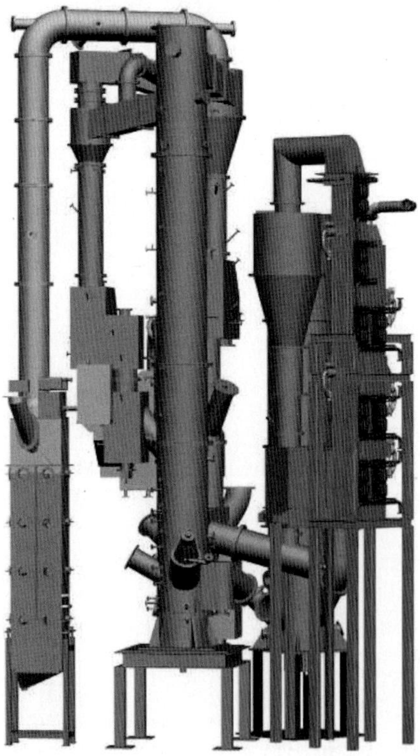

Fig. 18.6 CAD view of the 1 MW_{th} CLC test plant at TU Darmstadt.

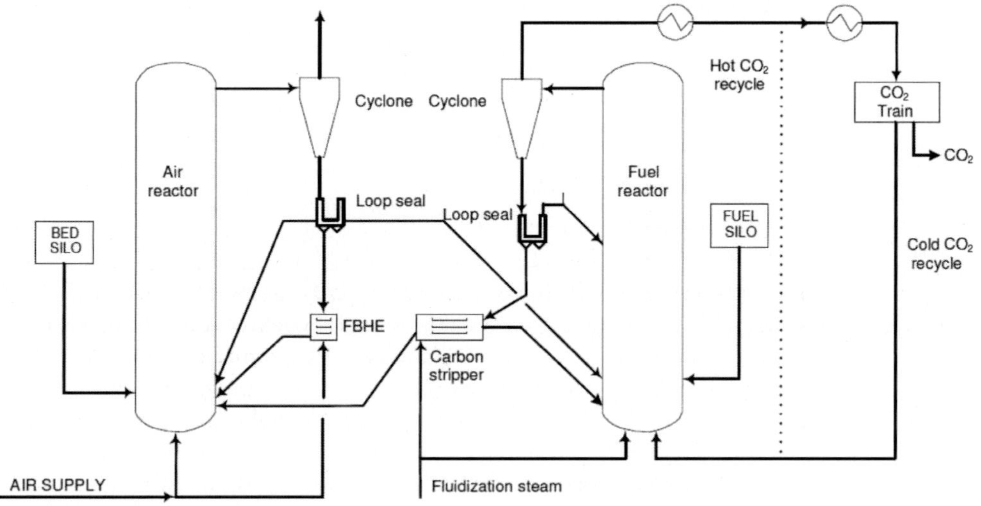

Fig. 18.7 Schematic diagram of a CLC reactor system for coal. Taken from Fillman *et al.* [26].

is shown in Fig. 18.7. Steam and/or recycled flue gas is used as the fluidization agent in the fuel reactor. A maximum electrical efficiency of 42% is reached with hot recycled flue gas only. The efficiency decreases linearly with the share of steam and is slightly lower with cold recycle. A high carbon stripper efficiency slightly lowers the electrical efficiency due to reduced carbon conversion, but significantly improves the CO_2 capture efficiency. A low overall air ratio has a positive effect on the electrical efficiency.

18.2.2.3 Release of Molecular Oxygen

Some Cu-, Mn, and Co-based oxygen carriers have an equilibrium O_2 partial pressure in the order of 3% at typical CLC temperatures, that is, around 900 °C. These oxygen carriers absorb O_2 under oxidizing conditions in the air reactor, whereas they release molecular O_2 under reducing conditions in the fuel reactor. The conversion rate of coal in presence of O_2 is several orders of magnitude faster than the gasification rate with H_2O or CO_2. This effect will significantly improve char conversion, so that the size/inventory of the fuel reactor can be reduced, and a carbon stripper could be omitted. Furthermore, an oxygen excess in the fuel reactor could lead to full conversion of the combustible gases, avoiding the supply of additional oxygen in a post-oxidation unit. This process, also named "chemical looping with oxygen uncoupling" (CLOU), has so far only been tested in small-scale units [27].

The CLOU process will significantly reduce the investment costs for the reactor system, since the carbon stripper and post-oxidation unit can be avoided and the fuel reactor can be built smaller. The operating costs on the one hand decrease due to the avoidance of oxygen supply and the lower pressure drop in the fuel reactor. On the other hand, the cost of Cu-, Mn-, and Co-based oxygen carriers is orders of magnitude higher than that of natural ores or industrial waste materials. Hence the lifetime of the oxygen carrier in the reactor system will be decisive for the success of the CLOU process.

18.3
Carbonate Looping Process

Carbonate looping involves post-combustion CO_2 capture by means of a CaO-based solid sorbent. In the carbonator, the CO_2 contained in the flue gas of a fossil fuel-fired power plant reacts with CaO to form $CaCO_3$, which is transferred to the calciner, where the CO_2 is released for subsequent storage by addition of heat. The CaO formed thereby is fed back to the carbonator, closing the loop. A schematic diagram of the process is shown in Fig. 18.8.

The reversible carbonation–calcination reaction is

$$CaO + CO_2 \rightleftharpoons CaCO_3 \tag{18.4}$$

The equilibrium partial pressure of CO_2 increases with increase in temperature. Under atmospheric conditions, the partial pressure of CO_2 in the calciner is close to 1 bar. The temperature in the calciner must be well above 850 °C to convert $CaCO_3$ rapidly and completely. Hence the endothermic calcination reaction is operated at around 900 °C.

The CO_2 partial pressure in the flue gas of a coal-fired power plant is around 0.15 bar. The equilibrium partial pressure in the carbonator must be below 0.03 bar to achieve 80% CO_2 capture. However, the temperature must be high enough to provide a sufficient reaction rate. Suitable temperatures for a fast exothermic carbonation reaction are in the range 600–700 °C [28]. The carbonation reaction occurs at the surface until a $CaCO_3$ layer is formed, preventing further contact of CO_2 with CaO. Hence a CaO particle will not react complete-

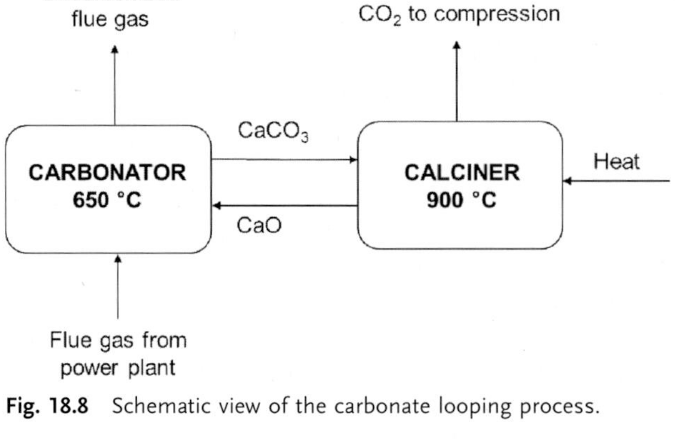

Fig. 18.8 Schematic view of the carbonate looping process.

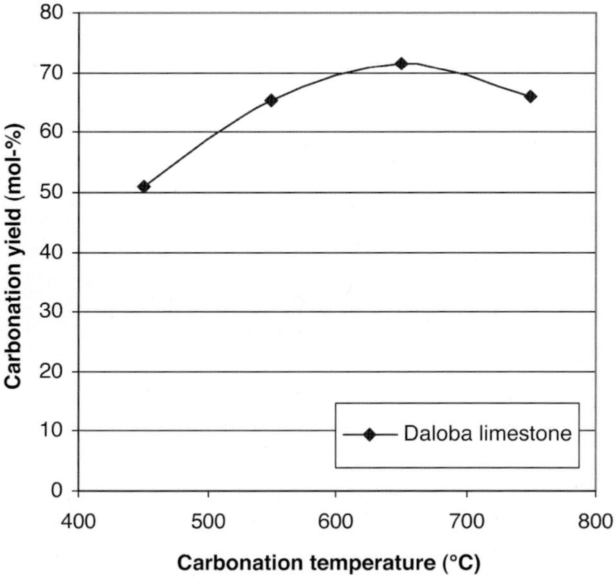

Fig. 18.9 Maximum carbonation yield as a function of temperature for "Daloba" limestone. Taken from Epple and Ströhle [29].

ly. The carbonation yield is a function of temperature, and the maximum is observed at 650 °C [28, 29], as illustrated in Fig. 18.9.

With increasing number of carbonation–calcination cycles, the ability of the CaO to capture CO_2 gradually decreases [29–31], as shown in Fig. 18.10. The internal surface area of the CaO particles decreases due to sintering of the pores [32]. However, the material does not deactivate completely. A remaining reactivity after a large number of cycles of around 10–20% was determined in laboratory tests [32–34]. In order to ensure high CO_2 absorption rates, a stream of fresh CaO or $CaCO_3$ (make-up) must be steadily supplied to the process. $CaCO_3$ is available at low cost in the form of natural limestone. The reactivity of dolomite stabilizes at a higher level than that of limestone [29, 35], which is also shown in Fig. 18.10. However, magnesium does not participate in the reaction and thus constitutes additional ballast for the cycling solids.

Apart from the carbonation reaction, CaO also reacts with SO_2 to form $CaSO_4$ (gypsum):

$$CaO + SO_2 + \tfrac{1}{2}O_2 \rightarrow CaSO_4 \tag{18.5}$$

This is an irreversible reaction in the relevant temperature range, so that the demand of make-up increases. SO_2 and CO_2 affect the ability of CaO to capture the other species. SO_2 reduces the CO_2 capture capability of lime after a large number of cycles, although the SO_2 concentration is orders of magnitude lower than the CO_2 concentration in the flue gas [36]. Carbonization–calcination cy-

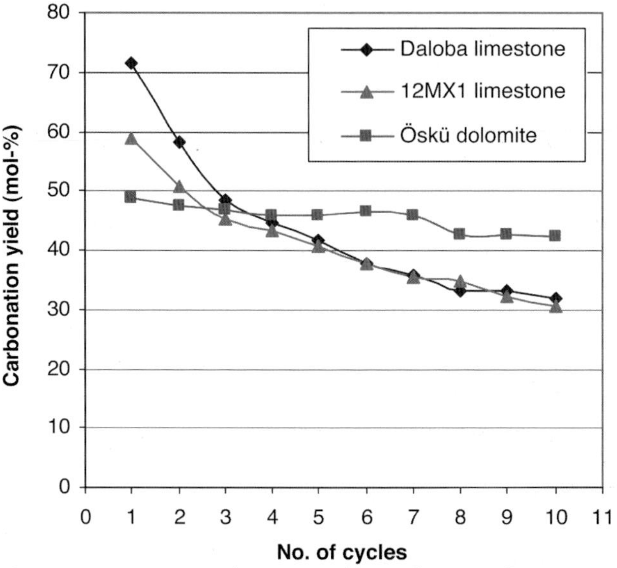

Fig. 18.10 Maximum carbonation yield as a function of number of cycles for "Daloba" and "12MX1" limestones and for "Öksü" dolomite. Taken from Epple and Ströhle [29].

cles, however, lead to an increase in macropores, enhancing the diffusivity with respect to sulfurization, so that SO_2 capture increases.

A fairly large heat input is required to increase the temperature of solids entering the calciner and to sustain the endothermic calcination reaction in the calciner. This heat can be provided either by supplementary combustion of fuel and oxygen or by heat transfer from an external energy source, for example, the combustor of the upstream power plant. The different options are discussed below.

18.3.1
Carbonate Looping with an Oxyfuel Calciner

The heat for the calciner can be provided by burning additional fuel in the calciner. The fuel must be fired with oxygen since air would dilute the CO_2 stream with N_2. A schematic diagram of the process is shown in Fig. 18.11.

The process setup of a carbonate looping unit for the retrofit of a coal-fired power plant is presented in Fig. 18.12. After the flue gas desulfurization (FGD) unit, a fan raises the pressure of the flue gas that is subsequently internally preheated. The CO_2 and remaining SO_2 are absorbed in the carbonator that is operated as a CFB. A steam generator extracts the excess heat from the carbonator resulting from the exothermic CO_2 capture reaction, maintaining a temperature of 650 °C. The solids are separated from the flue gas in a cyclone separator and are partly transferred to the calciner that is also a CFB reactor. There, the temperature of the solids is increased to around 900 °C. The heat for the endother-

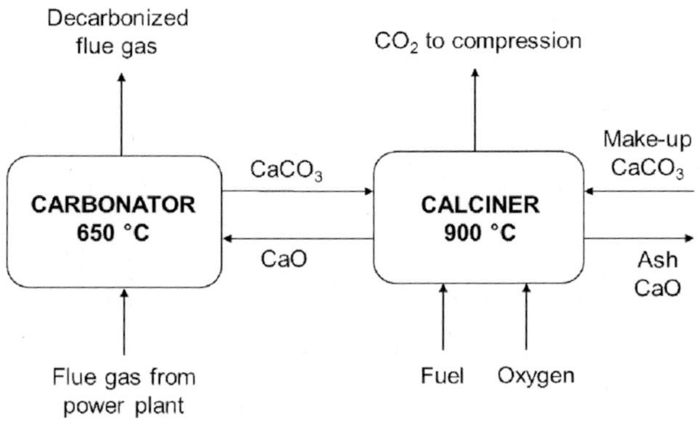

Fig. 18.11 Schematic view of carbonate looping with an oxyfuel calciner.

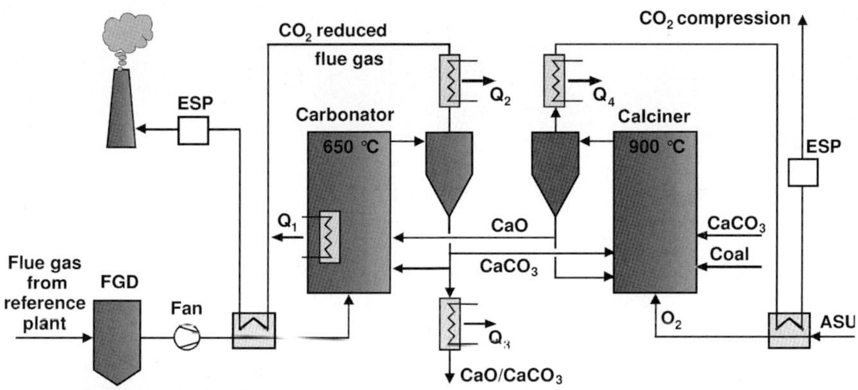

Fig. 18.12 Process scheme of a carbonate looping unit as a retrofit of a coal fired power plant. ASU = air separation unit. Taken from Epple and Ströhle [29].

mic calcination reaction is supplied by combustion of coal with oxygen. Furthermore, fresh limestone (make-up) is added to the calciner. In a cyclone, the "CO_2 stream," consisting of the released CO_2 and the products of coal combustion, is separated from the solids that are transferred back to the carbonator. The CO_2 stream is cooled, dedusted, and prepared for compression and geological storage. A solids stream is continuously discharged from the process. The heat extracted from these solids and the gases leaving the two reactors is used for steam generation and process internal preheating of flue gas and oxygen.

Considering a coal-fired power plant integrating a carbonate looping process for CO_2 capture, around one-third of the total coal feed must be raised for the calciner [37]. The heat input of the fuel is not lost, since the heat extracted from the carbonator and the heat in the off-gases from the carbonator and the calciner can be efficiently used to produce supercritical steam for power generation [38], which will increase the electrical capacity of the power plant by around 50%. An example of a heat recovery steam generator of a carbonate looping unit

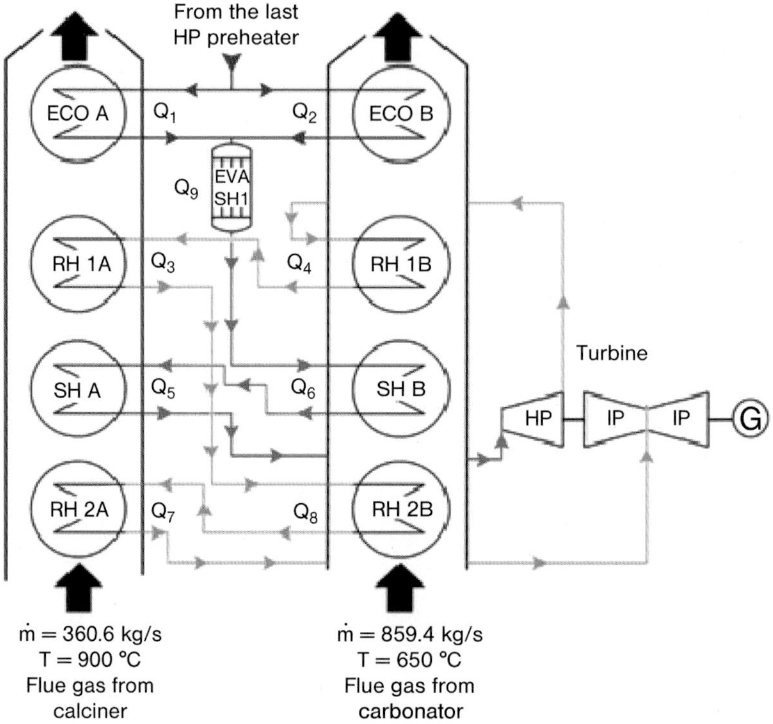

Fig. 18.13 Heat recovery steam generator of a carbonate looping unit for a 1052 MW_{el} coal-fired power plant. ECO = economizer; SH = superheater; RH = reheater; EVA = evaporator; HP = high pressure; IP = intermediate pressure. Taken from Epple *et al.* [38].

is illustrated in Fig. 18.13, where feed water enters the steam generator at 307 °C, live steam at 600 °C and 285 bar is produced in the economizers (ECO), evaporator (EVA), and superheaters (SH), and steam is heated from 356 to 620 °C at 59 bar in the reheaters (RH).

The supply of oxygen for the operation of the calciner is responsible for the major energy loss in the process. The carbonate looping process for CO_2 capture from a coal-fired power plant requires around one-third of the total coal feed that must be raised for the calciner, which has to be burnt with technical pure oxygen. Assuming an efficiency decrease of 9% for oxygen supply in an oxyfuel plant, the corresponding loss for the carbonate looping process can be estimated to be around 3% since only one-third of the oxygen amount is required. The demand for compression of the CO_2 gas stream needs to be added, as is the case for all CO_2 capture technologies. Choosing a reference pressure of 100 bar (temperature 30 °C), the efficiency further decreases by 2.5–4% depending on the process type.

Calculations by Abanades *et al.* [39] for a power plant with 100 MW thermal power and a reference efficiency of 46% resulted in a total net efficiency of

38.8%, including CO_2 compression. Ströhle *et al.* [40] reported net efficiencies of 41–42% and CO_2 capture rates in the range 70–95%, depending on make-up and circulating mass flow, without CO_2 compression for a plant of similar size with a reference efficiency of 45%. Romano [41] calculated a net efficiency of 37.6% and an overall CO_2 capture rate of 97% for a coal-fired power plant with a reference efficiency of 44.2%. Epple and Ströhle [42] calculated an efficiency decrease of 2.75% for 87% CO_2 capture for a 1052 MW_{el} coal-fired power plant with a reference efficiency of 45.6%, not including CO_2 compression. The calculated efficiency decrease of the same unit is 3% when the hydrodynamics in the carbonator are considered using an appropriate CFB model [43].

The process is associated with fairly low efficiency penalties in comparison with other technological solutions. The low price and high availability of limestone at various geographical locations is a further advantage of the carbonate looping process. These features lead to relatively low CO_2 avoidance costs that have been estimated to € 14.5 t^{-1} CO_2 [44] and € 18 t^{-1} CO_2 [45].

The carbonate looping process is currently being investigated in the same 1 MW_{th} test plant as for CLC, shown in Fig. 18.6. The major issues are the determination of solids circulation rates and make-up flow rates under realistic conditions.

18.3.2
Carbonate Looping with an Indirectly Heated Calciner

A significant improvement in electrical efficiency can be achieved if the need for oxygen from an air separation unit is avoided. One option is to transfer heat from an external source to the calciner. This energy source must be operated at temperatures significantly higher than in the calciner. Materials withstanding temperatures around 1000 °C have to be applied. The external energy source could be an additional combustor in the case of a retrofit solution, as illustrated in Fig. 18.14, or the combustion chamber of the upstream power plant. However, this will affect the water–steam cycle of the power plant significantly.

Abanades *et al.* [39] suggested two options to transfer heat from an air-fired combustor to the calciner:(1) through metallic walls and (2) using a solid heat carrier, as shown in Fig. 18.15. The CO_2 partial pressure in the calciner is reduced to 0.4 bar by steam injection and vacuum, respectively, in order to permit a reduced calciner temperature of 850 °C. The net efficiency of a power plant with 100 MW thermal power mentioned above increased to 39.4 and 40.0% for these two options, respectively.

Another option is heat transfer by means of heat pipes from an external fluidized bed heat exchanger of a CFB combustor. A schematic diagram of this process is illustrated in Fig. 18.16. This process is currently being investigated within an EU-funded research project coordinated by TU Darmstadt. A test rig on the scale of 1 MW_{th} has been erected at TU Darmstadt for further investigations regarding the technical implementation of the two processes.

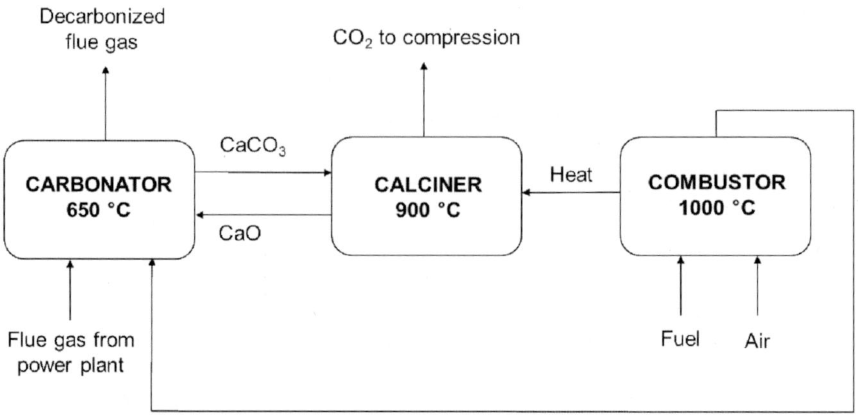

Fig. 18.14 Schematic diagram of a carbonate looping processes with indirectly heated calciner.

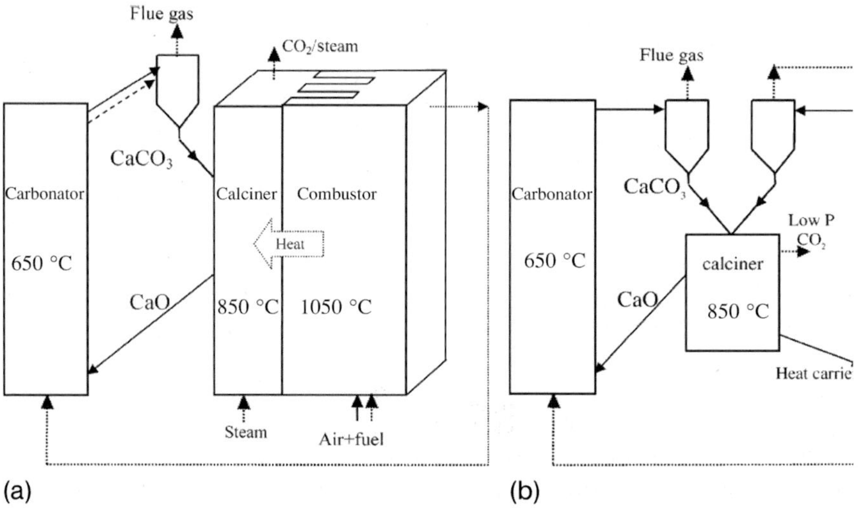

Fig. 18.15 Carbonate looping processes with indirect heat transfer to the calciner though metallic walls (a) and using a solid heat carrier (b). Taken from Abanades *et al.* [39].

18.4 Conclusion

Chemical and carbonate looping are two efficient and environmentally sound processes for CO_2 capture from power plants. The chemical looping process implies potentially the lowest efficiency drops since no energy is required for the separation of O_2 or CO_2. For natural gas firing, a pressurized reactor system in-

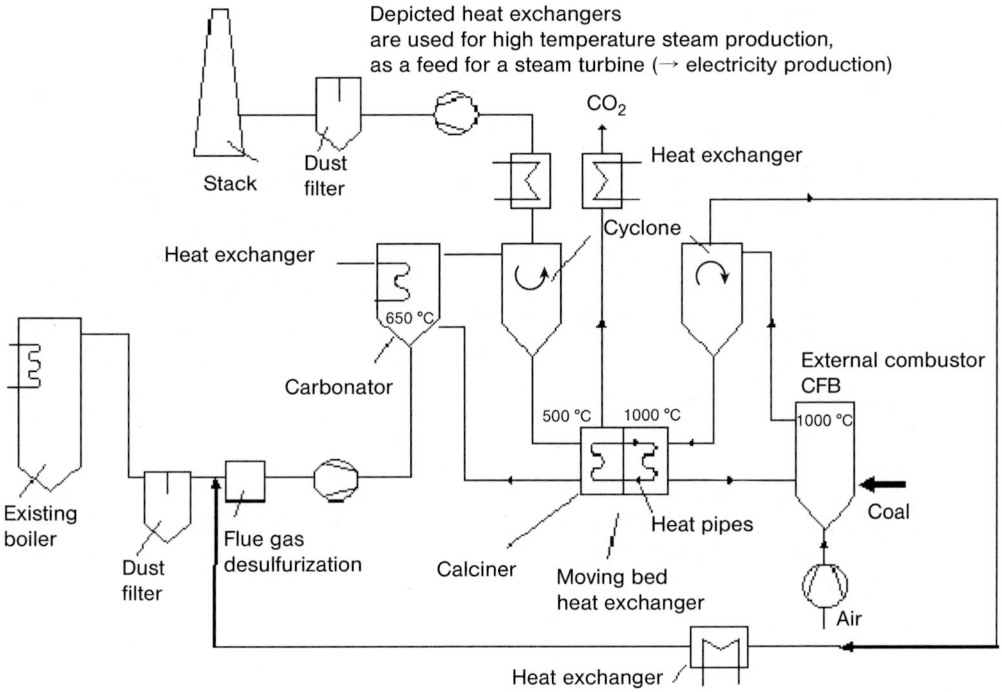

Fig. 18.16 Carbonate looping processes with indirect heat transfer from an external CFB combustor to the calciner by means of heat pipes.

tegrated into a combined cycle power plant can achieve electrical efficiencies of 53% if the air reactor can be operated at 1200 °C. For coal firing, the potential electrical efficiency can reach values up to 42% depending on the steam flow for fluidization of the fuel reactor. The CLOU process might be an option to improve char conversion and minimize the fraction of unburnt gases at the fuel reactor exit. The suitability of chemical looping for a retrofit of existing plants is limited, so that it is a promising option for new power plants with CO_2 capture. The carbonate looping process is particularly suited for the retrofit of existing plants, but suffers from a slightly higher efficiency decrease if the calciner is fired with oxygen. The electrical efficiency of a state-of-the-art coal-fired power plant integrating post-combustion CO_2 capture by carbonate looping lies in the range 39–40%. The process efficiency can be improved by integrating an indirectly heated calciner. The efficiency of both chemical and carbonate looping processes is significantly higher than for oxyfuel combustion or monoethanolamine scrubbing, but significant research work is needed to reach commercial application.

References

1 Strömberg, L. Discussion on the potential and cost of different CO_2 emission control options. *VGB PowerTech*, 2001, (10), 92–97.

2 Panesar, R., Lord, M., Simpson, S., White, V., Gibbins, J., and Reddy, S. Coal-fired advanced supercritical boiler/turbine retrofit with CO_2 capture. Presented at the 8th International Conference on Greenhouse Gas Control Technologies, Trondheim, 2006.

3 Gibbins, J., Lucquiaud, M., Li, J., Lord, M., Liang, X., Reiner, D., and Sun, S. Capture ready fossil fuel plants: definitions, technology options and economics. Presented at the 8th International Conference on Greenhouse Gas Control Technologies, Trondheim, 2006.

4 Abu-Zahra, M.R.M., Niederer, J.P.M., Feron, P.H.M., and Versteeg, G.F. The technical and economical analysis of CO_2 capture based on an absorption/desorption process using mono-ethanolamine. Presented at the 8th International Conference on Greenhouse Gas Control Technologies, Trondheim, 2006.

5 Bouillon, P.-A., Hennes, S., and Mahieux , C. ECO2: post-combustion or oxyfuel – a comparison between coal power plants with integrated CO_2 capture. *Energy Procedia*, 2009, **1**, 4015–4022.

6 Ekström, C., Schwendig, F., Biede, O., Franco, F., Haupt, G., de Koeijer, G., Papapavlou, C., and Røkke, P.E. Techno-economic evaluations and benchmarking of pre-combustion CO_2 capture and oxy-fuel processes developed in the European ENCAP project. *Energy Procedia*, 2009, **1**, 4233–4240.

7 Martelli, E., Kreutz, T., and Consonni, S. Comparison of coal IGCC with and without CO2 capture and storage: Shell gasification with standard vs. partial water quench. *Energy Procedia*, 2009, **1**, 607–614.

8 Johansson, M., Mattisson, T., and Lyngfelt, A. Use of $NiO/NiAl_2O_4$ particles in a 10 kW chemical-looping combustor. *Ind. Eng. Chem. Res.*, 2006, **45**(17), 5911–5919.

9 Johansson, E., Matthison, T., Lyngfelt, A., and Thunman, H. A 300 W laboratory reactor system for chemical-looping combustion with particle circulation. *Fuel*, 2006, **85**(10–11), 1428–1438.

10 Abad, A., Mattisson, T., Lyngfelt, A., and Rydén, M. Chemical-looping combustion in a 300 W continuously operating reactor system using a manganese-based oxygen carrier. *Fuel*, 2006, **85**(9), 1174–1185.

11 Abad, A., Mattisson, T., Lyngfelt, A., and Johansson, M. The use of iron oxide as oxygen carrier in a chemical-looping reactor. *Fuel*, 2007, **86**(7–8), 1021–1035.

12 de Diego, L.F., Garcia-Labiano, F., Gayán, P., Celaya, J., Palacios, J.M., and Adánez, J. Operation of a 10 kWth chemical-looping combustor during 200 h with a $CuO–Al_2O_3$ oxygen carrier. *Fuel*, 2007, **86**(7–8), 1036–1045.

13 Kolbitsch, P., Bolhàr-Nordenkampf, J., Pröll, T., and Hofbauer, H. Operating experience with chemical looping combustion in a 120 kW dual circulating fluidized bed (DCFB). *Int. J. Greenhouse Gas Control*, 2010, **4**, 180–185.

14 Ishida, M., Yamamoto, M., and Ohba, T. Experimental results of chemical-looping combustion with $NiO/NiAl_2O_4$ particle circulation at 1200 °C. *Energy Convers. Manage.*, 2002, **43**(9–12), 1469–1478.

15 Wolf, J., Anheden, M., and Yan, J. Comparison of nickel- and iron-based oxygen carriers in chemical looping combustion for CO_2 capture in power generation. *Fuel*, 2005, **84**(7–8), 993–1006.

16 Naqvi, R., Wolf, J., and Bolland, O. Part-load analysis of a chemical looping combustion (CLC) combined cycle with CO_2 capture. *Energy*, 2007, **32**(4), 360–370.

17 Zhang, X., Han, W., Hong, H., and Jin, H. A chemical intercooling gas turbine cycle with chemical-looping combustion. *Energy*, 2009, **34**(12), 2131–2136.

18 Mattisson, T., Garcia-Labiano, F., Kronberger, B., Lyngfelt, A., Adanez, J., and Hofbauer, H. Chemical looping combustion using syngas as fuel, *Int. J.*

Greenhouse Gas Control, 2007, **1**, 158–169.

19 Berguerand N. and Lyngfelt A. Design and operation of a 10 kWth chemical-looping combustor for solid fuels – testing with South African coal. *Fuel*, 2008, **87**(12), 2713–2726.

20 Ströhle J., Orth M., and Epple B. Simulation of the fuel reactor of a 1 MW_{th} chemical looping plant for coal. Les Rencontres Scientifiques de l'IFP – 1st International Conference on Chemical Looping, Lyon, 2010.

21 Leion, H., Mattisson, T., and Lyngfelt, A. Solid fuels in chemical-looping combustion. *Int. J. Greenhouse Gas Control*, 2008, **2**, 180–193.

22 Leion, H., Jerndal, E., Steenari, B.-M., Hermansson, S., Israelsson, M., Jansson, E., Johnsson, M., Thunberg, R., Vadenbo, A., Mattisson, T., and Lyngfelt, A. Solid fuels in chemical-looping combustion using oxide scale and unprocessed iron ore as oxygen carriers. *Fuel*, 2009, **88**(10), 1945–1954.

23 Cao, Y., Liu, K., Riley, J., and Pan, W. Application of a circulating fluidized bed process for the chemical looping combustion of solid fuels. *Prepr. Pap. Am. Chem. Soc. Div. Fuel Chem.*, 2004, **49**(2), 815.

24 Wall, T., and Liu, Y. *Chemical Looping Combustion and CO_2 Capture: Status and Developments, Technology Note 32*, QCAT Technology Transfer Centre, Pullenvale, Qld, 2008.

25 Beal, C., Epple, B., Lyngfelt, A., Adanez, J., Larring, Y., Guillemont, A., and Anheden, M. Development of metal oxides chemical looping process for coal-fired power plants. Les Rencontres Scientifiques de l'IFP – 1st International Conference on Chemical Looping, Lyon, 2010.

26 Fillman, B., Anheden, M., and Wolf, J. Parameter study in order to reveal critical design issues in the design for a CLC power plant using solid carbon as fuel. Les Rencontres Scientifiques de l'IFP – 1st International Conference on Chemical Looping, Lyon, 2010.

27 Mattisson, T., Lyngfelt, A., and Leion, H. Chemical-looping with oxygen uncoupling for combustion of solid fuels. *Int. J. Greenhouse Gas Control*, 2009, **3**(1), 11–19.

28 Bhatia, S.K. and Perlmutter, D.D. Effect of the product layer on the kinetic of the CO_2–lime reaction. *AIChE J.*, 1983, **29**, 79–86.

29 Epple, B. and Ströhle, J. CO_2 capture for the retrofit of power plants based on limestone. Presented at the VGB Conference "Power Plants in Competition," Prague, 2009.

30 Siliban, A. and Harrison, D.P. High temperature capture of carbon dioxide: characteristics of the reversible reaction between CaO(s) and CO_2(g). *Chem. Eng. Commun.*, 1995, **137**, 177–190.

31 Abanades, J.C. The maximum capture efficiency of CO_2 using a carbonation/calcination cycle of $CaO/CaCO_3$. *Chem. Eng. J.*, 2002, **90**, 303–306.

32 Abanades, J.C. and Alvarez, D. Conversion limits in the reaction of CO_2 with lime. *Energy Fuels*, 2003, **17**, 308–315.

33 Grasa, G.S. and Abanades, J.C. CO_2 capture capacity of CaO in long series of carbonation/calcination cycles. *Ind. Eng. Chem. Res.*, 2006, **45**, 8846–8851.

34 Grasa, G.S., Abanades, J.C., Alonso, M., and González, B. Reactivity of highly cycled particles of CaO in a carbonation/calcination loop. *Chem. Eng. J.*, 2008, **137**, 561–567.

35 Silaban, A., Narcida, M., and Harrison, D.P. Characteristics of the reversible reaction between CO_2(g) and calcined dolomite. *Chem. Eng. Commun.*, 1996, **146**, 149–162.

36 Sun, P., Grace, J.R., Lim, C.J., and Anthony, E.J. Removal of CO_2 by calcium-based sorbents in the presence of SO_2. *Energy Fuels*, 2007, **21**, 163–170.

37 Hughes, R.W., Lu, D.Y., Anthony, E.J., and Macchi, A. Design, process simulation and construction of an atmospheric dual fluidized bed combustion system for *in situ* CO_2 capture using high-temperature sorbents. *Fuel Proc. Technol.*, 2005, **86**, 1523–1531.

38 Epple, B., Ströhle, J., Priesmeier, U., Kluger, F., Radgen, P., and Marquardt, R. Post-combustion CO_2 capture from coal-fired power plants using limestone

sorbents. Presented at Power Gen Europe, Cologne, 2009.

39 Abanades, J.C., Anthony, E.J., Wang, J., and Oakey, J.E. Fluidized bed combustion systems integrating CO_2 capture with CaO. *Environ. Sci. Technol.*, 2005, **39**, 2861–2866.

40 Ströhle, J., Orth, M., Galloy, A., and Epple, B. Simulation of the carbonate looping process for CO_2 capture with interconnected fluidized bed reactors. Presented at the 9th International Conference on Circulating Fluidized Beds, Hamburg, 2008.

41 Romano, M. Coal-fired power plant with calcium oxide carbonation for postcombustion CO_2 capture. *Energy Procedia*, 2009, **1**, 1099–1106.

42 Epple, B. and Ströhle, J. CO_2 capture based on chemical and carbonate looping. *VGB PowerTech*, 2008, (11), 85–89.

43 Ströhle, J., Lasheras, A., Galloy, A. and Epple, B. Simulation of the carbonate looping process for post-combustion CO_2 capture from a coal-fired power plant. *Chem. Eng. Technol.*, 2009, **32**, 435–444.

44 Romeo, L.M., Abanades, J.C., Valero, A., Ballesteros, J.C., Escosa, J.M., Giménez, A., Cortes, C., and Pao, J. Process optimization in post-combustion CO_2-capture by means of repowering and reversible carbonation/calcination cycle. Presented at the 8th International Conference on Greenhouse Gas Control Technologies, Trondheim, 2006.

45 Poboss, N. and Scheffknecht, G. *BMWi Project 0327771B, Final Report*, Universität Stuttgart, 2008.

Part 6
Transportation and Storage of CO_2

Process Engineering for CCS Power Plants, First Edition. Edited by Detlef Stolten and V. Scherer

19
CO_2 Compression

Mark A. Gray

Abstract

Considerable industry resources are being applied on a global scale to accelerate the development of the carbon capture process for utility boilers. Clearly, the development of a cost-effective technology is critical to bring commercial viability to the industry if carbon capture and storage (CCS) is to be applied in any meaningful way. Similar focus must also be given to the technologies and practice of carbon dioxide compression, heat recovery, pipeline transportation, and storage. Given that each ton of coal consumed yields approximately *two* tons of CO_2 and that transport and storage require compression to at least 1500 psia (100 bar), the energy, infrastructure, and equipment requirements are significant. Economical compression technologies with useful heat energy recovery combined with prudent geologic storage practices will be critical for CCS to be accepted by the public. This chapter explores the issues and developing solutions to address the compression and storage component of CCS. Effective use of capital expenditures and reduced energy consumption are discussed and future technical and social-political challenges and possible solutions are explored.

Keywords: carbon dioxide, compression, storage, energy consumption

19.1
CO_2 Compression and Storage – Magnitude of the Issue

The United States today uses approximately 980 million short tons of coal for electric power production, which represents about 50% of all electric power consumption annually. This coal consumption results in CO_2 emissions of nearly 1900 Mt per year, and this does not account for the additional CO_2 emissions of natural gas-fired generation of approximately 420 Mt per year [1]. Clearly, the quantity and numerous sources of CO_2 emissions (over 6900 coal and natural gas units at more than 1500 plant sites in the United States [2]) provide a technical and infrastructure challenge to capture and store permanently this quantity of carbon dioxide. The trend of total CO_2 emissions in the United States

Process Engineering for CCS Power Plants, First Edition. Edited by Detlef Stolten and V. Scherer

has been flat to downwards due to the impact of the global recession and to a reduction in carbon intensity within the US economy. Power generation emissions are now at 2002 levels and are projected to increase only slightly, if at all, through 2030 [3].

Specifically, the electric power generation sector has seen emissions range from 2300 Mt in 2003 to a peak of 2409 Mt in 2007 and down-trending to below 2300 Mt in 2009. Current forecasts by the US Energy Information Agency (EIA) show CO_2 emissions remaining at these levels through 2015 [4]. The recent reductions in the electric generation sector are the result of both reduced demand and a shift to natural gas due to favorable pricing trends. Natural gas remains at very competitive pricing compared with coal and this situation will likely continue for the next 2–3 years as the economy regains its footing and new eastern US shale gas reserves are brought into production.

The quantity of CO_2 represented by the generation sector production of approximately 2300 Mt is worth consideration. Today, the United States consumes approximately 20 000 000 barrels (800 million gallons) of petroleum per day [5]. The quantity of compressed CO_2 represented by the electric generation sector alone is greater than 1.8 billion gallons per day of liquid (supercritical) CO_2 and at a 90% capture level is *over 2.25 times* the total daily petroleum consumption of the United States. The magnitude of the issue comes into focus when one considers the infrastructure of wellheads, piping, and control systems within the petroleum industry and the time and capital resources required to develop those systems.

The cost to capture CO_2 in the power sector will greatly determine the timing and penetration of carbon capture and storage (CCS) within the United States. The average cost of residential electricity in the United States is \$ 0.115 kWh^{-1} [6]. The projected cost of CCS could raise these rates by 30–50% over the next 8–12 years due to a combination of (1) added capital equipment, (2) reduced power plant net output due to the capture and compression process, (3) costs of CCS reagents and maintenance, (4) retirement of older generating units to be replaced with new more expensive generation technologies, and (5) the cost associated with CO_2 storage well field development, operation, and long-term management.

These are dramatic changes that will affect the landscape of power generation in the future. The industry is working to assess the appropriate direction and best technologies to meet the complex demands of CCS and assure a positive public perception of the technologies.

19.2
CO_2 Compression Energy Consumption – Heat Integration

The available CO_2 capture processes today using advanced amine- or ammonia-based chemistry produce a high-purity CO_2 stream at a compressor suction condition typically between 15 and 300 psia (1–20 bar). The supercritical pressure for liquid-phase CO_2 is approximately 1108 psia (75 bar) and represents the minimum required pipeline transport pressure to move the CO_2 to the wellhead for final storage. Given the normal pipeline losses and other system requirements to assure proper injection control, the compressor discharge pressure will be between 1500 and 2500 psia (102–170 bar). The compressor motor drive horsepower to achieve this discharge pressure will be approximately 55–65 horsepower per megawatt of unit generating capacity, assuming 90% CO_2 capture for a typical coal unit (assuming a typical multi-stage, intercooled compressor and a generating unit producing 1900 lb MWh^{-1} of CO_2). The projection of equipping up to 250 000 MW of coal power in the United States represents over 14 million operating horsepower of compressor capacity using nearly 11 000 MW of electrical power. The majority of the input power to a typical multi-stage compressor results in heat which is rejected to the atmosphere through the intercoolers. Clearly, the industry must find a method to recover at least part of this heat energy.

It should be kept in mind that for CCS to be applied on a coal unit, the flue gas must be very clean and typically contain less than 20 ppmv of SO_2, necessitating the use of high-efficiency flue gas desulfurization (FGDS) (chilled ammonia and advanced amine CCS system technologies currently require clean flue gas with SO_2 below 20 ppm for sustained operation). The additional auxiliary load associated with the FGDS is *not included* in the above numbers, nor is the typical steam consumption required for the CCS process. The total loss of net effective capacity in the United States as a result of CCS would be greater than 20% of the net coal generating capacity or approximately 50 000 MW. This loss of generating capacity and energy must then be replaced with some form of dispatchable renewable (wind or solar with energy storage) or nuclear power to assure that the CO_2 reductions from the applied CCS controls are fully realized.

The use of the heat from compression of the CO_2 will be largely driven by the available temperature of the compressor discharge stream. Typical multi-stage intercooled compressors do not provide sufficiently high rejection temperatures to use heat effectively in either the CCS or the generating unit heat cycles. Modifications to a multi-stage compressor can be made to achieve a high discharge temperature (210–260 °C) but at the expense of higher power consumption, which offsets the value of the heat energy. Figure 19.1 shows a typical multi-stage compressor (MAN Turbo) with intercooling and Fig. 19.2 represents the typical temperature–entropy diagram.

The development of new supersonic compression technology may serve this need by achieving single- or two-stage compression with high temperature (210–260 °C) discharge conditions and similar power consumption to traditional compressor designs. The Ramgen supersonic compressor (partially financed by the US Department of Energy) can provide 10 : 1 compression ratios within a single stage while providing a 230–250 °C discharge temperature [7]. This machine would allow more advanced heat integration within the CCS (reduced reboiler steam demand, for example), thus reducing the impact on the net generating output of the unit. The small footprint of the machine also reduces in-

Fig. 19.1 MAN 10-stage intercooled, multi-stage compressor with 200 bar discharge pressure.

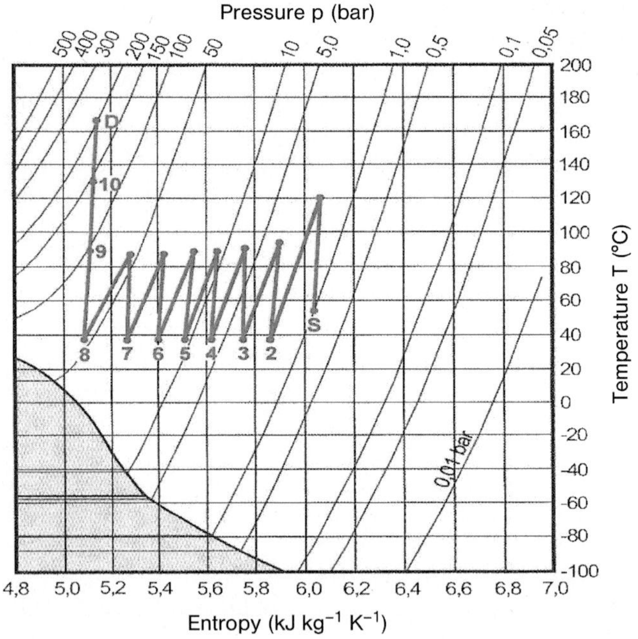

Fig. 19.2 Temperature–entropy diagram for multi-stage compressor on pure CO_2. Product line catalog data.

stalled costs and the use of only one or two compression stages should also provide for lower maintenance costs. Figure 19.3 is a computer-aided design (CAD) rendering of a single-stage Ramgen compressor capable of a 101 compression ratio and Fig. 19.4 represents the entropy diagram for the compressor. A prototype machine has been built and successfully tested, and full-scale compressors are now being considered for deployment for CO_2 compression.

Fig. 19.3 CAD rendering of a Ramgen compressor.

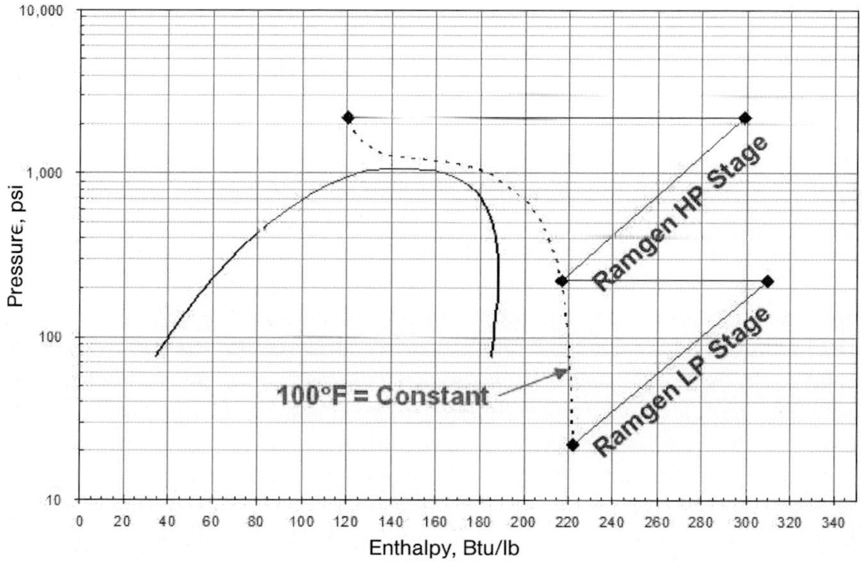

Fig. 19.4 Entropy diagram for Ramgen two-stage compressor. 1 Btu lb^{-1} = 1 kJ kg^{-1}.

19.3
Heat Recovery Opportunities

Given the amount of heat available from the compressed CO_2 liquid used for pipeline transportation and eventual deep well storage, a heat recovery approach is clearly needed. The CO_2 must be maintained as a liquid (supercritical fluid) with pressures greater than 1108 psia (75 bar) and temperatures near ambient conditions for final injection. The maximum available heat energy lies between the compressor discharge temperature of 210–260 °C and the typical pipeline temperature of 20–30 °C. Realistically, the opportunity for heat recovery lies between the compressor discharge temperature and the CCS process regeneration liquor inlet temperature or the lowest stage condensate heater discharge temperature within a typical steam power cycle (typically about 60 °C). Thus a driving temperature of 155–205 °C is available for useful heat recovery. Considering that each ton of coal burned produces nearly 2 t of CO_2, the energy recovery opportunity is significant.

The potential use of this available heat energy is given by the following example. A modern supercritical 750 MW coal unit will operate with a gross unit efficiency of about 42% LHV and produce about 500 t h^{-1} of CO_2, assuming a CCS with 90% design removal. The condensate system by comparison will pump about 2800 t h^{-1} of water through the steam cycle. The lowest condensate heater steam extraction plus above heater drains is about 260 t h^{-1} with an available energy of approximately 135 000 000 Btu h^{-1}. The CO_2 liquid stream leaving the compressor at 230 °C at 100 bar will have an available energy of over 350 000 000 Btu h^{-1} assuming a 65 °C terminal temperature and a specific heat capacity of about 5.0 kJ kg^{-1} K^{-1} or about 1.2 times that of water. The stream of CO_2 liquid could accommodate the regenerative heat applied within the first and second stage of heaters by the normal steam extraction and drains flow and thus allow the steam extraction from the turbine to be used for useful work to provide added electric generation. Liquid CO_2 undergoes a wide change in specific heat capacity as it moves into the supercritical range (Fig. 19.5). This wide range of thermal capacity is unique to supercritical fluids and provides both opportunities and design complexities to any heat recovery system.

The direct use of the high-temperature CO_2 stream from the compressor for condensate heating has several operational risks that must be considered. The piping of high-pressure CO_2 within the power plant poses significant human risk of asphyxiation should a substantial leak occur. Also, the leak of CO_2 into the condensate would result in an immediate decrease in condensate pH from the formation of carbonic acid and require a rapid controlled shutdown of the unit to avoid corrosion damage to the steam generator or turbine components. These risks could be reduced by using a secondary heat transfer fluid but this would add considerable cost and operational complexity to the energy recovery system.

A more reasonable use of the heat energy is within the carbon capture system reagent regeneration–CO_2 recovery cycle. Use of the available heat within the

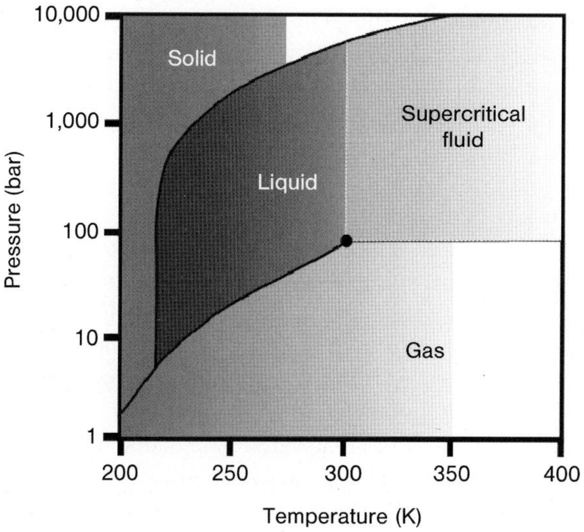

Fig. 19.5 CO_2 phase diagram.

"rich" reagent loop reboiler or similar components would reduce normal steam extraction demand from the steam turbine, reduce the parasitic losses, and reduce the high-pressure CO_2 piping distance (and considerable cost). Should a high-pressure leak of CO_2 occur, it would remain within the CCS system or be vented to the outside air with a much reduced risk of human injury. Most CCS designs today use open structural support steel with rain shields to provide immediate atmospheric venting of any CO_2 gas leaks. The use of the available heat from the compressed CO_2 gas would also reduce the sizing and related cost of any steam extraction piping needed to support the CCS heat demand. Clearly, energy recovery from the compression of the CO_2 will be a significant focus of attention as these systems evolve to improve the overall energy efficiency, reduce the impact on the unit net output, and reduce the costs of construction and operation.

19.4 CO_2 Purity and Pipeline Transport Issues

The purity of CO_2 liquid from the carbon capture and compression systems is of particular concern. CO_2 liquid is commonly transported today within carbon steel pipeline systems for enhanced oil recovery (EOR) operations and various other chemical processes. The typical pipeline quality purity specifications [8] include limits on water (to avoid carbonic acid corrosion), CO, H_2S, O_2, N_2, and other inert and hydrocarbon trace gases. There is much debate about this issue as no clear standards have been set for the CCS future market. Within the United States, a commercial purity specification has been established by Kinder-Morgan to help assure a consistent product for the EOR industry. The final use of the CO_2 liquid/gas plays a major role in determining the purity requirements. Obviously, if the CO_2 is to be used in the food production industry, the purity requirements are stringent. The chemical industry and EOR applications establish another, lower, purity level. If the CO_2 is captured strictly for well field storage, yet another purity level could be used.

The primary concern with purity is to assure a minimum of 95% CO_2 within the total liquid stream followed by limits on water (no free water within CO_2 liquid), H_2S (<1500 ppm), O_2 (<10 ppm), sulfur (<1450 ppm) and other inert or hydrocarbon gases (<4%). These purity values are to assure low pipeline corrosion rates and consistent measurement accuracy within the metering devices.

The coal-fired utility industry will likely find the water and O_2 the most challenging purity control issues. Some discussion of water limits below 30 lb per MCF (million cubic feet) have been considered to assure low pipeline corrosion. The use of FGDS should reduce sulfur and H_2S contamination to a minimum. Inert gases do not appear to be an issue with the current amine- or ammonia-based carbon capture processes as CO_2 purity of over 98% is common. Process control will need to be carefully monitored to assure low water and oxygen contents in the final discharge stream.

Finally, a pipeline transport pressure will need to be established to guide the design of the piping, valving, metering, and pumping systems. The CO_2 must be maintained above the critical pressure of 1108 psia (75 bar) to assure a liquid state within the systems. Delivery pressures to the wellhead for final storage will vary considerably based on the characteristics of the individual wells and rock strata that are used for final storage. Once the CO_2 is in a liquid state, booster pump equipment (as opposed to compressors) can be used to maintain pipeline pressure or to increase the wellhead delivery pressure as needed. Most of the development and commercial systems today are operating the pipelines at between 100 and 170 bar. This pressure is well within the compressor and pumping booster technology available today.

19.5
CO_2 Storage Development – Prudent Practices

The use of well field storage for vast amounts of CO_2 will be an intensely debated subject in the years ahead. Industry, academia, environmentalists, and the public at large will provide compelling arguments both for and against the use of deep saline reservoirs or other similar geologic storage opportunities – often citing the same data as the basis for their opinion. This will no doubt cause confusion and distrust of the 'scientific' argument by the public. If the industry is to succeed it must be able to relate clearly the experience of the EOR industry and the natural formations of CO_2 that exist as part of a compelling discussion of the merits and safety of well field storage. The industry must also provide clear evidence that it is using the most prudent practices in keeping with the "safety first" position. This will not be easy. The recent events in the Gulf of Mexico and other man-made calamities have raised legitimate concern about both the technical understanding and the appropriate regulation and monitoring of these complex operations. The industry must provide clear evidence of its prudence and move at an appropriate pace to build confidence in the technology and operation.

EOR has a long and successful history of natural CO_2 extraction, compression, transportation, injection, and monitoring for over 20 years in the United States. The existence of large natural formations of underground CO_2 also shows that the gas can be trapped in geologic formations for very long periods of time. The injection of CO_2 into deep saline reservoirs is less understood both on a predictive modeling basis and from a practical operating viewpoint. Experience to date shows successful injection when several prudent steps are taken during well field development and operation. These steps include

- Extensive geologic mapping and modeling of the proposed well field to confirm appropriate selection and understanding of the geology and robust nature of the confinement layers.
- Well casing and cementing designs that assure multiple barriers to avoid the possibility of fluid migration into potable water resources.
- Exploratory wells are drilled to evaluate the injection zones and layers of confinement rock. This exploration helps confirm the injectability of the saline formation (porosity and permeability) through the use of wireline technologies and core samples and to confirm the structural characteristics of the confinement rock formations (strength, fracture mechanics, discontinuities).
- Use of monitoring wells to track the injection plume pressure wave and movement within the target rock strata.
- A clear understanding of injection well pressure limits and a reasonable safety margin to avoid structural fracturing of the injection strata and cap rock.
- Emergency action plans to address failure scenarios and timely communication programs with the community to assure the public of a "safety first" approach to daily operations.
- Sharing of experiences throughout the industry to help promote best practices.

Figure 19.6 shows a typical multiple casing injection well for CO_2 used in the United States [9]. The well is nearly 8000 ft (2400 m) deep and has been receiving CO_2 for over a year into a saline reservoir without incident.

It is important to understand that each well casing requires a unique design based upon the findings of the exploratory work and well field overall design. The cost of developing a well field for long-term CO_2 will typically be about $ 50–80 million based on a 100 square mile well field needed for a 500–700 MW unit operating for 25–30 years. These costs do not include ongoing operation, final closure, and long-term monitoring. Current estimates of the total costs to inject CO_2 safely over multiple decades are between $ 5 and 10 per ton (2009 US $) [10]. Considering that a single 700 MW coal unit will produce over

100 million tons of CO_2 over a 30 year operating span, the costs of storage are significant.

Protection of the local environment during the well field development and injection period is a key focus. Careful analysis of the area being considered must include a thorough investigation of past underground mining activities, oil or gas exploration and production, potable water well usage, area seismic activity history, ground water impacts, and mineral rights. The example of the well casing in Fig. 19.6 uses more than one cap rock stratum to help assure long-term

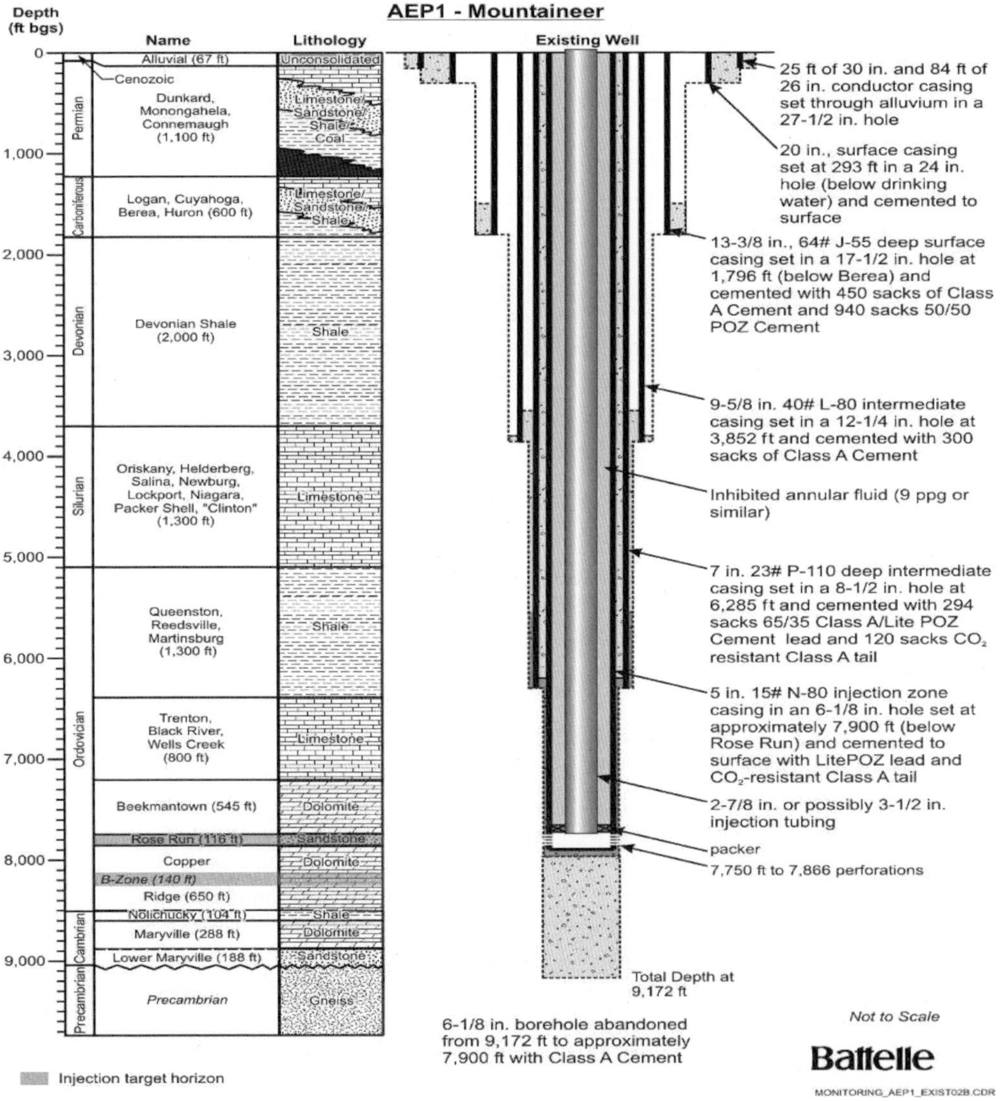

Fig. 19.6 Typical injection well configuration.

containment and a margin of safety within the overall design. The area needed for the CO_2 storage of large units is vast and the migration of the plume over time must be considered. Conservative assumption in the design of the well field will help assure successful permanent storage and tolerance to typical geologic activity. Many prime storage sites indicate that the CO_2 will, over time (50–500 years), convert to solid carbonates within the rock, thus assuring permanent fail-safe storage. All these technical issues must be considered in the initial site selection and final well field development.

The pipeline distribution network to the wellheads must also be carefully considered. The typical 700 MW coal unit will require many wellheads within the 100 square miles to accommodate the injection stream from the unit. The pipe distribution system must be designed to assure very low probabilities of failure and the release of any CO_2 gas is minimized. Appropriate automatic valves and metering control will be needed to assure rapid isolation of a possible pipeline failure. The ability to inspect the pipelines and effect repairs in a cost-effective and timely manner must also be considered in the overall design.

Finally, thoughtful, timely, and open communication with landowners, communities, environmental groups, and emergency response organizations must occur early and often as the well field development progresses. There is a great deal of public distrust of industry and the technical complexity of CCS will be the focus of much debate by many differing views. Open communication will be the bridge to allow meaningful dialog and agreement.

19.6 Public Policy and Long-Term Liability

The goal of large-scale CCS deployment will have important policy implications for the United States, the European Union, China, India, South Africa, and other large consumers of coal and natural gas for use in power generation. The United States alone could capture and store 50 000 Mt (5.0×10^{10} t) of CO_2 over the next 30 years. China might well double that amount depending upon international agreements and the development of capture technologies. Clearly, these are enormous quantities of material that must be monitored for many decades in the future. The policies, regulations, and business risks that drive both the timing and custodial responsibility of the stored CO_2 will be very important. If policy and regulation require an unacceptable level of business risk, the cost to the world and national economies could be traumatic. Rapid increases in energy costs combined with reduced reliability of generation due to the complex equipment operation and maintenance requirements will have far-reaching implications. The need for national support of large-scale CCS installations so that the industry can begin the important process of ‘learning by doing’ is critical. This support is needed for both the financial and technical aspects of these projects and also policy and regulatory support at the national and local levels. Custodial responsibility for the well field after injection operations have ceased and the

Fig. 19.7 Mountaineer wellhead.

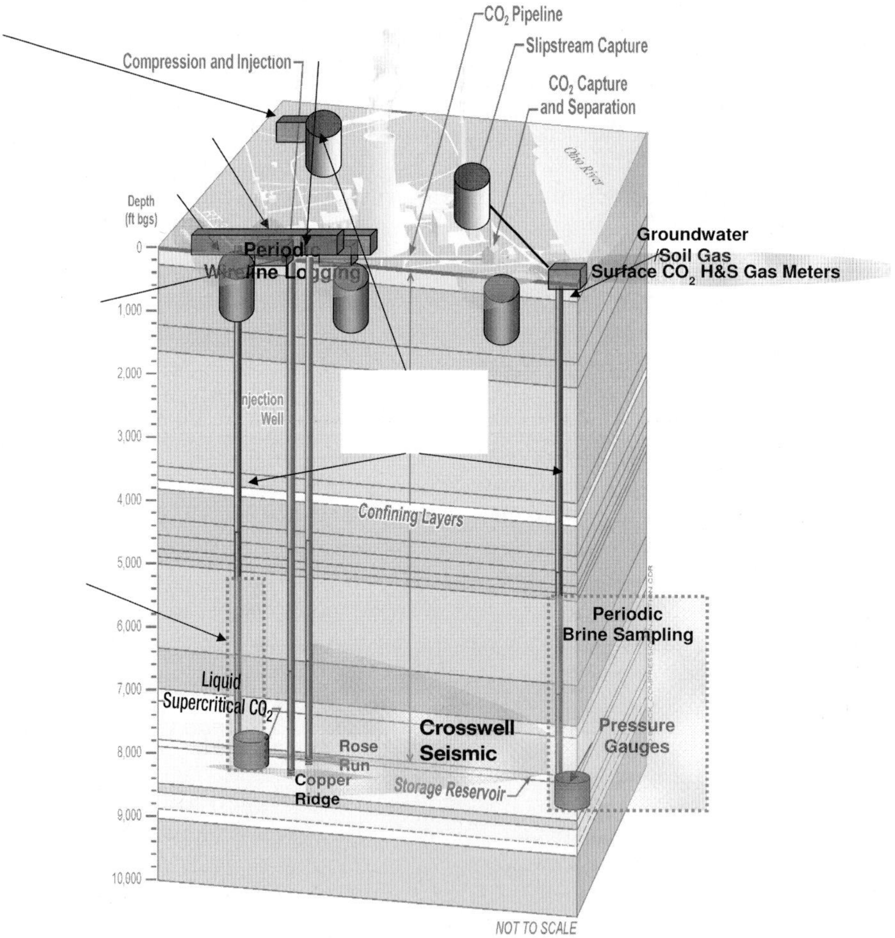

Fig. 19.8 Mountaineer well field configuration.

well heads have been capped cannot reasonably reside at a corporate level. There will likely need to be national government involvement for the decades following the closure of a large well field. These and other similar policy considerations require candid, open dialog to reach meaningful conclusions and agreements. The time to begin is now. A photograph of the injection wells at the American Electric Power CCS project located at the Mountaineer plant in New Haven, WV, USA is shown in Fig. 19.7. An artist rendering of the various injection and monitoring wells and instruments at the Mountaineer site is shown in Fig. 19.8.

19.7 Conclusion

- The potential impact to the coal fleet from the broad implementation of CCS both in terms of cost and operational efficiency is enormous.
- Heat integration from CO_2 compression is vital to minimize the impact on unit efficiency and net generating output.
- Advanced compression technologies will be needed to assure meaningful heat integration and minimize the cost of equipment and energy consumption to operate.
- Heat integration can be achieved by several means and should be an integral part of the design studies for each unique unit where CCS is considered.
- CO_2 purity standards need to be established that allow a broad array of CCS technologies to enter the marketplace while assuring reliable pipeline operation.
- Prudent well field design and development practices with appropriate regulatory oversight must be used to assure safe, long-term storage of CO_2.
- Regulatory oversight must consider the emerging knowledge of CCS and allow for a "learning by doing" approach to facilitate large-scale demonstration projects.

References

1 US Energy Information Agency. *Coal Consumption Data*, www.eia.doe.gov/, Table 2.1.A (accessed 15 November 2010).

2 US Energy Information Agency. *EIA Existing Capacity by Energy Source*, www.eia.doe.gov/, Table 1.2 (accessed 23 November 2010).

3 US Energy Information Agency. *EIA Annual Energy Outlook 2010*, www.eia.doe.gov/.

4 US Energy Information Agency. *EIA Emissions from Energy Consumption*, www.eia.doe.gov/, Table 3.9 (accessed 23 November 2010).

5 US Energy Information Agency. *EIA Annual Energy Outlook 2010 for Natural Gas and Oil Consumption*, www.eia.doe.gov/, Figure 79 (accessed 11 May 2010).

6 US Energy Information Agency. *EIA US Data Retail Prices to Consumers*, www.eia.doe.gov/, Table 5.3 (accessed 15 November 2010).

7 Ramgen Power Systems. *Ramgen Public Information Release*, www.Ramgen.com/apps_comp_unique, last accessed 28 Feb. 2011.

8 Kinder-Morgan. *CO_2 Purity Specifications for Commercial Pipelines*, www.sequestration.org/publish/MGSC_capture, last accessed 28 Feb. 2011.

9 American Electric Power. *Battelle Well Casing Design Features Used at AEP Mountaineer Site, 2009*, www.aep.com/investors/present/documents/MTCCSInvestorPresentation, last accessed 28 Feb. 2011.

10 Schlumberger Carbon Service. Taylorville Energy Center Cost Report.

20
CO_2 Transport – The Missing Link for CCS

Chris A. Hendriks, Erika de Visser, and Joris Koornneef

Abstract

Transport is an essential element in the carbon capture and storage (CCS) chain. Attached to the transport are technical issues, safety and regulation, investment and operational costs, long-term planning, and public involvement. In various regions worldwide, but especially in the United States, there is experience with large-scale transport of CO_2. If CCS has to contribute substantially to the long-term strategic emission reduction goal for greenhouse gases, effective and efficient transport networks have to be established worldwide. Pipelines will form the core of such system, but may be supplemented by ship transport under certain conditions, as this can add flexibility to the system. However, much is still unknown about technical specifications for pipelines and CO_2 in a network design, and integration of ships and pipeline networks for CO_2 transport has never been investigated in great detail. The role of the governments needs to be large to create an economically efficient system, to coordinate the availability of storage capacity and supply of CO_2, and to guarantee access for all parties on a competitive basis. Long-term vision and (governmental) commitment will be essential for the successful and efficient development of a large-scale transport network, including common and well-established procedures for performing environmental impact assessments and obtaining permits.

Keywords: carbon dioxide transport, pipelines, ship transport

20.1
Introduction

Transport of the captured CO_2 to the storage location is an essential element in the carbon capture and storage (CCS) chain. Although transportation will ultimately represent only a small portion of the total CCS costs, it requires both careful long-term planning and substantial upfront investments. Also, the guarantee of flow down the pipeline is essential for the capture operator, and possi-

Process Engineering for CCS Power Plants, First Edition. Edited by Detlef Stolten and V. Scherer

bly the injector operator. Discontinuity of the CO_2 flow will force capturing to be stopped unless there is intermediate storage available.

When CCS comes to be applied on a large scale, networks of CO_2 pipelines complemented with ship infrastructure will be required to form a cost-effective and efficient transportation system. In the demonstration phase and early commercial phase, dedicated pipeline or ship connections will be acceptable. Nevertheless, already in this stage careful planning is required as they ultimately will be part of the larger CO_2 transport network.

Without long-term contracts, transport will bear high financial risks. In the demonstration and early commercial phase, project developers need to design and control the full CCS project as it is unlikely that a separate transport company will take the investment risk. In the longer term, dedicated companies can facilitate transport on a commercial fee-based system.

If in 2050 10 Gt of CO_2 needs to be transported – resulting from the Blue Map scenarios from the IEA roadmap [1] – between 200 000 and 600 000 km of pipeline will be required. Various issues need to be resolved before the infrastructure of pipelines and ships can be developed and constructed effectively. In this chapter, we address the main issues of transport of CO_2.

In the next section, a description is given of the modes of transportation and considerations for application. In Sections 20.3 and 20.4, pipeline and ship aspects, respectively, are discussed in more detail. Section 20.5 focuses on CO_2 transport networks and Section 20.6 concentrates on the organizational, strategic planning, and safety aspects of CO_2 transport. The chapter concludes with an outlook on transport of CO_2 in the coming 40–50 years.

20.2 Experience with CO_2 Transport

20.2.1 Pipelines

CO_2 can be transported by pipeline, ship, truck, and train. The last two are typically good alternatives for smaller volumes of CO_2 but are not of interest for commercial-scale CCS operations. The most common way to transport large volumes of CO_2 is by means of pipelines. The operational experience with this mode of transport is significant, mainly in the oil and gas industry, and covers over 40 years. The largest CO_2 pipeline network is found in the United States, with a total length of ~6300 km [2]. These CO_2 pipelines primarily operate with the purpose of CO_2-driven enhanced oil recovery (EOR). Currently, over 100 CO_2-driven EOR projects are running in the United States, producing almost 300 000 barrels of oil per day [3]. The CO_2 originates mostly from naturally occurring CO_2 in underground reservoirs. Other sources of CO_2 are natural gas processing facilities, ammonia manufacturing plants, and a large coal gasification project in North Dakota, which provides CO_2 for the Weyburn EOR project in Canada. About 90 km of CO_2 pipelines can be found in Turkey and some others in Algeria and Hungary.

The first large CO_2 pipeline in the United States was the Canyon Reef Carriers, built in 1970. This pipeline of 352 km transports annually 4.4 Mt of CO_2 produced by Shell Oil Company gas processing plants in the Texas Val Verde basin to EOR sites elsewhere. The first project to use man-made CO_2 for EOR operations is the Weyburn–Midale CO_2 project. The carbon dioxide comes from the Dakota Gasification Plant in Beulah, ND, and is pipelined 320 km north to oil fields in Weyburn. Along with water, the CO_2 is injected 1500 m deep into a partially depleted oil field to boost oil production [4]. CO_2 injection activities started in 2000 and during 2010 18 Mt CO_2 has been stored in the Weyburn and Midale fields.

To date, most experience with transporting CO_2 is gained with transportation via land pipelines. The only subsea CO_2 pipeline that has been laid is the Snøhvit pipeline in Norway. This 153 km pipeline transports 0.7 Mt of CO_2 per year. Table 20.1 provides an overview of existing long-distance CO_2 pipelines. Figure 20.1 presents an impression of the pipeline diameters used by distance.

20.2.2 Ships

CO_2 is also transported commercially by ship. In Europe, a ship logistic system operates to distribute industrial and food-grade CO_2. Currently, there are about 10 relatively small ships operating worldwide transporting CO_2. Vessel sizes

Table 20.1 Overview of long-distance CO_2 pipelines.

Name of pipeline	Operator	Length (km)	Diameter (in)	Capacity (Mt a^{-1})	Country
NEJD Pipeline	Denbury Resources	295	20		USA
Cortez Pipeline	Kinder Morgan	808	30	19.3	USA
Bravo Pipeline	BP	350	20	7.3	USA
Transpetco Bravo Pipeline	Transpetco	193	12.75	3.3	USA
Sheep Mountain Part 1	BP	296	20	6.3	USA
Sheep Mountain Part 2	BP	360	24	9.2	USA
Central Basin Pipeline	Kinder Morgan	–	26 and 16	11.5	USA
Este Pipeline	ExxonMobil	191	12 and 14	4.8	USA
West Texas Pipeline	Trinity	204	8–12	1.9	USA
SACROC Pipeline	354	16	4.2		USA
Weyburn Pipeline	Dakota Gasification	330	12–14	4.6	USA
Canyon Reef Carriers	Kinder Morgan	225	16	4.6	USA
BatiRaman	Turkish Petroleum	90	1.1		Turkey
Snøhvit	Statoil Hydro	153	8	0.7	Norway

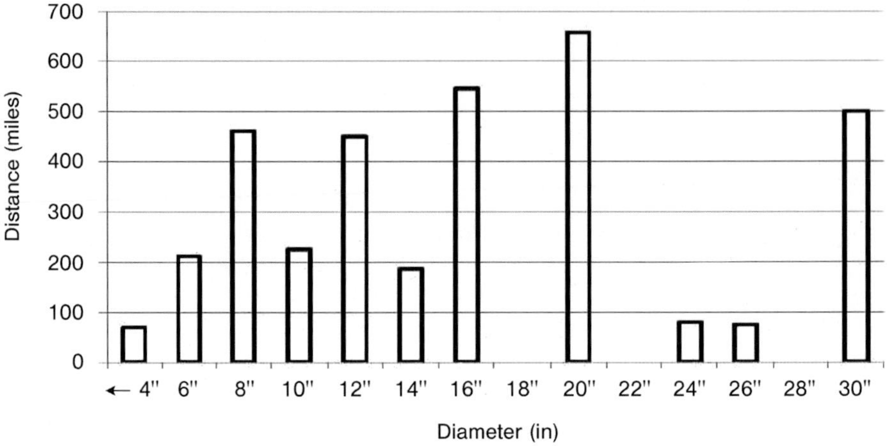

Fig. 20.1 Diameters versus mileage of CO_2 pipelines [5].

vary between 1000 and 1500 m^3 and the transport pressure is about 14–20 bar. Yara is currently operating three small-scale semi-pressurized CO_2 ships. These vessels transport liquid CO_2 to distribution terminals in a number of countries around the North Sea basin. In addition, at present six ships in the range 8500–10 000 m^3 are certified for carrying CO_2. The CoralCarbonic is the first purpose-built CO_2 tanker in the world with a capacity of 1250 m^3. The ship distributes CO_2 for Linde from Finland to import terminals in Sweden and Latvia [6]. New ships for dedicated CO_2 transport with capacities up to about 40 000 m^3 are currently under design [6–8].

20.3
CO_2 Transport by Pipeline

The previous section showed that there is already wide operating experience with CO_2 pipelines worldwide. The design of CO_2 pipelines is current practice, but designing CO_2 pipelines with the purpose of transporting CO_2 in CCS projects adds some more elements to the design process, for example, conditions of the CO_2 stream, CO_2 quality, requirements for pipeline materials, and costs. In the following sections, we elaborate on these issues. Further detailed description of design issues that apply to CO_2 pipelines is outside the scope of this chapter and for this we refer to, for instance, the report of the Energy Institute [6].

20.3.1 Pipeline Design Considerations

The design of pipelines is well established practice and a mature engineering subject. It relies to a large extent on the design of hydrocarbon pipelines. The sections that follow examine CO_2-specific additions to normal pipeline design.

The design of pipelines is an optimization process. The pipeline design can be optimized with regard to pipeline diameter, inlet pressure, pressure loss, and pipeline wall thickness, depending on flow rate and distance, and also on whether the pipeline is situated onshore or offshore [9]. The main issues that should be addressed when CO_2 pipeline transport is considered are as follows:

- transport conditions of CO_2
- CO_2 quality with respect to potential corrosive effects of CO_2 in combination with water and impurities
- pipeline materials
- safety and environmental issues.

20.3.1.1 **Transport Conditions**

For efficient and cost-effective pipeline transportation, the CO_2 needs to be compressed. Preferably, the CO_2 is transported in the dense phase, that is, in the liquid or supercritical phase. Under these conditions, the density of the CO_2 is high and the pipeline volume is used efficiently. Figure 20.2 shows a phase diagram for pure CO_2 at various pressures and temperatures. From this graph, it can be seen that at a transport temperature of 10 °C, typically the underground temperature in Europe, CO_2 is a liquid at pressures above 45 bar. CO_2 pipelines should be operated to avoid two-phase flow, that is, to avoid the coexistence of gaseous and liquid CO_2. Two-phase transport may lead to abrupt pressure drops, can cause material damage to the pipe by cavitation, decreases transport capacity, and increases energy requirements. To avoid two-phase flow, it is generally recommended that CO_2 is transported well above the critical pressure of 73.8 bar. Above this pressure, CO_2 exists as a single dense or supercritical phase over a wide range of temperatures (see Fig. 20.2). Lower pressures can be applied when CO_2 is transported at stable low temperatures.

Two-phase flow may also be induced by the presence of impurities in the CO_2. The presence of impurities might require higher pressures for the CO_2 stream in order to be in a single phase, either liquid or supercritical. Depending on the amount and type of impurities, this may lead to several (tens of) bar increased pressure (see also the subsection "Impact of Impurities and Water" in Section 20.3.1.2).

20.3.1.2 **CO_2 Stream Composition**

The composition of the CO_2 stream from the capture depends on the composition of the treated gases and the capture process applied. The most relevant

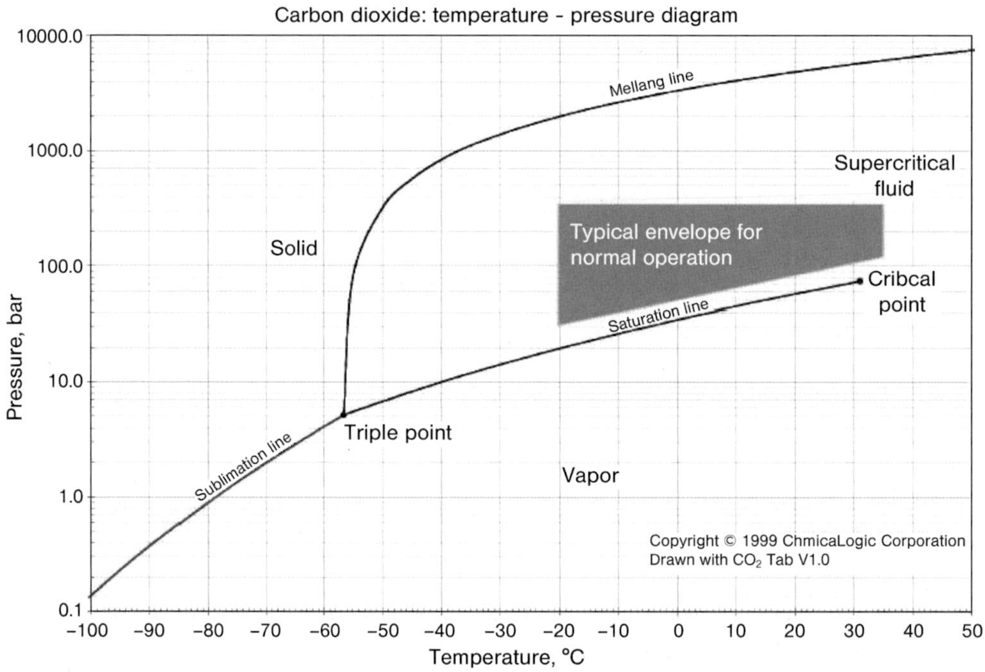

Fig. 20.2 Thermodynamic phases of pure CO_2 at various pressures and temperatures.

components in the captured CO_2 are water, SO_2, H_2S, O_2, and non-condensable gases such as N_2, CH_4, Ar, and H_2, and in some cases also particulates. The amounts and types of impurity may differ considerably between post-combustion, pre-combustion, and oxyfuel capture processes. Depending on the requirements for transportation and possibly for storage, additional purification steps might be needed after the capture unit [10].

Impact of Impurities and Water

The properties and behavior of CO_2 will change when impurities are present. Impurities can have a large effect on the thermodynamic properties of the CO_2 stream transported. Properties affected are, for example, the critical point, saturation content of water, density, and temperature/pressure phase envelope. Small amounts of methane and hydrogen have a large impact on the behavior of the CO_2 stream. The impact of argon, nitrogen, and oxygen is smaller. The effect is largely reduced if the fluid being transported remains in a gaseous or supercritical state. However, keeping the fluid in a supercritical state typically implies heating it above the critical temperature, which requires additional operating costs for the transportation network. Impurities in the CCS stream have two main negative effects on transport:

- Higher pressures might be required to avoid two-phase flow and free water formation and the consequent danger of hydrate formation and corrosion.

- The transport capacity might be reduced; as impurities occupy space and influence the compressibility of CO_2 negatively.

Free Water Formation, Hydrate Formation and Corrosion The water content in CO_2 is critical for transport. It should be controlled, because of risks of corrosion and hydrate formation in the pipeline. The water solubility limit is significantly lower when CO_2 is in the gaseous phase than the liquid phase. The water solubility also decreases with increase in temperature. The water content should be controlled more strictly when CO_2 is transported at low temperatures and at relatively low pressures.

The kinetics of hydrate formation of CCS streams are not as well understood as for natural gas. Experience with pipeline transportation of CO_2 shows that carbon dioxide hydrates form at temperatures of up to 10 °C at current transportation pressures [11]. The accumulation of hydrates in pipelines can cause problems such as plugging and equipment damage, as described by Carroll [12] and Bachu and Gunter [13]. Although hydrate formation is a serious threat for safe pipeline operation, there are good methods to avoid it, largely by avoiding the presence of free water. Carroll [12] mentioned, however, the small risk of hydrate formation with dissolved water.

Experience from existing pipelines shows that the corrosion rates are very low if the CO_2 is sufficiently dry. If free water is present, CO_2 can dissolve in water and form carbonic acid, which is corrosive. Corrosion rates can be in the order of millimeters per year when free water is present and in the order of micrometers per year when dry CO_2 is transported [14]. The discussion regarding risk of corrosion is therefore closely linked to the formation of free water.

In addition to corrosion from CO_2, corrosion effects could also occur from hydrogen sulfide, which forms a weak acid when dissolved in water [15]. The severity of the corrosion due to H_2S depends on factors such as CO_2 levels, temperature, gas velocity, pH, presence of water, internal or external stresses, and the concentration of H_2S.

Pipeline Capacity and Compression Work Modeling has indicated that impurities within the CO_2 will affect the phase envelope, the critical point, the viscosity, and the hydrate stability zone [16]. The presence of impurities shifts the boundary of the two-phase region towards higher pressures, so that higher operating pressures are required to keep CO_2 in the dense phase [9]. Impurities of only 2% may require an increased transport pressure of up to 25 bar to avoid two-phase formation.

The pipeline capacity can be significantly reduced with increasing levels of impurities. For instance, the decrease in volumetric capacity of the pipeline is 27% for a mixture of CO_2 with 10% of hydrogen compared with pure CO_2 [15]. From this perspective, the level of contaminants should be limited as much as possible.

Reducing the level of impurities after separation will increase capture costs, but will lower transport costs and increase operational reliability. Further research is required to determine the cost-optimal level of impurities in the CO_2.

20.3.1.3 Materials

Another important aspect of the design of CO_2 pipelines is the pipeline material. In the United States, pipelines used transporting CO_2 are commonly made of carbon steel. The selection of the material for CO_2 pipelines needs to take into account with the corrosive nature of CO_2 in the presence of water. Provided that the CO_2 is pure and dry (no free water), the risk of corrosion is low and carbon steel can be used without problems. A good example is the Snøhvit subsea pipeline that uses carbon steel without any internal coating, because dry CO_2 is transported.

In the case of very short transport distances, it may be economical to transport wet CO_2 and use more expensive stainless steel. In the Sleipner field, where wet CO_2 is injected into the Utsira aquifer, corrosion-resistant stainless steel has been justified because of the very short transport distance [14]. To reduce the risk of corrosion, inhibitors (e.g., glycol dehydrator) can be added to the CO_2 flow.

Both land and subsea pipelines need to be protected for corrosion from the outside environment. Often, an anti-corrosion protective layer is added by external coating of the pipeline or using cathodic protection.

20.3.2 Safety of CO_2 Pipeline Transport

The main safety issue for pipeline transport of large volumes of CO_2 is the risk of sudden leakages of large quantities of CO_2. This may lead to high concentrations of CO_2 in the vicinity of the pipeline, depending on, for instance, the nature of the blow-out and the meteorological conditions.

The design and operation of CO_2 pipelines presents some unique challenges that are different from traditional oil and gas transmission pipelines. Important differences are the specific properties such as toxicity, corrosiveness, and thermo-physical properties of the CO_2. Various quantitative risk assessments have been performed for CO_2 pipelines worldwide taking into account or omitting these specific characteristics [17–27]. Important variables – some of which may be influenced by the pipeline design – in such quantitative risk assessment are failure rates, operating pressure, pipeline and exit temperature, section length, diameter, orifice size (full rupture versus impingement), type and direction of release, meteorological conditions, jet diameter, vapor mass fraction in the release, and the toxicity of CO_2 [28]. We will not tackle all of these items here, but focus on some key aspects when assessing the risk of CO_2 pipelines, namely toxicity, release and dispersion in the case of a pipeline failure, the effect of impurities, and possibilities for risk mitigation.

20.3.2.1 Toxicity of CO_2

CO_2 is an asphyxiant and has toxicological effects. The impact of exposure to CO_2 on human health depends strongly on the concentration and duration of exposure. The short-term exposure limit of CO_2 is set at 3%, which is the normal exhale concentration. A level of 7% of CO_2 results in serious discomfort and reduced blood supply to the brain. A concentration of 10% (for 3–5 min) and 30% (for less than 1 min) may result in rapid loss of consciousness that ultimately may result in coma or death. Several CO_2 concentration thresholds are used worldwide in risk assessments and it has proven difficult to derive an accurate dose–response function for CO_2 that can be used in risk assessments. When reviewing risk assessments this is an important aspect to consider as it affects the outcome of safety assessments considerably.

20.3.2.2 Release and Dispersion in the Case of a Failing CO_2 Pipeline

Typically, a dispersing heavy gas without momentum will form a cloud that moves close to the ground, its progress being influenced heavily by local topography and obstructions. The dispersion of a heavy gas is different from that of gases that are lighter than air, such as natural gas, and is strongly influenced by meteorological conditions. Under stable weather conditions, elevated concentrations can be expected further away from the source in the case of low-momentum releases. High-momentum releases tend to disperse faster, which decreases the possibility of the formation of a dense cloud of CO_2.

Failures of CO_2 pipelines are typically divided into full ruptures and punctures. The type of failure has important influences on physical phenomena during the release and dispersion of CO_2. These phenomena comprise the dimensions and aspects of the source (area, height, direction), amount, velocity and duration of CO_2 release, and the phase of the CO_2 (solid, liquid or gas) [29].

The maximum CO_2 release rate from a failing pipeline has been estimated [28] to range between 0.001 and 22 t s^{-1}, depending mainly on the diameter of the pipeline and the size of the puncture, the operating pressure has less impact[1)] on the maximum release. Release characteristics such as the direction (vertical or horizontal) and momentum (impinged or unimpinged jet or instantaneous release) have a significant impact on the outcomes of a quantitative risk assessment (QRA).

The probability of failures assumed in QRAs for CO_2 pipelines ranges between 0.7×10^{-4} and 6.1×10^{-4} a^{-1} km^{-1} and is often based on experience with natural gas pipelines. The track record for failures with CO_2 pipelines in the United States shows failure rates of about $(3–4) \times 10^{-4}$ a^{-1} km^{-1} and did not result in fatalities [28, 30].

An important limitation when assessing risks accurately is the absence of validated release and dispersion models for high-pressure CO_2 pipeline failures, although significant research efforts are being deployed to fill this knowledge

1) With increasing operating pressure, the flow rate of the release is initially higher, but diminishes steeply with time.

gap. Validation of release and dispersion models is necessary for high-pressure CO_2 pipelines by performing experiments and field tests. Such tests should include CO_2 streams with impurities to analyze the effects of release and dispersion.

20.3.2.3 **Effect of Impurities**

The presence of impurities may influence phase transitions during sudden release or leakage from the pipeline [11, 30–33]. Impurities might also result in impacts on human safety as some of these substances are highly toxic (predominantly H_2S and CO).

Turner *et al.* [24] examined a CO_2 stream containing H_2S (0.2%) and CO (0.7%). The impurities had a small effect on the risk profile that was still dominated by the toxicity of CO_2. Another study showed a significantly greater distance to the exposure threshold for H_2S (up to 1180 m) than for the CO_2 content (up to 210 m) of the released CO_2 stream [34] implicating that H_2S (0.8%) was in that case the dominant toxic substance. Another study also indicated a higher safety risk due to possible exposure to H_2S (0.01%) compared with that of CO_2 [35]. Impurities may therefore dominate the risk profile depending on the concentration of impurities and the risk assessment methodology applied. From this perspective, the concentration of impurities should be limited. In the Dynamis project, for instance, a maximum H_2S concentration of 200 ppm is recommended [10]. Finally, as impurities may be of such high importance in the risk assessment, the assessment of the effects of impurities on operation, failure rates, and health impacts requires further attention.

20.3.2.4 **Possibilities for Risk Mitigation**

Risk-mitigation measures are typically aimed at (1) reducing the exposure to a failure mechanism, (2) increasing the resistance to a failure mechanism, (3) mitigating the effect of a failure, and (4) limiting the impact of a failure on the environment. An assessment of risk-mitigating efforts – such as block valves, protective materials, pipeline material selection, physical barriers, route selection, and management agreements – shows that mitigating the relevant risks can be considered part of current practice, making the risks of CO_2 transport by pipelines controllable. Thereby it is possible to meet stringent risk norms if required. Options for mitigating these risks are therefore available but will increase the cost of the pipeline infrastructure. Clearly an optimum has to be found between risk (decreasing failure rate, shorter duration, and less mass released) and economic aspects [30], also taking into account public risk perception [36].

Overall, quantifying the risks of CO_2 pipelines accurately is at present difficult. The way to mitigate these risks is, however, part of existing knowledge and existing risk-mitigation measures can be applied to CO_2 pipelines.

20.3.3
Cost of Pipelines

Costs of pipelines are determined by three main variables:

- construction costs (material/equipment costs and installation costs)
- operation and maintenance costs (energy costs, monitoring, operation, maintenance)
- other costs (design, project management, regulatory filing fees, insurance costs, right-of-way costs).

Figure 20.3 shows typical capital investment costs for pipelines under uncomplicated terrain conditions [9]. Extra investment is also required for artworks, that is, crossings of waterways and freeways. Costs can be up to € 2–3 million for each artwork.

Complex terrain conditions such as hilly areas and soggy or unstable soil may increase the investment costs considerably. Anther important factor is land use. Populated areas or nature reserves increase costs considerably. Onshore pipeline costs may increase by 50–100% or more when the pipeline route is congested and heavily populated. Costs increase in heavily urbanized areas because of accessibility to construction and additional required safety measures. Offshore pipelines generally operate typically at higher pressures and lower temperatures than onshore pipelines and may be 40–70% more expensive.

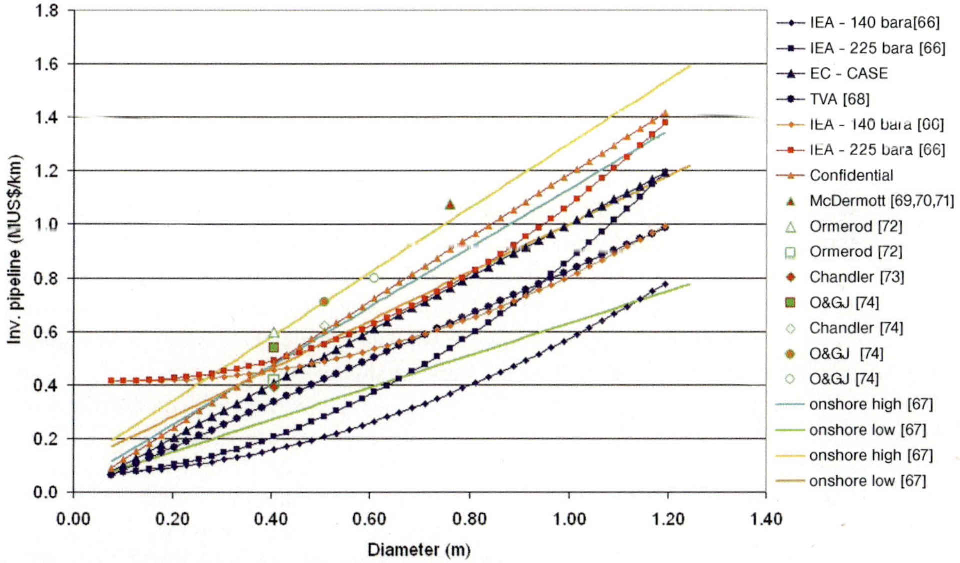

Fig. 20.3 Total investment costs for pipelines from various information sources for offshore and onshore pipelines. Costs exclude possible booster stations (IEA GHG, 2002; Hendriks et al., 2005; Bock, 2003; Sarv, 2000, 2001 a, 2001 b; Ormerod, 1994; Chandler, 2000; O&GJ, 2000).

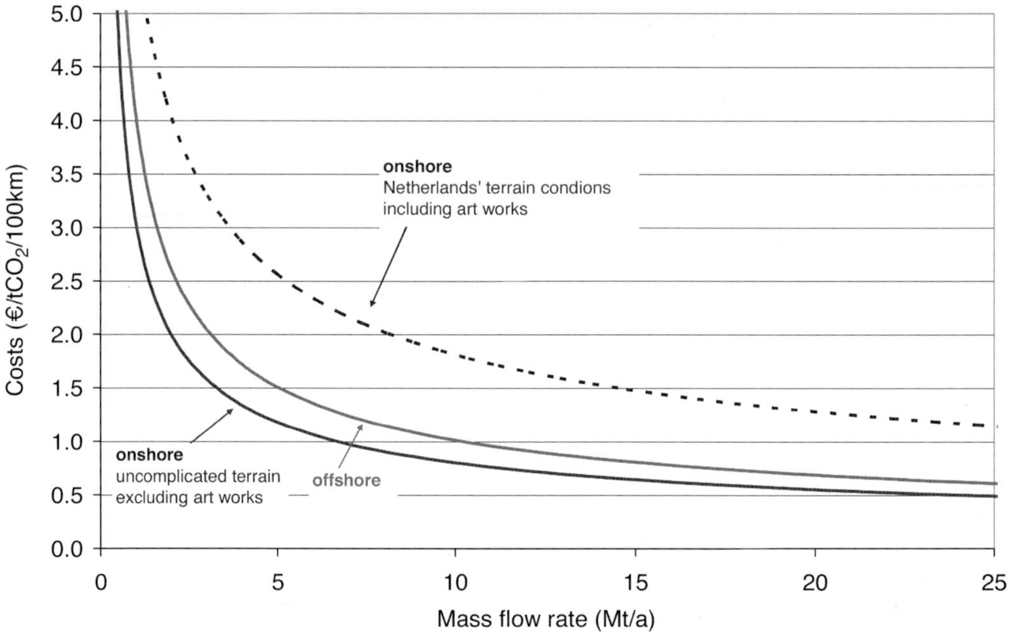

Fig. 20.4 Indication of transport costs for onshore and offshore pipelines per 100 km. A capital charge rate of 10% is assumed.

Figure 20.4 shows a range of transport costs for onshore and offshore pipelines per 100 km for "uncomplicated" onshore and offshore terrain conditions. The graph shows that transport costs are very sensitive to volume. For relatively difficult terrain conditions and including artworks, costs range from about € 1.8 t^{-1} for each 100 km for a large network transporting 10 Mt per year to € 1.2 t^{-1} for each 100 km for 20 Mt per year. They are about twice as high as those for onshore pipelines in uncomplicated terrains, excluding artworks.

Figure 20.5 shows the result of a cost optimization for the transportation of 170 kg s^{-1}. The discontinuity in the line is caused by the fact that a maximum of 40 bar pressure loss is considered acceptable. A higher pressure loss requires a booster station, which adds to the costs of capital and electricity.

20.4
CO_2 Transport by Ship

Large-scale shipping of CO_2 may be used to increase flexibility in the CO_2 transport system and to permit early CCS industry development.

Key differences exist between shipping and piping of CO_2. Ship transport distinguishes itself by different infrastructural requirements regarding the preparation (liquefaction, pressurization), loading , and unloading (injection) of CO_2 in

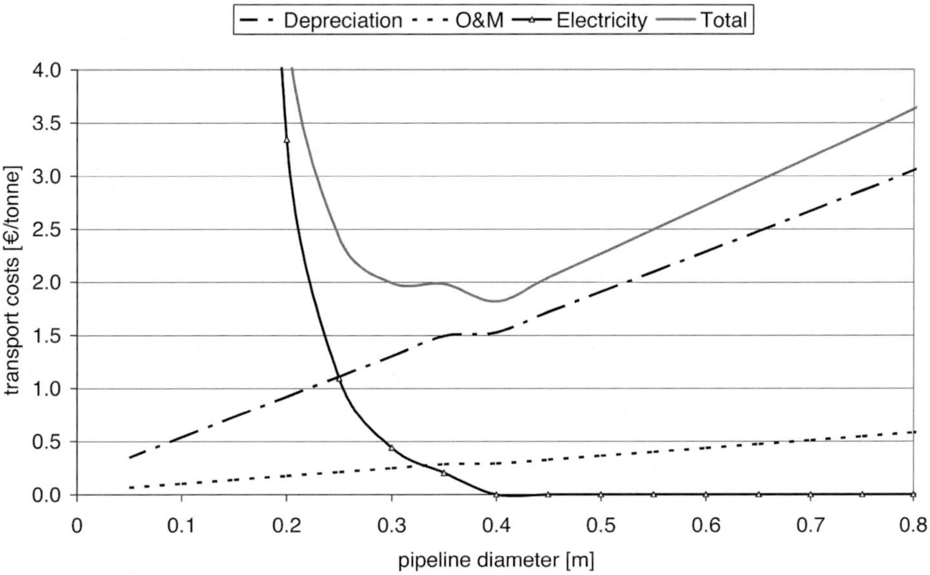

Fig. 20.5 Indication of optimal pipeline diameter for the onshore case. The discontinuity in the graph is caused by need for a booster station. Assumed required capacity is 170 kg s^{-1}, which corresponds to about 5 Mt per year.

the value chain. Ship transport has a different cost structure (capital and operating costs) to transport by pipeline. Capital expenditures are typically lower for ship transport, but operational expenditures are typically higher. Flexibility is also a distinct property of ship transport in contrast to the more rigid transport by pipeline. The flexibility of ship transport allows temporally servicing multiple sources and sinks and as such can be a key mode of transport in the start-up of large-scale CO_2 infrastructure. The start-up time, or lead time, of developing an infrastructure for CO_2 transport is also an advantageous aspect of shipping. This section briefly discusses the key requirements for and anticipated performance of CO_2 shipping and highlights the most important differences between ship and pipeline transport.

20.4.1 Typical Aspects of CO_2 Transport by Ship

CO_2 transport by ship is currently being conducted on a small scale with a limited number of ships transporting mainly food-grade CO_2 [37, 38]. The optimum phase for shipping was found to be the liquid phase in which the density is sufficiently high and the transport conditions could be maintained constant without requiring too costly materials and insulations. Also, solid CO_2 could be transported as dry cargo on ships, but seems not to be economically feasible mainly due to complex loading and unloading procedures [38, 39]. The most

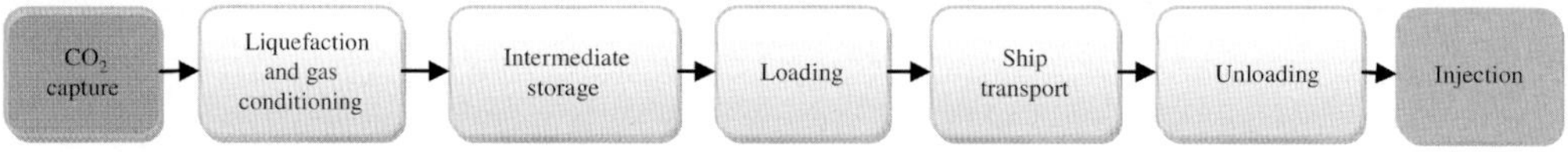

Fig. 20.6 Ship transport value chain.

likely options that remain are fully pressurized vessels and semi-pressurized/semi-refrigerated vessels in which the CO_2 is transported at approximately –55 °C and 6.5 bar. As depicted in Fig. 20.6, the ship transport value chain typically comprises five main processes: liquefaction and gas conditioning, intermediate storage, loading, ship transport, and offshore/onshore unloading. These processes are described below in more detail.

20.4.1.1 **Liquefaction**

Ship transport requires the energy-intensive liquefaction of CO_2, which entails the compression, purification, and cooling of the CO_2. The configuration of the liquefaction facility depends mainly on the quality and quantity of the CO_2 stream. Impurities have to be removed as they influence the properties of the stream, such as density, dew point, and boiling point, and consequently the energy requirement of different processes of the transport chain [38, 39].

20.4.1.2 **Intermediate Storage**

A key stage of the shipping value chain is intermediate storage, as the capture process of carbon dioxide is (semi)continuous whereas the cycle of ship transport is a discrete process.

The optimum capacity of intermediate storage tanks is determined by the cost of high-tensile steel versus the capacity, as the wall thickness (and hence the steel requirement) is proportional to the tank diameter and storage pressure. Hegerland *et al.* [40] reported an optimum capacity of about 20 kt for cylindrical storage tanks. Spherical tanks typically require relatively less steel due to a lesser wall thickness, but are more expensive to manufacture.

Multiple intermediate storage tanks would be required depending on the amount and capacity of ships utilized in the project or network. Typically, the intermediate storage capacity needs to be 1–1.5 times the capacity of a ship to allow for operational margins [41, 42].

20.4.1.3 **Loading**

In the loading stage, the CO_2 is pumped from the intermediate storage tanks to the ship using a loading arm. Boiled-off CO_2 vapor from the vessel is returned to the intermediate storage tank. The selection of materials and equipment, such as pumps and loading arms, should take into account possible corrosion, the low temperature, and the high pressure during operation.

20.4.1.4 **Ship Transport**

Ships for CO_2 transport have similar design considerations as vessels for transporting ammonia and liquefied petroleum gas (LPG). It is possible to rebuild LPG and ethylene vessels into CO_2 ships. According to van den Noort and Lemmes [43], investment costs for CO_2-carrying ships are 30–50% higher than for a similar sized semi-refrigerated LPG ship. Modification of LPG ships would most likely require investments of the same order of magnitude.

Ship designs are also being developed that may transport multiple cargos, for instance LPG and CO_2. These multipurpose carriers could mitigate investment risks but should meet different specifications with increased investment costs [8, 42, 44].

Important equipment on board the CO_2 vessel includes suitable tanks for stable storage of liquid/pressurized CO_2, a re-condensation unit (optional for long trips), and loading/offloading equipment such as pumps, heat exchangers, compressors, and heaters. Heating and compression of CO_2 are required to prevent dry-ice formation and to meet injection specifications. This equipment can also be placed on the offloading facility.

20.4.1.5 **Unloading**

Several offshore unloading concepts are currently being used in the oil and gas industry and may also be used for CO_2 ships; these include catenary anchor leg mooring (CALM), single-anchor leg mooring (SALM), tower mooring system (TMS), and submerged turret loading system (STL). These concepts are shown in Fig. 20.7.

According to Aspelund and Gundersen [46, 47], the STL variant is the most advantageous transfer system in terms of adaptability for different offshore installations. The choice for one of these concepts depends on the average weather conditions (i.e., wave height) at the injection site. Cost considerations and design specifications for the ship are also important elements in decision making. The TMS system does not use a buoy system, but a fixed rotating installation, and the hoses are always kept out of the water. STL tends to have fewer problems due to weather exposure/swell but higher installation costs. In contrast, SALM has higher weather exposure but lower installation costs. The TMS system has higher capital costs than the others but permits higher waves and thus higher availability. The system allows for additional space for the machinery required for unloading (compressors, pump, and heaters) and has in general lower operating and maintenance costs. In all concepts the CO_2 is transferred via a flexible hose, either submerged or not. The hose and mechanical equipment should be able to cope with low-temperature CO_2 if the CO_2 is not heated at the ship side of the transfer system [8, 42, 45–47].

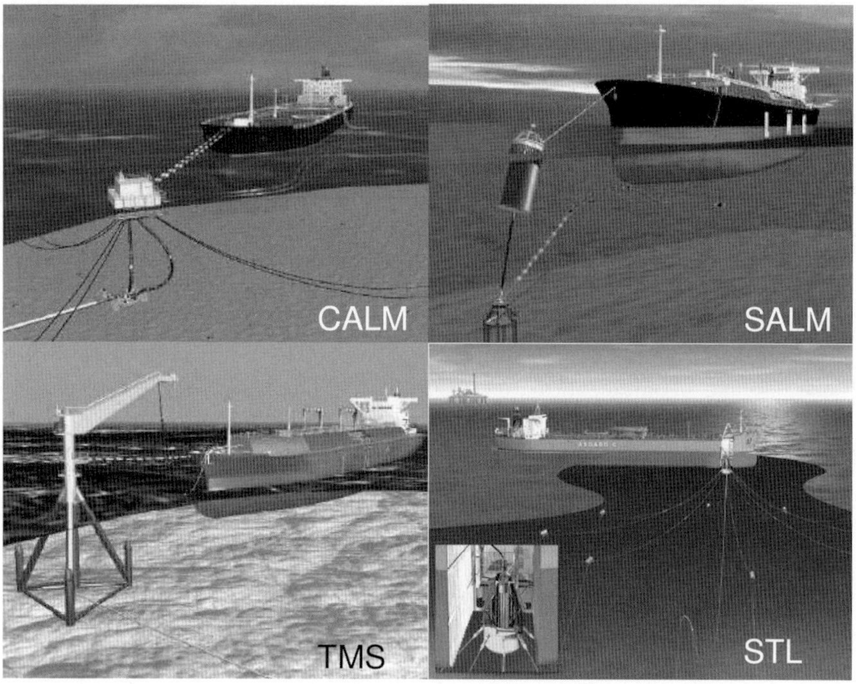

Fig. 20.7 Offshore mooring and offloading concepts [45].

20.4.2 Costs

Costs of a marine transport system comprise investment costs for liquefaction, intermediate storage, ships, and unloading system. Further costs are for operation (e.g., labor, ship fuel costs, energy costs, harbor fees) and maintenance. Important variables that influence the specific transport cost are the distance from source (harbor) to sink, site-specific injection characteristics, transported volume and ship capacity, and the project lifetime. The generic costs for marine transportation systems cannot be assessed in detail since no system has been implemented on a scale that handles several million tonnes of CO_2 per year and optimal value chain costs are very much case specific.

Figure 20.8 shows the breakdown of costs along the value chain for shipping CO_2 in the North Sea region. Various cost breakdowns are presented for different transport volumes, distance and ship capacity. Results by Nardon *et al.* [42] show that costs of liquefaction and unloading are fairly constant at about € 10 and € 3 t^{-1}, respectively. The costs of shipping and intermediate storage increase with longer distances traveled. Figure 20.8 shows that larger ship capacities are preferred when increasing the transport distance and annual transport volumes. Smaller ships are preferred for shorter distances and relative low ($\sim$1 Mt a^{-1})

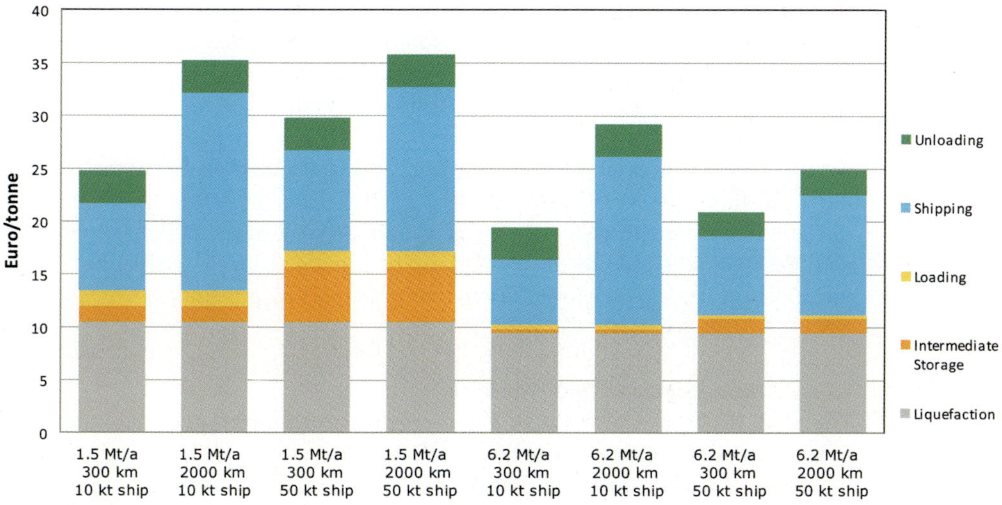

Fig. 20.8 Breakdown of CO_2 transport costs by ship when varying the annual CO_2 flow, transport distance, and ship capacity. Adapted from [42].

annual transport volumes. Overall cost estimates range between € 20 and € 35 t^{-1} transported.

20.5 Ships Compared with Pipelines

Key differences between pipeline and ship transport are found when comparing the two transport modes with respect to flexibility, lead time, and costs. Ships may provide a complementary mode of transport that can contribute to the flexibility of the transport infrastructure. Ships can be competitive in cases of long distances over sea, in cases of smaller storage reservoirs, and when the supply may show intermittent behavior or is only for a shorter period. Ships can also be put into operation faster than pipelines. Ships do not necessarily have to transport from the CO_2 supply side to the storage site, but can also be supplied by and/or fee into pipelines. In a larger CCS transport network, it is likely that a combination of ships and pipelines will be used, combining the advantages of each transport modes. Integration of ships and pipeline networks for CO_2 transport has never been investigated in great detail.

20.5.1.1 Flexibility and Speed of Implementation

A major advantage of marine transportation over pipelines is the increased flexibility of delivery and the possibility of shifting more easily from loading site and delivery site. Ships may serve multiple sources and sinks and also smaller storage reservoirs may be served that are uneconomical when in a pipeline infrastructure. Ships may also be deployed covering more markets than just CO_2 transport by combining CO_2 transport with LPG transport. Another advantage is that ship transport can scale more easily with demand and supply. Increasing the number of ships on one route or within a network of sources and sinks with increasing demand is relatively easier and faster than planning for and laying a pipeline [7].

The construction time for a CO_2 carrier is about 2 years [8, 9, 44] compared with the planning and construction phase of about 4 years for a pipeline 2). Ship transport can thus be deployed faster than transport by pipeline.

20.5.1.2 Cost Compared with Pipelines

The main difference between ship and pipeline transport can be found in share of capital and operation cost in specific transport cost. Typically, capital costs for the ship transport value chain is lower and operational costs are higher. For comparison, in the shipping phase capital cost represent about 50% [42] of the cost, whereas capital requirements for transport pipelines sum up to 60–80% [48] of specific transport cost. In addition, risks in the capital cost of ships that are designed to handle both CO_2 and other liquefied gases are lower as multiple markets can be served.

Pipeline transport costs increase considerably with transport distance, whereas this less adversely impacts ship transport. This means that high transport volumes for short distances are typically more economically transported by pipeline. Longer distances and lower volumes are transported more cost-effectively using ships.

The high capital costs and thus upfront investment costs are a main point of attention when developing a pipeline infrastructure. This can be mitigated by using ship transport in the start-up phase. In the start-up phase, smaller volumes are transported. When the supply of CO_2 increases and large volumes make pipelines more attractive, some transport routes may be served with pipelines. In such a configuration, the costly over-dimensioning of pipelines is prevented.

20.6 CO_2 Infrastructure Networks

If CCS wants to play a substantial role in the reduction of the emission of CO_2 towards 2050 and beyond, substantial amounts of CO_2 have to be captured and transported to the storage location, preferably in the most economical way. Timely development of transport infrastructure plays a crucial role in the successful deployment of CCS. Following the IEA roadmap [1], worldwide about 2.5 Gt of CO_2 will need to be captured and transported in 2030, increasing to

2) Planning, permissions, and construction for an onshore pipeline may even be longer, up to 6 years.

10 Gt in 2050. Ultimately, this will require a network of transport where supply and storage injection capacity are met and a system is established that can adequately balance changes in supply or storage capacity.

Principally two transport approaches can be distinguished. A first concept is to connect each capture project by dedicated pipelines to a storage location or to a cluster of storage structures. This approach will often be taken in the demonstration and early commercial phase of CCS. Gradually the stand-alone pipelines can be integrated into a grid or network of pipelines when the supply of CO_2 increases over time and multiple storage locations have to be used. An important question is how such a transition should be facilitated and what will be the most efficient and cost-effective approach.

A network can be realized, for instance, by one or more central backbone systems with satellite connections to individual (clusters of) sources and (clusters of) storage sites. A possible advantage of such backbone systems might be that remote, but large, storage sites, which would be prohibitively expensive for single sources, could be connected to become economically more attractive. Several studies have shown that large trunk lines are only cost-effective with large supplies of CO_2 (e.g., when CO_2 prices are high) and/or when storage availability is limited and clustered [49, 50]. Other advantages of networks are the increased flexibility in supply and injection capacity and centralized cross-border arrangements. Disadvantages are the need for strictly defined CO_2 specifications and for organizational institutes.

The timely availability of transport (networks) for CO_2 is essential for the success of CCS. However, (private) investment in transport will only be made if there is sufficient certainty regarding the required transport capacity. A strong political commitment will be required before a cost-effective and efficient network can be developed.

20.6.1
Development of Networks

Currently, CCS is hardly applied worldwide, and the design of future implementation schemes for CCS will need to be made almost from scratch. In an ideal world with perfect foresight into future demand and location of supply and storage capacity, a cost-optimal implementation trajectory for a network could be made. However, as CCS is a technology fully driven by (climate) policies, the future role of CCS in emission reduction policies is still unclear. This will impact the willingness to make large upfront investments which will only pay off in the longer term.

The development of a CO_2 transport network is a major challenge. Although transport costs may ultimately contribute only marginally to the total costs of CCS – say in the order of 5–15% – the development approach may well determine the success of CCS. Without secured transport capacity, operators may delay or call off projects to capture CO_2 as the associated (financial) risks may be perceived to be too high to implement the project.

Currently in Europe, six CCS projects are being developed, each capturing and storing about 1 Mt a^{-1} of CO_2. To make optimal use of these projects for future developments in CCS infrastructure, sufficient knowledge and understanding are needed about the commercial role of CCS. For instance, adequate insight may provide a basis for investing in oversized pipelines to accommodate future demand. The IEA roadmap [1] recognized the need for long-term infrastructural plans, to optimize the development CO_2 transport and to avoid conflicts of interest. This applies to land use at the surface and also to applications of the subsurface, where CO_2 storage is only one of several possible uses of the subsurface pore volume.

The main factors that will determine the layout of the network are how much CCS will be applied in the future and to which sources, and the availability of storage capacity. With respect to the latter, one can distinguish the geographic spread in storage capacity, the technical means to determine storage capacity – especially storage estimates for aquifers are highly uncertain – and the political position to allow access to the storage reservoirs. In some countries, such as The Netherlands and Germany, the public – especially people living in the vicinity of (planned) projects – have severe concerns about the transport of CO_2. Local opposition may reduce opportunities for onshore storage of CO_2 or increase the cost of the infrastructure through re-routing and/or the application of additional safety measures. In addition, several studies [49–51] have indicated that it is (very) unlikely that all CO_2 can be stored nationally, and networks will be more effective when they cross borders and use international waters. International cooperation and agreement allowing CO_2 transport, and associated liability transfer, between the countries involved would be required [51, 52].

In the development of networks, in principle use can also be made of existing (offshore) infrastructure, which may reduce costs and construction time. The BERR [52] concluded that existing trunk pipelines can be used, but at lower pressures and only when CO_2 is rigorously cleaned. Also, the timing of the availability is complicated as closure of the gas fields that they serve is difficult to predict precisely and subject to economic conditions, such as the price of fuels and new extraction technologies.

20.6.2
Cost Implications of Networks

In recent years, various studies have been carried out on the feasibility and costs of the CO_2 transport infrastructure. Studies commissioned by the IEAGHG focused on cost curves of transport and storage for North America and Europe [51, 53]. The European study examined the advantages of using a backbone compared with pipelines linking individual sources with individual

3) Storage data were taken from the EC GESTCO project completed by an additional inventory of seven additional countries not present in the GESTCO project. They comprise data on aquifers, oil and gas fields, and coal seams.

(clusters of) sinks. It concluded that when all storage capacity[3)] is taken into account, the construction of backbone pipelines does not reduce costs significantly. However, when storage capacity is scarce or concentrated in certain areas, using backbones may become financially attractive. For large-scale transport, average costs amount to about € 5 per tonne of CO_2. When storage capacity is limited (e.g., by excluding certain types of storage), costs may double [50]. In the UK, many feasibility and infrastructure studies have been performed for the North Sea region. In the BERR report [52] a one-to-one matching approach was used between sources and sinks for the UK and Norway. The study considered storing CO_2 from sources in the UK and Norway and the resulting infrastructure was limited to those countries. It was concluded that transport and storage costs amounts on average to about € 7 per tonne of CO_2. If EOR were to be applied, the revenues from oil should at least be € 60–80 per barrel to make EOR in the North Sea economically attractive, according to the BERR [52].

Figure 20.9 shows the outcome of a scenario where all main CO_2 sources in Europe are connected to the identified storage locations [50]. It should be noted that in this scenario a backbone pipeline has been predefined and was not a result of optimization of the system. Total investment costs amounts to about € 120 billion and more than 95% of the CO_2 can be transported and stored at less than € 20 t^{-1}. When storage is only possible in hydrocarbon fields, only one-third of the CO_2 can be stored at costs lower than € 20 t^{-1}.

According to Europipe [51], in 2050 1.2 Gt of CO_2 may be captured and transported through Europe. Assuming that storage can take place both onshore and offshore, about 22 000 km of pipelines will be required. About 18% of the CO_2 needs to be transported across borders. If only offshore reservoirs are used, the total pipe length increases by 50% to 33 000 km and 70% of the total CO_2 is transported across borders. This is not surprising as the majority of offshore storage capacity is located in only two countries (Norway and the UK) or in international waters.

20.7 Regulation and Investment Decisions

Effective planning and construction of CCS transport networks requires careful organization and coordination of the design, construction, and operation of the network. The role of the stakeholders involved – the public, industry, investors, governments – needs to be clear. Organization structures such as regulated monopoly or complete open market are possible forms to organize the competition in the market. In this section, we discuss various aspects of the organization of infrastructure development and the role of the government.

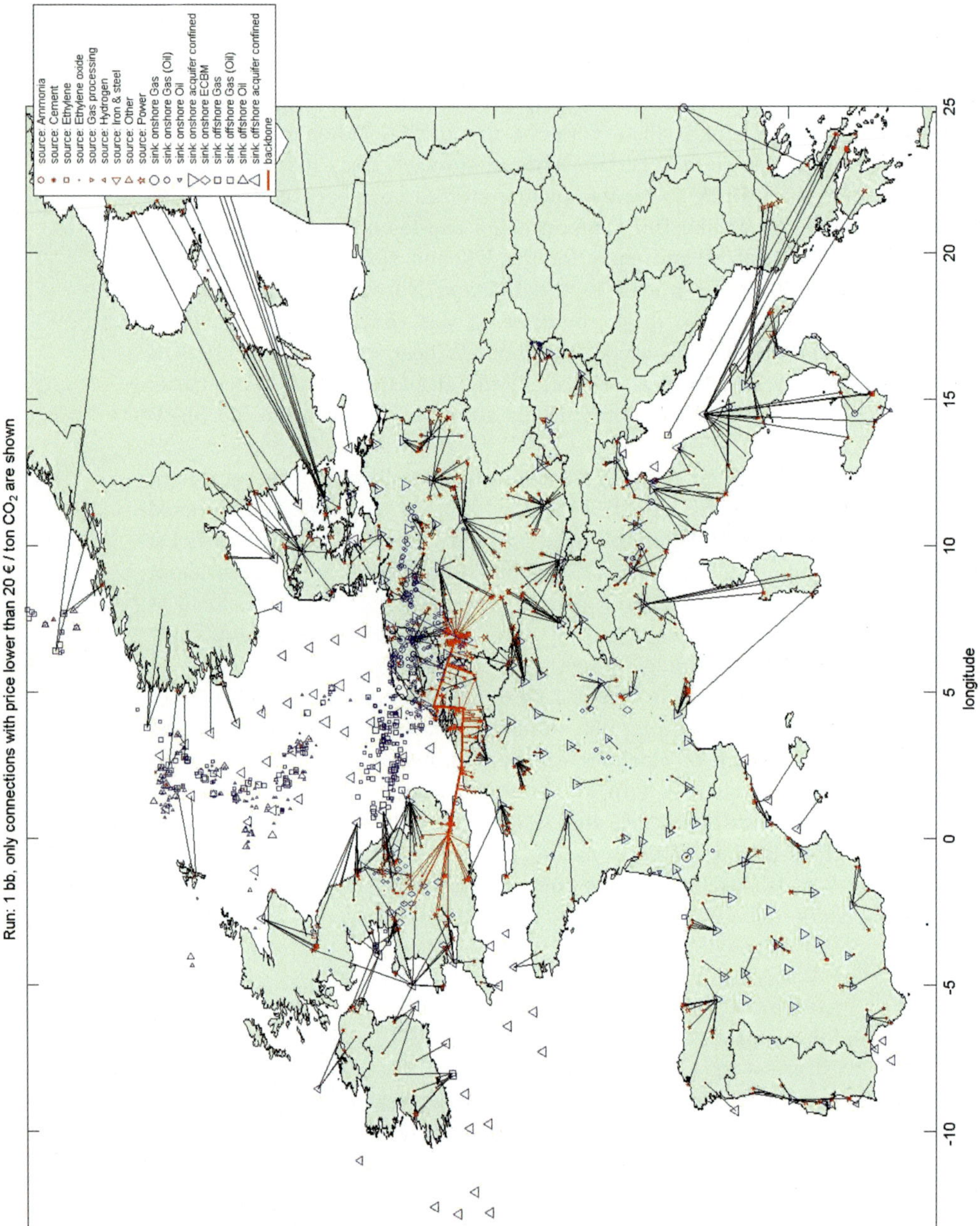

Fig. 20.9 Example of a map that shows the results of a future scenario for CO_2 sources.

20.7.1 Regulating CO_2 Transport Infrastructure

Infrastructures are systems that provide the physical connection for an economic transaction. This basic property makes CCS a partner in a transaction dependent on the infrastructure. Connected to this notion is the natural monopoly aspect of CO_2 infrastructure. A natural monopoly is characterized by strong economies of scale. The costs of investment are so large that it is an economically very unlikely that one would have two networks operating next to each other. Also, once the initial investments are made, the costs of operation are relatively small.

In the European CCS Directive (2009/31/EC), it is established that Member States are obliged to give any potential user access to transport and storage facilities (so-called third party access). This may be motivated by the fact that especially in the demonstration phase and early commercial phase a considerable contribution of public money is required. Exclusive rights by any of the private participants are therefore not preferable or possible. Moreover, it is important to create a level playing field between emitters of CO_2. Competitive advantage may otherwise be created with the support of governmental finance.

20.7.1.1 Regulation of Design Parameters

In determining the organization of CCS transport infrastructure, essential variables in the designs for the economic regulation of the network are ownership, competition regulation, and the degree of vertical integration of the transport into the CCS chain.

The effectiveness – or payback time – of the relatively high upfront investments in infrastructure is uncertain because it is subject to possible changes in (climate) policies. A reduced focus on climate change may result in a much lower demand for CO_2 transport than originally predicted. Therefore, following the rule of allocating a risk to a party that can best manage that particular risk, it follows that the government should take that part of the investment risk. Moreover, investment of public funds in the infrastructure signals a commitment to CCS by the government, and as such provides more certainty for private investors to invest in capture and storage equipment.

Another argument for (partly) public investment is the public interest, which is inherent in a well-functioning CCS system. Public interests provided by CCS are regional economic development and reduction of emissions, leading to the achievement of committed CO_2 reduction goals. Benefits in terms of these public values are not perceived by private companies, substantiating the need for the participation of a government. Another argument for government participation is related to initial overdimensioning of the transport capacity to anticipate future increases in demand. Constructing larger or stronger pipelines than needed for the initial amount of CO_2 to transport requires additional upfront investment with little certainty of return on investment. The time horizon of private companies is normally not long enough to take these capacity growth issues into account. Potential investors in transport facilities should have the right balance between investment risks and revenues. Coordination by a public authority in the design could justify higher upfront investments. Potential transporters of CO_2 have indicated that they see CO_2 transport as an activity with low technical risk and with low associated margins. They do not see themselves as a financial risk-bearing party in the CCS chain but will act upon transport contracts for a fixed (long-term) period of time. Infrastructure will not be developed or transport tariffs may increase substantially when such contracts may not be established.

Vertical integration considers the participation of companies investing in CO_2 capture and/or storage in transport. The effect of vertical integration is twofold. On the one hand, vertically integrated companies can efficiently coordinate the supply and demand of subsystems in the value chain. This would lead to cost reductions. On the other hand, vertically integrated companies can use their ownership of (part of) the network to strengthen their position. They can exclude their competitors from the network or charge high access prices. This trade-off is similar to the trade-off present in the unbundling discussion in other infrastructure sectors such as electricity and gas. The natural monopoly elements of the network have been separated from the competitive elements of the network. A similar line of reasoning can be followed for CO_2 infrastructures.

20.7.2
Long-Term Vision on CO_2 Infrastructure

To minimize the financial risk to private transport companies, the government should carefully guide the CCS implementation process in order to profit optimally from the CCS potential. The government can generally take three positions in the implementation of transport infrastructure: supervisor, driver, or participant/owner. The government should focus on those aspects that are not (naturally) performed by the market parties. Based on the current market structure, the available financial incentives, and existing uncertainties in the market, the government should take or facilitate initiatives especially in the demonstration and early commercial phase.

To avoid the transport network being be a major obstacle to the large-scale deployment of CCS, governments should create long-term visions on CCS implementation, preferably in an international context. In such a vision, a strategy should be developed in which the government prioritizes storage reservoirs and indicates what infrastructure can be re-used. Based on such a vision, a CO_2 transport strategy should be set up. This strategy can serve for industrial partners as a reference for future investments. In situations where an extensive network is required in a relatively short period, it may be effective to have the responsibility in the hands of only one party to stimulate fast decision-making processes and to create an efficient transport system. It will be clear that suffi-

Table 20.2 Estimated duration of legal and construction periods for CO_2 pipelines in The Netherlands, onshore and offshore [7].

Item	Network pipeline	Dedicated pipeline
Onshore	4–6 years	3–4 years
Planning and permits	2–4 years	2–3 years
Construction	2 years	1 year
Offshore	3–4 years	
Planning and permits	2–3 years	
Construction	1 year	

cient guarantees should be created to have competitive prices for CO_2 transport. However, such business models are not unusual in large infrastructures, for example, in natural gas transport. The government should remain responsible for the overall management.

20.8 Strategic Planning for Pipelines

Planning and permissions dominates the lead time of CO_2 pipelines, especially onshore (see Table 20.2). The most important legal requirements affecting the timeline of deploying pipelines are obtaining concessions from landowners (rights of way) and the environmental impact assessment (EIA/SEA) procedure. The latter procedure is required in many countries worldwide, although in different forms. Because this procedure is open to public involvement, it may potentially delay the construction of a pipeline for several years if there is opposition from the public or local governments [54]. Oversight by multiple authorities that regulate different aspects of siting CO_2 pipelines could also cause delays and increase costs for pipeline developers [55, 56].

20.8.1 Environmental Impact Assessment

The main goal of the environmental impact assessment procedure, on a project level (EIA) or strategic level (SEA), is to assess and evaluate the impacts on the environment of an initiative and possible alternatives. One benefit of the procedure is that all relevant information on environmental impacts and safety, required for various administrative decisions, is gathered in a single report which is open to third-party consultation. Possibilities for participation of the public in both the SEA and EIA procedures may play an important role in the public acceptance of CCS plans and projects. Given that plans are currently being drafted for CCS (pilot) projects, there is a growing need for clarity on administrative, juridical, and environmental implications of these projects (see also [57]).

Environmental impact assessments conducted in the past on comparable infrastructural projects (natural gas pipelines, power plants, and underground gas storage) may provide lessons for developing long-term CCS planning without many delays in formal procedures.

An example is the EIA procedure for underground gas storage (UGS) in an empty gas field at Langelo, The Netherlands. For this project, the storage location at Langelo was the subject of the assessment and nominated as the location of preference by the initiators. The procedure was significantly delayed[4] due to the demand for more information on the strategic value of the UGS facility for the Dutch natural gas supply. Furthermore, before the regional spatial plan could be revised to implement and formally allow the UGS activity, the location at Langelo had to be proven to be the most suitable one and therefore alternatives had to be assessed on their suitability [58, 59]. This led to a broadening of the scope of the assessment to include strategic alternatives not directly related to that UGS project, such as national energy-saving options.

Another EIA procedure for a natural gas pipeline includes the assessment of alternative locations for a compressor station, route alternatives, construction methods for the pipeline, and crossings with other infrastructure [60]. Interesting in this case is the advice by an independent EIA advisory body for that procedure, which stated that when describing the need for developing a pipeline it should be argued why other alternatives, such as transport by LNG tankers, are excluded [61]. A strategic and long-term vision could deal with these types of assessments and arguments for CCS activities and the linking infrastructure in advance, and with it preventing delays in formal procedures on the project level.

In a (currently cancelled) CCS transport and storage project in Barendrecht, The Netherlands, safety turned out to be a primary concern in the public debate [62]. The content of the EIA procedure, specifically the uncertainties surrounding risk assessments for transport and storage, turned out to be of very high importance for the governmental bodies involved in the decision-making process for that project. Delay occurred for the same reasons as for the UGS project described above. Third-party objections were submitted in the EIA procedure regarding the content and scope of the procedure. The scope of the procedure was deemed too narrow as no comprehensive assessment was made of alternative locations. Also, a discussion on the need and necessity for CCS was optioned to be assessed in more detail. With hindsight, a strategic vision accompanied by an SEA including allocated CO_2 storage sites and possibly transport routes would have eased the (legal) procedure for this specific project.

4) The EIA procedure started in June 1991 and the EIA report was finished at the end of 1992. In April 1993, the application for the revision of the local zoning plan was rejected. In December 1994, the Minister of VROM issued a designation order to revise the regional spatial plan of the province of Drenthe. The construction of the facility eventually started in 1995 and it was operational in 1997.

An initiative, the AMESCO[5] project, taken by the Dutch Oil Company (NAM) and supported by four provinces[6] also added to the knowledge base. The goal of that project was to produce a paper that can be used as a reference guide for future EIAs for underground CO_2 storage activities. This "generic" EIA is being drafted to "prevent identical environmental studies being performed for each location" [63] and thus speed up the implementation of individual projects. This document has been used in the EIA procedure for the Barendrecht project, but proved to have less authority than an official long-term vision on CCS that would be the responsibility of the national government.

20.8.2
Strategic Planning and Formal Procedures

Long-term strategic planning is deemed necessary to allow for a streamlined and cost-effective deployment of a CCS infrastructure. The strategic planning of a CO_2 infrastructure entails predominantly timing and spatial considerations. Spatial considerations are related to the locations of sources and sinks and timing considerations depend on the availability of sources and sinks over time [7, 49, 64].

The most important organizational and legal aspects of strategic planning for a CO_2 infrastructure arise when such a plan is to be embedded in existing long-term spatial planning by (local, national) governmental bodies and an SEA procedure is to be conducted [56]. Long-term planning may include deciding locations for storage and CO_2 sources (e.g., industrial activities and power plants) and deciding/reserving preferred routes for CO_2 pipelines and shipping. Planning for a CCS infrastructure then integrates this long-term vision for the development and allocation of locations for both CO_2 point sources, such as large-scale industry and the power sector, and also a long-term vision for the use of underground sites, including preferences for locations and reservoirs for, amongst, others CO_2 storage, geothermal heat, and hydrocarbon activities (exploration, production, and seasonal storage). However, a general problem is that national spatial plans do not, or only in a limited way, acknowledge the linkages between CO_2 capture, transport, and storage and that these policy plans may incorporate merely the individual elements of the CCS chain. This, together with the possible size and required deployment rate of CCS infrastructures, complicates the timing and scope of formal procedures compared with, for instance, the power and oil and gas infrastructure, which are not so strongly linked without CCS and for which infrastructure build-up took several decades.

An example of such a long-term vision is the structure vision for pipelines in The Netherlands that includes possible CCS infrastructure development [65]. Important legal and organizational aspects are that long-term spatial plans have

5) Generic Environmental Impact Study on CO_2 Storage (AMESCO in Dutch).

6) The Netherlands is divided administratively into 12 provinces. The provincial government is responsible for the formulation of policy at a regional level.

to be revised, which includes governmental authorities on several regional levels and is open to public consultation. This type of procedure typically takes several years (at least 5 years) to complete, due to the complex nature of "predicting" future infrastructure and due to the involvement of a large number of stakeholders (multi-level), governments, and the public.

20.8.3 Strategic Planning and Safety

Safety of CO_2 pipelines is an important aspect of the deployment of CO_2 infrastructure, especially in densely populated areas such as north-west Europe. In the IPCC report [9], it is noted that as "the consequences of CO_2 pipeline accidents potentially are of significant concern, stricter regulations for CO_2 pipelines than those for natural gas pipelines currently are in force in the United States." A recent study shows that uncertainties exist when assessing the risk of CO_2 pipelines [28], but also that the existing CO_2 infrastructure has a good safety track record [30]. In addition, ample technologies for the mitigation of risk are applicable to CO_2 pipelines. The risks of CO_2 pipelines should therefore be taken into account when deciding pipeline routes for projects or within long-term spatial planning, but is manageable and likely to be of the same order of magnitude as in other industrial activities for which risk assessments are required.

References

1 IEA. *Technology Roadmap Carbon Capture and Storage, International Energy Agency,* Paris, 2009.
2 Dooley, J.J, Dahowski, R.T and Davidson, C.L *Comparing Existing Pipeline Networks with the Potential Scale of Future U.S CO_2 Pipeline Networks.* In *Greenhouse Gas Control Technologies,* Energy Procedia, Washington, DC, 2008.
3 Ferguson, R. Presented at the SPE International Conference on CO_2 Capture, Storage, and Utilization, Advanced Resources International, 2010.
4 Baksht, D., *et al. Viability of Carbon Capture and Sequestration in Existing Coal-Fired Power Plants in the United States,* 2008, Chicago, University of Chicago.
5 Kadnar, J.O *CO_2 transportation via pipeline,* US Department of Transportation, NIST Workshop, 2009.
6 Energy Institute. *Good Plant Design and Operation for Onshore Carbon Capture Installations and Onshore Pipelines,* Energy Institute, Global Carbon Capture and Storage Institute, London, 2010.
7 Hendriks, C., Hagedoorn, S., and Warmenhoven, H. *Transportation of Carbon Dioxide and Organisational Issues of CCS in The Netherlands – Report Prepared by Order of EnergieNed, the Ministry of Economic Affairs and the Ministry of Housing Spatial Planning and the Environment,* Ecofys, SpinConsult, Utrecht, 2007.
8 Schulze, A. Maersk Tankers – a pioneer in CO_2, *Carbon Capture J.*, 2010, (31 March), 19–20.
9 IPCC. *IPCC Special Report on Carbon Dioxide Capture and Storage, Prepared by Working Group III of the Intergovernmental Panel on Climate Change* (ed. B. Metz, *et al.*), Cambridge University Press, Cambridge, 2005, p. 442.
10 DYNAMIS. DYNAMIS CO_2 quality recommendations. In *Towards Hydrogen and Electricity Production with Carbon Dioxide Capture and Storage,* Ecofys, Statoil, SINTEF-ER, PEL, Vattenfall, 2007.

11 Odru, P., *et al.* Technical and economical assessment of CO_2 transportation. Presented at Greenhouse Gas Control Technologies 8. Trondheim, 2006.

12 Carroll, J. J Problem is the result of industry's move to use higher pressures. *Pipeline Gas J.*, 2003, **230**(6), 60–61.

13 Bachu, S. and Gunter, W. D Overview of acid-gas operations in Western Canada. In *Proceedings of the 7th International Conference on Greenhouse Gas Control Technologies. Volume 1: Peer-Reviewed Papers and Plenary Presentations* (ed. Rubin, E. S, Keith, D. W and Gilboy, C. F), IEA Greenhouse Gas Programme, Cheltenham, pp. 443–448, 2004.

14 Seiersten, M. Material selection for separation, transportation and disposal of CO_2. In *Proceedings, Corrosion 2001*, National Association of Corrosion Engineers, Houston, TX, 2001, paper 01402.

15 Mohitpour, M., Golshan, H., and Murray, A. *Pipeline Design and Construction, a Practical Approach*, American Society of Mechanical Engineers, New York, 2003.

16 DYNAMIS. Assessment of technical experience and engineering practice for CO_2 pipelines. In *Towards Hydrogen and Electricity Production with Carbon Dioxide Capture and Storage*, Ecofys, British Geological Survey, 2009.

17 Golomb, D. Transport systems for ocean disposal of CO_2 and their environmental effects. *Energy Convers. Manage.*, 1997 **38**(Suppl. 1), S279–S286.

18 Vendrig, M., *et al. Risk Analysis of the Geological Sequestration of Carbon Dioxide*, DNV Consulting for Department of Trade and Industry, London, 2003.

19 National Energy Board. *Reasons for Decision Souris Valley Pipeline Limited, Report MH-1-98*, National Energy Board, Calgary, 1998.

20 Cameron-Cole. Air dispersion modelling of well blowout and pipeline rupture scenarios Salt Creek field. In *Environmental Assessment Howell Petroleum Phase III/IV CO_2 Enhanced Oil Recovery Project: Salt Creek Oil Field*, US Department of the Interior; Bureau of Land Management, Casper Field Office, 2005.

21 Molag, M. and Raben, I. ME. *Externe Veiligheid Onderzoek CO_2 Buisleiding bij Zoetermeer*, TNO, Apeldoorn, 2006, p. 46.

22 Heijne, M. AM. and Kaman, F. JH. *Veiligheidsanalyse Ondergrondse Opslag van CO_2 in Barendrecht – Appendix 6*, in *MER Ondergrondse opslag van CO_2 in Barendrecht*, Haskoning Nederland, Tebodin, The Hague, 2008.

23 Tetra Tech. *Final Risk Assessment Report for the FutureGen Project. Environmental Impact Statement*, Tetra Tech, Lafayette, CA, 2007.

24 Turner, R., Hardy, N., and Hooper, B. Quantifying the Risks to the public associated with a CO_2 sequestration pipeline: a methodology and case study. In *Abstracts of Greenhouse Gas Control Technologies 8*, Trondheim, 2006.

25 Kruse, H. and Tekiela, M. *calculating the consequences of a CO_2-pipeline rupture. Energy Convers. Manage.*, 1996, **37**(6–8), 1013.

26 Hooper, B., Murray, L., and Gibson-Poole, C. (eds.) *Latrobe Valley CO_2 Storage Assessment*. CO2CRC Publication No. RPT05-0220, Cooperative Research Centre for Greenhouse Gas Technologies, Canberra, 2005.

27 Lievense. *OCAP CO_2 v.of. CO_2 GREENGAS PROJECT Risico Analyse NPM-leiding Document 042282 rev. 3*, Raadgevend Ingenieursbureau Lievense, Breda, 2005.

28 Koornneef, J., *et al.* Quantitative risk assessment of CO_2 transport by pipelines – a review of uncertainties and their impacts. *J. Hazard. Mater.*, 2010, **177**, 12–27.

29 Uijt de Haag, P. AM. and Ale, B. JM. (eds.) *Guidelines for Quantitative Risk Assessment – the 'Purple Book'. Publication Series on Dangerous Substances (PGS 3)*, Committee for the Prevention of Disasters (CPR), VROM, RIVM, The Hague, 1999.

30 Gale, J. and Davison, J. *Transmission of CO_2 – -safety and economic considerations*, 2004. Energy 29(9–10), pp. 1319–1328.

31 Eldevik, F., *et al.* Development of a guideline for safe, reliable and cost efficient transmission of CO_2 in pipelines. *Energy Procedia*, 2009, 1(1), 1579–1585.

32 de Visser, E., *et al.* Dynamis CO_2 quality recommendations. *Int. J. Greenhouse Gas Control*, 2008, 2(4), pp. 478–484.
33 Oosterkamp, A., *et al.* Pipeline transport of CO_2 with impurities – a state of the art review. Presented at the 4th Trondheim Conference on CO_2 Capture, Transport and Storage, 16–17 October 2007, Trondheim.
34 National Energy Board. *Reasons for Decision Souris Valley Pipeline Limited, Report MH-1-98*, National Energy Board, Calgary, 1998.
35 Tetra Tech. *Final Risk Assessment Report for the FutureGen Project. Environmental Impact Statement*, Tetra Tech, Lafayette, CA, 2007.
36 Vlek, C. Environmental versus individual risk taking: perception, decision, behavior. In *Encyclopedia of Applied Psychology* (ed. Spielberger, C.), Elsevier, Amsterdam, 2004, Vol. 1, pp. 825–840.
37 IEAGHG. *Ship Transport of CO_2*, International Energy Agency Greenhouse Gas R&D Programme, Cheltenham, 2004.
38 Aspelund, A., Molnvik, M. J, and Koeijer, G. D Ship transport of CO_2 – technical solutions and analysis of costs, energy utilization, exergy efficiency and CO_2 emissions. *Chem. Eng. Res. Des.*, 2006, *84*(A9), 847–855.
39 Aspelund, A. and Jordal, K. A study of the interface between CO_2 capture and transport. Presented at Greenhouse Gas Control Technologies 8, Trondheim, 2006.
40 Hegerland, G., Jorgensen, T., and Pande, J. Liquefaction and handling of large amounts of CO_2 for EOR. In *Proceedings of the 7th International Conference on Greenhouse Gas Control Technologies. Volume 1: Peer-Reviewed Papers and Plenary Presentations* (ed. Rubin, E. S, Keith, D. W and Gilboy, C. F), IEA Greenhouse Gas Programme, Cheltenham, pp. 2541–2544, 2004.
41 Berger, B., Kaarstad, O., and Haugen, H. A Creating a large-scale CO_2 infrastructure for enhanced oil recovery. In *Proceedings of the 7th International Conference on Greenhouse Gas Control Technologies. Volume 1: Peer-Reviewed Papers and Plenary Presentations* (ed. Rubin, E. S, Keith, D. W and Gilboy, C. F), IEA Greenhouse Gas Programme, Cheltenham, pp. 321–328, 2004.
42 Nardon, L. *Techno-Economic Assessment of CO_2 Transport by Ship. A Case Study for The Netherlands*, Cá Foscari University, Utrecht University, Ecofys, Vicenza, 100 pages, 2010.
43 van den Noort, A. and Lemmens, H. Technical aspects of ship transport. Evaluation of the technical possibilities of ship transport into CO_2 transport infrastructure. In *CATO-2 WP 2.1*, Publ.: Gasunie, Inst.: KEMA, 2010.
44 Tetteroo, M. *CO_2 Transport by sea going ships*, in *Carbon Capture & Storage Conference 2010*, London, 19 March 2010.
45 Bluewater Energy Services. *North Sea CO_2 Injection CO_2 Transfer Options – Technical Feasibility Screening Study*, Ecofys, Gaz de France Suez, Skaugen, Bluewater Energy Services, 2009.
46 Aspelund, A. and Gundersen, T. A liquefied energy chain for transport and utilization of natural gas for power production with CO_2 capture and storage – Part 2. *Appl. Energy*, 2007, 781–792.
47 Aspelund, A. and Gundersen, T. A. liquefied energy chain for transport and utilization of natural gas for power production with CO_2 capture and storage – Part 3. *Appl. Energy*, 2007, 805–814.
48 EBN/Gasunie. *CO_2 Transport and Storage Strategy*, Energie Beheer Nederland, Gasunie, 2010.
49 van den Broek, M. A Modelling approaches to assess and design the deployment of CO_2 capture, transport, and storage. In *Science, Technology and Society*, Utrecht University, Utrecht, 2010, p. 273.
50 IEAGHG. *Building the Cost Curves for CO_2 Storage: European Sector*, International Energy Agency Greenhouse Gas R&D Programme, Cheltenham, 2005.
51 TNO, ECN, and Anthony Veder. Development of a large-scale CO_2 transport infrastructure in Europe: matching captured volumes and storage availability. In *CO_2Europipe Project – Towards a Transport Infrastructure for Large-Scale CCS in Europe*, TNO, ECN, Anthony Veder, 2010.

52 BERR. *Development of a CO_2 Transport and Storage Network in the North Sea: Report to the North Sea Basin Task Force.* Department for Business, Enterprise and Regulatory Reform, Element Energy Limited, Pöyry Energy, British Geological Survey, 2007.

53 IEAGHG. *Building the Cost Curves for CO_2 Storage: North America,* International Energy Agency Greenhouse Gas R&D Programme, Cheltenham, 2005.

54 Wood, C. *Environmental Impact Assessment – a Comparative Review,* 2nd edn., Pearson Education, Harlow, 2003.

55 WRI. *CCS Guidelines: Guidelines for Carbon Dioxide Capture, Transport, and Storage,* World Resources Institute (WRI), Washington, DC, 2008.

56 Koornneef, J., Faaij, A., and Turkenburg, W. The screening and scoping of environmental impact assessment and strategic environmental assessment of carbon capture and storage in The Netherlands. *Environ. Impact Assess. Rev.,* 2008, **28**(6), 392.

57 Mace, M.J, Hendriks, C., and Coenraads, R. Regulatory challenges to the implementation of carbon capture and geological storage within the European Union under EU and international law. *Int. J. Greenhouse Gas Control,* 2007, **1**(2), 253.

58 ECN. Gasopslag. In *Energie Verslag Nederland, 1994,* ECN, Petten, 1994.

59 NAM/Gasunie. *Startnotitie Milieueffectrapportage: Opslag van Aardgas in het Gasveld Norg,* NAM/Gasunie, 1991.

60 Arcadis. *Startnotitie/MER aardgasleiding Zuid-Beveland,* Delta, Middelburg, 2006.

61 Commissie MER. *Aardgasleiding Zuid-Beveland – Advies voor Richtlijnen voor het Milieueffectrapport,* Commissie MER, Utrecht, 2006.

62 Haskoning Nederland. *MER Ondergrondse opslag van CO_2 in Barendrecht – Rapport 1: Samenvattend hoofdrapport,* Shell CO_2 Storage, The Hahue, 2008.

63 Croezen, H., *et al. AMESCO Generic Environmental Impact Study on CO_2 Storage,* NAM, SEQ, Nogepa, Essent, Electrabel, Eneco, Province Groningen, Friesland, Drenthe, South-Holland, Ministry of VROM, State Supervision of Mines, 2007.

64 van den Broek, M., *et al.* An integrated GIS-MARKAL toolbox for designing a CO_2 infrastructure network in The Netherlands. *Energy Procedia,* 2009, **1**(1), 4071–4078.

65 Ministerie van VROM. *Notitie Reikwijdte en Detailniveau Planmer structuurvisie Buisleidingen,* Ministry of Environmental Affairs (VROM), Arcadis, 2009.

66 IEA GHG, 2002: Transmission of CO_2 and Energy, IEA Greenhouse Gas R&D Programme, Report PH4/6, IEA GHG, Cheltenham, UK (March).

67 Hendriks, C.A., Wildenborg, T., Feron, P., and Graus, W. 2005: Capture and Storage, prepared for EC, DG-ENV, Ecofys Energy and Environment, report nr. M70066.

68 Bock, B.R., Rhudy, R., Herzog, H., Klett, M., Davison, J., de la Torre Ugarte D.G., and Simbeck, D., 2003: Economic Evaluation of CO_2 Storage and Sink Enhancement Options. TVA Public Power Institute, February 2003.

69 Sarv, H. and John, J., 2000: Deep ocean sequestration of captured CO_2. Technology, 7S, 125–135.

70 Sarv, H., 2001 a: Further Technological Evaluation of CO_2 Storage in Deep Oceans. Presented at the 26th International Technical Conference on Coal Utilisation & Fuel Systems, March 5–8, 2001, Clearwater, Florida.

71 Sarv, H., 2001 b: Large-scale CO_2 transportation and deep ocean sequestration – Phase II final report. McDermott Technology Inc., Ohio. Technology Report DE-AC26-98FT40412, 2001.

72 Ormerod, B., 1994: The disposal of carbon dioxide from fossil fuel fired power stations. IEA Greenhouse Gas R&D Programme, Cheltenham, Technical Rep. IEAGHG/SR3, June 1994.

73 Chandler, H.M. 2000: Heavy Construction Cost Data – 14th Annual Editions. R.S. Means Company, Inc. Kingston, MA, USA.

74 O&GJ, 2000: Pipeline Economics. Oil and Gas Journal, **98**(36), 68–86.

21
Storage of Fossil Carbon

Klaus S. Lackner

Abstract

The use of fossil fuels will be limited by concerns over climate change, which is caused in large part by the associated carbon dioxide emissions. While the oxidation of carbon is the unavoidable consequence of extracting energy from fossil carbon, the subsequent emission could be avoided by capturing the carbon dioxide and storing it permanently. However, permanent storage represents an enormous challenge, mainly because of the sheer volume of emissions. This chapter considers various storage technologies and puts them in perspective with regard to technical difficulties, storage capacity, permanence of storage, storage safety, and the ability to account for the stored carbon.

Keywords: carbon capture and storage, geological storage, mineral carbonation

21.1
Introduction

Carbon capture and storage has been proposed as an important tool for meeting growing energy demands while avoiding climate change [1]. Capture of CO_2 may be the more expensive aspect of the technology, but it follows standard industrial practice of gas separation. By contrast, CO_2 storage is moving into uncharted territory because of the huge quantities that would need to be stored safely and permanently. Capture of CO_2 at power plants or other sources only makes sense if there is a viable course of action for the subsequent use or disposal of the gas. This requires a commercial use for the captured CO_2 or an affordable technology for its safe and permanent disposal.

This chapter summarizes a number of CO_2 storage options, considering various aspects of carbon storage ranging from storage capacity to the cost of storage and future monitoring, and from environmental constraints to potential safety issues.

Process Engineering for CCS Power Plants, First Edition. Edited by Detlef Stolten and V. Scherer

21.2 Summary of Storage Options

Carbon storage technology aims to isolate the carbon remaining after fossil fuel consumption in well-defined reservoirs or carbon sinks that have little or no interaction with the atmosphere. Waste carbon could be diluted into large natural carbon reservoirs, or new reservoirs could be created and filled with anthropogenic carbon. Carbon could be stored as CO_2, the form in which it is captured, or it could be converted to other more convenient forms prior to storage. Technology must provide sufficient storage space, assure the safety and permanence of storage, and minimize its long-term environmental impact.

The size of storage operations required is truly large. If the CO_2 concentration in the air is to be stabilized, the rate at which carbon needs to be stored will approach the rate of fossil carbon consumption, and the total storage capacity will have to match all the carbon removed from the ground [2].

This section provides a summary of the wide array of options that have been suggested for carbon storage. They range from simply injecting CO_2 into underground reservoirs to fostering algae growth in the ocean in order to increase ocean CO_2 uptake and drive the natural sequestration of CO_2 into deeper levels of the ocean. Even though there is a wide range of options, most of them are based on a few concepts.

There is no reason why these different technologies could not be developed and deployed simultaneously. However, they will differ in cost, in total storage capacity, and in environmental and safety risks. In addition, the ability to verify the storage and account accurately for the stored carbon will differ. The ability to monitor the stored carbon and to account for its whereabouts is critical for public acceptance of any carbon storage technology [3].

21.2.1 Underground, Geological Storage

At present the most widely studied and most mature storage technology is based on the injection of CO_2 into underground reservoirs [4, 5]. This is often referred to as geological storage. In many discussions, CO_2 storage has become nearly synonymous with geological storage, as for example in [6].

Underground geological storage sites can be below land, or the injection could be into a reservoir below the ocean floor. For shallow ocean depths, there is little difference between these two options. For very deep ocean sites, the combination of low reservoir temperature and high pressure results in a qualitative difference, as in such sites CO_2 is denser than the surrounding pore water [7, 8]. This implies that the CO_2 is gravitationally trapped and will not escape upwards toward the ocean floor. An exception to this might occur in very low-permeability formations, where injection-induced cracks may move upwards because a fluid-filled crack has buoyancy relative to the rock [8].

In some formations, the injection of CO_2 can support the extraction of a valuable product such as oil or gas. CO_2 has been used for this purpose even in the absence of economic incentives for carbon storage [9, 10, 11]. Injection into oil fields helps in maintaining pressure, and reduces the viscosity of the oil. CO_2 can also maintain pressure in a natural gas field and thus facilitate production. The injection of CO_2 into deep, unmineable coal seams can result in the production of methane, as the CO_2 successfully displaces the methane adsorbed on the coal [12, 13].

Existing oil and gas fields, once depleted, can also be used as a reservoir to store CO_2 [14]. A concern in some of these fields is the large number of holes that have been drilled into these formations, each of which represents a potential conduit for gas leaking back out [4].

Outside oil and gas fields and coal seams, the injection of CO_2 typically displaces water, often saline water, in a formation. Saline aquifers in sedimentary rock, or in basalts, are ubiquitous. The prevailing view suggests that storage capacities are very large [4, 15] and storage lifetimes could be thousands of years or longer.

Injection costs are fairly low, estimated at a few dollars per tonne of CO_2 [4, 16].

21.2.2 Mineral Sequestration

Also discussed in the IPCC report on carbon dioxide capture and storage [17] is the formation of carbonate minerals from CO_2 and silicates. CO_2 can react with various minerals to form carbonates. Carbonates typically are thermodynamically lower than CO_2. As a result, carbonation reactions are normally exothermic and thermodynamically downhill.

Carbonates can be solids such as calcium and magnesium carbonates, or they could be highly water soluble such as sodium and potassium carbonates. In the latter case, it is also possible to form hydrogencarbonates that typically remain in solution. Therefore, carbonates themselves can be used as the base which neutralizes carbonic acid. Once formed, carbonates are stable and it would require substantial inputs of energy to release the CO_2 again. Hydrogencarbonates, on the other hand, are only slightly bound and thus could be re-released again over time.

Carbonation processes can occur in underground reservoirs *in situ* and thus change the long-term stability of the CO_2 stored in an underground reservoir [18, 19]. Alternatively, carbonation can be performed *ex situ* in an industrial process that relies on mined minerals that react chemically with CO_2 to produce carbonate tailings, which need to be disposed of [20].

Ex situ mineral carbonation would most likely involve large mining operations and a large processing effort in crushing and grinding the rock. A large amount of rock would need to be processed, roughly twice the mass of the CO_2 or six times the mass of the coal [20]. Since it is easier to transport coal, CO_2, or electricity than the mineral rock, it is unlikely that the rock would be transported to the coal plant. Instead, it is more likely that the CO_2 will be pipelined to the mine, or future coal plants will be co-located with the mine. The cost of *ex situ* mineral sequestration is likely always to remain larger than that of underground injection as, by necessity, it involves a large amount of processing. Whereas ultrafine grinding can be prohibitively expensive in terms of cost and energy, typical grinding in mining operations, milling ores down to about 0.1 mm particle size, is usually very affordable, and it only requires a small percentage of the energy that was released in the production of the CO_2 [21, 22]. The most chemically active minerals are serpentine and olivine.

The advantage of mineral sequestration is the nearly unlimited capacity of the base mineral and a high degree of permanence [17]. The disadvantage is the extensive processing necessary that results in a high cost, currently estimated in the vicinity of $100 per tonne. The cost would have to decrease substantially before this method could be considered competitive [17].

Mineral sequestration that produces dissolved carbonate or hydrogencarbonate salts that are not held in an underground reservoir can be viewed as a dilution approach to storage. After treating the carbonic acid to form a more harmless carbonate or hydrogencarbonate salt, the salt is released to the environment and is likely to migrate to the ocean. Such dispersion could be deemed acceptable because the natural background concentrations of these substances completely dwarf the anthropogenic component.

Schuiling and Krijgsman [23] suggested a process that simply exposes finely ground olivine and serpentine to ambient weathering conditions. Particularly in a tropical environment, carbonation is expected to proceed rapidly; in this case, it may result in hydrogencarbonates that will be dissolved.

Most solid minerals would be stored at specific sites, but their gradual dispersal would not create problems as the materials are inert and environmentally safe. The advantage of materials in the thermodynamic ground state is that they tend to be inert, and there is no need to maintain a barrier indefinitely.

The most difficult technical hurdle for mineral sequestration is the naturally slow reaction rate, which needs to be greatly accelerated. The dominant cost of the process is associated with this acceleration, and it is still unknown. The cost of mining the raw minerals and disposing of the tailings is considered minimal, and would be less than $10 per tonne of CO_2.

21.2.3 Ocean Disposal

Dilution of CO_2 into the ocean, and the formation of pools of CO_2 with partial or full hydrate formation on the ocean floor, have been considered as options for CO_2 storage. They are discussed in detail in the special report on carbon capture and storage by the IPCC [24]. More recently, there has been very little activity on ocean disposal options because of a general perception of adverse environmental consequences. Ocean disposal is mentioned here only in passing, as the focus of this chapter is on underground storage options. Ocean disposal can overlap with mineral sequestration if the CO_2 is delivered as a carbonate or hydrogencarbonate [25, 26].

In the absence of providing alkalinity, the ocean could absorb roughly 1200–1500 Pg (1 Pg $=10^{12}$ kg or 1 Gt) of carbon before the pH would change by 0.3 and the carbonate ion concentration would decrease by a factor of two [27]. The capacity for neutral carbonate/hydrogencarbonate disposal would be an order of magnitude larger.

21.2.4 Biomass Options

Afforestation and ocean fertilization to foster biomass production in the ocean are both options that have been proposed and promoted for carbon storage. Like ocean disposal options, biomass production in effect is storing carbon inside the mobile carbon pool. Terrestrial biomass absorbs and releases about 100 Pg of carbon every year. As a result, the long-term stability of this reservoir for carbon storage is not easily established.

The US Department of Energy has referred to land-based biomass production for carbon storage as terrestrial carbon storage, which should not be confused with other land-based carbon storage options, such as geological storage under land and mineral sequestration in above-ground mines [28].

The biomass reservoir is not large enough to absorb the impact of storing large amounts of carbon. The entire biomass (mainly forests) contains about 600 Pg of carbon, which is close to the preindustrial carbon content of the atmosphere [29] (1 ppm of CO_2 in the atmosphere represents $\sim$0.5 Pg of carbon). For comparison, the annual consumption of fossil fuels releases 8 Pg of carbon. Hence on a century scale, carbon consumption exceeds the entire stock of biomass carbon.

21.2.5 Soil Carbon

Biomass production will take carbon from the atmospheric reservoir and convert it into reduced carbon. A small fraction of this carbon can be converted into very stable carbon-rich organic compounds that can persist in the soil for many centuries. Producing such materials as a means of storing carbon has therefore been suggested.

The natural carbon pool in soil, apart from living biomass, is about 1400 Pg [29], and managing and increasing this pool is another option for storing carbon. For example, a particularly stable form of biomass is biochar, charcoal produced from biomass. This could represent a potentially large sink [30].

21.3 Current Activities

Over the last decade, a number of carbon storage activities have been developed. Some of the earliest were initially intended not for carbon storage but for enhanced oil recovery. This section lists some of the more prominent and larger efforts.

21.3.1 West Texas Oil Fields

The oldest active injection sites are old oil fields in West Texas that have been using CO_2 for tertiary oil recovery since the mid-1970s [9]. Here, the economic driver is the value of the oil extracted from the ground plus tax incentives for enhanced domestic oil production that reduces the dependence on foreign oil. Originally, the CO_2 was produced at industrial sites, but soon after this process started, the CO_2 itself was produced from natural, underground sources. Today, CO_2 is pipelined from southern Colorado, across New Mexico to West Texas, a distance of 800 km. When the enhanced oil recovery projects in West Texas were developed, concerns over CO_2 storage played no role in their design. Since most of the CO_2 is derived from underground CO_2 wells, the net impact of these operations on the world carbon balance is small but results in increased carbon mobilization. In effect, fossil CO_2 is taken from the ground in one place and reinjected at another location. Associated CO_2 emissions result from CO_2 losses in the system, and from energy consumption related to the operation and construction of the infrastructure.

21.3.2 LaBarge

Some natural gas reservoirs contain a significant amount of CO_2 that is co-produced with the methane. Since 1986, Exxon has been separating CO_2 from methane obtained from gas wells in the LaBarge Field in Wyoming. The input to the Shute Creek Treating Facility is a methane, CO_2, H_2S, and He mixture, with 66% CO_2 [10]. Today, the H_2S is returned to the formation, whereas most of the CO_2 is stripped out and sold for enhanced oil recovery. The operation is large by CCS standards and is now approaching 7 Mt of CO_2 per year [31]. The sale of CO_2 for enhanced oil recovery (EOR) was planned from the beginning and provided CO_2 for a number of tertiary recovery operations.

21.3.3 Sleipner

The first large project strictly for CO_2 capture and storage to eliminate a waste product was the Sleipner Project owned by Statoil [4, 32]. Statoil injects CO_2

that is separated from natural gas produced at the Sleipner Field in the North Sea into a sandstone formation (the Utsira formation), which lies 800 m below the North Sea floor but still above the natural gas reservoir. The formation is a saline aquifer filled with a seawater-like brine. The formation is large and the permeability of the sandstone is high. Injection has been ongoing since 1996. Statoil has injected on the order of 10^6 t of CO_2 per year. The economic driver behind this effort is a CO_2 tax levied by Norway on such types of emissions, which is on the order of $ 50 per tonne.

The Utsira formation represents an extremely large reservoir and it has been suggested that this formation alone could take up most of the CO_2 that is produced at northern European power plants [32].

In Norway, Statoil is also operating a smaller injection site at Snohvit in the Barents Sea (0.7 Mt per year) storing CO_2 that had to be removed from liquefied natural gas. It is also planning to store CO_2 from a gas-fired combined heat and power plant at Mongstad near Bergen, which is expected to become operational after 2014 when decisions on the final size and operational details will be made.

21.3.4
Weyburn

Another early and large storage project is the Weyburn Project. At Weyburn in Saskatchewan, Canada, CO_2 is delivered for EOR. Approximately 10^6 t of CO_2 are shipped per year from the Dakota Gasification Company's Great Plains Synfuels Plant to the Weyburn field [33]. The CO_2 shipped to the Weyburn field contains ~2.5% of H_2S. Tolerating the H_2S improves the economics of the process and increases the miscibility of the gas with the oil [34]. Injection of H_2S is not new: there are a large number of projects in Canada that inject sour gases, such as CO_2 and H_2S, into formations in order to manage the sulfur emissions associated with many Canadian gas fields. These should be considered predominantly H_2S management projects which happen to co-sequester a relatively small amount of CO_2 [35]. The Weyburn field, in contrast, tolerates some H_2S in a CO_2 stream.

21.3.5
In Salah

A large storage project that has been going on for several years is geological storage of CO_2 in an abandoned gas field at In Salah, Algeria. This project, which involves Sonatrach, BP, and Statoil, removes CO_2 from natural gas before it is pipelined to the Mediterranean shore [4]. The CO_2, which on average comprises 5.5% of the gas, is stripped out to achieve a pipeline specification of 0.3%. The resulting CO_2 is then injected into an old gas field. The field is well monitored with regard to injection, CO_2 leakage, and other consequences of the injection of about 1.2×10^6 t of CO_2 per year. For example, satellite-based radar interferometry has shown that the ground rises near the injection in response to the additional volume added [36]. The rise is local, and it happens even though in the larger area there is a net removal of gas. The volume of the CO_2 injected is only a small fraction of that of the natural gas that is removed from nearby formations.

21.3.6
Future Large Projects

The IEA has compiled a list of 80 projects that are planned or already under way [37]. The list clearly indicates the dominance of geological storage in the present thinking about carbon dioxide storage. Nearly all projects that consider storage aim for geological storage. However, in the past there were several efforts to perform tests on ocean disposal, but all of these projects were stopped before they started to inject CO_2 because of public protests. One was an international project involving the United States, Japan, and Norway to inject CO_2 into the ocean off the coast of Hawaii. The project was fairly small, planning total injections of 20–40 t [38]. The project ran into a strong protests and never received the necessary permits. Norwegian authorities gave permission for a relocated version of the experiment but revoked it at the last minute, again because of strong objections by the public [39].

A similar fate seems likely also with some of the geological storage projects. An example is the Shell project at Barendrecht, which was first allowed to move forward but now has been cancelled because of strong local opposition [40]. These developments point to the need for clearly communicating with the public regarding the need for and the purpose of CO_2 storage. It also suggests that there is an inherent advantage to more remote locations, as it does remove the concerns and risks.

The projects that have moved forward are those in remote locations and all have some economic incentives associated with them. Some, such as LaBarge and the oil fields in West Texas, initially did not even consider CO_2 storage as a potential benefit of the project. The In Salah and Sleipner natural gas operations need to clean up the gas produced in any case. Sleipner avoids a substantial carbon tax, and the In Salah project has applied for credits under the Clean Development Mechanism (CDM). Both projects also have the advantage that they are associated with large gas extraction operations that are much bigger than the CO_2 injection operation, which deals with volumes more than an order of magnitude smaller. The North Dakota Gasification Plant has the advantage that it can create a concentrated stream of CO_2 with relatively little effort, and the Weyburn field has a significant economic interest in the EOR that results from this.

The next generation of efforts, in Norway, the United States, and China, are likely to take more CO_2 from power plants. For example, Monkstad takes CO_2 from a natural gas-fired power plant. Projects in China are taking power plant CO_2. The United States is working on FutureGen 2 and several other power plant scrubbers. Gorgon, on the other hand, one of the largest projects planned, is another natural gas project like Sleipner and In Salah [37].

The IEA has made the point that the commitments for the necessary number of projects to see an 80% reduction in emissions specified by the OECD by 2050 have not yet materialized. Many of the projects listed are too small or too uncertain to really matter.

Smaller projects are designed to address specific questions about the behavior of the stored carbon. For example, at Ketzin in Germany, in a project known as CO2sink, a detailed comparison between computational simulations and actual behavior will strengthen the understanding of the reservoir dynamics of CO_2 in an underground storage reservoir [41].

The US Department of Energy has a substantial ongoing effort in carbon capture and storage, supporting regional partnerships to develop CO_2 management strategies reflecting the needs and geological circumstances of a region. On the storage side, the focus is on geological storage, on terrestrial sequestration (carbon storage in terrestrial biomass), and on verification and monitoring [42, 43, 44].

21.4 Utilization Versus Disposal

Some storage technologies take advantage of other commercial activities such as EOR that nearly coincidentally store CO_2. Storage resulting from a practical application is preferable to waste disposal. Unfortunately, there is a huge mismatch between industrial and commercial demand for CO_2 and the CO_2 emissions created by the large demand for fossil fuels. Furthermore, not all uses of CO_2 result in storage; many simply re-emit the CO_2 after a short time of use. For utilization to have an impact on the size of the mobile carbon pool, the use has to result in carbon storage or at least eliminate the use of fossil carbon.

The scale of emissions is formidable; a single gigawatt coal plant, producing approximately 1 kg of CO_2 per kWh of electricity, could supply the entire commercial market for CO_2 in the United States, which is about 8×10^6 t per year [45]. CO_2 emissions from fossil fuels, which are about 30 Gt per year [46], dwarf all commercial and industrial uses of CO_2. Even the current use in EOR (about 40×10^6 t per year) is a tiny fraction of the world's emissions.

The requirement of a net zero carbon balance will likely drive CO_2 disposal, unless vast new opportunities for making use of CO_2 are developed or fossil fuels are abandoned in favor of other energy options.

21.4.1 Anthropogenic Carbon Flows

Whether or not a particular use of CO_2 will actually lower total CO_2 emissions is not always straightforward to answer. The special report by the IPCC raised a number of concerns over CO_2 utilization [17]. In many lifecycle analyses, one draws the box around the system too small, resulting in dubious claims about carbon reductions. For example, replacing methane with CO_2 in a chemical synthesis may reduce emissions, but it also may leave the process short of energy and thus require additional fuel to complete the process. A conceptually easy way of keeping track of net carbon accounting is to demand that the extraction of fossil fuels from the ground be matched by carbon storage. This follows from a simple mass flow analysis. Whatever enters the mobile carbon pool has to be actively removed and stored away from it.

Matching storage to extraction could most easily be accomplished with a certificate of sequestration that is issued for storage and required for extraction or loss from storage. Until certificates are available in sufficient quantities, a carbon permit that allows the extraction of fossil carbon without matching storage could be issued instead of a certificate. The number of carbon permits would set the allowable net emission [47]. Permits would gradually have to be phased out in order to stabilize the CO_2 concentration in the air.

21.4.2 Transient Use of CO_2

Some uses simply delay emissions. For example, CO_2 used for making dry-ice will be released within days. Similarly, CO_2 fed to plants growing inside a greenhouse is only temporarily fixed. The transfer of CO_2 from a fossil fuel-based power plant to such applications does not remove it from the mobile carbon pool and therefore would not warrant a certificate of sequestration. The coal mine operator would need to obtain certificates of sequestration from unrelated activities in order to continue mining, while the CO_2 from the coal itself would end up in the atmosphere. However, the requirement that fossil fuel extraction is properly permitted would discourage the use of additional fossil carbon – possibly from a CO_2 well – for such transient applications.

21.4.3 Synthetic Fuels

Synthetic fuels are in a separate category because fuel consumption is large enough to channel a large carbon stream. However, CO_2 as feedstock for synthetic fuels made with renewable or nuclear energy input would also be just a pass-through of carbon since the carbon is released as soon as the fuel is consumed. On the other hand, the use of synthetic fuels (presumably produced with non-fossil energy input) eliminates the extraction of additional fossil fuels. For every gallon of synthetic gasoline, a gallon of petroleum-derived gasoline has been avoided. The power plant operator may not be issued a certificate of sequestration for the transaction, but the fuel producer avoids the need to buy a second one. In effect, the carbon was used twice, cutting the demand for certificates of sequestration in half. Fuel use would disperse the CO_2 into the atmosphere and the only way to satisfy the demand for certificates of sequestration would be to reclaim the CO_2 emitted to the air. This would require either the use of biomass in power generation or the capture of CO_2 from the air [48].

Substituting synthetic fuels for petroleum fuels may not eliminate the use of petroleum altogether but may simply postpone it. The entire petroleum reserve may still be consumed eventually. Since carbon will remain mobile for thousands of years, the climatic impact would only be delayed and not avoided. If, however, one matches fossil fuel extraction with carbon storage, then the delay in use also allows a delay in committing carbon to storage, which is correctly taken into account by using certificates of sequestration in balancing carbon accounts.

21.4.4 Carbon Dioxide for Chemicals

CO_2 incorporation into the production of carbon-containing chemicals such as polycarbonates and urea could lead to substantial storage times [17]. This has led to the suggestion that these should be counted as carbon offsets. However, often these storage times are limited to decades, which would be too short to provide adequate carbon storage. Furthermore, the net impact could be negated by higher fuel consumption. For example, the substitution of CO_2 for CH_4 in a chemical synthesis leaves the process with less available energy, which may now need to be introduced elsewhere in the process flow. One could consider the fixation of carbon in synthetic materials as a temporary storage. If it appears economically useful, one could issue certificates of sequestration for temporary storage, provided that the owner of the storage agrees to be treated as a potential emitter, if the carbon is remobilized or simply lost.

21.4.5 Carbonate Minerals

In contrast to synthesizing organic materials from CO_2, which requires substantial amounts of energy, the formation of carbonates from alkaline minerals is thermodynamically favored. Ideally, the process releases energy, but most industrial processes will consume some energy in the preparation of the base. As a result, CO_2 storage can be combined with creating soil enhancers, for example by adding olivine to acidic soils, through the production of certain construction materials such as cement, or the stabilization of waste materials such as steel slags. Finally, there are a number of alkaline industrial waste streams, such as the red mud in bauxite processing, that could be neutralized with carbonic acid and hence be stabilized.

21.4.6
Enhanced Oil Recovery

Underground injection for EOR would likely result in net long-term storage of CO_2. It is generally recognized as a carbon emission reduction, but it is negated by the additional production of oil. If the oil were produced in any case, the injection of CO_2 would lead to a real reduction in total CO_2 emissions. This case is easily handled with certificates of sequestration. The extraction of petroleum would require certificates, some of which would be generated at the same site by the injection of CO_2. In most cases, additional certificates would be required to balance out the net carbon flow.

The same logic would apply to the use of CO_2 in flushing natural gas out of tight formations, shales, or deep unmineable coal beds. In the latter, if the coal were ever mined in the future, its carbon content would include the CO_2 that is absorbed on the coal, which would add a few percent to the total carbon content of the coal.

21.4.7
Summary

Unless the world abandons fossil fuels, there is a need for CO_2 storage. At any given time, the CO_2 storage rate must match the rate at which carbon is extracted from under the ground plus the rate at which carbon is lost from temporary storage. Reuse of carbon either remains small or it has to rely on CO_2 recycled from the air. As long as synthetic fuels are made from fossil CO_2, the carbon flux remains out of balance. Some storage will have additional commercial benefits, but most likely the bulk of all storage will need to put CO_2 away even if there is little or no additional benefit. Many geological storage options or mineral sequestration options fall into this category.

21.5
Different Forms of Stored Carbon

Carbon can take many different forms and some are easier or safer to store than others. Other factors include cost and accountability of storage. In this section, we follow through the list, from the thermodynamically lowest to the thermodynamically highest state.

21.5.1
Solid Carbonates

Near the surface of the Earth in the presence of oxygen and crustal minerals, the natural thermodynamic ground state of carbon is a carbonate. CO_2 is an acid that will react with a base to form a carbonate. The most abundant form of crustal carbon is therefore a mineral carbonate. Near the surface of the Earth, limestones are the largest carbon reservoirs.

Ex situ and *in situ* mineral sequestration aim to create these energetically low states of carbon. In this form of solid carbonate, the carbon will naturally keep out of the atmosphere and its ability to migrate from one carbon reservoir to another is very limited.

Solid carbonates are abundant on Earth; they are thermodynamically stable and hence almost inert and environmentally benign. Once formed, they are stable on geological time scales. As a result, they can be easily disposed of and need not be monitored indefinitely.

Even though the formation of solid carbonates from silicates is thermodynamically favored, the reactions are usually fairly slow. Since CO_2 is readily available, minerals that would react rapidly are naturally depleted. The only abundant minerals with reaction kinetics fast enough to make above-ground processing feasible are magnesium silicates, namely olivine and serpentine [49, 50]. The availability of olivine and serpentine far exceeds the availability of fossil fuels. Although there are some other minerals such as wollastonite that could be readily transformed, the availability of these minerals is insufficient for large-scale implementation [20].

Although most approaches envision an industrial above-ground process for *ex situ* mineral sequestration, Schuiling and Krijgsman's suggestion of exposing finely ground olivine and serpentine to weathering conditions [23] straddles the line between forming solid carbonates and forming dissolved CO_2. Particularly in a tropical environment, carbonation is expected to proceed rapidly; in this case, it may result in hydrogencarbonates that will be dissolved.

Other options which have not been studied in detail are biological process that can form carbonates from dissolved carbonates or hydrogencarbonates. Calcite and aragonite are continuously generated by biological processes in the ocean. However, for such a process to result in net carbon storage, the alkalinity that was present in the water must be replaced artificially by adding alkaline minerals. One calcium ion in the ocean is neutralized by nearly two hydrogencarbonate ions. After it has been precipitated as solid calcium carbonate, it only holds on to one carbonate ion, in effect stranding one CO_2 [51].

21.5.2
Dissolved Carbonates and Hydrogencarbonates

Weathering processes can also produce dissolved carbonates and hydrogencarbonates. In the ocean, about 90% of all dissolved inorganic carbon is in the form of hydrogencarbonate. Most dissolved carbonates or hydrogencarbonates eventually find their way into the ocean where they would remain as hydrogencarbonates, unless they are safely confined in an underground reservoir. This pool is as stable as the alkalinity of the ocean, which over long time scales can change. There are many alkaline minerals that could be dissolved in the ocean water. For example, Rau and Caldeira suggested dissolving limestone with carbonic acid and letting the brine be diluted into the ocean – this would indeed result in net carbon storage [26].

The formation of carbonates in the pore space during geological sequestration is another option of interest. Usually the process of carbonate formation, so-called mineral trapping, is very slow. However, there are some minerals in basalts that react sufficiently fast to make *in situ* sequestration a viable option. As a practical matter, it is important that the transition to carbonates happens on a time scale in which human institutions can comfortably operate. If reactions are reasonably fast, one could come to closure on a reservoir in 25–50 years from now, by establishing that virtually all of the injected carbon has successfully gone through the transition to carbonates. This is still five orders of magnitude slower than what one would need to accomplish in *ex situ* mineral sequestration, but it would make it possible to develop a regulatory framework that actually takes advantage of the fact that the reservoir will gradually transition into a carbonate storage.

This is the goal of experiments currently performed at Iceland [52]. The basalts in the formations studied have shown very high permeability and also are fairly reactive. Unfortunately for these lava flows, there is no competent seal on the formation. It is not possible simply to inject CO_2 in the formation, because it would likely escape to the surface. For gaseous CO_2, the time to escape would likely be shorter than the time to neutralize the carbonic acid by leaching alkalinity out of the rock. As a result, the operators decided to inject the CO_2 at 2.5 MPa pressure into a water stream that is taken from the reservoir and sent back down saturated with CO_2. Because of the large hydrostatic pressure at reservoir depth (about 800 m), the CO_2 remains dissolved in the water and does not come back out. Geochemical modeling suggests that it eventually precipitates out as magnesium carbonate.

21.5.3 Carbonic Acid

Dissolution of CO_2 in water is the simplest transformation and the basic premise for many ocean disposal options. Dissolution of CO_2 in water adds a weak acid to the water. Dissolution by its nature dilutes the carbon dioxide greatly, making storage reservoirs inherently large. As a result, the capacity of the entire ocean is limited; it takes roughly 1200 Pg of carbon to double the proton concentration in the ocean. Furthermore, as we will discuss below, dilution into the ocean may not be sufficiently permanent as the ocean turns over in less than 1000 years, resulting in an exchange with the atmosphere.

In underground injection, it is generally assumed that after a sufficient time, the CO_2 will dissolve in the surrounding pore water. After pore trapping, this is the next step in improving the quality of carbon containment. CO_2 in the reservoir is thought to move from geological trapping, that is, being confined by a physical barrier, to residual trapping, that is, being dispersed into small pores from which only a fraction can be removed again, to solution trapping, to chem-

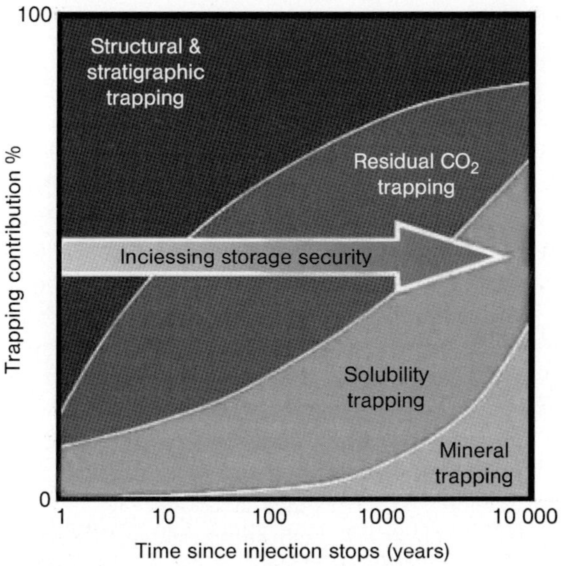

Fig. 21.1 Carbon dioxide trapping mechanisms in geological storage. Source: IPCC Report [4].

ical trapping, which removes alkalinity from the rock to neutralize the carbonic acid [4] (Fig. 21.1). It has been shown, however, that in many reservoirs dissolution of the CO_2 can take a very long time [53]. Wherever this is the case, it does not really affect the safety and security of the reservoir on a time scale relevant to human decision making. However, with a well-sealed cap rock geological trapping under a competent seal may well be sufficient to assure the safety of the reservoir.

The greatest challenge to dissolution is the sheer volume involved. A gigawatt coal-fired power plant would produce on the order of 8×10^6 t of CO_2 per year, or 400×10^6 t over a lifetime of 50 years. A single storage facility taking on the CO_2 from a few plants could thus easily end up storing 10^9 t of CO_2 over its lifetime, which needs to be at least on the order of 30 years, so as to justify the expenditure on large capital equipment to support the injection process. In rough numbers this is 1–2 km^3 of liquid CO_2 and an injection rate approaching 1 t s^{-1}.

This fluid needs to be spread into a pore space, which may range from a few percent to as high as 30–40% of the rock volume. The Rose Run formation, which is under consideration for CO_2 storage in Ohio, has a porosity of about 9.5%. By contrast, for the Utsira formation at Sleipner it is about 25%. The thickness of a formation can vary from a few meters to a few hundred meters. The Utsira formation is about 250 m thick at the injection point and the Rose Run formation has a net thickness of about 6 m. Given the limitations of pore space availability, the reservoir volume that needs to be filled is thus 3–10 times larger than the CO_2 volume. Dissolving the CO_2 in a brine at maybe a 2 M concentration would multiply the volume again by a factor of 25. Hence dissolved CO_2 in a pore volume could easily occupy a volume 100 times larger than that of liquid CO_2. Hence 1 Gt of CO_2 could therefore spread over 1000 km^2 or more.

Nevertheless, induced solution trapping opens up the possibility of using other types of reservoirs that may lack a strong cap over the reservoir. For example, as noted above, the injection into basalts in Iceland [52] aims to take advantage of the relatively fast chemical fixation in the strongly basic basalt.

21.5.4
Carbon Dioxide (Gas, Solid, or Liquid)

The simplest approaches and certainly the most cost-effective in the initial disposal steps store CO_2 directly. This is the most direct approach because the energy extraction from fossil fuels will naturally result in CO_2. Underground storage usually starts with the injection of CO_2 into a reservoir. In many reservoirs, CO_2 will remain chemically stable for centuries or millennia. The physical form will depend on the temperature and pressure at which it is stored. In order to achieve a reasonable density, it is necessary to compress the CO_2. Depending on pressure and temperature, the CO_2 can remain gaseous, turn into a liquid, or become supercritical. Typically one avoids gaseous CO_2 as the storage volume required tends to be too large. At depth one has sufficient hydrostatic pressure to liquefy CO_2; however, in most reservoirs the geothermal gradient is steep enough that at the requisite pressure the CO_2 is already supercritical (Fig. 21.2). The critical temperature and pressure of CO_2 are 31.1 °C and 7.3 MPa.

For most underground conditions of temperature and pressure, CO_2 is significantly less dense than water, hence it has a propensity to rise. The exception is storage below the deep ocean floor, because such reservoirs can combine high

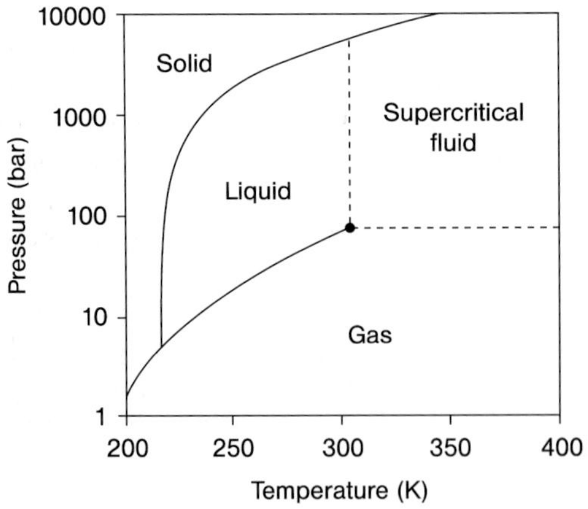

Fig. 21.2 Carbon dioxide phase diagram.

pressure with low temperature [7, 8]. At a depth of ~2700 m, the ocean is near freezing temperature and the ambient pressure will result in liquid CO_2 with a density that exceeds that of seawater. Below the ocean floor, the temperature of the formation increases with depth due to the geothermal gradient. Since CO_2 is far more compressible than water, and also has a far larger thermal expansion coefficient, eventually with increasing depth the CO_2 becomes buoyant again. As a result, there is a layer at which CO_2 is neutrally buoyant. If it is driven away from this level it will naturally return. Therefore, storage of CO_2 in permeable formations below the ocean floor at depths of ~3000 m would have no propensity to migrate upwards, back to the ocean floor. The CO_2 is gravitationally trapped.

Liquid storage in nearly all applications has been considered in pore spaces, where CO_2 is in close contact with rock. The only storage option that would take bulk liquid or supercritical CO_2 that has been suggested in the past would be in salt caverns. However, this approach has not been investigated, even though storage of natural gas in salt caverns is fairly common. Solid storage, in a large block of dry-ice, has also been suggested [54], but ultimately the cost of storing dry-ice and the need to maintain it cold have discouraged this avenue.

21.5.5 Carbon Dioxide Hydrates

Under conditions of high pressure and low temperature and in the presence of water, CO_2 can form a clathrate compound with water, a gas hydrate. The hydrate $(CO_2)_8(H_2O)_{46}$ represents a different modification of ice, which is stabilized by the guest molecules filling cages formed in the crystal structure [55]. The CO_2 forms a type I clathrate, which involves 46 water molecules in a unit cell that can hold up to eight CO_2 molecules. Clathrates are stable at high CO_2 pressure and water temperatures near but above freezing.

Clathrates offer an approach to stabilizing CO_2 either on the ocean floor below 450 m depth and under near freezing conditions [56] or in an underground reservoir that provides sufficiently high pressure (exceeding ~4.5 MPa) and sufficiently low temperatures (a few degrees above freezing). Forming hydrates by letting CO_2 flow on to the ocean floor releases a large amount of heat, which needs to be dissipated. The CO_2 output of a gigawatt coal plant would produce about 360 MW of low-grade heat [3]. Once deposited on the ocean floor, clathrates are not entirely stable. In the absence of liquid CO_2, the clathrate gradually dissolves in the ocean water, since the clathrate is only stable if the CO_2 partial pressure exceeds approximately 4.5 MPa (Fig. 21.3).

Most of these reservoirs would be below the ocean floor, but there are opportunities for storing CO_2 as a gas hydrate under land in regions with permafrost. Hydrate formation may be of particular interest in places where the CO_2 could displace methane in an existing hydrate [57, 58] and thus provides a means of extracting a commercially valuable product.

A more exotic approach would be to form hydrates under thick ice sheets in Antarctica where the hydrate, which is denser than ordinary ice, would be

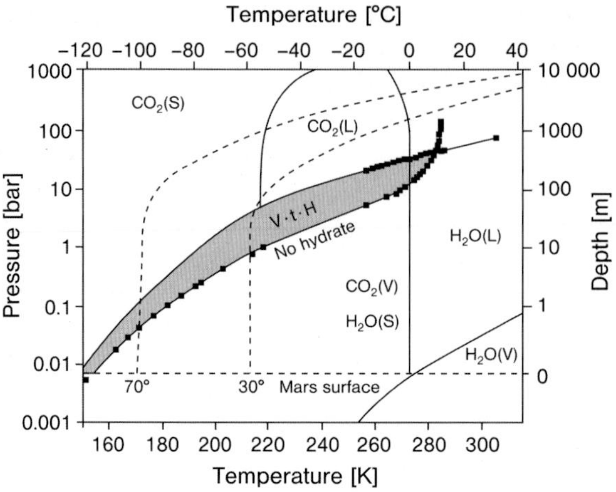

Fig. 21.3 Carbon dioxide clathrate phase diagram.

trapped at the bottom of a large depression. Unlike ordinary ice, which at the bottom of the ice sheet is often at the verge of melting, the clathrate has a higher melting point and would remain solidly frozen to the ground. Clathrate formation would produce sufficient heat to provide meltwater for the continuous formation of clathrate.

Hydrate formation also needs to be considered in the dissolution of CO_2 into ocean water.

A small droplet of CO_2 rising in the water column will cover itself with a CO_2 hydrate and thus resist further dilution. This has been studied in the context of CO_2 injection for dilution, because it helps define the time constants of dissolution [59]. Hydrate formation also plays a role on the ocean floor as large volumes of CO_2 can be stored under those conditions. It is worth noting that the formation of hydrates is exothermic and thus releases heat on the ocean floor.

21.5.6 Organic Forms of Carbon

Forming reduced carbon such as biomass is thermodynamically uphill, which suggests that the stored carbon is not entirely stable. Storing reduced carbon is certainly feasible, but natural reservoirs are either well protected, such as coal seams and gas reservoirs deep underground, or simply not large enough to take significant amounts of carbon. The deforestation of the planet was not a major contributor to the change in atmospheric CO_2 concentration. Therefore, reforestation and afforestation are not likely to absorb the CO_2 from fossil fuels.

Furthermore, any storage of reduced carbon suggests that the energy extracted from the fossil carbon has been returned to the carbon. Hence in an overall energy budget, no energy was extracted from the fossil carbon. This raises the question of why one would want to store organic carbon rather than leaving the fossil carbon in the ground. Rather than sequestering biomass, one should consider it a fuel, and store the CO_2 that is produced in the combustion process.

A possible exception is a suggestion by Steinberg to convert methane into hydrogen and carbon, use the hydrogen as a fuel, and dispose of the carbon for sequestration [60]. This would leave approximately half of the energy available in methane behind in the disposed carbon. Although certainly possible, it is not likely to happen while coal, which is a much lower quality of carbon, is still being mined.

Finally, storing CO_2 in a thermodynamically unstable form raises the question of pathways for its decay. Methane, a likely product, is a far more potent greenhouse gas than CO_2. Nominally it is considered to be 23 times stronger than CO_2. Even if only a small fraction of the stored carbon escapes as methane, the climatic impact could be worse than not storing the CO_2 at all.

21.6 Storage Lifetime

The size of the carbon emissions and the scope of future demand interact to define the minimum required storage time. A simplified but instructive way of looking at the problem is that the storage time defines the time during which the carbon is kept out of the mobile carbon pool. In effect, release from storage becomes another form of emission and must be controlled in the same way as any other emission.

There are two ways of controlling these emissions. One is to recapture an amount of CO_2 that is equal to what is released and store it again somewhere else. This would require the ability to create negative emissions or directly capture CO_2 from air [48]. The second is to demand a storage life time that is long enough to keep leakage small. Since the first approach implies a continuous and ever-increasing burden on future generations, it should be avoided. Hence storage times will have to be long enough that normal rates of release can safely be ignored. Demands on future generations would be limited to dealing with unexpected contingencies.

For storage to accomplish a reduction in CO_2 concentration in the air and not just delay the peak in time, the storage time needs to be comparable to the time it takes natural processes to absorb the CO_2. This absorption involves a number of different time scales, which range from a few hundred years to tens of thousands of years. The shorter time scales are due to ocean uptake, which by its nature can only redistribute the CO_2 in the larger combined ocean/atmosphere reservoir. Therefore, unless the anticipated emissions in the future are small enough that dilution into the combined reservoir is still acceptable, storage times need to be measured in thousands if not tens of thousands of years.

Enlarging the pool into which fossil carbon is diluted to include the biomass and the oceans enlarges the reservoir by a factor of $\sim$4–5. This additional dilution capacity may be sufficient to accommodate a rapid phasing out of fossil fuels, but it would not make it possible to consider fossil fuels in the longer term energy mix.

21.7
Storage Capacity Requirements

Storage capacity requirements are large. The scale of annual storage could easily exceed current annual CO_2 emissions (30 Pg per year) and the total amount could eventually match the fossil fuel reserves.

It is clear that the ocean does not have the capacity to contain all the CO_2 that could be made from fossil fuels. Ocean acidification needs to be avoided, and the ocean most likely could absorb large enough quantities of carbon if they were neutralized prior to injection or deposited in a fashion such that they would become neutralized on dissolution in the ocean. The necessary alkalinity could be brought in with the CO_2 or it could be derived from the ocean floor. Carbonate deposits that cover larger parts of the ocean floor could be dissolved to form hydrogencarbonates, which could be added in substantial quantities to the ocean.

Generally, the pore volume considered in geological sequestration seems sufficient to deal with this century's carbon dioxide emissions and possibly more. Estimates vary but move up from 1000 Pg of CO_2. There have been many capacity assessments that are very optimistic. However, there is a large uncertainty attached to these numbers. Not all storage sites will be suitable, and the fraction of the reservoirs that is actually acceptable is not yet known. Volume injections would be very large. The 4000 Pg of CO_2 injected under Alberta would add enough volume to raise the entire Canadian province by roughly 6 m even if the CO_2 dissolves and is present as carbonate or hydrogencarbonate ions. A way around this problem would be to extract water, which in turn would have to be processed (i.e., desalinated). However, such processing for most reservoirs would be cheap compared with the overall cost of carbon sequestration and thus could easily be added.

There is clearly a large amount of capacity in geological storage, and there is no reason why one could not start with the sites that are promising. If in the future a shortage develops, then one can consider additional options.

Finally, mineral sequestration would have a very large capacity in particular if one considers *ex situ* operations. The peridotite rocks in Oman, rich in olivine and serpentine, by themselves would be sufficient to absorb centuries of CO_2 emissions [19, 20]. The availability of serpentine and olivine far exceeds that of oil, coal, and gas. The constraint here is the obvious environmental impact of large surface mining operations and the high costs of the operation. However, the footprint of a serpentine mine to compensate for the CO_2 emissions associated with the coal from a specific coal mine would have a much smaller footprint than the coal mine, because serpentine occurs in large blocks as opposed to the thin seams in which coal occurs [61].

21.8 Closing Natural Carbon Cycles

In dealing with the large-scale disposal of a waste product, one needs to consider how the anthropogenic contributions are included in the natural material cycles of which they will become a part. The natural carbon cycle is complex and involves a hierarchy of different intertwined cycles into which many of the carbon storage options are interwoven. Although there are certainly natural underground CO_2 reservoirs, and they do provide a natural analog for geological sequestration, the total amount of carbon that naturally ends up in these reservoirs is small, and it appears that geological sequestration would rapidly create far more CO_2 reservoirs than there are naturally. In that sense, geological sequestration is trying its best to keep the captured CO_2 out of the natural cycle.

By contrast, ocean disposal can be viewed as an acceleration of the natural cycle, as is the production of biomass. In both cases, the process is linked to a very fast and very large natural cycle. At issue is whether sequestration could overwhelm this cycle or pass through the cycle so fast that the impact of CO_2 rise is simply delayed rather than avoided. At the very least, however, it would reduce the peak level of CO_2 in the atmosphere.

The entire living biomass adds up to about 600 Pg of carbon, that is, it has incorporated about 2200 Pg of CO_2 into this reservoir. Roughly 100 Pg of carbon are added and removed every year. Hence a large fraction of the reservoir operates on a very short time scale and is therefore unsuitable to long-term storage. On the other hand, some of the carbon incorporated in trees will be stored on a century scale, but a few hundred petagrams of storage capacity are clearly too small to find room for 30 Pg of CO_2 per year. Furthermore, a century-long residence time is clearly too short to provide a sufficient sink for a carbon storage effort that needs to store carbon for millennia. Similarly, the addition of CO_2 to the ocean would lead to ocean acidification, and it would not take much more than 1000 Pg of carbon, or 3700 Pg of CO_2. Most of the ocean would return the carbon in a matter of 800 years.

21.9 The Role of Alkalinity

In combusting fossil carbon, humans are producing large quantities of a weak acid. The majority of the acid is carbon dioxide and a few percent is delivered in the form of sulfur oxides. Hence apart from climate change, one should also consider the introduction of that much acidity as a significant environmental change. This suggests that the optimal means of storage involves the neutralization of all this acid. This, of course, implies some form of mineral sequestration as one would have to provide the base to neutralize the acid.

An interesting variation of this approach is to obtain alkalinity by removing chloride ions from brine or from seawater. The former was proposed by Dunsmore in 1992 [62]. The latter has been suggested by House but also been described in a patent by Lackner [63] In effect, Calera's approach to mineral sequestration also produces hydrochloric acid as a waste product.

Although it was suggested by Dunsmore [62] that HCl makes a valuable by-product, this is not the case on the scales considered here. One would easily overwhelm any plausible market for HCl. However, HCl may be an expensive but more readily disposed of waste product. The major difficulty in making magnesium carbonate from silicate is due to the weakness of carbonic acid. It is fairly easy to form magnesium chloride from serpentine and olivine and hydrochloric acid. Indeed, dissolution of the mineral in HCl (in the absence of CO_2 outgassing) was in the past considered a field test for olivine and serpentine minerals. Hence mineral sequestration of chlorides would be straightforward.

21.10
Storage Safety

Safety issues vary widely with different storage technologies. Each technology poses its own set of risks. In all cases, one must consider the safety of the process of bringing carbon into storage and the safety of the long-term storage. In general, the safety of the process is easier to manage and assess as it involves conventional industrial safety issues. The risks of long-term storage are more difficult to understand and to predict.

Injecting CO_2 underground is generally considered safe, and it is a well-understood technology. Injection technologies exist, but they have never been tried on the enormously large scales that need to be considered for CCS. Generally, new issues arise when processes are scaled up by several orders of magnitude.

For underground injection, it has generally been difficult to convince the public that underground injection should be done in populated areas. Concerns range from outright accidents with catastrophic release of CO_2 to gradual shifting of water reservoirs leading to the intrusion of salty brines into freshwater aquifers. Large-scale, sudden releases seem in general to be highly unlikely, and one could imagine a large number of safeguards that could prevent such disasters, or at least give ample warning about an impending instability of a reservoir. In geological storage, as a last resort, one would always have the option of releasing the CO_2 again in a controlled fashion. Lackner and Brennan suggested that such a planned release should be built into the contingency planning of every large reservoir [3].

On the other hand, a large-scale unplanned release of CO_2 would indeed be hazardous. The natural release at Lake Nyos in 1986 [64] showed that large volumes of CO_2 can flow downhill like a liquid and inundate a depression with high concentration of CO_2. Exposure to more than 5% CO_2 in the air is hazardous and often lethal, as it prevents the lungs from discharging the CO_2 produced in the body.

Probably one of the most spectacular analogs of a serious accident with an underground storage facility was the release of approximately 1000 t of natural gas from a storage facility near Hutchinson, Kansas in 2001 [65, 66]. It came as a surprise that the released methane did not dissipate itself near the release point but traveled up an incline above the salt stock from which it had escaped to blow out of holes drilled nearly a century earlier in the small town of Hutchinson.

On the other hand, ubiquitous natural gas operations have shown how safe and routine underground storage can be made. Facilities in Germany store natural gas in urban areas; well-known example are in Ketzin and Spandau near Berlin.

By contrast, the concerns with mineral sequestration are nearly entirely on the processing step. Mining operations create many hazards that need to be managed well. A specific hazard relevant to mining serpentine, or olivine rich in serpentine, is that the mineral-rich rocks often contain asbestos. Although safe mining methods for these materials exist, it is important that they are used in this context [17].

21.11 Storage Accountability

Accountability is challenging for two reasons: one is the long-term liability associated with storage times of centuries and even millennia, and the other has to do with the nature of the process. It is difficult to verify whether the CO_2 really has been disposed of, and whether it remains in the storage site.

21.11.1 Storage Monitoring

It is very likely that the scope and effort in monitoring storage sites over time will grow and raise the cost of storage operations. For comparison, one may consider the large effort that goes into surveying land in order to verify and assure the boundaries of private properties. A large institutional infrastructure has been created to deal with this effort. It is not motivated by the possibility of spontaneous changes in the landscape, but it addresses concerns over the transparency of ownership and the need to assure all parties that all transactions are properly monitored.

Storage will need to be monitored and verified as carbon is put into storage. For storage that relies on barriers that keep CO_2 away from the atmosphere, it is also important that there are accounting methods that can verify that the carbon has not been lost after it has been put into storage. Ideally, this is based on a positive accounting of the material still left, rather than the failure to see carbon escape.

Ex situ mineral sequestration has a clear advantage when it comes to monitoring. The measurement of the total amount of carbonate produced is fairly straightforward, hence it should be easy to certify the initial storage. Furthermore, if one has produced solid carbonates, it is easy to survey them, and if there is a demand for a re-evaluation of the size of the carbonate pile, it is possible to obtain a very good estimate with relatively little effort. Methods that ultimately disperse the carbon cannot, of course, verify past measurements, but since carbonates are stable, there is little concern that they will convert back to CO_2.

Geological storage provides a more difficult scenario. Current technologies may be able to show the location of the CO_2 underground but they are at best qualitative in assessing the amount of anthropogenic carbon present in the formation. Most verification methods therefore rely on observing fluxes from the reservoir to the surface. It does not appear likely that the public will settle for monitoring of potential leaks. The argument that nothing was leaking while one was looking is far less convincing than a positive accounting method that can measure the amount of anthropogenic carbon left in the formation.

21.11.2 Storage Liability

Ultimately, a question that transcends this chapter is the long-term liability issue for the storage. As the longevity of these storage sites gradually is sinking in, policy makers have begun to realize that liability issues need to be considered carefully.

In these early discussions, it is often the State which takes on the long-term liability. Other models also need to be considered.

21.12 Conclusion

CCS remains a necessity as long as the world depends on fossil carbon as a source of energy. Capture technologies may dominate the cost of CCS, but they essentially apply standard industrial methodologies to a fairly standard problem. The more difficult challenge is the storage of all the captured CO_2. Storing small amounts of CO_2 would not be particularly difficult; what makes this problem so difficult is the enormous volume of CO_2 that needs to be stored, and the length of time this storage has to be guaranteed.

As has been pointed out by Smil [67], the scale of CO_2 storage is truly formidable, as it exceeds the amount of oil extracted in terms of volume or mass by a good order of magnitude. However, it is not at all clear that Smil's pessimism is justified. The total amount of fossil fuel extracted from the ground is approximately one-third of the CO_2 that is produced from their use. The amount of material processed in coal mining exceeds the amount of CO_2 that is produced from coal. Ultimately, on a ton-for-ton comparison, the effort of CO_2 capture and storage appears to be manageable. This is what really matters. There are many large-scale processes required to support a sustainable infrastructure underpinning a modern society.

The cost of managing carbon is not insignificant, but the Stern report [68] seems to be correct when it estimates the cost at the order of 1% of world GDP. Estimating the cost of potential damage and economic implications of the small probability of truly catastrophic outcomes may be difficult, it is clear that damage and risks will grow as long as the CO_2 level in the atmosphere keeps rising. Hence it is virtually certain that in the end carbon storage will have to match fossil carbon production.

The advocates of storage of CO_2 in deep saline aquifers argue that the available storage space is plentiful and exceeds what will likely be needed. Only time will tell whether this is indeed the case. The major concern is that various constraints on reservoirs will rule out the vast majority of reservoirs. Rejection could vary from concerns over potential leakage paths, to constraints on developing reservoirs in populated areas. The NUMBY principle (*N*ot *U*nder *M*y *B*ack *Y*ard), may rule out all but the most remote storage sites.

Nevertheless, geological storage clearly works and can be implemented in some locations. There is an urgent need to develop more such sites, and the regulatory and economic frameworks for such actions need to be put in place.

Of all the options available, the basic geological storage is the only form that is ready to start operations at the scale of individual power plants. Therefore, it should be started as soon as possible. There are many doubts that the necessary scale of operation can be reached and that the storage capacity is really available. It has been pointed out that the world will need thousands of operations like the Statoil plant in the North Sea, but then the world operates thousands of power plants. There is enough capacity to get started, and other more exotic options that are still in the research stage should also be advanced.

Taking all the various options together, it seems clear that there are opportunities that are sufficiently large, and that could keep the carbon out of the mobile pool for many millennia. The cost of most of these processes is still highly uncertain, but the first ones have reached a level of maturity where they can be implemented.

Learning by doing requires doing, and therefore not much progress will be made until the political will for moving forward is there.

References

1 Metz, B., *et al.*, *IPCC Special Report on Carbon Dioxide Capture and Storage*, Cambridge University Press, New York, 2005.

2 Lackner, K. S. The conundrum of sustainable energy: clean coal as one possible answer. *Asian Econ. Pap.*, 2006, **4**(3), 30–58.

3 Lackner, K. and Brennan, S. Envisioning carbon capture and storage: expanded possibilities due to air capture, leakage insurance, and C-14 monitoring. *Climatic Change*, 2009, **96**(3), 357–378.

4 Benson, S., *et al.* Underground geological storage, in *IPCC Special Report on Carbon Dioxide Capture and Storage* (ed. Metz, B., *et al.*), Cambridge University Press, New York, 2005, pp. 195–276.

5 Holloway, S. Storage of fossil fuel-derived carbon dioxide beneath the surface of the Earth. *Annu. Rev. Energy Environ.*, 2001, **26**, 145–166.

6 International Energy Agency. *CO_2 Capture and Storage. A key carbon abatement opiton. Energy Technology Analysis.* OECD/IEA, Paris, 2008.

7 House, K. Z., *et al.* Permanent carbon dioxide storage in deep-sea carbonate sediments. *Proc. Natl. Acad. Sci. USA*, 2006, **103**(33), 12291–12295.

8 Levine, J. S., *et al.* Gravitational trapping of carbon dioxide in deep sea sediments: permeability, buoyancy, and geomechanical analysis. *Geophys. Res. Lett.*, 2007, **34**(24), 5.

9 Brock, W. R. and Bryan, L. A., *Summary Results of CO_2 EOR Field Tests, 1972–1987.* Presented at the Low Permeability Reservoirs Symposium, 6–8 March 1989, Denver, CO.

10 Wo, S., Whitman, L. D., and Steidtmann, J. R. Estimates of potential CO_2 demand for CO_2 EOR in Wyoming basins. Presented at the SPE Rocky Mountain Petroleum Technology Conference, 14–16 April 2009, Denver, CO.

11 Hunter, J. K. and Bryan, L. A. LaBarge project: availability of CO_2 for tertiary projects. *J. Pet. Technol.*, 1987, **39**(11), 1407–1410.

12 Gunter, W. D., *et al.* Deep coalbed methane in Alberta, Canada: a fuel resource with the potential of zero greenhouse gas emissions. *Energy Convers. Manage.*, 1997, **38**(Suppl. 1), S217–S222.

13 Damen, K., *et al.* Identification of early opportunities for CO_2 sequestration – worldwide screening for CO_2-EOR and CO_2-ECBM projects. *Energy*, 2005, **30**(10), 1931–1952.

14 Stevens, S. H., *et al.* CO_2 injection and sequestration in depleted oil and gas fields and deep coal seams: worldwide potential and costs. *Environ. Geosci.*, 2001, **8**(3), 200–209.

15 US Department of Energy National Energy Technology Laboratory. *Carbon Sequestration Atlas of the United States and Canada (Atlas III)*, 3rd edn., NETL, Pittsburgh, PA, 2010.

16 Eccles, J. K., *et al.* Physical and economic potential of geological CO_2 storage in saline aquifers. *Environ. Sci. Technol.*, 2009, **43**(6), 1962–1969.

17 Mazzotti, M., *et al.* Mineral carbonation and industrial uses of carbon dioxide. In *IPCC Special Report on Carbon Dioxide Capture and Storage* (ed. Metz, B., *et al.*), Cambridge University Press, New York, 2005, pp. 319–337.

18 Matter, M. J. *et al.* Permanent Cabon Dioxide Storage into Basalt: The CarbFix Pilot Project, Iceland. Energy Procedia, 2009, **1**(1), 3641–3646.

19 Kelemen, P. B. and Matter, J. *In situ* carbonation of peridotite for CO_2 storage. *Proc. Natl. Acad. Sci. USA*, 2008, **105**(45), 17295–17300.

20 Lackner, K. S., *et al.* Carbon dioxide disposal in carbonate minerals. *Energy*, 1995, **20**, 1153–1170.

21 Lackner, K. S., *et al. Carbon Dioxide Disposal in Mineral Form: Keeping Coal Competitive*, Los Alamos Report LA-UR-97-2094, Los Alamos National Laboratory, Los Alamos, NM, 1997.

22 Newall, P. S., Clark, S. J., *et al. CO_2 Storage as Carbonate Minerals*, IEA/ Greenhouse Gas Program Report PH3/

17IEA, Greenhouse Gas R&D Program and CSMA Consultants, 2000.

23 Schuiling, R. and Krijgsman, P. Enhanced weathering: an effective and cheap tool to sequester CO_2. *Climatic Change*, 2006, **74**(1), 349–354.

24 Caldeira, Ken, *et al.* Ocean storage. In *IPCC Special Report on Carbon Dioxide Capture and Storage* (ed. Metz, B., *et al.*), Cambridge University Press, New York, 2005, pp. 277–318.

25 Kheshgi, H. S. Sequestering atmospheric carbon dioxide by increasing ocean alkalinity. *Energy*, 1995, **20**, 915–922.

26 Rau, G. H. and Caldeira, K. Enhanced carbonate dissolution: a means of sequestering waste CO_2 as ocean bicarbonate. *Energy Convers. Manage.*, 1999, **40**(17),

27 Lackner, K. S. Carbonate chemistry for sequestering fossil carbon. *Annu. Rev. Energy Environ.*, 2002, **27**(1), 193–232.

28 Litynski, J., *et al.* An overview of terrestrial sequestration of carbon dioxide: the United States Department of Energys Fossil Energy R&D Program. *Climatic Change*, 2006, **74**(1), 81–95.

29 Schimel, D., *et al.* CO_2 and the carbon cycle. In *IPCC Special Report on Climate Change 1994: Radiative Forcing of Climate Change and an Evaluation of the IPCC IS92 Emission Scenarios* (ed. Houghton J. T., *et al.*.), Cambridge University Press, Cambridge, 1995, pp. 35–71.

30 Lehmann, J., Gaunt, J., Rondon, M. *Bio-char Sequestration in Terrestrial Ecosystems – A Review.* Mitigat. Adaptat. Strategies Global Change, 2006, **11**(2), 395–419.

31 Newsletter. ExxonMobil finishes Shute Creek expansion. *Oil Gas J.*, 2010, **108**(48), 13 December 2010.

32 Torp, T. A. and Gale, J. Demonstrating storage of CO_2 in geological reservoirs: the Sleipner and SACS projects. 2004, Energy, **29**(9–10), 1361–1369.

33 Preston, C., *et al.* IEA GHG Weyburn CO_2 monitoring and storage project. *Fuel Process. Technol.*, 2005, **86**(14–15), 1547–1568.

34 Schlumberger Excellence in Educational Development (SEED), 2010, http://www.seed.slb.com/subcontent.aspx?id=4182 (last accessed 6 Januqry, 2011.

35 Bachu, S. and Gunter, W. D. Acid-gas injection in the Alberta basin, Canada: a CO_2-storage experience. *Geol. Soc. Lond. Spec. Publ.*, 2004, **233**(1), 225.

36 Onuma, T. and Ohkawa, S. Detection of surface deformation related with CO_2 injection by DInSAR at In Salah, Algeria. *Energy Procedia*, 2009, **1**(1), 2177–2184.

37 International Energy Agency. *Carbon Capture and Storage, Progress and Next Steps.* IEA/CSFL Report to the Muskoka 2010 G8 Summit, Paris, 2010.

38 de Figueiredo, M. A. Sequestration field experiment: a case study in public perceptions and institutional effectiveness. PhD thesis, Massachusetts Institute of Technology, Cambridge, MA, 2003.

39 Haugan, P. M. Communicating scientific uncertainty for decision making about CO_2; storage. In *Uncertainties in Environmental Modelling and Consequences for Policy Making* (ed. Baveye, P. C., Laba, M., and Mysiak, J.), NATO Science for Peace and Security Series C: Environmental Security, Springer Netherlands, Dordrecht, 2009, pp. 253–264.

40 Pals, F. Shell's Barendrecht carbon-capture project canceled. *Business Week*, 4 November 2010, .

41 Förster, A., *et al.* Baseline characterization of the CO_2SINK geological storage site at Ketzin, Germany. *Environ. Geosci.*, 2006, **13**(3), 145–161.

42 Litynski, J., *et al.* US Department of Energy's Regional Carbon Sequestration Partnership Program: Overview. *Energy Procedia*, 2009, **1**(1), 3959–3967.

43 Figueroa, J. D., *et al.* Advances in CO_2 capture technology – the US Department of Energys Carbon Sequestration Program. *Int. J. Greenhouse Gas Control*, 2008, **2**(1), 9–20.

44 US Department of Energy National Energy Technology Laboratory. *Carbon Dioxide Capture and Storage R&D Roadmap*, DOE/NETL, Pittsburgh, PA, December 2010.

45 Rushing, S. A. *Biofuels Dig.*, 8 March 2010. Making Money from Carbon

Dioxide, published March 8th, 2010, available online at: http://chem.info/Articles/2010/08/Alternative-Energy-Making-Money-from-Carbon-Dioxide/, last accessed March 8th, 2011.

46 EIA data, http://www.eia.doe.gov. 47: p. 31–46.

47 Lackner, K.S., Wilson, R., and Ziock, H.-J. Free-market approach to controlling carbon dioxide emissions to the atmosphere: a discussion of the scientific basis. In *Conference on Global Warming and Energy Policy* (ed. Kursunoglu, B.N., Mintz, S.L., and Perlmutter, A., Kluwer Academic/Plenum Publishers, Fort Lauderdale, FL, 2000, p. 31–46.

48 Lackner, K.S. Capture of carbon dioxide from ambient air. *Eur. Phys. J. Spec. Top.*, 2009, **176**(1), 93–106.

49 Goff, F. and Lackner, K.S. Carbon dioxide sequestering using ultramafic rocks. *Environ. Geosci.*, 1998, **5**(3), 89–101.

50 Krevor, S.C., *et al.* Delineation of magnesium-rich ultramafic rocks available for mineral carbon sequestration in the United States. *Energy Procedia*, 2009, **1**, 4915–4920.

51 Butler, J. N. *Carbon Dioxide Equilibria and Their Applications*, Lewis, Chelsea, MI, 1991.

52 Oelkers, E.H., Gislason, S.R., *et al.* Mineral carbonation of CO_2. *Elements*, 2008, **4**(5), 333–337.

53 Orr, F.M., Jr. CO_2 capture and storage: are we ready?, *Energy Environ. Sci.*, 2009, **2**(5), 449–458.

54 Pearce, F. The high cost of carbon dioxide, *New Sci.*, **139**(1882), 26.

55 Sloan E.D., Jr. *Clathrate Hydrates of Natural Gases*, 2nd edn., Marcel Dekker, New York, 1998.

56 Koide, H., *et al.* Hydrate formation in sediments in the sub-seabed disposal of CO_2, *Energy*, **22** (2–3), 279–283, 1997.

57 Park, Y., *et al.* Sequestering carbon dioxide into complex structures of naturally occurring gas hydrates. *Proc. Natl. Acad. Sci. USA*, 2006, **103**(34), 12690–12694.

58 Castaldi, M.J., Zhou, Y., and Yegulalp, T.M. Down-hole combustion method for gas production from methane hydrates. *J. Pet. Sci. Eng.*, 2007, **56**(1–3), 176–185.

59 Brewer, PG, *et al.* Direct experiments on the ocean disposal of fossil fuel CO_2, *Science*, 1999, **284**(5416), 943–945.

60 Steinberg, M. The Hy-C process (thermal decomposition of natural gas) potentially the lowest cost source of hydrogen with the least CO_2 emission. *Energy Convers. Manage.* 1995, **36**(6–9), 791–796.

61 Ziock, H.-J., Lackner, K.S., Overview of the ZECA (Zero Emission Coal Alliance) Technology, 2000, Los Alamos National Laboratory, Technical Report, LA-UR-00-6002.

62 Dunsmore, H.E. A geological perspective on global warming and the possibility of carbon dioxide removal as calcium carbonate mineral. *Energy Convers. Manage.* 1992, **33**(5–8), 565–572.

63 Wright, A.B. *et al.* Removal of Carbon Dioxide from Air. US Patent Application No. US 11/996615 (Filed 28 July 2006). Publication No. US 2009/0294366A1. December 3, 2009.

64 Kling, G.W., *et al.* The 1986 Lake Nyos gas disaster in Cameroon, West Africa. *Science*, 1987, **236**, 169–175.

65 Allison, M.L. Hutchinson, Kansas: a geologic detective story. *Geotimes*, 2001, **46**(10), 14–18.

66 Evans, D.J. A review of underground fuel storage events and putting risk into perspective with other areas of the energy supply chain. *Geol. Soc. London Spec. Publ.*, 2009, **313**(1), 173–216.

67 Smil, V. Presented at the OECD Global Science Forum: Conference on Scientific Challenges for Energy Research, Paris, 2006.

68 Stern, N., Stern Review on the Economics of Climate Change. October 30, 2006, HM Treasury, London, United Kingdom.

Index

Process Engineering for CCS Power Plants, First Edition. Edited by Detlef Stolten and V. Scherer

f